Edited by
Manfred Wendisch and
Jean-Louis Brenguier

Airborne Measurements for Environmental Research

Related Titles

Bohren, C. F., Huffman, D. R., Clothiaux, E. E.

Absorption and Scattering of Light by Small Particles

2013
Softcover
ISBN: 978-3-527-40664-7

Camps-Valls, G., Bruzzone, L. (eds.)

Kernel Methods for Remote Sensing Data Analysis

2009
Hardcover
ISBN: 978-0-470-72211-4

Vincent, J. H.

Aerosol Sampling
Science, Standards, Instrumentation and Applications

2007
Hardcover
ISBN: 978-0-470-02725-7

Wendisch, M., Yang, P.

Theory of Atmospheric Radiative Transfer – A Comprehensive Introduction
Wiley-VCH Verlag GmbH & Co. KGaA, Weinheim, Germany

2012
ISBN: 978-3-527-40836-8

Bohren, C. F., Clothiaux, E. E.

Fundamentals of Atmospheric Radiation
An Introduction with 400 Problems

2006
Softcover
ISBN: 978-3-527-40503-9

Seinfeld, J. H., Pandis, S. N.

Atmospheric Chemistry and Physics
From Air Pollution to Climate Change

2006
Hardcover
ISBN: 978-0-471-72018-8

Lillesand, T. M., Kiefer, R. W., Chipman, J. W.

Remote Sensing and Image Interpretation

2008
Hardcover
ISBN: 978-0-470-05245-7

Edited by Manfred Wendisch and Jean-Louis Brenguier

Airborne Measurements for Environmental Research

Methods and Instruments

WILEY-VCH

WILEY-VCH Verlag GmbH & Co. KGaA

The Editors

Dr. Manfred Wendisch
Leipzig Institute for Meteorology (LIM)
Leipzig, Germany
m.wendisch@uni-leipzig.de

Dr. Jean-Louis Brenguier
Météo-France, CNRM, GMEI
EUFAR
Toulouse, France
jlb@meteo.fr

A book of the Wiley Series in Atmospheric Physics and Remote Sensing

The Series Editor

Dr. Alexander Kokhanovsky
University of Bremen, Germany
alexk@iup.physik.uni-bremen.de

Cover Picture

Photo taken near the Caribbean island of Antigua from the University of Wyoming King Air research aircraft during the Rain in Cumulus over the Ocean (RICO) project (funded by the US National Science Foundation). Courtesy of Gabor Vali.

Library of Congress Card No.: applied for

British Library Cataloguing-in-Publication Data
A catalogue record for this book is available from the British Library.

Bibliographic information published by the Deutsche Nationalbibliothek
The Deutsche Nationalbibliothek lists this publication in the Deutsche Nationalbibliografie; detailed bibliographic data are available on the Internet at <http://dnb.d-nb.de>.

Cover Design Grafik-Design Schulz, Fußgönheim, Germany
Typesetting Laserwords Private Limited, Chennai, India
Printing and Binding Markono Print Media Pte Ltd, Singapore

Print ISBN: 978-3-527-40996-9
ePDF ISBN: 978-3-527-65324-9
ePub ISBN: 978-3-527-65323-2
mobi ISBN: 978-3-527-65322-5
oBook ISBN: 978-3-527-65321-8

Contents

Preface

EUFAR
European Facility
For Airborne
Research

This book summarizes the knowledge of international experts in airborne measurements from 13 countries, which they have developed over many years of field experiments and application to environmental research. The book is produced within the framework of the European Facility for Airborne Research (EUFAR, *http://www.eufar.net/*). EUFAR is a research infrastructure network supported since 2000 by the European Commission, as part of the research infrastructures integration program; see the respective web page at *http://cordis.europa.eu/fp7/capacities/home_en.html*.

One of the EUFAR Networking Activities is dedicated to Expert Working Groups (EWGs), which facilitate cross-disciplinary fertilizations and a wider sharing of knowledge and technologies between academia and industry in the field of airborne research. Over the past 10 years, numerous workshops have been organized by the EWGs addressing technical, logistic, and scientific issues specific to airborne research for the environment. From the beginning, these workshops involved international experts; however, an ever increasing number of scientists from outside the European airborne science community became involved and played an active role. Thus, the EWGs within EUFAR have become a truly international collaborative effort and, as a consequence, the workshops had a continuously increasing impact on defining research foci of future international airborne research.

The EUFAR EWGs currently publish workshop reports and recommendations (i) to aircraft operators on best practice and common protocols for operation of airborne instruments, (ii) to scientific users on best usage and interpretation of the collected data, and (iii) to the research institutions on future challenges in airborne measurements. To ensure legacy of this accumulated knowledge, this book summarizes the major outcome of the EWG discussions on the current status of airborne instrumentation. The book has been designed to provide an extensive overview of existing and emerging airborne measurement principles and techniques. Furthermore, the book analyzes problems, limitations, and mitigation approaches specific to airborne research to explore the environment.

The target audience of the book is not only experienced researchers but also graduate students, the book intends to attract to this exciting scientific field. Also university teachers, scientists experienced in related fields and looking for additional airborne data, for example, for validation or analysis of their own measurements, modelers, and project managers will find a concise overview of airborne scientific instrumentation to explore atmospheric and Earth's surface properties in this book.

Chapter 1 examines the strengths and weaknesses of airborne measurements. The subsequent Chapter 2 deals with the description of instruments to measure aircraft state parameters and basic thermodynamic and dynamic variables of the atmosphere, such as static air pressure, temperature, water vapor, wind vector, turbulence, and fluxes. The next three chapters consider *in situ* measurements of gaseous and particulate atmospheric constituents (Chapters 3–5). Chemical instruments to measure gaseous atmospheric components are introduced in Chapter 3, whereas the instrumentation for particulate atmospheric constituents is described in Chapters 4 (aerosol particles) and 5 (cloud and precipitation particles). Special problems associated with airborne particle sampling (aerosol and cloud/precipitation particles) are discussed in Chapter 6. The following two chapters deal with airborne radiation measurements (Chapter 7) and with techniques for passive remote sensing of the Earth's surface (Chapter 8). The most commonly applied airborne active remote sensing techniques are introduced in Chapter 9. An extensive, albeit not complete, list of references the reader may consult for airborne instrumentation is given at the end of the book. Furthermore, some supplementary material has been compiled, which is not printed but available from the publisher's Web site.

We are very grateful to the European Commission, Research Infrastructure Unit, for its financial support to EUFAR, more specifically for the organization of expert workshops and the preparation of this book. We also acknowledge the support of the national research organizations from Europe and the United States, which are supporting the 91 scientific experts contributing to the book.

We particularly appreciate the considerable efforts of Ulrich Schumann and David W. Fahey to organize and steer the review process for the book; we also acknowledge their useful comments and suggestions. Before publication, the book was peer-reviewed by external experts, which has contributed to improve the quality of the book significantly. We explicitly thank the external reviewers of the book listed in alphabetic order: Charles Brock, Peter Gege, Jim Haywood, Dwayne E. Heard, Jost Heintzenberg, Robert L. Herman, Lutz Hirsch, Andreas

Hofzumahaus, Peter Hoor, Ruprecht Jaenicke, Greg McFarquhar, Matthew McGill, Ottmar Mühler, Daniel Lack, George Leblanc, Hanna Pawlowska, Tom Ryerson, Johannes Schneider, Patrick J. Sheridan, Geraint Vaughan, Peter Vörsmann, and Elliot Weinstock. Technical editor Dagmar Rosenow led many of the thankless but necessary tasks to pull this book together. We are grateful for her talents and dedication, without which the book could not have been completed. We also thank Matt Freer and Frank Werner for their help with editing the text and figures; the students Kathrin Gatzsche and Marcus Kundisch from the Leipzig Institute for Meteorology (LIM) of the University of Leipzig were of great help in compiling the extensive bibliography. Furthermore, we would like to list the leading authors of the chapters emphasizing their active role in writing this book.

Chapter 1: Ulrich Schumann, David W. Fahey, Manfred Wendisch, and Jean-Louis Brenguier
Chapter 2: Jens Bange, Marco Esposito, and Donald H. Lenschow
Chapter 3: Jim McQuaid and Hans Schlager
Chapter 4: Andreas Petzold and Paola Formenti
Chapter 5: Jean-Louis Brenguier
Chapter 6: Martina Krämer, Cynthia Twohy, and Markus Hermann
Chapter 7: Manfred Wendisch and Peter Pilewskie
Chapter 8: Eyal Ben-Dor
Chapter 9: Jacques Pelon and Gabor Vali

Leipzig and Toulouse
2012

Dr. Manfred Wendisch
Leader of EWGs
within EUFAR (Universität Leipzig)
and
Dr. Jean-Louis Brenguier
EUFAR coordinator (Météo-France)

A Tribute to Dr. Robert Knollenberg

There have been many important technologies that have been developed for the airborne measurement of atmospheric properties, and the scientific community owes its gratitude to the many distinguished researchers for their significant contributions to the measurement sciences.

One of these people who stands out, in particular, is Robert Knollenberg whose pioneering work in the 1970s led to the development of technology that is still employed in the majority of instruments that make airborne, real-time measurements of size distributions of atmospheric particles, in and out of clouds.

We are paying a special tribute to Robert Knollenberg in this book, the first to present a comprehensive overview of airborne instrumentation for atmospheric measurements, because of his many innovative and creative ideas that have allowed us to study the fine-scale structure of clouds and aerosol particles on a particle-by-particle basis. The development of the optical array probe (OAP), forward scattering spectrometer probe (FSSP), passive cavity aerosol spectrometer probe, single-particle soot photometer, and ultrahigh sensitivity aerosol spectrometer represent cutting-edge technology that has revolutionized how we look at clouds and aerosol particles. These groundbreaking instrument developments allowed the atmospheric science community to understand fundamental physical processes that, while theoretically predicted, could not be observed in the free atmosphere until Knollenberg's instruments provided the technology to measure the necessary particle characteristics to corroborate the theory.

After completing his PhD at the University of Wisconsin with continuous strong guidance from Dr. Robert Graham, Robert Knollenberg spent 3 years at the National Center for Atmospheric Research developing and field testing cloud particle spectrometers. In 1972, he decided to commercialize such instruments for use by the atmospheric community and formed Particle Measuring Systems (PMS) Inc. in Boulder, Colorado (USA). While under his direction, PMS developed a commercial version of the OAP, the 200-X and 200-Y, which were classified as "1D" probes, because they only measured the maximum size of particles with no shape information. This led, however, to the "2D" probes that measured the two-dimensional image of particles. In parallel to these developments, Knollenberg was also implementing single-particle light scattering to build the axially scattering spectrometer probe that evolved into the FSSP. Aside from the atmospheric community's use of these instruments, they became essential for quantifying the cloud structure when certifying aircraft for "flight into known icing conditions."

Knollenberg turned PMS over to other management in the early 1990s to concentrate on research and instrument development, while spinning out a PMS electro-optics division forming another company, Research Electro-Optics Inc., which concentrated on developing new lasers and related optical components needed for PMS instrumentation and more general use. PMS went on to become the industry leader in optical particle counters used by the semiconductor industry to assess clean room and process fluid microcontamination. Knollenberg built the multiangle aerosol spectrometer (MASP) for NASA that was used to derive the size and refractive index of stratospheric aerosol particles while flying on the NASA ER-2. The MASP is the precursor to the cloud and aerosol spectrometer that is currently in use and implements the same measurement approach. Most recently, Knollenberg discovered how to measure the mass concentration of black carbon in individual aerosol particles using the concept of incandescence in an infrared laser cavity, once again breaking new ground and giving researchers a new tool to dissect individual particles and better understand how black carbon affects health and climate.

It is difficult to assess the breadth and depth of knowledge that the community has gained with respect to the science of the atmosphere as a result of Knollenberg's technological contributions. His strong background as an atmospheric physicist allowed him to understand the limitations and uncertainties in the technology that he then removed or minimized with the development of new techniques. Knollenberg serves as an inspiration to all who are interested in understanding how the atmosphere works and in particular to young scientists with new ideas on how to measure the properties of the atmosphere.

It is with deep gratitude that we offer this tribute in his name.

List of Contributors

Armin Afchine
Institute of Energy and Climate Research
Stratosphere (IEK-7)
Forschungszentrum Jülich GmbH
52425 Jülich
Germany

Gérard Ancellet
Université Pierre et Marie Curie
Laboratoire Atmosphères
Milieux et Observations Spatiales (LATMOS)
4 Place Jussieu
75252 Paris Cedex 05
France

Maria Dolores Andrés-Hernández
Universität Bremen
Institut für Umweltphysik
Postfach 33 04 40
28359 Bremen
Germany

William D. Bachalo
Artium Technologies, Inc.
470 Lakeside Drive, Unit C
Sunnyvale
CA 94085
USA

Stephen Ball
University of Leicester
Department of Chemistry
Leicester LE1 7RH
UK

Jens Bange
Universität Tübingen
Umweltphysik
Hölderlinstr. 12
72074 Tübingen
Germany

Darrel Baumgardner
Droplet Measuring Technologies
2545 Central Avenue
Boulder, CO 80301
USA

Eyal Ben-Dor
Tel Aviv University
Department of Geography
Ramat Aviv PO Box 39040
69989 Tel Aviv
Israel

Jean-Louis Brenguier
Météo-France/CNRM/GMEI
42 Avenue Gaspard Coriolis
31057 Toulouse
Cedex 1
France

Birger Bohn
Institute of Energy and Climate Research
Troposphere (IEK-8)
Forschungszentrum Jülich GmbH
52425 Jülich
Germany

Agnès Borbon
Laboratoire Interuniversitaire des Systèmes Atmosphériques (LISA), IPSL
University of Paris Est Créteil (UPEC) and Paris Diderot (UPD)
UMR CNRS 7583
Créteil
France

Philip R. A. Brown
Observation-Based Research
Met Office
FitzRoy Road
Exeter, EX1 3PB
UK

Steven S. Brown
Chemical Sciences Division
National Oceanic and Atmospheric Administration (NOAA)
325 Broadway
R/CSD7
Boulder, CO 80305
USA

Anthony Bucholtz
Naval Research Laboratory
7 Grace Hopper Street
Stop 2, Monterey, CA 93943-5502
USA

Ulrich Bundke
Johann Wolfgang Goethe-Universität
Institut für Atmosphäre und Umwelt
Altenhöferallee 1
60438 Frankfurt/Main
Germany

Valery Catoire
LPC2E (UMR7328)
Universite Orleans
3A Avenue de la Recherche Scientifique
45071 Orleans cedex 2
France

Patrick Y. Chuang
University of California
Earth and Planetary Sciences
1156 High Street
Santa Cruz, CA 95064
USA

Hugh Coe
University of Manchester
School of Earth Atmospheric and Environmental Science
Sackville Street
Manchester M60 1QD
UK

Joachim Curtius
Johann Wolfgang
Goethe-Universität
Institut für Atmosphäre und
Umwelt
Altenhöferallee 1
60438 Frankfurt/Main
Germany

Thomas G. Custer
Max-Planck Institute for
Chemistry
Hahn-Meitner-Weg 1
55128 Mainz
Germany

Susanne Crewell
Institut für Geophysik und
Meteorologie der Universität
zu Köln
Albertus-Magnus-Platz
50923 Köln
Germany

Paul J. DeMott
Colorado State University
Department of Atmospheric
Science
1371 Campus Delivery
Fort Collins, CO 80523-1371
USA

Suresh Dhaniyala
Clarkson University
204 CAMP
Potsdam, NY 13699-5725
USA

Piero Di Carlo
Dipartimento di Scienze Fisiche e
Chimiche
Centro di Eccellenza CETEMPS
Universita' degli Studi di L'Aquila
Via Vetoio
67010 Coppito-L'Aquila
Italy

Volker Dreiling
Deutsches Zentrum für Luft- und
Raumfahrt (DLR)
Abteilung Flugbetrieb
Oberpfaffenhofen
82234 Wessling
Germany

Gerhard Ehret
Institut für Physik der
Atmosphäre
Deutsches Zentrum für Luft- und
Raumfahrt (DLR)
Oberpfaffenhofen
82234 Wessling
Germany

Marco Esposito
cosine Research BV
Niels Bohrweg 11
Leiden
CA 2333
The Netherlands

Biagio M. Esposito
Centro Italiano Ricerche
Aerospaziali
via Maiorise
81043 Capua
Caserta
Italy

David W. Fahey
National Oceanic and
Atmospheric Administration
(NOAA)
Earth System Research Laboratory
325 Broadway R/CSD6
Boulder, CO 80305
USA

Markus Fiebig
Department of Atmospheric and
Climate Research
Norwegian Institute for Air
Research
2027 Kjeller
Norway

Richard C. Flagan
California Institute of Technology
210-41
1200 E. California Blvd.
Pasadena, CA 91125
USA

Pierre H. Flamant
Laboratoire de Météorologie
Dynamique
Ecole Polytechnique
91128 Palaiseau
France

Paola Formenti
Laboratoire Interuniversitaire des
Systèmes Atmosphériques
(LISA), IPSL
University of Paris Est Créteil
(UPEC) and Paris Diderot (UPD)
61 avenue du Général de Gaulle
UMR CNRS 7583
Créteil
France

Jacob Fugal
Max-Planck Institute for
Chemistry
Hahn-Meitner-Weg 1
55128 Mainz
Germany

Timothy Garrett
University of Utah
Atmospheric Science Department
135 S 1460 East Rm 819 (WBB)
Salt Lake City
UT 84112-0110
USA

Jean-Francois Gayet
Université Blaise Pascal
LaMP UMR 6016 CNRS
24 avenue des Landais
BP80026, 63 171 Aubière Cedex
France

Hermann Gerber
Gerber Scientific Inc.
1643 Bentana Way
Reston, VA 20190
USA

Andreas Giez
Deutsches Zentrum für Luft- und
Raumfahrt (DLR)
Abteilung Flugbetrieb
Oberpfaffenhofen
82234 Wessling
Germany

Chawn Harlow
Observation Based Research
The Met Office
Cordouan 2 W007
FitzRoy Road
Devon, Exeter EX1 3PB
UK

Samuel Haimov
University of Wyoming
Atmospheric Science Dept. 3038
1000 E. University Ave.
Laramie, WY 82071
USA

Andy Heymsfield
MMM Division
National Center for Atmospheric Research (NCAR)
Boulder, CO 80301
USA

Markus Hermann
Leibniz Institute for Tropospheric Research
Department of Physics
Permoserstraße 15
04318 Leipzig
Germany

James Hopkins
National Center for Atmospheric Sciences
Department of Chemistry
Heslington
York YO10 5DD
UK

James G. Hudson
Desert Research Institute (DRI)
Nevada System of Higher Education
Division of Atmospheric Sciences
Reno, NV 89512-1095
USA

Evelyn Jäkel
Universität Leipzig
Leipzig Institute for Meteorology (LIM)
Stephanstr. 3
04103 Leipzig
Germany

Alexander Kokhanovsky
Institute of Environmental Physics
Universität Bremen
Otto-Hahn-Allee 1
28359 Bremen
Germany

Alexei Korolev
Cloud Physics Research
Meteorological Service of Canada
4905 Dufferin Street
Ontario M3H 5T4
Canada

Martina Krämer
Institute of Energy and Climate Research
Stratosphere (IEK-7)
Forschungszentrum Jülich GmbH
52425 Jülich
Germany

R. Paul Lawson
Stratton Park Engineering Company
3022 Sterling Circle
Suite 200
Boulder, CO 80301
USA

Donald H. Lenschow
University Corporation for
Atmospheric Research (UCAR)
3450 Mitchell Lane
Boulder, CO 80307-3000
USA

David Leon
University of Wyoming
Atmospheric Science Dept. 3038
1000 E. University Ave.
Laramie, WY 82071
USA

Larry Mahrt
Oregon State University
College of Oceanic and
Atmospheric Sciences
104 COAS Administration
Building
Corvallis, OR 97331-5503
USA

Tim Malthus
Christian Laboratory
CSIRO Land and Water
GPO Box 1666
Canberra
Australia

Szymon P. Malinowski
University of Warsaw
Faculty of Physics
Institute of Geophysics
Pasteura 7
02-093 Warsaw
Poland

Jim McQuaid
University of Leeds
National Centre for Atmospheric
Science
School of Earth and Environment
Leeds LS2 9JT
UK

James B. Mead
ProSensing
107 Sunderland Road
Amherst, MA 01002
USA

Andreas Minikin
Institut für Physik der
Atmosphäre
Deutsches Zentrum für Luft- und
Raumfahrt (DLR)
Oberpfaffenhofen
82234 Wessling
Germany

Andrew L. Pazmany
ProSensing
107 Sunderland Road
Amherst, MA 01002
USA

Jacques Pelon
Université Pierre et Marie Curie
Laboratoire Atmosphères
Milieux et Observations Spatiales
(LATMOS)
4 Place Jussieu
75252 Paris Cedex 05
France

Andreas Petzold
Institute of Energy and Climate Research
Troposphere (IEK-8)
Forschungszentrum Jülich GmbH
52425 Jülich
Germany

Klaus Pfeilsticker
Ruprecht-Karls-Universität
Institute of Environmental Physics
69120 Heidelberg
Germany

Peter Pilewskie
University of Colorado
Laboratory for Atmospheric and Space Physics
Department of Atmospheric and Oceanic Science
Boulder, CO 80309-0392
USA

Antonio J. Plaza
Remote Sensing
Signal and Image Processing Laboratory
University of Maryland
1000 Hilltop Circle
Baltimore, MD 21250
USA

Alain Protat
Centre for Australian Weather and Climate Research (CAWCR)
Australian Bureau of Meteorology
700 Collins Street
Docklands
VIC3008 Melbourne
Australia

Gregory C. Roberts
University of California
San Diego
Scripps Institution of Oceanography
9500 Gilman Drive
La Jolla, CA 92093-0221
USA
and
Centre National de Recherches Météorologiques
42 av. G. Coriolis
31057 Toulouse
France

Thomas Röckmann
Institute for Marine and Atmospheric Research Utrecht
Princetonplein 5
3584 CC Utrecht
The Netherlands

Alfred R. Rodi
University of Wyoming
Atmospheric Science Dept. 3038
1000 E. University Ave.
Laramie, WY 82071
USA

David C. Rogers
Research Aviation Facility (RAF)
National Center for Atmospheric Research (NCAR)
Broomfield, CO 80021
USA

Anke Roiger
Institut für Physik der
Atmosphäre
Deutsches Zentrum für Luft- und
Raumfahrt (DLR)
Oberpfaffenhofen
82234 Wessling
Germany

Hans Schlager
Institut für Physik der
Atmosphäre
Deutsches Zentrum für Luft- und
Raumfahrt (DLR)
Oberpfaffenhofen
82234 Wessling
Germany

Daniel Schläpfer
ReSe Applications
Langeggweg 3
CH-9500 Wil SG
Switzerland

Ulrich Schumann
Institut für Physik der
Atmosphäre
Deutsches Zentrum für Luft- und
Raumfahrt (DLR)
Oberpfaffenhofen
82234 Wessling
Germany

K. Sebastian Schmidt
University of Colorado
Laboratory for Atmospheric and
Space Physics
Space Science Building
3665 Discovery Drive
Boulder, CO 80303
USA

Raymond A. Shaw
Michigan Technological
University
Department of Physics
1400 Townsend Drive
Houghton, MI 49931-1295
USA

Rick Shetter
Atmospheric Chemistry Division
National Center for Atmospheric
Research (NCAR)
1850 Table Mesa Dr
Boulder, CO 80305
USA

Holger Siebert
Leibniz Institute for Tropospheric
Research
Department of Physics
Permoserstraße 15
04318 Leipzig
Germany

Herman Smit
Institute of Energy and Climate
Research
Troposphere (IEK-8)
Forschungszentrum Jülich
GmbH
52425 Jülich
Germany

Fred Stroh
Institute of Energy and Climate
Research
Stratosphere (IEK-7)
Forschungszentrum Jülich
GmbH
52425 Jülich
Germany

Walter Strapp
Cloud Physics Research
Meteorological Service of Canada
4905 Dufferin Street
Ontario M3H 5T4
Canada

Jonathan Taylor
Observation Based Research
The Met Office
Cordouan 2 W007
FitzRoy Road
Devon, Exeter EX1 3PB
UK

David D. Turner
National Severe Storms Laboratory
National Oceanic and Atmospheric Administration (NOAA)
120 David L. Boren Blvd.
Norman, OK 73072
USA

Cynthia Twohy
Oregon State University
College of Earth, Ocean and Atmospheric Sciences
104 CEOAS Administration Building
Corvallis, OR 97331-5503
USA

Gabor Vali
University of Wyoming
Atmospheric Science Dept. 3038
1000 E. University Ave.
Laramie, WY 82071
USA

Marc von Hobe
Institute of Energy and Climate Research
Stratosphere (IEK-7)
Forschungszentrum Jülich GmbH
52425 Jülich
Germany

Jian Wang
Brookhaven National Laboratory
Atmospheric Sciences Division
Building 815E, 75 Rutherford Drive
Upton
NY 11973--5000
USA

Zhien Wang
University of Wyoming
Atmospheric Science Dept. 3038
1000 E. University Ave.
Laramie, WY 82071
USA

Manfred Wendisch
Universität Leipzig
Leipzig Institute for Meteorology (LIM)
Stephanstr. 3
04103 Leipzig
Germany

Jonathan Williams
Max-Planck Institute for Chemistry
Hahn-Meitner-Weg 1
55128 Mainz
Germany

Mengistu Wolde
Institute for Aerospace Research
National Research
Council Canada
U-61, 1200 Montreal Road
Ottawa, ON K1A 0R6
Canada

Helmut Ziereis
Institut für Physik der
Atmosphäre
Deutsches Zentrum für Luft- und
Raumfahrt (DLR)
Oberpfaffenhofen
82234 Wessling
Germany

Martin Zöger
Deutsches Zentrum für Luft- und
Raumfahrt (DLR)
Abteilung Flugbetrieb
Oberpfaffenhofen
82234 Wessling
Germany

1
Introduction to Airborne Measurements of the Earth Atmosphere and Surface

Ulrich Schumann, David W. Fahey, Manfred Wendisch, and Jean-Louis Brenguier

Aircraft have been applied very effectively in many aspects of environmental research. They are widely used to investigate the atmosphere and observe the ground visually and by measurements with instruments on board the aircraft. They allow for

(i) *in situ* measurement of atmospheric properties from the ground to altitudes of 20 km;
(ii) remote sensing measurements of the ground and atmosphere and even extraterrestrial properties;
(iii) targeted measurements along a selectable flight path at controllable times and places (e.g., in the atmospheric boundary layer over ocean and ice, in the free troposphere, in the tropical and polar stratosphere, or over a specific landscape);
(iv) exploration of atmospheric phenomena or events and process studies for basic understanding (e.g., thunderstorm anvils, volcanic eruption plumes, tropical cyclones, cloud microphysics, and contrails);
(v) measurements at high temporal and spatial resolutions (e.g., 1 s and 100 m are typical, and smaller scales are reached for some turbulence and cloud microphysics measurements);
(vi) acquisition of data for model development and parameterization and validation (e.g., chemical composition and kinetics, cloud microphysics, and aerosol properties);
(vii) testing and validation of remote sensing measurements (e.g., trace gas measurements from satellites); and
(viii) comprehensive airborne measurements as part of campaigns or long-term investigations within an extensive observation and modeling system approach to explore complex and fundamental Earth–atmosphere system properties (e.g., aerosol/cloud/radiation interactions with atmospheric dynamics, and climate change).

The high maneuverability of aircraft allows researchers to chase atmospheric phenomena, follow their evolution, and explore their chemistry and physics from small spatial scales up to thousands of kilometers and over time scales of fractions

Airborne Measurements for Environmental Research: Methods and Instruments, First Edition.
Edited by Manfred Wendisch and Jean-Louis Brenguier.

of seconds to many hours or even days. Aircraft instruments uniquely complement remote sensing instruments by measuring many parameters that are currently not available from space- or ground-based sensors (e.g., turbulence, nanometer-sized particles, and gases without or with only low radiation absorption efficiencies, such as nitrogen monoxide). Aircraft can reach remote locations and can carry *in situ* as well as active and passive remote sensing instruments. Observations may be performed along streamlines or in a fully Lagrangian manner with repeated sampling of the same air mass over extended periods. Instrumented aircraft enable remote observations of the Earth surface with very high resolution and with minimum disturbances by the atmosphere between sensor and object.

Worldwide, an impressive fleet of research aircraft is available with airborne instruments designed for many applications, although further demand still exists. Traditionally, research has been performed with manned aircraft, but increasingly, unmanned aircraft are also being used. Research aircraft include stratospheric aircraft (e.g., the Russian Geophysica and NASA ER-2 and NASA Global Hawk Unmanned Aerial System), high-level jets (e.g., Gulfstream-505 aircraft at National Center for Atmospheric Research (NCAR), the High-Altitude and Long-Range Research Aircraft in Germany, and two Falcon 20 in France and Germany), large and mid-sized aircraft operating between near ground and in the lower stratosphere (e.g., the NCAR C130 in the United States, a BAe-146 in the United Kingdom, and an ATR-42 in France), and several smaller (e.g., CASA-212 in Spain and Do-228 in the United Kingdom) and low-level aircraft (e.g., Sky-Arrow in Italy and Ultra-light at the University of Karlsruhe). In addition to specialized research aircraft, commercial airliners have been equipped with inlets and instruments and have participated in measurement programs (e.g., Swiss Nitrogen Oxides and Ozone along Air Routes Project, Measurement of Ozone and Water Vapor by Airbus In-Service Aircraft, Civil Aircraft for Regular Investigation of the Atmosphere Based on an Instrument Container, and the forthcoming In-Service Aircraft for a Global Observing System) that have produced large amounts of climatologically relevant data on atmospheric composition and properties during many long-distance flights.

Improving access to these powerful research platforms and providing professional support and user training are objectives of leading international research institutions and agencies and they have been supported by European Facility For Airborne Research (EUFAR), in particular.

The design, integration, and operation of *in situ* and remote instrumentation on aircraft platforms require a number of special considerations to achieve desired performance. The first is the rapid motion of an aircraft through the atmosphere, which in contrast to other airborne platforms, such as balloons, is required for continuous aerodynamic lift. The motion changes the pressure, temperature, and flow fields near the aircraft surfaces that generally contain air sampling inlets and other openings. *In situ* instruments, in particular, require inlets and suitable sampling strategies. An important example is the measurement of ambient air temperature using a probe that involves contact of a temperature sensor with ambient air. In this case, the deceleration (relative to the aircraft reference frame) of gas molecules from the speed of the aircraft to stagnation conditions in the probe induces a compression heating of the order of 4 K (at 100 m/s) to 25 K (at 230 m/s). Accurate temperature measurements require careful probe calibrations and the availability of accurate airspeed and ambient pressure measurements. In clouds, phase changes reduce the temperature changes and cause strong humidity changes. Short-range, remote (e.g., infrared temperature) measurements offer an option that avoids airspeed effects.

Many aerosol instruments have been successfully designed to mount inside or outside the aircraft fuselages or on the underside of the wings. In aerosol sampling, particles below a certain size follow the flow lines around the curved fuselage and wing surfaces and make sampling inlets straightforward. Larger particles cross the flow lines due to their greater inertia, thereby complicating or simplifying sampling strategies depending on the objective. Particles approach sampling inlets with the airspeed of the aircraft. In clouds, droplets and ice particles affect inlet probe surfaces, and larger particles may shatter into many smaller ones. This effect may invalidate the intended measurements if the additional particles cannot be properly accounted for. New inlet designs have minimized these shattering effects.

Many gas and aerosol instruments are too large to be installed outside the aircraft. Therefore, an atmospheric sample must be continuously transported into the pressurized cabin or other payload area. As some gases react with or are absorbed on the walls of the sample inlet lines, special inlet materials or fast sample flow rates must be used to acquire representative samples. Similarly, aerosol particles can be lost by turbulent and inertial deposition in the inlet lines. Specialized inlet systems have been designed and implemented to provide ambient air, from outside the possibly polluted aircraft boundary layer, to a suite of sampling instruments inside the aircraft, while minimizing the loss of particles on the inlet walls.

Another consideration in the design of airborne instruments is accommodation of environmental conditions that sometimes rapidly change. For instance, some instruments are located in unpressurized payload areas and exposed to ambient pressure and temperature conditions during flight. In some cases, in descent from high altitudes and low temperatures, instruments experience rapid changes in temperature or humidity, which may lead to condensation of water vapor on optical, electronic, and other components. Pressure and temperature problems are typically avoided by the use of pressurized enclosures for critical components and heaters that control temperatures throughout a flight. In many *in situ* sampling instruments, special provisions are required to maintain ambient sample flows or other parameters constant in response to changes in ambient pressure between the ground and cruise altitudes. Inlet systems and instrument sampling volumes are typically sealed to avoid contamination from cabin air.

Another important consideration is aircraft turbulence encountered both in clear air and in convective cloud systems and lightning strikes. In turbulence, aircraft instruments (and crew members) are exposed to rapid and often large accelerations in all three dimensions. Vibration from turbulence or engine operation is also a concern. With the application of good materials and structural engineering principles in the design and construction phases, instruments are generally able to maintain high measurement quality under these conditions. Lightning strikes are a physical threat when sampling near convective cloud systems. These systems are of significant scientific interest because of their chemical and dynamical properties. Although lightning strikes often cause minor structural damage and can be unnerving to the crew and passengers, the aircraft systems and instrument payload are generally unaffected. After a strike, flight directors often end the scientific portion of a flight in the interests of safety and direct the aircraft to return to base for inspection.

For many measurements, precise geographical positions and three-axis orientations are needed along the flight track. Examples are measurements of the wind vector, of upward and downward irradiances, and all types of active or passive remote sensing. Accurate information on aircraft position, velocity, and translational and rotational accelerations can be provided accurately by advanced inertial systems at high frequencies (up to 100 Hz) and by global positioning systems at frequencies up to 1 Hz.

Airborne measurements require careful consideration of aviation safety. In recent years, the effort required for aircraft and instrument safety certification has increased. Aviation safety and airworthiness certification regulations have a significant impact on the development of airborne instrumentation and the planning and execution of field experiments. Like other structural components of the aircraft, airborne instruments must withstand extreme accelerations as in the case of severe turbulence or unexpected airframe loads. This requires special attention to allow installation of a comprehensive measurement system, especially in small aircraft. Research instruments may contain radioactive, explosive, flammable, toxic, or chemically active constituents that carry additional safety and regulatory requirements. Moreover, instruments mounted outside the aircraft must be able to withstand bird strikes, icing, and lightning. Less constraining, but still crucial for aircraft integration, are instrument weight, volume, and electrical power consumption. Limits for total payload weight and payload center of gravity are necessary considerations, which are often a challenge when the payload contains a number of large and heavy instruments.

An important part of carrying out airborne measurements is campaign planning. Planning includes identification of scientific objectives and key scientific questions, site selection, aircraft preparation, and preparation of instrumentation, flight templates, time lines, and the on-site decision process. Planning activities and strategies, which are usually not well represented in the scientific literature, are generally undertaken by the scientific leaders of a campaign. It is important to recognize that planning may extend many months before a campaign and that a high-quality planning effort greatly increases the likelihood of a successful campaign.

During or shortly after campaign science flights, preliminary "quick look" results and analyses often become available on board or on the ground. These analyses can be used for "in flight" modification of flight objectives or planning for subsequent flights. Quality data processing often requires postflight instrument calibration or corrections that depend on aircraft state parameters or other variables and may take months of work effort. Intercomparison flights of two or more instrumented aircraft operating in the same or equivalent air masses, sometimes wing by wing, have been found to be important for data quality checks and instrument improvements in many campaigns. Finally, instrument data sets generally must be made available to other investigators and the public following a campaign and archived in data banks. These data sets, alone or in combination with model and other observational results, provide the basis for subsequently addressing the scientific objectives and key scientific questions of a campaign in the scientific literature.

The preface introduces the objectives and chapters of this book, which address issues specific to airborne measurements. The book serves, in part, as a handbook to guide engineers and researchers involved in airborne research in the integration of airborne instrumentation, its operation in flight, and processing of acquired data. It also provides recommendations for the development of novel instrumentation and examples of successful projects to help researchers in the design of future flight campaigns. The substantial success of instrumentation on board aircraft platforms in the past decades suggests that instrumented aircraft will continue to play an important role in meeting the ongoing challenge of understanding the processes in our complex Earth system.

2
Measurement of Aircraft State and Thermodynamic and Dynamic Variables

Jens Bange, Marco Esposito, Donald H. Lenschow, Philip R. A. Brown, Volker Dreiling, Andreas Giez, Larry Mahrt, Szymon P. Malinowski, Alfred R. Rodi, Raymond A. Shaw, Holger Siebert, Herman Smit, and Martin Zöger

2.1
Introduction

Insofar as the atmosphere is part of a giant heat engine, the most fundamental variables that must be quantified are those describing its thermodynamic state and the air motions (wind). Therefore, this chapter focuses on describing methods for measuring basic thermodynamic and dynamic variables of the atmosphere, including aspects and calibration strategies that are unique to performing such measurements from airborne platforms. However, in order to be able to analyze airborne thermodynamic and dynamic measurements, aircraft motion and attitude have to be measured as well, both for the purpose of placing measurements in an Earth coordinate system and for making corrections that depend on those factors. Therefore, this chapter starts by describing techniques to measure these aircraft state parameters.

The chapter begins with some historical context (Section 2.2), immediately followed by a description of methods for measuring the motion, position, and attitude of the airborne measurement platform itself (Section 2.3). The structure of the remainder of the chapter is organized with the following train of logic: scalar properties of the atmosphere are dealt with first, followed by vector properties, and finally, the two properties are combined in the discussion of flux measurements. The scalar properties that are of primary relevance to the thermodynamic state of the atmosphere are static air pressure (Section 2.4), atmospheric temperature (Section 2.5), and water vapor (Section 2.6). Water vapor is one of several trace gases of atmospheric relevance, but it is particularly highlighted here because of its profoundly important coupling to the atmospheric thermodynamic state (e.g., through latent heating/cooling and through infrared (IR) absorption) and the fact that water is common in the atmosphere in all three phases (gaseous, liquid, and solid, i.e., ice), with respective phase transitions. Water vapor by itself could be the subject of its own chapter, but we have chosen to keep it in the context of the other thermodynamic and dynamic variables, for example, temperature, that

Airborne Measurements for Environmental Research: Methods and Instruments, First Edition.
Edited by Manfred Wendisch and Jean-Louis Brenguier.

combine to give critical thermodynamic variables such as relative humidity and supersaturation. The treatment of the dynamic motions of the atmosphere is divided into measurement of the large-scale, three-dimensional wind vector (Section 2.7) and the measurement of smaller-scale turbulent motions (Section 2.8). The chapter culminates with a treatment of flux measurements (Section 2.9), which ultimately are responsible for the changing state of the atmosphere itself.

2.2 Historical

The history of airborne measurements for atmospheric research can be traced back to free air balloon sounding of the atmosphere. The first meteorological ascent was reported by the French physicist, Jacques Charles, on 1 December 1783 in a hydrogen balloon equipped with a barometer and a thermometer. He recorded a decrease in temperature with height and estimated the atmospheric lapse rate. Joseph Louis Gay-Lussac and Jean-Baptiste Biot made a hot-air balloon ascent in 1804 to a height of 6.4 km in an early investigation of the Earth's atmosphere and measured temperature and moisture at different heights. They reported that the composition of the atmosphere does not change with decreasing pressure (increasing altitude). Manned balloons continued to be used throughout the next couple of centuries with the obvious advantage of being able to follow an air mass and thus allowing very detailed measurements in a small volume of air, but with the disadvantage of limited sampling statistics. In the early 1930s, Heinz Lettau and Werner Schwerdtfeger made direct measurements of vertical wind velocity in the lowest 4 km of the troposphere from a balloon using a combination of a rate-of-climb meter to keep the balloon height constant and a sensitive anemometer to measure the vertical air velocity relative to the balloon. They estimated that the accuracy of their technique was better than $0.2\ m\ s^{-1}$ (Lewis, 1997).

The use of powered aircraft for airborne measurements of atmospheric parameters goes back to at least 1911 when in Germany, Richard Assmann, the inventor of the aspirated psychrometer, motivated the aircraft designer, August Euler, to modify one of his aircraft to make upper-air soundings. The following year a meteorograph was installed in an Euler monoplane and it recorded pressure and temperature up to 1100 m altitude. Aircraft continued to be used for temperature soundings, in some cases on a daily basis, from the 1920s through the World War II. These measurements played a role in the major advances that occurred in synoptic meteorology during these years. Eventually, their routine sounding role diminished as pilot balloons and radiosondes became the standard tools for atmospheric sounding.

Thermodynamic and turbulence measurements were performed in 1936 with a Potez 540 aircraft from the French Air Force in the Puy de Sancy Mountain area (Dupont, 1938). The aircraft was equipped with an "anémoclinomètre" for the airspeed and attack and drift angles measurements, an accelerometer with three piezoelectrical channels for the vertical acceleration component, and a "météograph" for the pressure, temperature, and hygrometry measurements. Several flights were performed over the National Glider School Center to characterize turbulence and dynamic properties over the mountain site.

The use of aircraft for intensive research programs continued to expand. For example, a series of temperature and humidity soundings from aircraft in the lowest 300 m over the ocean in the fall of 1944 was used to study modification of stably stratified air along its trajectory as it passed from land to a relatively cold ocean offshore of Massachusetts, USA (Craig, 1949).

Turbulence measurements from aircraft date back to at least the early 1950s when a US Navy PBY-6A instrumented with a vertical accelerometer was used by Joanne Malkus and Andrew Bunker to estimate a "turbulence index" for cloud dynamics observations (Malkus, 1954). Later, an anemometer was combined with the vertical acceleration measurements to estimate vertical and longitudinal air velocity fluctuations, and thus to calculate vertical momentum flux (Bunker, 1955).

In the mid-1950s, a more complete turbulence measuring system was used on a McDonnell FH-1 (the first all-jet aircraft) to measure vertical velocity spectra in the planetary boundary layer. This system used either a rotating vane or a differential pressure probe mounted on a nose boom to measure the aircraft attack angle, an integrating accelerometer to measure aircraft velocity fluctuations relative to the Earth, and an integrating rate gyroscope to measure pitch angle fluctuations. By combining these measurements, fluctuations of vertical wind velocity were estimated (Lappe and Davidson, 1963).

A different approach to measuring turbulence intensity was used by MacCready (1964) starting in the early 1960s, who disregarded the long wavelength contributions to the longitudinal air velocity fluctuations by band-pass filtering the output of an airspeed sensor to estimate the turbulence dissipation from the Kolmogorov hypothesis. This provided a simple easily implemented system to provide a standardized measure of turbulence, albeit over a limited wavelength region, as well as a measure of the total turbulence energy production by equating it to the turbulence dissipation.

The next step in improving the complexity and accuracy for vertical wind velocity measurements was taken in the early 1960s in Australia, with the development of a system on a Douglas DC-3 by Telford and Warner (1962). They combined a nose-boom-mounted vane with a free gyroscope and a vertically stabilized (using signals from the free gyroscope) accelerometer. This reduced errors present in previous systems due to the varying contribution of gravity to the measured acceleration resulting from attitude angle variations. They also incorporated a fast temperature sensor and wet-bulb thermometer to measure heat and water vapor fluxes.

Afterward, an inertial navigation system (INS) was integrated, with improved accuracy and reduced drift rates, to measure the translational and rotational aircraft motions, as well as the absolute location of the aircraft. Today, GPS-based instruments are also utilized in combination with Inertial Measuring Units (IMUs) to provide a lighter and less expensive alternative to INS. In contrast, the air motion sensing systems have changed little in the past few decades and are now the limiting factor in measuring air motion.

At present, there is a remarkable variety of instrumented airborne platforms for atmospheric and environmental measurements, including high-performance jet aircraft for high-altitude and long-range measurements, smaller turboprop aircraft for intensive boundary layer measurements, armored aircraft for thunderstorm penetration, slow-moving helicopter-towed platforms for high-resolution measurements, and an emerging fleet of relatively small, remotely piloted vehicles carrying miniaturized but still highly capable instrument packages. Indeed, the airborne platforms are as varied and innovative as the instruments they carry, all matched to the specialized research objectives that drive the continuing innovation.

2.3
Aircraft State Variables

In order to place measurements into a proper geographical reference frame it is necessary to precisely measure the position and attitude of the aircraft from which measurements are made. These variables, including aircraft height or altitude, attitude (e.g., yaw, pitch, roll angles), position, and velocity, are collectively defined as the aircraft state.

2.3.1
Barometric Measurement of Aircraft Height

Hypsometric (or pressure) altitude can be estimated by an integration of the hydrostatic equation using measurements of virtual temperature T_{vir} and static air pressure p, assuming that the sounding is invariant as follows:

$$z - z_0 = -\int_{p_0}^{p} \frac{R_{dry} \cdot T_{vir}}{g} \, \mathrm{d} \ln p \qquad (2.1)$$

where $R_{dry} = 287.05\,\mathrm{J\,kg^{-1}\,K^{-1}}$ is the specific gas constant of dry air. The gravitational acceleration g varies with height z and location (geographic latitude). Errors result from horizontal temperature gradients, and also when nonhydrostatic conditions exist, for example, in strong atmospheric motions. Further errors are introduced when neglecting the effect of humidity and vertical variation of the gravitational acceleration g.

Alternatively, standard atmosphere models can be used to estimate the temperature from the pressure, which can then be integrated to obtain pressure altitude. The International Standard Atmosphere (ISA) sets the international standard (ISO, 1975). Below 30 km altitude, the ISA model is identical to that of the International Civil Aviation Organization (ICAO) and the US Standard Atmosphere, with variables as shown in Table 2.1. These standard atmospheres assume dry atmospheric conditions.

Table 2.1 ISA standard atmosphere properties (base values) in the troposphere and stratosphere.

Layer	Geopotential height h_0 (gpkm)	Geometric height z_0 (km)	Lapse Rate γ_0 (°C gpkm^{-1})	Temperature T_0 (°C)	Pressure p_0 (Pa)
0	0	0.0	−6.5	+15.0	101 325
1	11	11.019	+0.0	−56.5	22 632
2	20	20.063	+1.0	−56.5	5 474.9
3	32	32.162	+2.8	−44.5	868.02

Variation in the value of gravitational acceleration g is small. To account for this, instead of the geometric altitude, atmospheric models use geopotential height measured in geopotential meters (gpm), defined as

$$h - h_0 = \frac{1}{g_n} \int_{z_0}^{z} g(z')\, dz' = -R_{dry} \cdot \int_{p_0}^{p} T_{vir}\, d\ln p \tag{2.2}$$

where the subscript 0 refers to a reference state for each atmospheric layer as defined in Table 2.1, and $g_n = 9.80665$. After integration for each layer with constant lapse rate γ_0 (also given in Table 2.1), it can be shown that

$$p = p_0 \cdot \left(1 - \frac{\gamma_0 \cdot h}{T_0}\right)^{g_n/(R_{dry} \cdot \gamma_0)} \tag{2.3}$$

and

$$h = h_0 + \frac{T_0}{\gamma} \cdot \left[1 - \left(\frac{p}{p_0}\right)^{R_{dry} \cdot \gamma_0 / g_n}\right] \tag{2.4}$$

see Iribarne and Godson (1981). For the dry, tropospheric layer, using the ISA constants, we obtain

$$h = 44\,331 \cdot \left[1 - \left(\frac{p}{p_0}\right)^{0.19026}\right] \tag{2.5}$$

where h is obtained in gpm. In Eq. (2.5), p_0 is 1013.25 hPa, corresponding to the lowest atmospheric layer in the ISA (Table 2.1). An aircraft pressure altimeter in this lowest atmospheric layer indicates the ISA altitude when the altimeter setting is 1013.25 hPa. Typically, the altimeter setting is adjusted so that the altimeter reads exactly the airport altitude on landing. The details of how altimeter setting is mechanized in an aircraft pressure altimeter can be found in Iribarne and Godson (1981).

Both the hypsometric altitude from Eq. (2.1) and the pressure altitude from Eq. (2.5) assume that there are no horizontal pressure gradients. Height measurements based on RADAR are not covered here. The sum of RADAR altitude plus the height of the terrain above sea level approximates hypsometric or pressure altitude measurements, but accurate terrain data is not available at very fine scale, and surface artifacts such as buildings can complicate that determination except, of course, over the sea. Neither of these altitude estimates is as inherently accurate as those from the Global Navigation Satellite System, as described in Section 2.3.3. For use in comparing airborne measurements with atmospheric model output, pressure or potential temperature could be less ambiguous measures of height.

2.3.2
Inertial Attitude, Velocity, and Position

2.3.2.1 System Concepts

IMUs using Newton's laws, applied to motion on a rotating planet, integrate a triad of linear accelerations to determine aircraft velocity and position. Detailed theory of operation and design criteria for navigation units based on IMUs are presented in the studies by Broxmeyer (1964) and O'Donnell (1964). The accelerometer orientation must be known to accommodate accelerations due to gravity, and this is accomplished by mounting on a stable platform.

Two main approaches are in general use: gimballed and strapdown systems. The gimballed system is typically mechanized to keep the stabilized platform containing the accelerometers level with respect to the Earths gravity, rotating as necessary to maintain verticality as the aircraft moves, incorporating the effect of changes in the gravity vector as the aircraft changes latitude and altitude. The gimbals are a set of three rings that let the platform keep the same orientation while the vehicle rotates around it. Attitude angles can then be measured directly from the gimbal orientation. The big disadvantage of this approach is the relatively high cost and mechanical complexity causing reliability challenges related to the many precision mechanical parts. The coordinate transformation between Earth-fixed and aircraft body axis systems is described by Axford (1968) and Lenschow (1972). Figure 2.1 (Axford, 1968) shows the arrangement of a gimballed system and defines the coordinate transformation variables for its use.

In strapdown systems, which comprise most of the IMUs used in atmospheric research at present, accelerometers are fixed to the aircraft, and linear and angular acceleration measurements are integrated using a model to continually compute the orientation of gravity to the vehicle axis, creating a virtual stabilized platform. Compared to the gimballed systems, strapdown systems offer lower cost and higher reliability but require higher maximum angular rate capability and higher sampling rate capability to sufficiently capture aircraft motion on a maneuvering aircraft (Barbour, 2010). IMUs integrated into an INS with a gyroscope error of $0.01^\circ\,\mathrm{h}^{-1}$ will result in a navigation error of $\sim 2\,\mathrm{km\,h}^{-1}$ of operation.

2.3.2.2 Attitude Angle Definitions

Standard definitions for the attitude angles and other motion variables can be found in ISO 1151-1 (ISO, 1985) and ISO 1151-2 (ISO, 1988). A conventional INS defines the Earth-based coordinate system to be north-east-down (NED) and the aircraft body axis system to be forward-right-down (XYZ). The transformation matrix from the body axis XYZ to the Earth-based NED system has three successive

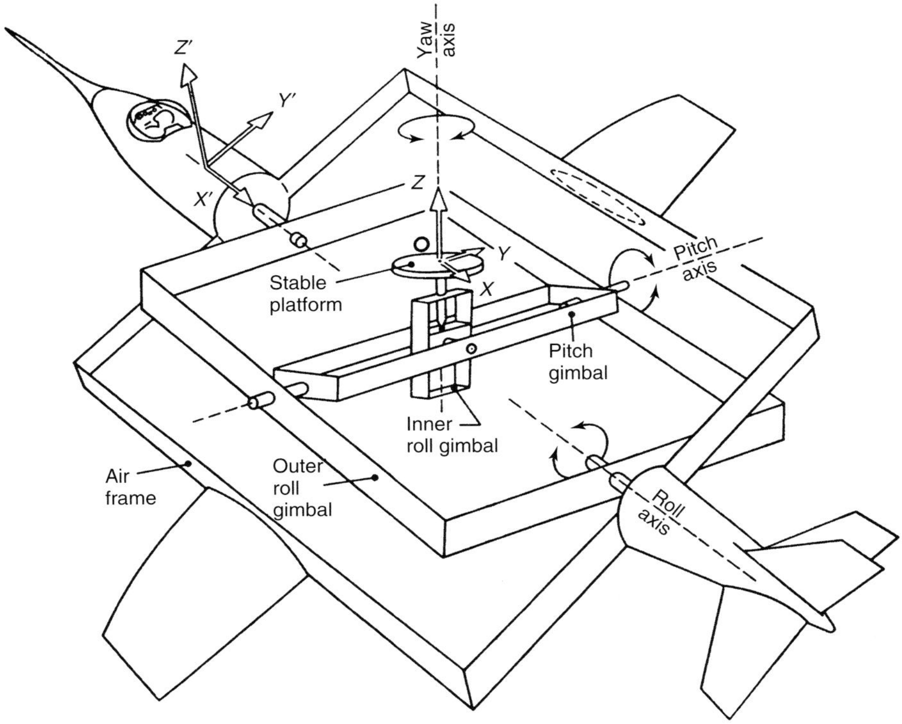

Figure 2.1 Sketch of gimbal system. (Source: Redrawn from Axford (1968). Copyright 1968 American Meteorological Society. Reprinted with permission.)

rotations that are prescribed by the order of the gimbals (roll innermost). For strapdown systems, the equations are written to emulate this gimbal order that defines the attitude angles using the Tait-Bryan sequence of rotations: (i) rotate to wings horizontal around body X (forward)-axis by roll angle (ϕ, right wing down positive); (ii) rotate to X-axis horizontal about body Y (right)-axis by pitch angle (θ, nose up positive); and (iii) rotate about Z (down)-axis to north by heading (ψ, true heading, positive from north toward east). Transforming a vector in the XYZ body axis to Earth-based NED coordinates requires the roll **R**, pitch **P**, and heading **H** rotation matrices

$$\begin{aligned}
\mathbf{H} &= \begin{pmatrix} \cos\psi & -\sin\psi & 0 \\ \sin\psi & \cos\psi & 0 \\ 0 & 0 & 1 \end{pmatrix}, \\
\mathbf{P} &= \begin{pmatrix} \cos\theta & 0 & \sin\theta \\ 0 & 1 & 0 \\ -\sin\theta & 0 & \cos\theta \end{pmatrix}, \\
\mathbf{R} &= \begin{pmatrix} 1 & 0 & 0 \\ 0 & \cos\phi & -\sin\phi \\ 0 & \sin\phi & \cos\phi \end{pmatrix}
\end{aligned} \tag{2.6}$$

Table 2.2 Performance of classes of unaided INS.

Class	Position performance	Gyro technology	Accelerometer technology	Gyro bias	Acceleration bias
Military grade	1 nmi $(24\,h)^{-1}$	ESG, RLG FOG	Servo Accelerometer	$<0.005\,^\circ\,(h^{-1})$	30 μg
Navigation grade	1 nmi h^{-1}	RLG FOG	Servo Accelerometer Vibrating Beam	$0.01^\circ\,(h^{-1})$	50 μg
Tactical grade	>10 nmi h^{-1}	RLG FOG	Servo Accelerometer Vibrating beam MEMS	$1^\circ\,(h^{-1})$	1 mg
AHRS		MEMS, RLG FOG, Coriolis	MEMS	$1\text{–}10^\circ\,(h^{-1})$	1 mg
Control system		Coriolis	MEMS	$10\text{–}1000^\circ\,(h^{-1})$	10 mg

to be applied to the vector in the following order:

$$\begin{aligned} \mathbf{T}_N^B &= \mathbf{H}\,(\mathbf{P}\,\mathbf{R}), \\ \begin{pmatrix} N \\ E \\ D \end{pmatrix} &= \mathbf{T}_N^B \begin{pmatrix} X \\ Y \\ Z \end{pmatrix} \end{aligned} \tag{2.7}$$

2.3.2.3 Gyroscopes and Accelerometers

Doebelin (1990) presents an overview of linear accelerometers and 3D gyroscopic angular displacement and angular velocity (rate) sensors. Barbour (2010) and Schmidt and Phillips (2010); Schmidt (2010) survey current inertial sensor issues and trends. Spinning electrically suspended gyroscopes (ESGs) offer the highest accuracy and stability, with the rotor supported in vacuum by an electric field, thus nearly eliminating errors caused by friction. Currently, there is an upsurge in solid-state sensors that include microelectromechanical systems (MEMS) devices, ring laser gyros (RLGs), fiber-optic gyros (FOGs), and interferometric gyros (IFOGs), which have significant cost, size, and weight advantages over spinning devices. Accelerometers are pendulous servo accelerometers, resonant vibrating beam accelerometers (VBAs), or MEMS implementations of either of these.

Table 2.2 indicates the gyro bias and accelerometer bias requirements of each class of application. In an unaided INS, initial alignment must be accomplished carefully so that the initial tilt of the system does not put a component of gravity into the horizontal accelerometers. Alignment is accomplished by tilting to zero the horizontal acceleration to establish level, and the initial heading is accomplished by establishing north by alignment with the Earth's rotation rate ($0.002\,^\circ\,s^{-1}$).

Table 2.3 shows the expected uncertainties from unaided navigation-grade INS.

Table 2.3 Accuracy of unaided navigation-grade INS (Honeywell LaserRef2 SM after 6 h).

Variable	Accuracy
Position	$1.5\,\mathrm{km\,h^{-1}}$
Ground velocity	$4.10\,\mathrm{m\,s^{-1}}$
Vertical velocity	$0.15\,\mathrm{m\,s^{-1}}$ (baro-damped)
Pitch and roll angles	0.05°
True heading	0.2°

Source: From Honeywell (1988).

2.3.2.4 **Inertial-Barometric Corrections**

Unaided INS does not have sufficient information available to damp errors in the Earth-vertical coordinate. Barometric pressure can be used to limit errors in the vertical acceleration that cause unbounded drift. A third-order baro-inertial loop described by Blanchard (1971) can be used for this. Lenschow (1986) discusses the considerations for choosing the time constant for the mechanization of the loop, being a trade-off among minimizing the effect of high-frequency noise in the pressure measurement, minimizing the recovery time from errors, and improving long-term stability. A time constant of 60 s has been used for the National Center for Atmospheric Research (NCAR) aircraft.

2.3.3 **Satellite Navigation by Global Navigation Satellite Systems**

Global Navigation Satellite Systems (GNSS) are constellations of satellites in medium Earth orbit at heights of about 2.5×10^7 m, corresponding to an orbital period of roughly 12 s. Gleason and Gebre-Egziabher (2009) provide detailed information about GNSS methodology and expected errors. Table 2.4 lists the status of GNSS as of 2010. Receivers compatible with multiple constellations benefit from the larger number of satellites in view.

Each satellite vehicle (SV) broadcasts a precise time measurement along with its ephemeris. GPS receivers use this to determine the transit time of the signal, which is then converted to distance (called *pseudorange*). Satellite positions can be obtained from either the broadcast ephemeris or the more accurate ephemeris published within hours or days, which can be incorporated into postprocessing of the GNSS signals. Four (or more) pseudorange measurements are used to unambiguously compute the receiver position using triangulation. Adding more SV signals increases the accuracy of the position estimate.

2.3.3.1 **GNSS Signals**

The details of the coding and decoding of GNSS signals are discussed in the texts of Gleason and Gebre-Egziabher (2009); Bevly and Cobb (2010); Hofmann-Wellenhof,

Table 2.4 Overview of operational and planned Global Navigation Satellite Systems.

System	Country	Number of satellites	Frequencies	Status
GPS	USA	24	1.57542 GHz (L1 signal) 1.2276 GHz (L2 signal)	Operational
GLONASS[a]	Russia	21	Around 1.602 GHz (SP) Around 1.246 GHz (SP)	Operational, six satellites in maintenance
Galileo[b]	EU	30	1.164–1.215 GHz (E5a and E5b) 1.215–1.300 GHz (E6) 1.559–1.592 GHz (E2–L1–E11)	Operational in 2014
COMPASS[c]	China	30	B1: 1.561098 GHz B1–2: 1.589742 GHz B2: 1.20714 GHz B3: 1.26852 GHz	Operational in 2015

[a]GLONASS stems from the Russian words Globalnaja Navigaziona Systema, which means in English, the Global Navigation System.
[b]Galileo is not an acronym, the system is named after the famous Italian astronomer Galileo Galilei (1564–1642).
[c]COMPASS means BeiDou (Compass) Navigation Satellite System. Compass is the translation from Chinese

Lichtenegger, and Collins (2001). The US GPS provides two precision positioning signals (military-accessible P-code) on frequencies L1 (1575.42 MHz) and L2 (1227.6 MHz) and a clear acquisition signal (C/A code) on L1. A third frequency L5 (1176.5 MHz) was added in 2009. Selective availability (SA) – the intentional addition of time varying errors of up to 100 m (328 ft) to the publicly available navigation signals – was discontinued in 2001.

In addition to the pseudorange-only triangulation techniques, carrier phase (CP) tracking on multiple frequencies can produce centimeter accuracies. CP techniques were developed for precise surveying and geodesy (including surface motions presaging earthquakes, continental drift), but they have also been recently successfully applied to moving vehicles such as aircraft. Each cycle of the L1 and L2 carrier frequency is about 19 cm long, and phase can be measured to better than 1% so that millimeter accuracy can be obtained under optimal conditions.

2.3.3.2 Differential GNSS

Differential GNSS (DGNSS), commonly termed *differential GPS (DGPS)*, is a means of removing almost all navigation errors (discussed later). GNSS receivers are installed at strategically placed ground locations. Several networks of these ground stations exist. Using the precisely known location of the ground-based stations, the simultaneous measurements of the SV location can be used to solve for satellite clock and ephemeris errors, and ionosphere and troposphere delays (see below). When reference stations are closely located, errors are highly correlated, and this assumption is used to solve the navigation equations simultaneously to estimate the errors in the moving platform. Also, wide area corrections can be made by estimating corrections over an extended area. Oceanic areas void of stations are an obstacle to these approaches.

2.3.3.3 Position Errors and Accuracy of Satellite Navigation

The true location of the satellite is known very accurately because the orbits of the GPS satellites are precisely determined by continuous error checking and computation. The broadcast ephemerides have 1–2 m accuracy; however, postprocessing centers provide ± 15 cm (available every 2 h), ± 5 cm (next day), and $\pm$ 2 cm (about two weeks) accuracies.

The speed of the GPS signals varies as they pass through the Earth's atmosphere, with the ionosphere having the largest effect. Consequently, the errors become greater for longer paths through the atmosphere, and the geometry of the triangulation of the receiver location then needs to be considered in estimating errors.

Ionospheric delay (dispersion) effects depend on the total electron content (TEC) in the path at altitudes of 50–100 km and can be very large. Ionospheric models, such as the one by Klobuchar (1996), can provide useful estimates of the effect. Measurement of delays for two or more frequency bands (L1 and L2 CP for the GPS) allows a more precise correction, and under optimal conditions, "ionosphere-free" solutions can be obtained.

Propagation errors caused by refractivity variations in the troposphere, which result primarily from variations in temperature and water vapor mixing ratio, are smaller than those caused by the ionosphere. Atmospheric models can be used here also.

GPS signals can reflect from external reflectors near the receiver antenna, and mixing the direct and reflected signals distorts the received signal tracking. Multipath effects are much less severe in aircraft at altitudes above reflecting surfaces, but low-level flights near buildings and/or topography can result in errors. For very precise positioning, these effects can be mitigated or eliminated using DGPS.

When visible GPS satellites are close together in the sky (i.e., small angular separation), the navigation solution becomes less precise. A figure of merit termed *dilution of precision* (*DOP*) provides a measure of how errors propagate through the geometry of multiple satellite position determination from the pseudorange measurements. The components of DOP-HDOP, VDOP, PDOP, and TDOP give horizontal, vertical, 3D position, and time dilutions, respectively. Hofmann-Wellenhof, Lichtenegger, and Collins (2001) provide a complete description of the DOP determination. DOP values of 1 are optimal, with increasing DOPs indicating increasing degradation of the solution, while DOPs of 6 are considered unacceptable.

2.3.4
Integrated IMU/GNSS Systems for Position and Attitude Determination

INS systems with GNSS aiding to minimize errors have been available since the 1970s. Recent advances in the GNSS technology and real-time and postprocessing software have provided combined systems with smaller than 1 m real-time error. Kalman filtering is widely used to merge the continuous IMU sensor data with the intermittent GNSS data (Brown and Hwang, 1997). Multiple GNSS antenna configurations are available that increase the accuracy of heading, but the three-antenna configuration can also be used to solve for the three attitude angles. DGPS error correction is available in real-time through terrestrial or satellite-based communication to the aircraft.

The biggest issue with the integrated IMU/GNSS systems is performance during periods of GNSS denial, that is, when satellite coverage is poor or unavailable, or when the number of satellites in view drops, as frequently occurs during aircraft maneuvers resulting in very poor DOP. Tactical-grade IMU technology has been shown to provide adequate stability during those periods.

2.3.5
Summary, Gaps, Emerging Technologies

In just a short period, the integrated IMU/GNSS technology has bridged the gap between relatively inaccurate navigation aids and expensive navigation-grade INS, providing small low-cost systems for attitude, position, and velocity determination. The anticipation is that this trend will continue with advances in FOG and MEMS gyroscope technology to reduce size and cost even further. The addition of the Galileo and the COMPASS constellations will bring further advantages in terms of the number of satellites in view. Additional frequencies on the GNSS are coming online, which will allow further reduction of GNSS propagation errors to make 1 cm accuracy routine on a maneuvering aircraft.

2.4
Static Air Pressure

A knowledge of static air pressure is critical for a wide range of atmospheric applications. It is needed to understand atmospheric measurements onboard an aircraft, it is used as a vertical coordinate, and it is an important parameter for physical and chemical atmospheric processes. Measurements of static air pressure onboard an aircraft also provide information on the actual flight altitude. This parameter influences the calculation of aircraft speed and other performance data as well as the cabin pressure regulation. Historically, with the increasing speed of aircraft, distortion of flow by the fuselage became more important. In the 1950s, Gracey (1956) published a fundamental work dealing with the shape of fuselage, shape of probes, and influence of speed on the measurement of static air pressure at various positions on aircraft. Today, we use models to calculate air stream around arbitrarily shaped bodies, but the results of this paper still give profound insight into this subject.

The concepts of measuring static air pressure have not changed much since then. Along the fuselage, a number of positions can be found where static pressure is found to be close to $p_{s\infty}$, the undisturbed static air pressure at the same altitude. Virtually, all the positions show cross-sensitivity to maneuvers. So tools such as booms in different positions are used to shift static air pressure probes out of the influence of the aircraft.

Instead of static air pressure, generally, some expression of flight altitude is given in avionic data sets. But the most useful coordinate for vertical position within the atmosphere is static air pressure, which is an expression for the weight of the atmosphere above the actual flight level. This is expressed by the hydrostatic equation.

$$\frac{\mathrm{d}p}{\mathrm{d}z} = -g \cdot \rho \tag{2.8}$$

where p is the air pressure (in units of Pa) at a given altitude z (given in units of m), g is the acceleration due to Earth gravity (gravitational acceleration) in units of $\mathrm{m\,s^{-2}}$, and ρ is the density of air (in $\mathrm{kg\,m^{-3}}$). Using the relation for an ideal gas,

$$\frac{p}{\rho} = R_{\mathrm{dry}} \cdot T \tag{2.9}$$

where $R_{\mathrm{dry}} = 287.05\,\mathrm{J\,kg^{-1}\,K^{-1}}$ is the individual gas constant for (dry) air and T is the air temperature in Kelvin, we have the differential form of the hydrostatic equation:

$$\frac{\mathrm{d}p}{p} = -\frac{g}{R_{\mathrm{dry}}} \cdot \frac{\mathrm{d}z}{T} \tag{2.10}$$

For a given local position, air pressure is steadily decreasing for growing altitude z. Although air temperature is decreasing with a gradient of $-0.0065\ \mathrm{K\,m^{-1}}$ as a global mean for tropospheric standard day conditions, the local deviation may be large. To express the air temperature for a given atmospheric layer $\Delta z = z_2 - z_1$, the mean temperature $\overline{T}$ is defined by

$$\overline{T} = \Delta z \cdot \left(\int_{z_1}^{z_2} \frac{\mathrm{d}z}{T} \right)^{-1} \tag{2.11}$$

Equation (2.11) can be applied to humid air by replacing the air temperature T with the virtual temperature T_{vir}. At this point, a problem arises that is typical for aviation: we can measure all atmospheric parameters *in situ*, but we also need to know the temperature (and humidity) profile below the aircraft, which normally is not known. This limits the practical application of Eq. (2.10) or enforces assumptions on the vertical temperature distribution. The density effect of humidity, which is accounted for by the virtual temperature, plays a major role at lower flight altitudes, whereas the low temperatures of the upper troposphere allow only relatively low water vapor pressures and hence small density effects.

This uncertainty is a very important problem in aviation, so a fundamental solution had to be found. This was done by defining an atmospheric standard. Since 1976, the last revision of the US Standard Atmosphere, the standard is based on a mean sea level temperature of 288.15 K, in which the pressure at mean sea level for a "standard day" is set to 101 325 Pa. The vertical temperature gradient for troposphere is set to a mean value of $-0.0065\,\mathrm{K\,m^{-1}}$. At the mean altitude of the tropopause (defined as 11 000 m) and above, up to an altitude of 20 000 m, the temperature is set constant at 231.65 K. This is the atmospheric layer important for normal aviation, although the tabulated standards continue further up to 86 000 m.

Applying this temperature distribution, we can calculate the altitude for a standard day from a pressure measurement alone in any given flight level. In aviation, this is exactly what happened to the avionic instrument standards. An altimeter that is set to standard conditions assumes a sea level pressure of 101 325 Pa and a temperature change according to $-0.0065\,\mathrm{K\,m^{-1}}$. Then, the instrument shows the pressure altitude. The importance of this standard procedure in aviation is the reliable vertical separation of traffic in the air. Safety assessment requests a classification of the measurement to a defined accuracy.

On the other hand, the given pressure altitude may deviate hundreds of meters from geometric altitude, which is nowadays easily reported from satellite-based measurements or calculated from independent atmospheric profile data. Also, the standard setting cannot be used in the vicinity of ground-based obstacles. There is a different procedure for low altitudes, where "low" depends on regional topography as well. In case of low flight altitudes (e.g., takeoff and landing), the real ground pressure will be set as a reference. Then, the instrument shows barometric altitude, where the actual measured ground pressure (e.g., taken at an airport tower at the actual field height) is extrapolated down to sea level for a reference.

2.4.1 Position Error

The air pressure p is approximated by the static air pressure indicated by p_s in the following discussion. p_s is the measured pressure corrected for the static defect so it is the best estimate of p. Methods for reliable measurement of static air pressure on fast-moving platforms are discussed in this section, with specific attention to the dependence of pressure measurements on their position on the aircraft; these deviations from background static air pressure are denoted as position errors. Sensor elements and conditioners are not discussed in detail. As far as these parts are mounted in unpressurized bays with widely changing environmental conditions, it is at least important to make sure that the instrument's signal processing is insensitive to extreme temperatures and pressures. All pressure sensors show more or less strong sensitivity to temperature changes, which, for best results, can be treated by stabilizing the sensors' temperature within a narrow interval. Pressure measurements often use some length of tubing, so it is necessary to care for the total volume involved and to keep in mind that the pressure signal can be damped and delayed. Neglecting measurement errors

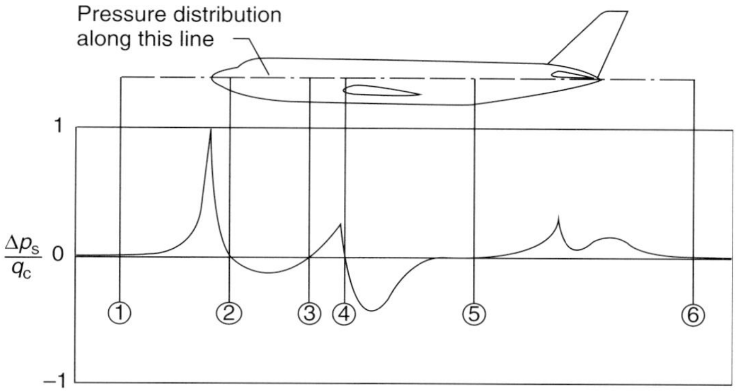

Figure 2.2 Pressure deviation around fuselage. (Source: Adapted from Haering (1995). Reprinted with permission. Courtesy of Tom Anderson.)

produced by tubing and sensor elements for now, the error of the static air pressure p_s is mainly the position error Δp_s, as defined in more detail in the next paragraph.

The fuselage of an aircraft distorts the flow field in a typical way, and the local pressure at the surface of the fuselage may deviate by some 1000 Pa from $p_{s\infty}$, the undisturbed static air pressure. Figure 2.2 (Haering, 1995) shows complex structures of positive and negative deviations from undisturbed $p_{s\infty}$, but moreover, these patterns of pressure deviation will vary with maneuvers such as turns, low or high speeds, and aircraft weight. This will result in errors for static air pressure that depend on the measurement location, called *position errors.*

For a standard aircraft, a number of different orifices along both sides of the fuselage are combined to provide a mean value over a number of sensors, which serve as an important reference. The positions of the orifices are carefully selected for each aircraft model : by modeling and testing, a manufacturer defines those locations where pressure deformation is close to zero. This holds usually for a "clean" aircraft, that is, without further devices mounted outside the fuselage or wings. Any change in the outer shape has the capability to change the quality of a pressure port.

Errors in static air pressure measurement also influence the measurement of the aircraft speed taken from a measurement of dynamic pressure q_c, which is the result of a difference of two large numbers:

$$q_c = p_t - p_{s\infty} \tag{2.12}$$

where p_t means total pressure, measured as the ram pressure of a pitot tube. An error in static air pressure will also produce an error in dynamic pressure and a correction will lead to a term of correction for speed. Although q_c can be measured directly as differential pressure between the two ports, the error can be expressed as a correcting term for the indicated static air pressure p_i.

$$\Delta p_s = p_{s\infty} - p_i \tag{2.13}$$

which leads to

$$q_c = p_t - (p_i + \Delta p_s) \tag{2.14}$$

The ram pressure p_t for speed reference is measured in a definite position in the nose area using a pitot tube, and as mentioned, the static air pressure can be measured through distributed orifices along the fuselage. It is also possible to use special tools such as a nose boom to put a flow sensor as far as possible out of the disturbed neighborhood of the fuselage.

The widely used five-hole probe (FHP) mounted at the tip of a boom reaches out into undisturbed flow fields and has better results for both dynamic ram pressure and static pressure. In addition, differential pressure measurements allow for determination of flow angles along vertical and cross axes. The blunt-tipped FHP for subsonic aircraft is a hemisphere on a cylindrical shaft. It combines a pitot port at the tip with a number of static bore holes around the shaft at a distance of about 4 shaft diameters behind the tip. Within the hemisphere, two pairs of ports for vertical (α, angle of attack) and crosswise (β, angle of sideslip) deviation from axial flow are positioned 45° off the axes for heading. The q_c signal shows some sensitivity to nonaxial flow, hitting the FHP with angles of more than 6° (de Leo and Hagen, 1976).

Some distance ahead of the nose, the influence of the fuselage on the pressure field may be very low, but still there is an effect on pressure generated by the boom itself. The deviation Δp_s at the boom tip is small but cannot be neglected. This is true for most other devices that may be used instead. An exception is the Flush Airdata Sensing (FADS) system that uses a system of many pressure ports distributed over the normal nose cap. This device needs no change in outer shape and hence produces no further influence on the pressure field. But for both versions, boom and FADS, adaptions might be required to aircraft systems, such as RADAR. Both systems require further investigation of position errors by flight testing. Then, the most important question for test flights is to find a reliable measurement for $p_{s\infty}$, that is, a pressure sensor has to be put out into the undisturbed air.

2.4.1.1 Tower Flyby

Near the ground it is possible to use stationary ground-based pressure sensors to serve as reference. In a procedure called *tower flyby*, the pressure reference of the airport tower or other instruments can be used. Here it is necessary to know the exact altitude of the aircraft relative to the ground-based instrument. This can be achieved through (i) a known flight path, for example, the centerline of the runway; (ii) a flight altitude just high enough to avoid disturbance by ground effects; (iii) some optical device to get an independent measurement of the altitude (which, e.g., can be done by photography); or (iv) tracking of flight altitude by high-precision DGPS.

The procedure should include data taken on the runway by the aircraft system. In this case, the ground-based instrument has mainly to record the pressure changes during the flights. For highest accuracy, temperature and humidity will also be recorded. The altitude of the aircraft above the ground can be obtained from a photograph (e.g., by applying a self-scaling procedure using the length of the fuselage as a known constant to scale the distance between the aircraft and the runway). A sequence of tower flybys are carried out to cover the applicable range of speeds and attitudes (e.g., flap settings). The correction is simply given by Eq. (2.13), where $p_{s\infty}$ is calculated from Eq. (2.10).

2.4.1.2 **Trailing Sonde**

The characterization of airborne static air pressure measurements needs more combinations of speed and attitude than tower flyby can cover. Test flights have to be carried out at higher altitudes (higher speeds, lower density) as well. The independent measurement of $p_{s\infty}$ can then be delivered by a calibrated chase aircraft, flying in more or less close formation (but at least at the same flight level).

Without a chase aircraft, autonomous methods can be used to install calibration sensors on board of the tested aircraft. Means to eliminate the pressure deviation around the fuselage use probes put out to a distance under or behind the aircraft (the "trailing bomb" or the "trailing cone," respectively) connected by a long tube to the fuselage. Given that the sensors are positioned at distances where aircraft disturbances are minimal, the sensors read the undisturbed pressure directly. The method of the trailing bomb is more useful for slower aircraft or Mach numbers below about 0.5. The trailing cone probe works above Mach 0.3 and up into the supersonic speed range. This probe consists of a cone of resin with 35° opening angle and some circular openings, stabilizing the tubing against rotation. The cone produces drag that is needed to keep a long tube and a static air pressure probe lifted up behind the tail of the aircraft. In general, the turbulence field behind the aircraft moves downward. At the distance of about one to two wingspans, the probe should be free of turbulence. At high air speeds, the drag of the cone keeps the tubing and probe above the turbulent wake and the measurements are very close to $p_{s\infty}$.

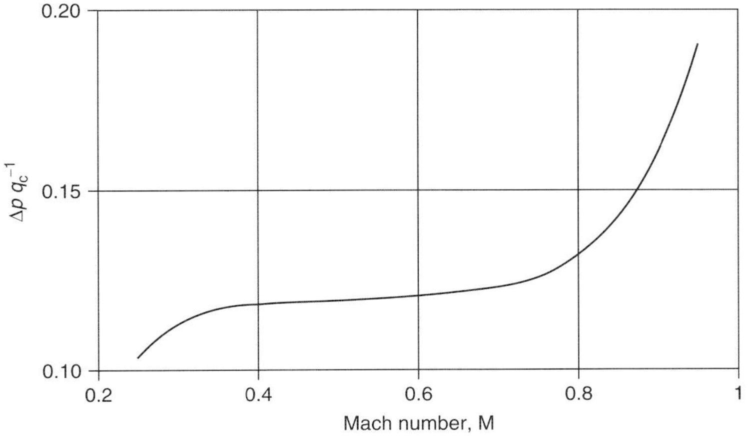

Figure 2.3 Schematic correction of static pressure by trailing cone for a jet aircraft over a wide range of Mach numbers.

Figure 2.3 shows that the correction of the indicated pressure, defined as

$$\frac{\Delta p_s}{q_c} = \frac{p_i - p_{tc}}{q_c} \tag{2.15}$$

where p_{tc} is the pressure reading of the trailing cone probe, approaches a constant value. It then remains almost constant in the speed range of Mach numbers between 0.35 and 0.75 and increases very strongly above. Furthermore, the correction clearly depends on the configuration of the aircraft (e.g., flap and gear positions). The data were taken at several flight levels and are valid only for a certain aircraft and for a defined configuration. With devices such as the FHP, it is also possible to investigate the dependence of Δp_s on the angles of attack, α, and sideslip, β. In this case, the error $\Delta p_{s\alpha\beta}$ of static air pressure can be expressed as a linear superposition:

$$\Delta p_{s\alpha\beta} = \Delta p_s + \Delta p_{s\alpha} + \Delta p_{s\beta} \tag{2.16}$$

which has been shown for a Falcon 20 by Boegel and Baumann (1991), analyzing a series of pitching or yawing oscillations.

2.4.2
Summary

Static air pressure measurement onboard an aircraft depends on the configuration of the aircraft (flaps, landing gear, etc.), on the position of the pressure probes, and on maneuvers, speed, and aircraft attitude. Errors of static air pressure will also propagate to speed calculation. A careful investigation of the errors will result in correction for position error Δp_s, and for errors of attitudes in angles of attack and sideslip. For best results, research aircraft use routinely special devices such as booms or FADS to improve the measurement of important flow parameters. By flight testing, corrections can be achieved for a given configuration of the aircraft. Flight tests apply additional probes or procedures (e.g., tower flyby, trailing cone) to obtain better measurements for static air pressure of the undisturbed air $p_{s\infty}$.

2.5
Static Air Temperature

Temperature is an important variable for describing physical and chemical processes in the atmosphere. An exact knowledge of atmospheric temperature is necessary to understand energy and heat transfer within the Earth system driving the climate and weather systems, to characterize the rate of chemical reactions, and to know the state of aggregation of atmospheric compounds.

2.5.1
Aeronautic Definitions of Temperatures

There are a number of subtleties involved with the measurement of temperature from a fast-moving platform because the actual air temperature is disturbed by the aircraft. From the technical point of view, the following aeronautic definitions of temperature are in common use (Stickney, Shedlov, and Thompson 1994):

(i) *Static air temperature* T_s is the temperature of the undisturbed air through which the aircraft is about to fly. Ideally, this should correspond to the usual air temperature T. This is similar to the distinction between common air pressure p and static air pressure p_s.
(ii) *Total air temperature* T_t is the maximum air temperature that can be attained by 100% conversion of the kinetic energy (per unit mass of air) of the flight.
(iii) *Recovery temperature* (T_r) is the adiabatic value of local air temperature on each portion of the aircraft surface due to incomplete recovery of the kinetic energy.
(iv) *Measured temperature* (T_m) is the actual temperature as measured, which differs from T_r because of heat transfer effects due to imposed environments.

All temperature sensors have to be calibrated to a defined temperature scale. More detailed calibration standards to produce the Kelvin or Celsius scale are described in the International Temperature Scale (ITS), 1990 version (ITS-90).

2.5.2
Challenges of Airborne Temperature Measurements

Beside the common problems of high-precision laboratory temperature measurements, airborne temperature measurements involve additional challenges. Temperature probes are exposed to a harsh environment if installed onboard an aircraft. Vibrations can change material properties or even destroy a sensor if no precautions are taken. Vibration of wires or connector contacts can cause additional signal noise in signal lines. Temperature probes and electronics are often exposed to strong temperature gradients or variations of temperature (typically -60 to $+50\,^\circ$C). Temperature gradients can produce a hard-to-detect offset to the signal by inducing a thermoelectric voltage at points where different types of metal are connected with each other (e.g., soldered joint, connector contact), and strong variations of temperature are harmful to most electronics since nearly all electronic components show a temperature dependence in their properties. Pressure changes and the possibility of condensation at the sensor or electronics can cause additional problems.

The most challenging aspect of airborne temperature measurements is the aerodynamic effect of the flow distortion caused by the aircraft and the sensor itself. Each solid inserted into a flow will cause a distortion, changing the pressure, temperature, and density field of the flow. Bernoulli's theorem states that the sum of all forms of energy in a fluid flowing along an enclosed path (a streamline) is the same at any two points in that path. Considering a compressible flow and provided that the steady flow of the gas is adiabatic, we obtain Bernoulli's theorem for a compressible gas.

$$\left(\frac{\gamma}{\gamma - 1}\right) \cdot \frac{p_s}{\rho} + \frac{u^2}{2} = \text{constant} \tag{2.17}$$

with p_s, the static air pressure; ρ, the air density; u, the free-stream air velocity; and $\gamma = c_p/c_v$, the adiabatic exponent. $c_p = 1004\,\text{J kg}^{-1}\,\text{K}^{-1}$ and $c_v = 717\,\text{J kg}^{-1}\,\text{K}^{-1}$ are the specific heat capacities at constant pressure and volume (both for dry air), respectively. This equation can be rewritten as

$$\frac{u^2}{2} + c_p \cdot T = \text{constant} \tag{2.18}$$

using the ideal gas law for dry air

$$p = \rho \cdot R_{dry} \cdot T \tag{2.19}$$

Here we have the specific gas constant R_{dry} of dry air given by

$$R_{dry} = c_p - c_v = \left(\frac{\gamma - 1}{\gamma}\right) \cdot c_p \tag{2.20}$$

At the stagnation point where the flow comes to rest, there is a complete conversion of the energy of motion into thermal energy. The temperature rise, ΔT, in this case can be calculated using Eq. (2.19) from the free air stream temperature T and the free air stream velocity u; hence,

$$\Delta T = \frac{u^2}{2\,c_p} \tag{2.21}$$

Using the definition of the Mach number M as the ratio of the air stream velocity u and the speed of sound c,

$$M = \frac{u}{c} = \frac{u}{\sqrt{\gamma \cdot \frac{p_s}{\rho}}} \tag{2.22}$$

we can calculate the temperature rise as a function of the Mach number:

$$\Delta T = T \cdot \left(\frac{\gamma - 1}{2}\right) \cdot M^2 \tag{2.23}$$

As shown in Figure 2.4, the temperature rise at the stagnation point reaches around 1.3 K at an air stream velocity of $50\,\text{m s}^{-1}$ (~100 knots) and around 16 K at $180\,\text{m s}^{-1}$ (~350 knots).

To allow for the fact that, apart from at the stagnation point, the conversion of the kinetic energy of motion into thermal energy is incomplete, it is common to introduce a local recovery factor r in Eq. (2.21):

$$\Delta T = r \cdot \frac{u^2}{2\,c_p} = T \cdot r \cdot \left(\frac{\gamma - 1}{2}\right) \cdot M^2 \text{ for } r \leq 1 \tag{2.24}$$

Typical values of r measured for a cylindrical element with its axis orientated normal to the flow are between 0.6 and 0.7 and for a spherical element, around 0.75 (Lenschow and Pennell, 1974).

In bringing a temperature sensor onboard an aircraft, the flow distortion caused by both the aircraft and the sensor itself should be considered. Considering the aircraft as a solid surrounded by airflow with the air stream velocity identical to the true airspeed of the aircraft, the temperature field around the aircraft can be described by a field of local recovery factors. The local recovery factors of the temperature sensor and of the aircraft at the position of the sensor interact with each other such that at least the higher value of both will supersede the other. To avoid complex calibration of the combined aircraft/sensor system, either the position of the sensor has to be selected very carefully or the sensor should be designed with a recovery factor very close to 1. Therefore, most commercial aircraft use total air temperature T_t sensors with recovery factors between $r = 0.95$ and $r = 0.99$. For scientific temperature measurements, the highest absolute accuracy under normal conditions is still reached using these commercial total air temperature T_t sensors, which are commonly mounted close to the aircraft nose or on forward-extending booms. In addition to these total air temperature T_t sensors, a wide variety of different contact (immersion) and non-contact-type sensors are also in use.

2.5.3 Immersion Probe

Immersion type temperature sensors are fast-responding instruments (typically in the range of 1 Hz to 1 kHz) and are thus suited to measure the turbulent fluctuations. They are designed for use in a wide range of altitude, weather, and flow speed. The main component of an immersion probe is usually a platinum

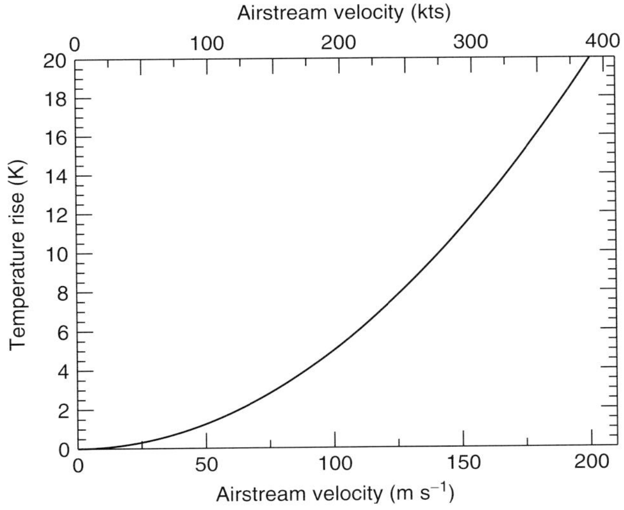

Figure 2.4 Maximum kinetic temperature rise as function of the airstream velocity (sea level, dry, 0 °C). (Source: Courtesy of Martin Zöger.)

or nickel resistance sensing element, such as an open wire or a thin coil around a ceramic element. The electrical resistance of the wire is usually assumed to be a linear function of the wire temperature. The sensor is more or less directly exposed to the airflow but the measured temperature is not the required static air temperature T_s. From the previous equations, the static air temperature T_s can be calculated from the measured temperature T_m by

$$T_s = \frac{T_m}{1 + \frac{r\cdot(\gamma-1)}{2} \cdot M^2} \tag{2.25}$$

with

$$r = \frac{T_m - T_s}{T_t - T_s} \leq 1 \tag{2.26}$$

If data of the Mach number M are not available with high precision, an additional pressure sensor very close to the temperature sensor, measuring total air pressure p_t, allows for the calculation of the static air temperature T_s:

$$T_s = \frac{T_m}{r \cdot \left[\left(\frac{p_t}{p_s}\right)^{\kappa} - 1\right] + 1} \tag{2.27}$$

with

$$\kappa = \left(\frac{\gamma - 1}{\gamma}\right) \tag{2.28}$$

Some manufacturers (e.g., Goodrich, formerly Rosemount) use the recovery correction η instead of the recovery factor r. The recovery correction is defined as

$$\eta = \frac{T_t - T_m}{T_t} \tag{2.29}$$

which is related to r by

$$r = 1 - \eta \cdot \left[1 + \frac{2}{(\gamma - 1) \cdot M^2}\right] \tag{2.30}$$

Depending on the design of the sensor, the recovery factor can depend not only on the Mach number but also on the mounting position and aircraft attitude. In the following, the most common design of immersion type sensors is discussed.

Total air temperature sensors (Figure 2.5a) are designed to achieve a high recovery factor ($r > 0.95$) to minimize the aerodynamic influence of the aircraft and the sensor housing on the temperature measurement. These sensors are commercially available and well characterized by intensive wind tunnel testing. Some designs are also optimized to separate particles such as water droplets or ice crystals from the air stream. The mounting position of these sensors is not relevant as long as the housing is aligned parallel to the local flow field and sources of turbulence (e.g., propellers, other inlets, or antennae) in front of the sensor are absent. Different sensing elements are available, such as open wire elements for fast time response or more robust encapsulated elements. To improve time response, modified sensing elements such as miniature thermistors have been investigated (Friehe and Khelif, 1992).

2.5.4 Reverse-Flow Sensor

Several investigations have shown that in heavy clouds, the particle separation of commercial total air temperature sensors may fail and sensor wetting may occur (Lenschow and Pennell, 1974; Lawson and Cooper, 1990; Sinkevich and Lawson 2005). A number of attempts at solving the *in situ* thermometer in-cloud wetting problem are described by Lawson and Cooper (1990) and are all based on some manner of inertially separating the cloud water from the airstream. One approach, initially developed in Canada and at the University of Chicago (Rodi and Spyers-Duran 1972), is the reverse-flow housing, one version of which is shown in Figure 2.5b. This uses exhaust ports that produce a negative pressure inducing the reverse flow through the housing. Lawson and Rodi (1992) and Lawson and Cooper (1990) report tests of the efficacy of the reverse-flow housing using a device at the sensing element location measuring conductivity on a surface that is very sensitive to the presence of liquid water. The conductivity tests clearly indicated that water reached the reverse-flow sensing element, although results from supercooled cloud penetrations, while not as definitive, indicate immunity from wetting. Lawson and Cooper (1990), using wind tunnel observations, suggested that water accumulated and streaming back on the housing may be the mechanism for ingestion of water into the reverse-flow housing, also explaining why the housing is much more effective in supercooled cloud.

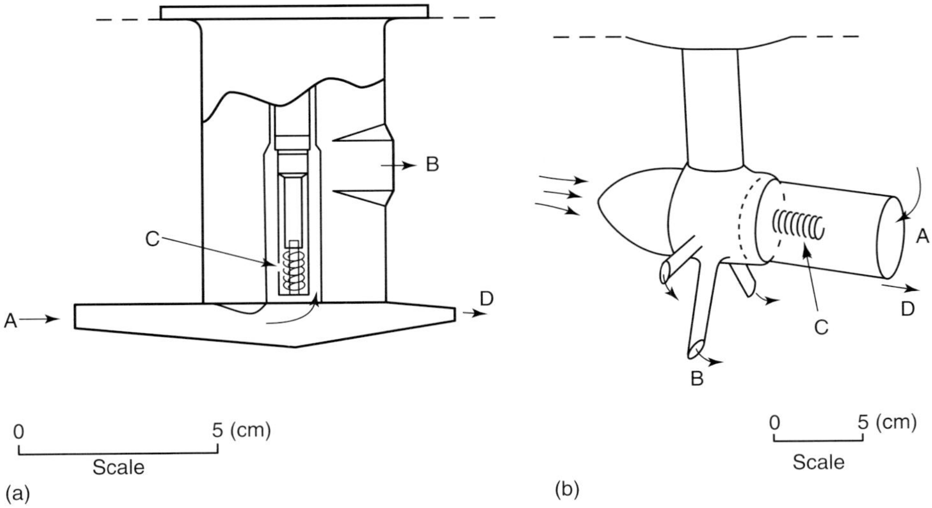

Figure 2.5 (a) Goodrich (formerly Rosemount) total temperature probe and (b) reverse-flow probe as used on the NSF/NCAR King Air. Air enters through port (A) and is exhausted through port (B) after coming in thermal contact with the platinum wire sensor (C). These probes are designed to separate cloud hydrometeors (D) from the airstream reaching the sensing element. (Source: From Lawson and Cooper (1990). Copyright 1990 American Meteorological Society. Reprinted with permission.)

2.5.5 Radiative Probe

Air temperature may be derived from measurements of the emitted radiance in the thermal infrared (TIR) spectral region. It is desirable that the weighting function of the detected radiation should be confined within a short distance (~10–100 m) of the detector. This reduces the sensitivity to changes in aircraft attitude, when the viewing path of the instrument may be shifted from the horizontal and may, therefore, view through the vertical temperature gradient of the atmosphere. Suitable wavelengths for measurement are, therefore, strongly absorbed in the atmosphere, and a typical choice is the 4.25 μm absorption band of CO_2 (Beaton, 2006).

The temperature may be determined by inversion of the Planck function that describes the radiance, $B_\lambda(T)$, emitted by a blackbody of temperature T (Chapter 7, Eq. (7.22)). The inversion of the Planck function gives

$$T^{-1} = \frac{k_B \cdot \lambda}{h \cdot c} \cdot \left\{ \ln \left[\frac{2\,h \cdot c^2}{\lambda^5 \cdot B_\lambda(T)} \right] + 1 \right\} \tag{2.31}$$

with the Boltzmann constant $k_B = 1.3806 \times 10^{-23}\,\mathrm{J\,K^{-1}}$; the Planck constant $h = 6.6262 \times 10^{-34}\,\mathrm{J\,s}$; T, the absolute temperature in Kelvin; and λ, the wavelength. When the atmospheric path is totally absorbing, and hence its emission is perfect (emissivity is unity), the brightness temperature is equal to the temperature of the air.

A recent implementation of this principle is described by Beaton (2006) (Figure 2.6). The instrument consists of a filter radiometer, with a passband width of ~0.05 μm. A rotating chopper wheel allows the detector to view alternately the atmospheric radiance and the emission from an internal temperature-controlled blackbody target. Measurement of the difference signal and the blackbody temperature allows the atmospheric brightness temperature to be determined.

The instrument housing has an external window that is transparent in the TIR. This allows the internal temperature and humidity of the instrument to be more easily stabilized. The window must be maintained free of any materials that are strongly absorbing at the detection wavelength. This includes liquid water that might form a thin film across the window when the instrument is in liquid-phase clouds or rain.

Liquid- and ice-phase clouds are both strongly absorbing at the 4.25 μm wavelength. The impact of this fact when making measurements in cloud is that the absorption within the wings of the passband of the filter is increased compared to that in clear air. This has the effect of decreasing the effective viewing path within cloud from 100 to ~20 m (Beaton, 2006).

In principle, the instrument can be radiometrically calibrated to give an absolute true air temperature measurement. In practice, however, the stability of such calibrations is insufficient and they are normally calibrated against an immersion temperature sensor using cloud-free in-flight data. Such a calibration will typically exclude data from periods when the aircraft roll and pitch angles

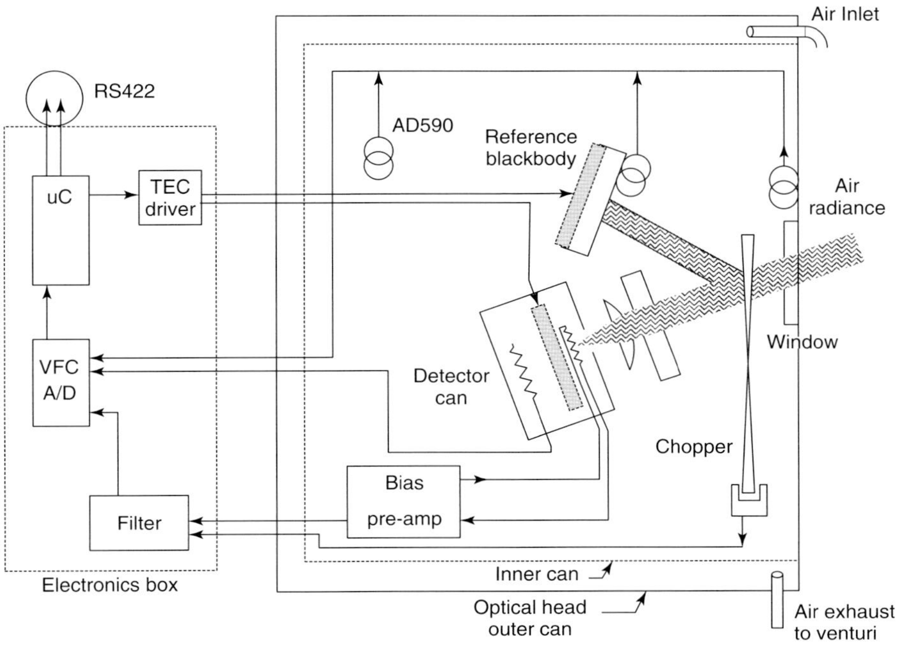

Figure 2.6 A block diagram of the Ophir air temperature radiometer (2006). The external window is at the right. Behind it is the chopper wheel, the 4.3 μm interference filter, the focusing lens, and then the detector can. Inside the detector can is the HgCdTe detector, the thermistor to monitor the detector temperature, and the thermoelectric cooler for the detector. The TEC driver supplies power to the thermoelectric coolers for the detector and controlled blackbody. The entire optical system is kept near the external air temperature by air circulating between the inner and outer cans of the optical head. (Source: Figure redrawn from Beaton (2006) and used by permission.)

exclude certain limits. This ensures the rejection of any data obtained when the instrument may be viewing up or down the atmospheric vertical gradient of temperature.

The sample rate of such a radiometric temperature sensor is typically around 1 Hz. At typical flight speeds of 70–100 m s^{-1}, this means that the along-track averaging length is comparable with the instrument viewing path length. Higher-frequency sampling is possible but will increase the noise level.

2.5.6 Ultrasonic Probe

Ultrasonic thermometry is based on the measurement of the speed of sound of the air that mainly is a function of temperature. The speed of sound is derived from the measurement of the transit time of a short sound pulse over a well-known distance. A relative movement of the air with respect to the emitter of the sound pulse (e.g., wind) will be superimposed on the speed of sound. Measuring the transit time back and forth along the same path allows extraction of the speed of sound as well as the wind vector component along the sound propagation path. This principle is widely used for ground-based measurements of 3D wind and temperature simultaneously (Section 2.8.3). Owing to the noncontact type of measurement, a high time resolution is possible, making the method useful for the measurement of temperature fluctuation. But its ability for absolute temperature measurement is strongly reduced by secondary effects in sound wave propagation theory based on the assumption that air is an ideal gas (Cruette *et al.* 2000). Up to now, only a few ultrasonic temperature probes have been used for airborne measurement, mainly on slow-flying aircraft or helicopters. Calculation of static temperature using this type of probe requires measurement of water vapor mixing ratio because this type of probe measures the sonic temperature, which is closely related to virtual temperature:

$$T_{\text{vir}} = T \cdot \frac{w + \epsilon}{\epsilon \cdot (1 + w)} \tag{2.32}$$

where w is the mixing ratio and $\epsilon = R_{\text{dry}}/R_{\text{wv}} \approx 0.622$ in the Earth's atmosphere. $R_{\text{dry}} = 287.05\,\text{J kg}^{-1}\,\text{K}^{-1}$ is the specific gas constant of dry air, and $R_{\text{vw}} = 461.7\,\text{J kg}^{-1}\,\text{K}^{-1}$ is the specific gas constant for water vapor.

2.5.7
Error Sources

The following error discussion mainly deals with immersion type total air temperature sensors because they are the most common and widely used static air temperature sensors onboard an aircraft. Nevertheless, most points also apply for other types of sensors or give at least a good guideline for a more specific error analysis.

2.5.7.1 Sensor

2.5.7.1.1 **Calibration Accuracy** The error discussion of immersion type temperature sensors starts with the basic temperature versus resistance calibration of the sensor element itself. Commercially available calibration systems based on the dry well technique achieve accuracy in the range of 0.2–0.5 K. With more sophisticated stirred immersion bath calibration systems, accuracies better than 0.1 K can be achieved for temperatures above $-70\,^{\circ}\text{C}$. This calibration accuracy includes contributions from many error sources such as bath homogeneity, temporal stability, accuracy of reference, accuracy of fit, accuracy of ohmmeter, and the contribution of contact voltages.

2.5.7.1.2 **Conduction and Radiation** The immersion type air temperature senor is based on the heat transfer from the air to the sensor by advection. Whenever temperature gradients between the sensor and its housing or mounting exist, heat transfer to the sensor element by conduction or radiation has to be considered.

Possible countermeasures are thermal insulation and radiation shields. Typical total air temperature sensors are constructed so that conduction and radiation errors can be neglected as long as a sufficient airflow across the sensor is guaranteed (Stickney, Shedlov, and Thompson 1994).

2.5.7.1.3 **Self-Heating** The typical resistance measurement of an immersion temperature sensor applies a constant current source to the resistor. Within the resistor, electric power is converted into heat influencing the temperature measurement itself. The self-heating effect can either be measured and corrected by, for example, wind tunnel tests or be minimized by keeping the electric power dissipated in the sensor as low as possible.

2.5.7.1.4 **Deicing** In order to be able to measure static air temperature T_s during icing conditions and to avoid aircraft damage by ice shedding, some total air temperature probes are equipped with deicing heaters. The deicing heats up the front part of the housing preventing the build up of ice. The heated housing often influences the temperature measurement inside the housing. If known, for example, as a result of wind tunnel testing, this deicing error can be corrected. In-flight calibration procedures can also be used to identify the deicing error. These typically involve making measurements in a region of uniform temperature with and without the deicing heating.

2.5.7.1.5 **Time Constant** If not properly addressed, the unknown time response of the sensor can introduce an additional error to the temperature measurement, especially if flying in varying temperature conditions, for example, during ascent and descent. Several approaches have been made to correct the temperature measurement for the known time lag mainly to improve the time response of turbulence measurement and to provide greater accuracy of sensible heat fluxes (Rodi and Spyers-Duran, 1972; McCarthy, 1973; Inverarity, 2000).

2.5.7.2 Dynamic Error Sources

As stated, the flow distortion caused by the aircraft and the sensor itself can strongly influence the static air temperature measurement. Figure 2.4 shows that this dynamically induced error becomes significant for airspeeds above $\sim$20 m s^{-1}. If not using a well-characterized total air temperature probe, extensive wind tunnel testing or airflow simulation is necessary to distinguish the dynamic correction. Commercially available total air temperature probes often provide all information necessary to correct the dynamic error. Stickney, Shedlov, and Thompson (1994) published a recovery correction for different types of Rosemount total air temperature probes derived from wind tunnel experiments. The remaining dynamic error resulting from production tolerance and the repeatability error of the wind tunnel tests are in the range of 0.2 K. The Mach number dependence of the recovery correction cannot be neglected for Mach numbers ≥ 0.4. Using a constant recovery factor instead of the Mach-number-dependent recovery correction would introduce an additional error of a few tenths of a Kelvin.

2.5.7.3 In-Cloud Measurements

Typical total air temperature probes as well as special reverse flow sensors are designed to separate particles from the airstream avoiding contamination of the sensor element. Nevertheless, erroneous temperature measurements in clouds in the range of a few Kelvin have been documented (Lenschow and Pennell, 1974; Lawson and Cooper, 1990; Sinkevich and Lawson, 2005). Lawson and Cooper (1990) identified evaporative cooling of the wetted sensor as a possible reason for the measurement error in clouds and quantified the evaporative cooling effect for a completely wetted sensor. By comparing in-cloud temperature measurements of a Rosemount total air temperature probe with a radiative probe, Lawson and Cooper (1990) found that the full evaporative cooling effect applies only at very high liquid water content. For lower liquid water contents, partial wetting of the sensor or housing will lead to a cooling effect that is difficult to quantify (Lawson and Rodi, 1992).

2.5.8 Calibration of Temperature Sensors

Laboratory calibration of temperature sensors is restricted to the calibration of the sensor element itself. As mentioned earlier, sophisticated stirred immersion bath calibration systems can achieve overall accuracies better than 0.1 K. Depending on the bath fluid in use, a temperature range of −70 to +50 °C is achievable. Extensive error analysis of the whole calibration chain as well as traceability to national standards is necessary to achieve the aforementioned accuracy.

To avoid extensive wind tunnel testing, most non-total-air-temperature-type sensors as well as radiative probes rely on in-flight calibration and are, therefore, used together with a well-characterized total air temperature probe. Under cloud-free and stable conditions, both instruments are compared against each other, yielding a correction factor valid for these particular flight conditions. A fully independent characterization of the temperature probe requires extensive flight testing to quantify the dependencies on Mach number, density, and attitude (pitch, roll, yaw). Nevertheless, the achievable accuracy of in-flight calibration is always limited to the accuracy of the reference sensor. If no reference sensor is available, special flight test maneuvers can be used to determine the recovery factor. From Eq. (2.25) it follows that plotting the measured temperature as a function of the Mach number will yield the static air temperature, T_S, as the coordinate intercept. From the slope of this curve, together with T_S, the recovery factor, r, can be calculated. However, the absolute accuracy of this method is limited by the temporal stability of T_S as well as by the fact that a constant recovery factor is assumed.

2.5.9 Summary, Gaps, Emerging Technologies

Huge efforts have been made in the past decades to improve airborne static air temperature measurements. Nevertheless, most reliable and accurate measurements still are based on standard commercial total air temperature probes. Currently, the achievable absolute accuracy is around half a Kelvin for cloud-free conditions. Future scientific demands mainly focus on higher absolute accuracy, especially within clouds where, for example, temperature accuracy is a limiting factor on the accuracy of estimates of supersaturation. Some research is underway to develop optical temperature probes, mainly for military purposes. These instruments are currently in the stage of proof of concept with bulky laboratory installations. Expected accuracies are in the range of 0.3 (optimal conditions) to 1.2 K (clouds).

2.6 Water Vapor Measurements

This section is dedicated to Cornelius Schiller. With the development of FISH, the "Fast In situ Stratospheric Hygrometer," in the 1990s, Cornelius Schiller started a new era of the measurement of water vapor under the challenging conditions of the upper troposphere and lower stratosphere (UT/LS). Accurate water vapor measurements under such conditions are exceedingly difficult, but Cornelius together with a few colleagues succeeded in pushing the boundaries of science and technology by taking measurements with this instrument on multiple platforms, from the tropics to high latitudes. FISH is considered to be one of the most accurate hygrometers in the world. Today, FISH represents a reference for water measurements around the world on a variety of platforms in the UT/LS. Cornelius strongly fostered utilization of high-flying research aircraft as an important platform to study atmospheric processes in the UT/LS region. As a result of this activity, he was strongly engaged in EUFAR (European Facility for Airborne Research) and particularly in what the future fleet of European research aircraft will look like. Cornelius' research resulted in a large number of high-quality publications with great impact on the scientific community. Although at times his work has necessarily been quite technical, his motivation remained always very clear: striving for scientific truth and a better understanding of the Earth's atmosphere and climate. He led large international measurement campaigns, and we all followed him because of his strong scientific integrity combined with his kind and congenial nature. Cornelius Schiller, our dear colleague and friend, passed away on 3 March 2012 in Neuss, Germany, after he lost his fight against cancer. We will not forget him, but continue to work in his spirit on the questions that he raised.

2.6.1 Importance of Atmospheric Water Vapor

Water vapor plays a key role in the atmospheric energy budget and greenhouse effect. Water vapor is one of the key drivers for and key tracers of atmospheric transport, and as a source of the hydroxyl radical, it has a strong influence on the chemistry (i.e., oxidative capacity) of the atmosphere. Water vapor is the source of clouds and precipitation and is crucial in the removal of aerosol particles or water-soluble gases through heterogeneous reactions with aerosol or cloud particles. Water vapor has a strong influence on the size distribution of hygroscopic aerosol particles and their optical properties, that is, atmospheric radiative effects.

Therefore, airborne measurements of ambient water vapor are important in many areas of atmospheric research.

Important considerations for an airborne hygrometer are fast time response for adequate spatial resolution, insensitivity to the presence of liquid water or ice particles, demonstrated insensitivity to contamination from instrument or platform surfaces, capability of measuring ice supersaturation, ease of calibration, and sensitivity over a wide dynamic range. Water vapor levels drop from a few parts per hundred at the ground to a few parts per million in the lower stratosphere. For this reason, many different measurement methods and sensors have been developed through the years, each having certain advantages and limitations and each being suitable for some but not all applications. There is a long history of water vapor measurements using various techniques.

This section describes the different types of humidity measurement techniques that are in airborne use today, their advantages and limitations, and the applications where certain instruments should or should not be used.

2.6.2 Humidity Variables

In the atmosphere, water generally can exist in three states (thermodynamic phases): as frozen ice crystals, as liquid water droplets, and as gaseous water vapor. Important variables for humidity measurements are the water vapor (partial) pressure and the saturation vapor pressures that can be in equilibrium with either the liquid or the ice phase. Only at the triple point of water, all three phases coexist in thermodynamic equilibrium, and then the (saturated) vapor pressure of both ice and liquid water is $e_{\mathrm{s,tri}} = 611.657$ Pa at the triple point temperature of $T_{\mathrm{tri}} = 273.16$ K (Guildner, Johnson, and Jones, 1976). The transition between gaseous phase and the liquid or ice phase increases in an exponential fashion with temperature T at the surface of the transition. The Clausius–Clapeyron equation that describes the nonlinear dependence of the saturated vapor pressure e_{s} on temperature T may be written as (Rogers and Yau, 1989)

$$\frac{\mathrm{d}e_{\mathrm{s}}}{\mathrm{d}T} = \frac{e_{\mathrm{s}} \cdot l^*}{R_{\mathrm{wv}} \cdot T^2} \tag{2.33}$$

where l^* represents either the specific latent heat of evaporation $l_{\mathrm{v}} \approx 2.501 \times 10^6\,\mathrm{J\,kg^{-1}}$ or sublimation $l_{\mathrm{s}} \approx 2.835 \times 10^6\,\mathrm{J\,kg^{-1}}$, and $R_{\mathrm{wv}} = 461.7\,\mathrm{J\,kg^{-1}\,K^{-1}}$ is the specific gas constant for water vapor. Although l_{v} and l_{s} are both temperature dependent, in the first approximation, the Clausius–Clapeyron equation can be integrated by regarding the latent heat as constant in order to give an expression for the saturation vapor pressure over liquid water ($T \geq T_{\mathrm{tri}}$) or an ice surface ($T \leq T_{\mathrm{tri}}$) as a function of temperature, that is,

$$e_{\mathrm{s}}(T) = e_{\mathrm{s0}} \cdot \exp\left[\frac{l^*}{R_{\mathrm{wv}}} \cdot \left(\frac{1}{T_0} - \frac{1}{T}\right)\right] \tag{2.34}$$

where e_{s0} is the value of saturation vapor pressure at temperature T_0. At $T_0 = 0\,^{\circ}\mathrm{C}$, $e_{\mathrm{s0}} = 611$ Pa.

Equation (2.34) shows that the saturation vapor pressure e_s is a strong function of temperature T and increases almost exponentially with increasing temperature. At temperatures below the triple point temperature T_{tri}, next to the stable ice phase, a transient metastable supercooled liquid phase may exist under certain conditions such that at the same temperature the saturation water vapor pressure is higher over the liquid than over the ice (Murphy and Koop, 2005). In this case, one can measure the dew point temperature, that is, saturation with respect to liquid water, or the frost point temperature, that is, saturation with respect to ice. Supercooled liquid water can be expected in clouds down to temperatures as low as ~235 K.

Equation (2.34) is not an exact fit because water vapor is not an ideal gas and the latent heats of evaporation and sublimation are temperature dependent. This finding is directly related to the temperature dependence of the specific heat capacity of water vapor, ice, and supercooled water. There are several saturation water vapor equations to calculate the equilibrium pressure of water vapor over a plane surface of liquid water or ice as a function of temperature (Sonntag, 1994; Murphy and Koop 2005).

Several measures can be used to express the water vapor abundance in the atmosphere (Rogers and Yau, 1989). The most important ones are the (partial) water vapor pressure, the molar or mass density of water vapor, specific humidity (the ratio of the mass of the water vapor to the mass of the moist air), the mass mixing ratio (the ratio of the mass of water vapor to the mass of dry air), and the volume mixing ratio (the ratio of the water vapor pressure to (total) ambient air pressure). Other measures often used are the dew point or the frost point (temperature), defined as the temperature at which the air parcel would be saturated with respect to liquid water or ice, respectively. Atmospheric water vapor abundances are also frequently reported as the relative humidity, RH (ratio of the actual water vapor pressure e to the saturation pressure, e_s, at the prevailing ambient air temperature, T, multiplied by 100%). RH can be expressed with respect to the liquid water or the ice phase.

The calculation of saturation ratios and relative humidity from water vapor partial pressures critically depends on the ambient gas temperature because of the saturation pressures being exponential functions of T. Relative changes with temperature T are about 6% K^{-1} at 300 K and increase to about 15% K^{-1} at 200 K. Therefore, not only accurate water vapor measurements but also accurate ambient air temperature measurements are needed.

2.6.3
Dew or Frost Point Hygrometer

The dew or frost point hygrometer is the most widely used instrument to measure atmospheric water vapor concentration onboard a research aircraft. The chilled mirror technique in its basic form (Figure 2.7) detects the dew or frost point by cooling a small reflective metal surface or mirror in contact with ambient air until a layer of dew (liquid droplets) or frost (ice crystals) begins to form. Formation of

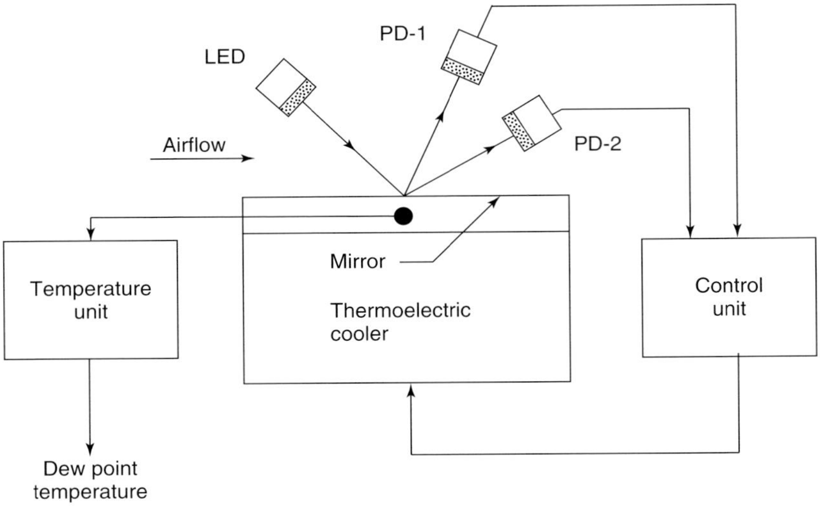

Figure 2.7 Block diagram of a chilled mirror hygrometer. The LED is the light source and the two photodetectors sense the scattered (PD-1) and reflected light (PD-2) from the mirror to determine if dew or frost is present or not. The ratio of the two detector signals determines if mirror heating or cooling is required.

dew or frost on the mirror is detected optically with a light-emitting diode (LED) illuminating the mirror and one or more photodetectors that sense the change in light reflecting or scattering when dew or frost forms on the mirror. The signal is fed into an electronic feedback control system to regulate the mirror temperature through cooling and heating, to the point where dew or frost just starts to form. A temperature sensor, usually a thermistor, is embedded within the mirror surface to measure its surface temperature. When a predetermined layer of dew or frost is maintained on the mirror surface, the measured mirror temperature corresponds to the dew or frost point of the ambient air flowing over the mirror. A comprehensive overview of chilled mirror hygrometers is given by Wiederhold (1997).

For airborne applications in the lower or middle troposphere, the thermoelectrically cooled mirror type is most widely used. Thermoelectric coolers are small and inexpensive and work over a modest range of dew or frost point temperatures down to approximately $-40\,^\circ$C for single-stage and $-60\,^\circ$C for two-stage thermoelectric coolers. However, their capabilities at low humidities (frost point temperatures below $-50\,^\circ$C) are limited. At the low end of their range, they become increasingly inefficient and slow to respond. At low humidities, that is, frost point temperatures below $-50\,^\circ$C, cryogenic cooling of the mirror is preferred.

In the cryogenic chilled mirror hygrometer, the mirror is cooled by a freely boiling cryogen, such as Freon or nitrogen, through an attached rod of high thermal conductivity. The mirror is permanently cooled to a temperature well below its measurement range. To make a measurement, the mirror temperature is raised to the dew or frost point and maintained at that point by a servo-controlled electrical heater. Most airborne cryogenic dew or frost point hygrometers in use nowadays are based on the original balloon-borne design made by Mastenbrook (1968) and have been modified for airborne use (Spyers-Duran, 1991; Buck, 1991; Ovarlez and Velthoven, 1997). Since the mid-1990s, commercial instruments became available (Busen and Buck, 1995). The efficient cooling allows operation down to frost point temperatures of $-90\,^{\circ}$C and faster response than feasible with thermoelectric cooling. In the more advanced instruments, a continuously operating stirling cycle cryogenic refrigerator or cryopump is used, allowing for continuous operation at very low frost points.

The chilled mirror method may seem to be a fundamental technique of measuring the dew or frost point temperature with a very accurate and well-calibrated temperature sensor (accuracy better than $\pm 0.1\,^{\circ}$C), but this type of hygrometer is influenced by several factors that can have a large and significant impact on the quality and reliability of the performance of the instrument (Wiederhold, 1997). Among the factors are instability of the feedback controller, air temperature variations, dew or frost point ambiguity between 0 and $-30\,^{\circ}$C, and the presence of water-soluble contaminants on the mirror (the Raoult effect).

The dew or frost point hygrometer is capable of quasi-continuous operation to measure dew or frost point temperature with an accuracy varying from 0.5 to 2 °C. It is often difficult to determine when or if a chilled mirror is operating correctly, particularly when ambient conditions are changing rapidly (e.g., when entering a cloud). Dew or frost point devices typically have slow responses.

A lightweight miniaturized application of a dew or frost point hygrometer is the surface acoustic wave (SAW) hygrometer that uses a tiny piezoelectric crystal to detect the onset of condensation. The intrinsic sensitivity of the SAW to mass loading changes is a result of the propagation of the acoustic wave occurring only on the surface of the crystal. This increased sensitivity means that theoretically, SAW hygrometers have fast response times and can operate in very dry environments (Hansford *et al.* 2006).

Another version of the dew or frost point hygrometer uses basic components similar to those shown in Figure 2.7, but differs in its ability to measure the water vapor mixing ratio in the high RH range of about 95–105% (Gerber, 1980). This version, termed *saturation hygrometer*, is based on the observation by Wylie, Davies, and Caw (1965). The accuracy of the saturation hygrometer was estimated to be ~0.02% RH near ambient RH = 100%, 0.1% at RH = 99%, and 0.6% at RH = 97% in radiation fog measurements (Gerber, 1991b).

2.6.4
Lyman-α Absorption Hygrometer

The Lyman-α absorption hygrometer uses the water vapor absorption of vacuum ultraviolet (VUV) light in a narrow optical band around the Lyman-α emission line of atomic hydrogen at a center wavelength of 121.56 nm. While for Lyman-α light water vapor absorption is very strong, oxygen absorption is uniquely low, and most other common gases are relatively transparent, for example, nitrogen. A significant

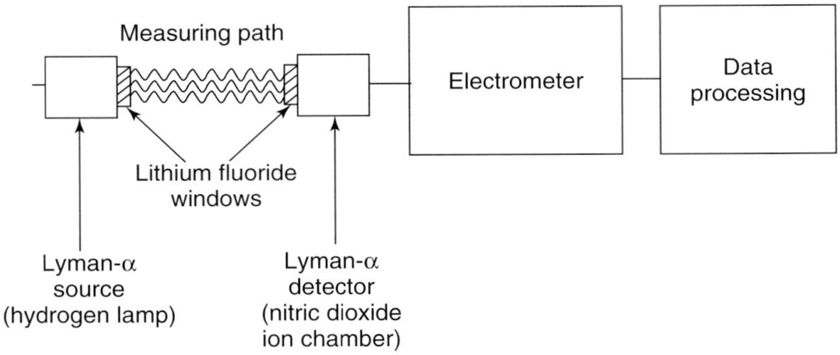

Figure 2.8 Single-beam Lyman-α absorption hygrometer.

fraction of radiation is absorbed over a few millimeters path length under normal conditions. The response is very fast, on the order of milliseconds. Developments in the Lyman-α absorption hygrometers for airborne use have been made by Buck (1985), Weinheimer and Schwiesow (1992), and others. The Lyman-α absorption hygrometer in its basic form is shown in Figure 2.8.

The Lyman-α absorption hygrometer is a secondary measurement device and must be regularly calibrated, usually in the laboratory in an enclosed airflow circuit with a chilled mirror hygrometer as calibration standard. When a reference hygrometer is not available, a variable path length self-calibration technique can be used (Buck, 1976). The Lyman-α absorption hygrometer for airborne use has been made commercially available by Buck Research Instruments (although production was stopped at end of 1990s). The device offers a very fast response (≈5 ms), and can measure water vapor densities of 0.1–25 $\mathrm{g\,m^{-3}}$ with a relative precision of 0.2% and an accuracy of 5%. The fast response makes it a suitable instrument for ultrafast hygrometry (sample rates of 10–100 Hz), water vapor flux measurements, or micrometeorological measurements inside and outside clouds.

It is not possible to use the Lyman-α absorption hygrometer alone to measure absolute water vapor density; the Lyman-α light source aging and optical window contamination are the main factors that prevent a stable predictable calibration. Therefore, in practice, a chilled mirror hygrometer is simultaneously used for slow accurate dew point measurements (Friehe, Grossman, and Pann, 1986).

2.6.5 Lyman-α Fluorescence Hygrometer

This type of hygrometer uses the Lyman-α light absorption in conjunction with the photodissociation of H_2O molecules, whereby fluorescence light is emitted and used as a measure of the H_2O abundance. The method was developed by Kley and Stone (1978) and Bertaux and Delan (1978). Figure 2.9 shows the schematics of the Lyman-α fluorescence hygrometer that consists of a monochromatic Lyman-α light source, two VUV detectors (VUV_A and VUV_B) to measure the Lyman-α light intensities, and a photomultiplier to measure OH-fluorescence light intensity.

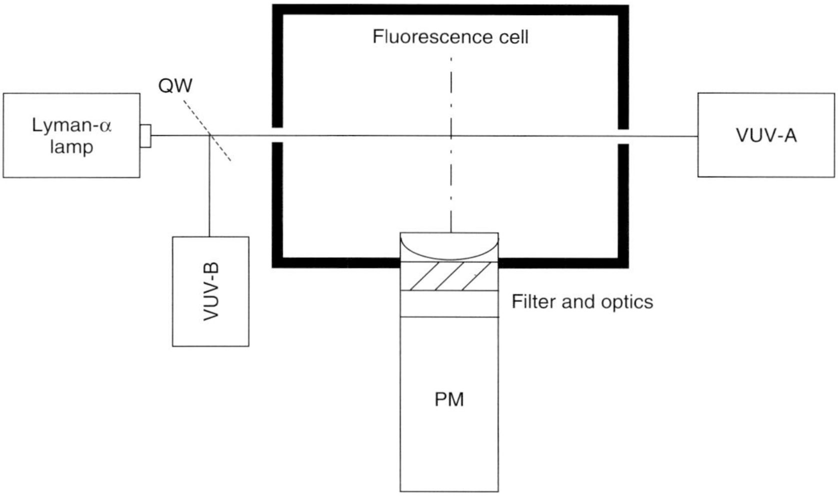

Figure 2.9 Lyman-α fluorescence hygrometer after geometry by Kley *et al.* (1979). QW, quartz window; VUV-A, NO cell A; VUV-B, NO cell B; PM, photo multiplier.

The Lyman-α fluorescence technique can achieve a large dynamic range for measurements from the middle and upper troposphere at about 1000 μmol mol^{-1} into the dry stratosphere with only 2–5 μmol mol^{-1}, where changes on the order of 0.1 μmol mol^{-1} can be detected with a relative uncertainty of ±5%. Although it is more usual practice for the volume mixing ratio to be expressed in terms of ppm, or ppmv, these are nonstandard units, and for the purpose of this text, we have used μmol mol^{-1} for ppm or ppmv. Large flow rates through the hygrometers together with integration times on the order of 1 s enable the measurement of small-scale features in the atmosphere.

Only a few well-established Lyman-α fluorescence hygrometers for use on research aircraft exist, such as the NOAA Aeronomy instrument developed by Kley and Stone (1978) and Kley *et al.* (1979), the Harvard instrument (Weinstock *et al.* 1994, 2009), the FISH (Fast In Situ Stratospheric Hygrometer) instrument (Zöger *et al.* 1999; Schiller *et al.* 2008), and the UK Met Office instrument (Keramitsoglou *et al.* 2002). Although the principle of operation is the same, the instrumental layout of each instrument is different. The fluorescence technique needs laboratory calibration, but in-flight calibration can be achieved by combining Lyman-α fluorescence with direct Lyman-α absorption measurements of water vapor (Kley *et al.* 1979).

2.6.6 Infrared Absorption Hygrometer

The IR absorption hygrometer uses the absorption of IR radiation by water vapor at certain distinct wavelength bands (e.g., Hyson and Hicks, 1975). Usually a

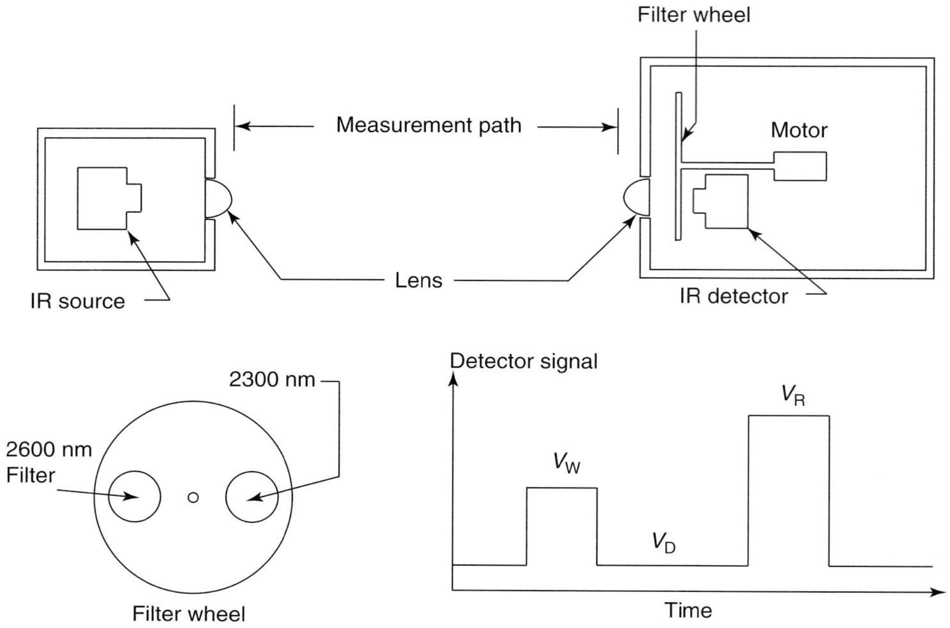

Figure 2.10 Schematic of a typical implementation of a single-beam IR absorption hygrometer.

dual-wavelength differential absorption technique is applied, whereby one (primary) wavelength is subject to strong water vapor absorption and the other (reference) is not. An example of a single-beam absorption hygrometer with one IR source and one IR detector is shown in Figure 2.10.

Two IR band filters, one in the reference band (e.g., 2.3 μm wavelength) and the other in the absorbing band (e.g., 2.6 μm wavelength) are mounted on a chopper wheel that rotates into the IR light beam in front of the IR detector. The detector signal would be composed of three components (Figure 2.10) sampled when the absorbing filter, no filter, and the reference filter are in the beam, respectively. This yields a normalized signal that is a direct measure of the abundance of water vapor, and at the same time is insensitive to drifts, deposits on the optical windows, and haze or fog within the sample volume. Through the use of lock-in (i.e., phase-sensitive) detection techniques, a high sensitivity of $0.01\,g\,m^{-3}$ and sample rates of 20 Hz can be obtained (e.g., Ohtaki and Matsui, 1982; Cerni, 1994). Nowadays, a suite of extractive as well as open-path instruments have become commercially available, mostly designed for fast and simultaneous measurements of CO_2 and H_2O concentrations. The measurement range is thereby about $0.3–30\,g\,m^{-3}$, such that airborne use is limited to the lower part of the troposphere.

2.6.7
Tunable Laser Absorption Spectroscopy Hygrometer

Tunable laser absorption spectroscopy (TLAS) is based on the use of a narrow-band, wavelength-tunable diode laser source to scan one or more characteristic spectral absorption lines of the target trace gas, here water vapor, in the path of the laser beam. A basic TLAS hygrometer (Figure 2.11) consists of a tunable diode laser, optical absorption cell with sample gas, and a photodiode as detector. The transmitted light intensity can be related to the concentration of the absorbing water vapor by the Lambert–Bouguer law (also called *Beer's law*); see also Section 7.2.5. A comprehensive overview of TLAS is given by Heard (2006).

In the past two decades, development of tunable diode lasers has yielded devices operating at around room temperature, particularly in the 1–2 μm wavelength band where water vapor has strong absorption lines. Communication laser diodes have become available, which allow the use of very compact, robust, and lightweight instruments. Furthermore, through the use of a multipass absorption cell, where the optical beam is reflected back and forth between a set of mirrors, a larger optical path length is obtained that offers higher sensitivity in a fairly compact system (Heard, 2006; May, 1998; Diskin *et al.* 2002; Zondlo *et al.* 2010). In conjunction, often a 2f-modulation technique is used to increase the sensitivity even further. However, this requires regular and precise calibrations in the laboratory.

If regularly calibrated, a TLAS hygrometer system can achieve relative accuracies of about 5–10% with a precision of ~2–3%. Most advanced instruments can even achieve this in the sub-μmol mol^{-1} range of 1–10 μmol mol^{-1}. Developments in the new TLAS hygrometer systems, which use direct absorption, are in progress (Gurlit *et al.* 2005). These new instruments measure water vapor in the lower range

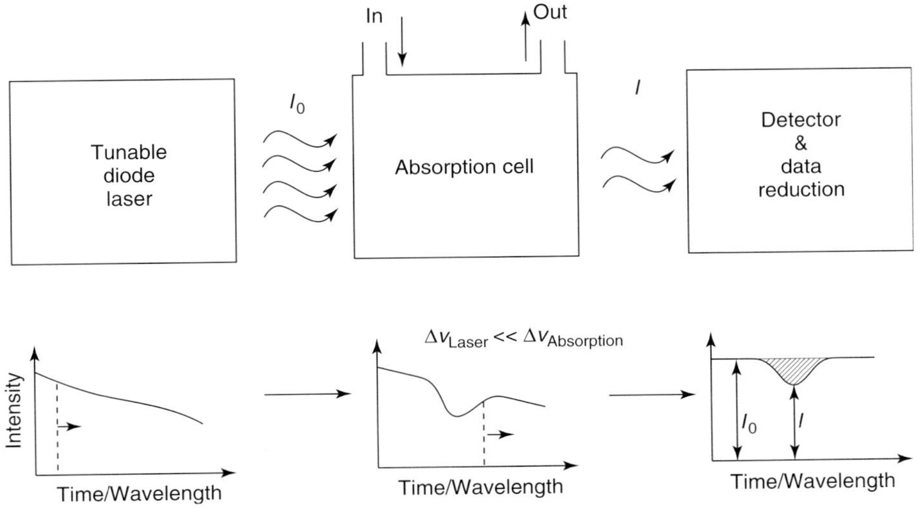

Figure 2.11 Basic setup of tunable laser absorption spectrometer (TLAS).

of 1000 μmol mol^{-1} down to a few μmol mol^{-1} or even lower. They use a new self-calibrating data evaluation strategy based on the first principles approach and known parameters such as the absorption line strength, pressure, gas temperature, and absorption path length. This strategy may provide a very robust, compact, lightweight, highly accurate, and absolute laser hygrometer without the need for regular recalibration.

2.6.8 Thin Film Capacitance Hygrometer

Capacitive humidity sensors are miniaturized sensors that measure dielectric changes of thin films resulting from water vapor uptake as they come into equilibrium with the water vapor pressure in the surrounding air. In the scope of the Measurement of Ozone and Water Vapor by Airbus In-Service Aircraft (MOZAIC) project, this type of humidity sensor is deployed on board commercial aircraft to measure relative humidity in the troposphere (Helten *et al.* 1998). The humidity sensing element (Humicap) together with a PT100 resistor to measure temperature is mounted in a total air temperature housing (Figure 2.12a), which protects the sensors against particles and avoids any wall contact of the sampled air. Although capacitive humidity sensors are widely deployed on radiosondes, their use on aircraft in the middle and upper troposphere requires careful and regular calibrations.

The air entering the total air temperature housing is subject to adiabatic compression caused by the strong speed reduction in the inlet part of the housing. The adiabatic compression produces an appreciable temperature rise relative to the ambient static air temperature T_s if the aircraft speed is comparable to the speed of sound. For a fast high-flying aircraft, the resulting difference between total and static air temperature ($T_t - T_s$) increases from 2 K near ground to ~30 K at 10–12 km cruise altitude (Figure 2.12b). Because of the strong temperature increase, the dynamic relative humidity RH_D detected by the sensing element in the total air temperature housing is appreciably lower than the static relative humidity of the ambient air, RH_S (Helten *et al.* 1998). We thus obtain

$$RH_s = RH_D \cdot \left(\frac{T_s}{T_t}\right)^{\frac{c_p}{c_p - c_v}} \cdot \frac{e_s(T_t)}{e_s(T_s)} \tag{2.35}$$

where $e_s(T_s)$ and $e_s(T_t)$ are the water vapor saturation pressures of liquid water at static air temperature T_s and total air temperature T_t, respectively. $c_p = 1004\,\mathrm{J\,kg^{-1}\,K^{-1}}$ and $c_v = 717\,\mathrm{J\,kg^{-1}\,K^{-1}}$ are the specific heats of dry air at constant pressure and volume, respectively. Therefore, for a fast high-flying aircraft, the sensor operates in the lowest 10% of its full dynamic range ($RH_S/RH_D > 10$), and it is obvious that individual calibrations of each sensor are necessary. This fact is not adequately covered by the factory calibration provided with the transmitter unit and hence requires regular individual recalibration of each sensor.

The response time of the humidity sensor is dependent on the polymer's ability to adsorb and desorb water vapor and on the sensor design, and it is strongly

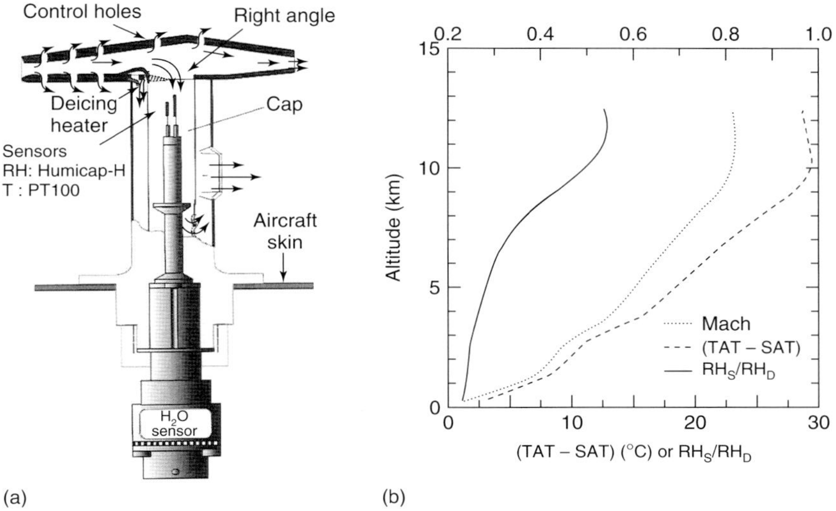

Figure 2.12 (a) Cross-sectional view of the airborne capacitive sensing element in air sampling total air temperature housing; control holes are for boundary layer and right angle causes particle separation. (b) Mean vertical profiles of Mach number, difference between total air temperature (TAT) and static air temperature (SAT): $T_t - T_s$, and ratio of static and dynamic relative humidity (RH_S/RH_D) for subsonic high-flying aircraft as obtained from MOZAIC measurements (1998).

dependent on the temperature of the sensor. The time response of the sensor in the lower or middle troposphere is good (1–10 s) but increases at lower temperatures to values of about 1 min at ambient air temperatures of $-60\,^{\circ}$C. The sensor is sensitive to chemical contamination by either additional bonding of the nonwater molecules or reduction in the ability of the polymer to adsorb water molecules, which may cause either a dry bias or reduce the sensitivity of the sensor, respectively. Regular calibration and cleaning of the capacitive sensor is a prerequisite for proper performance.

Long-term experience in MOZAIC has demonstrated that if the capacitive sensors are carefully calibrated every 500 h of flight operation, uncertainties better than $\pm$(4–6)% RH for measurements between the surface and 12 km can be obtained (Helten *et al.* 1998, 1999; Smit *et al.* 2008). For measuring stratospheric humidity, where relative humidities well below 5% prevail, the uncertainty of the capacitive humidity device is insufficient for quantitative water vapor measurements.

2.6.9
Total Water Vapor and Isotopic Abundances of ^{18}O and ^{2}H

A special airborne application is the measurement of total water vapor, that is, the sum of gaseous phase and vaporized liquid or ice phase. The air is sampled by a forward-facing inlet tube mounted outside the aircraft, while the contribution of liquid or ice phase is forced to evaporate by heating before detection. The gas and particle sampling characteristics for the inlets of the different research aircraft are approximated by computational fluid dynamics modeling (Chapter 6). Usually, gaseous water vapor content is measured simultaneously and independently in order to derive from the difference of both measurements the liquid or ice water content of clouds (Weinstock *et al.* (2006); Schiller *et al.* (2008)). A more detailed overview is given in Chapter 6, Section 6.4.

Measurements of relative isotopic abundances of ^{17}O, ^{18}O, and ^{2}H in atmospheric water vapor constitute a complementary and powerful proxy to study various processes in which atmospheric water vapor is involved (Moyer *et al.* 1996). The traditional way is to use cryogenic techniques to trap water vapor in the atmosphere into samples that are measured subsequently off-line with laboratory-based isotope-ratio determination by mass spectroscopy (Zahn, 2001; Franz and Röckmann, 2005). A detailed introduction to water sampling for isotopic analysis, including scientific examples, are given in Chapter 3. While mass spectroscopy can provide high-precision measurements, cryogenic trapping requires long sample times, particularly in dry conditions. This reduces the spatial resolution of the measurement with the speed of the aircraft, such that small spatial structures such as isolated clouds cannot be resolved. The use of tunable diode laser spectroscopy techniques such as TLAS (Webster and Heymsfield, 2003; Dyroff, Fuetterer, and Zahn, 2010) and cavity ring-down spectroscopy (CRDS) (Kerstel *et al.* 2006; Sayres *et al.* 2009a) enables *in situ* measurements of water vapor isotope ratios also to be performed. Another challenging technique to detect isotopic ^{2}H abundance at low humidities is using the photodissociation of water vapor followed by laser-induced fluorescence detection of the OH fragment (HOxtope) (St Clair *et al.* 2008). With these new *in situ* measuring devices, sampling times can be reduced by more than a factor 20 to achieve similar performance compared to the conventional off-line sampling techniques (Dyroff, Fuetterer, and Zahn, 2010).

2.6.10
Factors Influencing In-Flight Performance

Numerous factors can influence the in-flight performance of airborne water vapor measurements. To mention them all is beyond the scope of this chapter. Of crucial importance are the sticking of water vapor at surfaces and the appropriate use of sampling systems.

2.6.10.1 Sticking of Water Vapor at Surfaces

Water molecules are highly polar, such that water molecules attach themselves tenaciously to surfaces. Particularly, at low temperatures, this can lead to large memory effects of the water vapor measurements. Additional heating may eliminate any such memory effects. The selection of hydrophobic materials is an important part of the sampling and measuring system design of the hygrometer. In general, to avoid any water vapor contamination or memory effect from the aircraft skin, it is most favorable that the air inlets are sampling air outside the aerodynamic boundary layer at the aircraft skin. In addition, moisture must not be allowed to leak into the measurement system or interfere with the measurements. This is most critical in dry regions or at low humidities in the upper troposphere and stratosphere.

2.6.10.2 **Sampling Systems**

Most airborne hygrometers use extractive sampling systems that force the sampled ambient air through an appropriate inlet system into a closed-path detector system installed inside the aircraft. Usually the sideward or backward facing type of air inlets are deployed by using a pump, and the forward directed type of air inlets use the ram pressure caused the moving aircraft. Some special hygrometer designs use measuring systems located outside the aircraft, such as fast Lyman-α or IR absorption hygrometers, or open-path TLAS systems. These open-path systems have the advantage of virtually eliminating contamination issues and removing the requirement for a reference instrument for in-flight calibration, but they require ambient air temperature and pressure measurements. Furthermore, at high speed, this type of measurement is sensitive to any flow disturbances that may lead to poorly defined pressure and temperature conditions within the measuring section of the open-path system.

2.6.11 Humidity Measurements with Dropsondes

A dropsonde, or dropwindsonde, is a compact meteorological device that is released from high-flying aircraft. While descending (speed of $\sim 10\,m\,s^{-1}$) through the atmosphere on a special balloon-like parachute, continuous measurements of pressure, temperature, relative humidity, and horizontal wind velocity and direction are made and transmitted by radiotelemetry to the aircraft for further onboard data processing (Figure 2.13a).

Dropsondes, an adaption from radiosonde technology (Dabberdt *et al.* 2002), were first developed in the 1960s for hurricane reconnaissance and forecasting purposes. Since then the dropsonde technology has been further developed at the NCAR (Boulder, USA), see for example Govind (1975). The latest major achievement in dropsonde technology is the NCAR GPS dropwindsonde (Hock and Franklin, 1999). The wind-finding capability of the dropsonde is based on the GPS satellite navigation, while pressure, temperature, and relative humidity sensors (all capacitive) are the same as used in radiosondes (model type RS92) manufactured by Vaisala (Finland).

The dropwindsonde consists of four major components: the pressure, temperature, relative humidity (PTU) sensor module; the digital microprocessor circuitry; the GPS receiver module; and the 400 MHz radio transmitter (Figure 2.13b). Relative humidity is measured by the H-Humicap, the same type of capacitive thin film sensor as deployed in the capacitance hygrometer (Section 2.6.8). Artifacts in the measurements caused by condensation and icing can be avoided through the use of two sensors operating on a preprogrammed heating cycle: while one sensor

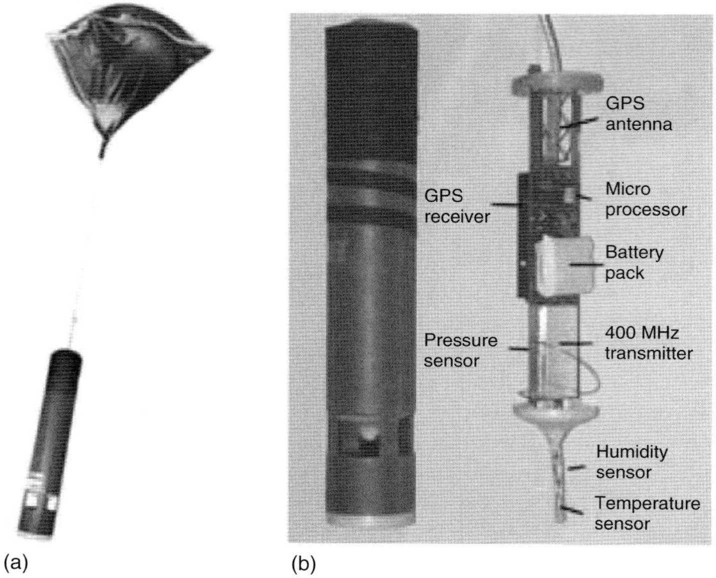

Figure 2.13 (a) GPS dropsonde descending on its parachute. (b) GPS dropsonde and internal view.

measures ambient relative humidity, the second sensor is heated and allowed to recover. The sondes are manufactured in license by Vaisala (Finland), and thousands of them are flown every year for hurricane reconnaissance and other atmospheric research purposes.

2.6.12
Calibration and In-Flight Validation

Essential for an accurate and reliable humidity measurement is regular calibration against an accurate reference instrument, sometimes called a *transfer standard*, that operates on fundamental principles and is capable of providing stable and accurate results. Most widely used transfer standards are chilled mirror hygrometers. Thereby, it is important that each transfer standard is traceable to a primary standard. The primary standard, which relies on fundamental principles and base units of measurements, is very accurate, but cumbersome, expensive, time consuming, and thus not applicable in practice. Only a few national standard laboratories have the availability of a primary standard. A detailed description of different calibration techniques and procedures used by national standard laboratories are given by Wiederhold (1997). General guidelines for calibration of atmospheric humidity instruments are given by the World Meteorological Organization (1983).

While the accuracy of these standards has been well established under atmospheric pressure, they have not been validated at the low pressures under which the most controversial water vapor measurements have to be made, in the UT/LS. Also, instruments that might be checked and validated under ideal laboratory conditions might be subject to unknown biases and systematic errors under flight conditions. The AquaVIT water vapor intercomparison, discussed below, made a significant effort in addressing this issue.

In addition, it is essential to validate the performance of different airborne hygrometers through intercomparison with other airborne hygrometers under realistic measurement conditions. This is especially important for those instruments that measure low humidities in the UT/LS. Particularly, at the lowest range of 1–10 μmol mol^{-1}, water vapor observations show large uncertainties, as was noted by the comprehensive 2000 SPARC Assessment of Upper Tropospheric and Stratospheric Water Vapor (2000). It includes intercomparisons of satellites, aircraft, balloon-borne, and ground-based water vapor instruments. Since the report of the World Meteorological Organization (1983), discrepancies remained between key instruments such as Harvard–Lyman-α (Weinstock *et al.* 2009), FISH–Lyman-α (Zöger *et al.* 1999; Schiller *et al.* 2008), and (Jet Propulsion Laboratory) JPL–TLAS (May, 1998). For example, from aircraft intercomparisons during the AURA–MLS (Microwave Limb Sounder) satellite validation (Read *et al.* 2007), the key instruments showed discrepancies of 10–20% or more at low water vapor values ($\leq$ 10 μmol mol^{-1}). Particularly large differences were observed at temperatures below 190 K. A major laboratory intercomparison was the AquaVIT Water Vapor Intercomparison campaign (https://aquavit.icg.kfa-juelich.de/WhitePaper/AquaVITWhitePaper_Final_23Oct2009_6 MB.pdf), where a large number of hygrometers were compared to each other under controlled pressure, temperature, and humidity conditions typical of the UT/LS. Generally, the UT/LS hygrometers such as Harvard–Lyman-α, FISH–Lyman-α, and JPL–TLAS showed agreement within ±10% in the water vapor range of 1–150 μmol mol^{-1}, whereas the Harvard–Lyman-α tended to larger readings at lower water vapor values. These differences observed at low water mixing ratios were of the same character but significantly smaller than those exhibited in flight intercomparison campaigns. However, AquaVIT does not address atmospheric sampling issues, which primarily affect the in-flight performance and have to be addressed separately.

2.6.13
Summary and Emerging Technologies

Airborne measurement of atmospheric humidity has been and continues to be a challenging task. Airborne humidity measurements require continuous care, regular maintenance, and intensive calibration. There is no airborne sensor available that can cover the full dynamic range of water vapor levels from a few percentage near the surface down to a few μmol mol^{-1} in 15–20 km altitude. A survey of different techniques and their performance in terms of time response, precision and accuracy, and specific airborne applications is presented in Table 2.5. While a broad spectrum of instruments exists today, all of them have their limitations, making it necessary to combine several instruments to achieve the required data quality.

Substantial improvements in the performance of airborne water vapor measuring systems or new aircraft applications are in progress, particularly through new developments in the field of tunable diode laser spectroscopy. New developments of tunable diode laser spectroscopic techniques are on their way, such as CRDS, a direct absorption technique based upon the gradual decrease of light intensity as a tunable diode laser light pulse undergoes multiple reflections between two highly reflective mirrors in an optical cavity (Berden and Engeln, 2009). An introduction to CRDS and related methods is given in Chapter 3, Section 3.5.3. In the next decade, good performance at low atmospheric humidities (1–100 μmol mol^{-1}) can be expected. Also new airborne applications using CRDS hygrometry to measure, for example, isotopic abundances of ^{17}O, ^{18}O, and ^{2}H in atmospheric water vapor to identify their sources and sinks are becoming feasible (Berden and Engeln, 2009).

Laser-induced photoacoustic spectrometry (LPAS) uses the physical effect that the absorption of periodically modulated laser light can generate a sound wave at the frequency of the light modulation and is proportional to the concentration of the absorbing compound. The dynamic range (or the measurement range) of a photoacoustic instrument is rather large, so that five to six orders of magnitude change in concentration can be achieved (Bozoki, Pogány, and Szabó, 2011). In general, in order to achieve stable relative accuracies of 10% or better, a photoacoustic hygrometer needs to be regularly calibrated.

Advanced instrumental developments of the CIMS (chemical ionization mass spectroscopy) detection techniques enable water vapor measurements at extremely low stratospheric values.

2.7 Three-Dimensional Wind Vector

The three-dimensional (3D) wind vector needs to be measured to characterize large-scale atmospheric motions. We distinguish here between motions on scales on the order of 10 or 100 m and above, and the velocity fluctuations at smaller scales that can be considered turbulence; the latter is covered in Section 2.8. The use of research aircraft to measure mean wind components and vertical eddies has been evolving quite greatly in the past few decades. The techniques to measure wind and large-scale turbulence by means of airborne instrumentation are (i) the aircraft's response to wind (Lenschow, 1976), (ii) remotely sensed wind measurement using Doppler wind Light Detection and Ranging (LIDAR) (Bilbro *et al.* 1984a; Bilbro *et al.* 1986), and (iii) *in situ* wind measurement. The last technique remains the most accurate method and is the subject of this section.

Table 2.5 Survey of different water vapor sensing techniques and their specifications of performance for airborne use.

Sensor type	Altitude range	Measured quantity	Measured range	Time response	Accuracy	Reliability	Limitations
Dew–frost point [thermoelectric cooling]	LT, MT, and UT	Dew–frost temperature	230–300 K	LT: few seconds UT: few minutes	0.2–0.5 K	Good	Ambiguity dew and frost point 233–273 K
Dew–frost point [cryogenic cooling]	LT, MT, UT, and LS	Dew–frost temperature	185–300 K	10–20 s	0.2 K	Good	Ambiguity dew and frost point 233–273 K
Lyman-α absorption	LT and MT (open path)	Density	0.1–25 g m^{-3}	5 ms	5%	Light source short lifetime; regular cleaning of optic windows needed	Requires in-flight calibration
Lyman-α fluorescence	MT, UT, and LS	Volume mixing ratio	1–1000 µmol mol^{-1}	1 s	5%	Very good	Only for dedicated mission; requires high expertise
IR absorption	LT and MT	Density	0.3–30 g m^{-3}	5 ms	1 g m^{-3}	Sensitive to aircraft vibrations and fast pressure changes	Interferences with large aerosol loadings
IR absorption (closed path)	LT and MT	Density	0.3–30 g m^{-3}	1 s	0.1 g m^{-3}	Good	Interferences with large aerosol loadings
TLAS (open path)	MT, UT, and LS	Density	0.005–1 g m^{-3}	5 ms (20 Hz)	5% (1 Hz)	Good	Only for dedicated mission; requires high expertise
TLAS (closed path)	LT and MT	Density	0.05–30 g m^{-3}	2 s	5–10%	—	For dedicated and in-service operation
CRDS	LT and MT	Density	0.1–30 g m^{-3}	10–30 s	5–10%	—	—
Thin film capacitance [Humicap]	LT, MT, and UT	Relative humidity	0–100% RH	LT: 1 s MT: 10 s UT: 1 min	5% RH	Good, when regularly calibrated	Not suitable for LS
Dropsonde [Humicap]	LT, MT, and UT	Relative humidity	0–100% RH	LT: 1 s MT: 30 s UT: 5 min	5–10% RH	Good	Slow response in UT; not suitable for LS

[a] LT, lower troposphere; MT, middle troposphere; UT, upper troposphere; LS, lower stratosphere.

2.7.1 Airborne Wind Measurement Using Gust Probes

Measuring the wind components from moving platforms is a challenge since both ground speed and airspeed of the aircraft are about one order of magnitude larger than the meteorological wind, and errors arise from compressibility, adiabatic heating, and flow distortion. Basic principles of wind measurement from aircraft have been given by Axford (1968) and Lenschow and Stankov (1986).

2.7.1.1 True Airspeed (TAS) and Aircraft Attitude

The horizontal wind speed and direction are computed by the vector sum of the true airspeed (TAS) and the ground relative velocity of the aircraft. The TAS from instrumented aircraft is usually measured by means of gust probes, which can be either installed on nose or wing-tip booms or directly radome mounted. Gust probes are equipped with dynamic, static, and differential pressure ports from which the TAS and the flow angles can be calculated, allowing for the computation of the wind components.

The most common gust probe in use today is the so-called FHP (Figure 2.14). In the FHP, the local wind vector in the aircraft coordinate system is determined from the dynamic pressure increment Δp_{q} and the pressure differences between four opposite pressure holes in the FHP, that is, the pressure difference in the horizontal plane $\Delta p_\beta = p_2 - p_4$, and in the vertical plane $\Delta p_\alpha = p_1 - p_3$, where p_j, with $j = 1, 2, 3, 4, 5$, denotes the individual holes of the FHP, with p_5 being the central hole. The pressure differences Δp_α and Δp_β increase when the angle of attack α and the angle of sideslip β increase. But the pressure differences also depend on the airspeed (and therefore on both the dynamic pressure increment Δp_{q} and the Mach number) and on the air density ρ (and therefore on the altitude z). Therefore, these pressure differences can be used for both the TAS and aircraft

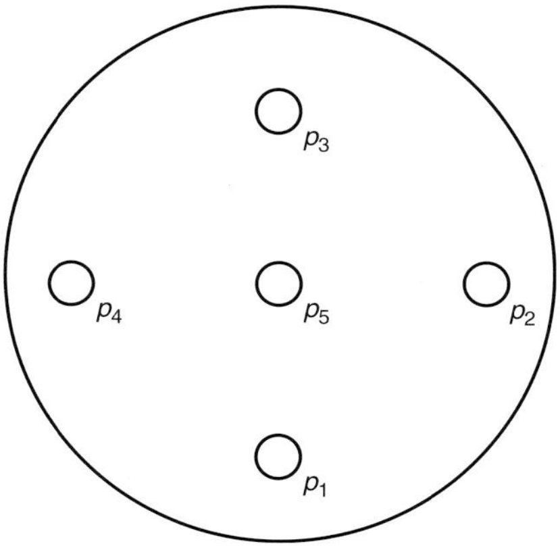

Figure 2.14 Schematic illustration of an FHP showing the pressure ports p_1 to p_5 (head-on perspective, i.e., starboard is on the left side from this point of view).

attitude measurements. Calibration routines, both for wind tunnel experiments and flight maneuvers can be found in the literature by Haering (1990); Wörrlein (1990); Haering (1995); Barrick *et al.* (1996); Friehe *et al.* (1996); Khelif, Burns, and Friehe (1999); Weiß, Thielecke, and Harders (1999); Williams and Marcotte (2000); van de Kroonenberg *et al.* (2008); and van de Kroonenberg (2009). For details of the principle of the FHP, see Section A.1, given in the Supplementary Online Material provided on the publisher's web site.

The angles of attack and sideslip (α and β, respectively) are defined as the flow angles with respect to the longitudinal axis of the aircraft in the lateral and vertical directions. These angles along with the TAS define the velocity vector relative to the aircraft. The calculation of airspeed and flow angles is based on the measurement of the surface pressure distribution (Brown, Friehe, and Lenschow, 1983) on the nose of the aircraft itself, from which the angles of attack and sideslip, and the dynamic pressure are obtained. Examples of such instruments are the Best Aircraft Turbulence (BAT) probe, an effort by the NOAA Atmospheric Turbulence and Diffusion Division (ATDD) and that can be installed on slow aircraft; the Aventech Aircraft Integrated Meteorological Measurement System (AIMMS-20) (Foster, 2003); and the FHP by Rosemount that can also be used on fast aircraft. The relative ground speed and the aircraft attitude determination (Euler angles) are described in Section 2.3.

2.7.1.2 Wind Vector Determination

The wind vector can be determined through the following logical progression:

(i) TAS calculation,
(ii) flow angles calculation,
(iii) rotation matrices, and
(iv) ground speed calculation.

At each stage, the measured variables must be corrected for undesired airflow distortion as explained later in this section.

The meteorological wind vector $\mathbf{v}$ (in the Earth coordinate system) is the difference between the velocity vector of the instrument or sensor $\mathbf{v}_{\text{sensor}}$ in the Earth-fixed coordinate system and the TAS vector $\mathbf{v}_{\text{TAS}}$ (Lenschow and Stankov, 1986; Lenschow and Spyers–Duran, 1989):

$$\mathbf{v} = \mathbf{v}_{\text{sensor}} - \mathbf{v}_{\text{TAS}} = \mathbf{v}_{\text{aircraft}} + \mathbf{v}_{\text{AS}} - \mathbf{v}_{\text{TAS}} \tag{2.36}$$

$\mathbf{v}_{\text{sensor}}$ can be expressed as the sum of two velocity components, namely, the speed $\mathbf{v}_{\text{aircraft}}$ of the aircraft reference point (the location of the inertial reference system, IRS) and a relative speed $\mathbf{v}_{\text{AS}}$ of the sensor with respect to this point caused by changes in the aircraft attitude. $\mathbf{v}_{\text{AS}}$ can be calculated from the aircraft rotational velocity vector ω and the distance vector $\mathbf{r}_{\text{AS}}$ between aircraft reference point and the flow sensor in an aircraft-fixed coordinate system:

$$\mathbf{v}_{\text{AS}} = \omega \times \mathbf{r}_{\text{AS}} \tag{2.37}$$

Different coordinate systems are used in the wind calculation since some of the data measurements are referenced to the aircraft (gust probe data) and some to an Earth-based coordinate system (IRS data). Therefore, two main coordinate systems are important in the measurement of wind speed, namely, the Earth-fixed (geodetic) coordinate system (x_g, y_g, z_g) and the aircraft-fixed (body) coordinate system (x,y,z). Three Euler angles are used to describe the relative orientation of these two coordinate systems to each other: aircraft yaw (ψ), pitch (θ), and roll (ϕ) angles. The definition of the coordinate systems and the transformation matrices between them are subject to national and international standards, which are commonly available (ANSI, 1992; DIN, 1990).

The TAS as the speed of an aircraft with respect to the air in which it is flying is completely based on pressure data from the gust probe and meteorological parameters such as air temperature, pressure, and humidity. The respective formula can be directly derived from the Bernoulli equation, the ideal gas law, and the adiabatic equation. We thus obtain

$$\mathrm{TAS} = \sqrt{2 \cdot \left(\frac{k}{k-1}\right) \cdot R \cdot T_s \cdot \left[\left(\frac{p_t}{p_s}\right)^{\frac{k-1}{k}} - 1\right]} \tag{2.38}$$

with adiabatic index k, gas constant R for humid air, static air temperature T_s, total air pressure p_t, and static air pressure p_s of the undisturbed air.

The TAS wind vector $\mathbf{v}_{\mathrm{TAS}}$ is usually measured in two steps. First, the absolute value $|\mathbf{v}_{\mathrm{TAS}}| = \mathrm{TAS}$ is determined using Eq. (2.38). It is important to note that the TAS wind vector is normally not aligned with the gust probe due to changes in the aircraft angle of attack and the influence of wind. Especially for slow aircraft, this deviation can be significant. Therefore, the accuracy of the TAS calculation depends strongly on the proper parameterization of the angular dependency of the total and static air pressure measurement (see later discussion). The second step is the measurement of the angles of attack and sideslip between the airflow vector and the gust probe axis. A typical TAS probe is the FHP, see Section A.1, given in the Supplementary Online Material provided on the publisher's web site.

2.7.1.3 **Baseline Instrumentation**

Two main sensor configurations are commonly used for flow angle measurements. The first one uses the aircraft radome as a sensor, equipped with appropriate flush-mounted pressure ports. This method has the advantage of short tubing between pressure sensors and the gust probe, which guarantees fast response and avoids damping and resonance effects along these lines. This design is also very insensitive to mechanical vibration. However, the aircraft weather RADAR, which is typically located aft of this installation, puts some restrictions on the pressure port design. The determination of the inlet location on the typically nonsymmetric radome and the disturbed pressure field in this area require extensive airflow simulation and in-flight calibration. Another disadvantage is the lack of a proper static pressure port close to the probe.

The second configuration uses a nose boom to locate a flow sensor ahead of the aircraft in order to minimize the effects of aircraft-induced pressure disturbance. If the probe is equipped with a static air pressure port, static source calibration becomes easier and no delay effects between the different pressure sources have to be accounted for. The boom is usually inclined in order to compensate the mean aircraft angle of attack. However, there are serious restrictions to the installation. The intention of bringing the sensor as far ahead of the aircraft as possible (in order to leave the aircraft-induced pressure disturbance) is limited by aeroelastic considerations and the boom natural frequency.

Recent developments have shown that the advantages of both methods can be combined in a boom solution where the pressure sensors are located directly behind the flow sensor at the tip of the boom. This requires a very stiff boom construction as well as specially modified compact pressure sensors (Crawford and Dobosy, 1992; Cremer, 1999).

Pressure sensors for airborne flow measurements must be of high accuracy, small, light, fast and inert to the environmental conditions on the aircraft. The desirable absolute accuracy is 0.1 hPa, which requires regular calibration and an accurate reference. The optimum acquisition rate for these measurements is given by the pneumatic response time of the system. A typical value is 100 Hz, which requires response times of some milliseconds for the sensor itself. Since the most critical parameter of a pressure measurement is temperature, the sensors should be actively temperature controlled in order to withstand an environment that can range from +70 to $-70\,^{\circ}$C (e.g., on a jet). Thermal error corrections based on measured sensor temperature are subject to systematic errors when the outside temperature changes. The sensitivity to aircraft accelerations can be minimized by choosing an appropriate orientation of the sensor with respect to the aircraft axes.

The second important measurement for the calculation of wind speed is the position and attitude of the gust probe in an Earth-fixed coordinate system. This is usually accomplished using an IRS. Owing to the high costs of these systems, data from the aircraft IRS is often used for this. Time delays and data steps due to internal processing and correction schemes, the Schuler effect, and data drift are possible effects that have to be treated when processing the data (Lenschow and Spyers-Duran, 1989; Matejka and Lewis, 1997). At present, off-the-shelf stand-alone systems are available, which combine accurate low-speed GPS information with data from a fast IMU by applying sophisticated filter techniques. Some systems are even able to use real-time DGPS.

2.7.1.4 **Angles of Attack and Sideslip**

As mentioned, the two flow angles measured by the gust probe are used to determine the orientation of the TAS vector, one of the two vectors needed to directly determine wind speed. The angle of attack α is the angle between the projection of the airflow vector onto the x, z-plane of the aircraft coordinate system and the aircraft x-axis itself. The angle of sideslip β is defined as the angle between the airflow vector and the aircraft x, z-plane (Boiffier, 1998; Luftfahrtnorm, 1970). Since the free-stream flow lines are bent due to the influence of the aircraft, the local flow angles at the gust probe location are biased (Figure 2.15).

This means that the measurement of α and β require a detailed characterization of the airflow ahead of the aircraft and a precise knowledge about the position and

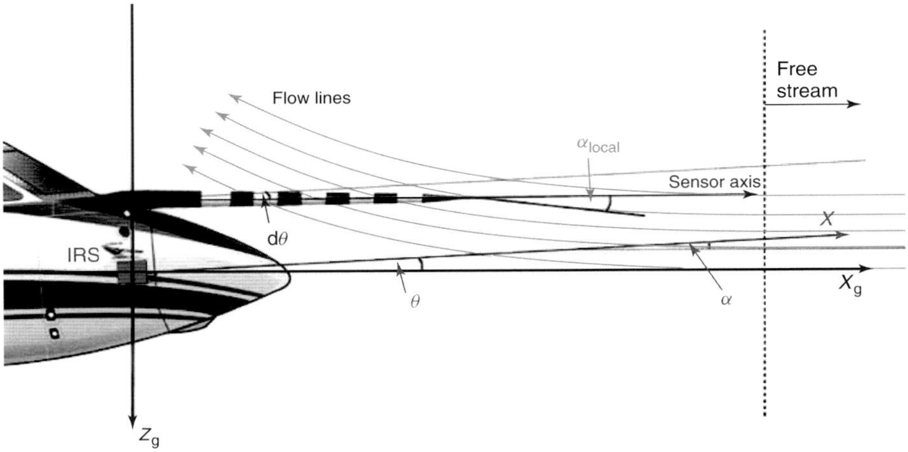

Figure 2.15 The angle of attack can usually not be measured directly by a gust probe, even if the sensor is perfectly characterized. This is due to the flow disturbance caused by the aircraft itself and a possible angular offset between the axes of the aircraft and the flow sensor. When flying at a constant level in a stable atmosphere, the angle of attack is identical to the aircraft pitch as measured by the IRS.

orientation of IRS and the gust probe relative to each other. The aircraft-induced vertical deflection of the flow lines is also known as upwash. When the gust probe is not installed on the aircraft nose or a centered nose boom (e.g., a wing-mounted boom), there will be a lateral deflection of the flow lines, which is called *sidewash*.

2.7.2 Errors and Flow Distortion

2.7.2.1 Parameterization Errors

The calculation of atmospheric parameters such as pressure, temperature, or flow angles from an initial measurement always involves correction terms or specific sensitivity coefficients. These terms account for aerodynamic effects or certain sensor properties and they usually depend on flight parameters such as the Mach number and the flow angle. While some of them are provided by the manufacturer of the respective sensor, many others are specific to the aircraft and the sensor configuration. They have to be determined by the aircraft operator using in-flight calibration procedures or airflow analysis (Lenschow and Spyers–Duran, 1989; Boegel and Baumann, 1991; Crawford, Dobosy, and Dumas, 1996; Kalogiros and Wang, 2002).

2.7.2.2 Measurement Errors

A proper calibration must cover the whole logical chain between the sensor and the data acquisition. Depending on the aircraft being used, the required data range can be significant: $-70\,^{\circ}$C, 200 hPa, and stratospheric humidity require sophisticated calibration equipment. The impact of environmental conditions on a sensor can be significant and is hard to detect during flight. An example is the sensitivity of a pressure sensor to low temperatures or accelerations. A very special sensor is the IRS, one must investigate the instrument during flight in order to learn about its characteristics. In some cases, GPS data and sophisticated filter methods are used to improve the accuracy of aircraft position or velocity data (Lenschow and Spyers-Duran, 1989; Khelif, Burns, and Friehe, 1999).

2.7.2.3 Timing Errors

The processing of wind speed involves many different data sources each having its own characteristic temporal behavior, that is, different response time constants or delays, caused by physical properties of the sensing element or electronic (processing) delays. For humidity and temperature, these response times are not constant but depend on parameters such as the aircraft speed and static pressure. It is immediately clear that the wind calculation will have possibly large errors if one uses data sources that are not exactly synchronized (Lenschow and Spyers-Duran, 1989). Any time shift between the different time series will lead to artificial wind signals because the contributions caused by aircraft motion will not completely cancel each other out. In-flight calibration techniques are necessary to identify these delays, which have to be applied during data processing (Lenschow and Spyers-Duran, 1989; Boegel and Baumann, 1991).

2.7.2.4 Errors due to Incorrect Sensor Configuration

The wind signal can be degraded by vibrations of a "soft" nose boom and by long pressure lines between the gust probe and the pressure sensor (Whitmore *et al.* 1990). Especially on large aircraft, the fuselage can no longer be seen as rigid and a large distance between flow angle sensor and IRS will cause errors due to fuselage bending. Misalignment of sensors is another problem that concerns the IRS and the nose boom (i.e., flow sensor) orientation.

2.7.3 In-Flight Calibration

The calibration of wind sensing systems is a complex task that begins with the ground calibration of each instrument in the system (pressure, temperature, humidity), includes the characterization of the probe in wind tunnel, and ends with in-flight calibration of the whole system. Two different ways of performing in-flight calibration are common: the Lenschow and Rodi maneuvers. More advanced techniques such as measurement of divergence (Lenschow, Savic–Jovcic, and Stevens, 2007) can provide a very accurate calibration/verification of the horizontal wind measuring system. The in-flight calibration can be found in Section A.1.3, given in the Supplementary Online Material provided on the publisher's web site.

2.8 Small-Scale Turbulence

Turbulence in atmospheric flows plays a dominant role for many processes such as mixing, turbulent transport, and collisions of particles (Wyngaard, 2010). In principle, the contributing scales range from the largest scale L that has the dimension of the considered phenomenon itself (e.g., the boundary layer height for convective plumes or the cloud diameter for cloud turbulence) down to the dissipation scale, also called the *Kolmogorov microscale* η given by

$$\eta = \left(\frac{\nu^3}{\overline{\varepsilon}}\right)^{1/4} \tag{2.39}$$

with the air viscosity $\nu \approx 1.5 \times 10^{-5}\ \mathrm{m^2\,s^{-1}}$ and the mean turbulence energy dissipation rate per unit mass of

$$\overline{\varepsilon} \sim \frac{u_{\mathrm{rms}}^3}{L} \tag{2.40}$$

where u_{rms} is the root-mean-square value of the flow velocity. For typical atmospheric conditions, η is on the order of millimeters; that is, atmospheric turbulence spans a huge range of spatial scales resulting in high Reynolds' number:

$$\mathrm{Re} = \frac{L \cdot u_{\mathrm{rms}}}{\nu} \sim \left(\frac{L}{\eta}\right)^{4/3} \tag{2.41}$$

Therefore, the atmosphere provides a natural laboratory for high Reynolds' number turbulence, characterized by strong small-scale intermittency. This makes small-scale turbulence measurement in the atmosphere also appealing for fundamental turbulence research.

This section focuses on sensors and devices suitable for airborne small-scale turbulence measurements. With some arbitrariness, the "small scales" are defined as scales within and below the inertial range, typically below the order of tens of meters or so where standard aircraft instrumentation has its limitation. An introduction of a few basic concepts of sampling requirements can be found in Section A.2, given in the Supplementary Online Material provided on the publisher's web site.

2.8.1 Hot-Wire/Hot-Film Probes for High-Resolution Flow Measurements

If flow measurements with a resolution down to the dissipation scale are required, there is no alternative to hot-wire anemometry. Note that if a more robust sensor is required for airborne applications, thin hot-film probes might be used instead of the fragile thin wires with a slight reduction of temporal resolution. The basic principle of hot-wire anemometry is the forced convection of a heated sensing wire in a fluid. "Forced" convection means that the bulk relative velocity between sensor and fluid is due to external forces, whereas "natural" convection is due to buoyancy. The heat transfer power (measured in Watts) can be described as

$$W = U \cdot A \cdot (T_{\mathrm{sen}} - T_{\mathrm{f}}) \tag{2.42}$$

with the sensor temperature T_{sen}, the fluid temperature T_f (all temperatures in K), the surface area of the sensor A, and the convective heat transfer coefficient U (in units of W m^{-2} K^{-1}). Here, the conduction heat transfer between the heated portion of the sensor and its support is ignored for simplicity.

The dimensionless Nusselt number is defined as

$$\mathrm{Nu} = \frac{L \cdot U}{\kappa} \tag{2.43}$$

where L is a typical dimension and κ is the thermal heat conductivity. The dependence of the dimensionless Nusselt number Nu on the convection heat transfer coefficient U is used to relate the heat transfer to the flow velocity u. For forced convection from a cylinder, the Nusselt number is semiempirically found as

$$\mathrm{Nu} \approx 0.24 + 0.56 \cdot \mathrm{Re}^{0.5} \tag{2.44}$$

where $\mathrm{Re} = u \cdot d/\nu$ is the Reynolds number of a flow around a wire with diameter d and ν is the fluid viscosity, often referred to as *King's law* (King, 1914). This finally results in

$$U = \frac{0.24\,\kappa}{d} + 0.56\,\kappa \cdot \sqrt{\frac{u \cdot d}{\nu}} = a + b \cdot \sqrt{u} \tag{2.45}$$

Thus, the convection heat transfer coefficient U depends on the geometrical sensor, physical fluid properties, and the flow velocity u.

The most common type of hot-wire anemometer is the constant temperature anemometer (CTA). Here, the sensor temperature T_{sen} and, therefore, the resistance of the sensing wire (R_{sen}) are maintained at a constant value. With the heating current I, we find that for steady-state conditions,

$$W = I^2 \cdot R_{sen} = U \cdot A \cdot (T_{sen} - T_f) \tag{2.46}$$

If the flow velocity u increases, U will increase and the system has to increase I through the sensor to restore equilibrium. Since R_{sen} is constant, the voltage drop $E_{out} = I \cdot R_{sen}$ over the sensor increases with current, thus giving a voltage signal proportional to $u^{1/4}$.

The main advantage of a CTA is its high bandwidth of up to 100 kHz (or higher) over a huge range of flow velocities. With multisensor probes, the two-dimensional (2D) or 3D velocity vector can also be measured. Sensor calibration can be performed with the help of special pressure nozzles in a free jet created by compressed air up to Mach 1. The disadvantage of hot wires, especially on aircraft, is their fragility and that each sensor has to be carefully calibrated individually, including its supports, connectors, and cables. Alternatively, the sensor can be calibrated by comparing with a standard anemometer (e.g., gust probe) as in-flight calibration or postprocessing. The hot-wire signal is highly nonlinear in u, and calibration depends on the temperature difference between the sensor and the environment. More sophisticated probes are temperature compensated. The influence of natural convection around the sensing wire can be neglected for airborne measurements because of the high TAS. Although robust hot-film probes are available for high

TAS (with some degradation of the bandwidth), only a few airborne measurements are reported (Sheih, Tennekes, and Lumley, 1971; Merceret, 1976a; Merceret, 1976b; Lenschow, Friehe, and Larue, 1978; Payne and Lumley, 1965). Most hazards to the wires are due to impacting aerosol particles (in particular, during takeoff and landing) or cloud droplets (Siebert, Lehmann, and Shaw, 2007; Siebert, Shaw, and Warhaft, 2010), and aircraft vibration. Electromagnetic noise can also be troublesome, but it can be minimized with shielded cables and by placing signal amplifiers and data recording devices as close to the sensors as possible.

A more detailed introduction of all aspects of hot-wire anemometry can be found in the works by Comte-Bellot (1976), Bruun (1995), and Goldstein (1996).

2.8.2
Laser Doppler Anemometers

Laser Doppler velocimetry (LDV) provides a measurement of the speed of particles in a flow, via heterodyne detection of Doppler-shifted light scattered by individual particles. The method has been refined and widely adopted for the study of engineering flows, and detailed reviews are available (Adrian, 1996; Buchhave, George, and Lumley, 1979). The method has been further modified to enable determination of the diameter of spherical particles (Bachalo, 1980). It is relatively recent that LDV has become sufficiently robust to be commonly used for atmospheric measurements (Chuang *et al.* 2008). For example, it has been used in the measurement of turbulence statistics in clouds where other methods can be troublesome due to the multiphase environment (Siebert *et al.* 2006a). Because the method measures the speed of particles, rather than the flow itself, it can be of considerable use in studying dynamics of inertial particles in turbulence (Saw *et al.* 2008).

The physical principle underlying LDV is the Doppler shifting of light scattered by particles such as cloud droplets or aerosol particles that are moving relative to a light source and detector. The Doppler shift in the detected radiation is exceedingly small in practical terms. Therefore, the measurement is made by mixing two slightly different Doppler-shifted signals from the same particle and measuring the resulting beat frequency. The difference in frequencies arises from the geometrical arrangement of the two crossing laser beams. An additional user-imposed frequency shift from a Bragg cell is sometimes imposed to minimize directional ambiguity. The beat frequency is proportional to the component of the particle velocity vector that lies in the plane of the crossing laser beams perpendicular to the optical axis; see Section A.3 (given in the Supplementary Online Material provided on the publisher's web site) for a derivation of the frequency dependence on particle motion and system geometry. Additional velocity components can be measured with more complex, multilaser, and detector systems. As a particle moves through the beam-crossing region, the measured signal has the form of a Gaussian envelope, as a result of the Gaussian laser beam profile modulating the Doppler beat frequency.

By its very nature, LDV is ideally suited for measurements in clouds. The very droplets that tend to disturb other high-resolution methods, such as hot-wire anemometry (Section 2.8.1), are the source of the signal. The ability of an LDV system to accurately sample a turbulent flow field in a cloud depends on its spatial resolution and its velocity resolution. The spatial resolution is determined by the cloud particle number density and the instrument sample cross section (modern sampling systems are designed such that sampling frequency is not a limiting factor). In practice, because of the nonuniform spatial sampling, the effective resolution can be degraded to as much as 10 times the average distance between sampled cloud particles. The velocity resolution must be sufficiently fine to capture typical velocity fluctuations corresponding to a desired spatial scale r. For a turbulent flow with kinetic energy dissipation rate ε, the magnitude of the velocity fluctuations can be estimated from

$$v_\mathrm{r} = (\varepsilon \cdot r)^{1/3} \tag{2.47}$$

The velocity resolution scales with the resolution with which the Doppler beat frequency can be measured. Ultimately, this depends on instrument parameters such as signal-to-noise ratio (SNR), sampling frequency, and sample volume size via the Cramer–Rao error estimate (Chuang *et al.* 2008). Finally, it is again emphasized that particle, not fluid, speeds are measured. If the fluid speed is desired then a further source of error is the inability of large cloud particles to follow high-frequency velocity fluctuations. Typical cloud droplets of diameters up to $\sim$30 μm can be safely approximated as flow tracers for the moderate energy dissipation rates of most clouds and for the purposes of estimates of average turbulence properties. Details on the response of cloud droplets to turbulent fluctuations are given in Section 4 of the work by Chuang *et al.* (2008).

An example of a turbulent velocity spectrum from a cumulus cloud is shown in Figure 2.16. The solid curve is for data obtained from an LDV instrument (Chuang *et al.* 2008), and the dashed curve is for data obtained simultaneously from a

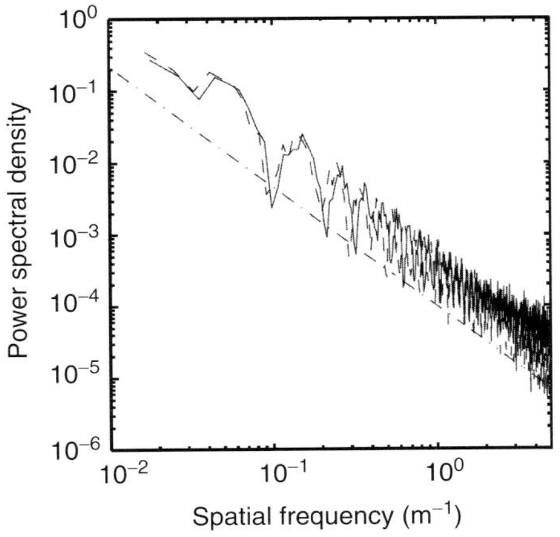

Figure 2.16 Turbulent velocity energy spectrum measured with an LDV system (solid) and a sonic anemometer (dashed); a line with a slope of −5/3 is included for reference (dashed–dotted). (Source: Adapted from Figure 12 in Chuang *et al.* (2008).)

sonic anemometer. The measurements were made aboard the Airborne Cloud Turbulence Observation System (ACTOS) deployed via a helicopter (Siebert *et al.* 2006a). The agreement between the two instruments is reasonable throughout the resolvable subset of the inertial range. Furthermore, both power spectra match, at least to within the sampling uncertainty, the expected $-5/3$ power law dependence (exemplified by the dotted–dashed line) for the energy spectrum of the longitudinal velocity component within the inertial range. The energy spectra are plotted up to a spatial resolution of 20 cm (spatial frequency, 5 m^{-1}), which is approximately the limit of the sample-and-hold method used for the selected segment of LDV data, based on the average cloud droplet arrival frequency, as well as the spatial resolution of the sonic anemometer. The slight flattening of the LDV power spectrum at high frequencies is characteristic of the sample-and-hold method.

The focus here has been on Doppler measurements from light scattered by single particles (e.g., cloud droplets) at spatial and temporal resolution suitable for turbulence characterization. The essential elements of LDV have also been applied in a variety of systems, however. Doppler LIDARs typically measure light scattered from aerosol particles contained in a relatively large volume of air using a similar heterodyne detection method as LDV, for example, see review by Huffaker and Hardesty (1996). Furthermore, LIDAR systems operate with time (range) gates. Therefore, they provide velocity information over a region of space rather than in a single measurement volume. LIDAR-type laser Doppler techniques have been developed for airborne measurements of airspeed as well, and in principle, they offer several advantages over traditional measurement methods such as differential pressure. For example, the measurement volume can be located upwind of the disturbance generated by the moving aircraft and systems can be configured to offer spatial information such as vertical wind shear. The system described by Keeler *et al.* (1987) provided turbulence spectra up to ~10 Hz, allowing the large scales of the turbulence inertial subrange to be observed. A new system known as the *Laser Air Motion Sensor* developed at the NCAR (Spuler *et al.* 2011) provides an absolute measurement of airspeed independent of flight maneuvers and atmospheric conditions and is being extended to include a single forward-pointing beam plus three 30° off-axis beams in a single wing-pod canister to allow measurement of the full wind vector.

2.8.3
Ultrasonic Anemometers/Thermometers

Ultrasonic anemometers/thermometers (hereafter called *sonics*) are standard devices for tower-based atmospheric boundary layer studies and turbulent flux measurements (Section 2.9). Even though such instruments are not widely used on aircraft, they have a few advantages, making them attractive for at least slow-flying aircraft and helicopter-borne applications at low TAS.

The basic principle of a simplified 1D sonic is transit time measurements of two subsequent sound pulses traveling with and against the wind component along two transducers T1 and T2, which are separated by a distance L. The first sound pulse is sent from T1 to T2 with pulse velocity $u_1 = c + U$ (where c is the speed of sound), and the second pulse is immediately sent back with a velocity $u_2 = c - U$ after the first pulse is received by transducer T2. The measured transit times t_1 and t_2 are $t_1 = L/(c+U)$ and $t_2 = L/(c-U)$, and a combination of both equations yields

$$U = \frac{L}{2} \cdot \left(\frac{1}{t_1} - \frac{1}{t_2}\right) \tag{2.48}$$

and for the speed of sound,

$$c = \frac{L}{2} \cdot \left(\frac{1}{t_1} + \frac{1}{t_2}\right) \tag{2.49}$$

Note that in principle, the measurement of U depends only on the constant distance L and the two measured transit times but is independent of c, and the measurement of c is independent of U.

Under adiabatic conditions, c is given by

$$c = \sqrt{\gamma \cdot R_{\text{dry}} \cdot T_{\text{vir}}} \tag{2.50}$$

where $\gamma = c_p/c_v = 1.4$ is the adiabatic exponent (c_p and c_v are the heat capacities of dry air for constant pressure and constant volume, respectively), $R_{\text{dry}} = 287\ \text{J kg}^{-1}\ \text{K}^{-1}$ is the gas constant for dry air, and:

$$T_{\text{vir}} = T \cdot \left(1 + 0.38\,\frac{p_w}{p}\right) \tag{2.51}$$

is the virtual temperature, with actual temperature T, air pressure p, and water vapor pressure p_w. With the same shot, the wind velocity component and the virtual temperature can be measured within the same volume, making this device quite attractive (Section 2.5.6). It is straightforward to design a configuration of three transducer pairs to measure the three-component wind vector. Typical resolution is about $1\ \text{cm s}^{-1}$ for the velocity components and 10 mK for temperature measurements.

A few general design issues have to be considered. The ratio of the diameter of the transducer d and the distance L has to be small ($d/L \ll 0.1$ with L typically $\sim$0.1 m) to keep transducer shadowing effects in an acceptable limit (Wyngaard and Zhang, 1985). The effect of line averaging over the path L is discussed by Kaimal, Wyngaard, and Haugen (1968).

Usually, standard ultrasonic anemometers used for ground-based studies cannot be applied directly to fast-flying aircraft without serious modifications of the framework and electronics. The maximum wind speed (approximate TAS on an aircraft) that can be measured with a sonic is limited to about $50\ \text{m s}^{-1}$, and the framework is usually not stiff enough to avoid vibrations (Siebert and Muschinski, 2001). However, a few special developments of sonics for aircraft and helicopter are in use, and TAS up to $100\ \text{m s}^{-1}$ has been reported (Cruette *et al.* 2000; Avissar *et al.* 2009).

2.8.4 Measurements of Atmospheric Temperature Fluctuations with Resistance Wires

This section focuses on measurements of small-scale temperature fluctuations in turbulent atmosphere and clouds. Further description of airborne temperature measurements can be found in Section 2.5.

Generally, temperature readings from aircraft are strongly affected by temperature fluctuations due to dynamic pressure variations (depending on TAS and attitude angles of the aircraft), the effects of heat transport/thermal inertia of sensor supports and housings, and even the effects of viscous heating. Altogether these effects introduce bias and noise (in the case of unsteady flow around the sensor) and limit effective time constants to $\sim 10^{-1}$ s (Friehe and Khelif, 1992; Mayer *et al.* 2009). All these effects make measurement of small-scale temperature fluctuations difficult, especially in clouds and rain where wetting may cause problems (Lawson and Cooper, 1990; Sinkevich and Lawson, 2005).

The only sensors of resolution capable to detect small-scale temperature fluctuations from the aircraft described so far are fine resistive wires (2.5 μm diameter) with minimum shielding mounted on a vane that adjusts to the local flow. These ultrafast thermometers, UFTs, (Haman *et al.* 1997, 2001) are developed on the basis of similar or even finer unshielded sensors used in laboratory turbulence research. Specific airborne requirements mean that even recent versions of UFTs are of limited applicability. They are unstable over long times, require laborious and frequent replacements of sensing elements, and have to be accompanied by a stable, calibrated, slow-response thermometer as a reference. Nevertheless, they provide high-resolution information not available with other techniques and are well suited for in-cloud measurements (Haman and Malinowski, 1996; Siebert *et al.* 2006a; Haman *et al.* 2007).

The Prandtl number of air (the ratio of momentum and thermal diffusivities) is $\mathrm{Pr} \approx 0.72$. This means that smallest scales of temperature fluctuations η_t in turbulent airflow are close to the Kolmogorov microscale η:

$$\eta_t = \eta \cdot \mathrm{Pr}^{-3/4} \approx 1.28\,\eta \qquad (2.52)$$

Assuming $100\,\mathrm{m\,s^{-1}}$ as a typical TAS of the aircraft and η on the order of 10^{-3} m, a required response time of the sensor resolving turbulent temperature fluctuations should be $\sim 10^{-5}$ s with the corresponding size of the temperature sensor of $\sim \eta_\mathrm{t}$ an order of magnitude less than the capabilities of UFT estimated from a heat balance equation (Haman *et al.* 1997):

$$\frac{\mathrm{d}T_\mathrm{sen}}{\mathrm{d}t} = \frac{(T_\mathrm{sen} - T_\mathrm{a})}{\tau} \qquad (2.53)$$

where T_sen is the temperature of the sensor; T_a, temperature of the air; and τ, the time constant. Assuming that the resistive wire is long (length-to-diameter ratio, ≥ 1000), τ can be estimated as

$$\tau = \frac{c_\mathrm{sen} \cdot d^2 \cdot \rho_\mathrm{sen}}{4\kappa \cdot \mathrm{Nu}} \qquad (2.54)$$

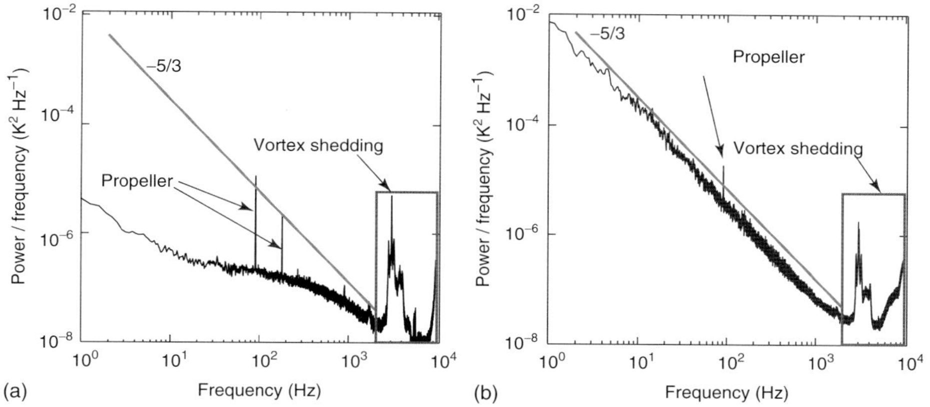

Figure 2.17 Power spectral density of temperature fluctuations close to the top of a stratocumulus cloud recorded by a UFT thermometer at TAS of 55 m s^{-1} (a) above clouds in calm air with no significant temperature fluctuations, (b) in the topmost part of a stratocumulus cloud undergoing mixing with dry environment.

where c_{sen} and ρ_{sen} are, respectively, the specific heat and density of the sensor material and d is the diameter of the cylinder (Nu is defined in Section 2.8.1). For a UFT sensing element consisting of a platinum-coated tungsten wire (length/diameter ratio, $\approx$2000) in the range of TAS of 25–100 m s^{-1} and viscosities of $(1.46 - 2.03) \times 10^{-5}\,\text{m}^2\,\text{s}^{-1}$ gives $\tau = (0.71 - 1.37) \cdot 10^{-4}$ s. This is slightly less than the experimentally measured response of UFT wire ($1.68 \pm 0.17 \times 10^{-4}$ s at 40 ms^{-1}).

Figure 2.17 presents power spectral density of temperature fluctuations demonstrating the performance of UFTs. In Figure 2.17b (data recorded in turbulence of high-temperature contrasts), power spectral density follows $-5/3$ power law at a wide range of frequencies of 1–1000 Hz (wavelengths of 55 m–5.5 cm). In calm thermally homogeneous air (Figure 2.17a), power spectral density shows only very weak natural variance of temperature. In both panels, a peak at 90 Hz due to acoustic waves from the propellers can been seen, with the second harmonic obscured by natural temperature fluctuations in Figure 2.17b. Above $\approx$2 kHz, temperature fluctuations in vortices shedding from the protecting rod dominate. Amplitude/frequency of these vortices depends on TAS, and details of these effects were investigated experimentally by Haman *et al.* (2001) and numerically by Rosa *et al.* (2005). Vortex shedding limits the effective resolution of the UFT sensor down to $\approx$ 1 kHz cutoff frequency as illustrated by Kumala *et al.* (2010). Temperature fluctuations at distances less than $\approx$ 5 cm are not properly resolved with UFT.

The relatively high-frequency response of the UFT creates many technical problems. Usually research aircraft are not equipped to record signals with sampling rates of 2×10^4 samples per second or more, and hence, special data acquisition systems are necessary. The sensor (bare wire) acts as an antenna and is sensitive to electromagnetic noise from radio, RADAR, and avionics. All these effects limit our abilities to measure small-scale processes in the atmosphere.

2.8.5 Calibration of Fast-Response Sensors

Many turbulence sensors are characterized by a high-frequency response with high sensitivity but low absolute accuracy and/or long-time drift. Such behavior is typical for fine-wire sensors, which have to be individually calibrated for each sensor element. This calibration can be performed in different ways, the most common procedures are (i) complementary filtering (e.g., Kálmán filters) or, more simply, (ii) calibration against other more accurate sensors by applying a regression.

Complementary filters perform low-pass filtering to a data set $s(t)$ measured by a highly accurate but slowly responding sensor, and high-pass filtering to a data set $f(t)$ measured by a fast sensor with poor long-term accuracy. Both filtered time series are then merged resulting in a highly accurate signal $n(t)$ with high temporal resolution. Such filters can be applied in real time (e.g., Kálmán filter, often used for navigation) or after the measurement as postprocessing. Two simple and fast methods for postprocessing are described in the following. For both methods, if the time series were not sampled at the same frequency, the slower one has to be interpolated before applying the filter.

The first method is based on Fourier transformations. Both time series $s(t)$ and $f(t)$ are Fourier transformed to complex series $\widetilde{s}(\nu)$ and $\widetilde{f}(\nu)$ (with frequency ν). After definition of a certain cutoff or merging frequency ν_c, all elements of $\widetilde{s}(\nu)$ with $\nu > \nu_c$ and all elements of $\widetilde{f}(\nu)$ with $\nu < \nu_c$ have to be removed. Then the two complex Fourier series are merged to a new complex Fourier series:

$$\widetilde{g}(\nu) = [\widetilde{s}(\nu_{min}) \ldots \widetilde{s}(\nu_c), \widetilde{f}(\nu_c) \ldots \widetilde{f}(\nu_{max})] \tag{2.55}$$

which is finally transformed back to a new time series $g(t)$ using the inverse Fourier transformation. While this method is fast and easy, the two numerical Fourier transformations of finite data sets cause unwanted modification of the measured data due to imperfect data window functions (e.g., Hanning). Also, Fourier transformation is not very suited to time series of intermittent and nonperiodic signals (e.g., turbulence).

An alternative to the Fourier method is a filter in time like the Savitzky–Golay filter (Savitzky and Golay, 1964) that acts similar to a low-pass filter. It replaces each data point $s(t_i) \equiv s_i$ by a linear combination $\widehat{s}$ of itself and a number $(n + m + 1)$ of nearby neighbors:

$$\widehat{s_i} = \sum_{j=-n}^{m} c_j \cdot s_{i+j} \tag{2.56}$$

Using a symmetric window ($n = m$) with constant weights $c_j = (2n + 1)^{-1}$, the filter equals a moving-average window. More sophisticated Savitzky–Golay filters define c_j as a polynomial to preserve features of the time series such as local maxima and minima that are usually flattened by moving averages. Then, Eq. (2.56) represents the convolution of time series s with c.

A symmetric filter window ($n = m$) is still suitable for most measurements. The width of the window is usually a function of the cutoff frequency ν_c and the data sampling frequency ν_s, for example, $n = \nu_s/\nu_c$ floored to the nearest integer. The high-frequency part of time series $f(t)$ (see above) is obtained by subtracting the low-frequency part $\widehat{f}(t)$ (Eq. (2.56)) from the original time series $f(t)$. Finally, the new time series (high accuracy and high resolution) is obtained from

$$g(t) = \widehat{s}(t) + f(t) - \widehat{f}(t) \tag{2.57}$$

2.8.6
Summary, Gaps, and Emerging Technologies

Besides the design of the sensor itself, the biggest challenge of airborne turbulence measurements is likely the location of the sensor on the aircraft to keep the influence of flow distortions below an acceptable threshold or to quantify the influence to compensate for. On fixed-wing aircraft, nose booms or wing pods are preferred locations for high-resolution measurements (see examples in Figure 2.18). Such integrated systems have the further advantage of closely collocated measurements of the different parameters (wind vector, temperature, and humidity), which allows a more detailed interpretation of covariances and correlation of different parameters.

An alternative to fast-flying fixed-wing aircraft are integrated pods to be carried as external cargo by slow-flying helicopters. The influence of the rotor downwash can be overcome by an appropriate relation between tether length and minimum TAS. These setups are characterized by high flexibility and can be operated on board ships or in remote areas. Furthermore, due to the comparably low TAS, many technical limitations and sampling problems of fast-flying aircraft, such as adiabatic sensor heating or inlet problems, are significantly reduced. Another technical advantage of such compact systems is the possibility of being maintained and calibrated in whole without any influence of the aircraft such as electromagnetic noise or similar hazards. At present, two such systems are in use and are shown as an example in Figure 2.19.

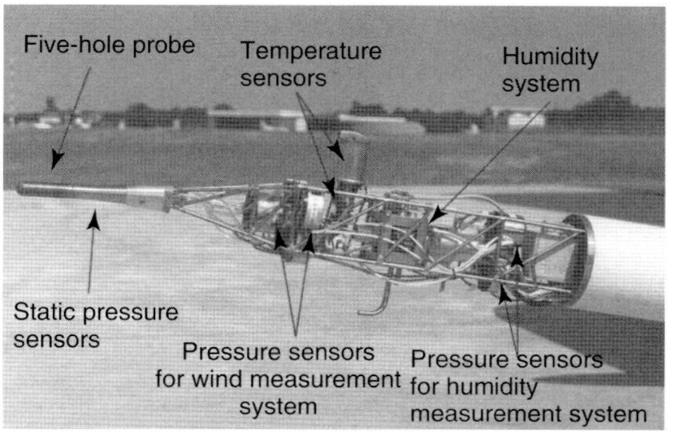

Figure 2.18 Integrated nose boom for aircraft use to measure the wind vector, temperature, and humidity of the DO 128–6 of the Technische Universität Braunschweig. (Source: Reprinted with permission of TU Braunschweig.)

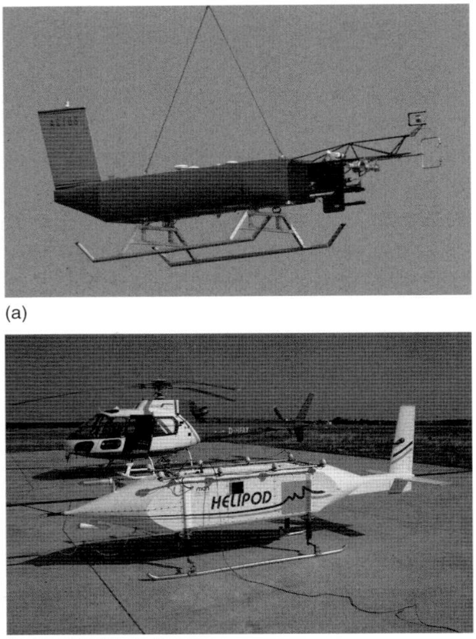

(a)

(b)

Figure 2.19 Two measurement payloads carried by helicopters as external cargo: (a) the "Airborne Cloud Turbulence Observation System ACTOS" and (b) the "Helipod." Both payloads contain various sensors to measure turbulent quantities. (Source: Courtesy of Holger Siebert and Jens Bange.)

Another critical issue is the ability to perform high-resolution turbulence measurements in atmospheric clouds. The presence of droplets is critical for many turbulence probes. In particular, fine-wire sensors can be operated only under certain conditions (Siebert, Lehmann, and Shaw, 2007) or need special shielding to avoid droplet impaction (Section 2.8.4). Fast-response humidity measurements in cloudy environments are biased by possible evaporation of cloud droplet, which is difficult to quantify, and fast measurements of supersaturation with an absolute accuracy of a tenth of a percentage probably remain impossible for the near future.

2.9 Flux Measurements

2.9.1 Basics

Turbulent quantities in the Earth's atmosphere include the air pressure p, the air density ρ, the air temperature T, as well as the potential temperature θ, the mixing ratios m of the individual gases in the air (of special interests in meteorology are, e.g., water vapor, carbon dioxide, and ozone), and the wind vector $\mathbf{v} = (u, v, w)$. The calculation of the vertical turbulent fluxes of these quantities is the eddy covariance method (Montgomery, 1948). However, a covariance computed from a certain measured data set is only an estimator for the required ensemble-averaged covariance (Lenschow and Stankov, 1986; Crawford *et al.* 1993). This estimator is defined via spatial averaging, assuming homogeneous turbulence. Since airborne measured data sets are usually time series, the spatial average is often substituted by a time average, assuming that (i)Taylor's hypothesis is fulfilled, (ii) the atmosphere is in a quasi-stationary state, and (iii) the aircraft's ground speed is constant. The last point is not true in real flight experiments, but then, no significant difference between fluxes calculated via spatial and temporal averaging has been found in experimental data (Crawford *et al.* 1993). Also, nonevenly distributed spatial data makes Fourier analysis, for instance, more complicated. Therefore, time series and time averaging are used in the following.

To obtain the turbulent fluctuations $\phi'(t)$ of a measured time series $\phi(t)$, its mean value $\overline{\phi}$ has to be removed (the Reynolds decomposition). In practice, it is often recommended to remove a certain trend of the time series, especially for nonstationary situations (e.g., measurements around solar noon) and over heterogeneous terrain. This trend can be approximated by a linear regression, the second or third order of a polynomial regression, or the low-frequency modes of a Fourier series (i.e., removed by a high-pass filter). The detrending has to be carried out carefully because the resulting flux may be very sensitive to the definition of the trends and the mean values (Caramori *et al.* 1994). The most commonly measured vertical turbulent fluxes in the lower atmosphere are

$$H = \rho \cdot c_p \cdot \overline{w' \cdot \theta'} \quad \text{Sensible heat flux} \tag{2.58}$$

$$LE = \rho \cdot L \cdot \overline{w' \cdot m'} \quad \text{Latent heat flux} \tag{2.59}$$

$$F_c = \overline{w' \cdot \rho_c'} \quad CO_2 \text{ flux} \tag{2.60}$$

$$\tau_x = -\rho \cdot \overline{w' \cdot u'} \quad \text{Horizontal momentum flux} \tag{2.61}$$

$$\tau_y = -\rho \cdot \overline{w' \cdot v'} \quad \text{Horizontal momentum flux} \tag{2.62}$$

where $c_p = 1006\,\mathrm{J\,kg^{-1}\,K^{-1}}$ is the dry air isobaric specific heat, $L = 2.5 \times 10^6\,\mathrm{J\,kg^{-1}\,K^{-1}}$ is the latent heat of water vaporization, ρ is the mean air density at the measurement altitude, m is the mixing ratio of water vapor, and ρ_c is the CO_2 density.

2.9.2 Measurement Errors

In airborne atmospheric experiments, measurement errors may result from non-ideal measurement equipment or disturbing effects that cannot be eliminated. These are, for instance, dynamic heating (Heinemann, 2002), oscillations of the instrument (e.g., organ pipe resonance in pressure tubes, nose boom oscillation, engine vibration, or pendulum oscillation; (Hauf (1984), Bange and Roth (1999)), or inexact static air pressure and geometric attitude measurement (Khelif, Burns, and Friehe, 1999). Additional sources of measurement errors are effects that limit the spectral range of the instrument, such as noise and slowly responding sensors (McCarthy, 1973; Isaac *et al.* 2004), both of which mainly affect the high frequencies of the measured spectrum.

Many errors can be identified in a Fourier spectral analysis, for instance, noise in the power supply within a certain frequency range or sensor oscillation. If only one of the two turbulent quantities in each Eqs. (2.58)–(2.62) is affected by, for example, an oscillation, the corresponding flux error might be negligible because the oscillation and the turbulence are not correlated. This can be inspected in a Fourier cross-spectrum.

Broader effects such as high-frequency noise and comparison with the Kolmogorov hypothesis of locally isotropic turbulence within the inertial subrange can be analyzed using structure functions or spectra (i.e., the $k^{-5/3}$ or the $r^{2/3}$ law, respectively). The ability to measure small-scale turbulent fluctuations is limited by the sensor with the slowest response time when applying the eddy covariance method (Eqs. (2.58)–(2.62)). For example, if the minimum response time of the vertical wind measurement unit is 0.1 s but the temperature sensor response time is 1 s, only fluctuations slower than 1 Hz contribute to the measured sensible heat flux. The sensor response time should not be confused with the sampling rate of the system. A covariance can be measured by sampling at a rate much less than the time response of the slowest sensor if the measurement is made fast enough to resolve all the scales that contribute to the covariance. This is the basis for disjunct eddy sampling (Lenschow, Mann, and Kristensen, 1994; Rinne *et al.* 2000; Rinne *et al.* 2001; Karl *et al.* 2002), which has been used to measure fluxes of trace species; the sample is collected quickly and the species subsequently measured by a slow-responding instrument. If the separation between samples is less than the integral scale (discussed in the next section), the random error of the covariance estimate is only slightly increased over that obtained from a higher sample rate measurement.

Measurement errors can also be caused by the experimental setup (i.e., the meteorological boundary condition and the flight strategy) and the data analysis method. For instance, mesoscale fluctuations usually mix with turbulent motions at the low-frequency transition of the measured spectrum. Under some meteorological conditions (especially under stable thermal stratification, see van den Kroonenberg *et al.* (2008)) these can be separated from the turbulent flow during the data analysis. In a convective boundary layer (CBL), this is much more complicated (Howell and Mahrt, 1997). Here, convection is the main source of turbulence, and the transition between mesoscale structures and turbulent eddies may be smooth, so the mesoscale cannot be decoupled from the turbulent range.

This discussion reveals the variety of possible errors. In the framework of this book, it is impossible to give a universal recipe on how to avoid such errors for each and every instrument, scientific mission, and atmospheric condition.

2.9.3 Flux Sampling Errors

The following errors do not primarily involve sensor characteristics but are caused by the measurement strategy. In order to achieve a certain accuracy, a minimum flight distance has to be flown under quasi-stationary conditions, as the sampling error is a direct function of the sampling length (Lenschow and Stankov, 1986; Grossman, 1992; Lenschow, Mann, and Kristensen, 1994; Mann and Lenschow, 1994). Flight legs that are not large compared to the largest energy-transporting turbulent eddies cause systematic errors since they lead to a systematic underestimation- (or overestimation) of the turbulent fluxes (Grossman, 1984). In this regard, it should be noted that, for instance, a 10 km flight leg cannot be substituted by 10 legs of 1 km length because the coherence of the turbulent flow is destroyed by the fragmentation. The systematic error in an ensemble of 10 measurement flights that each underestimate the turbulent heat flux cannot be corrected by increased averaging.

A measured turbulent flux, that is, the vertical sensible heat flux H, is only an estimation of the ensemble-averaged vertical sensible heat flux H_E. The ensemble average is an average taken over many different flow realizations that have the same initial and boundary conditions. In the limit of the sample size going to infinity, the ensemble average approaches the ensemble mean and may be a function of both time and position. When the flow is steady and homogeneous, the ensemble, space, and time means are equal (under identical experimental conditions). The atmosphere is normally nonstationary and heterogeneous to some degree, in which case the concept of an ensemble average, and the following error estimates, are only approximations.

2.9.3.1 Systematic Flux Error

The absolute deviation of a certain flux measurement H from the (unknown) ensemble average is

$$\Delta H = |H - H_\mathrm{E}| \tag{2.63}$$

This systematic flux error can be estimated by a simple expression (Lenschow, Mann, and Kristensen, 1994; Mann and Lenschow, 1994):

$$\Delta H \approx 2\frac{I_\mathrm{H}}{P_\mathrm{m}} \cdot |H| \tag{2.64}$$

as long as the averaging time P_m is large compared to the integral timescale I_H of the flux.

The integral scale I_ϕ is the outer scale or macroscale of a turbulent quantity ϕ (Rotta, 1972). The associated integral timescale can be interpreted as the correlation time, the persistence or memory of the turbulent flow (Kaimal and Finnigan, 1994). The transformation into the integral length scale is carried out by multiplication of the integral timescale by the aircraft's ground speed, assuming that Taylor's hypothesis of frozen turbulence is valid and that the ground speed variations are not too large (Crawford *et al.* 1993). The length scale can then be interpreted as the typical size of the largest or most energy-transporting eddies.

The integral timescale of a measured quantity ϕ is defined by

$$I_\phi = \int_0^{\tau_1} \frac{\overline{\phi'(t+\tau)\cdot\phi'(t)}}{\overline{\phi'^2}}\, d\tau = \int_0^{\tau_1} \frac{\mathrm{cov}_{\phi\phi}(\tau)}{var_\phi}\, d\tau \tag{2.65}$$

In Eq. (2.65), ϕ represents turbulent quantities such as temperature, humidity, wind components, and combinations of these. Hence, an integral scale of a turbulent flux can be defined. For instance,

$$\phi(t) = \rho \cdot c_p \cdot w'(t) \cdot \theta'(t) \tag{2.66}$$

yields the integral timescale I_H of the vertical turbulent flux of sensible heat. In practice, I_H is calculated by integration from zero lag to the first crossing with zero at τ_1 (Lenschow and Stankov, 1986). In order to make quantitative estimates of statistical properties of a turbulence time series, the integral scale has to exist. Often it was reported that I_ϕ was difficult to calculate since the autocorrelation function – the integrand in Eq. (2.65) – behaved unpredictably (Mann and Lenschow, 1994; Lenschow, Mann, and Kristensen, 1994) or did not cross zero at reasonable lags (Lumley and Panofsky, 1964). Then an upper limit of integral scale (and thus an upper limit of the error) can be estimated by

$$I_H \le \frac{\rho \cdot c_p}{|H|} \cdot \sigma_w \cdot \sigma_\theta \cdot \sqrt{I_w \cdot I_\theta} \tag{2.67}$$

with the integral timescales I_θ and I_w of θ and w, respectively, and their standard deviations,

$$\sigma_\theta = \sqrt{\mathrm{Var}_\theta} \tag{2.68}$$

and

$$\sigma_w = \sqrt{\mathrm{Var}_w} \tag{2.69}$$

Combining Eq. (2.67) with Eq. (2.64), the upper limit of the systematic flux error is

$$\Delta H \le 2\rho \cdot c_p \cdot \sigma_w \cdot \sigma_\theta \cdot \frac{\sqrt{I_w \cdot I_\theta}}{P_m} \tag{2.70}$$

2.9.3.2 **Random Flux Error**

In general, different measurements of finite duration P_m under identical boundary conditions will lead to different fluxes and different deviations from the ensemble mean. The random flux error is defined as the averaged squared differences between the ensemble and the actually measured fluxes:

$$\sigma_H^2 = \overline{(H_E - H)^2} \tag{2.71}$$

thus σ_H can be interpreted as the standard deviation of the measured flux H. For any turbulent quantity ϕ, and for measurement (averaging) periods P_m much larger than the integral timescale I_ϕ, an estimate for the random error is given by the variance (Lumley and Panofsky, 1964; Lenschow and Stankov, 1986):

$$\sigma_\phi^2 = 2\,\frac{I_\phi}{P_\mathrm{m}} \cdot \overline{\phi'^2} = 2\,\frac{I_\phi}{P_\mathrm{m}} \cdot \left(\overline{\phi^2} - \overline{\phi}^2\right) \tag{2.72}$$

For instance, the random error of the vertical flux of sensible heat is then defined by (with $\phi = \rho \cdot c_\mathrm{p} \cdot w' \cdot \theta'$)

$$\begin{aligned} \sigma_H^2 &= 2\,(\rho \cdot c_\mathrm{p})^2 \cdot \frac{I_\mathrm{H}}{P_\mathrm{m}} \cdot \overline{(w' \cdot \theta')'^2} \\ &= 2\,(\rho \cdot c_\mathrm{p})^2 \cdot \frac{I_\mathrm{H}}{P_\mathrm{m}} \cdot \left(\overline{w'^2 \cdot \theta'^2} - \overline{w' \cdot \theta'}^2\right) \end{aligned} \tag{2.73}$$

The ratio of the systematic to the random flux error decreases slowly toward zero for large averaging time or measurement duration P_m:

$$\frac{\Delta H}{\sigma_H} \sim \frac{1}{\sqrt{P_\mathrm{m}}} \tag{2.74}$$

In other words, an increasing measurement duration leads to a systematic error that becomes a decreasing fraction of the random error. Field campaigns with the Helipod and the Do 128 demonstrated that in a typical mid-European summer afternoon (heterogeneous terrain and a moderately CBL with some cumulus clouds), flight legs of 10 km length are sufficient to neglect the systematic flux error (Bange, Beyrich, and Engelbart, 2002). Furthermore, simultaneous flights of the Helipod and the Do 128 confirmed the dependence of σ_H on the flight distance. The applicability of this error analysis was also analyzed in numerical flight experiments (Schröter, Bange, and Raasch, 2000).

2.9.4 Area-Averaged Turbulent Flux

The turbulent fluxes of momentum and heat representative of an area or region can be achieved using area-covering flight patterns such as simple horizontal squares or horizontal grids (Bange *et al.* 2006b). First, the individual fluxes are calculated by averaging continuous data for each straight and level flight section (leg). Then, for example, for a simple square and assuming homogeneity, all four legs at constant altitude z_f are averaged to obtain an estimator for the area-representative flux (in this example, the sensible heat flux):

$$H_{z_\mathrm{f}} = \rho(z_\mathrm{f}) \cdot c_\mathrm{p} \cdot \frac{1}{4} \sum_{j=1}^{4} \langle w' \cdot \theta' \rangle_j \tag{2.75}$$

The random error of H measured on a square flight can be obtained from the Gaussian error reproduction calculated from the individual errors of the four flight legs:

$$\sigma_H = \frac{1}{4} \sqrt{\sum_{j=1}^{4} \sigma_{H_j}^2} \tag{2.76}$$

Of course, four samples give only a measure of the random error.

The total vertical flux of horizontal momentum is defined by combination of Eqs. (2.61) and (2.62) as

$$\begin{aligned} \tau_{xy} &= \sqrt{\tau_x^2 + \tau_y^2} \\ &= \rho(z_f) \cdot \sqrt{\overline{w' \cdot u'}^2 + \overline{w' \cdot v'}^2} \end{aligned} \tag{2.77}$$

The corresponding random flux error is again calculated using the Gaussian error propagation (Bange, Beyrich, and Engelbart, 2002):

$$\begin{aligned} \sigma_{\tau_{xy}}^2 &= \left(\frac{\partial \tau_{xy}}{\partial \tau_x} \cdot \sigma_{\tau_x}\right)^2 + \left(\frac{\partial \tau_{xy}}{\partial \tau_y} \cdot \sigma_{\tau_y}\right)^2 \\ &= \left(\frac{\tau_x}{\tau_{xy}} \cdot \sigma_{\tau_x}\right)^2 + \left(\frac{\tau_y}{\tau_{xy}} \cdot \sigma_{\tau_y}\right)^2 \end{aligned} \tag{2.78}$$

For many applications the surface flux is of interest. Besides flying extremely low, there are also methods to extrapolate airborne measured fluxes to surface level from area-covering flight patterns (Bange *et al.* 2006b). At first, it seems reasonable to perform area-representative flux measurements above the atmospheric boundary layer flux blending height (Wieringa, 1986; Cantrell, 1991; Mahrt, 2000; Mahrt, Vickers, and Sun, 2001) since there the influence of the individual surface patches within a heterogeneous land surface vanish. But then this blending height can be poorly defined or quite high, up to the middle of the atmospheric boundary layer (Bange *et al.* 2006a), where the fluxes are small or not definable and difficult to be extrapolated to the ground.

2.9.5
Preparation for Airborne Flux Measurement

Finally, this section provides a rough guide on how to prepare for airborne measurements of vertical turbulent fluxes. In order to measure turbulent fluxes of thermodynamic quantities and others, the corresponding sensors should be installed in front of disturbing aircraft elements such as wings and propulsion, ideally mounted on a nose boom. All sensors should be installed within a small volume in order to avoid significant phase shifting between the measurements. If there are significant longitudinal displacements of sensors, or time lags introduced by, for example, sampling ducts, the fluxes have to be corrected by shifting the time series of one variable relative to the other before calculating the flux.

The quantities to measure are ground speed vector (in the inertial coordinate system), altitude above ground and above mean sea level, the Euler angles and angle rates, TAS vector, static air pressure or air density, air temperature, humidity, and other scalars of interest. Of course, all sensors have to be calibrated before flight.

Sensors that have a short response time (that allow for fast sampling rate and thus a high temporal and spatial resolution) usually are not long-time stable and vice versa. To achieve high-resolution measurements that do not drift in time due to changing sensor physics (for instance, contamination of a very thin resistance thermometer), a second, slower but more stable, sensor can be used complementarily (Muschinski and Wode, 1998; van den Kroonenberg *et al.* 2008).

Most sensors are only calibrated for small angles of attack, sideslip, and acceleration. Also the flow around the aircraft can become quite complicated and disturbed during flight maneuvers. Thus, turbulent fluxes should be measured during level flight sections that are straight (legs) or with only small turn rates (i.e., $<0.3^{\circ}\,s^{-1}$).

The legs should be as long as possible although nonstationarity of the atmosphere and heterogeneity of the experimental site have to be taken into account. With surface heterogeneity, flight tracks are often partitioned to provide more meaningful flux estimates and restore applicability of the random error estimates. This procedure limits the length of the flight track.

3
In Situ Trace Gas Measurements

Jim McQuaid, Hans Schlager, Maria Dolores Andrés-Hernández, Stephen Ball, Agnès Borbon, Steven S. Brown, Valery Catoire, Piero Di Carlo, Thomas G. Custer, Marc von Hobe, James Hopkins, Klaus Pfeilsticker, Thomas Röckmann, Anke Roiger, Fred Stroh, Jonathan Williams, and Helmut Ziereis

3.1
Introduction

There are a great many trace gases, many of which can be grouped together, generally by their chemical property and structure. The chemical reactivity of these species can span as much as 10 orders of magnitude. These range from the highly reactive hydroxyl radical, OH, which has an atmospheric residence time of less than one second, to highly stable chlorofluorocarbons, which are highly resistant to destruction via chemical or physical processes such as photolysis, which can have atmospheric lifetimes in excess of 100 years. This temporal range means that gas-phase measurements can be used to probe oxidative capacity on a very local scale and cloud processing and vertical transport dynamics on a larger scale through to large-scale or long-range transport across intercontinental dimensions.

Due to of the nature of a turbulent atmosphere, it is necessary to approach any chemical interpretation of the composition of the atmosphere from a dynamically biased point of view. This is because of the constant production and destruction of trace atmospheric compounds coupled to atmospheric motion, mixing air of differing composition together. Despite these continual concentration changes, the atmosphere is relatively stable over long timescales. The lifetime of a specific species can be related to the distance it can potentially travel in the atmosphere before it is lost. This has been elegantly expressed in a single illustration that considers the spatial and temporal variability (Figure 3.1).

This chapter describes, with specific relevance to airborne measurements, the properties used to determine the presence and abundance of gas-phase species, including interactions with electromagnetic radiation (spectroscopy), chemical reagents (chemical conversion), and a direct measure of the mass-to-charge ratio of the ionized species of interest. The final technique covered here is that of chromatographic separation of ambient air coupled to a detection technique suited to the target analyte(s). Many of the techniques described here derive from

Airborne Measurements for Environmental Research: Methods and Instruments, First Edition.
Edited by Manfred Wendisch and Jean-Louis Brenguier.

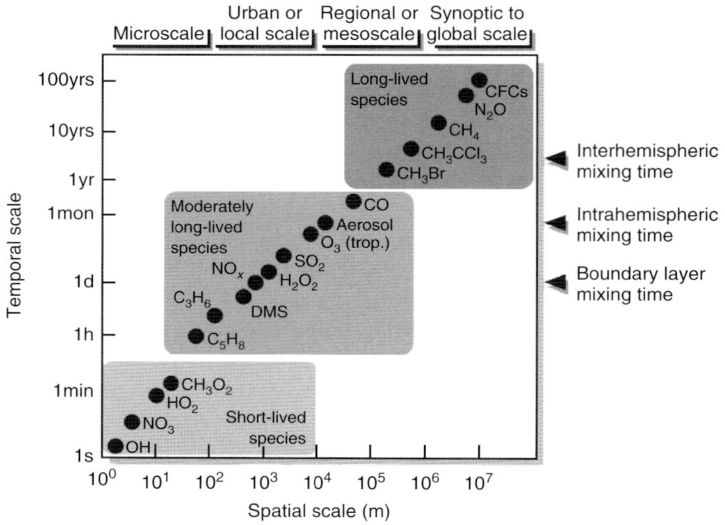

Figure 3.1 Residence times of a range of trace gases illustrating their capacity to be transported in the atmosphere, based on average lifetimes over a 24-h period, using daytime lifetimes. (Source: Adapted from Seinfeld and Pandis (1998). Copyright 1998 by Blackwell Publishing Ltd. Reproduced with permission of John Wiley and Sons.)

laboratory instrumentation with suitable modifications to enable their transfer to the aircraft environment. The book by Heard (2006) has been described as the "first book to bring together all the various instrumental techniques for making atmospheric measurements, with all the significant information is in one place." This is a key reference text that could be used to provide the ground work for this chapter. Most recently, Warneck and Williams (2012) have also produced a very useful text. Other excellent reviews are available, including Parrish and Fehsenfeld (2000); Clemitshaw (2004); Hoffmann, Huang, and Kalberer (2011); Finlayson-Pitts and Pitts (1999). For the basic underpinning principles of the analytical techniques, the reader is directed toward these in addition to more standard analytical chemistry textbooks (Skoog *et al.* 2003).

Gas-phase chemical measurements benefit significantly from the fast pace of the commercial sector, many airborne instruments are derived from commercial advances. There are a few airborne instruments that are modified "off-the-shelf" units; the level of modification required does depend on the target species as well as the expected abundance to be detected. Perhaps the most common of these is that of ozone measurement using ultraviolet (UV) photometry. Furthermore, the increased level of public funding toward environmental and atmospheric science has meant that there are now commercial versions of several instruments, which originally hailed from research laboratories. An identifying feature of instruments for airborne observations is enhanced data rates, for example, an aircraft flying at 100 m/s will require at least 10 s data rates from instruments to achieve a spatial resolution of 1 km; for flux studies, data must be collected at a rate higher than

Table 3.1 Details of the key gas-phase species measured on research aircraft.

Species/parameter	Reference	Technique	Averaging time	Accuracy	Precision	Detection limit
Ozone (O_3)	Nedelec *et al.* (2003)	UV photometer	4 s	5%	2 nmol/mol	1 nmol/mol
Ozone (O_3)	Yushkov *et al.* (1999)	Fast-Response Chemiluminescent Airborne Ozone Analyzer (FOZAN) dye chemiluminescence	1 s	0.001 µmol/mol	8%	1 nmol/mol
Ozone (O_3)	Ryerson *et al.* (1998)	NO/O_3 chemiluminescence	1 s	3%	0.1 nmol/mol	0.2 nmol/mol
Carbon monoxide (CO)	Gerbig *et al.* (1999)	Resonance fluorescence	1 s	2.50%	1 nmol/mol	2 nmol/mol
Carbon monoxide (CO)	Viciani *et al.* (2008)	Tunable diode laser absorption spectroscopy (TDLAS)	4 s	9%	1%	3 nmol/mol
Nitric oxide (NO)	Ziereis *et al.* (2000)	NO/O_3 chemiluminescence	1 s	7%	4%	5 pmol/mol
Nitrogen dioxide (NO_2)	Ziereis *et al.* (2000)	UV photolysis–chemiluminescence	1 s	10%	10%	15 pmol/mol
Total reactive nitrogen (NO_y)	Ziereis *et al.* (2000)	Au converter + Chemiluminescence	1 s	12%	7%	15 pmol/mol
Nitric acid (HNO_3)	Schlager *et al.* (1997)	Ion-trap chemical ionization mass spectrometer	2 s	15%	10%	50 pmol/mol
Nitrate radical (NO_3), dinitrogen pentoxide (N_2O_5)	Dubé *et al.* (2006)	CRDS	1 s	25%	2%	1 pmol/mol
Nitric acid (HNO_3), ammonia (NH_3)	Neumann *et al.* (2002)	Chemical ionization mass spectrometer	1 s	15%	25 pmol/mol	50 pmol/mol
PAN	Whalley *et al.* (2004)	GC/ECD (electron capture detection)	90 s	5 pmol/mol	5%	20 pmol/mol
PAN, PPN, PiBN & MPAN	Slusher *et al.* (2004)	Chemical ionization mass spectrometer	1 s	15–30%	2%	5 pmol/mol (30 pmol/mol MPAN)

(*continued overleaf*)

Table 3.1 *(continued)*

Species/parameter	Reference	Technique	Averaging time	Accuracy	Precision	Detection limit
Real-time VOC	de Gouw *et al.* (2003)	Proton transfer reaction mass spectrometer	1 s every 15 s	10–20%	5–30%	50–250 pmol/mol
VOCs and Oxygenated volatile organic compound (OVOCs)	Lewis *et al.* (2007)	Whole air sampler (WAS) and ground analysis with GC/FID	60 s	5–10%	1–3%	1–10 pmol/mol
Halocarbons	Schauffler *et al.* (1998)	WASs and ground analysis with GC/MS	60 s	5–10%	1–5%	0.1 pmol/mol
Speciated peroxides (inorganic/organic)	Penkett *et al.* (1993)	Fluorometric	10 s	20%	2 pmol/mol	5 pmol/mol
Formaldehyde (HCHO)	Cárdenas *et al.* (2000)	Hantzsch synthesis, fluorimetry	10 s	30%	12%	50 pmol/mol
Formaldehyde (HCHO)	Jimenez *et al.* (2003)	TDLAS	1 s	7%	300 pmol/mol	140 pmol/mol
Peroxy radicals (RO_2 + HO_2)	Green *et al.* (2003b)	Chemical amplifier – PeRCA	30–60 s	40%	6%	2 pmol/mol
Hydroxyl radical (OH)	Commane *et al.* (2010)	FAGE (fluorescence assay by gas expansion) laser-induced fluorescence	60 s	35%	25%	2.3×10^6 molecule/cm^3
Hydroperoxy radical (HO_2)	Commane *et al.* (2010)	FAGE laser-induced fluorescence	60 s	50%	26%	2.0×10^6 molecule/cm^3
CO_2	Schulte and Schlager (1996)	ND–IR photometer	1 s	0.3 µmol/mol	0.1 µmol/mol	0.1 µmol/mol
Halogen oxides (ClO & BrO)	Von Hobe *et al.* (2005)	HALOX chemical-conversion resonance fluorescence + thermal dissociation	60 s & 600 s	17% and 35%	5% and 15%	5 and 2 pmol/mol*
SO_2	Schlager *et al.* (1997)	Ion-trap chemical ionization mass spectrometer	2 s	10 pmol/mol	5 pmol/mol	20 pmol/mol
SO_2	Ryerson *et al.* (1998)	Pulsed UV fluorescence	3 s	10%	0.35 nmol/mol	1 nmol/mol

(*) The detection limit refers to the averaging time given in the middle column.

1 Hz. Table 3.1 details the key gas-phase species measured on research aircraft. It includes key parameters such as time resolution, accuracy, and precision. Please consider that the most common unit for the concentration of gas-phase species in atmospheric science is that of a volume mixing ratio such as parts per million (ppm), parts per billion (ppb), and parts per trillion (ppt). The advantage of using these units is that the volume mixing ratio of a gas is remaining constant, while the air density changes; thus, the volume mixing ratio is a robust measure of atmospheric composition. Although it is a more usual practice for ambient concentrations to be expressed in terms ppm, ppb, or ppt, these are nonstandard units (non-SI units), and for the purposes of this text, we have used μmol/mol for ppm, nmol/mol for ppb, and pmol/mol for ppt.

3.2 Historical and Rationale

Perhaps the earliest airborne measurements of atmospheric composition were of water vapor. These observations were not from the perspective of a chemical species present in the atmosphere but of precipitation precursors. Not forgetting that microphysical processes are tied up by the latent energy associated with phase changes in water, making water alone an important physical measurement. Section 2.6 provides the reader with much greater detail regarding measurements of water vapor. In this section, however, we focus on the trace gases.

To resolve the underlying dynamical processes, which distribute constituents in the atmosphere, measurements of trace gases are the only direct access to the slow overturning circulation; (Brewer, 1949; and Dobson and Lawrence, 1929) deduced stratospheric circulation alone on the basis of ozone and water vapor measurements. The crucial role of tracers in this field today is reflected by the current stratospheric age-of-air discussion, which was stimulated by *in situ* sulfur hexafluoride (SF_6) measurements (Engel *et al.* 2009), contrasting with model results. Furthermore, airborne (or balloon-borne) *in situ* tracer measurements are still needed to resolve the spatial dynamical scales beyond the restrictions of satellite and model resolution and still play a dominant role for understanding dynamical processes (Engel *et al.* 2006). High-resolution observations still play a key role in this field and can only be provided from *in situ* techniques. The validation of numerical models is also a very significant beneficiary of airborne observations. This is often the driver for major observational programs to investigate whether we can adequately simulate the atmosphere in a regional and, indeed, global context. Other important scientific goals of gas-phase measurements onboard aircraft include the calibration and validation of satellite observations as well as test-beds during the development of satellite instruments.

The trace gases form a very minor fraction of the total atmospheric bulk ($>0.1\%$), but their distribution, abundance, and reactivity are of critical importance to control what is commonly referred to as the oxidative capacity, that is, the atmospheric potential to "cleanse itself." There are a number of species that contribute to this,

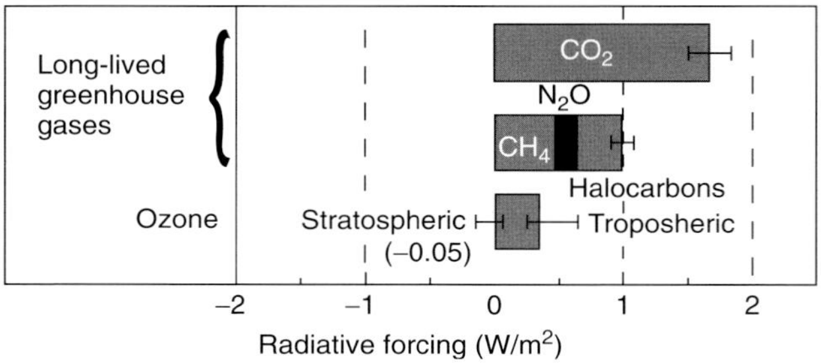

Figure 3.2 Summary of the major trace gas components of climatic radiative forcing. The values represent the forcings in 2005 relative to the start of the industrial era (about 1750). Positive forcings lead to warming of the climate and negative forcings lead to cooling. The thin black line attached to each bar represents the range of uncertainty for the respective value. Adapted from Forster *et al.* (2007).

perhaps the main oxidant species being the hydroxyl (OH) radical, a radical that controls the lifetime of most atmospheric trace gas species. The measurement of the OH radical is one of the major challenges for atmospheric chemists, and the comparison of observations against numerical models is, perhaps, the test of our success in understanding the atmospheric system.

Closely coupled to the hydroxyl radical is the hydroperoxy (HO_2) radical, these two species exist in equilibrium in the atmosphere, the precise balance of which is controlled by other atmospheric species such as nitric oxide (NO) and nitrogen dioxide (NO_2). Ultimately, it is the hydroxyl radical that controls the abundance of these non–carbon dioxide greenhouse gases (GHGs) because it controls the lifetime of these species, be they the very long-lived halocarbons such as methyl bromide or the much shorter lived anthropogenic and biogenic hydrocarbons.

In addition to the long-lived "climate gases" such as carbon dioxide (CO_2) and nitrous oxide (N_2O), there are a number of other trace gases that have a climate impact owing to their ability to absorb outgoing terrestrial radiation and are also considered as important GHGs in their own right (Figure 3.2). These include ozone (O_3), methane (CH_4), and many halogenated species. In fact, the global warming potentials of some of these gases are significantly higher than CO_2. CH_4, for example, has a global warming potential 70 times greater than CO_2 (based on a 20-year time horizon) Forster *et al.* (2007).

Furthermore, trace gases are known precursors for atmospheric aerosol particles (Pöschl, 2005; Jacobson *et al.* 2000), which also affect the climate in a number of ways (Haywood and Boucher, 2000).

From an air quality perspective, it is often long-range transport of pollutants, which provides a significant contribution toward regional and local enhancements in ground-level ozone and particular matter. This clearly demonstrates that air quality must be considered in a wider context than just local observations. Airborne observations are often used to target long-range transport. There have been a number of field campaigns in recent times to study this (Lewis *et al.* 2007; Parrish *et al.* 2009). Through the variation in atmospheric lifetimes across the many different species that can be measured, it becomes possible to provide a time dimension to atmospheric transport processes.

3.3 Aircraft Inlets for Trace Gases

The method of transmitting external air to cabin-mounted instrumentation/sampling equipment is very much governed by the characteristics of the target analyte. Ambient air will undergo rapid and extreme changes in temperature, pressure, and relative humidity. Compression heating as the air passes into the sample inlet can elicit phase changes in water and can disturb chemical equilibria such as $HO_2NO_2 \longleftrightarrow HO_2 + NO_2$. For least reactive species of interest such as carbon dioxide and chlorofluorocarbons, a very simple inlet may be sufficient. However, there are a range of more increasingly specialized inlets required as compounds become more reactive or susceptible to being "lost" to or sourced from sampling lines (SLs), the former more often being termed as "sticky" compounds. Inlets may require heating or be manufactured from other materials such as Polytetrafluoroethylene (PTFE) and perfluoroalkoxy (PFA) or even glass or PTFE liners inserted as opposed to the basic stainless steel construction, which is the common default for the more inert species. Some species such as nitric acid (HNO_3) and ammonia (NH_3) exhibit reversible uptake on inlet lines, which require careful consideration when designing a sampling system. In many cases, specially designed inlets are required with additional surface passivation and heating capabilities because there may be as much as a 100 K temperature difference between the ambient air external to the aircraft and the cabin environment. Extensive tests have been conducted as to the applicability of different materials for HNO_3, concluding heated Teflon to be the best candidate (Neuman *et al.* 1999). Other studies have addressed materials for a range of the more challenging species (Eisele *et al.* 1997; Ryerson *et al.* 1999).

Sample inlets must extend beyond the boundary layer of the aircraft, thus robust knowledge of the depth of this boundary layer is required. Often this is available from aircraft manufacturers, but careful consideration of other external factors such as inlet fairings should be included as these may be beyond the design of the baseline airframe. The orientation of the sample inlet to the bulk airflow around the aircraft leads to advantages and disadvantages. Rearward-facing inlets avoid water ingress when flying in clouds and also prevent larger aerosol particles from entering the sample stream, while forward-facing inlets can have a ram effect that enhances the mass flow into the inlet. Figure 3.3 presents a general inlet cluster, which has been designed for the high-altitude and long-range research aircraft (HALO). It includes both forward- and rearward-facing inlets. Such a structure also provides capacity to heat the inlets if required, and, for simplicity, it is mounted on a standard window blanking plate. This is a common place and these provide a relatively straightforward component to add/remove from the aircraft pressure skin.

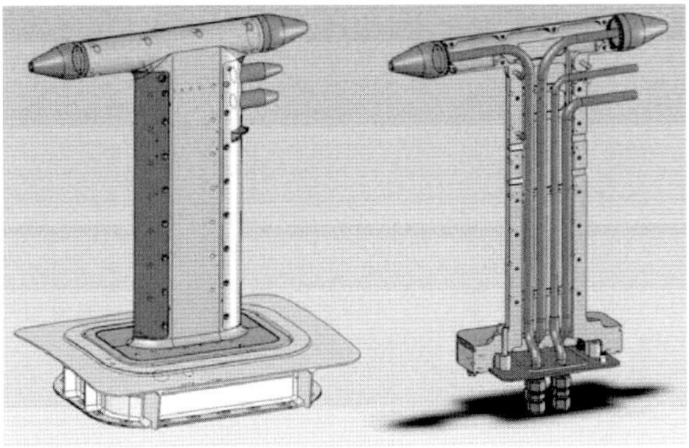

Figure 3.3 Gas sampling inlet for HALO, the cluster includes a single forward-facing and three rearward-facing inlets (or vice versa), it can accommodate tubes of a range of diameters (1/2″, 3/8″, and 1/4″) and also different materials (stainless steel, PTFE, or PFA), as per the requirement. (Design by enviscope GmbH/DLR, 2009)

Transmission inside the aircraft cabin also needs careful consideration. The altitude will determine the pumping requirements to draw the sample into the instrument. Furthermore, the inlets for instruments carrying out fast radical measurements such as that described in Section 3.5.3 are designed so that the probe/excitation laser axis is close to the inlet itself and, furthermore, large pumping systems bring the sample to this position in microseconds.

Finally, the outlet and exhaust flows from these instruments need to be considered carefully. When sampling from the reduced pressures outside the cabin, it is preferred to exhaust the instrument back to outside the cabin, this reduces pumping requirements and also provides a means to exclude any noxious gas that may have been used in the analytical determination, an example being excess concentrations of ozone used in the chemiluminescent detection of nitric oxide. Instruments connected to a common exhaust must be complimentary and organized in such a way that there can be no contamination pathways between instruments. Individual instruments may require some level of humidity control or particle filtration before analysis. Other factors such as adiabatic heating and air flow deceleration are included in accompanying sections. Such is the range of complexity for aircraft inlets that the reader is directed toward the relevant measurement technique and references therein.

3.4 Examples of Recent Airborne Missions

Recent times have seen the exploitation of a great deal of the new technology described here across multiple platforms. The Intercontinental Consortium for Atmospheric research on Transport and Transformation (ICARTT) brought together several US aircraft alongside the UK's BAe-146 and Germany's Falcon-20 to investigate intercontinental transport across the Atlantic (Fehsenfeld *et al.* 2006). ICARTT addressed a range of topics, including the role of different oxidizing species, partitioning of key tracers during transport, and photochemical ozone and secondary aerosol formation during long-range transport. Finally, the wealth of data collected under the auspices of ICARTT provided a vast database by which the measurements could be compared and model uncertainties could also be assessed both against the observations and the model–model comparisons. In 2008, several of the European fleet (DLR Falcon-20, French ATR42, and UK BAe-146) took part in a series of joint missions as part of the European Aerosol Cloud Climate and Air Quality Interactions project to study interactions of climate and air pollution during which a wide range of measurements of aerosol and gas phase were made across Europe (Hamburger *et al.* 2003).

Projects such as the High-Performance Instrumented Airborne Platform for Environmental Research (HIAPER), Pole-to-Pole Observations (HIPPO) of carbon cycle, and GHGs study (Wofsy, S. C. and the HIPPO Science Team and Cooperating Modellers and Satellite Teams 2011) open up a new dimension on airborne observations. HIPPO is the first comprehensive, global survey of atmospheric trace gases. It approximately measures the cross sections of atmospheric concentrations from one pole to the other (generally transecting the Pacific Basin), from the surface to the tropopause, 4–6 times during different seasons over a 2.5- to 3-year period. Using a range of established techniques and recent developments, it will perform a comprehensive observational suite of tracers of the carbon cycle and related species, including CO_2, CH_4, CO, N_2O, $^{13}CO_2$, $^{12}CO_2$, H_2, SF_6, Carbon oxide sulfide (COS), CFCs, Halogenatedfluorocarbon (HFCs), Hydrochlorofluorohydrocarbon (HCFCs), and selected hydrocarbons, many of which will have multiple techniques being applied to them, which clearly will increase confidence in the results.

Exciting times are ahead with the increasing use of unmanned aerial vehicles (UAVs), such as NASA's Global Hawk, although operational restrictions for this aircraft limit it to 43 000–65 000 ft. There is, however, still an astounding capability offered by such a platform, Global Hawk is able to remain aloft for 30 h and has a range of over 18 000 km, while carrying a moderate scientific payload (700 kg). As aeronautical command and control technology matures, it may be that such well-developed unmanned technologies are ideally placed to venture below 43 000 ft. The first science mission for the Global Hawk was under the Global Hawk Pacific Mission (GloPac) (GloPac, 2010). GloPac aimed at "directly measure and sample greenhouse gases, ozone-depleting substances, aerosol particles, and constituents of air quality in the upper troposphere and lower stratosphere." The campaign succeeded in many of these objectives, laying a solid foundation for the future of using such platforms for atmospheric observations. GloPac included deployments of the HIAPER aircraft, which covered the surface to tropopause below the flight track of the Global Hawk, this illustrates the huge potential in combining two such aircraft to cover the lower most 20 km of the atmosphere in the remote environment.

3.5 Optical *In Situ* Techniques

There are a number of mature spectroscopic techniques used in trace gas observations, notably used for ozone, carbon dioxide, and water vapor. There are numerous undergraduate- and graduate-level textbooks on the subject. The reader is directed toward those to grasp the basics. In the past decades, airborne ultraviolet/visible/near-infrared (UV/VIS/NIR) spectroscopy emerged as a powerful tool to remotely sense the sources and the atmospheric distribution of a larger suite of important primary pollutants and their decay products, as well as photochemically formed secondary trace gases and radicals (Wahner *et al.* 1990; Brandtjen *et al.* 1994; Pfeilsticker and Platt, 1994; Prados–Roman *et al.* 2010) and others. The rapidly growing market for optical technology driven predominantly by telecommunications has resulted in significant steps forward, including size, cost, wavelength selection, and stability to name a few. Atmospheric spectroscopy has been a significant beneficiary of such advances and it is perhaps this area where some of the largest steps forward have been achieved in the past decade. For example, highly stable quantum cascade lasers (QCLs) have transformed the observation of a number of species, including CO_2, CO, CH_4, and N_2O. Spectroscopic measurement techniques use the absorption, (re-)emission, or scattering of electromagnetic waves by atoms or molecules (or atomic or molecular ions) to measure their concentration.

The absorption technique provides a direct measurement of the target species by comparing the absorption of electromagnetic radiation due to its interaction with the sample. Absorption spectroscopy can be used to observe a number of molecules, including ozone, carbon dioxide, carbon monoxide, and also water vapor. There are a number of methods for generating the source radiation and these will be dependent on the absorbing wavelength in question.

3.5.1 UV Photometry

If one was to measure just a single trace gas in the atmosphere, it would probably be ozone. It is not only an important climate gas but also a useful tracer for a number of atmospheric processes, stratospheric/tropospheric exchange (Homan *et al.* 2010), long-range transport (Lewis *et al.* 2007), as well as surface emissions. Far and away the most common technique for measuring ozone in the atmosphere is UV photometry. Ozone mixing ratio is calculated using the ozone absorption cross section at 253.7 nm and the Beer-Lambert–Bouguer (also called the Beer-Lambert or Beer's) law. UV photometry provides a robust absolute measurement, and response times are 5–20 s. Modern instruments adopt a dual-cell configuration that results in a more dependable observation (improved precision, accuracy, and response time). Obviously should one cell become contaminated, then the advantage of dual cells quickly becomes a disadvantage.

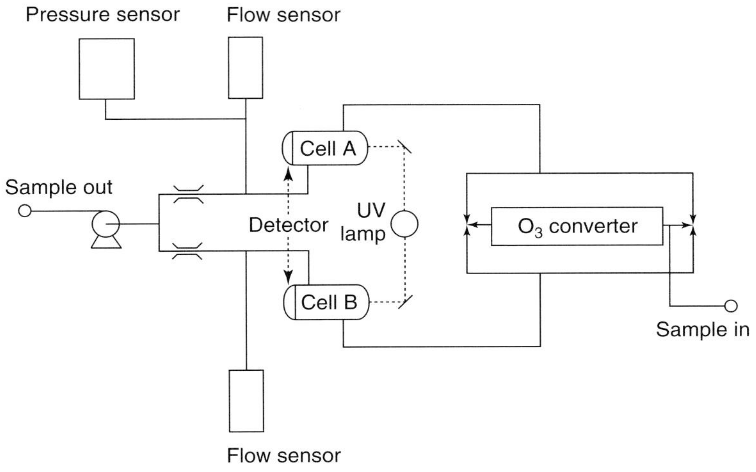

Figure 3.4 A commercial UV photometer, this design employs dual cells.

Proffitt and McLaughlin (1983) further refined the UV photometry technique and produced faster response times. This instrument has long and successful career having been fitted to a number of US airborne platforms, including the NASA ER-2 and WB-57 as well as the new HIAPER aircraft.

Figure 3.4 illustrates a commercial dual-cell UV photometer for measuring ozone. In recent years, a miniaturized UV photometer has become available (Bognar and Birks, 1996) and deployed successfully on small balloons (Helmig *et al.* 2002). This paves the way for installation onto miniature platforms such as UAVs.

A number of modifications to the basic construction of a UV photometer may be considered for airborne use. The altitude range of the planned usage may require additional pumping. Another consideration that must be borne in mind when measuring ozone using UV photometry is the effect of rapid changes in ambient relative humidity. While no water vapor interference is expected for UV absorbance measurements of ozone because water vapor does not absorb appreciably at 253.7 nm, a number of reports have been published (Kleindienst *et al.* 1993; Wilson and Birks, 2006). Many operators now install a NAFION^{TM} membrane drier in the sample stream ahead of the instrument. NAFION^{TM} is highly permeable to water vapor, and by operating with a counterflow of dry sheath air, it is possible to remove significant levels of ambient water vapor very quickly. The residence times required for water removal is calculated as being in the region of 50–100 ms for a standard instrumental setup (Robinson, Bollinger and Birks, 1999). Wilson and Birks (2006) have proposed that the ozone scrubber used for the zero measurement can act as a reservoir for water. Any water coating on the internal surfaces will modify the total internal reflection, thus, modifying the transmission efficiency within the cell. This effect is only seen during rapid changes in humidity, as are observed when profiling in the atmosphere through layers of different composition. This is very apparent when crossing between the boundary layer and the free troposphere.

One must consider potential coabsorbers within the sample matrix when using UV spectroscopy, this is especially important when only a single wavelength is used (differential optical absorption spectroscopy relies on algorithms to deduce a spectral fit). In particular, aromatic hydrocarbons such as toluene have a known absorption at 253.7 nm (Kleindienst *et al.* 1993). Williams (2006) demonstrated, however, that this was not a problem even in heavily urbanized and industrial areas in Texas.

There are other techniques used to measure ozone, but only chemiluminescence with either nitric oxide (Ridley, Grahek and Walega, 1992) or organic dyes (Yushkov *et al.* 1999) has been deployed on anything approaching a regular basis. Chemiluminescence is a faster technique but unlike the absolute measurement of O_3 by UV absorbance, it requires external calibration and is usually operated with the support of the slower, absolute technique of UV photometry. A number of studies have investigated the performance of UV O_3 monitors in comparison to NO chemiluminescence (Ryerson *et al.* 1998). Chemiluminescence is covered in detail in Section 3.7.2.

3.5.2
Differential Optical Absorption Spectroscopy

Absorption spectroscopy can also be deployed for remote sensing over extended path lengths (up to 20 km). For airborne differential optical absorption spectroscopic (DOAS) observations, natural ambient light is used as a passive light source. Using detailed knowledge of absorption spectra of target compounds allows the retrieval of integrated column density.

3.5.2.1 Measurement Principle

Remote sensing of atmospheric trace gases and optical parameters by spectroscopy in the UV/VIS/NIR spectral range involves three major elements:

(i) the spectroscopic measurements and the retrieval of slant column densities (SCDs) from measured atmospheric spectra using the DOAS technique (Platt and Stutz, 2008);
(ii) forward modeling of the radiative transfer for each individual observation; and
(iii) mathematical regularization and inversion techniques, which invert measured SCDs into absolute concentrations of the target gases (Rodgers, 2000).

Passive DOAS, element (i) of the outlined technique, relies on evaluating spectroscopic measurements of foreground spectra against a background spectrum by inspecting either direct sun-, moon-, or starlight or scattered skylight. In most cases, it is sufficient to record the light at medium spectral resolution (typical spectral resolutions at full width and half maximum are several nanometers depending on the application), because the DOAS retrieval mainly exploits differential signatures of electronic/vibrational UV/VIS/NIR absorption bands.

Table 3.2 Targeted trace gases and their indicative detection limits in airborne applications.

	2D IDOAS (2 km, AMF = 2)		**Mini-DOAS nadir (10 km, AMF = 1)**	**Mini-DOAS (10 km)***	**Mini-DOAS Limb (15 km)***
O_3	$3{\cdot}10^{17}/cm^2$	35 nmol/mol	$1\times10^{18}/cm^2$	35 nmol/mol	80 nmol/mol
NO_2	$5.7\times10^{15}/cm^2$	0.6 nmol/mol	$2\times10^{15}/cm^2$	13 pmol/mol	20 pmol/mol
HONO	$1\times10^{16}/cm^2$	1 nmol/mol	$1\times10^{16}/cm^2$	35 pmol/mol	45 pmol/mol
BrO	$1.2\times10^{14}/cm^2$	13 pmol/mol	$5\times10^{13}/cm^2$	0.7 pmol/mol	0.9 pmol/mol
OClO	$1.4\times10^{14}/cm^2$	15 pmol/mol	$4\times10^{13}/cm^2$	3 pmol/mol	4.5 pmol/mol
IO	$5.4\times10^{13}/cm^2$	6 pmol/mol	$7\times10^{13}/cm^2$	0.2 pmol/mol	0.4 pmol/mol
OIO	$4\times10^{13}/cm^2$	0.2 pmol/mol	$4\times10^{13}/cm^2$	0.2 pmol/mol	0.4 pmol/mol
CH_2O	$2.2\times10^{16}/cm^2$	2.4 nmol/mol	$3\times10^{16}/cm^2$	600 pmol/mol	1 nmol/mol
SO_2	$2.8\times10^{15}/cm^2$	0.3 nmol/mol	$2.8\times10^{15}/cm^2$	NA	NA
CHOCHO	$2.5\times10^{15}/cm^2$	0.3 nmol/mol	$5\times10^{15}/cm^2$	NA	NA
O_4*	$3.6\times10^{42}/cm^2$		$3.6{\cdot}10^{42}/cm^2$	±6.5%	±9%
CH_4	$2.5\times17/cm^2$	17 nmol/mol	$2.5\times17/cm^2$	1.7 nmol/mol	2 nmol/mol
CO_2	$7.1\times19/cm^2$	5 µmol/mol	1 µmol/mol	0.5 µmol/mol	0.8 µmol/mol
H_2O (vapor)	$9.4\times10^{21}/cm^2$	N/A	$2.5\times10^{20}/cm^2$	1 µmol/mol	1.5 µmol/mol
H_2O (liquid)	N/A	N/A	$3.5\ g/m^2$	$0.1\ g/m^3$	$0.1\ g/m^3$
H_2O (ice)	N/A	N/A	$2.5\ g/m^2$	$0.15\ g/m^3$	$0.15\ g/m^3$

(*) for observations from an aircraft flying at 10 or 15 km.

In terms of comparing foreground and background spectra, DOAS is a relative measurement of two optical states of the atmosphere, whereby the background state needs either to be known from a priori information or to be determined from the measurements (e.g., by Langley method) or to be eliminated using differential inversion schemes (Kritten *et al.* 2010).

Using modern optical instrumentation, the two optical states of the atmosphere can be very precisely compared, and detection limits as low as 10^{-4} in terms of optical thickness are reported in the literature. Furthermore, because atmospheric observations often offer extremely long light paths as compared with the laboratory results, for example, several kilometers to several 100 km, and typical UV/VIS/NIR absorption cross sections are as large as $10^{-17}\ cm^2$, atmospheric DOAS measurements eventually provide detection limits as low as $10^4\ cm^{-3}$ for individual gases (Table 3.2).

Many atmospheric DOAS applications today use a variety of different viewing directions either by simultaneously monitoring the atmosphere with imaging spectrometers or by subsequently scanning it with movable telescopes.

Element (ii) of the discussed technique addresses the interpretation of the atmospheric measurements with respect to the detection sensitivity and probed atmospheric region. Evidently, it involves radiative transfer modeling to answer the question which part of the atmosphere was probed with what optical path length. Atmospheric radiative transfer is largely dominated by Rayleigh and varying amounts of Mie scattering. Thus, the accuracies to which air mass factors can be calculated and thus of the DOAS technique itself may largely depend on the knowledge of the spatial distribution of Mie scatterers (e.g., aerosol particles and cloud particles) and their optical properties but not the radiative code in use. As DOAS spectroscopy does not use broadband extinction features caused by absorption and scattering of aerosol particles and cloud particles, radiative transfer modeling does not need to describe all radiative properties and processes to the greatest detail. Rather it attempts to provide a measure on photon path lengths and their distribution (frequently expressed by air mass factors) as a function of the observation geometry, solar, moon, or stellar zenith and azimuth angles, wavelength, and any other relevant parameters of the atmosphere's optical state. In effect, the involved radiative transfer modeling requires treating accurately the three-dimensional nature of the atmosphere, the Earth's sphericity, atmospheric refraction, major optical absorption and scattering properties, the field of view (FOV), and the direction and position of the detector. Therefore, for tropospheric applications, frequently either efficient discrete ordinate schemes or Monte Carlo codes are used to fulfill this task.

In return, atmospheric DOAS measurements inherently offer means to constrain the radiative transfer for each individual measurement. Such constraints can be inferred from observed relative radiances or from other measured quantities, for example, the atmospheric absorption of gases with known concentration (e.g., O_2 and O_4).

On the whole, numerical errors in the DOAS technique largely rely on specific features of the individual measurements, for example, on the targeted trace gas, and type of atmospheric observations. While the former is mostly due to the knowledge of the absorption cross section and potential interferences with other absorbers, the latter largely determines to what degree the optical properties of the atmosphere can be simulated. For example, while for direct sun, moon, and star measurements the air mass factors can be calculated to better then 1%, or 5% for limb measurement in a Rayleigh scattering–dominated atmosphere (e.g., the stratosphere), errors might be much larger for measurements (up to 50%) where effects of the ground albedo, aerosol, and cloud particles play a prominent role. As all these factors largely depend on the situation of the individual measurement, quantitative statement on the error or uncertainty is usually made by the individual authors when detailing their studies.

Element (iii) of the outlined technique combines elements (i) and (ii) with mathematical inversion techniques. Mathematical inversion is necessary to infer vertical concentration profiles of the target absorbers from the measured SCDs and the radiative transfer–modeled observation geometry. Thereby, an appropriate choice of the number and nature of the retrieval parameters is evidently problem dependent. In particular, it depends on the observation geometry, the optical state of the atmosphere, the frequency of (independent) observations, the assumed spatial and temporal distribution of the targeted parameters, and finally the measurement errors (Rodgers, 2000).

However, in most cases, the magnitude of the measurement errors and the uncertainties and errors of the forward model are crucial for the interpretation of the measurements. Accordingly, modern inversion algorithms need to provide some metrics for information content and a reliable assessment of error propagation based on estimates for the quality of the measurement and the forward model. To maximize the information content, the measurements require to be as mathematically independent as possible, which can be achieved by reducing the measurement errors and increasing the "independence" of individual observations (e.g., by significantly varying the observation geometry between the individual measurements). Therefore, UV/VIS/NIR remote sensing, as any other remote sensing technique, attempts to use a variety of spatial arrangements assembled among the position of the light source, the targeted region, and the detector (e.g., by off-axis, limb solar, lunar, or stellar occultation observations, or by using moveable observing platforms). In practice, the overall error is often determined by the uncertainty and systematic error of the forward model, rather than errors due to the spectroscopic measurement or inversion schemes.

3.5.2.2 **Examples of Measurement**

Three applications are briefly discussed in the following to illustrate some strengths and weaknesses of the UV/VIS/NIR atmospheric remote sensing technique.

Figure 3.5 shows a latitudinal cross section of tropospheric NO_2 monitored by an airborne multiaxis differential optical absorption spectrometry (AMAX-DOAS) instrument during a DLR Falcon-20 flight from Bale/Switzerland to Tozeur/Tunisia on 19 February 2003 (Bruns *et al.* 2004; Wang *et al.* 2005a). A particular strength of the AMAX-DOAS observations is to monitor simultaneously the whole atmosphere in up to 10 different viewing directions (five upward and five downward looking). In this case, the AMAX-DOAS observations provided NO_2 concentrations every

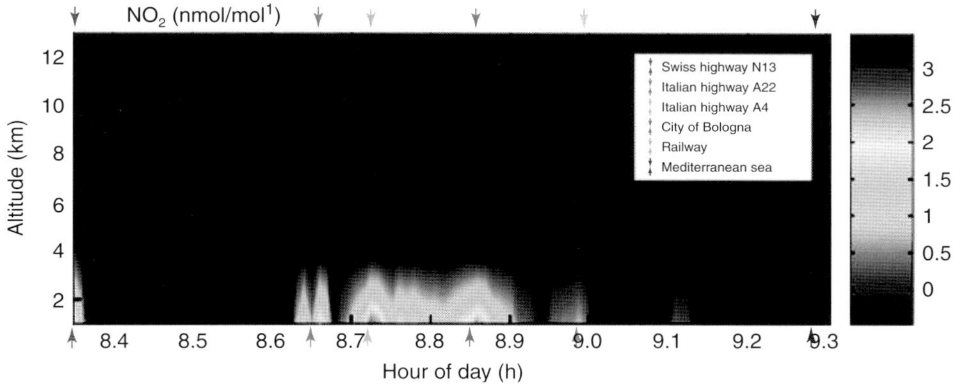

Figure 3.5 Tropospheric NO_2 profiles measured by the AMAX-DOAS instrument during a DLR Falcon-20 flight from Bale (Switzerland) to Tozeur (Tunisia) on 19 February 2003 (Bruns *et al.* 2004). The NO_2 concentrations are given in nmol/mol. The arrows indicate the locations of elevated NO_x emissions, as detailed by the legend.

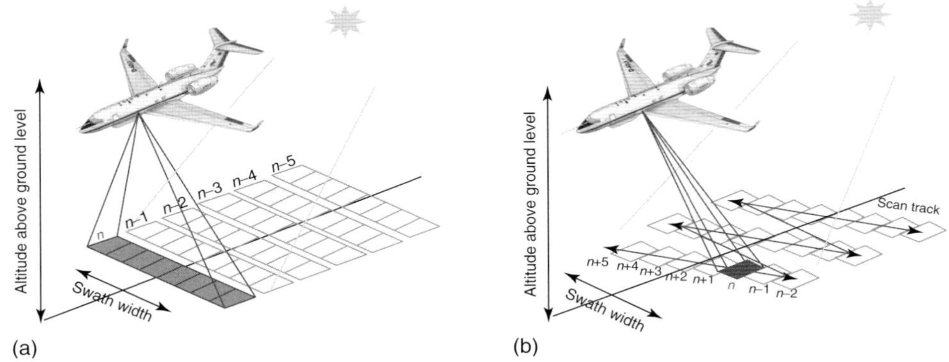

Figure 3.6 Different types of observation geometries used for aircraft-based 2D imaging DOAS measurements: push-broom technique that simultaneously images the red row of pixels (perpendicular to the flight direction) into an imaging spectrometer (a) and whisk-broom technique during which the ground pixels are subsequently scanned (b).

minute (or every 10 km along the flight track for typical DLR Falcon-20 cruise speed) for 3.5 height layers (i.e., with 3.5 pieces of independent information for each hemisphere) with a precision of ± 0.3 nmol/mol. The height versus time/latitude cross section indicates the location of larger NO_x emissions overflown by the aircraft, that is, highways in Switzerland and Italy or larger cities such as Bologna and Pisa/Florence (Italy). Limitations of these AMAX-DOAS measurements are moderate height resolution and detection precision for the targeted gases.

Another approach to identify sources of pollutants and to monitor the near-surface air quality offers the two-dimensional (2D) imaging DOAS (for the observation geometry of the 2D imaging technique, see Figure 3.6, left panel). Figure 3.7 shows the example of NO_2 observations performed by a 2D imaging DOAS instrument that was deployed on a South African Partenavia aircraft (Figure 3.7). On 5 October 2006, the aircraft flew over the Majuba Power Plant/South Africa (27.1 °S, 29.8 °E) roughly 3.5 km above the ground. Beside other species, the vertical column amounts of NO_2 and SO_2 were recorded with a swath width of 80 m in cross track and 90–200 m along the track, the latter depending on the aircraft speed (Heue *et al.* 2008).

NO_2 and SO_2 are known as primary pollutants emitted by coal-fired power plants, which do not apply flue gas denoxification and desulfurization techniques to clean the exhaust. Monitoring of the exhaust of several power plants located in the South African Highveld region indicated SO_2/NO_x emission ratios in the range of 3–7, mostly because of large sulfur content (<3%) of the coal used to fire the plants and to lesser degree due to the details of the combustion process.

The last example reports on aircraft-borne UV/VIS/NIR limb profiling of tropospheric BrO. Major advantages of UV/VIS/NIR limb observations are the high detection sensitivity for important atmospheric radicals and trace gases (e.g., NO_2, BrO, IO, CH_2O, and HONO) owing to the extremely long light paths (some 10–100 km) achievable when viewing through the limb.

Figure 3.7 Measured NO_2 vertical column over the Majuba Power Plant/South Africa (27.1 °S, 29.8 °E) on 4 October 2006, overlaid onto a local map.

The UV/VIS/NIR limb measurements were performed during the Arctic Study of Tropospheric Aerosol, Clouds, and Radiation (ASTAR) campaign, which took place in the Arctic during spring 2007; see the following web site: *http://www.pa.op.dlr.de/aerosol/astar2007/*. For the campaign, a limb scanning mini-DOAS instrument was deployed on the DLR Falcon-20 aircraft (Prados–Roman *et al.* 2010). One DLR Falcon-20 sortie was particularly devoted to the detection of BrO profiles within air masses potentially affected by ozone depletion events (ODEs). ODEs are known to occur within the planetary boundary layer (PBL) during Arctic spring, caused by unique bromine-mediated photochemistry. It converts inert halide salt ions (e.g., Br^-) into reactive halogen species (e.g., Br atoms and BrO) that deplete ozone in the boundary layer to near-zero levels (e.g. Simpson *et al.* (2007)).

Figure 3.8 shows simultaneously measured O_3 (left panel) and CO (middle panel) concentrations, both measured by *in situ* techniques, and BrO (right panel) concentrations measured by the mini-DOAS instrument in limb viewing direction during the DLR Falcon-20 sorties on 1 and 8 April 2007. The O_3, CO, and BrO profiles indicate that within the PBL, BrO was enhanced with typical mixing ratios

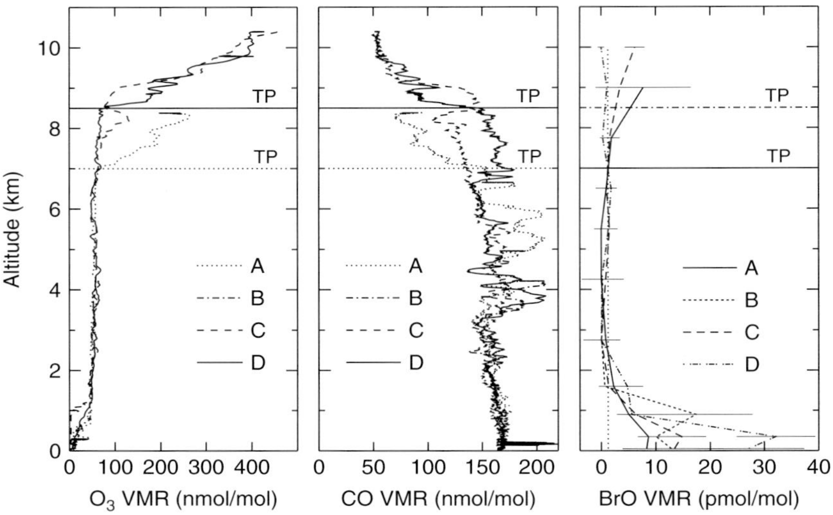

Figure 3.8 Measured profiles of O_3 (left panel), CO (middle panel), and BrO (right panel) in the Arctic during the ASTAR campaign in spring 2007 (VMR = volume mixing ratio, TP = tropopause). A = 1 April 2007 (descent 11:25 UTC); B = 8 April 2007 (descent 13:00 UTC); C = 8 April 2007 (ascent 14:30 UTC); D = 8 April 2007 (descent 15:20 UTC). While O_3 and CO were measured by *in situ* instrumentation from the DLR (Hans Schlager, personal communication), the BrO concentration profile was inferred from aircraft-borne limb scanning DOAS observations. For the latter, measured BrO slant column absorptions are inverted into BrO concentration profiles by a Bayesian inversion scheme. O_4 absorption bands where used to constrain the radiative transfer–modeled light path. This procedure appears to be necessary to properly account for the radiative transfer during the observations, which was largely affected by a varying degree of Mie scattering by aerosol and clouds (Prados–Roman *et al.* 2010).

between 5 and 30 pmol/mol, whereas ozone (<1 nmol/mol) was largely depleted as compared with unperturbed or pristine tropospheric conditions.

This finding underlines that the Arctic spring ODEs are due to bromine-mediated ozone destruction within the PBL. These BrO profile measurements were the first of their kind, demonstrating the potential of limb scanning UV/VIS/NIR spectroscopy for radical and trace gas detection.

The three major elements involved in airborne UV/VIS/NIR spectroscopy were described. Advantages of the technique are its high sensitivity for the detection of atmospheric radicals and trace gases, the flexibility in choosing the observation geometry, for example, by varying the viewing direction and by using movable platforms, or both, and thus its potential to access and study most atmospheric regions. Accordingly, UV/VIS/NIR spectroscopy has already been applied on any kind of ground-based, airborne, or spaceborne vehicle. Modern applications clearly advance to use the inherent tomographic skills of remote sensing, for example, by mathematically combining sets of measurements to provide one-dimensional plus time-dependent or, in specific cases, 2D plus time-dependent information on the target.

3.5.3
Cavity Ring-Down Spectroscopy

Cavity ring-down spectroscopy (CRDS) is a type of optical spectroscopy that provides extremely long (multikilometer) absorption path lengths inside a compact instrument. Several reviews exist of CRDS and related cavity-based techniques (Berden, Peeters, and Meijer, 2000; Berden and Engeln, 2009), with some reviews specifically focusing on applications in atmospheric research (Atkinson, 2003; Brown, 2003; Hancock and Orr-Ewing, 2009).

3.5.3.1 Measurement Principle

The central component of any cavity-based spectrometer is a high-finesse optical resonator formed, most usually, between two highly reflective concave mirrors. The very many reflections experienced by photons inside a high-finesse cavity enable the cavity to act as a short-term storage device for light.

CRDS was first demonstrated by O'Keefe and Deacon (1988), who noticed that an absorbing (or scattering) sample present inside a cavity reduces the cavity's ability to store light and hence shortens the 1/e decay time constant of a laser light pulse circulating within the cavity; this decay time is commonly called the ring-down time (τ). Measuring the light's ring-down time also means that CRDS is more immune to intensity fluctuations in the light source than most other types of spectroscopy. O'Keefe and Deacon (1988) demonstrated their new technique by measuring the 690 and 630 nm B$\leftarrow$X absorption bands of oxygen in ambient air. Modern CRDS experiments follow essentially the same approach: the sample's absorption is computed from ring-down times measured when the cavity contains the gas sample (τ) and when the sample is excluded from the cavity (τ_0), for example, by flushing the cavity with a non-absorbing gas such as nitrogen; see also Table 3.3. Typically, the sample's absorption spectrum is obtained from two sets of wavelength-resolved ring-down times recorded by scanning a pulsed tunable laser over the spectral region of interest with and without the sample inside the cavity. A target absorber's concentration is retrieved from the cavity ring-down spectrum using analogous spectral fitting methods to those employed with other types of absorption spectroscopy. In favorable cases, the absorption due to a specific molecule can be inferred from τ measurements at a single wavelength allied to the selective removal of the target absorber, for example, by chemical titration – see the National Oceanic and Atmospheric Administration (NOAA) cavity ring-down instrument for airborne measurements of NO_3 and N_2O_5 described in the following section.

Two notable variations on CRDS exist. Phase-shift cavity ring-down spectroscopy pumps a cavity with a sine wave-modulated continuous wave (CW) laser (Engeln *et al.* 1996). As noted earlier, the light's journey from source to detector is delayed by its passage through the cavity, thereby creating a phase delay in the

Table 3.3 A summary of the main features of cavity-based spectroscopic techniques.

Technique	CRDS	CEAS	Phase-shift cavity ring-down spectroscopy
Light source	Pulsed laser or CW laser (interrupted or deflected)	CW laser	CW laser (modulate at angular frequency ω)
Quantity measured	1/e time constant of the decay in light intensity: the ring-down time, τ	Light intensity transmitted through cavity at steady state, I	Phase delay of light exiting cavity relative to its source, ϕ
Relationship to ring-down time	–	I is proportional to τ	$\tan\phi = -\omega \cdot \tau$
Absorption coefficient of sample	$\alpha = \frac{1}{c} \cdot \left(\frac{1}{\tau} - \frac{1}{\tau_0}\right)$ c = Speed of light	$\alpha = \frac{1-R}{L} \cdot \left(\frac{I_0-I}{I}\right)$ R = Mirror reflectivity, L = Cavity length	$\alpha = \frac{\omega}{c} \cdot \left(\frac{1}{\tan\phi} - \frac{1}{\tan\phi_0}\right)$
Absolute absorber concentrations	Yes	Yes, if mirror reflectivity known versus wavelength	Yes

detected light's modulation, which is directly related to the cavity's ring-down time. Hence, the absorption coefficient of the sample can be found from comparison of phase shifts measured with and without the sample present in cavity (Table 3.3). Cavity-enhanced absorption spectroscopy (CEAS) measures the steady-state light intensity transmitted through a cavity pumped with a CW laser (Engeln *et al.* 1998). The closely related technique of integrated cavity output spectroscopy measures the time-integrated light intensity from a cavity pumped by a pulsed laser. The transmitted light intensity scales inversely with the total optical losses within the cavity and so is directly proportional to the ring-down time. Thus, the sample's absorption can be calculated from the fractional difference in the transmitted light intensities with and without the sample inside the cavity. In CEAS, the cavity enhances the distance the light travels through the intracavity sample by a factor of $1/(1 - R)$ compared with a single-pass measurement, where R is the mirror reflectivity. Cavity enhancement factors of the order of 1000 to 10 000 are typical. The simplicity of CEAS instrumentation (does not require a pulsed laser system or a high-speed detector) means CEAS and its variants are now widely used.

A key advantage of regular (e.g., single pass) absorption spectroscopy is that, via the Lambert–Bouguer law, it delivers an absolute measure of an absorber's concentration $[x]$ from measurements of relative light intensity before and after passing through the sample. The only additional information required are molecular absorption cross sections (σ) that are usually known to within a few percent accuracy from laboratory studies. In a similar way, CRDS yields absolute absorber concentrations from its measurements of ring-down times and hence the sample's absorption coefficient ($\alpha = \sigma[x]$). Thus, CRDS can have advantages over techniques of comparable sensitivity, but that requires a separate independent calibration, for example, laser-induced fluorescence (LIF), (resonance-enhanced) multiphoton ionization, intracavity laser absorption spectroscopy, and photoacoustic spectroscopy. CEAS and phase-shift CRDS also measure absorber concentrations directly, albeit a separate measure of the mirror reflectivity is required in CEAS.

There is no intrinsic limit on the spectral region accessible to cavity-based spectroscopy, provided optics and suitable light sources are available. In the laboratory, CRDS and related methods have been applied from 200 nm in the UV to millimeter wave spectroscopy (Berden and Engeln, 2009). A number of optics manufacturers now offer cavity mirrors "off the shelf" with reflection efficiencies of 99.99% or better. A cavity constructed of $R = 99.99\%$ mirrors separated by $L = 50$ cm (a length that comfortably fits into a standard aircraft rack) has an empty cavity ring-down time of $\tau_0 = 16.7\,\mu s$, equivalent to an absorption path of 5 km. The minimum detectable absorption in CRDS is limited by being able to accurately determine the fractional change in ring-down time ($\Delta\tau$) owing to the absorber's presence in the cavity:

$$\alpha_{\min} = \frac{1-R}{L} \cdot \left(\frac{\Delta\tau}{\tau_0}\right)_{\min} \tag{3.1}$$

A well-designed instrument can determine $\Delta\tau/\tau_0$ to 0.1%, and thus the minimum absorption detectable using the example cavity above is $\alpha_{\min} = 2\times10^{-9}\ \text{cm}^{-1}$. For a strong electronic absorption band, for example, that of the NO_3 radical at 662 nm, an absorption of $2\times10^{-9}\ \text{cm}^{-1}$ is equivalent to an absorber concentration of 10^8 molecules/cm^3 (or $[NO_3] = 4\ \text{cm}^{-1}$ pmol/mol at 1 bar pressure). In principle, the detection limit can be improved by using better mirrors: more highly reflective mirrors yield greater absorption path lengths with concomitantly more opportunity for photons to interact with the sample. But this comes at the cost of fewer photons being transmitted into the cavity through the entrance mirror, which adversely affects the accuracy of the ring-down time determination. CRDS instruments typically have sensitivities of the order of 10^{-7} to $10^{-10}\ \text{cm}^{-1}$ (the limiting sensitivity in cavity-enhanced absorption and phase-shift instruments is similar for analogous reasons). Higher sensitivity has been achieved in specialist instruments where a narrowband CW laser has been actively locked to excite a single Fabry–Perot mode of the cavity (Paldus *et al.* 1998; Ye and Hall, 1998).

An alternative restatement of Eq. (3.1) is that, for CRDS to provide an efficient measurement of a trace gas absorption, the per-pass loss due to the sample's absorption inside the cavity cannot be more than a couple of orders of magnitude less than the reflection loss at each mirror. In practice, the wavelength range accessible to a cavity-based spectrometer in atmospheric applications will be limited by the properties of the atmosphere itself, as well as the hardware. At short wavelengths and even in the absence of aerosol scattering, the per-pass loss will be dominated by Rayleigh scattering by the bulk atmospheric gases (and at very short wavelengths by efficient absorption by ozone), meaning there is no advantage to be gained in using very highly reflective mirrors (Brown, 2003). At infrared (IR) wavelengths, it is necessary to avoid the near-ubiquitous and often very strong absorption bands due to H_2O: in some cases, it may be possible to dry or pre-concentrate the gas samples or to find narrow absorption lines of a target species lying in micro-windows between individual rotation–vibration lines of water.

3.5.3.2 **Aircraft Implementation**

The recent advances in cavity-based spectroscopic methods have enabled nighttime aircraft measurements of NO_3 and N_2O_5. Before the advent of these techniques, the only viable method for quantifying atmospheric NO_3 radicals was based on DOAS (Platt, 1994) and electron spin resonance (ESR) spectroscopy (Geyer *et al.* 1999). In the either case, the temperature-dependent equilibrium mixing ratio of N_2O_5 must be calculated from the column average measurements of NO_2 and NO_3 (Atkinson, Winer, and Pitts, 1986).

In situ aircraft instrumentation for NO_3 and N_2O_5 has enabled detailed investigation of the nocturnal atmospheric chemistry of nitrogen oxides. Nighttime aircraft measurements are particularly important because of inefficient mixing and vertical stratification that typically leads to large vertical concentration gradients at night. Surface-level measurements do not necessarily characterize the bulk of the chemistry occurring within the shallow nocturnal boundary layer or the deeper residual daytime boundary layer above it. Because the residual layer is typically decoupled from the surface, aircraft measurements may be the only way to track the evolution of pollutants mixed into this layer during the previous day or to sample nighttime emissions from elevated stacks, such as power plants, that emit directly into this altitude range. Within the nocturnal boundary layer, aircraft sampling can provide direct measurements of vertical concentration gradients depending on the minimum altitude capability of the particular aircraft, for example, by missed approaches into airfields. Figure 3.9 illustrates an example of nighttime vertical profiles of NO_3, N_2O_5, and related species (Brown *et al.* 2007). The aircraft encountered a power plant plume just above the nocturnal boundary layer and sampled distinct layered structure in NO_3 and N_2O_5 within both the residual daytime and nocturnal boundary layers (the latter marked by the sharp increase in potential temperature up to approximately 150 m). The presence of SO_2 in the large plume identifies its source as a coal-fired electricity generating plant. The anticorrelation between NO_2 and O_3 is a characteristic of the NO_x plumes emitted after dark in which O_3 is lost to titration by NO and the further, slower reaction with NO_2, but is no longer replenished by photochemical NO_x reactions. The right graph shows that the N_2O_5 lifetime, or the ratio of $[N_2O_5]$ to its production rate from the NO_2+O_3 reaction (a common measure of reactivity for NO_3 or N_2O_5), was long enough for significant overnight transport of NO_x in the form of N_2O_5 to occur.

Currently, the two examples of aircraft instruments for NO_3 and N_2O_5 measurements are the NOAA cavity ring-down instrument (Dubé *et al.* 2006; Fuchs *et al.*

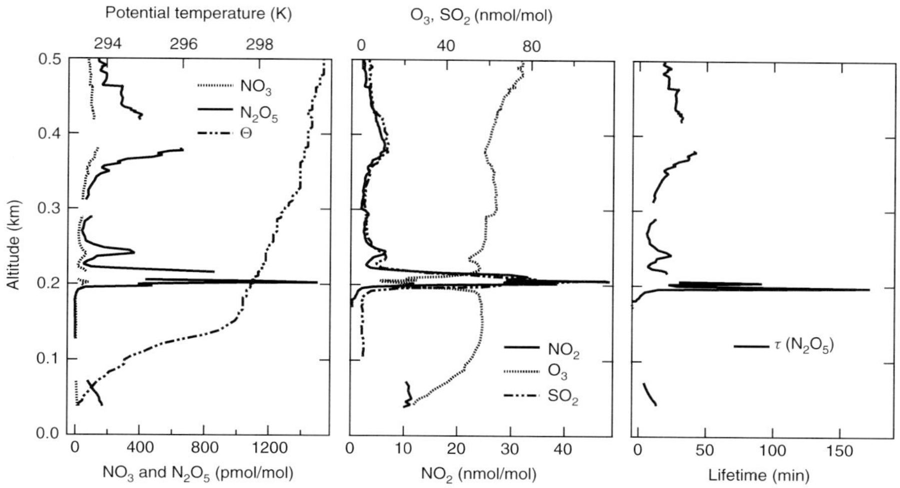

Figure 3.9 Aircraft vertical profile from surface level to 500 m showing potential temperature (top axis, left), NO_3 and N_2O_5 (bottom axis, left), O_3 and SO_2 (top axis, center), NO_2 (bottom axis, center), and N_2O_5 lifetime (right graph). Data were acquired during the ICARTT campaign in August 2004 from instruments on board the NOAA P-3 aircraft. (Source: Adapted from Brown *et al.* (2007). Copyright 2007 American Geophysical Union (AGU). Reproduced/modified by permission of AGU.)

2008) that operates on the NOAA P3 aircraft in the United States and the University of Cambridge broadband cavity-enhanced absorption instrument that operates from the UK's BAe-146 Facility for Airborne Atmospheric Measurements (FAAM) aircraft (see the following section). The NOAA cavity ring-down instrument has flown during three separate aircraft campaigns in 2004 (ICARTT) (Fehsenfeld *et al.* 2006), 2006 (Texas Air Quality Study) (Parrish *et al.* 2009), and 2010 (California Nexus (CalNex); *http://www.esrl.noaa.gov/csd/calnex/*). In the more recent CalNex campaign, the light source was a 100-mW tunable diode laser near 662 nm, modulated at 500 Hz; this replaced a bulkier Nd:YAG pumped dye laser system (pulse energy 1 mJ at 662 nm, pulse width 6 ns, repetition rate 50 Hz) used in the earlier campaigns.

A schematic of the NOAA instrument for the CalNex campaign is shown in Figure 3.10. The optical cavities consist of two 1-m radius of curvature mirrors with a reflectivity of 99.9995% separated by 95 cm. Ring-down time constants in the absence of NO_3 or N_2O_5 are in the order of 400 μs (equivalent to a path length of 120 km). The instrument's response time is 1 Hz or better, typically achieved by coadding a series of individual ring-down transients and fitting the sum to a single-exponential decay to acquire the ring-down time constant.

3.5.3.3 **Calibration and Uncertainty**

Aside from mounting the instrument frame on a set of vibration isolators, there are no special requirements for operation of the NOAA cavity ring-down

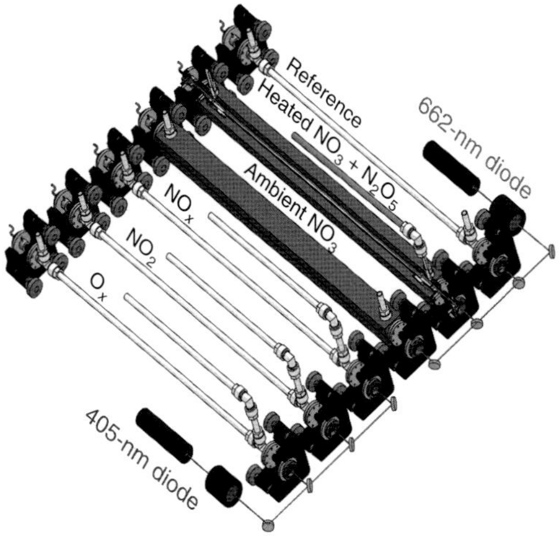

Figure 3.10 Schematic of the NOAA cavity ring-down instrument. (Source: Adapted from Brown *et al.* (2007). Copyright 2007 American Geophysical Union (AGU). Reproduced/modified by permission of AGU.)

instrument on an aircraft relative to a laboratory or ground-based setting. It has similar intrinsic noise statistics in most environments. The inherent detection sensitivity of this instrument is equivalent to $[NO_3] = 0.2$ pmol/mol when it is not under flow; however, turbulence at the sample rates required for reactive trace gases such as NO_3 and N_2O_5 leads to optical noise that degrades the sensitivity to the range 0.5–2 pmol/mol (2σ, 1 s). The instrument is zeroed by addition of NO to the inlet to chemically destroy NO_3 via its rapid conversion to NO_2 (i.e., $NO + NO_3 \rightarrow 2\ NO_2$). The combination of 662 nm optical extinction and chemical titration with NO provides a measurement that is highly specific to NO_3. During aircraft sampling, the optical extinction at 662 nm due to various other trace species, including water vapor, NO_2, and O_3, as well as the pressure-dependent Rayleigh scattering background, may vary rapidly between acquisitions. The latter can be particularly large because the instrument's sample pressure changes with the external aircraft pressure. Background ring-down time constants vary in the range 400–450 μs as the aircraft ascends and descends through its normal altitude range; however, correction for this Rayleigh scattering–induced pressure dependence is straightforward to implement. The much smaller corrections for absorption due to O_3 and NO_2 at 662 nm can be straightforwardly subtracted if sufficiently accurate measurements of these compounds are available from other instruments on the aircraft. Lastly, the sensitivity to water vapor with the pulsed dye laser instrument is generally very small because its 0.5 to 1 cm^{-1} bandwidth allows for spectral discrimination against water vapor absorption lines with no loss in sensitivity for NO_3.

Accurate sampling of NO_3 and N_2O_5 from ambient air is a tractable but difficult task owing to their reactivity on inlet surfaces. The problem is compounded in the NOAA cavity ring-down instrument by the use of an inlet filter, which is required in this instrument to eliminate the large and variable background extinction signal due to ambient aerosol. Several steps mitigate the sampling accuracy issues associated with surface losses. First, filters are housed in a pneumatically actuated automatic changer and are typically changed each hour (or more often, if needed) to reduce aerosol accumulation that may lead to heterogeneous loss of N_2O_5. Second, the instrument samples at flow rates and residence times that are rapid relative to the measured inlet wall loss rate for NO_3. Operation at reduced pressure (approximately half of ambient pressure, achieved with a short inlet restriction) significantly reduces the residence time and the rate at which aerosol accumulates on the inlet filter. Lastly, periodic calibrations are preformed via introduction of an N_2O_5 or NO_3 source to the inlet followed by conversion to NO_2, which is measured at either 532 or 405 nm in a separate cavity ring-down channel downstream of the N_2O_5 and NO_3 cavities. The overall instrument accuracy is therefore estimated at $-9/+12\%$ for NO_3 and $-8/+11\%$ for N_2O_5, limited in part by the transmission efficiency and in part by uncertainty in the absorption cross section (Fuchs *et al.* 2008).

3.5.3.4 **Broadband Cavity Spectroscopic Methods**

Cavity-based spectroscopic techniques employing spectrally broad light sources offer certain advantages for atmospheric applications (Ball and Jones, 2003, 2009). As the emission bandwidth of the light source is typically wider than the molecular absorption features of interest, an extended portion of the sample's absorption spectrum can be obtained in a single observation. This is an important advantage in analytical applications where the sample's composition changes rapidly, such as when sampling from a moving aircraft.

3.5.3.4.1 **Measurement Principle** Broadband methods are particularly useful when targeting atmospheric absorbers with broad but structured electronic absorption bands at VIS and near-UV wavelengths. Methods borrowed from DOAS (Platt, 1994) can be applied to positively identify a target absorber and quantify its concentration in the presence of other absorbing/scattering species. Indeed, there is generally no requirement to filter aerosol particles from the sample flow (except where one wishes to preserve the cavity's optimum ring-down time), and in favorable cases, broadband methods can measure the aerosol extinction from the broad continuum in the sample's absorption spectrum underlying the structured molecular absorption features.

At present, almost every variant of CRDS has its broadband equivalent. Broadband cavity ring-down spectroscopy (BBCRDS) simultaneously records ring-down traces at multiple wavelengths using a spectrograph and a clocked charge-coupled Device (CCD) camera. BBCRDS has been used in ground-based measurements of NO_3 (Ball and Jones, 2003), NO_2 (Fuchs *et al.* 2010), and iodine in the marine boundary layer. The size of the BBCRDS hardware currently precludes its operation on aircraft. Broadband cavity-enhanced absorption spectroscopy (BBCEAS) uses the output from arc lamps or light-emitting diodes (LEDs). Several ground-based BBCEAS instruments exist targeting, for example, NO_3, N_2O_5, NO_2, HONO, I_2, and glyoxal (Ball, Langridge, and Jones, 2004; Gherman *et al.* 2008; Langridge *et al.* 2008a; Washenfelder *et al.* 2008; Varma *et al.* 2009). Light from a modulated LED has also been used to perform phase-shift CRDS for ambient NO_2 (Kebabian *et al.* 2008). Further into the future, supercontinuum laser light sources and prism cavities offer the potential for extremely broadband cavity systems (Johnston and Lehmann, 2008; Langridge *et al.* 2008b).

3.5.3.4.2 **Aircraft Instrumentation** A BBCEAS instrument has recently been flown on the UK's FAAM BAe-146 aircraft as part of the Role of Nighttime Chemistry in Controlling the Oxidizing Capacity of the Atmosphere (RONOCO) consortium. Like the NOAA cavity ring-down instrument, the BBCEAS instrument has three channels to measure: (i) NO_3 at ambient temperature, (ii) N_2O_5 by thermal decomposition to NO_3 in a heated channel, and (iii) NO_2 (Kennedy *et al.* 2011). The detection limits for NO_3, N_2O_5, and NO_2 are 0.7, 1.0, and 25 pmol/mol, respectively (1σ, 1 s), when operating the instrument in the laboratory, with only a modest degradation in the NO_3 and N_2O_5 channels' detection limits observed in-flight (1.1 and 2.4 pmol/mol, respectively). Ambient air is sampled into the BBCEAS instrument through two rear-facing inlets approximately 4 m behind the aircraft's nose and 10 cm away from the aircraft's body. The first inlet (thermally insulated PFA tube, 60 cm in length and 16 mm in internal diameter) supplies the three cavities at a combined volumetric flow rate of 50 l/min, such that the residence time inside the inlet and instrument is short compared with the NO_3 and N_2O_5 wall loss rates. A second inlet is used to draw ambient air through a sheath surrounding the NO_3 channel to maintain its temperature close to that outside the aircraft and thus to minimize any perturbation of the NO_3/N_2O_5 equilibrium.

The NO_3 and N_2O_5 channels use light from red LEDs and operate with a bandwidth of 640–680 nm, whereas the NO_2 channel uses light from a blue LED and a bandwidth of 442–480 nm. Fiber optic cables convey light from the LEDs to their cavities and out of each cavity into its own dedicated mini-spectrometer. Spectra of the steady-state light intensity transmitted through the cavity in the absence of the atmospheric sample are obtained periodically by back-flushing the cavities with dry nitrogen. In addition, the mirror reflectivity of each cavity is determined by measuring the phase shift of the transmitted light intensity during off-line calibrations performed by stepping through 2-nm portions of the LED's output selected by a monochromator (Langridge *et al.* 2008b).

Initial results from the RONOCO flights are encouraging. Highly structured time series for NO_3 and N_2O_5 were observed during most nighttime flights, indicative of the aircraft passing through multiple pollution plumes (Figure 3.11). In addition, both the NO_2 measured in the blue channel and the water vapor measured via an absorption band that overlaps NO_3's absorption in the red channels showed excellent correspondence with NO_2 and humidity measured by the aircraft's core instruments.

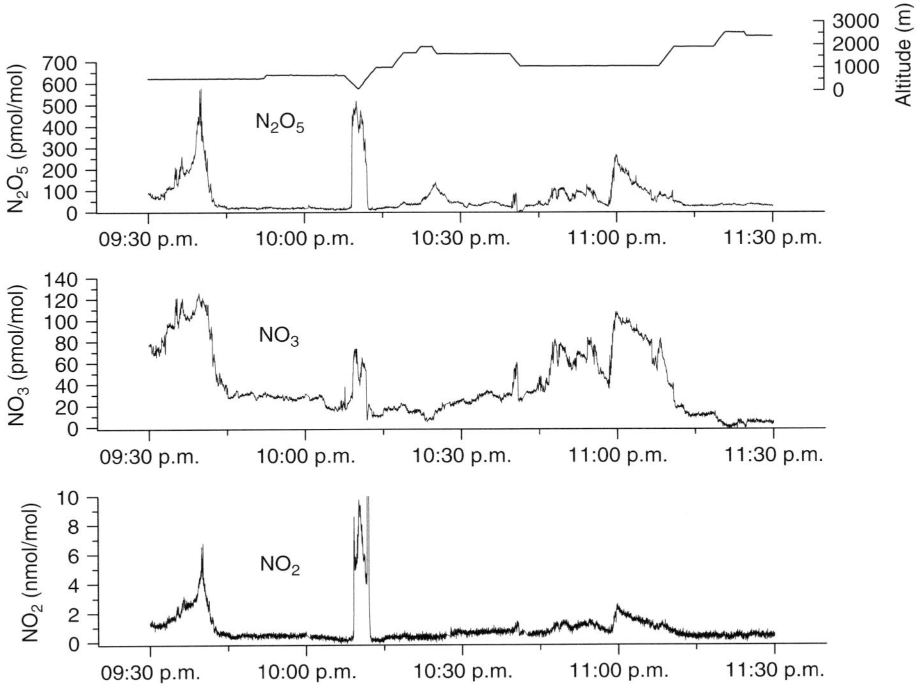

Figure 3.11 Time series of N_2O_5, NO_3, and NO_2 volume mixing ratios (1-s data) recorded by the Cambridge University BBCEAS instrument during a RONOCO flight over southern England on the night of 20 July 2010 (Kennedy *et al.* 2011). The large spike in all three time series around 10:10 p.m. is because of a missed approach into London Southend Airport. The elevated concentrations around 9:40 and 11:00 p.m. correspond with the aircraft flying across and subsequently recrossing a polluted air mass (probably outflow from London).

3.5.4
Gas Filter Correlation Spectroscopy

Gas filter correlation (GFC) spectroscopy is an enhanced version of nondispersive infrared (NDIR) spectroscopy. The NDIR technique utilizes a broadband light source, usually a heated filament. The signal from the sample is compared with that obtained by passing the IR beam through a cell containing a known concentration of the target analyte. GFC improves this technique by incorporating a multipass cell (usually a white cell) that increases the overall path length and thus sensitivity; it also uses a chopper wheel to modulate the light signal (Figure 3.12).

GFC can be applied to a number of trace gases, including carbon dioxide, carbon monoxide, nitrous oxide, methane, and water vapor. Several groups have modified commercial units designed to measure urban levels of carbon monoxide for more pristine environments. They replaced low-cost aluminum mirrors with gold or silver ones and CaF_2 lenses were used to focus the light beam onto a hand-picked

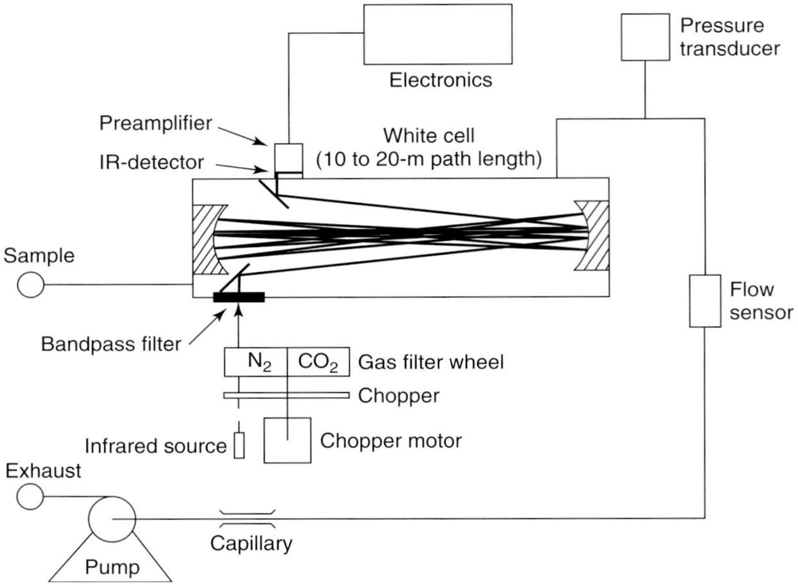

Figure 3.12 Internal schematic of a GFC instrument (in this case, CO_2).

high-sensitivity detector (Dickerson and Delany, 1988). Frequent zeroing improves the overall measurement accuracy, and additional pumping capacity is often used to increase the response times.

GFC instruments continue to be used on some research aircraft, compared with other more sensitive techniques, they are still a robust measurement system and have considerable advantages as they can be made autonomous relatively easily, they require no gas cylinders also, which makes them very attractive for installation onto commercial aircraft such as that done as part of the Measurement of Ozone and Water Vapor by Airbus In-Service Aircraft (MOZAIC) (Nedelec *et al.* 2003) and this will continue into the In-Service Aircraft for a Global Observing System project.

3.5.5
Tunable Laser Absorption Spectroscopy

The basics of the tunable laser absorption spectroscopy (TLAS) technique have been described in Section 2.6.7, with the example of water vapor measurements. A schematic of the measurement principle is shown in Figure 2.11. It is based on the absorption of an IR laser beam by the trace gas to measure, leading to a transition between a rotational level of a lower vibrational state and a different rotational level of a higher vibrational state. The laser beam is tuned over a spectral microwindow (1/cm) to a wavenumber matching one (or sometimes a few) vibrational line of the molecule. The Lambert–Bouguer law relationship is used, providing quantitative results thus based on fundamental molecular parameters and measured quantities (transmittance signal, optical path length, pressure, and temperature), which are independent of instrumental factors. Therefore, absorption measurements are considered absolute and do not require calibration.

The essential components of a tunable laser spectrometer are the laser source, the optical absorption cell containing the sample gas, the photodetector, and the control–acquisition electronics. Overall, this leads to a highly compact instrument with high frequency (in a few seconds or less) and precise (<0.2%) and accurate (a few percent for absolute concentrations) measurements. A comprehensive overview of this technique has been given by Heard (2006). Hence, only the latest developments are presented, in particular, concerning the laser sources, for which significant advances have been made.

3.5.5.1 Tunable Diode Versus QCLs

High-performance airborne TLAS instruments currently employ solid-state semiconductor laser sources. Until the 2000s, TLAS using lead–salt diode lasers were widely used in the mid-infrared (MIR, 3–30 μm) domain, where molecules are active for the fundamental transitions, leading to the strongest absorptions. TLAS has been successfully applied to airborne measurements, still recently, for instance, for CH_4, N_2O, CO, and HCHO (Scott *et al.* 1999; Bartlett and Webb, 2000; Kormann *et al.* 2002; Fischer *et al.* 2002; de Reus *et al.* 2003; Wert *et al.* 2003; Viciani *et al.* 2008).

However, their use is decreasing at present, given the emergence of a new type of lasers, the QCLs, with more advantages. The difference in the technology of these two kinds of laser results in better intrinsic properties for the QCLs (Heard, 2006; Werle *et al.* 2002; Beck *et al.* 2002):

(i) reproducibility in the wavelength emission domain and the optical output power, whereas the lead–salt diode characteristics may change with temperature cycling;
(ii) pure single-mode and single-frequency operation by incorporation of a distributed feedback structure;
(iii) wider range of wavenumber tunability (typically 10/cm) compared with lead–salt diodes (1–2/cm), potentially leading to versatility in the choice of the species to analyze; and
(iv) much higher optical output power, usually >10 mW, that is, around two orders of magnitude higher than the lead–salt ones, which is interesting for long optical path operation.

Finally, the QCLs' final drawbacks have been very recently overcome:

(i) operation near room temperature, in the range 240–300 K where small Peltier thermoelectric coolers can replace large cryogenic liquid N_2 dewars in use for the lead–salt diodes (operating at 80–120 K);
(ii) operation in CW mode instead of pulsed mode, leading to negligible line widths (HWHM < 6 MHz, i.e., 2×10^{-4}/cm) with respect to the molecular

absorption line, which is not always the case for lead–salt diodes, giving measurements of higher accuracy;

(iii) extension of the wavelength emission down to 3 μm, instead of initially 4 μm. QCLs can now scan an MIR (3–20 μm) region similar to the one of the lead–salt diodes.

Thus, leading in addition to more compact instruments, QCLs are more widely used for trace gas measurements, for instance, for HCHO and HCOOH (Herndon *et al.* 2007), for N_2O (Kort *et al.* 2011), and for CO and CH_4 (Wunch *et al.* 2010; Klippel *et al.* 2011; Wecht *et al.* 2012).

In the NIR domain (1–2 μm), TLAS usually employs InGaAs diode laser sources developed for telecommunication industry, extending the species measurement possibilities. These lasers have similar properties as the QCLs, for example, narrow line widths (<10 MHz), high output powers (>10 m W), single-mode emission, and no mode-hops in a long continuous tunability range (>2/cm). Substantial advantages are in addition their lower cost and their real room-temperature operation. However, the molecular vibrational transitions taking place are overtones or combinations and so, the absorptions are much less intense (at least by a factor of 100) than the fundamental transitions. This drawback has no consequences for measurements of the more abundant trace gases, for example, H_2O, CO_2, and CH_4, which are classically performed in this domain (Zondlo *et al.* 2010; D'Amato, Mazzinghi, and Castagnoli, 2002). Again, significant progress in this type of laser design has been very recently made, allowing for the extension of the NIR region toward 3 μm, with the same advantages. The new generation of InGaAsSb diode lasers operating at room temperature has been marketed, which covers the 2- to 3-μm domain, making possible the scanning of other, sometimes more intense, transitions and even fundamental ones, for example, at 2.63 μm of H_2O (Durry *et al.* 2007). Note that TLAS (NIR and MIR) has also been applied to airborne measurements of water vapor and carbon dioxide isotope ratios (Kerstel and Gianfrani, 2008; Dyroff, Fuetterer, and Zahn, 2010).

3.5.5.2 **Further Progress**

Finally, mentioned briefly are other essential parts of TLAS in which advances have arose. Depending on the wavelength domain, InAs, InGaAs, InSb, or HgCdTe photodetectors are used, and again great progress has been realized in moving from cryogenically cooling to thermoelectrically cooling by Peltier effect or by Stirling cycle, making the instruments completely free of cryogenic liquid. Concerning the absorption cell, the atmospheric gas to analyze is continuously drawn through a closed cell at reduced pressure (10–50 hPa) or probed through an air-opened optical path, the latter case being well suited for reactive and highly adsorbing gases. It is worth to note the recent patented invention of a new type of optical multipass cell, the Robert cell made of three spherical mirrors. This cell combines the design of the Herriott and white cells, behaving as a multiplier of Herriott cell (Robert, 2007). Its main advantages are that it can be made of standard mirrors without particular tolerance (therefore, greatly lowering the cost), its operating conditions are very easy, and the path length is greatly versatile and easily determined, while allowing for very long optical paths in a very compact volume. For instance, more than 150-m path length has been easily reached with a cell of 7.4-cm diameter and 64-cm length, recently built for the airborne spectromètre infrarouge in situ (SPIRIT) instrument associating three QCLs sequentially operating in the same Robert cell (Guimbaud *et al.* 2011). This compact instrument (0.36 m^3) enabled *in situ* measurements of CO, CH_4, CO_2, and N_2O at four time resolutions on board the DLR Falcon-20 within the frame of the European Stratospheric Ozone: Halogen Impacts in a Varying Atmosphere campaign in Malaysia in November–December 2011.

3.5.6 Fluorescence Techniques

3.5.6.1 Resonance Fluorescence

As has already been demonstrated, absorption spectroscopy is not sensitive enough to provide access to the low concentrations that are frequently observed in both remote environments and also outside of the PBL. Such a gap in the "measurement market" is often the driver for advances such as the application of resonance fluorescence as an alternative method for measuring carbon monoxide. Originally developed for stratospheric balloons (Volz and Kley, 1985), the instrument achieved a detection limit of 1 nmol/mol for an integration time of 10 s from ground level up to an altitude of 34 km. Using a resonance lamp excited by an Resonance Fluorescence (RF) discharge in combination with an optical filter produces photons between 145 and 151 nm, and fluorescence occurs across a wavelength range in the vacuum ultraviolet (VUV) of 160–190 nm, this is in turn detected by a photomultiplier followed by a fast counter, as with the modified GFC instrument, CaF_2 windows and lenses are used (Figure 3.13). The prototype instrument was then modified for use on board the UK Meteorological Office C-130 Hercules aircraft (Gerbig *et al.* 1996, 1999) and has subsequently been commercialized and VUV resonance fluorescence has become the standard technique for airborne observations of CO. It also has an increased time resolution and is capable of operating at up to 10 Hz.

Comparisons of resonance fluorescence and a TLAS system resulted in excellent agreement (within 11% with systematic offsets of $<$1 nmol/mol) between the two methods (Holloway *et al.* 2000).

3.5.6.2 LIF Techniques

The earliest attempts to use LIF technique on aircraft platforms dates back to the middle of the 1970s, but only a decade later, owing to improvements on pulsed laser efficiency and the reduction in their dimensions and weight, more reliable airborne LIF systems were developed. The LIF technique is based on the molecule fluorescence stimulated by laser and its detection to determine the molecules concentration. On the aircraft, LIF systems are used to observe directly reactive radicals such as OH, precursors of ozone such as NO_2, and after thermal or chemical conversion, HO_2, peroxy nitrates (RO_2NO_2), alkyl nitrates ($RONO_2$), and nitric acid (HNO_3). There are a number of other species, such as peroxy radical (RO_2) and

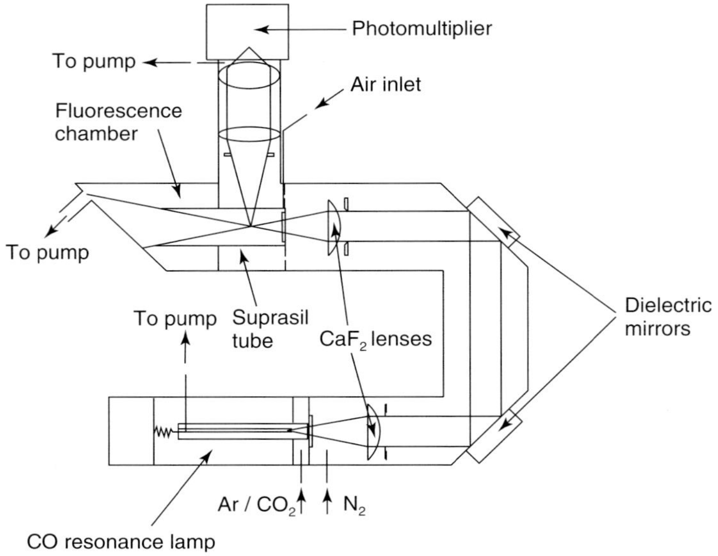

Figure 3.13 Optical layout of a commercial VUV resonance fluorescence carbon monoxide analyzer. (Source: Adapted from Gerbig *et al.* (1999). Copyright 1999 American Geophysical Union (AGU). Reproduced/modified by permission of AGU.)

halogen monoxide (IO), which can also be detected using the LIF technique. ROxLIF can be used for measurement of atmospheric peroxy radicals using a two-step chemical conversion scheme and LIF for radical detection (Fuchs *et al.* 2008).

The scheme of an LIF system is as follows: A laser emits radiation exactly at one of the wavelength absorbed by the molecules and populates their excited states, then a detector measures the following fluorescence emitted during molecule relaxation to the lower states.

Technical and instrumental complexities arise because the fluorescence is in competition with other relaxation processes such as electronic quenching. The quenching is a nonradiative relaxation of the molecule to a lower energy state by giving up the energy in collision with bath molecules, mainly N_2, O_2, and H_2O. Owing to quenching, the number of photons emitted by fluorescence is lower than the number of the excited ones. The fluorescence signal S (photon counts per second) includes factors such as laser power, absorption cross section, quenching and relaxation factors, as well as the optical geometries inside the detection cell. It is not easy to calculate all the individual parameters, and for this reason, in the LIF technique, it is usual to calibrate the system instead of finding the values of all the parameters. The equation used is

$$S = \alpha \cdot C \tag{3.2}$$

where S is the fluorescence signal in counts/s, α the calibration constant, and C the concentration of the species to be detected. The calibration constant (counts/s/nmol/mol) is determined by injecting known amounts of the target

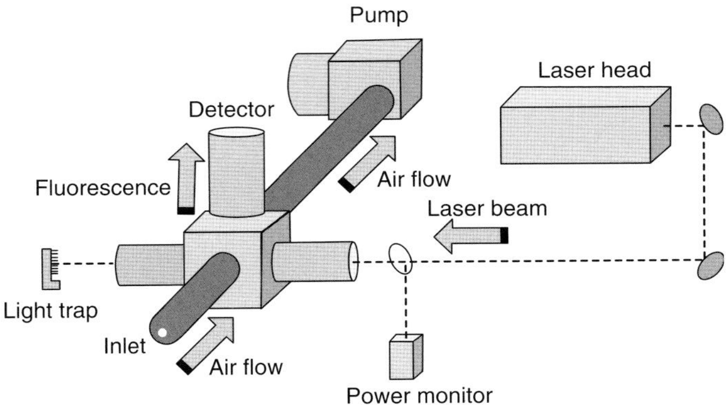

Figure 3.14 Sketch of the main parts of a LIF system.

molecule, in the detection cell of the instrument, and measuring the fluorescence as a function of the molecule concentrations. Knowing α, Eq. (3.2) is used to determine the concentration of the molecule in the atmosphere, because this equation gives a relation between fluorescence in photon counts/s measured by the LIF detector and the species concentrations.

The heart of the LIF system is the detection cell shown in Figure 3.14 where there is the intersection of three perpendicular axes:

(i) the air flow of the ambient air to be sampled;
(ii) the laser beam; and
(iii) the detector that collects the fluorescence emitted by the molecules.

The scheme reported in Figure 3.14 is the basic configuration of almost all the LIF systems. Differences among LIFs include the laser source, the detector, filters, design of the fluorescence cell, and other minor parts, which basically depends on the molecule to be detected. Sometimes, a single-beam configuration is replaced by a multipass setup where the laser light is reflected several times (20–30 times) between three mirrors (white cell) or two mirrors (Herriott cell).

When a molecule is excited by a laser and it fluoresces, there are other radiative processes (Rayleigh, Mie, and Raman scattering) and walls scattering, which can be mixed with fluorescence photons. To distinguish fluorescence photons from others, optical filters and temporal filters are used in the detection scheme of LIF system. For molecules that fluoresce at the same wavelength of the laser light used for their excitation, such as OH (excitation and fluorescence at 308 nm), the difficulties in distinguishing fluorescence photons from other scattered photons are overcome by taking advantage of the different time duration of the fluorescence compared with other radiative processes as function of the pressure inside the detection cell. When a pulsed laser is used in an LIF technique, the Rayleigh and Mie scatters have the same temporal length of the laser pulse (usually tens of nanoseconds), whereas the fluorescence can be longer than hundreds of nanoseconds when the pressure inside the detection cell is reduced to a few hPa because the quenching is less efficient. This feature is utilized by the use of a delay gate, which turns on the detector, usually a photomultiplier tube (PMT) or a multichannel plate, after a delay of a few nanoseconds after the end of the laser pulse and it is kept on for hundreds of nanoseconds until the fluorescence ends (Thornton, Wooldridge, and Cohen, 2000; Faloona *et al.* 2004).

The measurement sequence of an LIF system includes a tuning of the laser wavelength to one of the absorption lines of the molecule, followed by a period of acquisition of the fluorescence signal and a changing of the wavelength of the laser to one side of the absorption line, where the molecule absorption is absent or weak to acquire for a while the background signal. Usually, the background takes into account all the nonfluorescence signals as the molecule to be detected, including fluorescence of other molecules, is measured by injecting a compound such as perfluoropropene (C_3F_6) that selectively removes only OH (Faloona *et al.* 2004; Heard and Pilling, 2003), or for other molecules, a compound such as NO_2, using the sample ambient air that travels through a tube coated with TiO_2, that selectively removes NO_2 (Matsumoto and Kajii, 2003). High-purity zero air as well is usually sent into the detection cell to observe the system background.

For OH, a complex calibration system is needed to generate a known amount of OH to be sent into the detection cell for the calibration. There are few different techniques; one of them uses the photolysis of water vapor by the mercury lamp line at 185 nm to make OH. This calibration system is quite big so none of the LIF systems for OH measurements installed on aircraft uses on-flight calibrations. However, most of them include a relative in-flight calibration that uses an Hg lamp to monitor the changes in the instrument sensitivity during different flight conditions (Martinez *et al.* 2008; Commane *et al.* 2010).

The uncertainty of LIF measurements depends mainly on the calibration uncertainties. For systems that use standard gas, the main sources of errors are

(i) the uncertainty of the concentration of the standard gas in the cylinder (usually for NO_2 between 3% and 6%) and
(ii) the uncertainty of the mass-flow controllers, which is usually 0.5% of the reading plus 0.2% of the full range and globally account between 2% and 5%.

Systems, such as OH-LIF, that use a calibration system to produce OH usually suffer from big uncertainty in the determination of the actinic flux of the mercury lamp that implies an uncertainty on the OH measurements, which can be between 25% and 32% (Faloona *et al.* 2004; Commane *et al.* 2010).

Another parameter that characterizes an LIF system is the detection limit that is the minimum detectable mixing ratio of a species. The detection limit of OH-LIFs used on aircraft range between 0.003 pmol/mol of the Pennsylvania State University (PSU) system (Faloona *et al.* 2004) and 0.029 pmol/mol of the Leeds system (Commane *et al.* 2010). For NO_2, a detection limit of 8 pmol/mol is reported by Bertram *et al.* (2007).

In LIF systems, as mentioned earlier, the detection cell is kept at a low pressure using vacuum pumps. The pressure inside the cell is a function of the ambient pressure. Therefore, when the aircraft changes the flight altitude, the internal pressure of the cell changes and consequently the sensitivity of the LIF. This means that a calibration as a function of the altitude is needed, and for LIF systems not using in-flight calibration systems, a correction for the altitude changes in the calibration constant is required.

LIF systems that measure very reactive species such as OH require special inlets to reduce the air speed, to minimize the loss of the compound to be measured. There are two inlet designs, at the moment: a nacelle described in the study by Faloona *et al.* (2004) and used by the PSU-LIF and one adapted from the design reported in the study by Eisele *et al.* (1997) and mounted by the European OH-LIF (Martinez *et al.* 2008; Commane *et al.* 2010).

The first LIF installed on aircraft was the two-photon system for NO observations developed by the Georgia Tech (USA) and was used in 1983 (Bradshaw *et al.* 1985). This system used two wavelengths. Therefore, two lasers were used, one at 226 nm and another in the IR, at 1.1 μm, to excite NO molecules and detect fluorescence at about 190 nm. At the beginning of the 1990s, an LIF system for OH and HO_2 measurements was developed at the Harvard University (USA) and installed on the ER-2 aircraft used for stratospheric studies. This system uses the transition near 282 nm to excite the fluorescence of the OH molecules at 308 nm. HO_2 is measured after chemical conversion into OH, injecting NO into the detection cell. More details on the instrument design, performance, and campaign results could be found in the studies by Wennberg *et al.* (1994a,b, 1995).

A tropospheric LIF for OH and HO_2 measurements was developed by the PSU (USA). As the water vapor content is higher in the troposphere than in the stratosphere, to reduce the self-production of OH below the detection limit, the PSU-LIF reduces the pressure in the detection cell compared with ambient and excites the OH molecules at 308 nm (where the quantum yield of the production of $O(^1D)$ from O_3 photolysis is much less than that at 282 nm), and the fluorescence is detected at the same 308-nm wavelength. HO_2 is detected after conversion into OH as in the Harvard system. The PSU-LIF has been used in several campaigns in the past 15 years. More details on the PSU-LIF and campaign results can be found in the studies by Brune *et al.* (1998), Jaegle *et al.* (2000), Faloona *et al.* (2004), and Ren *et al.* (2008). An LIF system based on the PSU design was developed at the Max Planck Institute (MPI, Germany) and installed on a Learjet aircraft. The major difference between the PSU-LIF and the MPI system is the inlet. The first uses a nacelle, the latter an inlet similar to the one described in the study by Eisele *et al.* (1997). More details on the MPI LIF design and campaign results can be found in the studies by Lelieveld *et al.* (2008) and Martinez *et al.* (2010b).

Another LIF system for OH and HO_2 observations is installed on the British BAe-146 aircraft and the system was developed at the University of Leeds (UK). It applies an inlet similar to that described in the study by Eisele *et al.* (1997) and uses a single-beam configuration, whereas PSU and MPI LIFs use a multipass laser beam configuration. The first campaign involving the Leeds system was in 2004, details on the instrument and campaign results can be found in the study by Commane *et al.* (2010).

An LIF system for NO_2, total peroxy nitrates, total alkyl nitrates, and HNO_3 was developed at the University of California, Berkeley (USA), and uses a multipass configuration, at 585 nm, to excite NO_2 molecules. It has two cells that alternatively measure the four species mentioned earlier. The first campaign where this LIF was used is the Tropospheric Ozone Production about the Spring Equinox in 2000 on board the C-130. More details on this LIF system can be found in the studies by Thornton, Wooldridge, and Cohen (2000) and Bertram *et al.* (2007).

Recently, another LIF system for NO_2, total peroxy nitrates, total alkyl nitrates, and HNO_3, developed at the University of L'Aquila (Italy), has been installed on the BAe-146 aircraft. This LIF has a single-pass configuration and excites NO_2 molecules at 532 nm. This system uses four cells, one for each species above that are observed simultaneously, more details can be found in the studies by Dari-Salisburgo *et al.* (2009) and Di Carlo *et al.* (2011).

Finally, there is a system that uses the LIF technique as the detector, but it is a flow reactor (FR) to measure the OH reactivity (inverse of the OH lifetime), developed by the PSU and installed on the DC8. This system is becoming very popular and almost all the OH-LIF around the world have developed one of them, but at the moment, only the PSU system is on aircraft, specifically on the DC8 (Mao *et al.* 2009). At the Forschungszentrum Jülich (Germany), another OH-LIF system was developed and used on the Zeppelin in the ZEPTER-2 campaign in 2008 (Oebel *et al.* 2010). This will be installed on the DLR HALO aircraft together with the MPI OH-LIF, so in the near future, there will be an aircraft with two OH-LIFs on board. This will be a good opportunity for an in-flight intercomparison that is very rare for radicals. To date, the only intercomparison was made during the TRACE-P campaign between an LIF and a chemical ionization mass spectrometer instrument (Section 3.6).

During the past decades, the LIF technique has been used to make observations in several aircraft campaigns and is contributing to understand the radical and nitrogen chemistry from boundary layer up to the stratosphere. It is a very selective and sensitive technique, the main issues being the need of large pumps and the fact that lasers are usually expensive. Compact pumps and low-cost lasers, such as diode laser, could give a big impetus to the development of more LIF systems for aircraft. Future development could be the use of LIF systems for in-flight flux observations, because they are very fast (up to 10 Hz) and the development of other LIF systems for the observations of species not yet observed on aircraft with this technique such as formaldehyde and glyoxal.

3.5.6.3 Chemical Conversion Resonance Fluorescence Technique

Halogen compounds are important trace species throughout the troposphere and stratosphere. The technique described here is not well suitable for airborne measurements in the lower troposphere. Therefore, we focus on the upper troposphere and lower stratosphere (UT/LS). Their total stratospheric abundance currently corresponds to roughly 3300 pmol/mol of chlorine, 21 pmol/mol of bromine, and sub-pmol/mol amounts of iodine. The direct airborne measurement of the appropriate species at such low mixing ratios, especially at reduced ambient pressures as low as 50 hPa, and low stratospheric concentrations either requires extremely sensitive detection techniques or measurement over long path lengths. However, to improve the understanding of the partitioning and chemical conversion of the halogen species *in situ*, measurements probing well-defined, compact air volumes are preferable.

Until recently the only established *in situ* technique for the halogen oxides ClO and BrO has been the chemical conversion resonance fluorescence (CCRF), which was developed in the 1970s from similar techniques for the detection of oxygen atoms and OH radicals (Anderson, 1975, 1976). Chemical ionization mass spectrometers have now been developed to measure BrO in the troposphere, but CCRF remains the only established *in situ* technique capable of measuring ClO and BrO at stratospheric pressures. While initially the technique was employed to probe chlorine atoms and ClO in the middle and upper stratosphere on parachute sondes, it became the work horse for analyzing the chemical processes generating the ozone hole (Anderson, Toohey, and Brune, 1991). In the late 1990s, the technique was extended to measure not only the halogen oxides but also chlorine nitrate, $ClONO_2$ (Stimpfle *et al.* 1999) and the ClO dimer, and ClOOCl (Stimpfle *et al.* 2004; Von Hobe *et al.* 2005) by first thermally dissociating the appropriate compounds into ClO (TD-CCRF). Only three airborne TD-CCRF instruments are operated worldwide – the instrument of the Harvard University (Stimpfle *et al.* 2004), the HALogen OXide monitor (HALOX) instrument of Forschungszentrum Jülich (Von Hobe *et al.* 2005), and the instrument of the Colorado University Boulder (Thornton *et al.* 2005).

3.5.6.3.1 **Scientific Background** Halogen species resulting from the breakdown of both anthropogenic and natural organic halogen compounds (mainly CFCs, halons, and methyl bromide) are the primary cause of severe ozone loss observed in the polar stratosphere above Antarctica and less often in the Arctic during late winter and early spring. Once appreciable amounts of ClO are formed, ClO and BrO react in very efficient catalytic cycles, thereby destroying ozone. The two major cycles active in the polar winter stratosphere have been widely described and involve the ClO dimer cycle (Molina and Molina, 1987) and the ClO–BrO cycle (McElroy *et al.* 1986).

The main goal of the current measurements is the establishment of a good chemical model representation of the processes governing the ozone loss cycles to provide a basis for sound predictions of the future development of the ozone layer. Highest uncertainties being the absorption cross section of chlorine peroxide, ClOOCl, the branching of various photochemical reactions, and the amount of the total stratospheric bromine. This is under discussion due to the fact that very short-lived substances may contribute up to 8 pmol/mol to total bromine due to their fast transport to the stratosphere or even through transport of their product gases (Salawitch *et al.* 2010). *In situ* measurements can provide a good means to validate measurements of satellite-borne sensors that have to rely on complicated retrievals to calculate trace gas concentrations (Dorf *et al.* 2006).

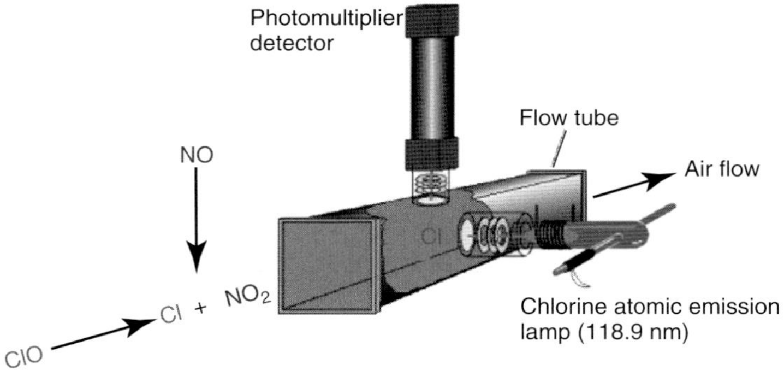

Figure 3.15 Schematic diagram of the CCRF detection technique for ClO.

3.5.6.3.2 **Measurement Principle** Initial descriptions of the implementation of the CCRF technique to measure ClO and BrO on board of aircraft have been given by Brune *et al.* (1990) and Brune, Anderson, and Chan (1989). Examples given correspond to the Jülich HALOX instrument and may be more or less different for the other instruments.

The CCRF technique is a multistage measurement as obvious from its name. Figure 3.15 presents a schematic diagram of the technique, while Figure 3.16 shows an actual instrumental setup.

A fast air flow is generated through a duct either by using the aircraft ram pressure or by sucking air into a nozzle by means of a suitable pump. The duct walls should be as inert as possible (PTFE or halocarbon wax coating). At a location where laminar flow is established, an excess of nitric oxide, NO, in a carrier stream of synthetic air is injected into the flow. The NO facilitates the fast chemical conversion of the oxides into the respective halogen atoms, thereby forming NO_2.

The atoms are detected downstream by means of atomic resonance fluorescence in the VUV at 118.9 nm for Cl and 131.7 nm for Br. Atoms in the flow tube absorb the radiation and fluoresce isotropically within a short time interval (about 10^{-8} s) or are quenched by collisions. Sensitive solar-blind VUV PMTs are arranged perpendicularly to the lamps to detect the resonance fluorescence radiation, S_{res} (also termed S_{Cl} or S_{Br}), on top of the background signal, S_{bgr}, consisting of Rayleigh scatter from air molecules (S_{Rayl}) and radiation reflected off the walls of the flow tube (chamber scatter, S_{ch}). Therefore, the total signal (S_{tot}) detected, including possible dark signal of the PMT (S_{dark}), is given by

$$S_{tot} = S_{res} + S_{bgr} = S_{res} + S_{ray} + S_{ch} + S_{dark} \tag{3.3}$$

To keep the fraction of chamber scatter low, the walls opposite the lamp and PMT are equipped with arrays of stacked razor blades, minimizing the reflection. The lamp beam as well as the FOV of the PMT are confined by apertures as indicated in Figure 3.16, thereby constraining the detection volume to the center region of the duct. Generally, the allowable total duct length enables stacking of at least two detection modules as shown in Figure 3.15.

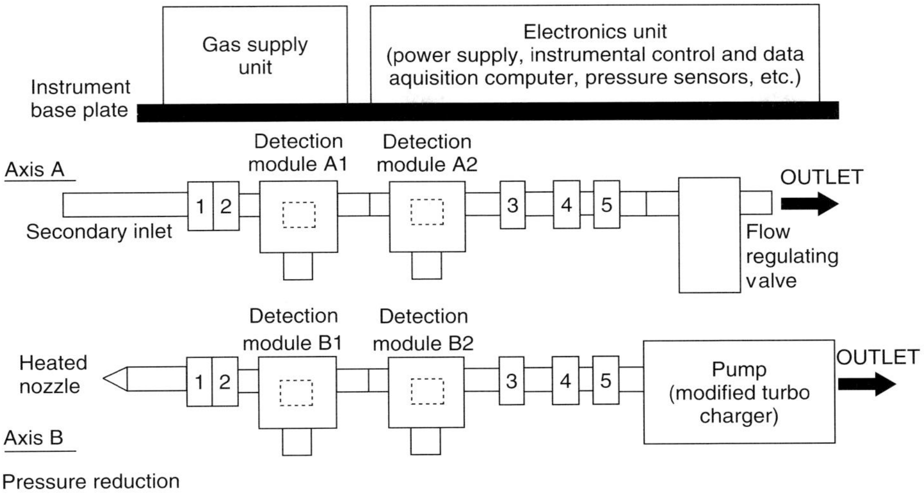

Figure 3.16 Schematic setup of the HALOX instrument. Basic components are the base plate with a gas supply unit and the electronics unit mounted on top of it. Below the plate, two parallel measurement ducts are supported. Each duct consists of an inlet, thermistor array for flow diagnostics, and NO injector followed by two stacked resonance fluorescence detection modules. Each of these comprises a flow tube element with a lamp housing and detector unit mounted to it. Downstream is a second thermistor array, a thermoanemometer and a pressure sensor followed by a butterfly valve for flow regulation for the ram-fed duct A. Duct B has a pump for flow generation in its back, while the inlet is equipped with a heatable nozzle. 1: Thermistor array A1 resp. B1; 2: NO injector A resp. B; 3: thermistor array A2 resp. B2; 4: thermoanemometer A resp. B; 5: pressure sensor A resp. B.

Owing to the higher reactivity of chlorine atoms as compared with bromine atoms, the first module usually detects chlorine, while the second module is set up for bromine detection. However, to better corroborate the conversion reaction kinetics (see further for details), double chlorine or bromine detection can also be employed. The parallel detection of additional species such as ClO dimer and $ClONO_2$ can be achieved by employing two parallel measurement ducts, one of them equipped with a heating device for the air flow.

By periodically switching the NO addition on and off, the resonance fluorescence signal S_{res} can be modulated, as shown in Figure 3.16. This facilitates separation from the other signal sources Eq. (3.3). NO on–off time intervals between about five and a few tens of seconds are usually employed.

The resonance fluorescence measurement provides halogen atom concentrations [Cl] (Cl is synonymous for Br here), while actually initial halogen oxide mixing concentrations $[ClO]_0$ are required. Therefore, the chemical conversion efficiency ($[ClO]_0/[Cl]$) has to be calculated employing an experimentally validated kinetic model.

Where available, temperature-dependent reaction rates are used from the study by Sander *et al.* (2006); for the bromine conversion system, additionally data are used from the study by Dolson and Klingshirn (1993). For a wide range of pressures and temperatures, maximum conversion efficiencies of generally better than 85% for ClO and better than 95% for BrO are calculated with a plateau for reaction times of about 2–5 ms. After the plateau, the conversion efficiency drops slowly as a result of the back reactions.

Stimpfle *et al.* (2004) have reported that under extremely cold conditions, the conversion of ClOO to ClO by NO addition at the room-temperature reaction rate is needed to explain the kinetics diagnostics within the Harvard instrument, reducing the conversion efficiency by up to 15%. No similar strong anomalies have been found for the HALOX instrument under cold conditions where this reaction is neglected in the calculation of the conversion efficiency.

Besides the halogen oxides, species such as ClOOCl, $ClONO_2$, and $BrONO_2$ may also be measured with the CCRF technique by thermally dissociating them and then measuring the oxides. So far, these measurements have only been implemented for ClOOCl (Stimpfle *et al.* 2004; Von Hobe *et al.* 2005) and $ClONO_2$ (Stimpfle *et al.* 1999) but not for $BrONO_2$.

The different bond dissociation energies allow for the selection of three distinct temperature regimes: at temperatures $<30\,^{\circ}C$, neither ClOOCl nor $ClONO_2$ dissociates, at $\sim 100\,^{\circ}C$, only ClOOCl is thermolyzed, and at temperatures above $250\,^{\circ}C$, both ClOOCl and $ClONO_2$ are quantitatively converted.

The two airborne instruments using this thermal dissociation step are slightly different with respect to the position and nature of the heating element and the derivation of ClOOCl and $ClONO_2$ concentrations. In HALOX, a wire heater coil is placed in the inlet nozzle of one measurement duct (Von Hobe *et al.* 2005) so that the thermal dissociation step takes place before NO addition and detection of any Cl atoms. In the Harvard instrument, heating grids are placed behind the first detection axis in both measurement ducts, so that both NO addition and the detection of Cl atoms released from background ClO take place before the thermal dissociation step. Von Hobe *et al.* (2005) describe laboratory tests of the conversion efficiency for ClOOCl.

Generally, the TD-CCRF technique relies on the following assumptions:

(i) ClOOCl as well as $ClONO_2$ almost quantitatively dissociate to ClO fragments when heated. For $ClONO_2$, this has been shown by lab calibrations (Stimpfle *et al.* 1999).
(ii) There are no significant other chlorine species producing ClO on heating: Possible candidates for ClO sources could be higher chlorine oxides such as Cl_2O_3. These, however, are not expected to play an important role in stratospheric chlorine chemistry. Abundant species such as HCl and OClO do not efficiently decompose at the employed temperatures.

For these reasons, no major cross sensitivities are expected for the TD-CCRF technique and it is deemed a reliable technique for the *in situ* detection of $ClONO_2$ and ClOOCl.

With the exception of $ClONO_2$, the species measured by the CCRF instruments are very reactive and wall contact of the probed air during the measurement should

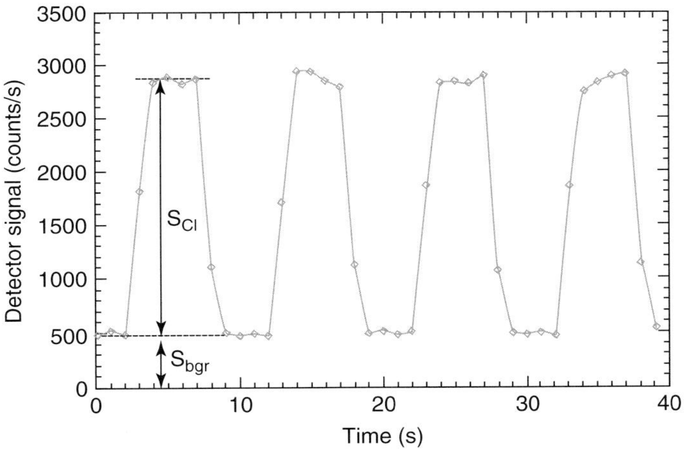

Figure 3.17 A 40-s PMT signal from a chlorine detection module as measured by the HALOX instrument onboard Geophysica on a flight in the Arctic polar vortex on 7 March 2005 during level flight on an altitude of 18.9 km. The signal corresponds to a ClO mixing ratio of 1.6 nmol/mol.

be minimized. Therefore, sensible duct flow speeds are in the range of 5–20 m/s. As common aircraft speeds are around 200 m/s, the air flow has to be slowed down by at least one order of magnitude without causing turbulence in the probed air mass. This is achieved with a double-stage inlet where a secondary inlet (start of the measurement duct) samples the core from a primary inlet that reduces flow speeds by a factor of about 4 (Soderman, Hazen, and Brune, 1991). In the Jülich HALOX instrument inlet, the primary inlet is of circular cross section with a diameter of about 40 cm, the measurement duct is quadratic with a dimension of $5 \times 5\,cm^2$.

Alternatively, air can be sampled by means of an inert nozzle into a duct that is pumped. This will induce turbulence but NO injection is placed downstream where a laminar flow has been reestablished (Von Hobe *et al.* 2005). This technique and the much more involved technique of a perpendicular sampling device (Thornton *et al.* 2003; Brune *et al.* 1998) allow to work at reduced pressure inside the flow tube.

The resonance fluorescence signal S_{res} (equivalent to S_{Cl} in Figure 3.17) can be translated into atom concentrations if the overall intensity of the excitation radiation, an optical quality factor for the measurement setup, and some intrinsic factors such as the quenching efficiency for excited state atoms and the Doppler–Doppler overlap of excitation and absorption line profiles are known. Descriptions of the practical and mathematical procedures involved are given in the studies by Brune *et al.* (1990) and Toohey *et al.* (1993) for chlorine.

The chlorine calibration is based on the absorption of the 118.9-nm atomic doublet as measured by Schwab and Anderson (1982). Detection modules are integrated into a flow system (two or even three modules can be inserted), as shown in Figure 3.18.

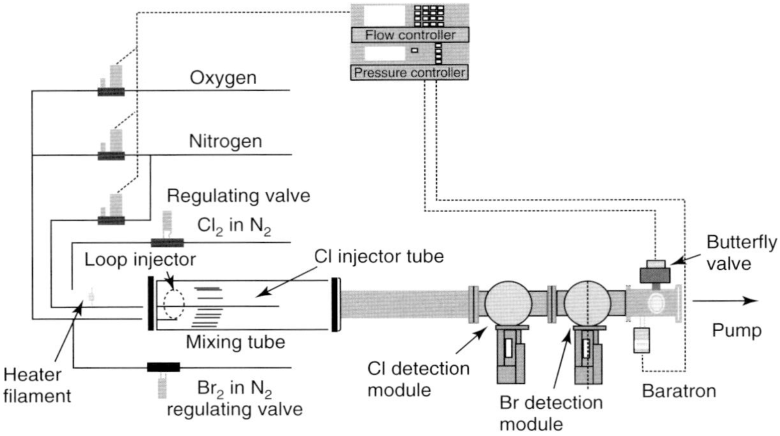

Figure 3.18 Schematic setup employed for the laboratory calibration of CCRF detection modules.

A fast flow of nitrogen or air can be generated through the system, while the pressure is regulated by a downstream butterfly valve. Chlorine atoms generated from the thermolysis of a nitrogen/Cl_2 mixture on a hot tungsten filament are homogeneously mixed into the main flow. Chlorine concentration is quantified by means of a vacuum UV monochromator monitoring the absorption of the 118.9 nm doublet emitted from the first chlorine lamp module across the flow tube. The calibration constant is defined as the chlorine resonance fluorescence signal divided by the chlorine atom concentration: $C_{Cl} = S_{Cl}/[Cl]$. The monochromator also enables the determination of the Lyman$-\alpha$ ratio for the Cl lamp and quality checks of the filtered emission spectrum.

Absorption path lengths of a few centimeters enable the detection of Cl atoms in the range of tens of nmol/mol.

The bromine calibration is based on the quantification of a given Cl atom concentration by a calibrated chlorine module (module #1 in Figure 3.18) and then quantitatively titrating the Cl atoms to Br via the fast reaction with excess Br_2. This way the resonance fluorescence signal of the bromine module (module #2) can be calibrated and then referenced to the appropriate scatter signal.

3.5.6.3.3 **Measurement Precision, Accuracy, and Limitations** For aircraft measurements of ClO, an accuracy of around 17% can be obtained where the calibration contributes about 14% and uncertainty of in-flight parameters and conversion efficiency around 10% (all on the 2σ confidence level). Precision can be as low as 5% or a few pmol/mol, whichever is higher. For the TD-CCRF measurements of ClOOCl, the accuracy is around 25%. Detection limits for ClO are about 5 pmol/mol (with signal integration of 1 min); with very high ClO (and ClOOCl), the detection limit for $ClONO_2$ may be as large as 100 pmol/mol, while for low ClO_x amounts, limits as low as 10 pmol/mol may be detected.

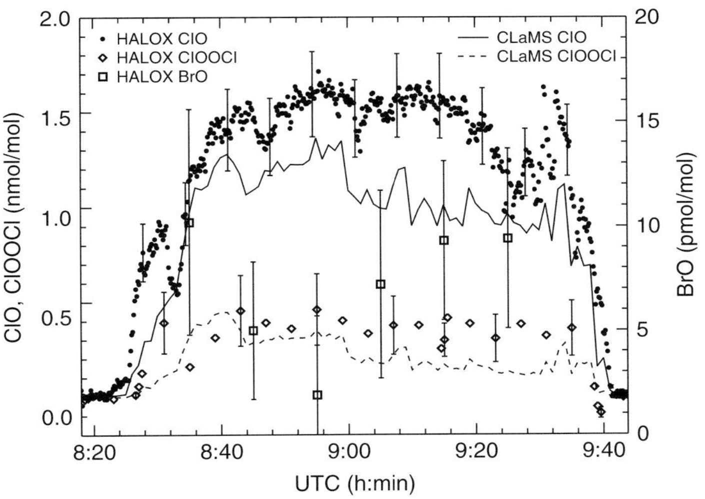

Figure 3.19 Time series of the HALOX measurements (symbols) of ClO, ClOOCl and BrO carried out on 7 March 2005. For ClO and ClOOCl, simulations by the CLaMS chemical transport model of Forschungszentrum Jülich (lines) are shown for comparison.

For BrO, owing to the indirect calibration procedure and more frequently observed drifts in the Br lamp calibration, the total accuracy is only around 20%, limiting the total measurement accuracy to 30%. Owing to the low abundance and the associated low Signal-to-noise Ratio (SNR), the precision is of the order of 15%. The detection limit is 2 pmol/mol with integration times of 10 min at pressures around 60 hPa.

3.5.6.3.4 **Measurement Example** Figure 3.19 exhibits a ClO and ClOOCl measurement of the HALOX instrument on the inbound leg of a flight of M55-Geophysica carried out on 7 March 2005 into the cold Arctic vortex extending as far south as Rome on that particular day (Von Hobe *et al.* 2006). ClO, BrO, and ClOOCl were measured inside a highly activated polar vortex under conditions where about 60% of the available ozone had been destroyed on the flight level.

The recorded data enabled a detailed study of the ClO_x chemistry and the available ozone depletion potential by the major catalytic halogen cycles. The CLaMS model (McKenna *et al.* 2002a,b) reproduces the relative time series quite well, while the absolute values are slightly underestimated.

The CCRF and TD-CCRF measurement techniques offer very sensitive tools for fast airborne measurement of ClO, BrO, ClOOCl, and $ClONO_2$. The measurements are important for studying processes highly relevant to stratospheric ozone as there are ClO_x photochemistry and quantification of stratospheric total bromine. Also, validation of satellite-borne sensors is an important application. However, the high instrumental and logistical efforts of the technique limit the number of available aircraft instruments to just three worldwide.

3.6
Chemical Ionization Mass Spectrometry

Chemical ionization mass spectrometry (CIMS) is a powerful tool for the detection of trace gases in the atmosphere owing to its high selectivity and sensitivity, fast response time, and low detection limit. It was first introduced in airborne research using rocket- and balloon-borne mass spectrometric measurements (Arnold, Krankowsky, and Marien, 1977; Arnold *et al.* 1980; Arnold and Henschen, 1978; Arijs, Nevejans, and Ingels, 1983; Schlager and Arnold, 1987a,b). In the past decades, the CIMS technique was significantly advanced and utilized for a large variety of trace gases. Recent reviews on CIMS for atmospheric trace gas measurements include the studies by Viggiano (1993), de Gouw *et al.* (2006), and Huey (2007). CIMS instrumentation has also been reviewed by Clemitshaw (2004) and Williams (2006).

An airborne CIMS instrument consists basically of an ion source (IS) and a flow reactor (FR) coupled to a differentially pumped mass spectrometer (quadrupole, ion trap, and time-of-flight (ToF) mass spectrometers are used). The trace gas of interest is selectively ionized in the sample air followed by the detection of the reagent and characteristic product ions in the mass spectrometer. In principle, the abundance of the trace gas can be inferred from the measured ratio of the reagent and product ion signal, the effective rate constant of the ion–molecule reaction, and the effective residence time in the FR. However, sensitivity changes due to variations in temperature and humidity and inlet issues require online calibrations for airborne CIMS instruments. In principle, CIMS utilizes negative (e.g., CO_3^-) and positive reagent ions (e.g., H_3O^+) depending on the specific application. Negative reagent ions, for example, can be used to detect a wide range of inorganic acids and a series of peroxyacyl nitrates (Huey, 2007) and H_3O^+ ions to measure trace gases with a higher proton affinity than H_2O, including a large number of volatile organic compounds (VOCs) (de Gouw *et al.* 2006). In the following, we discuss aircraft CIMS instruments using negative-ion chemistry and proton transfer reactions in separate sections.

3.6.1
Negative-Ion CIMS

A number of aircraft-borne CIMS instruments using negative-ion chemistry have been developed during the recent years (Crounse *et al.* 2006; Curtius and Arnold, 2001; Fortner, Zhao, and Zhang, 2004; Hanke *et al.* 2002a; Huey *et al.* 1998; Hunton *et al.* 2000; Leibrock and Huey, 2000; Mauldin, Tanner, and Eisele, 1998, 2001; Miller *et al.* 2000; Neumann *et al.* 2000; Roiger *et al.* 2011a; Slusher *et al.* 2004; Speidel *et al.* 2007; Zondlo *et al.* 2003). Reagent ions used include CO_3^-, NO_3^-, HSO_4^-, SiF_6^-, SF_5^-, I^-, and others allowing the detection of a range of trace gases, including important inorganic acids (HNO_3, HONO, HO_2NO_2, N_2O_5, $ClONO_2$, HCN, HCl, SO_2, and H_2SO_4) and peroxyacyl nitrates (peroxyacetylnitrate (PAN), Peroxypropionyl Nitrate (PPN), and Peroxymethacryloyl Nitrate (MPAN))

(Huey (2007) and references therein). CIMS coupled with chemical conversion and amplification in a dilution region added to the air intake is utilized for measurements of the hydroxyl radical (OH) and peroxy radicals (HO_2 and RO_2). OH measurements using CIMS include conversion of OH to $H_2{}^{34}SO_4$ by adding sufficient $^{34}SO_2$ and subsequent charge-transfer reaction of $H_2{}^{34}SO_4$ with NO_3^- of (Eisele *et al.* 1997). Peroxy radical detection by CIMS uses chemical amplification that converts HO_2 and RO_2 to H_2SO_4 by addition of NO and SO_2 (Hanke *et al.* 2002a; Hornbrook *et al.* 2011). In the following, we describe, as an example, the CIMS technique and instrumental setup of an aircraft chemical ionization ion-trap mass spectrometer (CI-ITMS) developed for the new German research aircraft HALO in a cooperation of DLR, Oberpfaffenhofen and MPI, Heidelberg (Roiger *et al.* 2011a). This instrument has recently been successfully deployed on the DLR Falcon-20 aircraft (Roiger *et al.* 2011b) and first HALO flights.

3.6.1.1 Measurement Principle and Aircraft Implementation

Figure 3.20 shows a schematic of the CI-ITMS as employed for PAN measurements. The instrument includes several major modules: a gas inlet system, a peroxyacetylnitrate calibration source (PCS), a tubular FR, an IS, and an ion-trap mass spectrometer (ITMS). Atmospheric air is sampled by a rotary vane pump through a backward-oriented air inlet (AI) mounted on top of the aircraft fuselage. The sample air passes through the sampling line (SL) via a back-pressure controller into a thermal decomposition region (TDR). Here, atmospheric PAN undergoes thermal decomposition leading to the formation of peroxy acetyl (PA) radicals $CH_3C(O)O_2$ and NO_2. Subsequently, the sample air enters the FR and reacts with I^- ions generated by an IS. After passage through the FR, the sample air is pumped out of the instrument via an air exhaust. A small fraction of the sample air containing the reagent and product ions enters the ITMS via a small circular entrance orifice and is analyzed.

The reagent I^- ions are produced by adding trace amounts of a CH_3I/N_2 gas mixture to an N_2 carrier gas, which is passed through a radioactive ^{210}Po inline ionizer. The I^- ions react rapidly with atmospheric water vapor molecules leading to hydrated $I^-(H_2O)_n$ ions. These serve as effective reagent ions for the detection of the PA radicals formed in the TDR leading to product ions $CH_3C(O)O(H_2O)_n$.

$$CH_3C(O)O_2 + I(H_2O)_n \rightarrow CH_3C(O)O(H_2O)_n + IO \quad (3.4)$$

The product ions are efficiently dehydrated in the ion trap through collisions with He as damping gas. Thereby mainly the unhydrated product ion is present in the mass spectra resulting in an enhanced SNR. The thermal decomposition products of other PAN homologues (e.g., PPN and MPAN) react in the same way with I^-. It should be noted that the sensitivity for the MPAN detection is much lower than that for PPN, and this makes the MPAN quantification with I-CIMS more difficult.

3.6.1.2 Calibration and Uncertainties

In-flight calibration of CIMS instruments can be performed in two ways: addition of a well-known amount of the trace gas of interest several times during the flight or

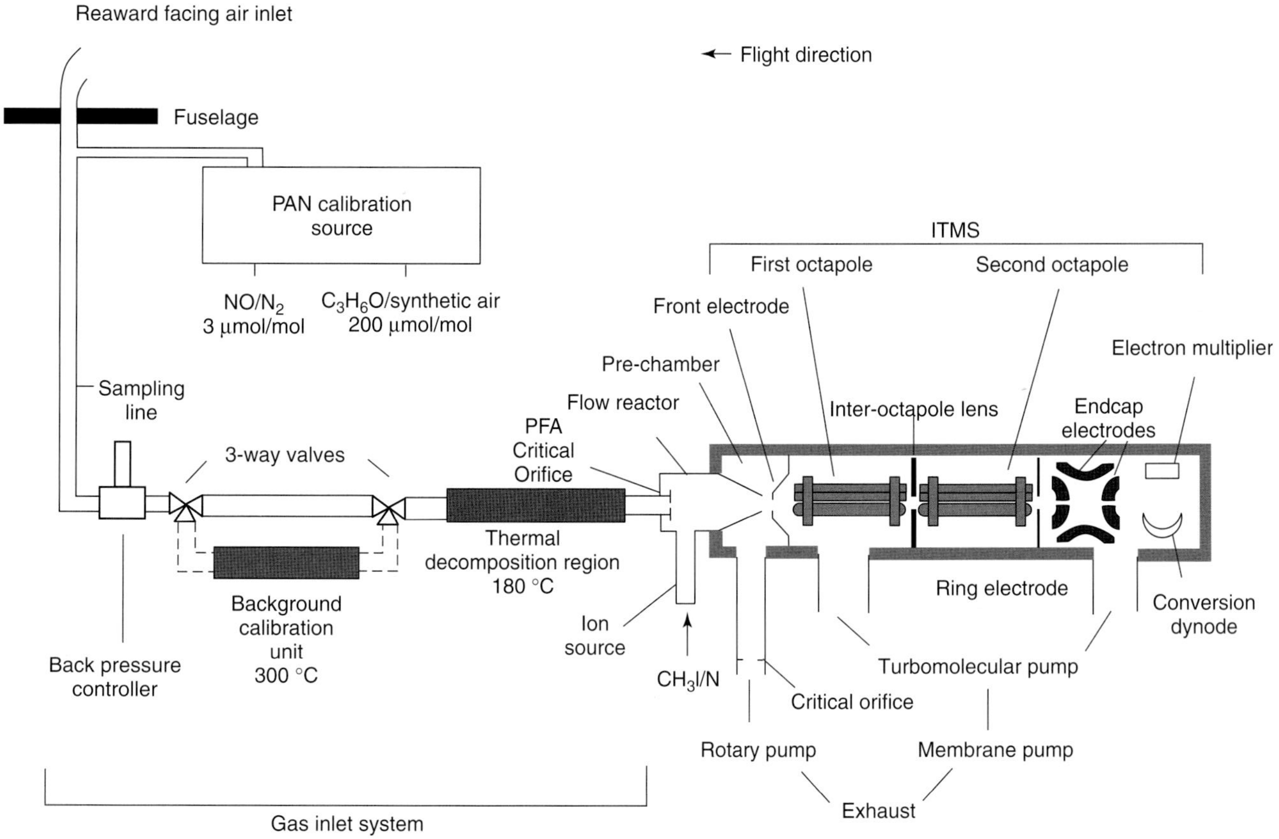

Figure 3.20 Experimental setup of the DLR/MPI CI-ITMS aircraft instrument for PAN measurements, including a gas inlet system, a PCS, a tubular FR, an IS, and an ITMS. (Roiger *et al.* 2011a)

continuous addition of an isotopically labeled standard. The latter is to be preferred, because it provides calibration throughout the flight (each mass spectrum) and it does not interfere with the product ion of the measured atmospheric trace gas. Isotopically labeled PAN can be produced on board using a photochemical method (Warneck and Zerbach, 1992). A mixture of isotopically labeled acetone ($^{13}C_3H_6O$) in synthetic air is photolyzed and the formed product CH_3CO reacts with abundant oxygen, thereby forming PA radicals ($^{13}CH_3{}^{13}C(O)O_2$). These react with added NO_2 with known efficiency to PAN; for details see the study by Roiger *et al.* (2011a).

Also, the in-flight calibration with an isotopically labeled standard significantly reduces the uncertainties arising from variable temperature and humidity conditions during the flight. Reagent ions may also react with other atmospheric trace gases than the target gas (cross sensitivity). For I^- chemistry, cross sensitivities with abundant trace gases such as HNO_3, O_3, NO_2, acetone, and acetic acid were not found (Huey, Hanson, and Howard, 1995; Slusher *et al.* 2004).

The accuracy of the CI-ITMS measurements described here is largely determined by the uncertainty of the isotopically labeled standard. For PAN mixing ratios >200 pmol/mol, the nominal accuracy is estimated to $\pm10\%$. The detection limit is defined as the concentration corresponding to the 2σ standard deviation at background conditions. At a high time resolution of 2 s, the 2σ detection limit is calculated as 25 pmol/mol.

3.6.1.3 Measurement Example

We present as a measurement example PAN observations with the CI-ITMS instrument described earlier together with other chemical data performed during a flight with the DLR Falcon-20 on 10 July 2008 during the Greenland Aerosol and Chemistry Experiment (GRACE) campaign, a subproject of the Polar Study using Aircraft, Remote Sensing, Surface Measurements, and Models, of Climate, Chemistry, Aerosol particles, and Transport (POLARCAT) initiative. Figure 3.21 presents a time series of several trace gases for a part of the flight between 15:30 and 17:00 UTC over northern Greenland. The Falcon encountered twice a pollution plume from East Asia at an altitude of 11.3 km, $\sim$1000 m above the dynamical tropopause. This anthropogenic pollution plume experienced strong up-lift in a warm conveyor belt located over the Russian east coast. Subsequently, the Asian air mass was transported across the North Pole into the sampling area where it was already mixed with stratospheric air (Roiger *et al.* 2011b).

3.6.2 The Proton Transfer Reaction Mass Spectrometer

Proton transfer reaction mass spectrometry (PTR-MS), introduced by Lindinger and coworkers at the University of Innsbruck (Lindinger, Hansel, and Jordan, 1998), is a form of CIMS (Munson and Field, 1966), exploiting aspects of flow/drift tube technology and chemical kinetics to enable fast, sensitive, and portable measurements of VOCs. Its speed, portability, and broad VOC detection capabilities are particularly well matched for use on fast moving aircraft in studies of the Earth's

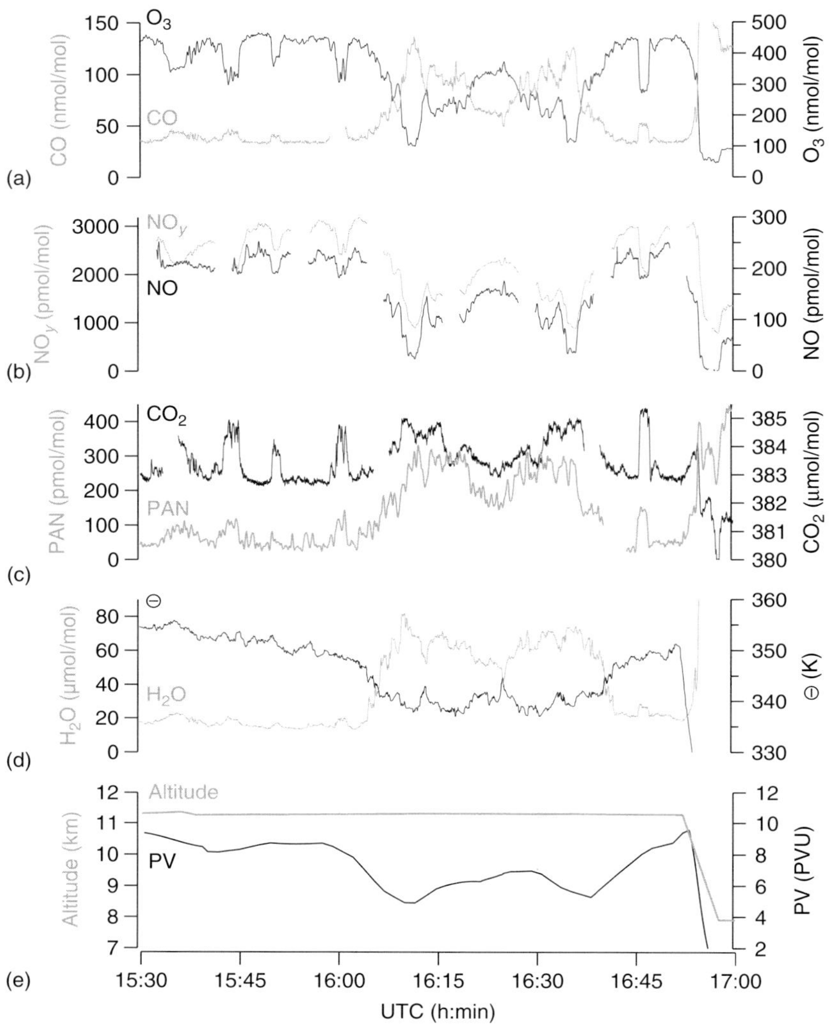

Figure 3.21 Time series of chemical measurements in the lowermost stratosphere during POLARCAT-GRACE on 10 July 2008, between 15:30 and 17:00 UTC. (a) O_3 (black) and CO (gray), (b) NO_y (gray) and NO (black), (c) PAN (gray) and CO_2 (black), (d) H_2O (gray) and potential temperature (black), and (e) flight altitude (gray) and potential vorticity (black), as interpolated from the European Centre for Medium Range Forecasting (ECMWF) analysis at 18:00 UTC. The flight segment between 16:05 and 16:41 includes measurements in an East Asian pollution plume transported over the North Pole to the lowermost stratosphere over Greenland. Adapted from Roiger *et al.* (2011b).

Table 3.4 Proton affinities, data from the study by Hunter and Lias (1998).

Compound	Formula	Proton affinity (kJ/mol)
Nitrogen	N_2	464.6
Oxygen, ozone	O_2, O_3	396.3, 595.9
Noble gases	Ar, Ne, He	346.3, 174.4, 148.5
Carbon mon-/dioxide	CO, CO_2	562.8, 515.8
Methane	CH_4	544
Water	H_2O	660.0
Methanol	CH_3OH	725.5
Acetonitrile	CH_3CN	779.2
Acetone	CH_3COCH_3	782.1
Dimethylsulfide	CH_3SCH_3	801.2
Isoprene	$CH_2C(CH_3)CHCH_2$	797.6

atmosphere. In the following, the key components of the PTR-MS and its principle of operation are briefly summarized. In addition, several examples are given of studies undertaken with airborne PTR-MS systems.

PTR-MS instruments exploit proton transfer reactions between H_3O^+ ions and a species of interest, R, to continuously monitor its mixing ratio in the sampled air:

$$H_3O^+ + R \rightarrow H_2O + HR^+ , \quad k_1 \sim 2 \times 10^{-9}\,cm^3\,s^{-1} \tag{3.5}$$

This reaction proceeds at or near the collision rate for compounds having a proton affinity larger than that of H_2O. Use of this reaction is convenient for atmospheric scientists as it is essentially "blind" to the major air constituents N_2, O_2, Ar, and CO_2 (Table 3.4) but sensitive to many important atmospheric species such as acetone, methanol, benzene, dimethyl sulfide, and isoprene. In comparison to instruments employing more common electron impact ionization, ionizing reactions with H_3O^+ occur at relatively low energies so that product fragmentation is minimal, considerably simplifying interpretation of mass spectra in the absence of any preseparation step.

Although from one instrument to the next there can be significant variety in design, all PTR-MS comprise an IS, a reaction drift tube, and a mass analysis/detection system. The PTR-MS in Figure 3.22 might be considered typical today and will be used for the following description. Water vapor is passed through a hollow cathode IS, which produces high number densities of $H_3O^+(H_2O)_n$ ions. In a small pumped region directly following the IS, much of the water is pumped away, while $H_3O^+(H_2O)_n$ ions are declustered into mainly H_3O^+ as they are accelerated toward the reaction drift tube. Ions and residual water are then sampled into the reaction drift tube where they are combined with sample air maintained at a constant pressure and temperature; see Table 3.5 for typical conditions. Gases traverse the reaction drift tube in less than a second, while ions drift through more than 1000 times faster. During their time in the reaction region, ions undergo

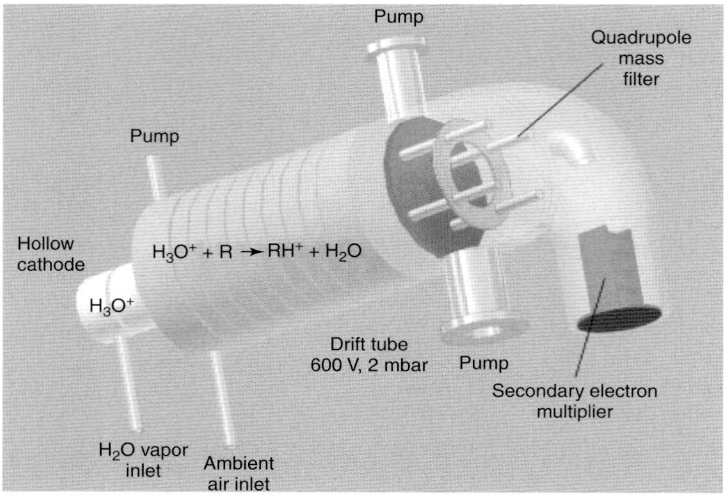

Figure 3.22 A schematic diagram of the PTR-MS. (Source: Adapted from Heard (2006). Reproduced with permission of John Wiley and Sons. Copyright 2006 by Blackwell Publishing Ltd.)

Table 3.5 Typical reaction conditions for PTR-MS measurement.

Parameter	Approximate value
IS water flow	8 STP cm^3/min
Reaction drift tube length	9.3 cm
Reaction drift tube diameter	0.98 cm
Sample air flow rate	15 STP cm^3/min
Gas residence time	0.46 s
Drift field	$600\ V/(9.3\ cm) = 64.5\ V/cm$
Temperature	50 °C
Pressure	2.2 hPa
E/N	130 Td ($1\ Td = 10^{-17}\ cm^2/V/s$)
Ion mobility (H_3O^+ in N_2)	2.8 $cm^2/V/s$ (1976)
Ion/molecule reaction time	88 μs

numerous collisions, some of which result in a reaction. The ionizing reaction is effectively quenched, long before the reaction proceeds to completion (all primary ions reacted away), by sampling the mixture of primary and products ions into the vacuum of a mass spectrometer. Here, remaining neutrals are pumped away and ions are separated from one another using a quadrupole mass filter and sensitively detected using a secondary electron multiplier.

Quantification is generally based on periodic measurements of certified chemical standards containing individual species or mixtures of species diluted to span mixing ratios relevant to the measurement at hand. The measured relationship between ion signals and introduced mixing ratios is then used to convert between the two. Background for such measurements typically comes from measurement of VOC-free air generated either by passage of sample air through a catalytic converter (typically a heated, platinum powder-coated substrate) or from purchased high-purity air mixtures (a possibility for low-humidity sampling situations). The interpolated background signal before and after a measurement is then subtracted from the ambient signal to derive a corrected mass signal. This accounts for signals generated by instrumentation and surface production or uptake occurring downstream of the zero air addition. It does not account for chemicals in sample air, which produce indistinguishable or overlapping ion products or fragments. Detection limits depend heavily on reliable background measurement and vary by chemical and from one instrument to the other.

Optimization during one airborne measurement (Williams *et al.* 2001) resulted in detection limits for most species between 10 and 100 pmol/mol. Detection limits are also affected by the total parent ion number density that can be cleanly introduced into the reaction drift tube.

PTR-MS is operated under well-controlled conditions under which the reaction proceeds (Table 3.5). This control makes it possible to use chemical kinetics to estimate mixing ratios in the absence of a calibration gas standard. The following equation can be used to provide a remarkably accurate first approximation:

$$[\mathrm{R}] = \frac{1}{k \cdot t} \cdot \frac{[\mathrm{HR}^+]}{[\mathrm{H_3O}^+]} \tag{3.6}$$

Here, [R] is the number density of species R in the reaction drift tube (molec/cm^3), k is the reaction rate coefficient for reaction between R and H_3O^+ (cm^3/s), t is the reaction time (s), $[HR^+]$ and $[H_3O^+]$ are product and parent ion count rates observed in the mass spectrometer, which are proportional to the number densities of these species in the reaction drift tube. All exothermic proton transfer reactions proceed at or close to the collision rate. Where specific reaction rates have not been measured, they can be calculated (Su and Chesnavich, 1982; Zhao and Zhang, 2004). The reaction time t is determined by the instrument reaction drift tube geometry, drift field, pressure, and temperature or may be experimentally determined (Lindinger, Hansel, and Jordan, 1998). This equation is often modified in an attempt to account for variable ion transmission through a given instrument (Steinbacher *et al.* 2004), humidity (de Gouw *et al.* 2003; Warneke *et al.* 2001), differential mobilities and diffusion (Keck, Oeh, and Hoeschen, 2007), gas composition (Keck, Hoeschen, and Oeh, 2008), energetically accessible back reactions (Hansel *et al.* 1997; Knighton *et al.* 2009), or to generally automate quantification (Taipale *et al.* 2008).

The precision of the measurement (reflecting instrument stability or measurement reproducibility) is dominated by a statistical error associated with stable production and sampling of ions as well as the counting statistics related to measuring these ions with an electron multiplier (Hayward *et al.* 2002).

$$NS = \frac{\mathrm{meansignal(cps)}}{\sqrt{\mathrm{meansignal(cps)} \cdot \mathrm{dwell(s)}}} \tag{3.7}$$

In Eq. (3.7), *NS* is the noise statistic that is related to the standard deviation of the signal. This can be used to calculate the change in signal necessary to indicate an "event" or significant change in sampled air mixing ratio. The "mean signal" is the mean measured signal and the dwell is the time for which the ion in question was monitored. In one field measurement on board an aircraft, the estimated precision for protonated acetone at m/z 59 was 20%, the more abundant protonated isoprene ion at m/z 69 had an estimated precision of 10%, and the less abundant protonated acetonitrile at m/z 42 had an precision of 30% (Williams *et al.* 2001). In general, more abundant species produce more product ions, decreasing the scatter of the signal. Faster measurement (shorter dwell times per mass) produces lower signals with more scatter or less precision.

The accuracy of PTR-MS measurements depends very much on the standard gas used for calibration, whether and how ion/molecule reaction kinetics are used to determine mixing ratios, and appropriate background determination. Assuming an appropriate calibration is performed using a well-defined standard gas mixture (mixture contains no two chemicals that produce indistinguishable ions, and calibration is performed at the same humidity, gas composition, temperature, drift field, and pressure as measurements), accuracy can approach quoted values of the gas standard itself. Overall, assuming that background is correctly determined, total uncertainties (the sum of the precision and accuracy) will reflect mostly the precision and be on the order of 20–30%.

The first time a PTR-MS was used on board an aircraft was in February 1998 during the Cooperative LBA Airborne Regional Experiment (LBA-CLAIRE) field campaign. The instrument was installed on a Citation jet alongside several other atmospheric measurement systems and the aircraft was flown low over the tropical rainforest (Crutzen *et al.* 2000). The mass information provided by the PTR-MS was interpreted within the context of the rainforest location to deduce the likely predominant contributor to each mass and to screen for potentially important new compounds, in particular, those that may have been missed by other techniques (Williams *et al.* 2001). As expected, over the rainforest, isoprene (m/z 69) and its oxidation products methyl vinyl ketone and methacrolein (both m/z 71) were observed in relatively high abundance, with temporal profiles consistent with a primary emission and photochemical product (Warneke *et al.* 2001). It should be noted that m/z 71 corresponds to the sum of methyl vinyl ketone and methacrolein. Individually, these compounds cannot be distinguished. In a second follow-up campaign to the region in 2005, the emission rates of isoprene, acetone, and methanol were estimated from a combination of in-flight and ground-based measurements (Eerdekens *et al.* 2009). Recently, the speed of the PTR-MS measurement has been further exploited so that fluxes can now be directly determined by PTR-MS from aircraft both over forests and cities (Karl *et al.* 2007, 2009). This is a useful way of verifying emission inventories.

PTR-MS technology has been adapted for operation on a variety of aircraft platforms: a Lear jet (enviscope GmbH, Germany), a DC8 (NASA, USA), a P3 (NOAA, USA), a Citation jet (University of Delft, The Netherlands), and a C-130 (NSF, USA), as well as other research aircraft; PTR-MS technology has also been built for routine operation onto a specially adapted commercial passenger jet within the Civil Aircraft for the Regular Investigation of the atmosphere Based on an Instrument Container (CARIBIC) project (Brenninkmeijer *et al.* 2007). Airborne PTR-MS systems have been used to investigate anthropogenic pollution emerging from South East Asia and North America (Warneke *et al.* 2007; Wisthaler *et al.* 2002), in both cases serving to improve regional emission databases. The ability of the PTR-MS to measure acetonitrile (CH_3CN), a gas almost exclusively emitted from biomass burning, has proven very useful to atmospheric scientists. In conjunction with back-trajectories analysis, the high-frequency PTR-MS measurements of acetonitrile can define biomass burning plumes sufficiently well that sources and plume ages can be defined (de Gouw *et al.* 2006). Acetonitrile measurements have also been used to identify biomass burning plumes in flight and to examine *in situ* photochemistry of other species such as acetone (Holzinger *et al.* 2005). In some circumstances, ground-level pollution can be lifted to the upper troposphere through convection and such cases have also been investigated through PTR-MS measurements from jet aircraft (Colomb *et al.* 2006).

Airborne PTR-MS measurement data sets are growing rapidly in number, and by combining data from around the world, new insights can be derived for global budgets of VOC such as methanol (Millet *et al.* 2008) and acetone (de Reus *et al.* 2003).

3.6.3
Summary and Future Perspectives

Over the past 10 years, the application of CIMS and PTR-MS on airborne platforms has advanced our understanding of the distribution and chemistry of atmospheric trace gases considerably. Much more can be expected as these experimental techniques further mature, methodologies improve, and the number and diversity of flights employing such instruments increase. Improvements in instrumentation generally focus on one or more of the following: decreasing size and power, increasing chemical specificity, improving sensitivity, enabling measurement of new or challenging atmospheric species, and utilization of CIMS technique to measure aerosol chemical composition.

Minimizing size, weight, and power consumption of instrumentation without sacrificing performance is of paramount importance for aircraft experiments. Increased efforts are, for example, being made to develop dedicated miniature CIMS and PTR-MS systems that could be used with small or even unmanned aircraft. Efforts at miniaturization have so far manifested in the use of alternative, smaller mass filters, redesign of existing MS vacuum chambers using lighter materials and lighter, more efficient pumps, and reducing the size of existing commercial electronics.

In recent years, a variety of alternative mass filters have been introduced and tested to replace the relatively slow and low-resolution quadrupole mass filters with which most CIMS and PTR-MS instruments are currently equipped. Although not yet smaller or lighter in weight than a quadrupole, ToF instrumentation (Jordan *et al.* 2009b; Tanimoto *et al.* 2007; Wyche *et al.* 2007) has a much higher mass resolution (having $\frac{m}{\Delta m}$ of around 4000 or better; *http://www.ptrms.com/*), broader mass range (up to 50 000 amu), and higher data acquisition rate (many scans per second). Increased resolution aids in identification of those ions that are not structural isomers of one another but that lie very close to one another in mass, improving the specificity of the technique. Ion traps (Mielke *et al.* 2008; Prazeller *et al.* 2003; Steeghs *et al.* 2007; Warneke *et al.* 2005a,b; Roiger *et al.* 2011a) are generally comparable to or smaller in size than quadrupole mass spectrometers and aim at increasing specificity through use of collision-induced dissociation experiments or chemical-specific reactions within a trapping volume. Triple quadrupole instruments would provide essentially the same capabilities, and first versions of such instruments have been used.

Improvements in IS and drift tube design have also been investigated and are perhaps more important than the mass analyzer in increasing sensitivity of the measurement by increasing the number density of ions available for reaction, increasing the purity of these ions, changing the selection of ions that can be used for chemical ionization, or changing the reaction conditions (ion/molecule reaction time, center of mass collision energy, and influence of water clusters). IS designs that have been reported recently include hollow cathodes capable of switching between a small selection of ions (Jordan *et al.* 2009a), direct current discharge (Inomata *et al.* 2006), radioactive ISs (Leibrock and Huey, 2000), Molecular Weight (MW) plasma discharges (Spanel, Dryahina, and Smith, 2007), and other forms of discharge (Sellegri *et al.* 2005) to select just a few. High-pressure drift tubes (Hanson *et al.* 2009; Sellegri *et al.* 2005) have been tested to increase the sensitivity. Drift tubes have also been made from various materials that could simplify construction (Thornberry *et al.* 2009).

Another direction for future expansion is the use of new or novel ion chemistry. Viable candidate ions must be easy to generate in high purity and in high number densities, preferably without preseparation and undergo chemically specific reactions with trace molecules of interest but not with the main components of air. A variety of other positive and negative ions have been explored, a few of which have been adopted on a wider scale or actively used in field measurements (Marcy *et al.* 2005; Reiner, Hanke, and Arnold, 1997; Slusher *et al.* 2001; Veres, 2008). Recently, there have been a number of reports of use of O_2^+ and NO^+ ions, in part due to their ease of production in existing instrumentation as well as due to an abundance of available information concerning their reactions with VOCs published in support of closely related selected ion flow tube air monitoring instrumentation (Fortner and Knighton, 2008; Jordan *et al.* 2009a; Wyche *et al.* 2005). Also, inlet modifications such as online water control through cryotrapping (Jobson and McCoskey, 2009) may aid in observation of currently hard-to-detect species such as formaldehyde, hydrogen cyanide, and hydroperoxides. Finally, first test experiments have been performed using CIMS technique to analyze the chemical composition of aerosol particles (Curtius and Arnold, 2001; Curtius *et al.* 2001).

3.7
Chemical Conversion Techniques

3.7.1
Peroxy Radical Chemical Amplification

The radical chemistry reaches special significance in the free troposphere. In the presence of enough NO from lightning and aircraft emissions, peroxy radicals can be responsible for rapid and effective production of O_3 in the upper tropospheric layers and thus have a global impact (Jaegle *et al.* 2001; Wennberg *et al.* 1999). In addition, the upper tropospheric budget of radical oxidative species, which mainly define the processing of long-lived pollutants and global climatic effects, changes dramatically as a consequence of injection of biogenic and anthropogenic precursors from the boundary layer. Variations in VOC/NO_x ratios lead to a set of radical conversion and propagating reactions, which define the oxidative capacity of air masses and their potential to produce O_3. Analytical techniques for atmospheric measurements of radicals have recently been extensively reviewed and the theoretical background of the methods used for chemical conversion and amplification of peroxy radicals has been described in detail (Heard, 2006). This chapter therefore basically focuses on the most recent developments for the application of the peroxy radical chemical amplification (PeRCA) technique to airborne measurements.

Hydroperoxy (HO_2) and alkyl peroxy (RO_2; $R =$ organic chain) radicals are short-lived species that play an important role in the chemical processing of the troposphere (Lightfoot *et al.* 1992; Monks, 2005). They generally originate from photolytic and oxidation processes of atmospheric trace species such as carbon monoxide (CO) and VOCs. Their relevance in tropospheric reactions has been studied in some detail (Monks, 2005; Atkinson, 2000; Jenkin and Clemitshaw, 2000), especially in those initiated by the hydroxyl radical (OH) in the presence of nitrogen oxides (NO_x), which affect the formation mechanisms of ozone (O_3) and other important pollutants such as PAN, aldehydes, and acids. Since the chemical amplification was pioneered for the ground-based measurement of peroxy radicals by Cantrell, Stedman, and Wendel (1984), this low-cost and portable technique has been successively characterized and improved for its deployment in different platforms (Cantrell *et al.* 1993; Volz–Thomas *et al.* 1998; Andrés-Hernández *et al.* 2001; Sjostedt *et al.* 2007).

3.7.1.1 Measurement Principles

Briefly, PeRCA utilizes the amplified conversion of oxy, alkoxy, hydroxyl, and alkyl peroxy radicals into the more stable NO_2, via a chain reaction involving NO and CO. As OH and RO abundances in the troposphere are much lower than the rest, PeRCA measures to a good approximation the total sum of peroxy radicals defined as RO_2^*, that is, $RO_2^* = HO_2 + \Sigma\ RO_2$. The NO_2 produced by the radical conversion is discriminated from other NO_2 background sources by modulating the signal between the amplification and the background measurement modes. This is achieved by adding CO at the front (amplification mode) and at the back (background mode) of a reactor, which turns on and off the chain chemistry. The total radical concentration is calculated from the difference between the amplification and background signals, ΔNO_2, and the knowledge of the amplification factor or chain length (CL) in the reactor.

Owing to the high reactivity of the peroxy radicals, a PeRCA reactor must be so designed that the chain reaction initiates promptly after sampling. The extent of the chain reaction in the reactor, indicated by the CL, is the result of the competition between amplification, chemical loss reactions, and wall loses. On top of this, the CL depends on the relative humidity (Mihele and Hastie, 1998; Mihele, Mozurkewich, and Hastie, 1999; Salisbury *et al.* 2002; Reichert *et al.* 2003).

The accurate determination of radical losses in SLs before the amplification zone is extremely difficult, but the overall radical conversion for a particular setup, that is, the effective chain length (eCL), can be determined using adequate calibration procedures reproducing operating conditions. The eCL obtained experimentally does not correspond exactly with the actual NO_2 yield in the reactor as the radicals not reaching the amplification zone in the reactor cannot be discriminated from the total. Under some circumstances, this radical loss determines the value of the eCL of a particular setup, while the concentration of the reactants added for the chemical conversion, the residence time in the reactor, and/or the reactor material and geometry only play a secondary role.

In the original PeRCA, NO_2 is measured by detecting the chemiluminescence of its reaction with luminol (3-aminophthalhydrazide: $C_8H_7N_3O_2$). PeRCA has recently introduced other NO_2 detection techniques such as LIF (Sadanaga *et al.* 2004; Kanaya *et al.* 2001), cavity ring-down spectroscopy (Liu *et al.* 2009), and cavity-enhanced spectroscopy in an attempt to overcome shortcomings of luminol detectors such as instabilities derived from luminol flow, temperature, and pressure. In the past decade, Reiner *et al.* (1999) and Reiner, Hanke, and Arnold (1997) developed a similar technique for the measurement of peroxy radicals based on chemical amplification and conversion in gases other than NO_2, followed by detection of H_2SO_4. This technique called the RO_x chemical conversion mass spectrometry (ROxMAS) or peroxy radical chemical conversion ion molecule reaction mass spectrometry (PerCIMS) has been further developed for radical speciation (Hanke *et al.* 2002b; Edwards *et al.* 2003) and implemented in field studies.

3.7.1.2 **Airborne Measurements**

Although progressively increasing, the number of airborne peroxy radical measurements still remains very limited. Concerning chemical amplification measurement techniques, the rapid response and low detection limit of PerCIMS and ROxMAS makes these instruments well suited for airborne measurements despite their relatively high volume and power requirements (Hanke *et al.* 2002a; Cantrell *et al.* 2003a,b; Mauldin *et al.* 2004; Ren *et al.* 2005).

In contrast, PeRCA offers less complex instruments as those recently developed for the airborne *in situ* measurement of peroxy radicals and deployed in field measurements in different platforms (Green *et al.* 2003b; Kartal *et al.* 2010;

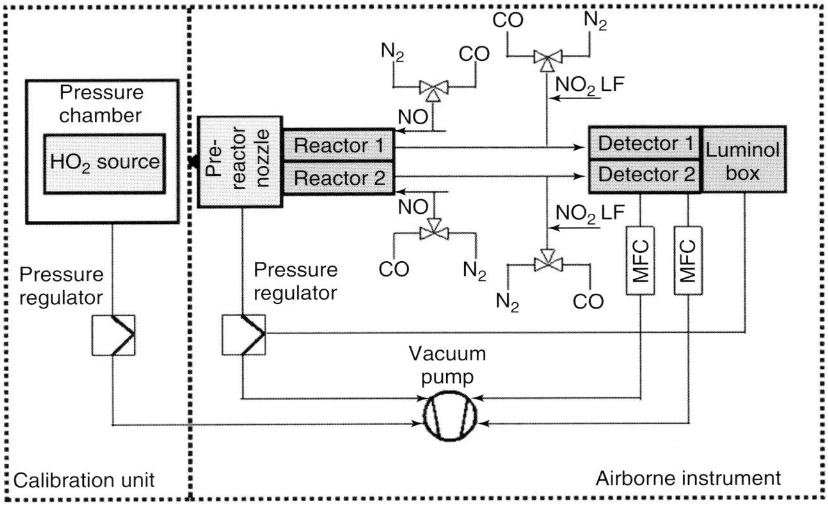

Figure 3.23 Schematic setup of a dual airborne instrument based on PeRCA. The calibration unit is used in the laboratory to determine the eCL at the constant instrumental pressure set during the flights. MFC, mass-flow controller; LF, linearization flow.

Andrés-Hernández *et al.* 2009; Brookes, 2009). The last generation of those devices consists of dual systems comprising two identical reactors coupled with two detectors, as first suggested by Cantrell *et al.* (1996) for environments with rapid changing background concentrations (Kartal *et al.* 2010; Green *et al.* 2006; Andrés-Hernández *et al.* 2010). A schematic is given in Figure 3.23. Similar to their approach, these instruments mainly differ in the way to handle with pressure variations during sampling. The use of two reactors that concurrently measure in the same or alternate modes (signal/background) enables to monitor short-term background variations and related interferences, and thereby to increase sensitivity and accuracy during the airborne sampling cycle.

3.7.1.3 Calibration and Uncertainties

The straightforward application of PeRCA to airborne radical measurements needs to address thoroughly some experimental drawbacks that are related to the following issues:

3.7.1.3.1 Variability of ambient conditions: pressure, temperature, and relative humidity

The stability of the chain chemistry at the reactor and/or the detection is very sensitive to variations in ambient conditions such as pressure, relative humidity, and temperature, during the operation.

The CL of PeRCA reactors has been observed to decrease when decreasing the pressure in laboratory experiments. Provided that the retention time in the reactor is sufficient for the chain reaction to be completed, the CL is expected to decrease at lower reactor pressures because the molecular collisions become less probable and the wall losses gain importance. The effect of the pressure seems to be less pronounced in Teflon than in glass reactors, which have shown a CL decrease up to 60% between 1000 and 200 hPa (Green *et al.* 2003b; Kartal *et al.* 2010).

There are two options to deal with variations in the pressure during airborne sampling: (i) to apply a correction of the CL with the pressure based on a laboratory characterization comprising potential ambient pressures or (ii) to keep the reactor pressure constant during operation. Both have been used in the field. To minimize disturbances in the reactor chemistry due to variations in the concentration ratios, the mass flows of sampling and addition gases (NO and CO) are kept constant during the flight. Therefore, option (i) is associated with changes in the retention time of the air sampled in the reactor, which possibly affect the CL. This potential perturbation must additionally be characterized under controlled conditions. On top of this, the decomposition of thermal labile peroxynitrates such as peroxy nitric acid can interfere depending on the retention time in the reactor.

The RH dependence must be characterized for each individual setup and operating conditions as it varies with the CO/NO ratio and the wall losses in the reactor (Mihele, Mozurkewich, and Hastie, 1999; Reichert *et al.* 2003). This effect becomes especially important for airborne measurements. Although the water content in upper tropospheric air is generally low, it is associated with high RH. Therefore, special care must be taken when applying the RH correction factor to the CL in the case of ΔNO_2 close to the detection limit, that is, low $[RO_2^*]$. Otherwise, the background noise of the luminol detector can be unrealistically amplified (Andrés-Hernández *et al.* 2010). Reducing the RH in the reactor is thus advantageous for minimizing uncertainties.

The impact of humidity on the CL can be additionally reduced if the reactors are kept at a constant pressure lower than the ambient owing to the associated decrease in $P_{\text{partial}}^{H_2O}$ (Kartal *et al.* 2010).

Concerning the detection, laboratory experiments with homemade luminol detectors have shown a degradation of the NO_2 detector sensitivity up to 45% when reducing the pressure from 500 to 200 hPa. These results indicate the effect of drying out of the luminol on the filter above a certain threshold in the volume flow through the detector (Kartal *et al.* 2010).

Furthermore, the sensitivity of luminol detectors decreases when increasing the temperature. The effect of temperature is related to the consequent variations in the vapor pressure of the alcohol, which is added to the luminol solution to enhance sensitivity and specificity for NO_2 (Wendel *et al.* 1983). As the temperature in the aircraft cabin can change dramatically during flight preparation and measurement phases, it is advisable to thermally stabilize luminol detectors.

In summary, keeping the PeRCA instrument at constant temperature and pressure contributes to improve the stability of its performance at changing atmospheric conditions. The latter has been recently achieved by introducing a prereactor nozzle in front of the reactors, which is continuously held at a settable constant pressure (Kartal *et al.* 2010).

3.7.1.3.2 **In-Flight Stability of the Detector Sensitivity** Apart from the above-mentioned variations in luminol sensitivity with the pressure and temperature, the overall performance of the luminol detectors is critically affected by instabilities in the luminol flow. It has been reported that the luminol feed is regularly disrupted by degassing when commercial detectors (Luminox LMA-3, Scintrex) are used at low pressures. A more stable performance up to 6000-m altitude can be achieved by detector upgrading (Green *et al.* 2003b). The luminol solution can also be supplied to LMA-3 detectors by pressure displacement of luminol with regulation of the flow by needle valves. This method increases the stability of the luminol flow but still requires some improvement for its application to airborne measurements (Green *et al.* 2006; Brookes, 2009). Despite pressure stabilization in upgraded homemade detectors, short-term instabilities in the luminol flow can occur during the flight and must be carefully taken into account in the data processing. These in-flight instabilities lead to short-term and not simultaneous variations in the sensitivities of the detectors, which might introduce uncertainties in the determination of ΔNO_2, that is, $NO_{2_{amplification}} - NO_{2_{background}}$, as it involves the signal of both the single detectors. Thus, regular NO_2 calibrations are highly desirable. However, there is always a fine balance between the number of in-flight calibrations and actual measurement time in a rapid changing environment. NO_2 calibrations with external cylinders before and after the flights are insufficient to monitor potential in-flight variations in the detectors sensitivity. Hence, additional methods are required to keep track of the relative performance of the detectors. In that context, simultaneous airborne O_3 measurements have been used to monitor the stability of the detector sensitivity.

Another important aspect is the linearity of the luminol detector, which is affected by the NO added at the reactor to the sampling air (Sadanaga *et al.* 2004). Although ambient O_3 background concentrations generally assure the linear response of the detector, an NO_2 offset should be added when sampling air masses with a $[NO_2 + O_3] < 35$ nmol/mol to prevent nonlinearities (Clemitshaw *et al.* 1997).

3.7.1.3.3 **Inlet Size Limitations** As the inlet is outside the fuselage and hence exposed to the free air stream, its shape, orientation, and position are constrained by external factors related to accessibility and safety regulations, which are particularly restrictive for the potential case of bird strike. In that context, loss of radicals before conversion is one of the most critical issues in the design of a suitable inlet. The high reactivity of peroxy radicals requires the immediate conversion into NO_2 after the entrance of the ambient air in the instrument. Consequently, the inlet has to act as a chemical reactor and must be able to sample the air outside the boundary layer of the aircraft fuselage. Pressure nozzles to regulate the reactor pressure can cause significant radical losses, which might then significantly reduce the eCL (Kartal *et al.* 2010). Furthermore, the valves used for the gas addition should be nearby the reactors as the length of tubing defines the time for signal stabilization after switching between amplification and background modes. Although reactors with small surface-to-volume ratio generally lead to higher CL owing to the consequent reduction in wall losses, low volume reactors that enable shorter modulation periods are more suitable for airborne measurements. Different inner materials have been used and Teflon or Teflon-coated surfaces represent an adequate compromise between matching sensitivity and safety requirements.

3.7.1.3.4 **Instrumental Time Resolution and Detection Limit** While the chain chemistry is usually modulated every 60 s, detector data are generally saved as 30-Hz averages. If the reactors are operating in alternate modes, total and background NO_2 are separately and simultaneously measured and the effect of atmospheric fluctuations in O_3 between terminating modes is minimized. Signal instabilities associated with switching modes reduce the number of useful data, which are usually provided as 1-min RO_2^* averages. The in-flight detection limit and accuracy of an airborne PeRCA instrument depend critically on the stability of the luminol flow, the variability of the air mass composition sampled, and the operating pressure. Improved sensitivity can be achieved by increasing the CO added to the reactor, which has a direct effect on the CL. This is, however, limited by safety regulations that restrain the CO permitted on board. The installation of secondary containments, which can instantaneously be emptied in the outer air if required, has notably improved the applicability of PeRCA instruments to long-range airborne measurements. RO_2^* detection limits between 2 and 5 pmol/mol for 20- to 60-s time resolution have been reported during airborne field measurements (Green *et al.* 2003b; Kartal *et al.* 2010; Andrés-Hernández *et al.* 2010). The total uncertainty is directly related to the in-flight detector performance and generally ranges between 25% and 45%.

3.7.1.3.5 **Calibration Procedures** The overall sensitivity of the DUal channel Airborne peroxy radicaL chemical amplifiER (DUALER) PeRCA instrument relies both on the performance of the NO_2 detector and on the efficiency of the conversion of radicals into NO_2 at the reactor. The instrumental calibration procedure must therefore comprise regular NO_2 and CL calibrations. These basically do not differ from calibrations used for ground-based measurements. The detector sensitivity is generally calibrated by directing air mixtures of known NO_2 concentrations from permeation tubes or calibrated gas cylinders through the inlet or before the detector.

The CL calibration of the reactor is based on the production of HO_2 or CH_3O_2 radicals from the UV photolysis of water at 184.9 nm (Schultz *et al.* 1995; Creasey, Heard, and Lee, 2000; Mihele and Hastie, 2000) and of CH_3I at 253.7 nm (Ashbourn, Jenkin, and Clemitshaw, 1998), respectively. In-flight CL calibrations are associated with high experimental complexity caused by the inherent difficulty to adequately introduce known radical concentrations into the inlet so that they experience identical conditions to those prevailing during the in-flight sampling. Losses related to steady variations of pressure and temperature can be taken into account by laboratory calibrations under controlled conditions. The actual in-flight sampling is, though, additionally affected by radical losses before entrance into the inlet, which might vary with flight speed, angle of flight path, or during the sampling through clouds, aerosol particles, or rain. Given the high reactivity of peroxy radicals, these additional losses might be significant and difficult to quantify. Their estimation can be nevertheless attempted by a thorough analysis of radical and ancillary measurement data.

Furthermore, as the calculation of the total sum of peroxy radicals generally bases on the CL calibration of HO_2 or CH_3O_2 radicals, the accuracy of the RO_2^* determination depends on the radical partitioning of the air sample. This can be very much variable along a flight track. The potential interference of active chlorine by the amplification of ClO radicals with NO and CO via $ClCO_3$ formation (Perner *et al.* 1999) should be evaluated as well for those environments with significant load of halogens. The sensitivity of the PeRCA to different mixtures of radicals has been demonstrated to be essentially defined by differences in heterogeneous and homogeneous processes limiting the yield of HO_2 reaching the amplification zone in the reactor (Ashbourn, Jenkin, and Clemitshaw, 1998; Kartal *et al.* 2010). This information is nevertheless insufficient to derive the radical partitioning of the air mass sampled. Therefore, the development of new reactors and measurement procedures for directly discriminating HO_2 from the rest of the sampled organic peroxy radicals by exploiting differences in wall losses is a very actual subject of research (Miyazaki *et al.* 2010). With this in mind, in-flight and laboratory intercomparison exercises are essential to assess airborne instrumental reliability and to identify potential artifacts or interferences in the quantification of peroxy radicals.

3.7.2 Chemiluminescence Techniques

Reactive nitrogen species play a key role in atmospheric photochemistry, particularly in controlling the cycling of OH and the production of ozone in the UT/LS and therefore have a strong impact on the radiative forcing of the atmosphere. The budget of nitrogen oxides in the UT/LS is controlled by a variety of different sources and processes, including transport from the polluted boundary layer, lightning, and air traffic emissions.

Reactive nitrogen species are usually released into the atmosphere as NO. By reaction with O_3 and photolysis, a photo stationary state between NO and NO_2 is established within a few minutes under daylight conditions. Further oxidation processes lead to the formation of other oxidized reactive nitrogen species. The family of all reactive nitrogen species NO_y includes NO, NO_2, NO_3, N_2O_5, HONO, HNO_3, HO_2NO_2, $ClONO_2$, PAN, and other organic nitrates.

3.7.2.1 Measurement Principle

The method predominantly used for the measurement of atmospheric NO is based on its chemiluminescence reaction with ozone (Clyne, Thrush, and Wayne, 1964; Clough and Thrush 1967):

$$NO + O_3 \rightarrow NO_2 + O_2 \tag{3.8}$$

$$NO + O_3 \rightarrow NO_2^* + O_2 \tag{3.9}$$

$$NO_2^* \rightarrow NO_2 + h \cdot \nu \tag{3.10}$$

$$NO_2^* + M \rightarrow NO_2 + M \tag{3.11}$$

The chemiluminescence signal ranges roughly between 0.5 and 3 μm. The maximum of the chemiluminescence signal peaks at about 1200 nm. A number of studies showed that this reaction could be used for atmospheric measurements (Fontijn, Sabadell, and Ronco, 1970; Stedman *et al.* 1972). For this purpose, the sample air containing NO is mixed with ozone in excess in a reaction chamber and the produced light is detected by a photomultiplier. The ozone is usually produced from a silent discharge using oxygen from a cylinder. Further studies have shown that the detector signal depends on the sample flow, the ozone flux, the pressure in the reaction chamber, and the choice of the photomultiplier (Steffenson and Stedman, 1974).

The signal recorded by the photomultiplier is not only caused by the chemiluminescence signal from NO but also from the dark current of the photomultiplier itself and the light arising from the chemiluminescence reaction of other species. To subtract these signals, most chemiluminescence detectors use a prereaction or zeroing volume. The sample flow is regularly directed to a prereaction chamber where the sample air is mixed with ozone. The dimensions of the zeroing volume and the flows are chosen in a way that the chemiluminescence signal from the NO reaction has already decayed before the air reaches the reaction chamber with the photomultiplier (Drummond, Volz, and Ehhalt, 1985). This of course is based on the assumption that the chemiluminescence signal from other reactions does not change significantly during the time of the sample flow through the zeroing volume.

3.7.2.2 **Measurement of Ozone Using Chemiluminescence**

Atmospheric ozone is usually measured by instruments using its UV absorption (Schlager *et al.* 1997). In these instruments, the sample air is passed through an absorption chamber that is irradiated by a mercury lamp.

However, ozone can be measured in the same way as NO using the O_3–NO chemiluminescence (Stedman *et al.* 1972). While for measuring NO, ozone is produced in excess by a silent discharge, for the ozone measurement, NO is added from a compressed gas cylinder. An ozone chemiluminescence detector was developed for the National Center for Atmospheric Research (NCAR) Sabreliner jet aircraft (Ridley, Grahek, and Walega, 1992). This instrument was part of a multichannel chemiluminescence instrument for the concurrent measurement of NO, NO_2, and NO_y.

Ridley, Grahek, and Walega (1992) have shown that using chemiluminescence small ozone detectors with low detection limit, high sensitivity, and fast response are feasible. While the benefits of faster data rates are obvious, there are some disadvantages with such chemiluminescence detectors for ozone on board of aircraft. NO is a highly toxic gas and thus requires special safety precautions.

3.7.2.3 NO_y and NO_2 Conversion

The conversion of total reactive nitrogen to NO offers the opportunity to measure NO_y with a chemiluminescence detector in combination with a reduction converter. Usually, gold converters are used for the conversion of NO_y species to NO. This technique was introduced by Bollinger *et al.* (1983) and Fahey *et al.* (1985). Laboratory studies have shown that gold converter heated to 300 °C converts the most important NO_y species with an efficiency of more than 90%. As a reducing agent, pure carbon monoxide is added. As an alternative reducing agent, H_2 (from a metal hydride storage device) is used to satisfy the requirements of the certification for civil in-service aircraft (Volz–Thomas *et al.* 2005). Usually, at least two-channel instruments are used to allow simultaneous measurements of NO and NO_y. Although catalytic conversion is widely used for the measurement of total reactive nitrogen in the atmosphere, this technique requires careful operation and regular checks in the laboratory (Kliner *et al.* 1997).

Furthermore, NO_2 can be measured by using a chemiluminescence detector in combination with a photolytic converter. Radiation from a UV source is used to convert NO_2 to NO that is subsequently measured as described earlier. For measurements on the NASA ER-2, for example, a metal halogen lamp was used (Gao *et al.* 1994). A high-pressure Hg arc lamp for NO_2 conversion was described by Ryerson, Williams, and Fehsenfeld (2000). The interference from the photo dissociation or thermal decomposition of other reactive nitrogen species has to be considered. More recently, LEDs have been used for the NO_2 photolytic conversion (Fuchs *et al.* 2010; Pollack, Lerner, and Ryerson, 2010).

NO and NO_2 are relatively easy to sample because their interactions with the inlet walls are small. On the other hand, trace gases such as nitric acid (or ammonia) are very "sticky" and are readily adsorbed to surfaces (Ryerson *et al.* 1999), especially if the surfaces are wet. As nitric acid is a main component of atmospheric reactive nitrogen, NO_y measurements can be perturbed by surface reaction in the inlet and inlet tubing. A further complication is that nitric acid is not adsorbed irreversibly but can be desorbed from the inlet surface at changing flight conditions and concentrations in the sample air. The adsorption of nitric acid strongly depends on the material of the surface and the temperature. Laboratory studies by Neuman *et al.* (1999) have shown that PFA Teflon tubing heated to temperatures above 10 °C shows the best behavior with respect to sampling of atmospheric nitric acid. To keep interferences from the inlet as small as possible, high sample flow rates and short inlets are a prerequisite (Figure 3.24).

3.7.2.4 Calibration and Uncertainties

Chemiluminescence detectors in combination with converters for the detection of NO_y and NO_2 offer low detection limits and fast response measurements. However, to assure a high accuracy and precision of the measurements, regular calibration of the instruments and quality checks are essential.

The detected signal of the chemiluminescence depends on the ozone concentration and therefore on the performance of the ozone generator and the quality of the photomultiplier. Although the sensitivity of a detector does not change rapidly

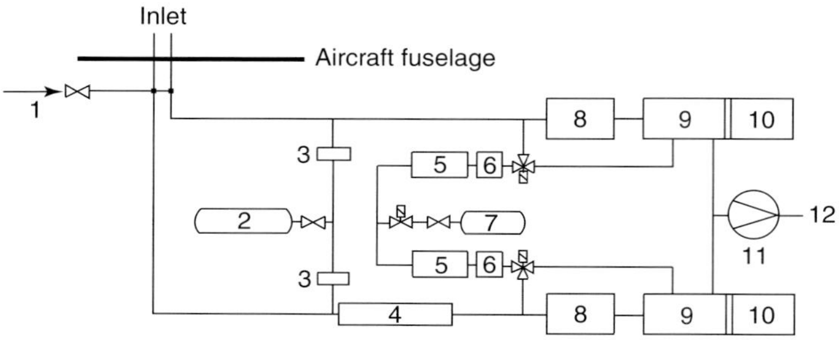

Figure 3.24 Schematic diagram of a two-channel NO detector with NO_y/NO_2 converter as used on the DLR research aircraft Falcon. 1: Calibration gas; 2: CO or metal hydride storage for H_2; 3: mass-flow controller; 4: NO_y or photolytic converter; 5: ozonizer; 6: humidifier; 7: O_2; 8: prereaction chamber; 9: reaction chamber; 10: photomultiplier; 11: vacuum pump; 12: exhaust.

with time, it has to be checked regularly by adding a small flow of a calibration gas, usually NO in pure N_2 blended with NO-free synthetic air. As mentioned earlier, the influence of the dark current of the photomultiplier and chemiluminescence resulting from O_3 reactions with other species is checked regularly by operating the instrument in the zero mode. Adding NO-free synthetic air to the measuring systems shows that even in this case the difference between measured and zero-mode signals is not equal to zero. This instrument background signal or zero air artifact is probably caused by the chemiluminescence of wall reactions or by impurities in the O_2 gas used for the production of ozone (Drummond, Volz, and Ehhalt, 1985) and has to be subtracted from the measured NO signal.

The measurement of NO_2 and NO_y using photolytic or gold converters, respectively, requires additional calibrations. The conversion efficiency of the converters for NO_2 can be checked by adding known amounts of NO_2 produced by gas-phase titration of NO with O_3. Alternatively, NO_2 can also be produced by using commercial permeation tubes. The conversion efficiency for HNO_3 can also be checked by using a permeation tube. However, it was found that the source strength of HNO_3 permeation cells may change with altitude and a pressure control is recommended (Talbot *et al.* 1997).

The conversion efficiency of photolytic converter depends on the pressure inside the converter. To avoid changing pressure with changing altitude in the atmosphere, the pressure can be kept constant by means of a Teflon valve. However, in this case, one has to take into account the potential loss at the valve. Catalytic converters usually do not show any pressure dependence as long as its surfaces are not contaminated (Fahey *et al.* 1985).

Of special interest is the conversion of species that are not considered to belong to the reactive nitrogen family. HCN is considered to be a major candidate for affecting the NO_y measurement. Different studies found conversion efficiency between a few and 100% depending on the temperature, reducing agent (CO or H_2), and ozone (Ziereis *et al.* 2000; Volz–Thomas *et al.* 2005).

Calibration also has to characterize the potential influence of the inlet on the measurement. Therefore, calibration gas should be added to the sample line as close to the inlet tip as possible.

Time response, sensitivity, detection limit, precision, and accuracy of the measurements depend on a variety of parameters and may change with time. A regular intercomparison in flight between aircraft instruments is highly desirable and provides confidence in the measurement quality (Brough *et al.* 2003). The typical time resolution of NO and NO_y measurements is about 1 s. To improve the detection limit and accuracy, data can be averaged over several seconds. The detection limit for NO and NO_y for the 10-s data can be as low as a few ppt.

3.7.2.5 **Measurement Examples**

The chemiluminescence technique for the detection of NO and NO_y in the atmosphere is widely used onboard research aircraft such as NASA ER-2 (Fahey *et al.* 1989), NASA DC-8 (Kondo *et al.* 1996), the Russian research aircraft M55-Geophysica (Huntrieser *et al.* 2009), the British FAAM (Stewart *et al.* 2008), the DLR Falcon-20 (Ziereis *et al.* 1999), a Learjet A35 (Pätz *et al.* 2006), and others. The chemiluminescence technique is also used for the measurement of NO and NO_y onboard civil in-service aircraft such as the MOZAIC (Volz–Thomas *et al.* 2005) and the CARIBIC program (Brenninkmeijer *et al.* 2007). In the following, two examples for measurements in the atmosphere are briefly discussed to illustrate which scientific questions can be addressed with these measurements.

Lightning is an important source for atmospheric nitrogen oxides. Nevertheless, the amount of reactive nitrogen globally produced by thunderstorms is still very uncertain (Schumann and Huntrieser, 2007). In recent years, a number of aircraft campaigns have been performed to reduce the uncertainty of the global NO_x production rate by lightning (Huntrieser *et al.* 2007). In spring 2005, the DLR Falcon-20 was used to probe the outflow of tropical and subtropical thunderstorms during the Tropical Convection, Cirrus, and Nitrogen Oxides Experiment (TROCCINOX) over Brazil.

Figure 3.25 presents the results of the NO and NO_y measurements during a particular flight. On 18 February, a flight with the DLR research aircraft Falcon was performed in the anvil of an active isolated thunderstorm over Brazil. The thunderstorm was penetrated several times at different altitudes. NO and NO_y volume mixing ratios were clearly enhanced with respect to the background, while the Falcon was passing through the outflow of the anvil. Maximum NO_y values up to 3 nmol/mol have been observed. High NO/NO_y ratios indicate that these emissions have been injected into the atmosphere only recently. For this particular case, it was found that 80% of the total anvil–NO_x was produced by lightning.

The uptake of atmospheric nitric acid on cirrus cloud particles has the potential to lead to a substantial removal of nitric acid from the gas phase and therefore may have an impact on the chemistry of the upper troposphere.

The chemiluminescence technique in combination with two NO_y converters can be used to study the uptake of reactive nitrogen species on ice crystals in cirrus clouds (Feigl *et al.* 1999).

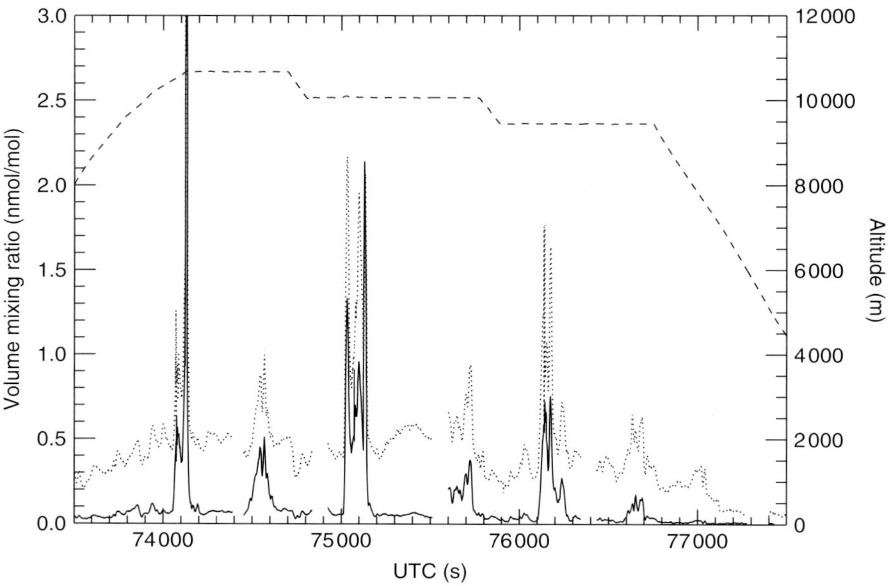

Figure 3.25 Time series of NO (solid line), NO_y (dotted line), and pressure altitude (dashed line) of a Falcon flight on 18 February 2005 during the TROCCINOX project. The anvil of a tropical thundercloud was penetrated six times.

In April 2000, NO_y measurements with the DLR Falcon-20 were performed during the interhemispheric differences in cirrus properties from anthropogenic emissions (INCA) experiment out of Punta Arenas (Chile) (Ziereis *et al.* 2004). Gas-phase total reactive nitrogen and particulate nitrate were measured by employing forward and aft-facing inlets with separate detection channels. The forward-facing inlet is subisokinetic and samples particles with an enhanced efficiency relative to the gas phase. Condensed-phase NO_y expressed as a gas-phase equivalent is inferred from the difference between forward- and aft-facing inlets' NO_y signals corrected by the effective enhancement factor. In Figure 3.26, particle NO_y as gas-phase equivalent is shown along with ice particle surface area measurements. The surface area was measured by means of a polar nephelometer. It was found that the uptake of nitric acid on ice particles depends on temperature, the nitric acid partial pressure, and the available surface area of the cirrus cloud particles. For the conditions encountered during INCA, $<1\%$ of the total reactive nitrogen was found as particulate nitrate.

3.7.2.6 **Summary**

Fast time response and a low detection limit make the chemiluminiscence technique most suitable for aircraft instruments. The total reactive nitrogen (NO_y) and NO_2 can be observed in combination with the gold or photolytic converter. Chemiluminescence instruments are widely used on different research and civil in-service aircraft.

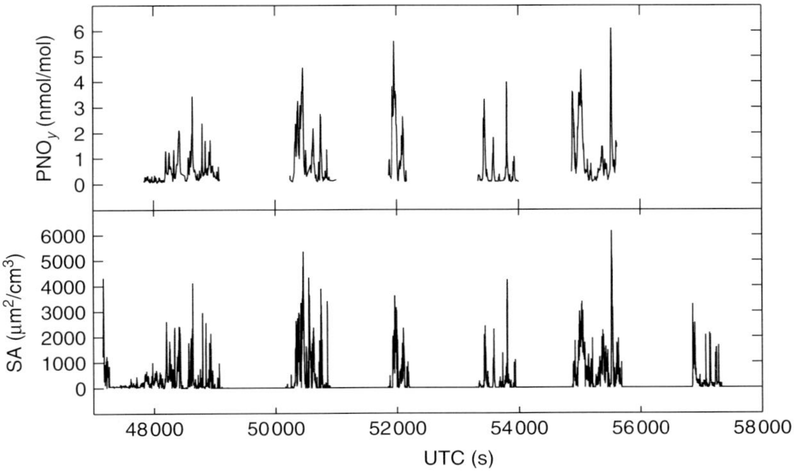

Figure 3.26 Particle NO_y as gas-phase equivalent (in pmol/mol) and surface area density (in $\mu m^2/cm^3$) for the flight on 12 April 2000 during the INCA field mission out of Punta Arenas (Chile). (Source: The figure was adapted from Ziereis *et al.* (2004). Copyright 2004 American Geophysical Union (AGU). Reproduced/modified by permission of AGU.)

Accuracy and precision of the measurement depend on a variety of different parameters. Interferences arising from the conversion or thermal composition of other species than total NO_y or NO_2, respectively, and inlet effects require regular pre- and in-flight calibrations of the instruments, including the inlets.

3.7.3 Liquid Conversion Techniques

The general basics of liquid conversion technique rely on the affinity with water of the target species present in the gas phase and as a consequence on their ability to partition between gas and water phase. As a consequence, atmospheric trace gases can be measured with this technique: (i) hydroperoxides (including hydrogen peroxide (H_2O_2) and organic hydroperoxides (ROOH)) and (ii) aldehydes: formaldehyde and other aldehydes. The Henry law coefficients (K_H) are 10^3 to 10^5/Matm.

Hydroperoxides are products of atmospheric photochemistry, which serve as reservoirs of odd-hydrogen radicals and as indicator of prior chemistry. Aldehydes and formaldehyde in particular are important intermediates of atmospheric photochemistry, being the by-products of most of the primary VOC. It is also a source of HO_x radicals.

This section focuses on the aircraft techniques that have produced most of the data in the literature for these two chemical groups of compounds.

3.7.3.1 Measurement Principles

All the liquid conversion techniques employ sampling devices to collect gaseous species continuously in an aqueous (or organic) solution before their chemical conversion (derivatization and catalyzed enzymatic reduction) and online or offline analysis (fluorimetry and UV–VIS detection). Among collection techniques, the scrubbing glass coil of Lazrus, Fong, and Lind (1988) is the most widely used. While air and collection solution flow together through the glass coil, soluble trace gases are partitioned and incorporated into the collection solution. Depending on the species, different chemical conversion reactants have been used and are summarized here. More details on measurement principles, conditions, and chemical reactions can be found in the studies by Lee, Heikes, and O'Sullivan (2000), Reeves and Penkett (2003), and Heard (2006).

3.7.3.1.1 **Hydrogen Peroxide (H_2O_2) and Hydroperoxides (ROOH)** In the aqueous phase, the peroxide-catalyzed reduction of H_2O_2 and other peroxides by *para*-hydroxy phenyl acetic acid (POPHA) is operated. A similar reaction occurs for organic peroxides, and in both the reactions, a stable dimer is formed, which fluoresces with a peak absorption at 320 nm and a peak emission at 400 nm. To discriminate between H_2O_2 and organic hydroperoxides, a second channel is treated with the enzyme catalase to destroy H_2O_2 before the POPHA reaction, so as to provide a measurement of total ROOH. A significant enhancement of this technique making it much more specific is the use of high-performance liquid chromatography (HPLC) to separate the hydroperoxides before being quantified (Lazrus, Fong, and Lind, 1988; Lee *et al.* 1995). More recently, Nunnermacker *et al.* (2008) have reported the use of three independent channels for the differential detection of H_2O_2, methylhydroperoxide, and hydroxymethylhydroperoxide during the Megacity Initiative: Local And Global Research Observations (MILAGRO) campaign in Mexico.

3.7.3.1.2 **Formaldehyde and Other Aldehydes** In the continuous mode, dissolved formaldehyde reacts with a dione and ammonium acetate to ultimately form a fluorescent dihydropyridine in a Hantzsch reaction (Kormann *et al.* 2003). In contrast, in an indirect mode, dissolved formaldehyde (and potentially other aldehydes) is derivatized with 2,4-dinitrophenylhydrazine and collected in a batch sample. Samples are analyzed after flight using HPLC (Lee *et al.* 1996). Six additional aldehydes could be analyzed by this technique by Lee *et al.* (1996).

3.7.3.2 **Aircraft Implementation**

The dual-channel enzymatic technique for hydroperoxides and the Hantschz reaction technique for formaldehyde have been widely used on aircraft platforms since two decades. Commercially, instruments based on these techniques are available (*http://www.aero-laser.de/html/home.html*; AL2021 for H_2O_2 and AL4021 for CH_2O). Lee, Heikes, and O'Sullivan (2000) reviewed in detail the airborne measurements of hydroperoxides up to year 2000.

To operate instruments on aircraft, one has to account for the pressure decrease in ambient air with increasing altitude, the inert character of the inlet line and interferences. To compensate the pressure decreasing with altitude, this can be done by means of a constant pressure inlet (Kormann *et al.* 2003). The constant pressure inlet guarantees a pressure-independent measurement between the boundary layer and the maximum ceiling of the aircraft.

The transmission of hydroperoxides and formaldehyde in aircraft inlets is a second critical point both considering adsorption during transfer or contamination by the inlet. Very few studies have reported on the SL effects and their quantitative evaluation under various environmental conditions. For hydroperoxides, Lee *et al.* (1995) did not see any alterations in the performances of the dual-enzymatic technique, while testing the thermo-regulated (35 °C) inlet with gas-phase standard additions during the flight. Other groups report that the inlet was designed to minimize contact of sampled air with dry surfaces before scrubbing (Nunnermacker *et al.* 2008). For CH_2O, the most comprehensive study with testing different sampling inlets under various conditions (temperature, relative humidity, concentration range, and pressure) has been reported for the TLAS technique by Wert *et al.* (2002). Observed sampling artifacts were limited to cases where liquid water was collected in the inlet and when heavily polluted air was sampled. Inlet shielding and heating and background substraction have prevented recurrence of this problem. Given the different techniques currently used to measure CH_2O, and the resultant variety of inlet designs and operational conditions, the applicability of this study to systems other than similar TLAS instruments must be carefully considered.

The evaluation of interferences is the third critical issue. Lee *et al.* (1995) and Kormann *et al.* (2003) have, respectively, reported on the different interferences for peroxides and CH_2O measurements (SO_2, higher aldehydes, and ketones), which can be either suppressed by adjusting pH stripping solution or neglected. Nevertheless, lower stratospheric measurements of CH_2O during the Mediterranean Intensive Oxidation capacity Study (MINOS) campaign indicated a significant increase in the instrument reading above the tropopause owing to a direct cross sensitivity of the Hantzsch reaction itself toward O_3 after laboratory tests as well as a sensitivity to temperature (Kormann *et al.* 2003).

3.7.3.3 **Data Processing**

The determination of the atmospheric concentrations $[X]$ collected by the liquid conversion techniques described here depends on several factors that have to be evaluated: calibration factor (A), background signal (S_b), collection efficiency (CE) into the glass coil, sampled air and scrubbing solution flows (fliq and fair), catalase destruction efficiency for H_2O_2, and correction factor for known interferences (F_{corr}). We propose here a generic expression for $[X]$:

$$[X] = A\,(S_X - S_b) \cdot \frac{f_{liq}}{f_{air}} \cdot \frac{1}{CE} \cdot F_{corr} \qquad (3.12)$$

X stands for CH_2O or total ROOH (including H_2O_2). One should note that H_2O_2 is derived from the difference in the two fluorescence channels accounting for the destruction efficiency, which is not expressed here. Destruction efficiency is determined by calibration and has been reported to be >95%. Each factor has to be thoroughly evaluated to provide accurate measurements. The calibration factor (A) and signals (S_{rmb} and S_X) are the parameters to be determined in the field. The flows depend on operating conditions during aircraft operations and are fixed. CE and F_{corr} are evaluated in the laboratory.

3.7.3.4 Limitations, Uncertainties, and Error Propagation

Limitations and uncertainties are related to the difficulty to evaluate some of the factors in Eq. (3.12). In particular, collection efficiency is one of the factors affecting the most the accuracy of organic hydroperoxide measurements (ROOH) and formaldehyde (Heikes *et al.* 2001). While the CE of H_2O_2 equals 99% due to its elevated K_H and is independent from temperature, the solubility of other hydroperoxides is much lower and CEs are comprised between 60% and 100%, depending on the temperature and the volume air flow (Lee, Heikes, and O'Sullivan, 2000). For measurements in remote areas or in the free troposphere, methylhydroperoxide is the main organic hydroperoxide and the uncertainty in the ROOH concentration is less important (Lazrus, Fong, Lind, 1988). A CE of 60% is generally admitted. Collection efficiency can be theoretically and empirically determined. While Lee *et al.* (1995) showed very good agreement within ±10%, an experimental determination appears more appropriate provided it takes into account the aircraft sampling conditions as, for instance, CE can vary with the volume flow rate of the sampled air, the pressure, and the temperature. To conclude, ROOH observations must be carefully interpreted. In particular, it is critical to test the collection efficiency of each hydroperoxide if one has to provide a quantitative measurement of these compounds.

3.7.3.5 Calibration and Maintenance

In the field, all instruments are calibrated against aqueous standard solutions, generally before and after the flight and occasionally during the flight. Lee, Heikes, and O'Sullivan (2000) reported in detail the calibration procedures. For hydroperoxide measurements, one critical issue is the synthesis of organic hydroperoxides, which cannot be purchased. Gas-phase calibration for aircraft measurements is essential to check interferences, inlet artifacts, and collection efficiency.

To prevent from contamination for H_2O_2, the inlet was periodically cleaned with methanol and water during PEM-West A (Heikes *et al.* 1996).

3.7.3.6 Measurement Examples

The interpretation of hydroperoxides and formaldehyde measurements derived from liquid conversion techniques has provided important insights on their spatial and temporal distribution and their role in atmospheric tropospheric chemistry. Reeves and Penkett (2003) reviewed in detail the main results deduced from hydroperoxide observations (2003 and references therein). Beyond providing the spatial distribution of formaldehyde, models–observations comparisons have been generally performed in remote marine and continental troposphere to test our understanding of formaldehyde budget. Evidence was found of direct production of formaldehyde in biomass burning plumes (Lee *et al.* 1997). Also, the potential role of long-range transport of formaldehyde precursors and chemical processing – nonmeasured intermediate products – in plumes was shown (Frost *et al.*

2002; Kormann *et al.* 2003). Furthermore, the importance of deep convection on formaldehyde budget in the upper troposphere, respectively, in clean marine background troposphere (Atlantic) and continental background troposphere (Europe) was shown (Fried *et al.* 2002; Stickler *et al.* 2006).

3.7.3.7 Summary and Emerging Technologies

The basics of liquid conversion techniques have not changed since the 1980s, showing the suitability and efficiency of the technique for aircraft measurements for hydrogen peroxide and formaldehyde. However, the measurement of total and speciated ROOH and other aldehydes remain more uncertain (see the preceding sections). Few alternative techniques based on optical detection or spectroscopy have been proposed. TLAS emerged for formaldehyde measurement (Fried *et al.* 2003). This technique worked fairly well compared with the coil-enzymatic technique during Intercontinental Chemical Transport Experiment (INTEX-NA) (Fried *et al.* 2003) in 70% of the cases. Chemical ionization mass spectroscopy developed by Crounse *et al.* (2006) appeared promising for H_2O_2 measurements (25 pmol/mol as detection limit and high time resolution (<1 s)) and compared well with the HPLC-fluorescence instrument.

Future development in using liquid conversion techniques are (i) the speciation of the organic peroxides and aldehydes (excluding formaldehyde). To do so, the systematic deployment of chromatographic separation or three channels have appeared suitable. (ii) A second future development is the extension to other soluble molecules such as nitrous acid (HONO) as reported recently by Zhang *et al.* (2009) and nitric acid (HNO_3).

3.8 Whole Air Sampler and Chromatographic Techniques

3.8.1 Rationale

Aircraft are important measurement platforms because they allow probing large parts of the atmosphere, almost all of which is not accessible from the ground. There are two principle ways of doing this: (i) deploying an instrument on the aircraft for *in situ* measurements and (ii) collecting air samples for subsequent analysis in the laboratory. When high-precision, fast, automatic, and reasonably small and light instruments are available, option (i) is preferred because the number of flasks that can be filled on a flight – and therefore the temporal and spatial resolution of the measurements – is limited. However, there are still numerous applications where instruments that fulfill these requirements are not available. Measurements of certain parameters can be carried out with higher precision under controlled conditions in the laboratory and/or require large and heavy instrumentation and a skilled operator. In other cases, the duration of a measurement is long, so the temporal/spatial resolution of *in situ* measurements may be even worse than what can be reached on air samples. Examples are the analysis of very-low-concentration hydrocarbons and halocarbons or the isotopic composition of many trace gases. For such applications, option (ii) is preferred. New analytical developments have led to analyzers that fulfill the above requirements for *in situ* measurements, which can lead to big advances because of the huge increase in temporal resolution (e.g., new spectroscopic analyzers for the isotopic composition of CO_2). For other applications, it is still necessary to sample air and bring it back to the laboratory for analysis on dedicated instruments. As air samples from aircraft campaigns or regular flights are rare and therefore of high value for atmospheric scientists, given the unique snapshot of the atmosphere that they provide, they are often analyzed by a number of different laboratories.

3.8.2
Whole Air Sampling Systems

Air sampling systems have been developed for different purposes and on different aircraft platforms using both scientific and commercial planes. Such systems for use in aircraft are not commercially available, and each sampler is custom-made, often for a specific aircraft (Heidt *et al.* 1989; Flocke *et al.* 1999; Brenninkmeijer *et al.* 2007; Cairo *et al.* 2010). Thus, each instrument and its individual components have to comply with safety regulations and must be certified for use in aircraft.

The most general type of sampling is to collect whole air samples in sampling containers. Owing to the decrease in pressure with altitude in the atmosphere, usually compressors are employed to pressurize the containers above surface ambient pressures (Heidt *et al.* 1989). In addition to increasing the sampled volume, this also reduces the possibility of contamination from air leaking into flasks when they come back to the surface and are often stored for longer times after the flight. For some dedicated applications, it is possible to perform the extraction of the target compounds already during the sampling. For example, for hydrocarbons, adsorption tubes that selectively adsorb carbon compounds can be employed (Brenninkmeijer *et al.* 2007). This limits the volume of the sampling unit and often also limits the number of compounds that can be analyzed.

3.8.2.1 Design of Air Samplers

Air samplers for use in aircraft can be separated into three subunits: the AI, the compression stage, and the sampling containers. In addition, the tubing connecting these parts and the valves used require consideration. For the AI and tubing, air sampling systems usually employ similar technology as used for *in situ* measurement systems. For example, silcosteel tubing (Restek, Bellefonte, PA, USA) can be used, which minimizes surface effects.

The market for aircraft-qualified compressors is small, and in many sampling units, currently operating metal bellows compressors (Model PWSC 28823-7, Senior Aerospace Inc., Sharon, Mass., USA) are in use. Processing of air through a compressor before taking a sample naturally invokes the risk of contamination from the compressor. In addition, the temperature change associated with the compression stage can affect some species (e.g., selected hydrocarbons). In particular, for unstable compounds and those with very low mixing ratios, contamination is always an issue, and the compressor unit needs to be tested for each target compound.

The sample canisters are likely the most critical point of air sampling systems, because the air samples are in contact with the canisters for usually weeks to many months or even years before analysis. In most sampling systems, electropolished or passivated stainless steel flasks have been used (Rudolph, Muller, and Koppmann, 1990; Kumar and Viden, 2007), there is also a well-described protocol from the US EPA as to the use of sample canisters. Glass flasks are used in the extensive ground air sampling programs (*http://www.esrl.noaa.gov/gmd/ccgg/flask.html*), but for safety reasons, they are rarely employed on aircraft. However, some recently developed aircraft sampling systems from NOAA/ESRL (Earth System Research Laboratory) (Bakwin *et al.* 2003) or in the new CARIBIC container (Brenninkmeijer *et al.* 2007) do make use of glass flasks.

When an air sampling system has been developed, it is necessary to test its suitability for the target compounds (Laube *et al.* 2010a,b; Flocke *et al.* 1998; Schauffler *et al.* 1998). Controlled moistening of the flasks after evacuation can reduce contaminations for some species and can lead to adverse effects on other target species, for example, isotope exchange between CO_2 and water (Lämmerzahl *et al.* 2002).

3.8.2.2 The M55-Geophysica Whole Air Sampler

The whole air sampler (WAS) for the high-altitude aircraft M55-Geophysica was developed by the MPI for Nuclear Physics (Heidelberg, Germany) and enviscope GmbH (Frankfurt, Germany). It has been deployed in several European measurement campaigns since 2003 and is now operated by the Institute for Marine and Atmospheric Research Utrecht (Utrecht University, The Netherlands). Similar to the systems used in the ER2 or DC8 aircraft (Heidt *et al.* 1989; Flocke *et al.* 1999), it uses custom-made, 2-l volume, electropolished, stainless steel canisters, equipped with Nupro SS-4H valves. The flask unit holds 20 flasks, which are connected to a manifold via modified electronic two-way valves (Clippard Minimatic, USA); an additional valve at the outlet of the manifold is used for flushing when no samples are collected. Nupro hand valves on the individual flasks are then opened and they are closed again when the sampler is accessible on ground after the flight. Because of the low pressures at Geophysica altitudes, two metal bellows compressors are connected in series. They permit the collection of sample amounts of 8 l STP (Standard Temperature and Pressure). The sampling lasts between 1 and 5 min depending on the altitude.

Three pressures are recorded at 0.3 Hz, at the inlet, between the two compressors, and after compressor 2; the latter gives the pressure in the sample flasks during sampling, temperatures are also recorded at three places. The WAS unit is located inside the aircraft, and despite the low ambient temperatures, it usually operates above 25 °C, the gas manifold can be heated with a heating wire if necessary. The compressors generate additional heat, and when temperatures on one of the compressors rise above 60 °C, the compressors are temporarily turned off to cool down. They can also be turned off in between while taking different samples.

The air sampler can be switched on by the pilot and compressors start when the pressure is below 200 hPa, as a precaution to avoid condensation of water vapor. In addition, a drying cartridge can be installed behind the inlet. In this configuration, the WAS instrument has been deployed in numerous past aircraft campaigns (Cairo *et al.* 2010).

3.8.3 Water Vapor Sampling for Isotope Analysis

Numerous optical techniques are available for *in situ* measurements of water vapor on aircraft, but until very recently, online isotope measurements on water vapor have not been possible with the desired precision. Therefore, several water vapor samplers have been constructed for aircraft. Although it is in principle straightforward to collect water cryogenically, clean sampling of water vapor is challenging because the polarity of the water molecule leads to unavoidable interactions with surfaces of the sampling system. In the lower troposphere, where high water vapor levels prevail, contamination can be minimized by sampling large volumes, but in the upper troposphere, sampling systems approach their limits, either because of the large amount of air that needs to be processed or because of the tremendous care that has to be taken to avoid contamination of water from the sampling equipment itself. Zahn *et al.* (1998) collected liquid water of the order of 1 ml from 15 m^3 of air using large cryogenic traps in the polar tropopause region. Unfortunately, the need for large amounts of liquid nitrogen has limited the applications because of safety considerations regarding the use of cryogens on aircraft. In the miniaturized sampling system described in the study by (Franz and Röckmann, 2005), special gold-plated containers and careful cleaning procedures were employed, and water samples of the order of 100 nl (liquid volume) could be collected and analyzed for isotopic composition. A version of this water vapor sampler has been integrated into the WAS for the M55-Geophysica aircraft during the African Monsoon Multidisciplinary Analysis (AMMA) campaign (Cairo *et al.* 2010). However, this work is presently not continuing, because for deuterium and ^{18}O analysis, the sampling approaches have in recent years been surpassed by *in situ* optical analyzers (Johnson *et al.* 2001; Sayres *et al.* 2009b). However, for ^{17}O and tritium analysis, still no adequate alternatives to sampling methods are available.

3.8.4 Measurement Examples

For illustrative purposes, we present two examples from measurements that were obtained during M55 test flights above Germany in 2009 and Kiruna (Sweden) in 2010, as part of the EU campaign Reconciliation of essential process parameters for an enhanced predictability of arctic stratospheric ozone loss and its climate interactions (RECONCILE).

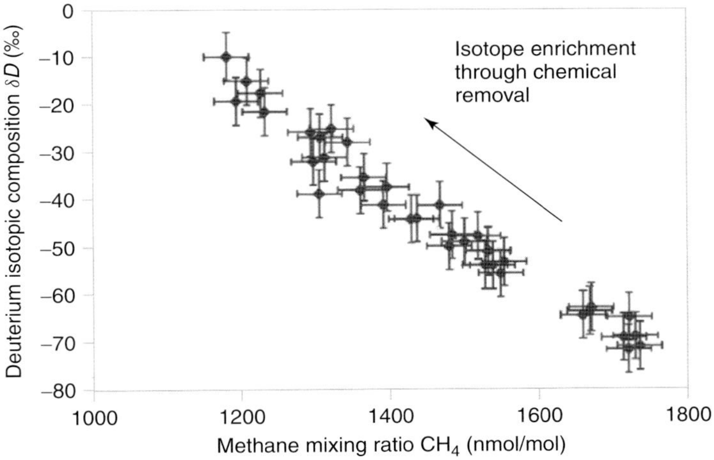

Figure 3.27 Deuterium isotopic composition (δD) versus methane mixing ratio as measured on M55-Geophysica samples collected with the WAS instrument of the Utrecht University during two flights from Oberpfaffenhofen, Germany, between 10 and 20 km altitude. Methane removal is accompanied by isotope enrichment (arrow). The isotopic composition is quantified as δD value, where $\delta D = [D/H]_{SA}/[D/H]_{VSMOW} - 1$ is the relative difference (usually expressed in permille) of the D/H atom number ratio in a methane sample (subscript SA) from the international isotope standard Vienna Standard Mean Ocean Water (subscript VSMOW, $[D/H]_{VSMOW} = 1.55678 \cdot 10^{-4}$). Error bars are estimated as $1 - \sigma$ errors of ±30 nmol/mol for CH_4 and ± 5‰ for δD (Brass and Röckmann, 2010).

Figure 3.27 shows the evolution of the heavy isotope (in this case deuterium) content of atmospheric methane, while it is destroyed in the stratosphere. Such high-precision isotope measurements are very difficult and presently cannot be performed *in situ*. However, with new methods (Brass and Röckmann, 2010), only small aliquots of air are required (40 ml for one measurement), and they can be performed on the samples from the WAS instrument with high precision.

Higher δD values (see the caption of Figure 3.27 for definition) mean that the sample is more enriched in deuterium, that is, the $[D/H]$ ratio is higher. This is because as methane is removed in the stratosphere, CH_4 is removed more quickly than CH_3D. We call such a difference in chemical removal rates due to isotopic substitution a kinetic isotope effect, and the ratio of the two removal rate coefficients is referred to as isotope fractionation factor $\alpha = k_{CH_3D}/k_{CH_4}$ (Brenninkmeijer *et al.* 2007). In the case of methane removal in the stratosphere, three different removal reactions are important, which have very different hydrogen fractionation factors. It is the relative importance of these three removal processes, plus mixing effects, which determine the isotopic composition of methane in the stratosphere (Rice *et al.* 2001).

As a second example, Figure 3.28 shows the mixing ratios of three halogenated hydrocarbons, the two well-known chlorofluorocarbons CFC-11 and CFC-12 and

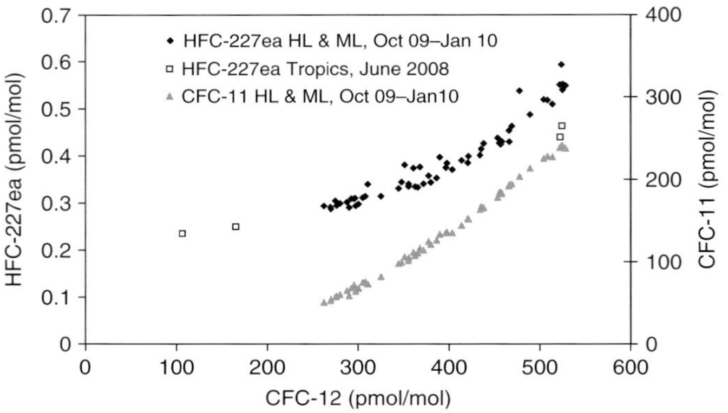

Figure 3.28 Scatter plot of the mixing ratios of HFC-227ea (left axis) and CFC-11 (right axis) versus CFC-12 (*x*-axis). HL and ML refer to high latitude and mid-latitude air, respectively, and black and gray points refer to samples collected in the tropics during a balloon flight in June 2008 (for details see (Laube *et al.* 2010b). Measurement uncertainties are less than the size of the symbols.

the recently discovered second-generation CFC-replacement compound HFC-227ea (Laube *et al.* 2010b).

These measurements also require a specialist operator and were carried out at the University of East Anglia on aliquots of the same sample flasks that were discussed in the first example, plus samples from the Kiruna flight. Note that the mixing ratios of the replacement compound HFC-227ea are about 1000 times smaller than that for CFC-12. The points at the top right corner of the scatter plots represent air that is freshly entering the stratosphere from the troposphere, and the more the three tracers are removed, the lower their mixing ratios become (toward the bottom left). The slopes of such tracer–tracer correlations can be used to assess the relative rates of removal, and thus the relative stratospheric lifetimes of the two tracers in the stratosphere.

These examples illustrate how highly specialized results can be obtained when clean samples from remote locations are collected without contamination and the air is then analyzed with dedicated instruments under controlled laboratory conditions.

3.8.5
Off-Line Analysis of VOCs

Gas chromatography has proven to be a powerful tool for atmospheric composition analysis. Ground-based studies on the atmosphere have used the technique for speciation of VOCs for many years (Roberts *et al.* 1985; Bonsang, Kanakidou, and Lambert, 1990; Hopkins *et al.* 2002; Helmig *et al.* 2008) and have been coupled to grab sampling methods for remote (Jobson *et al.* 1994; Solberg *et al.* (1996a); Saito, Yokouchi, and Kawamura, 2000) and airborne observations (Penkett *et al.* 1993; Blake *et al.* 1997; Purvis *et al.* 2003). *In situ* airborne gas chromatographs have been developed (Apel *et al.* 2003; Whalley *et al.* 2004) with some success, but the need for continuous, high-frequency, high-precision observations makes the technique less suitable for this type of analysis and so are yet to be made commercially viable.

The relatively recent development of PTR-MS allows fast resolution observations of a subset of VOCs to be made (Karl *et al.* 2004) and these have been successfully adopted for aircraft use (Crutzen *et al.* 2000; Sprung *et al.* 2001; Karl *et al.* 2009) (Section 3.6.1). These instruments, however, have known interferences from organic compounds of the same mass (such as methacrolein and methyl vinyl ketone, both of which are, in fact, oxidation products of isoprene) (Hewitt, Hayward, and Tani, 2003). The high degree of precision, reliability, and high level of speciation (there are some coelution issues such as some of the xylene isomers) achievable from chromatographic methods is unrivalled using current technology and can provide a powerful tool for the analysis of VOCs in the atmosphere. The major drawback with whole air samples or even in-flight GCs are the time required for the sample acquisition and separation. It is worth noting that the deployment of a PTR-MS instrument, for high-frequency observations, alongside a GC-based technique, for speciated data, is the most desirable scenario.

Speciation of VOCs is important not only for the reporting of mixing ratios for each compound and the calculation of emission fluxes from ground-based sources but also for use in atmospheric models. VOCs have a huge range of reactivities with many oxidants in the atmosphere and have a wide range of atmospheric lifetimes as a consequence. They can be present in the atmosphere in concentrations ranging from sub-ppt levels to hundreds of ppb.

Observations of VOCs can be used in a variety of ways to describe the state of the atmosphere.

The wealth of data available from even a small number of whole air samples makes them very useful to determine a number of atmospheric parameters. Knowledge of rates of reaction with respect to different atmospheric oxidants can provide clues as to the age and origin of the sampled air. The lifetime and dynamic range of nonmethane hydrocarbons makes them excellent tracers for pollution sources; furthermore, where compounds have a relatively short lifetime, their observed presence can indicate strong vertical transport from the surface (Purvis *et al.* 2003).

3.8.5.1 **Air Mass Ageing**

During transport from source regions, the VOC content of an air mass will deplete as a result of dilution and chemical processing within the atmosphere. For example, the OH rate coefficients for benzene and toluene differ greatly with reaction of toluene being more than four times faster than benzene. As reaction with OH is the major removal pathway for these compounds, it is assumed that their chemical lifetimes in the troposphere will be determined almost solely by this reaction; the ratio of benzene and toluene will change during exposure to OH radicals. The effect of dilution of an air mass by surrounding "nonzero VOC" air further complicates the matter. However, using a suite of VOCs, as can be determined from a single whole air sample, it is possible to extract information as to the history of the airmass (Arnold *et al.* 2007; Honrath *et al.* 2008). Benzene and toluene are most commonly used in these studies as both have a single major atmospheric sink in the OH radical as well as often more reliable source ratios compared with other VOC species, but the same methodology can be applied to any pair of VOCs with differing OH lifetimes.

If we assume that, for a given data set, all VOCs are emitted from a common source and are only removed from the atmosphere through reaction with hydroxyl radicals, then a clear linear relationship between the standard deviation of VOCs and their OH lifetimes, according to Eq. (3.13), will be revealed (Jobson *et al.* 1998).

$$S\left(\ln x_i\right) = A \cdot \tau^{-b} \tag{3.13}$$

$$\log_{10} S\left(\ln x_i\right) = \log_{10} A - b \cdot \log_{10}\left(\frac{1}{k_{\mathrm{OH}} + \mathrm{HC[OH]}}\right) \tag{3.14}$$

where S is the standard deviation, x_i the concentration of component i, τ the lifetime of the compound, A the proportionality coefficient, and b the exponent relating to the dominance of the sink terms.

The gradient (b) of such a plot reveals the extent of chemical processing within an air mass. Its lower limit, $b = 0$, signifies observations at a source where no chemical processing has occurred, whereas its upper limit, $b = 1$, can be considered as the chemical kinetic limit where variability is dominated by chemical loss alone, with all sources being at a sufficient distance from the sampling point to become negligible. Deviation from the linear relationship results from alternative reaction pathways for VOCs or from differing source regions of the air masses under investigation. VOCs whose standard deviations lie above the linear fit have undergone reaction with oxidants other than OH, whereas those that lie below the line have been influenced by additional or different source regions. This type of analysis is most effective for large data sets under differing air mass origin regimes (Bartenbach *et al.* 2007).

3.8.5.2 **Using VOC Observations to Probe Radical Chemistry**

While hydroxyl radicals provide the major chemical loss route for VOCs, other oxidants are present in the atmosphere and, under certain conditions, can have a major role in their atmospheric processing. Nitrate radicals are rapidly photolyzed and hence are present at very low concentrations during the daytime, but during hours of darkness, they can build up to more significant levels. Iso- and *n*-pentane prove to be useful compounds in identifying periods of nitrate radical activity due to their significantly different rates of reaction with respect to nitrate radicals (iso-pentane reacts much faster due to the presence of its tertiary carbon atom) yet almost identical reaction rates with respect to hydroxyl radicals. The ratio of iso-/*n*-pentane can therefore reveal information about the chemistry occurring in a particular environment (Penkett *et al.* 1993).

The rates of reaction of VOCs with halogens are, in general, considerably larger (up to two orders of magnitude) than those for OH, implying their importance even at low concentration. Halogen atoms can therefore play a significant role in the chemistry of the troposphere, especially in the marine environment (Jobson *et al.* 1994; Solberg *et al.* 1996b; Read *et al.* 2008) where oceanic wave action produces large amounts of water droplets, which can evaporate to produce suspended particles of compounds contained in the droplet, including halogen atoms (Wayne *et al.* 1999). The ratio of iso- to *n*-butane can be used to investigate halogen atom chemistry in a given air mass due to their almost identical OH reaction rate constants and significantly different Cl atom rate constants (Atkinson *et al.* 2006). The rate of reaction with respect to chlorine for *n*-butane is almost double that of iso-butane, this results in significant changes in their ratio if affected by Cl atom chemistry (Read *et al.* 2006). Halogen radical abundances have also been probed using similar approach in volcanic plumes (Baker *et al.* 2011). Interpretation of atmospheric data using these tools is used only to provide qualitative insight into the processing history of an air mass rather than for quantitative analysis or kinetic studies.

4
In Situ Measurements of Aerosol Particles

Andreas Petzold, Paola Formenti, Darrel Baumgardner, Ulrich Bundke, Hugh Coe, Joachim Curtius, Paul J. DeMott, Richard C. Flagan, Markus Fiebig, James G. Hudson, Jim McQuaid, Andreas Minikin, Gregory C. Roberts, and Jian Wang

4.1
Introduction

4.1.1
Historical Overview

The airborne observation of atmospheric properties such as temperature, pressure, and visibility are likely to date back to times when mankind first learned to fly. Airborne platforms for scientific observations developed since then include balloons, airships, airplanes, helicopters, and very recently unmanned airborne vehicles (UAVs). However, the measurement of atmospheric aerosol particles is still a major challenge for understanding the role of airborne particulate matter in the global climate system, although the knowledge about properties of interest has developed considerably.

Historically, the development of airborne aerosol particle observations was closely related to the availability of modern measurement methods for particle properties. First airborne sampling of biological particles and insects was reported by Felt (1928). Sticky flypaper was used on an aircraft for collecting insects at elevations above 300 m. Before World War II, aerosol observations on aircraft were limited to simple collection devices, mainly for biological particles to investigate the spread of crop diseases: Peturson (1931); Proctor and Parker 1938 in the United States; Hubert (1932), Rempe (1937) in Germany; Craigie (1945) in Canada; and van Overeem (1936) in the Netherlands. A comprehensive historical overview on transoceanic aircraft sampling for organisms and particles is given by Holzapfel (1978). In the 1950s, meteorological research flights focused on the measurement of charged clusters in the atmosphere (Sagalyn and Faucher, 1955) using World War II B-17 bomber aircraft as the measurement platform.

In the early years of coordinated observation of atmospheric properties, networks of ground-based or ship-borne stations were set up for monitoring the atmospheric

Airborne Measurements for Environmental Research: Methods and Instruments, First Edition.
Edited by Manfred Wendisch and Jean-Louis Brenguier.

chemical composition including atmospheric aerosol particles. The need for airborne atmospheric observations was first recognized in the field of stratospheric research where instrumented balloons were operated for collecting information about the nature of the lower and middle stratosphere, including the discovery of the stratospheric aerosol layer (or "Junge layer") by Junge, Chagnon, and Manson (1961) and the investigation of stratospheric aerosol particles by Hofmann *et al.* (1973). In those days, instrumented platforms were used for gathering information on atmospheric properties, chemical composition, and particulate matter from regions where no ground-based platforms existed. Among the first coordinated studies including airborne platforms, Arctic aerosol particles were studied intensively in 1983 in Alaska (Schnell, 1984).

The effect of aerosol particles on the radiative properties of liquid water clouds was stated as early as 1974 by Twomey (1974). In 1990, model results suggested an aerosol-related effect on global climate, which to a significant degree may counteract the effects of greenhouse gases; see Penner *et al.* (1994) for an overview. It was also recognized that the climate response, specifically the global mean temperature change caused by various atmospheric constituents, is sensitive to the altitude, latitude, and nature of the resulting forcing (Hansen, Sato, and Ruedy, 1997). Large-scale atmospheric studies integrating multiple platforms such as the series of aerosol characterization experiments (ACE-1, ACE-2; ACE-Asia) and studies targeting specific aerosol types (tropospheric aerosol radiative forcing observational experiment (TARFOX); Lindenberg aerosol characterization experiment (LACE); Indian ocean experiment, INDOEX; and many others) were set up to investigate the impact of the natural and anthropogenic aerosol on climate. Despite the substantial efforts undertaken in recent years for improving the understanding of global aerosol–climate interactions, the latest Intergovernmental Panel on Climate Change (IPCC) report (IPCC, 2007) still asks for detailed airborne observations of aerosol particles. The complexity and nonlinearity of feedback dominating aerosol–cloud interaction require large-scale and complex measurement resources to be brought to bear on this problem and *in situ* aircraft measurements will continue to be a vital component of this effort.

Today, ground-based networks for *in situ* measurements have developed into the Global Atmospheric Watch (GAW) program of the World Meteorological Organization (WMO; *www.wmo.int/pages/prog/arep/gaw*) or the IMPROVE network (*www.vista.cira.colostate.edu/improve*) in the United States. Remote sensing networks were set up successfully for sun photometers (AERONET; *www.aeronet.gsfc.nasa.gov*) and aerosol LIDAR (*www.earlinet.org*). More recently, the advent of modern instrumentation and miniaturization techniques means that a global airborne *in situ* observing system for atmospheric constituents including greenhouse gases, aerosol, and cloud particles is becoming a reality as the In-Service Aircraft for a Global Observing System (IAGOS) network (*www.iagos.org*), which uses a globally operating fleet of Airbus commercial aircraft. This airborne network complements the data obtained from coordinated field studies and ground-based networks by mapping the vertical distribution of aerosol particle properties.

Driven by improved optical methods for cloud and aerosol particle sizing such as the forward scattering spectrometer probes (FSSPs; Section 4.3), the airborne measurement of aerosol properties developed rapidly. Furthermore, groundbreaking steps in airborne aerosol measurement were the development of sensitive methods for measuring aerosol particle optical properties with instruments such as the particle soot absorption photometer (PSAP) and integrating nephelometer; see Section 4.5. Furthermore, sophisticated instruments for measurements of aerosol chemical composition (Section 4.4) and the cloud-forming potential of particles (Section 4.6) were developed. However, it has to be noted that well-designed sampling systems that are discussed in Chapter 6 are an indispensable prerequisite for quantitative aerosol measurements.

4.1.2 Typical Mode Structure of Aerosol Particle Size Distribution

One of the most important properties governing physicochemical behavior and atmospheric lifetime of airborne particles is their size given as particle diameter D_p or as typical dimension in the case of irregularly shaped particles. Atmospheric aerosol particles can be regarded as a superposition of four particle modes: the nucleation mode ($D_p < 20\,\text{nm}$) contains particles freshly formed from gaseous precursors by gas-to-particle conversion or homogeneous particle nucleation; most aerosol particles in the nucleation mode are composed of sulfuric compounds and hydrocarbon compounds. The Aitken mode ($D_p = 20-100\,\text{nm}$) includes particles formed by coagulation of nucleated particles and condensation of vapors on already existing particles or emitted directly into the atmosphere. The accumulation mode ($D_p = 0.1-1.0\,\mu\text{m}$) consists of particles formed from the Aitken mode by particle coagulation and particles emitted directly from primary sources such as combustion of vegetation or fossil fuels; for atmospheric aerosol particles, the accumulation mode forms the sink of particles growing from nucleation via the Aitken mode into the accumulation mode. The coarse mode contains all particles larger than 1.0 μm. These particles are generated mainly by mechanical processes such as wind-blown dust, sea spray, or plant debris or are emitted from volcanoes or large fires. Because of the different formation pathways, coarse mode particles are separated from smaller particles with respect to their chemical composition. Bioaerosol particles such as spores, bacteria, or pollen are also mainly part of the coarse mode, but may be part of the accumulation mode as well. Figure 4.1 shows a schematic representation of particle size distribution of typical boundary layer aerosol.

4.1.3 Quantitative Description of Aerosol Particles

Properties required for an assessment of aerosol–climate interactions are discussed by Penner *et al.* (1994) and Ogren (1995). The two major pathways to aerosol–climate interaction lead through the direct interaction of particles and radiation (direct effect) and the indirect interaction via the modification of cloud

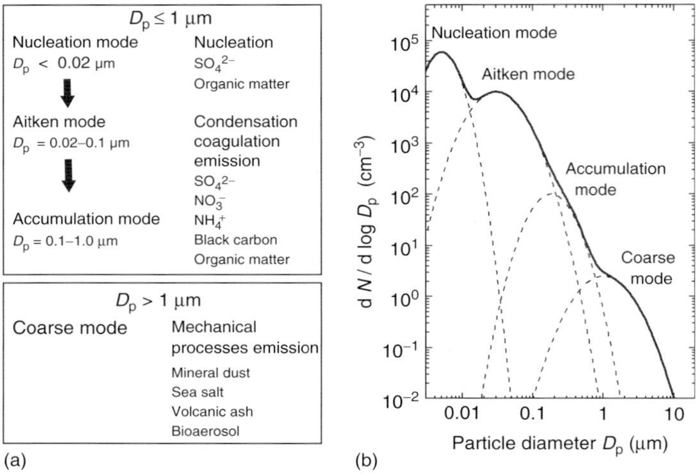

Figure 4.1 Schematic representation of the modes of a typical atmospheric aerosol particle population of the planetary boundary layer for different particle diameters D_p. (a) Key chemical compounds and particle generation processes. (b) Aerosol particle size distribution or the number of particles per logarithmic diameter interval plotted as a function of particle diameter. (Source: Reprinted in modified form from Schumann (2012), with permission of Springer Science + Business Media.)

properties by aerosol particles (indirect effects). Therefore, extensive (depending on particle concentration) and intensive (independent of particle concentration) properties to be known for the major types of natural and anthropogenic aerosol particles are summarized below.

Extensive aerosol particle properties are as follows: (i) the aerosol particle number concentration and size distribution (1 nm to < 100 μm); (ii) the particle scattering, absorption, and backscatter coefficients; and (iii) the aerosol particle extinction coefficient. Intensive aerosol particle quantities are as follows: (i) the particle single-scattering albedo; (ii) the particle shape and scattering phase function; (iii) the upscatter fraction; (iv) the aerosol particle chemical composition; (v) the particle hygroscopic growth; and (vi) the cloud condensation nuclei (CCN) or ice nuclei (IN) concentration.

The extensive properties, particle number concentration and number size distribution, characterize an aerosol population in terms of number of particles per unit volume such as cm^3 or m^3, and in terms of the number concentration of particles in a range of particle diameters $(D_p, D_p + \Delta D_p)$. As aircraft-borne measurements are conducted at a variety of pressures and temperatures, the aerosol number concentration measurements are often converted to standard temperature and pressure (STP)conditions, which are commonly defined as 273.15 K and 1013.25 hPa. The applied conversion is based on the ideal gas law and relates the volume sampled at ambient conditions to the respective volume that would have been sampled at STP conditions.

Absorption, scattering, and backscatter coefficients characterize the interaction of aerosol particles with electromagnetic radiation, given as an inverse length m^{-1} or $Mm^{-1} = 10^{-6} m^{-1}$. The extinction coefficient of the particles describes the combined effects of scattering and absorption of electromagnetic radiation by the particle population. The extinction, scattering, and absorption coefficients of the aerosol particles are referred to by the symbols $b_{ext,p}$, $b_{sca,p}$, and $b_{abs,p}$, respectively. The aerosol optical depth (AOD) $\tau_p(\lambda)$, which precisely should be termed aerosol particle optical depth, at a specific wavelength λ is calculated from the integral of the particle extinction coefficient $b_{ext,p}$ over the thickness z of an aerosol layer; see Section 7.2.3, Eqs. (7.4)–(7.5). Details on optical theories for the calculation of these properties, such as Mie theory valid for spherical particles (Bohren and Huffman, 1983), can be found in Chapter 10: Supplementary Online Material, Section 10.1.4.

Intensive properties are independent of particle concentration but describe properties characteristic for the investigated aerosol type. The particle single-scattering albedo $\tilde{\omega}_p$ defines the fraction of total extinction that is caused by scattering with the remainder being caused by absorption. The angular distribution of scattered light is related to the size and shape of the particles. The upscatter fraction β_{up} for a given solar zenith angle describes the part of incoming solar radiation that is scattered back into space. It is determined by the scattering phase function of the particles in the aerosol layer.

Knowledge of the aerosol particle chemical composition is required for a large variety of applications. It is characteristic of the aerosol source as well as for the transformation processes during the atmospheric lifecycle of a particle. Chemical composition is also closely linked to aerosol radiative properties via the complex index of refraction $m = n + i \cdot k$. The real part n defines the scattering properties, while the imaginary part k may equal zero for nonabsorbing particles and $k > 0$ for absorbing particles. Values for m are listed in the literature (Hess, Koepke, and Schult, 1998; Seinfeld and Pandis, 1998) for a variety of chemical substances and wavelengths. The refractive index of absorbing carbon, which is the most important particulate absorber in the atmosphere, is extensively reviewed by Bond and Bergstrom (2006). Refractive indices of dust are reviewed by Formenti *et al.* (2011), while respective information on other particulate absorbers such as brown carbon is still sparse. Mixing rules that apply for calculating properties of mixed particles are described by Bohren and Huffman (1983).

If an aerosol particle contains hygroscopic material such as sulfate, nitrate, or sodium chloride, then the particle takes up water vapor from the gas phase in a humid atmosphere. Depending on the substance, the hygroscopic growth is a continuous process, or it shows a steplike behavior at a certain humidity threshold. The uptake of water vapor causes a growth in size as well as a modification of the effective complex index of refraction for a mixed particle. Both effects impact the optical properties of aerosol particles, namely, $\tau_p(\lambda)$, $b_{ext,p}$, and $\tilde{\omega}_p$.

Treating the aerosol-loaded atmosphere by a multiple reflection model for an optically thin partially absorbing aerosol, the aerosol-related modification of the top-of-atmosphere (TOA) radiative flux density (irradiance, see Chapter 7) per unit

AOD can be estimated (Haywood and Shine, 1995). In combination with the solar constant S_0, the atmospheric transmission T_{at}, the fractional cloud coverage A_c, and the surface reflectance R_s, the TOA direct aerosol radiative forcing, or aerosol forcing term per unit optical depth, respectively, $\Delta F_{up}/\tau_p(\lambda)$, is

$$\frac{\Delta F_{up}}{\tau_p(\lambda)} \cong -\frac{1}{2} S_0 \cdot T_{at}^2 \cdot (1 - A_c) \cdot \widetilde{\omega}_p \cdot \beta_{up} \cdot \left[(1 - R_s)^2 - \left(\frac{2 R_s}{\beta_{up}} \right) \cdot \left(\frac{1}{\widetilde{\omega}_p} - 1 \right) \right] \tag{4.1}$$

With the exception of particle properties relevant for cloud formation, Eq. (4.1) combines all extensive and intensive properties introduced before, as optical and radiative properties are directly related to particle size distribution, chemical composition and resulting hygroscopic properties, and to particle shape. In addition, the cloud-forming potential of aerosol particles is an essential input to any assessment of aerosol indirect effects on climate.

4.1.4 Chapter Structure

One key prerequisite for modern aerosol research is the availability of sensitive methods for measuring aerosol properties. Because of moderate-to-high true air speed (TAS; see Section 2.7.1) of aircraft of 50–200 m s^{-1}, methods allowing online and in real-time measurements are preferred to methods of reduced time resolution and thus of limited spatial resolution.

This chapter summarizes respective methods and instruments for the measurement of particle number concentration (Section 4.2), particle size distribution (Section 4.3), chemical composition (Section 4.4), and optical properties (Section 4.5). The used terminologies are introduced in the respective sections. The need for better understanding the aerosol–cloud interactions over the last two decades has been the main driver for the development of specific instruments for measuring the ability of particles to form cloud droplets and ice crystals (Section 4.6) and significant progress has been made.

Each of these sections starts with an introduction identifying the scientific target of the presented methods. The physical principles are introduced, method uncertainties and limitations are described, and finally, procedures for data quality assurance and data quality control such as instrument calibration are identified. The chapter closes with an outlook to ongoing developments and future challenges for method development.

Table 4.1 provides a quick guide to the instruments measuring physical properties that are discussed in this chapter. The table includes the techniques that they implement and other useful information that gives the reader a broad perspective of the types of sensors available and their specifications. Techniques available for measuring the chemical composition are compiled in a separate table in Section 4.4.

Table 4.1 Overview of airborne microphysical and optical instruments for aerosol measurements.

Parameter measured	Measurement technique	Instrument	Measurement range	Time resolution	Calibration method	Chapter reference	Primary reference
Particle number	Condensation growth	Condensation particle counter (CPC)	< 0.01 – >100000 cm^3	5 s	Comparison with "Golden Instrument"	4.2	Hinds (1999)
Particle size	Light scattering	UHSAS	0.06 – 1.0 mm	Depending on aerosol concentration	Beads or drop generator for sizing, positioned scatterer for DOF and beam width	4.3.1	Cai *et al.* (2008)
		PCASP	0.1 – 10 mm			4.3.1	Liu *et al.* (1992)
		FSSP – 300	0.3 – 20 mm			4.3.1	Baumgardner *et al.* (1992)
		FSSP – 100	0.5 – 47 mm			4.3.1	Baumgardner *et al.* (1985)
		CAS	0.5c – 50 mm		Beads for sizing, analytical sample area		
		CDP	2 – 50 mm				Lance *et al.* (2010)
	Aerodynamic sizing	Aerodynamic particle sizer (APS)	> 0.9 mm (Aerodynamic diameter)	5 s	Beads or PSL spheres	4.3.2	Wang *et al.* (2002b)
	Electrical mobility	DMA/SMPS	0.005 –1.0 μm	>10 s Depending on operation mode	PSL spheres	4.3.3	Baron and Willeke (2005)
Scattering coefficient	Multiwavelength nephelometry	Integrating nephelometer	$>10^{-7}m^{-1}$	> 10 s	Calibration gas	4.5.1	Anderson *et al.* (1996)
Absorption coefficient	Multiwavelength absorption photometry	Particle soot absorption photometer (PSAP)	$>10^{-7}m^{-1}$	> 10 s	n/a	4.5.2	Bond *et al.* (1999)
	Photoacoustic spectroscopy	Photoacoustic spectrometer	$>10^{-6}m^{-1}$	5 s Depending on aerosol concentration	Calibration gas	4.5.2	Arnott *et al.* (2006)
Extinction coefficient	Cavity ring down spectroscopy	CRD	$>10^{-7}m^{-1}$	5 s Depending on aerosol concentration	Calibration gas	4.5.3	Strawa *et al.* (2003)
Nuclei for cloud drop formation	Cloud condensation nucleus activation chamber	CCN counter/ spectrometer	> 0.007% Supersaturation; 0.1 to 10^5cm^{-3}	1 s for single supersaturation; > 1 min for CCN spectra	Size-selected salt or sulfate particles	4.6.1	Robertsand Nenes (2005); Hudson (1989); Snider *et al.* (2003)
Nuclei for ice crystal formation	Ice nucleation counting chamber	IN counter	Depends on instrument design, IN concentration, averaging time (10^4 to $10^{-3}cm^{-3}$). uncertainties scale inversely with IN concentration.	5s Depending on IN concentration	No calibration standards available; artifact corrections necessary, depend on instrument method.	4.6.2	Rogers *et al.* (2001); Bundke *et al.*(2008); Stetzer *et al.* (2008); Klein *et al.* (2010b)

4.2
Aerosol Particle Number Concentration

The aerosol particle number concentration is defined as the total number of particles present in a volume of air. It is commonly given as the number of particles per cubic centimeter. Number concentration data are reported either for ambient temperature and pressure or for STP conditions. Investigating vertical transport phenomena or comparing number concentration data at different altitudes requires the conversion to STP conditions.

4.2.1
Condensation Particle Counters

The number concentration of aerosol particles is typically measured by using condensation particle counters (CPCs) and they are modified versions of those developed over many decades for ground-based applications. In general, a CPC used for aircraft-borne measurements operates in the same way as standard CPCs for ground-based measurements. The CPC determines the integral particle concentration for a large range of particle sizes, determined by the instrument's lower and upper size cut-offs. Typically, a CPC detects particles in the size range between a few nanometers up to a few micrometers, and therefore, the entire range of particle sizes that usually determines the atmospheric aerosol number concentration is covered by this instrument. Ultrafine particles smaller than 0.1 μm in diameter, which are difficult to detect by optical instruments, can be measured by CPCs. A comprehensive review of the operating principles and the historical development of CPCs in general is given by McMurry (2000); here, we focus on the special conditions and features required for airborne CPC operation.

CPCs are also used as standard particle detectors in combination with other instrumentation to determine various aerosol properties, for example, with differential mobility analyzers (DMAs) for measuring the particle size distribution or with cloud condensation nucleus counters (CCNC) for measuring the fraction of particles that act as CCN. Measuring size distributions of ultrafine particles using DMAs, however, generally results in poor temporal resolution as well as significant uncertainty due to poor counting efficiencies associated with DMAs. As a result, several CPCs with different lower cut-sizes have been operated in parallel as a "CPC battery" for fast, efficient, size-resolved determination of ultrafine particle concentration. These are particularly useful for the detection of aerosol nucleation events (Brock *et al.* 2000; Feldpausch *et al.* 2006; Petzold *et al.* 2005b). Adaptations of these instruments include a heated inlet line in front of the CPC to measure only the nonvolatile particles (Rosen, 1971; Clarke and Porter, 1991; Clarke, 1992).

Here, we describe briefly the principle of operation of the continuous flow diffusion CPC, which is the one most frequently used today for aircraft-borne operation (Figure 4.2). Ambient aerosol enters via a sampling inlet (Chapter 6) and is then exposed to a supersaturated vapor of the working substance. The working substance condenses rapidly onto the particles (activation) and the particles

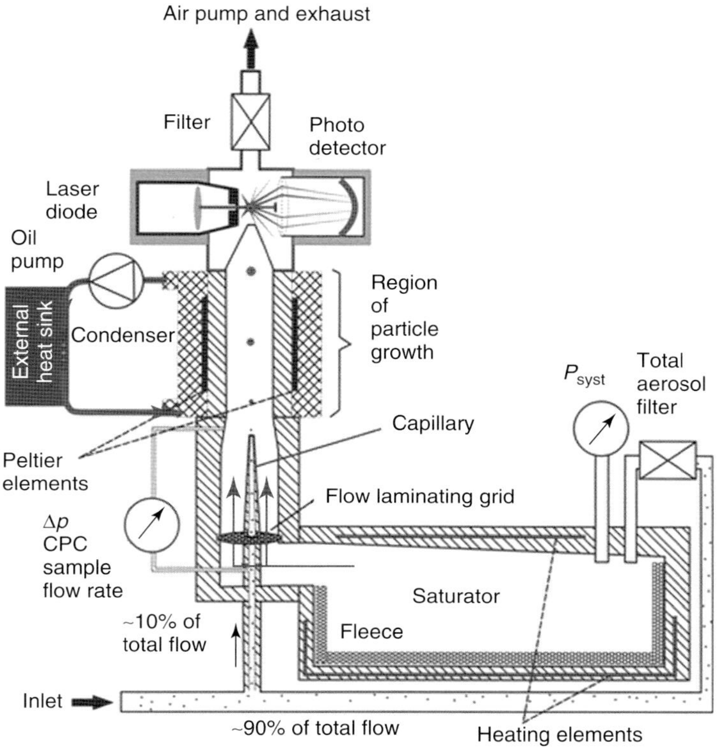

Figure 4.2 Schematics of an aircraft-borne CPC as originally developed by Wilson, Hyun, and Blackshear (1983). Figure adapted from Weigel *et al.* (2009). (Courtesy of R. Waigel.)

subsequently grow to sizes of several micrometers in the condenser. The condenser unit is maintained either at a fixed temperature of, for example, $T_{\text{cond}} = 10\,^{\circ}\text{C}$, or at a fixed temperature difference to the saturator, for example, $\Delta T_{\text{sat-cond}} = 27\,^{\circ}\text{C}$, with both temperatures floating and only the temperature difference being controlled. Thereafter, the particles pass through an aerodynamic nozzle and traverse a laser beam. The light scattered by the enlarged particles is then detected by a photodiode, and thereby, the individual particles are counted. For a known flow rate, the aerosol number concentration can then be calculated.

In detail, the flow from the inlet is divided into two subflows. A smaller subflow that contains the particles that are to be counted is passed through a capillary into the center of the flow in the condenser, while the larger flow is cleared from all particles by using a filter and is then passed into the saturator. The saturator contains a heated reservoir (e.g., 37 °C) of the working fluid, where gas phase saturation is attained in the flow above the liquid. The flow then passes from the saturator into the condenser where it is cooled and the vapor of the working fluid becomes supersaturated with a precisely controlled saturation ratio at the center line of the condenser. For airborne CPCs working fluids such as perfluorotributylamine ($C_{12}F_{27}N$ or FC-43, 3 M, St. Paul, MN), 1-butanol ($C_4H_{10}O$), or ethylene glycol ($C_2H_6O_2$) are used. For the continuous flow CPC setup presented here, it is important that the diffusion of the working fluid in air is slower than the heat diffusion to allow for sufficient supersaturation at well-defined conditions in the condenser. Therefore, if water is to be used as the working fluid, the measurement principle has to be changed (Hering and Stolzenburg, 2005). Continuous flow CPCs can measure in a very broad dynamic range of concentrations from < 0.01 to $> 100\,000\,cm^{-3}$. Measurement precision is typically around $\pm 10\,\%$ for a 1 Hz sampling frequency.

Schröder and Ström (1997) and Hermann and Wiedensohler (2001) describe in detail the necessary modifications of a commercial CPC to be adapted for aircraft-borne operation. However, attention has to be paid to data interpretation during aircraft maneuvering as steep ascents or descents or narrow turns affect sampling efficiency or can potentially cause spurious counts. Usually these sequences should be excluded from the data analysis (Petzold, Kramer, and Schönlinner, 2002).

Historically, the first airborne CPCs were flown by Junge (1961) on a balloon, using a CPC setup based on an adiabatic expansion principle. Continuous flow CPCs for regular operation on research aircraft were developed by Wilson, Hyun, and Blackshear (1983). Their CPC design led to the development of the broadly used laminar flow ultrafine aerosol-CPC (Stolzenburg and McMurry, 1991).

4.2.2
Calibration of Cut-Off and Low-Pressure Detection Efficiency

When operating aircraft-borne CPCs, two issues need special attention: (i) the pressure-dependent calibration of the lower size cut-off and (ii) the pressure-dependent particle detection efficiency. Besides pressure, both depend on the exact geometry and flow conditions of the instrument, the temperature difference between the saturator and condenser, the working fluid as well as the chemical composition, surface structure, and shape of the activated particles (Ankilov *et al.* 2002a, 2002b; Brock *et al.* 2000; Hering *et al.* 1990; Hermann and Wiedensohler, 2001; Hermann *et al.* 2005a; Magnusson *et al.* 2003; McMurry, 2000; Mertes, Schroder, and Wiedensohler, 1995; Seifert *et al.* 2004; Weigel *et al.* 2009; Wiedensohler *et al.* 1997; Zhang and Liu, 1991).

Counting efficiency curves defines the fraction of particles that are detected by the particle counter as a function of particle size (Figure 4.3). Because of diffusion losses of particles in the instrument and other loss mechanisms, not all particles are counted. Additionally, small particles are only activated with a certain probability, as the curvature of the particle surface raises the equilibrium vapor pressure above the surface (the Kelvin effect). The lower particle size cut-off $D_{p,50}$ is defined as the particle diameter at which 50 % of the particles are activated. The asymptotic particle counting efficiency is given by the maximum particle counting efficiency

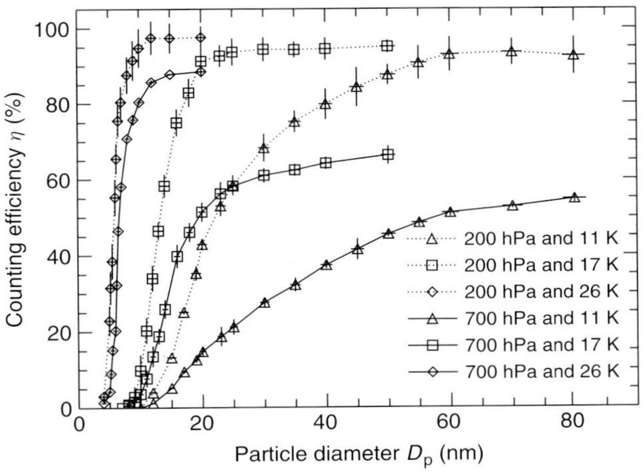

Figure 4.3 Efficiency curves for a TSI 7610 CPC operated with FC-43 as the working fluid at 200 and 700 hPa operating pressures and for temperature differences between the saturator and the condenser of 26, 17, and 11 K. Sulfuric acid particles are used as test aerosol. Lower size cut-offs $D_{p,50}$ of 5–6 nm, sharp cut-off curves, and high asymptotic counting efficiencies close to 100 % are reached for the largest temperature difference. (Source: Adapted from Hermann *et al.* (2005a). Copyright 2005 by Elsevier. Reproduced with permission by Elsevier.)

attained at sizes much larger than $D_{p,50}$. As one can see from Figure 4.3, the asymptotic particle counting efficiency changes with pressure.

Hermann *et al.* (2005a) shows for operation of a modified TSI model 7610 CPC (TSI Inc., St. Paul MN) that for low-pressure operation at or below 200 hPa using FC-43 as the working fluid leads to favorable particle detection efficiencies, while butanol is favorable for higher pressure operation. Figure 4.3 shows that a large temperature difference between the saturator and the condenser is favorable as it leads to a sharper cut-off curve with well-defined $D_{p,50}$ cut-off size and maximum asymptotic detection efficiency close to 100 %. Furthermore, because of the higher supersaturation reached in the condenser for larger temperature differences, the lower size cut-off shifts to smaller particle size.

In conclusion, it is necessary for aircraft-borne CPC operation to experimentally determine the cut-off size $D_{p,50}$ and the particle counting efficiency of the instrument as a function of pressure and working fluid and to correct the measured data accordingly. Data correction, however, requires an additional measurement of the particle size distribution. We also note that pressure-controlled inlets as such have been used for aerosol mass spectrometers might be effective in limiting measurement issues related to pressure changes and in maintaining a constant cut-point at different altitudes and during aircraft changes. The multichannel "CPC battery" developed by Brock *et al.* (2000) operates at a constant low-pressure downstream as critical orifice to maintain constant pressure conditions in the CPC units for the full flight envelope. Finally, careful instrument checks are required when commercial CPCs are put on pressurized aircraft. They usually do not have effectively sealed flow systems so that cabin air can leak into the instrument even if all other seals and fittings are tight.

4.3
Aerosol Particle Size Distribution

Since all methods presented in this chapter rely on different aerosol properties, the equivalent diameter approach is required for relating the results from the different methods to each other. A careful analysis of reported equivalent diameters is an essential prerequisite for the determination of reliable number size distributions of irregularly shaped particles by different methods. Optical methods measure the optical equivalent diameter referring to a sphere of given refractive index, which scatters the same amount of electromagnetic radiation into a given solid angle and which has the same cross-section as the irregularly shaped particle. Electrical mobility-based methods measure the electrical mobility diameter referring to a single-charged sphere of the same electrical mobility as the single-charged particle in question. Aerodynamic sizing methods measure the diameter of a sphere of unit density having the same terminal velocity in an accelerated airflow as the particle in question (Hinds, 1999). Geometric sizing methods such as scanning electron microscopy can measure properties such as the two-dimensional projection area, the surface-to-volume ratio (Sauter mean diameter), or the gyration diameter (Baron and Willeke, 2005).

4.3.1
Single-Particle Optical Spectrometers

The airborne, *in situ* instruments, mounted on the exterior of the aircraft (fuselage or wings) and that are presently used to measure size distributions of particles using optical techniques, are the ultrahigh sensitivity aerosol spectrometer (UHSAS), the passive cavity aerosol spectrometer probe (PCASP), the FSSP models 100 and 300, the cloud and aerosol spectrometer (CAS), and the cloud droplet probe (CDP). The size ranges of the UHSAS, PCASP, FSSP-300, FSSP-100, CAS, and CDP are 0.06–1.0 μm (UHSAS), 0.1–10 μm (PCASP), 0.3–20 μm (FSSP-300), 0.5–47 μm (range selectable, FSSP-100), 0.5–50 μm (CAS), and 2–50 μm (CDP), respectively. There are other optical particle counters (OPCs) that were developed for ground-based operation that have also been mounted in the cabin of aircraft, and sample air brought in with an external inlet system, but these will not be discussed here. As discussed in greater detail in Chapter 6, there is a number of issues with bringing aerosol particles into the cabin of the aircraft to sample them, for example, inertial losses to the walls of the inlet system. These types of issues may still be present but are minimized with exteriorly mounted probes.

As an example, Figure 4.4 shows a PCASP mounted on the wingtip pylon of a Hercules C-130. The UHSAS uses an identical inlet and canister. The FSSPs,

Figure 4.4 This photo shows a PCASP mounted on the wingtip pylon of the National Center for Atmospheric Research (NCAR) C-130.

CAS, and CDP are also used for measuring cloud particle size distributions and are discussed in greater detail in Chapter 5.

4.3.1.1 Measurement Principles and Implementation

Electro-optical spectrometers, for measuring the size of individual particles from aircraft, were introduced by Robert Knollenberg who developed the axial forward scattering spectrometer probe (ASSP) in 1972 and the FSSP in 1976, specifically for measuring cloud particles. At that same time, he also developed technology to measure aerosol particles using the active cavity technique whereby single particles scatter light within the resonant cavity of a gas laser (Schuster and Knollenberg, 1972). The airborne version, designated the active scattering aerosol spectrometer probe (ASASP), was flown regularly for research on aerosol particles until the early 1990s when it was replaced with the PCASP. The FSSP-100 was developed originally to measure cloud particles, but an aerosol version was developed in 1988 for measurements of stratospheric aerosol in the Arctic polar vortex (Baumgardner *et al.* 1992). Most recently, the UHSAS, originally developed for clean room measurements, was adapted for airborne research (Cai *et al.* 2008).

All of the instruments discussed here are single-particle, light-scattering instruments whose detection principles are based on the concept that the intensity of scattered light is a function of particle size and can be predicted theoretically if the shape and refractive index of a particle is known, as well as the wavelength of the incident light. The theoretical basis for this interaction is described in detail in Chapter 10: Supplementary Online Material, Section 10.1.4. However, the important aspect to understand is that the intensity of light scattered by a particle varies as the angle with respect to the incident illumination. This theory, known by its originator, Mie (1908), is applied in OPCs by collecting scattered light, over some range of angles that depend on the design of the instrument, from particles that pass through a light beam of controlled intensity and wavelength. This collected light is converted to an electrical signal whose amplitude is then associated with the size of the particle. We refer to this size as an optical diameter due to the manner in which it is derived.

The scattering cross-section is the property that defines how a particle interacts with the incident light and describes the relative intensity of scattered light within the solid angle defined by the collection optics of a particular instrument. The scattering cross-section is a function of the particle's physical cross-sectional area, shape, refractive index, and wavelength and polarization of the incident light. The UHSAS, PCASP, FSSPs, CAS, and CDP differ primarily in their optical configurations with respect to the angles over which the scattered light is collected (collection angles) and in their sampling sections.

The UHSAS and PCASP both collect a cone of side-scattered light, the UHSAS over a solid angle of 1.2 sr, or $90^\circ \pm 34^\circ$, and the PCASP over 2 sr, or $90^\circ \pm 60^\circ$. It has to be noted that recent detailed investigations of the UHSAS scattering geometry yielded scattering angles of 33.0–75.2° and 104.8–148.0°. The FSSP-100, FSSP-300, CAS, and CDP collect a forward scattered cone of light with a half-angle of approximately 4–12°. The UHSAS and PCASP can measure down to small diameters due to the implementation of optical systems that focus a high density of photons onto the particles.

As shown in Figure 4.5, the optical configuration of the UHSAS (Figure 4.5a), PCASP (Figure 4.5b), and FSSP-100/-300 (Figure 4.5c) are quite different. The UHSAS consists of a laser, a focused sample flow jet, and a high-efficiency detection system. Particles exit the sample jet, cross the laser principal intensity mode, and scatter light into the detection system. The laser is a high-quality factor, optical resonator built around an Nd^{3+}:YLF active laser crystal, pumped end-on by a semiconductor diode laser. The laser mirrors have reflectivities near 0.99999 at the lasing wavelength of 1053 nm. The imaging system is a pair of Mangin mirrors arranged to image the volume of space at which the flow intersects the laser mode. One Si avalanche photodiode (APD) is placed at the focus of the imaging system. The system size sensitivity is limited by a fundamental noise process: the photon shot noise on the detected molecular scatter from background gas.

In the PCASP, the method of particle input is also through an aerodynamically focused jet, which concentrates particle flow to a 150 μm diameter stream surrounded by a filtered and dried sheath flow. The sample stream is positioned at the focus of a 5 mm focal length parabolic mirror. The collected light is collimated by the parabolic mirror, reflected off a 45° flat mirror, and then refocused by an aspheric lens onto the APD. The laser source is a Gaussian mode, He–Ne gas laser operating at a wavelength of 632.8 nm. The high photon density is achieved with the external mirror that reflects the incident laser back through the sample volume, effectively illuminating the aerosol particles from the front and back.

Both FSSP-100 and FSSP-300 use the same optical system where the illumination of the sample volume is provided with a He–Ne gas laser that is focused to the center of the sample tube where aerosol particles pass unobstructed. The laser is projected onto a black, light-absorbing "dump spot" that also defines the inner

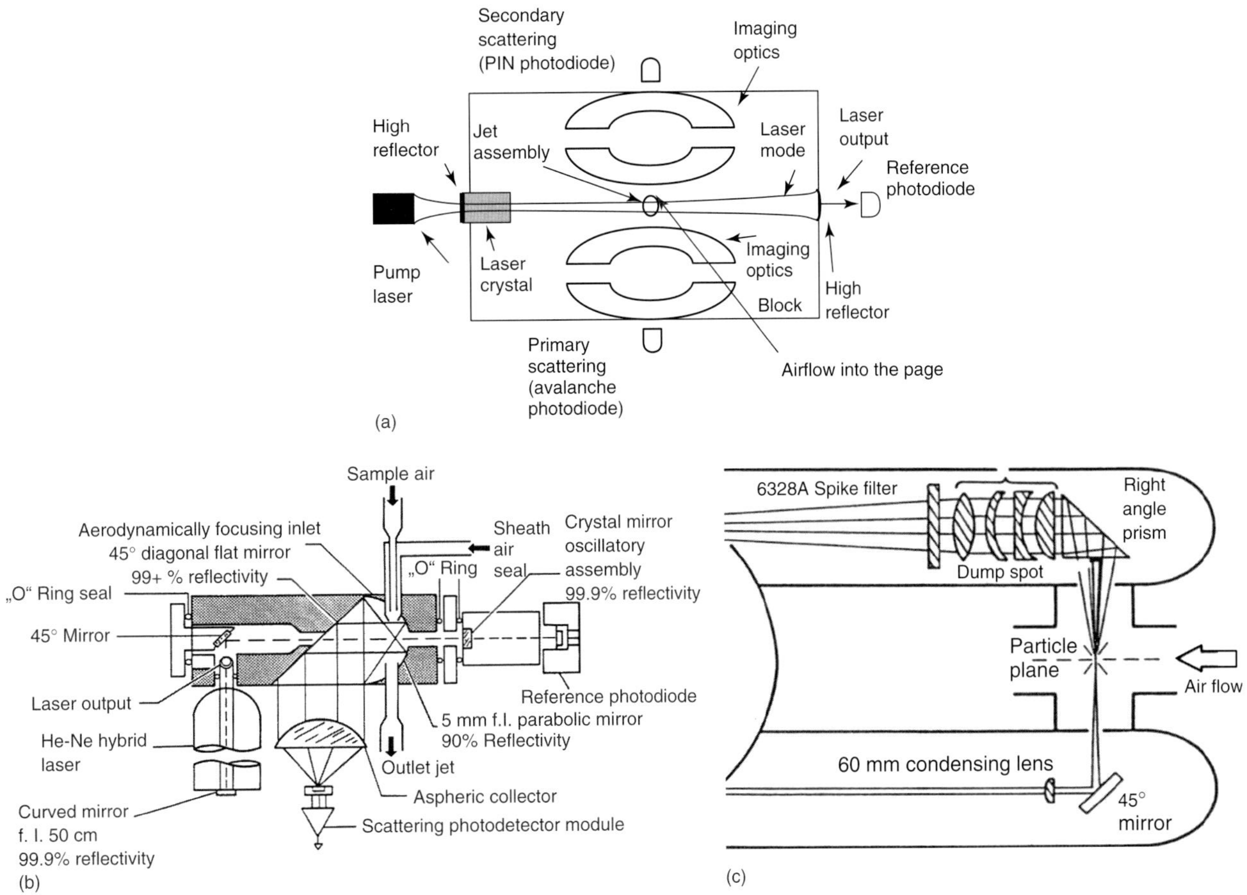

Figure 4.5 These three schematic diagrams show the optical configurations for the (a) UHSAS, (b) PCASP, and (c) FSSP-100/-300. Besides the differences in the optics, the FSSP-100/-300 does not use aerodynamic focusing and does not disturb the airflow to obtain the aerosol particles into the active sample volume. The figures are taken from the respective instrument manuals.

collection angle and the opening in the arm of the instrument restricts the forward scatter light to the outer collection angle. Additional information on the FSSPs is found in Chapter 5.

The particle size is derived from the measured intensity of scattered light using both application of the Mie theory and calibration with standard reference particles. The first step is to establish the relationship between the output of the photodetector onto which the scattered light is focused. This is done by measuring the reference particles, normally polystyrene latex (PSL) beads, that come with single sizes and a very small dispersion, normally less than 2%. The scattering cross-section for a specific size is calculated from the Mie theory for the laser wavelength, collection angle, particle diameter, and refractive index. Computer programs are readily available such as those that were developed by Bohren and Huffman (1983).

Once the scattering cross-section $C_{sca,p,c}$ is calculated for the reference particles, a calibration scale factor S_c is calculated based on the average peak voltage, V_c, measured from the detector when the reference particles are passed through the instrument. The scale factor is used to convert the peak voltages, V, measured for any particle, to a scattering cross-section, $C_{sca,p}$, that is, $C_{sca,p} = V \cdot S_c$ where $S_c = C_{sca,p,c}/V_c$. This new value of $C_{sca,p}$ is related to a particle diameter by using a lookup table between size and scattering cross-section for any assumed refractive index. An alternative approach is to calibrate the instruments using a (DMA) to produce monodispersed particles of known refractive index and size. Because of the upper size limit of DMA, this calibration can be done only up to about 1 μm.

4.3.1.2 **Measurement Issues**

The sizing accuracy of the single-particle light-scattering probes is limited by several factors: (i) the nature of how particles scatter light, (ii) the probability of multiple particles in the sample volume, and (iii) changes in particle size due to evaporation.

As previously discussed, the scattering cross-section of a particle is sensitive not only to the diameter but also to the refractive index. This has been investigated in detail by Garvey and Pinnick (1983); Jeung (1990); Kim and Boatman (1990); Pueschel *et al.* (1990); Liu *et al.* (1992); Kim (1995); Pinnick, Pendleton, and Videen (2000) who concluded that the manufacturer's calibration needs to be adjusted based on the composition of the particle. However, under normal, ambient conditions, the composition is not known *a prioris*. Hence, there will always be an uncertainty that is, on average, approximately ± 20% for diameters less than 1 μm.

Another source of uncertainty stems from particle asphericity such as for dust, soot, or biological aerosol particles whose refractive indices can vary over a wide range and whose shapes can be quite irregular. This uncertainty is difficult to quantify, although some studies have been done using particles of known shape (Pinnick and Auvermann, 1979; Pinnick and Rosen, 1979; Liu *et al.* 1992). For example, Liu *et al.* (1992), using NaCl compared to PSL, found that the lower threshold of the PCASP was 0.125 μm compared with the nominal value of 0.10 μm for PSL spheres.

The measurement of the OPCs is predicated on the assumption that only a single particle is in the sample volume at a time. Under high-concentration conditions, the probability increases that more than a single particle will be in the sample volume at the same time. This probability can be estimated by assuming that the particles are uniformly and randomly distributed so that Poisson statistics can be used to estimate the probability of more than a single particle in the sample volume. The UHSAS and the PCASP have approximately the same beam sample volume of $7 \times 10^{-6}\,\text{cm}^3$, which means that the concentration needs to exceed $1 \times 10^5\,\text{cm}^3$ before there would be a significant probability of two or more particles in the beam at the same time.

The sampling systems of the UHSAS and PCASP are anisokinetic, that is, the velocity of particles through the sampling volume is controlled with a pump at a speed of approximately $50\,\text{m s}^{-1}$, normally slower than the ambient airflow. As a result, the deceleration produces a heating of the airstream that is proportional to the square of the velocity (the Bernoulli effect) and volatile components on a particle may partially evaporate. This has been quantified by Strapp, Leaitch, and Liu (1992) who showed that the diameters of hydrated aerosol can be undersized by as much as a factor of two or more depending on the ambient humidity. In general, since the amount of heating and drying by anisokinetic sampling is a large uncertainty, the majority of users of these instruments choose to operate them with the inlet heaters activated to insure that any water is removed and the reported sizes are in dry diameter to remove this uncertainty.

Anisokinetic sampling can also lead to preferential size sorting, whereby larger particles will be more likely to be oversampled, that is, an increased concentration measured with respect to the ambient population. This measurement issue is discussed in greater detail in Chapter 5.

Figure 4.6 illustrates measurements made with a PCASP, FSSP-300, and FSSP-100 over the Mexican East Pacific. The size spectra are for two separate

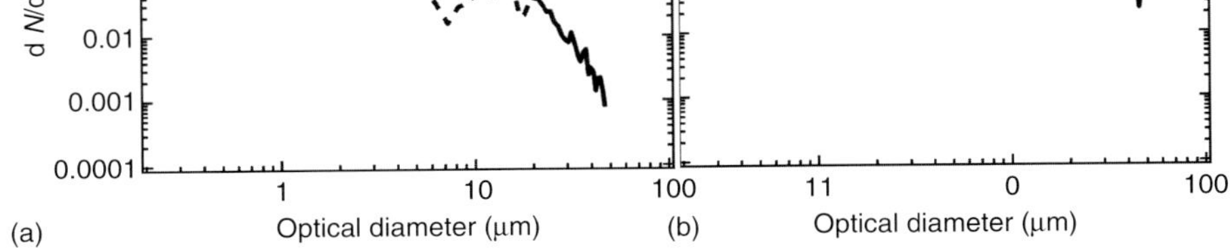

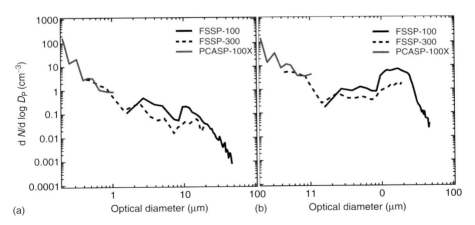

Figure 4.6 In this particle size distribution, measurements in the marine boundary layer are shown, taken with a PCASP, FSSP-300, and FSSP-100 over the Mexican East Pacific. The measurements were made on two different days under different wind conditions. (a) Research flight 9 was made under calm conditions and (b) research flight 13 under strong winds.

flights in the marine boundary layer, below 175 m. On one research flight, shown in Figure 4.6a, the winds near the surface were calm, with no white caps observed, whereas during the flight shown in Figure 4.6b, the winds were strong, more than 20 m s^{-1} at the altitude of the aircraft and large waves and white caps were observed on the ocean surface. These differences are reflected in the size distributions in the concentration of particles larger than an optical diameter of 5 μm, "giant nuclei." The strong wind conditions that generate breaking waves also loft the sea salt particles to the altitude of the aircraft.

4.3.2
Aerodynamic Separators

Measurement techniques based on aerodynamic separation of particles rely on the movement of particles in an accelerated airflow. This family of methods includes techniques such as multistage impactors and the aerodynamic particle sizer (APS). Although these methods are widely used in ground-based measurements (Baron and Willeke, 2005), their application on airborne platforms is not straightforward because the instrument response depends crucially on atmospheric state parameters such as temperature and pressure, which makes instrument calibration for realistic airborne applications a difficult task. Nevertheless, these methods have strong benefits in the field of size distribution measurements of super-micrometer-sized, irregularly shaped particles such as dust or volcanic ash. In particular, measuring particle size distributions of airborne dust is still a challenge.

Any particle moving with a speed U relative to the surrounding air is exposed to a drag force:

$$F_{\mathrm{D}} = \frac{3\pi \cdot \eta \cdot U \cdot D_{\mathrm{p}} \cdot \chi}{C_{\mathrm{c}}(Kn)} \tag{4.2}$$

with gas viscosity η. The dynamic shape factor χ considers the deviation from spherical shape, while the Cunningham factor $C_{\mathrm{c}}(Kn)$ corrects for the deviation from the continuum flow regime as the particle Knudsen number, $Kn = 2\lambda/D_{\mathrm{p}}$ (the ratio of the mean-free path of the gas molecules to the particle radius) becomes large. The slip correction factor is empirically derived, commonly fitted to a form:

$$C_{\mathrm{c}}(Kn) = 1 + Kn \cdot \left(\alpha + \beta \cdot \mathrm{e}^{-\gamma/Kn}\right) \tag{4.3}$$

where $\alpha = 1.142$, $\beta = 0.558$, and γ=0.999 (Allen and Raabe, 1985). Because of the pressure dependence of the mean-free path of gas molecules, the Knudsen number shows a distinct change with flight altitude, which has to be considered in the data analysis. For conditions close to ground, $C_{\mathrm{c}}(Kn)$ becomes relevant for particles <1 μm in diameter (Allen and Raabe, 1985; Hinds, 1999), while for lower pressures C_{c} is also relevant for super-micrometer-sized particles. While accurate only for spherical particles, C_{c} can be considered approximately correct for particles that are not grossly irregular. Values of the dynamic shape factor χ range from 1.0 for spheres to 1.8 for dustlike silicate particles (Reid *et al.* 2003).

If the particle is exposed to an external force F creating acceleration a_F, then a terminal velocity v_T will establish under equilibrium conditions with

$$v_T \cong \frac{\rho_p \cdot D_p^2 \cdot C_c(Kn)}{18\,\eta \cdot \chi} \cdot a_F \tag{4.4}$$

Aerodynamic sizing methods measure the aerodynamic equivalent diameter D_{ae}, that is, the diameter of a sphere of unit density ρ_0 (= $1000\,\mathrm{kg\,m^{-3}}$) having the same terminal velocity under an external aerodynamic force as the irregularly shaped particle of density ρ_p (Hinds, 1999). The APS in particular infers particle size by measuring the final drift velocity of particles inertially accelerated in an expanding air stream. For a constant external force, large particles undergo smaller acceleration and achieve lower velocity than small particles. Terminal particle drift velocities are measured and compared to calibration curves. This technique works for an aerodynamic equivalent diameter $D_{ae} > 0.9\,\mu\mathrm{m}$.

The applicability of an APS instrument for airborne measurements is carefully analyzed in Wang, Flagan, and Seinfeld (2002b). Equation (4.4) is valid only in the Stokes regime with low particle Reynolds numbers (Re < 0.05). While for Stokesian conditions, the particle drift velocity is a function of the aerodynamic size only (Hinds, 1999), under ultra-Stokesian conditions (Re > 0.05), the particle speed after acceleration is a function of particle aerodynamic size as well as of the density of the particle. Wang, Flagan, and Seinfeld (2002b) propose an ultra-Stokesian correction factor that includes the particle Reynolds number.

Airborne application of APS is not straightforward for mainly two reasons. The major source of uncertainty in APS-derived particle size spectra is the applicability of laboratory calibration to airborne measurements. For most cases, laboratory calibration is performed using PSL particles with known diameters. Applying this calibration to data from airborne measurements requires the evaluation of the slip correction factor and the Reynolds number at the APS sensing volume, which has a lower pressure and temperature than the ambient condition as a result of the expansion of the air stream. This evaluation has to be performed for the entire flight conditions. Knowing these conditions, the APS measurements can be analyzed using the ultra-Stokesian correction factor proposed by Wang, Flagan, and Seinfeld (2002b). The second major source of uncertainty is the transmission of super-micrometer-sized particles through an inlet to the instrument. The commercial APS instruments (e.g., TSI model 3320) have been engineered for laboratory measurement or at least for ground-based field measurement. For both applications, an isokinetic flow of the sample into the instrument without bends in the sampling line is easy to achieve. For an airborne application, special inlets are required. Furthermore, isokinetic conditions have to be achieved for all operation conditions, that is, for different ambient pressures and different TAS of the aircraft. Therefore, a careful analysis of the inlet transmission efficiency as a function of particle size and measurement conditions (TAS, atmospheric temperature and pressure) is required. One example is given by Wang, Flagan, and Seinfeld (2002b).

4.3.3
Electrical Mobility Measurements of Particle Size Distributions

The size distributions of particles smaller than 1 μm diameter are generally measured using the electrophoretic migration of charged particles in an electric field, commonly called *electrical mobility analysis*. This measurement determines the relationship between the aerodynamic drag on the particle and the applied electrostatic force. After a very short transient period (e.g., less than 10^{-5} s for sub-micrometer particles), the migration velocity approaches a steady state that is described by a force balance on the particle:

$$n_e \cdot e \cdot E = \frac{U_{mig}}{B} \tag{4.5}$$

where $B = C_c(Kn)/(3\pi \cdot \eta \cdot D_p)$ is the mechanical mobility of the particle, that is, the steady-state migration velocity, U_{mig}, per unit applied force. Here, η is the viscosity of the gas, D_p is the particle diameter, and $C_c(Kn)$ is the slip correction factor that accounts for noncontinuum effects as particles become of similar size as the mean-free path of gas molecules. Mobility analysis is generally focused on singly charged particles such that the net charge on the particle is $q = n_e \cdot e = \pm e$, where e is the elementary unit of charge; particles are, therefore, classified in terms of their electrical mobility:

$$Z = \frac{U_{mig}}{E} \tag{4.6}$$

Mobility analysis requires knowledge of, or assumptions about, the charge on the particles and the particle shape. When the particle shape is not known, particle size inferred from mobility analysis is reported as the mobility equivalent diameter that corresponds to the diameter of a singly charged spherical particle that has the same electrical mobility as the particle being measured; this sizing method characterizes singly charged particles in terms of their aerodynamic drag that, with suitable assumptions, can be used to deduce the particle size.

By making measurements of particle concentrations over a range of mobilities, the particle size distribution can be deduced, provided one knows the probability distribution for the number of charges carried by particles as a function of particle size. A highly stable charge distribution is generally achieved by exposing the aerosol to an electrically neutral cloud of positive and negative gas ions. The gas ions are generally produced by ionizing radiation from a low-level, shielded radioactive source in a device that is called a *bipolar diffusion charger* or, slightly confusingly, an *aerosol neutralizer*, or by a corona discharger or an X-ray neutralizer. The latter methods are of advantage for operation on an aircraft since using radioactive sources on an aircraft is subject to very strict regulations. Gas ions exchange charge with aerosol particles. After some time, a steady-state charge distribution is achieved that is approximately neutral, accounting for the latter name for the device.

Because the positive and negative gas ions consist of different chemical species, with different diffusivities and electrical mobilities, the final charge state of the aerosol is not perfectly neutral; instead, the rates of adding positive and negative charges are balanced in the steady-state charge distribution, which are often slightly negative due to higher diffusivity of negative ions. Moreover, the charge distribution is not the thermodynamic equilibrium state that is commonly referred to as the *Boltzmann equilibrium distribution*. To produce an equilibrium charge distribution, the charge transfer "reaction" would have to be fully reversible. At temperatures of atmospheric mobility analysis, charge desorption from an aerosol particle is an extremely unlikely event, and so the aerosol charge state is produced by reactions in which gas ions transfer charge to the particles; particle charge neutralization occurs only by transfer of charges of opposite polarity to the particle.

The steady-state charge distribution in bipolar diffusion charging can be predicted using a theory developed by Fuchs (1963) that accounts for the differences in molecular properties of the positive and negative ions and for noncontinuum effects encountered in fine particle measurements. While a detailed discussion of that theory is beyond the scope of this book, Wiedensohler (1988) corrected in a subsequent erratum a parametric fit, given in Table 4.2, that can be used to determine size distributions from mobility analyzer data.

Aerosol particle mobility is measured in a laminar flow apparatus in which charged particles that enter near one electrode migrate across a particle-free gas flow to a sample extraction point at a downstream location in an opposing electrode. In most DMAs, the electrodes are coaxial cylinders as illustrated in Figure 4.7. The aerosol is introduced in an azimuthal slot in the outer electrode of this cylindrical DMA (CDMA), while the classified aerosol is extracted through a slot near the opposite end of the inner electrode, although migration from the inner electrode toward the outer one has recently been shown to enable slightly higher resolution (Alonso, 2002). Mobility classification is also performed using parallel disk electrodes by introducing the aerosol through an annular slit in one of the electrodes, and inducing migration toward the opposite electrode as the aerosol and sheath flows carry the particles toward classified sample and exhaust ports at the centers of the two electrodes (Brunelli, Flagan, and Giapis, 2009; Zhang *et al.* 1995). DMAs with this geometry are referred to as *radial DMA* (RDMA). The operating principles of the two designs are the same, and the performances of ideal instruments differ only in geometry factors that are of order unity. Assuming ideal performance, the mobility of the particles that are transmitted through the different DMA designs is given by

$$
\begin{aligned}
Z^*_{\mathrm{CDMA}} &= \frac{(Q_{\mathrm{sh}} + Q_{\mathrm{e}})}{4\pi \cdot L \cdot \Phi} \cdot \ln\left(\frac{R_2}{R_1}\right) \\
Z^*_{\mathrm{RDMA}} &= \frac{(Q_{\mathrm{sh}} + Q_{\mathrm{e}}) \cdot b}{2\pi \cdot (R_2^2 - R_1^2) \cdot \Phi}
\end{aligned}
\tag{4.7}
$$

where Φ is the voltage difference between the two electrodes; R_1 and R_2 are the outer and inner radii of the classifier, respectively; Q_{sh} and Q_{e} are the volumetric flow rates of the sheath and excess flows, respectively; L is the length of the CDMA; and b is the gap between the electrodes of the RDMA. If the aerosol inflow and

Table 4.2 Coefficients $a_i(N)$ for the parametric fit proposed by Wiedensohler (1988).

	n_e				
$a_i(n_e)$	−2	−1	0	1	2
a_0	−26.3328	−2.3197	−0.0003	−2.3484	−44.4756
a_1	35.9044	0.6175	−0.1014	0.6044	79.3772
a_2	−21.4608	0.6201	0.3073	0.4800	−62.8900
a_3	7.0867	−0.1105	−0.3372	0.0013	26.4492
a_4	−1.3088	−0.1260	0.1023	−0.1553	−5.7480
a_5	0.1051	0.0297	−0.0105	0.0320	0.5049

Here $f(D_p, n_e)$ is the fraction of particles with diameter D_p carrying n_e elementary charges; the approximation formula is $f(N) = 10^{\left[\sum_{i=0}^{5} a_i(N)\times(\log D_p/\mathrm{nm})^i\right]}$.

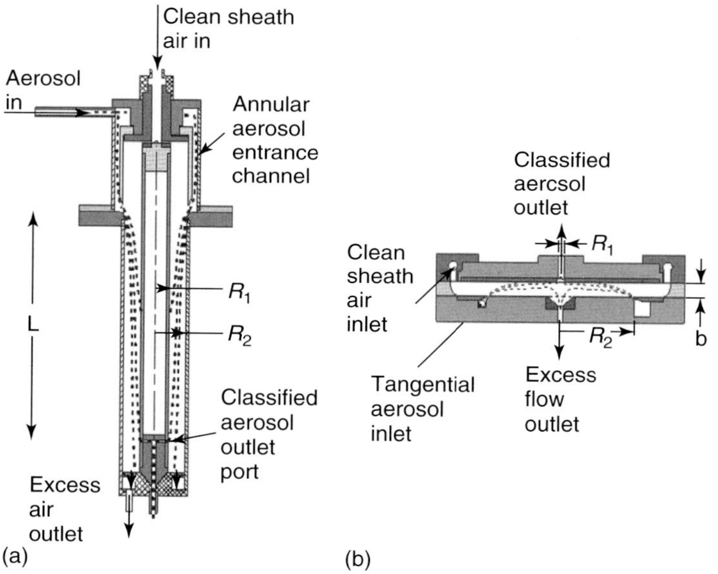

Figure 4.7 Schematic representations of (a) a cylindrical differential mobility analyzer (CDMA) and (b) a radial differential mobility analyzer (RDMA). (Courtesy of R. Flagan.)

classified aerosol outflow rates, Q_a and Q_c, are equal (balanced flow operation of the DMA), and if diffusion can be neglected, the probability that a particle of mobility Z will be transmitted through the DMA falls linearly from unity at Z^* to zero at $Z = Z^* \pm \beta \cdot Z^*$ where $\beta = (Q_a + Q_c)/(Q_{sh} + Q_e)$; DMAs are generally operated at $\beta \sim O(10)$, although lower values may be used to increase particle throughput, and larger values may be used to increase mobility resolution. The resolving power of

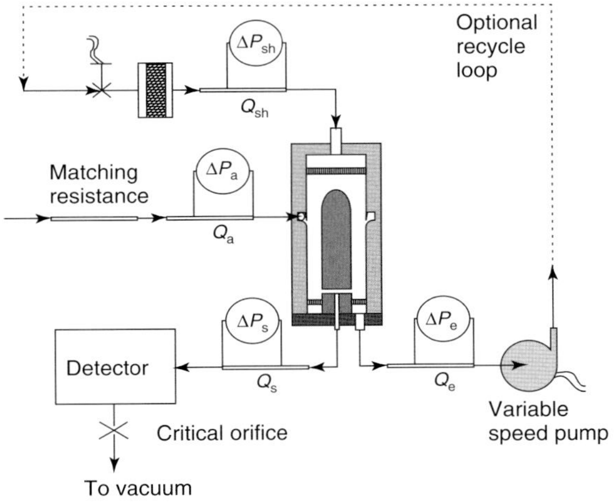

Figure 4.8 Schematic diagram of flow control system for DMA operation. The system shown includes laminar flow metering of all four DMA flows. Using a variable speed pump, the excess flow can either be discharged or recycled as indicated by the dashed lines. (Courtesy of R. Flagan.)

the DMA is generally expressed as the ratio of the mobility of the particles that are transmitted with the highest efficiency to the full width at half maximum of the transmission probability function, that is, $R = Z^*/\Delta Z_{1/2}$. In this kinematic limit, $R_K = \beta^{-1}$.

More generally, diffusion causes particles to migrate from their kinematic trajectories, reducing the resolving power of the DMA. At sufficiently high voltage, diffusion is unimportant, while at low voltage, diffusion dominates the classification process. Stolzenburg (1988) derived a semianalytical expression for the probability of transmission, also known as the *DMA transfer function*. Details are given there.

Differential mobility analysis requires precision control of the volumetric flow rates of the aerosol and sheath air entering the DMA and of the classified aerosol and exhaust flows exiting the DMA. There is little pressure drop within the DMA, but a number of flow elements impact the distribution of air between the different pathways in the flow networks on the inlet and outlet sides of the DMA. While simple manual valves have been used in many ground-based instruments, airborne measurements pose special challenges as the pressure in the aircraft varies in flight. For airborne measurements, the flows are driven by the pressure difference between the inlet to the DMA and pumps that are installed at the outlet of the instrument. Figure 4.8 shows the plumbing elements in a typical measurement scenario. The sheath air passes through a filter, flow control valve, and flow metering device, each of which introduces a pressure drop. The aerosol flow must experience a matching pressure drop without introducing flow elements that would cause excessive particle losses. Again, a flow metering element is required, but an additional pressure drop element is required to achieve the same pressure drop as in the sheath flow.

Since the mobility is selected within the DMA in terms of the volumetric flow rate, it is convenient to use laminar flow elements for metering the flows; the pressure drop in fully developed flow through a tubular laminar flow element is

$$\Delta p = -\frac{128\,\eta \cdot Q \cdot L_{\text{tube}}}{\pi\, d_{\text{tube}}^4} \tag{4.8}$$

where L_{tube} and d_{tube} are the length and diameter of the tube. The flow must pass through an entrance length, $L_{\text{entrance}} = 0.05\,\text{Re} \cdot d_{\text{tube}}$, before a fully developed velocity profile is established within the tube. $\text{Re} = \rho \cdot V \cdot d_{\text{tube}}/\eta$ is the Reynolds number of the tube flow (ρ is the fluid density). Metering elements may, therefore, include an entrance length that is often installed upstream of the actual metering pressure drop region to ensure a linear pressure drop/flow rate relation. Additional pressure drop elements may be added in series with the metering element to match impedances between the aerosol and sheath flows. One concern that must be addressed is particle losses within the passages that carry aerosol, especially metering elements that employ small diameters to achieve measurable pressure drop within reasonable lengths. The penetration of particles through laminar tube flow can be estimated using the solution to the Graetz problem (Friedlander, 1977) wherein

$$\frac{N_{\text{out}}}{N_{\text{in}}} = 8 \sum_{j=0}^{\infty} \frac{G_j}{\lambda_j} \cdot \exp\left[-\lambda_j^2 \cdot \left(\frac{2\,L_{\text{tube}}}{d_{\text{tube}}} \cdot \frac{D}{V}\right)\right] \tag{4.9}$$

where D is the particle diffusivity, $\lambda_0 = 7.312$, $\lambda_1 = 44.62$, $\lambda_2 = 113.8$ and for $j > 2$, $\lambda_j = 4j + 8/3$, and $G_0 = 0.749$, $G_1 = 0.544$, $G_2 = 0.463$, and for $j > 2$, $G_j = 1.01276\,\lambda_j^{1/3}$. This model can be applied to flows through other tubing within the instrumental system as well, although it must be recognized that bends and elbows induce secondary flows that significantly increase particle losses (Wang, Flagan, and Seinfeld, 2002b).

Size distribution measurements are performed by measuring the concentrations of particles transmitted through the DMA in a series of mobility intervals that span the range of mobilities of interest. Early DMA measurements employed the stepping mode in which the voltage is changed in discrete steps. Commercially, the combination of a stepping mode DMA and a CPC detector was known as the *differential mobility particle sizer* (DMPS). Because sufficient time must be allowed between measurements to ensure that the particles in each counting period have experienced a constant electric field throughout their migration time, DMPS measurements are slow, and particles are measured for only a small fraction of the time. Continuous particle measurements were enabled by the introduction of the scanning mode of operation (Wang and Flagan, 1990) in which the voltage is scanned continuously in an exponential ramp; particles are counted into consecutive time bins that can be related to mobility intervals. The scanning mode is labeled the *scanning electrical mobility spectrometer* (SEMS), *scanning mobility spectrometer* or *scanning mobility particle sizer* (SMPS). The latter name should not be confused with the sequential mobility particle sizer plus counter (or electrometer) from Grimm (Grimm Aerosol Technik, Germany), which is abbreviated as SMPS+C or SMPC+E. The latter instruments operate in a fast-stepping mode in which measurements are made too quickly to establish the steady state of the DMPS measurements. The mobilities of the particles that exit the DMA during a particular time interval are determined by the mean electric field strength during the migration time. Scanning-mode operation eliminates the wait between measurements and ensures that all particles with mobility within the scan range are counted, thereby accelerating the measurements significantly; however, DMPS measurements take many minutes to span the mobility range, SEMS/SMPS measurements can be performed in a few minutes or less. Using fast-response detectors, SEMS measurements have been performed in as little as 1 s.

Response time is a critical issue particularly for electrical mobility-based instruments operated on a fast moving aircraft. Considering, for example, an aircraft cruising with a speed of $100\,m\,s^{-1}$, a scan time, or stepping time, respectively, of 10 s results in a particle size distribution averaged over a distance of 1 km. This reduced lateral resolution may cause conflicts when small-scale phenomena are targeted.

Ideally, the SEMS/SMPS should give the same size distribution data as in DMPS measurements. Unfortunately, particles do not pass uniformly through CPC detectors due to mixing within the internal flow passages in these devices. This leads to a residence time distribution that delays particle detection by as much as several minutes. This mixing effect has been described by modeling the CPC as a continuously stirred tank reactor (Collins, Flagan, and Seinfeld, 2002; Russell, Flagan, and Seinfeld, 1995). Particles are, therefore, delayed to different degrees in their arrival at the detection point due to the mixing. A number of data inversion algorithms (Section 4.3.4) have been developed for tackling this problem, including nonnegative least squares matrix inversion (Lawson and Hanson, 1974) and one widely used simple approach proposed by Twomey (1975); see also Markowski (1987) and Collins, Flagan, and Seinfeld (2002).

4.3.4 Inversion Methods

Most methods for measuring the particle size distribution onboard aircraft use an indirect measurement principle. A DMA, for example, actually measures the particle number concentration as a function of electrical particle mobility Z. Particle electrical mobility is subsequently related to the geometric particle diameter D_p to obtain the particle size distribution (Section 4.3.3). The relationship between Z and D_p is not unique. Since the particles counted in the DMA may have one, two, or even three elementary charges, three different size classes of particles are counted at a given mobility Z, each with different probability.

Similarly, an OPC measures the particle number concentration as a function of particle scattering cross-section $C_{sca,p}$. $C_{sca,p}$, as a function of D_p, is specific for the probe geometry that determines the angular collection efficiency (CE) of light scattered by each particle passing through the spectrometer. To obtain the particle size distribution, $C_{\mathrm{sca,p}}$ needs to be related to D_p. Especially for probe geometries that focus on light scattered into the forward direction (as seen from the scattering particle) this relation is not unique; see Figure 4.9 for illustration.

Additionally, the instrument transfer functions or CEs usually deviate from the intended ideal. This is illustrated in Figure 4.10, which gives examples for CPCs, DMA, and OPC. In the DMA transfer function, that is, the probability a particle of diameter D_p has to pass the DMA and be counted, it is obvious that the transfer function is not a sharp peak, but has a certain width $\Delta D_\mathrm{p}(\Delta Z)$ for a given $D_\mathrm{p}(Z)$. In addition, the transfer function is asymmetric due to diffusion of smaller particles. In a CPC battery, several CPCs sampling the same aerosol are operated in parallel with their lower 50% counting efficiency at different particle diameters. The size information lies in the concentration difference measured by the CPCs that corresponds to the difference of transfer functions between the CPCs (Figure 4.10). It is obvious that it is not straightforward to attribute a diameter to such a class of particles.

Mathematically, the measurement process may be described by a Fredholm integral equation:

$$\int f^i(D'_\mathrm{p}) \cdot \frac{\mathrm{d}N(D'_\mathrm{p})}{\mathrm{d}\log D'_\mathrm{p}} \mathrm{d}\log D'_\mathrm{p} = b^i \tag{4.10}$$

where $f^i(D'_\mathrm{p})$ denotes the instrument CE or transfer function (known from the calibration), $\mathrm{d}N(D'_\mathrm{p})/\mathrm{d}\log D'_\mathrm{p}$ the particle size distribution (to be determined), and the particle concentration (measured). For describing a whole instrument,

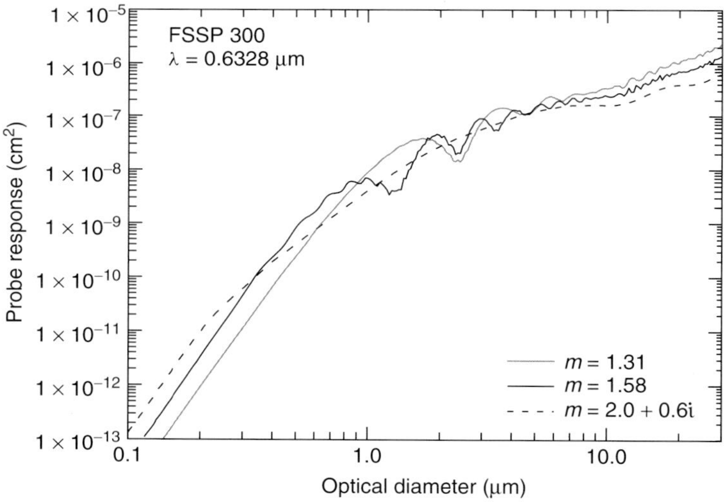

Figure 4.9 Scattering cross-section of particles seen by an FSSP-300 as a function of particle diameter. The threshold values between size bins in the instrument are marked with horizontal lines. Especially for particle diameters between 1 and 4 μm, it is apparent that the threshold values cannot be uniquely related to a particle diameter.

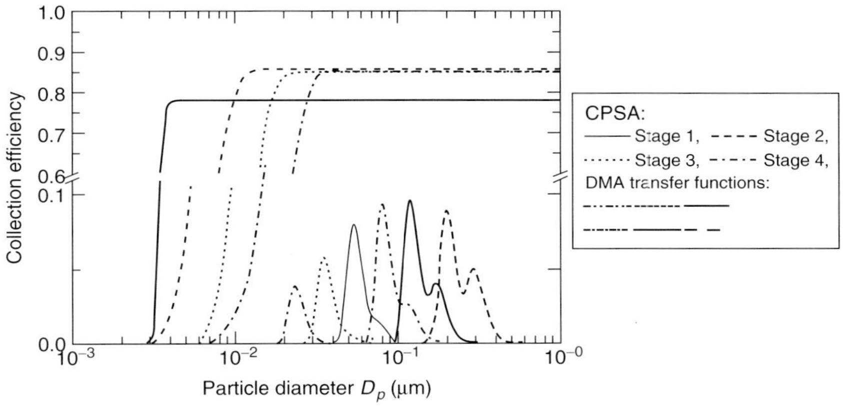

Figure 4.10 Examples of instrument transfer functions or CEs, that is, probability that a particle is registered as a function of particle size. The graph illustrates that a transfer function can result in a nonunique measurement or that the size attribution to a measurement can be diffused, requiring data analysis by inversion; adapted from Fiebig *et al.* (2005). (Source: Copyright 2005 by Elsevier. Reproduced with permission by Elsevier.)

one such equation is necessary for each DMA size channel, each optical particle spectrometer size bin, or each CPC of a CPC battery. This results in a set of m equations of type Eq. (4.10) that are numbered by the index i. The inversion process amounts to solving this equation system for the particle size distribution $\mathrm{d}N(D_\mathrm{p})/\mathrm{d}\log D_\mathrm{p}$.

To proceed with this task numerically, the integrals are usually discretized and converted into a sum:

$$\sum_j F_j^i \cdot x_j = b^i \tag{4.11}$$

where F_j^i denotes the discretized transfer or CE functions, and x_i the discretized particle size distribution. The discretization needs to be fine enough to resolve the relevant details of the transfer or CE functions. When the particle size distribution covers an extended size range, the spacing of the discretization is often logarithmic. With this step, all equations representing the individual DMA size channels, OPC size bins, or CPC battery CPCs can be assembled into one linear equation system:

$$F_j^i \cdot x_j = b^i \tag{4.12}$$

The origin of the term inversion now becomes obvious. With the transfer matrix or CE functions F_j^i and the vector of measured particle concentrations b_i known, solving Eq. (4.12) for the unknown particle size distribution amounts to finding the inverse of F_j^i. This is challenging due to two properties of F_j^i: (i) F_j^i is underdetermined, and (ii) the inversion is ill posed.

(i) The number of discretized particle size steps necessary to resolve the transfer or CE functions is normally larger than the available number of size bins or channels, that is, F_j^i has more columns than rows. The result to be expected can, therefore, only be the pseudo-inverse of F_j^i. This problem is usually solved by introducing an additional boundary condition, for example, that the resulting discretized particle size distribution x_j has to be smooth, that is, the absolute value of its second derivative minimized.

(ii) The column or row vectors of F_j^i are close to being linearly dependent. Graphically described, this is caused by the overlap of transfer or CE functions, for example, caused by the presence of multiply charged particles in a DMA.

Numerous algorithms for solving this type of problem have been described over the past decades, most of them optimized for a specific type of instrument. The algorithm of Twomey (1975) as modified by Markowski (1987) works iteratively, starting out with an initial guess of the solution. While being rather stable, the use of an initial guess always leaves a doubt as to whether the solution is independent of the initial guess. The algorithms of Stratmann *et al.* (1995) and Voutilainen, Kolehmainen, and Kaipio (2001) were specifically developed for inversion of DMA data. The algorithm of Fiebig *et al.* (2005) focuses on combining measurements of different instrument types, each optimized for a different size region, into one consistent particle size distribution, including propagation of the measurement uncertainty to the result.

4.4 Chemical Composition of Aerosol Particles

Chemical composition is a fundamental property of atmospheric aerosol particles ruling their climatic and environmental impacts. The chemical composition of aerosol particles depends on the process from which particles originate and varies with time after emission owing to deposition, cloud processing, mixing, and aging.

Aerosol particles contain organic and inorganic material. Organic matter originates from primary emission of carbonaceous particles not only from combustion (fossil fuel, domestic, and fire burning of vegetation and agricultural waste) and from emission of viruses, bacteria, fungal spores, and plant debris, but also from secondary conversion of volatile organic compounds (VOCs) from natural and anthropogenic sources. Inorganic matter originates not only from primary emission of soil crust species from aeolian erosion, sea salt but also from secondary conversion of reactive gaseous species as in the case of nonsea salt sulfate and nitrates.

The aerosol composition depends on particle size. Carbon (elemental and organic) and some secondary ionic species (sulfates and ammonium) tend to be found more on the fine fraction of particles smaller than 1 μm diameter, whereas wind-friction-produced particles (soil dust and sea salt), and also pollen and vegetation debris, are more concentrated in the coarse fraction of particles whose diameter is larger than 1 μm diameter. Some species (e.g., nitrates) are found in equivalent proportions in the fine and coarse fractions. Bacteria split these modes depending on desiccation. Viruses are clearly far below 1 μm; see Figure 4.1 for a summary.

Nonetheless, composition is a generic word. As a matter of fact, various types of composition (elemental, mineralogical, water soluble, isotopic) can be investigated according to the objectives of the respective study. To do so, different analytical techniques are employed, to which the sampling material and methodology need to be adapted.

Historically, composition has been studied by offline direct methods, coupling collection on a support medium and subsequent analysis. In this case, information on the bulk composition is obtained by filtration on a single-stage filter, whereas size-segregated composition can be studied by sampling on stack filter units or multistage impactors that select particles according to their aerodynamic size. Depending on the type of analysis, the information on composition will be related to the mass or to the number size distribution of the aerosol and to the geometric or aerodynamic diameter of the particle. In either case, the filtration/impaction support, filter or membrane, used for sampling has to be adapted to the consequent analytical technique. Offline direct methods are discussed in Section 4.4.1.

Because of analytical detection limits (by mass) and the counting statistics (by number), offline sampling requires minimum sample exposures of several minutes depending on atmospheric concentrations. At common aircraft speed ($50–200\,m\,s^{-1}$), the spatial resolution is low and often incompatible with the necessity of characterizing localized or low-concentration aerosol plumes or in pristine conditions. To achieve higher temporal resolution, the development of online methods based on mass spectrometry and particle solubility in water has started, and those methods are now readily available on research aircraft. Even online methods may have statistical sampling issues and may require a few minutes of data integration depending in the detection limit of the method and the concentration level of the investigated air sample. These online methods are described in Section 4.4.2.

Finally, indirect information on the aerosol composition can be achieved by measurements of extensive quantities such as the optical properties of scattering and absorption, or volatility. Because these quantities also depend on the concentration, size, and shape of the particles, concurrent measurements are needed to derive information on composition. These techniques are described in Section 4.4.3.

4.4.1
Direct Offline Methods

Direct offline methods consist of collecting aerosol samples on the aircraft and in analyzing them in the laboratory after the flight. General concerns are the need of sampling the whole range of tropospheric concentrations and to represent their spatial variability. Atmospheric concentrations vary from a few micrograms per cubic meter in pristine air (Quinn and Coffman, 1998) to thousands of micrograms per cubic meter in mineral dust storms (Marticorena *et al.* 2010). The scale of horizontal and vertical variability of aerosol layers ranges from tens of meters to thousands of kilometers (Abel *et al.* 2003; Esselborn *et al.* 2009).

As a consequence, particular care must be taken in pretreating and cleaning the collection media before exposure and in manipulating and storing the samples prior, during, and after the exposure, to avoid ambient contamination (losses of volatile compounds, adsorption of water vapor or gases). In addition, the exposure time must be kept to a minimum with respect to the detection limit of the analytical technique.

Aerosol sampling via particle collection is performed on straight and leveled runs at constant altitude. High-volume sampling (flow rate of $50\,l\,min^{-1}$ or higher) for bulk chemical analysis of the aerosol composition is used to limit the sample exposure and increase the spatial and temporal representativeness. As an example, this will result in a 20–30 min time resolution for elemental analysis by reference techniques such as X-ray fluorescence (XRF) or particle-induced X-ray emission (PIXE) to approximately one hour for analysis of the organic and elemental carbon. At the average flying speed of most tropospheric research aircraft ($\sim$100 m s^{-1}), the horizontal resolution ranges from approximately 100 to 360 km. Typical detection limits for some aerosol tracers as detected by some of most common analytical techniques, and the corresponding exposure times, volumetric flow rate are provided in Table 4.3.

Bulk aerosol sampling devices consist generally of in-house systems specially developed for aircraft sampling. Air enters the aircraft cabin via an inlet designed to reduce particle losses or at least whose transmission efficiencies as a function of particle size should be characterized (Chapter 6). The aerosol intake system has to be designed so that rain and large cloud water droplets are removed from the sampled air stream by inertial separation. A diffuser is then used to diverge and slow down the air flow. This is needed to ensure that (i) the particle deposition pattern is radially uniform on the collecting filter in case different subportions have to be analyzed by different techniques and (ii) to avoid particle break down on impaction with the filter. An unavoidable artifact of the deceleration of the airstream will be the evaporation of semivolatile compounds by ram heating, which is generally followed by additional heating as the air mass is transported to the aircraft cabin. These evaporation losses are mostly relevant to sampling from high-altitude, high-speed aircraft (Mach numbers $\sim$0.7), when the temperature differences between the outside and the cabin air might reach 50 K (Wilson *et al.* 1992; Hermann, Stratmann, and Wiedensohler, 1999). An example of a bulk filter sampling system is shown in Figure 4.11.

An example of special samplers is the Particle Concentrator–Brigham Young University Organic Sampling System (PC-BOSS) onboard the C-130 of NCAR, which has been designed for the quantification of organic aerosol particles (Eatough *et al.* 1999; Lewtas *et al.* 2001; Kawamura *et al.* 2003). This system is composed of a two-way sampler, one consisting in a "sideall" filter sampling undenuded air upstream of the sampler on a quartz filter for the study of organic gases such as dicarboxylic acids. Preconditioned (baked) quartz filters are used to sample organic aerosol particles in the main stream of the PC-BOSS sampler, in which a denuder

Table 4.3 Detection limits for some aerosol tracers as detected by some of most common analytical techniques.

Tracer	Analytical method	Range of atmospheric concentrations ($ng\,m^{-3}$ STP)	Range of quantification ($ng\,m^{-3}$ STP; flow rate $50\,l\,min^{-1}$; exposure time 30 min)
SO_4^{2-}	Ion chromatography (IC)	1 000-9 000[a]	>510[b]
NO_3^-	Ion chromatography (IC)	140–2 500[a]	>510[b]
Fe	X-ray fluorescence (XRF)	0.6 — 13 800[c]	>20[b]
Organic carbon	Thermal/optical reflectance (TOR)	500–9 900[a]	>2 240[b]
Elemental carbon	Thermal/optical reflectance (TOR)	10–2 300[a]	>2 240[b]
Levoglucosan	Gas chromatography–mass spectrometry (GC–MS)	1 200–40 000[d]	>8[c, e]

[a]Seinfeld and Pandis (1998).
[b]Assume detection limits reported in Chow (1995).
[c]Assume detection limits reported in Schkolnik and Rudich (2006).
[d]Mayol–Bracero *et al.* (2002); Simoneit *et al.* (2004); Puxbaum *et al.* (2007).
[e]Graham *et al.* (2002); Bin Abas *et al.* (2004).

has removed VOCs. The schematic diagram of this PC-BOSS sampler is shown in Figure 4.12.

Single or multijet inertial impactors are widely used for the size-selective collection of aerosol particles, for mass size distribution, or single-particle analysis purposes. In this case, a jet or an ensemble of jets of particle-laden air is directed at a flat impaction plate. Large particles are collected on the plate, while smaller particles follow the airflow around the impaction region. The primary parameter that governs the collection of particles in an inertial impactor is the Stokes number Stk representing the ratio of the particle stopping distance to the characteristic length of the sampling system, and is defined as

$$\mathrm{Stk} = \frac{\rho_\mathrm{p} \cdot C_\mathrm{c} \cdot U_0 \cdot D_\mathrm{ae}^2}{9\,\eta \cdot W} \tag{4.13}$$

where ρ_p represents the particle density, C_c is the slip correction, U_0 is the average air velocity at the nozzle exit, D_ae is the aerodynamic equivalent diameter, μ is the air viscosity, and W stands for the characteristic nozzle diameter. Note that the factor 18 is used instead of 9 when W refers to the nozzle radius (Seinfeld and Pandis, 1998).

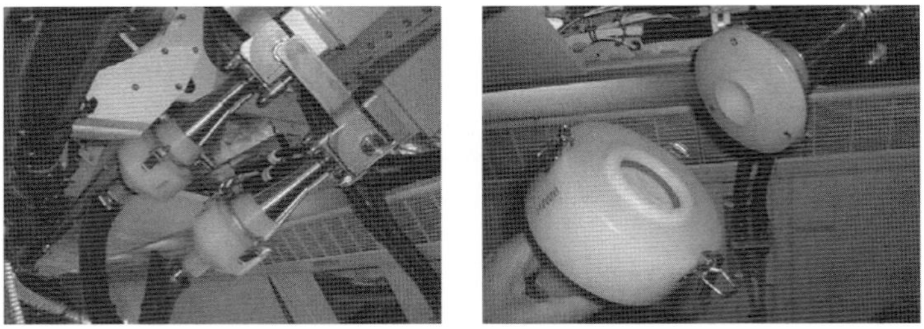

(a) (b)

Figure 4.11 Filter sampling system designed by the Max Planck Institute for Biogeochemistry in Mainz, Germany, and currently onboard the FAAM BAe-146. (a) The samples consist of two stacked-filter units (SFUs) mounted in parallel. (b) Each SFU can hold a maximum of three filters on sequential 47 or 90 mm diameter polyethylene supports (Andreae *et al.* 2000). Photo courtesy of Paola Formenti.

The Stokes number Stk relates the impactor geometry and flow rate to the particle CE. The value of Stokes number Stk_{50} for which the CE is equal to 50 % defines the value of the so-called size cut-off diameter (or radius) $D_{\mathrm{p},50}$ as

$$D_{\mathrm{p},50} \cdot \sqrt{C_\mathrm{c}} = \sqrt{\frac{9\,\eta \cdot \mathrm{Stk}_{50}}{\rho_\mathrm{p} \cdot U_0}} \tag{4.14}$$

The sampling efficiency curve of an ideal impactor stage would be a step function of the square root of the Stokes number Stk, separating particles that are collected (100 % sampling efficiency) from particles that pass the impactor stage (0 % sampling efficiency). In practical terms, the CE curve of an impactor stage is an S-shaped function with deviations from the ideal case mainly for high- and low-CEs (Baron and Willeke, 2005). The volume flow rate control is critical for impactor sampling as it determines the cut-off diameter (Eq. (4.13)). The effect of the deviation of the actual volumetric flow rate from the calibration one on the cut-off size can be estimated using Eq. (4.14).

Lists of commercially available impactors are given, for example, by Baron and Willeke (2005) and references therein. Some of them (e.g., the MSP manufactured by Dekati Ltd., Tampere, Finland, or the Micro-Orifice Impactor MOUDI from MSP Corporation, Minneapolis, MN) are also operated on aircraft at flow rates of 30–100 $\mathrm{l\,min^{-1}}$ at STP. In addition, some specific low-volume multiple-stage impactors such as the microinertial five-stage impactor by Kandler *et al.* (2007) and the two-stage impactor described by Matsuki (2005) have been specifically designed for single-particle analysis. These impactors have small collection surfaces (about 4 mm diameter) and operate at low flow rates (0.5–1 $\mathrm{l\,min^{-1}}$ at STP) to reduce particle bounce and break-up during impaction. These impactors have been designed to be used with collection substrates adapted to electron microscope analysis, for example, coated grids or nickel plates. An alternative device to collect samples for electron microscopy analysis used onboard research aircraft is the airborne streaker sampler, which is currently used onboard the NCAR C-130. This sampler was developed in the mid-1990s at the Max Planck Institute for Chemistry,

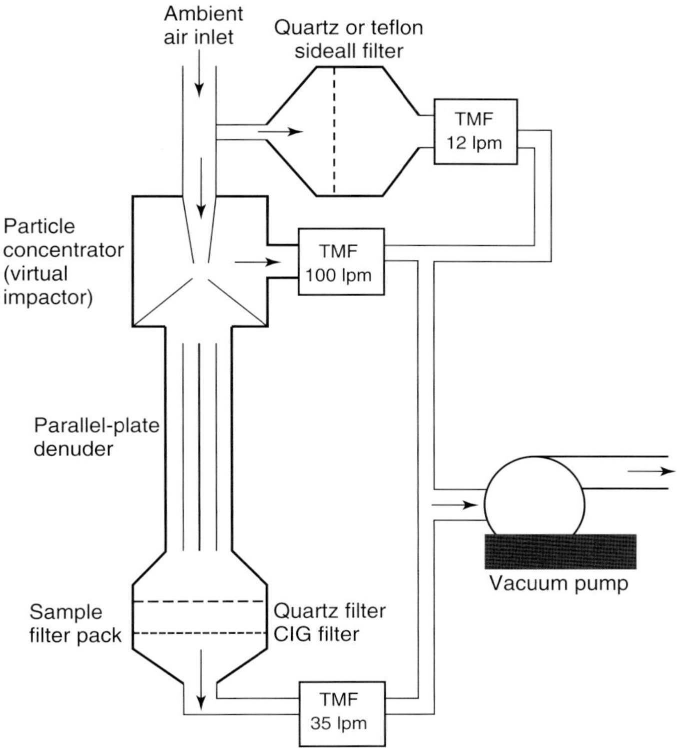

Figure 4.12 Schematics of the PC-BOSS onboard the NCAR C-130 aircraft (Kawamura *et al.* 2003).

Mainz, based on the commercial ground-based PIXE circular streaker sampler (Annegarn *et al.* 1996). It allows the collection of multiple samples on a single-filter disk of 85 mm diameter, which is rotated by a microprocessor-controlled stepping motor to a new section of the filter after a preprogrammed interval. The 8 mm diameter circular suction orifice allows up to 20 samples per filter frame and per flight. The maximum flow rate through a Nuclepore membrane at atmospheric pressure is $5\,l\,min^{-1}$, through a two-stage thermostatically controlled isokinetic inlet. The optimum exposure time for collections of samples for electron microscope analysis will be dimensioned such that it allows obtaining a uniform deposit of a monolayer of nonoverlapping particles. Typically, this will be achieved with a few second of sampling in urban polluted areas and less than one hour in remote areas.

The use of filter and impactor sampling devices onboard aircraft are subject to some special requirements. To achieve good spatial representativeness and to reduce all artifacts related to evaporation of semivolatile material and reaction during sampling, the exposure time of the aerosol samples should be kept to the minimum, which results on having low mass loading on the collected substrates. The sample preparation, handling, and storage prior, during, and after the flight are therefore crucial to reduce any possible occurrence of external contamination. Prewashing of collected substrates is at times performed for sampling at high altitudes or in remote regions. A good precaution consists in keeping the inlet covered, while the aircraft is parked and flashing the sampling line through "dummy" samples during taxing, take off and landing, and prior the beginning of scientific observations. The sampling lines and the working space on the aircraft should be regularly swept with kimwipes and deionized water. Furthermore, to avoid contamination from the aircraft cabin, several sample holders should be prepared in advance to limit the handling of substrates during the flight. It must be added that handling onboard an aircraft is particularly difficult because of the aircraft movement and operations. For the same reasons, samples cannot be conditioned and stored during the flight, but care should be taken in doing so immediately after the flight to avoid contamination, condensation, and losses by evaporation and chemical reactivity. Field sample blanks should be collected at various moments during each flight. A common procedure for field blank collection consists in handling the sample as a real sample by placing it in the collecting device but exposing it only few seconds to the air flow. For a given compound, the quantification limit will be estimated by the blank variability. Finally, unavoidable artifacts can arise from the evaporation of volatile material on the samples as the cabin temperature is often high, especially when flying in the boundary layer.

Lower blank levels are sometimes obtained by prewashing the sampling substrates before the exposure. An example of prewashing protocols is given in Kline *et al.* (2004) for Gelman Zefluor Teflon filters to reduce the nitrate variability in the filter blank values. The choice of the filter medium is determined by the design of the sampling device and by the compound to be measured, and therefore by the consequent analysis to be performed. The reader is referred to the paper of Chou *et al.* (2011) who present a general review of the substrate materials for filters and their characteristics. Typically, porous thin membranes such as Nuclepore polycarbonate or Zefluor Teflon are used for the analysis of inorganic components by X-ray and ion-beam techniques (XRF, PIXE, and instrumental neutron activation analysis, (INAA)) or water-extraction methods such as ion chromatography (IC). Quartz and glass-fiber filters are commonly used to dose organic and elemental carbon by thermal-evolution methods, and special conductive substrates on Cu- or Ni-grids are used for subsequent individual particle analysis by scanning and transmission electron microscopy. Impactor substrates should be greased to avoid losses due to bouncing material. It should be noted that, having comparable CE for particles larger than 0.3 μm, Zeflour Teflon is often preferred to Nuclepore polycarbonate because of its lower pressure drop, therefore allowing reducing the strength of the vacuum pump. Often Nuclepore and Zeflour filters are used in a tandem sequential arrangement consisting in a 8 μm pore size Nuclepore polycarbonate, followed by a 2 μm Zefluor Teflon downstream (Andreae *et al.* 2000; Formenti *et al.* 2003). At a given particle density and air velocity, filtration through large-pore size membranes can provide size segregation for multiple sampling, and it is often preferred to impactor sampling (John *et al.* 1983).

Samples should be preferentially collected on straight and leveled runs within a single air mass to ease the interpretation of the measurements. The use of volume flow controller or critical orifice should be preferred to be able to compare values measured at different altitudes. In addition, aerosol concentrations, in nanograms per cubic meter or micrograms per cubic meter, should be expressed at STP.

4.4.2
Direct Online Methods (Aerosol Mass Spectrometer, Single Particle Mass Spectrometer, and Particle-Into-Liquid Sampler)

The composition of atmospheric particles has also been measured using a range of online methods. There are broadly two types of approach: (i) the first technique is to collect the aerosol and subsequently perform an analysis on the material on board the aircraft and (ii) the second is to perform real-time online detection, which to date has been reliant on mass spectrometric methods or IC techniques.

There are two main classes of mass spectrometric methods that have been applied to airborne measurement: (i) single-particle detection and (ii) ablation based on laser methods and thermal vaporization techniques, which largely provide information on particle ensembles.

Online aerosol composition airborne measurements were pioneered in the late 1990s (Murphy and Schein, 1998; Thomson, Schein, and Murphy, 2000). In addition to the aforementioned methods for the investigation of a broad range of particle chemical characteristics, some online instruments have been designed to measure a single chemical component of atmospheric particles; most notable is a single-particle incandescence method. In the last decade, considerable development has taken place to apply the above range of methods widely across the atmospheric airborne research fleet.

4.4.2.1 Bulk Aerosol Collection and Analysis

These techniques typically use a combination of inertial removal of the particles from the ambient airstream and subsequent delivery of the particles to the detection system using an eluent flow system. The inertial removal of the particles is typically enhanced by exposing them to an environment that is supersaturated with respect to water and growing them to droplet size. This also has the benefit of providing the eluent for the delivery and detection of the components of interest.

One of the first measurements of this type was a mist chamber (Scheuer *et al.* 2003). A droplet mist is created by accelerating ambient air through a nozzle through which ultrapure water is aspirated. The water, with particulate and dissolved gaseous species, is removed from the chamber by a syringe pump and injected into an IC for the analysis of dissolved ionic species. Soluble gas species will also be sampled by such a system as no denuder or scrubber is used upstream of the mist generation nozzle. The mist chamber was positioned close to the aircraft inlet, providing a short residence time and the sample line heated to 40 °C. This approach was successfully used to study particle composition during airborne studies across the Pacific region (Dibb *et al.* 2003a, b) and provided some of the first measurements of vertical profiles of sub-micrometer particle component mass.

The particle-into-liquid sampler (PILS, Brechtel, USA) instrument has been used to date on several aircraft. The instrument, initially developed by Weber *et al.* (2001), was first installed on an aircraft by Orsini *et al.* (2003). While droplets are again used to dissolve aerosol components and deliver them for analysis close to real time, the instrument differs from the mist chamber in that the ambient particle-laden flow from outside the aircraft is mixed with a turbulent flow of steam held at 100 °C from an ultra pure water source. The cooling of the steam when mixed with ambient air causes a large supersaturation of sample flow to occur in which the ambient particles will rapidly grow to super-micrometer-sized solution droplets. The droplets are large enough to be separated from the gas flow and inertially impacted onto a vertical quartz surface via a single jet nozzle at the end of the sampler that focuses the droplets. Quartz is used as it is a highly wettable surface and does not have a high-sodium-ion background, which is important when sampling sea salt. A filtered flow of deionized water, spiked with a marker compound (typically LiF) to allow dilution to be determined, transfers the solution sample into a liquid flow to a detection system. Mostly, IC has been used to detect typical inorganic ions and simple organic acids observed atmospheric aerosol particles. Typical liquid flows of 0.03–0.1 ml min^{-1} have been used to deliver solution to sample loops on the timescale of 1–5 min.

The PILS particle collection system has the advantage that the particles are preferentially sampled relative to soluble gases that are separated rapidly from the liquid droplets, although upstream denuders are also used. The PILS has been further modified to allow the sample collection via up to 80 vials that can be subsequently analyzed using a wider range of analytical procedures postflight (Kleinman *et al.* 2007; Sorooshian *et al.* 2007). Sample efficiencies of greater than 96% are to be expected for all anions and cations except for ammonium, which had a sample efficiency of 88% due to greater volatilization (Sorooshian *et al.* 2006).

Particle CEs have been shown to be close to 100% between 30 nm and 10 μm diameter at the sample flow rates of 15–17 l min^{-1}, although of course diffusion losses of small particles and impaction and gravitational losses of super-micrometer particles can be substantial in the inlet systems of all airborne instruments sampling from within the cabin of an aircraft. As a result, PILS is typically run with a PM1 or PM2.5 inlet in front of the instrument.

The analytical method used dictates the duty cycle and detection limits of the method. IC has been extensively used since it is specific for a range of inorganic ions and common simple organic compounds in solution. Limits of detection of 10 and 50 ng m^{-3} have been achieved for anions and cations, respectively. The speciation of major inorganic aerosol components is possible on a timescale of about 4 min with longer timescales of 15 min or more necessary for organic components. Other analytical methods have also been employed. Peltier, Weber, and Sullivan (2007) used a total organic carbon (OC) analyzer to measure OC online. Comparisons with a National Institute for Occupational Safety and Health (NIOSH) method for water-soluble OC show agreement to within 5%, although the method showed limitations when sampling insoluble OC.

PILS and mist chamber methods have been compared with one another and with filter and impactor sampling during wing tip to wing tip comparisons during major airborne experimental studies (Ma *et al.* 2004) and showed that for most inorganic ions the different methods agreed to within 30–40%, comparable with similar results from ground-based experiments.

4.4.2.2 Mass Spectrometric Methods

Mass spectrometers have been developed for the analysis of ambient aerosol particles for a number of years. There are several thorough reviews covering both the historical development of these methods and providing an overview of current developments, and the reader is referred to these for details (Canagaratna *et al.* 2007; Coe, 2006; Murphy, 2007). Here, we provide a brief overview of the main measurement principles of particle mass spectrometry as applied to airborne measurements. There are two main types of aerosol mass spectrometry that have been developed to date. (i) The first of these uses laser-based methods to ablate and ionize the sampled particles before mass spectrometric detection. These methods are typically able to sample single particles with a wide range of compositions though are often not optimized for quantitative mass analysis. (ii) The second use thermal volatilization of the particles. To date only the aerodyne aerosol mass spectrometer (AMS) has applied this technique to airborne sampling. This method performs a separate ionization process using electron impact on the volatilized gaseous material and hence can deliver mass quantification. However, the AMS is not able to detect refractory components.

Key differences in instrument design result from the ablation and ionization methods used. Nevertheless, there are common features that both types of mass spectrometer need to incorporate. Ambient aerosol particles need to be focused into the detection region that is held under vacuum. The size of the particles should be measured in some manner, before the particles are volatilized and ionized and the mass fragments subsequently detected by mass spectrometry. However, the AMS can work without the size information, and a laser ablation instrument can work in an untriggered, high-frequency laser fire mode. There is no single method that is optimal for every application and particle composition, and the approaches to solving the above challenges to a large extent depend on the measurement needs.

4.4.2.2.1 Single-Particle Laser-Based Ablation Methods

Single-particle laser-based aerosol mass spectrometers were developed as research tools, and their detailed operation cannot be described generically. Murphy (2007) provides the most recent detailed review of the design of single-particle mass spectrometers. Aerosol particles are entrained into a vacuum through an inlet, which also acts to focus the particles toward the laser detection region. Various inlet systems have been used including pinholes, which allow a wide range of particle sizes to be transmitted but are highly divergent and are therefore inefficient and have particle-size-dependent transmission, capillaries, which improve transmission to some extent, and aerodynamic inlets that tightly focus the particles into a beam and hence have high throughput but only over a limited size range.

Once in the vacuum region, the particle size is detected by optical methods, either by optical scattering using a continuous laser (Section 4.3) or by aerodynamic sizing. The latter works by employing two continuous lasers set a known distance apart and calibrating the flight times between them for particles of known size and density. As sizing the particle, these systems are used to trigger a laser pulse with sufficient energy to ablate the particle and ionize the fragments. The ions are then transmitted to a mass spectrometer. The design ethos for the laser system also depends on the scientific aims of the instrument. Aerodynamic sizing provides better size resolution than light scattering but requires greater spacing between the inlet and the ablation laser and as a result increased losses may be incurred. Typically pulsed lasers are used to ablate and ionize the particles.

The choice of ablation laser is important and is influenced by several factors. These include the pulse power, duration and repetition rate, power consumption, cost, portability, and ionization threshold (minimum power flux required to generate ions). The latter is dictated by both the wavelength of the laser light used and the composition of the particles under analysis, as the absorption bands and ionization energies vary greatly between particle constituents. As the amount of energy absorbed by the particle increases, a greater fraction of the constituents become desorbed and/or ionized, and a greater amount of molecular fragmentation occurs. These phenomena continue to increase until a second threshold is reached, whereby a plasma is formed. The key technical parameter is the total amount of energy delivered per unit area during a laser pulse, known as the *fluence*.

Time-of-flight (ToF) mass spectrometers are the preferred mass spectrometric method in laser ablation instruments. They provide a simultaneous measurement of all the transmitted ions, which are separated by their acceleration toward the detector by an electric field. The method is ideally suited to pulsed laser ablation as the laser pulsing can also be used to trigger the timing of the mass spectrometer. The high energies involved in laser ablation supply the ions with a wide range of kinetic energies, this reduces the resolution of the mass spectrometer but electric fields can be used to reflect the ion beam and reduce this effect substantially, leading to mass resolutions ($\Delta m/m$) typically of the order of several hundreds. Both unipolar and bipolar ToF systems have been used on aircraft over the past decade.

The ion detection by multichannel plates is sufficiently sensitive that mass detection of ions from single particles is possible, allowing, for example, the discrimination of tropospheric versus stratospheric particles, and isolation of specific particles such as those composed of meteoritic material (Murphy and Schein, 1998). The strengths of ToF aerosol mass spectrometers are that they provide rapid, online analysis of the composition of single particles and that due to the high laser energies used they are able to ionize a wide range of components. These key characteristics have been used in a range of applications to deliver a number of unique insights into the composition of atmospheric particles. On the other hand, there are a number of limitations. Chief among these are as follows: first, inlets limit the size range of particles that can be sampled by the instrument and induce size-dependent transmission efficiencies. Secondly, quantification of the mass of components within particles is not readily possible using laser ionization methods, unless the laser fluences are increased to very high values, something that to date has not been carried out on aircraft. This is due to the complex interaction between the pulsed light from the laser and the chemical matrix of the particles as well as the ability to couple light energy into two identical particles in the same manner.

The first mass spectrometer to be flown on an aircraft (Thomson, Schein, and Murphy, 2000) was the particle analysis by laser mass spectrometer (PALMS) instrument, which used a 30 cm long capillary inlet, to sample from a 5 cm diameter duct in the nose of the NASA WB-57 aircraft. The instrument has been used to establish the presence of meteoritic material in the upper troposphere and lower stratosphere and to establish the presence of a large fraction of organic particles in the upper troposphere (Cziczo, Thomson, and Murphy, 2001; Froyd *et al.* 2009; Murphy and Schein, 1998; Murphy *et al.* 2006). More recently, a design of single-particle laser ablation aerosol mass spectrometry has incorporated an aerodynamic lens system with aerodynamic particle sizing and has incorporated a Z-shaped mass spectrometer to maximize the mass resolution in as small a volume as possible (Pratt *et al.* 2009a). This instrument has most recently been coupled to a counterflow virtual impactor (CVI) (Chapter 6) and used to investigate the IN efficiency of biological particles (Pratt *et al.* 2009b). Another airborne single-particle mass spectrometer has been developed and successfully operated by Zelenyuk *et al.* (2009); Zelenyuk and Imre (2009). Most recently, an instrument using similar principles, but with an even more compact design, has been developed and used to determine the aerosol mixing state around an urban airshed (Brands *et al.* 2011).

4.4.2.2.2 **Thermal Volatilization Methods** The philosophy of this approach is very different from that of the single-particle laser ablation instruments. The aim is to provide quantitative mass information. In doing so, only a limited number of chemical species can be measured and true single-particle information is not obtained. Nevertheless, high sensitivity and fast time response mean that this approach has recently seen widespread use on aircraft for the analysis of sub-micrometer sulfate, nitrate, organic material, and ammonium.

The only instrument that has operated on aircraft platforms using thermal methods to volatilize particles before subsequent mass spectrometric selection is the AMS. The method, first developed by Jayne *et al.* (2000) and described in detail by Canagaratna *et al.* (2007), uses an aerodynamic lens system (Zhang *et al.* 2002) with a transmission efficiency of close to 100 % for particle diameters between 60 and 600 nm to form a narrow beam of aerosol particles in vacuum from ambient air. The particle beam is directed at a tungsten heated surface held at approximately 600 °C, which leads to the flash vaporization of nonrefractory components. The tungsten heater is located close to an electron impact ionizer, which provides a source of 70 eV electrons and ionizes the gas plume resulting from the vaporization. The vacuum system is differentially pumped, giving a vacuum pressure in the ionization region of about 10^{-7} torr in the early versions of the instrument, although subsequent improvements have meant that pressures as low as 10^{-8} torr are now typical.

In the original version of the instrument, the ions were extracted into a quadrupole mass analyzer that is rapidly scanned repeatedly through 0–300 m z^{-1} and the ions are detected by an electron multiplier. Operating in this mode, known as *mass spectrum* or MS mode, the AMS delivers a mass spectrum of the nonvolatile component of the sub-micrometer particles. More recently, orthogonal pulsed ion extraction meant that it has been possible to gate the ions into a ToF mass spectrometer, a development that has led to significant improvements in sensitivity and the simultaneous collection of ions of all m z^{-1}. High-ion throughput using a C-shaped ion flight configuration, C-ToF-AMS (Drewnick *et al.* 2005), and high-ion resolution, HR-ToF-AMS (de Carlo *et al.* 2006) versions of the instrument have now been developed and have been flown on different aircraft. The latter has a mass resolution of about 4000 and can be used to probe atomic composition of the mass fragments, at the cost of sensitivity compared to the C-ToF AMS.

The acceleration of the particles through the aerodynamic lens system is size dependent and this provides a method for size determination. A second AMS operating mode is known as the *particle ToF* (PToF) mode. It uses a chopper wheel operating at 100 Hz with a 2% duty cycle mounted in the vacuum region close to the lens. When inserted into the particle beam, short packets of particles are transmitted to the heater. The vaporization process and transmission of ions through the instrument is rapid compared to the particle flight time to the heater and hence the ion detection as a function of the chopper wheel phase provides a measurement of the particle size. This is calibrated by a combination of size-selected polystyrene latex spheres and ammonium nitrate particles. When the AMS is fitted with a quadrupole mass analyzer, it is not possible to analyze a full mass spectrum in this mode and size-dependent information is only sampled at a few key preselected masses. The C-ToF and HR-ToF-AMS systems can deliver full mass spectral information as a function of particle size but usually alternate between MS and PToF modes as the sensitivity of the former is a factor 50 larger than the latter.

The conversion of ion signals to mass loadings has been described by Jimenez *et al.* (2003) and the associated error propagation discussed in Allan *et al.* (2003). Briefly the mass concentration of a chemical species is directly proportional to the detected ion rate assuming that the flow rate, ionization efficiency, and fraction of ions from a given compound at a given m z^{-1} is known. The ionization efficiency for ammonium nitrate is calibrated directly by sampling ammonium nitrate particles of a known size, typically 300–400 nm. Once the ionization efficiencies of other species are established from laboratory characterization, their relative ionization efficiencies compared to that of ammonium nitrate can be used. Clearly ions from different chemical constituents can be present at a given m/z. An inversion procedure to separate ion signals from different chemical species has been developed (Allan *et al.* 2004), which has been verified now that high resolution m z^{-1} data are available (Aiken *et al.* 2008). During operation, the sample flow may fluctuate due to changes in ambient pressure or material build up on the inlet orifice and the performance of the detector deteriorates with time. The signal strength at either N_2^+ or O_2^+ is monitored and this air beam is used to correct for such changes. This is important during aircraft observations as the data are automatically corrected for pressure changes and the data are reported as mass concentrations at standard pressure.

The first airborne measurements using an AMS were made with a quadrupole system during the ACE-Asia study (Bahreini *et al.* 2003). While the mass loading is unaffected by pressure due to the air beam correction, the performance of the lens is dependent on upstream pressure. Over a certain range of pressures, this can be corrected by laboratory calibrations (Bahreini *et al.* 2003). However, as the upstream pressure falls further, particle losses will occur as the sizes over which 100 % transmission occur change and accumulation particles are no longer sampled efficiently. This can be negated through the use of a constant pressure inlet (Bahreini *et al.* 2008); however, such inlets are only efficient over a limited range of pressures (Schmale *et al.* 2010).

The quadrupole version of the AMS had relatively poor detection limits (Bahreini *et al.* 2003) quote detection limits of 2.4, 3.1, 4.3, and 11.8 $\mu g\,m^{-3}$, respectively for nitrate, sulfate, ammonium, and organic mass for a 1 min average at ground level. Detection limits increase commensurately with reduced pressure. The detection capability of the quadrupole system for aircraft sampling was improved by only sampling selected masses when in MS mode (Crosier *et al.* 2007b). Most aircraft making AMS measurements now use either C-ToF-AMS (Morgan *et al.* 2009) or HR-ToF-AMS (de Carlo *et al.* 2008), which typically have sensitivities 30 and 10 times greater than the quadrupole systems.

The overall mass accuracy with the AMS is largely determined by the calibrations, with typical instrument comparisons showing agreement to within 20%. The main systematic uncertainty associated with the AMS is that the phase of the particles can affect the efficiency of the volatilization process. This is known as the *collection efficiency* (CE). Laboratory characterizations have shown that dry ammonium sulfate particles were sampled with 25 % efficiency, whereas liquid particles were sampled with nearly 100 % efficiency as a result of particles sticking to the heater rather than bouncing off (Matthew, Middlebrook, and Onasch, 2008). In the atmosphere, particles composed of ammonium nitrate have CE=1, whereas CE values as low as 0.3 have been reported. CE corrections to ambient data have used empirical comparisons with PILS (Kleinman *et al.* 2007) or filters (Crosier *et al.* 2007a), but then systematic biases can be propagated as has been shown by Matthew, Middlebrook, and Onasch (2008) who have used the measured nitrate fraction as a robust method of applying a CE to ambient particles.

4.4.2.3 **Incandescence Methods**

Previous measurements of black carbon (BC) from aircraft have been made either by offline analysis of filters or by measuring the absorption characteristics of the particles. The former of these has rather poor time resolution; the latter approaches are dependent on a known relationship between BC mass and absorption, the mass absorption coefficient that is known to be very variable for different types of BC.

A new, online approach to the measurement of BC mass and its size distribution, the single-particle soot photometer (SP-2) (Stephens, Turner, and Sandberg, 2003), has recently been used on research aircraft. The first deployment investigated Arctic lower stratospheric BC aerosol particles (Baumgardner, Kok, and Raga, 2004).

Schwarz *et al.* (2006) provide a detailed description of the SP-2 instrument. Briefly, the instrument uses light scattering to size particles within the laser cavity of a Nd:YAG laser at 1064 nm; if particles absorb at the wavelength of the laser, then they are rapidly heated to incandescence. The black body radiation emitted by the incandescence is proportional to the mass of BC within the particle and hence single-particle mass determination is possible. This technique provides a more direct measurement of BC mass than absorption instruments, although the instrument has some dependence on the type of soot being measured.

Two detectors are used to measure the incandescent light. These are fitted with optical filters to detect light over a broad and narrow range of VIS wavelengths (350–800 nm and 650–800 nm respectively). The ratio of light intensity from these two detectors provides information on the color temperature of the incandescence and hence is a means of identifying the absorbing species. While in laboratory studies metal particles and mineral iron-oxide such as hematite have been shown to incandesce in the SP-2, the only observed absorbing particle in ambient atmospheric measurements to date has been BC. Given this, some studies have chosen to maximize the range of BC masses detected by the SP-2 by altering the gain settings of the incandescence detectors, although this reduces the size range of particles over which the broad and narrow band signals can be compared.

Absorbing particles may lose mass through evaporation of a coating or partial incandescence during their passage through the laser beam, producing non-Gaussian scattering signals. These non-Gaussian pulse shapes have been exploited to provide an estimate of the thickness of nonrefractory coatings on the soot particles (Gao *et al.* 2007). At the present time, there is no universal BC calibration material and this provides a fundamental limitation on the absolute calibration of BC mass. However, most recently, calibration of the SP-2 has been carried out using several BC calibration particle types using a combination of a DMA and an aerodynamic particle mass analyzer (APM) to probe the directional dependence on thermal emission resulting from nonspherical particles (Moteki *et al.* 2009).

In addition to the size distributed BC mass, the SP-2 can also deliver some information on the coating of a BC particle. On exposure to the SP-2 laser, a BC particle with little or no coating will be flash heated and rapidly incandescence and there will be little or no time difference between the measured scattering and incandescence signals. However, if the particle has a significant coating, then incandescence will be delayed by the time taken for the absorbed laser energy to evaporate the semivolatile coating. This delay time has been used by multiple authors from aircraft platforms (Moteki *et al.* 2007; Schwarz *et al.* 2008) and has been used to investigate the radiative impact of the BC mixing state (Shiraiwa *et al.* 2008).

Aircraft studies using the SP-2 have been undertaken to investigate the abundance and source of BC in the Arctic upper troposphere and lower stratosphere

(Hendricks *et al.* 2004) and to investigate the evolution of urban plumes (Moteki *et al.* 2007) and provide tests of the vertical distribution of BC in global models (Schwarz *et al.* 2006; Schwarz *et al.* 2010).

4.4.3 Indirect Methods

Indirect methods are based on the inverse retrieval of information on composition based on measurements of the extensive properties such as volatility and absorption. Volatility analysis of aerosol relies on the principle that a heated aerosol will lose mass by either a phase transition or through mechanical breakdown (shattering). Publications on aerosol volatility date back to the 1960s, Goetz, Preining, and Kallai (1961) and several groups have used heated tube ("thermodenuder") volatility measurements to infer particle composition by, for example, measuring changes in size distributions (Clarke, 1991). This approach has been used extensively on balloon-borne systems to infer vertical profiles of aerosol composition (Deshler, Johnson, and Rozier, 1993). The standard system consists of a thermal denuder coupled to an OPC, and the ambient sample is passed to the denuder from a suitable aerosol inlet. Many measurement techniques have been coupled with heated volatilization tubes to indirectly determine chemical composition. Usually, these were used to infer aerosol sulfate (SO_4^{2-}) concentrations. Volatility analysis is particularly good for identifying sea salt, which has a phase transition occurring at 650 °C (Jennings and O'Dowd, 1990); many of the newer techniques such as the AMS do not provide such access to sea salt information. Since such instrumentation offers a degree of compositional information coupled to the size distribution of the population, it can provide an insight to the mixing state of the aerosol population. As a particular transition is passed, there will be a change in the total aerosol mass, indicating the presence of any associated chemical species. If one considers the aerosol population to be externally mixed, the loss of mass will be seen as a reduction in the number concentration; however, an internally mixed population will not lose aerosol by number, but the volatile component will be lost from the aerosol, resulting in a shift in the size spectrum with the affected aerosol shrinking in size. Overall, this technique was viewed as being very much qualitative, but recent studies have shown that there is significant confidence in the quantification possible through volatility analysis (Brooks *et al.* 2007).

Several recent thermodenuder designs have also been developed to address the main performance limitations of these instruments, namely, poor temporal response due to insufficient sample residence time (An *et al.* 2007; Wehner, Philippin, and Wiedensohler, 2002) and also vapor recondensation artifacts (Fierz, Vernooij, and Burtscher, 2007). Huffman *et al.* (2009) improved the design by Wehner, Philippin, and Wiedensohler (2002) so as to reduce the thermal inertia and improving the temperature control and stability, thus allowing for more rapid temperature scanning to allow measurement of particle volatilities across a much wider temperature range within a timescale of 1–3 h.

Twomey (1968) applied heated quartz tubes in front of a thermal diffusion cloud chamber to measure CCN as a function of temperature and concluded that CCN in the Northeastern USA were composed mainly of ammonium sulfate. Pinnick, Jennings, and Fernandez (1987) used a similar approach with an OPC to show that 60–98 % of sub-micrometer aerosol in New Mexico was composed of ammonium sulfate or ammonium bisulfate. Jennings and O'Dowd (1990) and Clarke (1991) each used a form of this heated tube design also upstream of an OPC to infer that fine aerosol in remote marine environments was also mostly composed of sulfates. Jennings *et al.* (1994) used the same idea, but operated the thermodenuder temperature up to 860 °C to measure evaporation of inferred elemental carbon. One of the most common ambient particle volatility instruments used incorporates a heated metal flow-tube placed between two DMAs to measure the particle size distribution changes as a function of temperature. This change is used to infer size-resolved aerosol chemical composition (Orsini *et al.* 1999; Villani *et al.* 2007).

Light absorption measurements are widely used for measuring the mass concentration of light-absorbing carbon. This approach assumes a constant factor for converting the light absorption coefficient (see the following section for a definition) into a mass concentration. In their comprehensive review, Bond and Bergstrom (2006) suggest mass absorption coefficients of 7.5–10 $m^2\,g^{-1}$ at a wavelength of 550 nm for aged and fresh light-absorbing carbon. Despite the large uncertainty in the conversion factor, the operation of multiple-wavelengths absorption photometers is one key method for measuring vertical profiles and lofted layers of light-absorbing carbon from airborne platforms; examples from recent airborne field studies are given in McMeeking *et al.* (2011) and Petzold *et al.* (2007). Aerosol absorption measurements at multiple wavelengths are used to identify particulate light absorbers by the spectral dependence of the respective light absorption coefficient. The fact that mineral dust absorbs strongly in the green but weakly in the red spectral region, while light-absorbing carbon absorbs strongly across the entire visible spectral range was recently used for the separation of mineral dust and light-absorbing carbon (Petzold *et al.* 2011).

4.5
Aerosol Optical Properties

Some of the fundamental aerosol parameters governing the direct aerosol impact on climate are the AOD and the ratio of particle scattering to extinction, the single-scattering albedo. Both quantities have been defined in Section 4.1.3. AOD is a common product delivered by space-borne sensors for the global mapping of atmospheric aerosol particles. Measurement of the single-scattering albedo requires reliable data at least for two of the three basic aerosol optical properties scattering, absorption, and extinction.

Airborne measurements of aerosol optical properties are a crucial prerequisite for the assessment of the aerosol impact on climate because satellite instruments only deliver a column-integrated value of the light extinction caused by particles, while ground-based *in situ* networks cannot detect, for example, light-absorbing aerosol layers in the free troposphere that may be transported between continents (Petzold *et al.* 2007). However, vertical profiles of the aerosol extinction coefficient and of the aerosol absorption coefficient are put high up on the agenda of observation programs such as GAW.

In the past decade, substantial improvements were achieved for measuring light extinction from airborne platforms. Schmid *et al.* (2006) conducted a comprehensive intercomparison of remote sensing and airborne *in situ* methods for measuring aerosol light extinction, scattering, and absorption. During the writing of this book, a new multiple-wavelength instrument for measuring light absorption was flown (Lack *et al.* 2012). Respective *in situ* measurement methods and techniques are discussed for light scattering in Section 4.5.1, for light absorption in Section 4.5.2, and for light extinction in Section 4.5.3. Inversion methods were further developed to provide comprehensive data sets on aerosol optical properties; these approaches are discussed in Section 4.5.4.

4.5.1
Scattering Due to Aerosol Particles

Integrating nephelometry is the direct method for measuring light scattering by aerosol particles. This technique, employed for the first time by Beuttell and Brewer (1949), is based on the geometrical integration of the angular distribution of the light scattered by gas molecules and aerosol particles in an enclosed volume. The incident light is provided by a near-Lambertian light source (generally an array of light-emitting diodes (LEDs), a flash lamp, or a quartz–halogen lamp) and measured by a photomultiplier detector, placed at 90° with respect to the lamp position to avoid stray light detection. A scheme of the general nephelometer geometry is reproduced in Figure 4.13 from Anderson *et al.* (1996).

This scheme illustrates the principle of the geometric integration of scattered light over a range of scattering angles θ by a finite volume dv illuminated by a light source of intensity I_0 and placed at a distance x and at an illumination angle Φ. The reader is referred to the original paper by Anderson *et al.* (1996) for the details of the formal derivation of the detected scattered intensity. The aerosol particle scattering coefficient $b_{\mathrm{sca,p}}$ is obtained by the geometrical integration of the light scattered between 0° and 180°. The aerosol particle backscatter coefficient $b_{\mathrm{bs,p}}$, used to approximate the aerosol upscatter fraction and the asymmetry parameter, is obtained by integrating the diffuse light between 90° and 180°, where the 90° truncation is obtained by blocking periodically the illumination at 90°. In reality, the range of sensed scattering angle is reduced to 7° in the forward direction to 170° in the backward direction because of the actual internal geometry of the nephelometer. This geometry applies to the TSI Model 3563 three-wavelength integrating nephelometer, which is described by Anderson *et al.* (1996). The Model Aurora 3000 (Ecotech Pty, Knoxfield, VIC, AUS) is another recently available three-wavelength integrating nephelometer with the option for measuring the

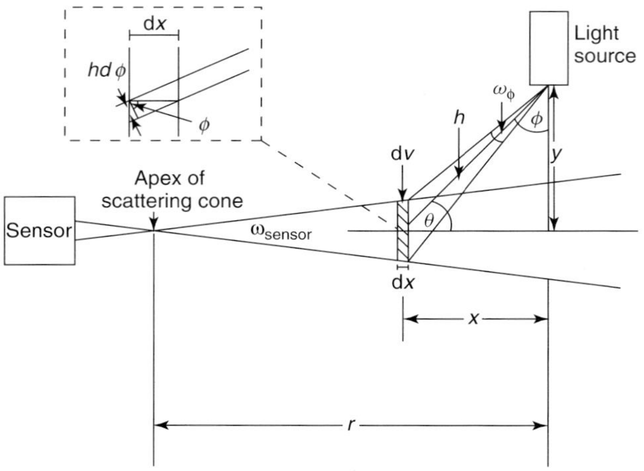

Figure 4.13 Schematic of the internal geometry of a nephelometer. The scattering finite volume is indicated as dv. The distance between the light source and the sensor axis is defined as h. The distance between the sensor and the light source position is r, whereas x indicates the distance between the finite volume and the light source. Angles are θ, the scattering angle, and ϕ, the illumination angle. (Source: Further details are given in Anderson *et al.* (1996). Copyright 1996 American Meteorological Society (AMS). Reprinted with permission by AMS.)

aerosol backscatter coefficient. Airborne applications, however, are published so far only for the TSI Model 3563.

The difference between the ideal and the measured $b_{sca,p}$ and $b_{bs,p}$ represents the truncation error. This is likely the most prominent source of error affecting nephelometer measurements. Because in the Mie regime, forward scattering becomes more prominent with increasing particle size, and the truncation error increases with particle size and might reach 50% for particles characterized by an important coarse mode, such as mineral dust or sea salt. If concurrent measurements of the aerosol composition and number size distribution are available, the truncation error can be estimated by optical calculations (Mie or others) of the full and truncated phase function. Otherwise, at least for the sub-micrometer fraction, Anderson and Ogren (1998) have proposed empirical parameterizations based on the spectral scattering dependence. Other sources of errors, such as wavelength and angular illumination nonidealities, calibration errors, and photon counting statistics, are reviewed by Anderson *et al.* (1996). Bond, Covert, and Mueller (2009) and Massoli *et al.* (2009) extended the discussion of truncation and angular scattering corrections for light-absorbing aerosol particles that are not treated adequately by the correction scheme from Anderson *et al.* (1996).

While airborne, the size and composition of aerosol particles might vary due to water uptake. As a consequence, the particle scattering efficiency might increase. The scattering particle growth factor $f_{sca,p}(RH)$ is a measure of this increase with respect to the value that the scattering coefficient would have at dry conditions. This parameter is measured by running two nephelometers in parallel, one maintained at dry conditions (e.g., RH = 30 %) and the second one maintained either at a fixed relative humidity $\geq$ 80 % or at relative humidity ramping from dry to wet conditions by a Nafion®-based humidifier tube. The humidifier generally consists of an annular design where a counterflow water streams around a water vapor permeable membrane (Teflon, Permapure®) stretched on a wire mesh tube. The aerosol air stream flows through the wire mesh tube. The controlled temperature of the water determines the relative humidity to which the air stream is exposed.

The dependence of scattering on relative humidity is described by Hänel (1976). Hegg, Larson, and Yuen (1993) have identified the processes contributing to the increase in light scattering by particles with increasing relative humidity (changes in size distribution, changes in refractive index, and coincidence of the particle size distribution with the effective light-scattering range). Various airborne studies have provided with experimental values of $f_{\text{sca,p}}(\text{RH})$ for atmospheric aerosol particles of different origins (sea salt, mineral dust, urban/industrial, and biomass burning), and linked it to their chemical composition (Carrico *et al.* 2003; Gasso *et al.* 2000; Kotchenruther, Hobbs, and Hegg, 1999). Examples of airborne humidifier systems are found in Carrico *et al.* (2003) and references therein. This chapter also provides a range of values of $f_{\text{sca,p}}(\text{RH})$ for different aerosol types.

Because of the importance of the particle size dependence of aerosol scattering, and the possibility of measuring both the scattering and backscatter coefficient, the three-wavelength nephelometer commercialized by TSI (Model 3563) is widely used in airborne research. A comparison of the performances of the commercially available nephelometers is presented by Heintzenberg *et al.* (2006).

The angular dependence of the light scattered by particle can be measured by polar nephelometry. Particles flow through a sampling volume where they intercept a laser beam. The scattered light is detected by a circular array of photodiodes sensing signals corresponding to a range of scattering angles. The size and position of the detectors determine the angular resolution of the differential scattering pattern. The polar scattering angle θ is 0° in the forward direction with respect to the propagation vector of the incident beam. A description of an airborne polar nephelometer is provided by Gayet *et al.* (1997) and Crépel *et al.* (1997). Currently polar nephelometer instruments cover the size range relevant for super-micrometer-sized particles and ice crystals. For the sub-micrometer-sized aerosol, no such instrument is available for airborne applications.

4.5.2
Absorption of Solar Radiation Due to Aerosol Particles

Measuring the aerosol particle absorption coefficient $b_{\text{abs,p}}$ is still a challenging task of high relevance. Compared to extinction and scattering, the absorption of solar electromagnetic radiation by aerosol particles is a much smaller effect. Except for particles emitted from anthropogenic or biogenic combustion processes, single-scattering albedo values of atmospheric aerosol particles are usually > 0.80, indicating that absorption contributes 20% and less to aerosol extinction. Reviews of available methods are given by Horvath (1993) for the status in the mid-1990s and by Moosmüller, Chakrabarty, and Arnott (2009) referring to the current status. A detailed evaluation study on aerosol particle absorption measurement methods is reported by Sheridan *et al.* (2005).

Approaches for measuring aerosol particle light absorption are manifold. In principle terms, they can be divided into one group of methods measuring the light absorption after particles have been deposited on an appropriate filter matrix, and the group of *in situ* methods that measure light absorption of airborne particles.

4.5.2.1 Filter-Based Methods

Early methods for aerosol absorption were based on the visibility of plumes emitted from incomplete combustion processes. Respective measurement approaches used as early as in the nineteenth century relied on the collection of the particulate matter on an appropriate filter matrix (Moosmüller, Chakrabarty, and Arnott, 2009). The darkening of the initially blank filter spot by the deposited matter was taken as a measurement of aerosol absorption. Improved approaches based on the collection of particulate matter on an appropriate filter matrix are still widely in use. Current commercially available instruments suitable for aircraft operation are the Aethalometer (Hansen, Rosen, and Novakov, 1984), the PSAP (Bond, Anderson, and Campbell, 1999), and the multiangle absorption photometer (MAAP) (Petzold and Schönlinner, 2004).

Filter transmission methods (Aethalometer, PSAP) measure filter transmissivity of a blank filter, T_0, and of the particle-loaded filter, T (Figure 4.14). The data analysis is based on the application of the Lambert–Beer law:

$$\ln\left(\frac{T_0}{T}\right) = b_{\text{meas}} \cdot \left(\frac{\text{Volume}}{\text{Area}}\right) \tag{4.15}$$

where the ratio of the sampled volume of air divided by the exposed filter area expresses the "length" of the probed atmospheric column, and b_{meas} refers to the measured signal that is a function of both the particle absorption and scattering coefficient, $b_{\text{abs,p}}$ and $b_{\text{sca,p}}$, respectively.

Filter transmittance methods have to assume that the optical properties of deposited particles do not deviate from those of the particles in their airborne state. Furthermore, it has to be assumed that the collected aerosol does not scatter ($b_{\text{sca,p}} = 0$) or the aerosol scattering coefficient is known so that their effect can be corrected. Finally, the effects of the filter matrix on the measurement in terms of multiple scattering have to be considered. Correction equations are of the form (Bond, Anderson, and Campbell, 1999):

$$b_{\text{abs,p}} = \left(\frac{b_{\text{meas}} - \text{K}_1 \cdot b_{\text{sca,p}}}{\text{K}_2}\right) \cdot f(T) \tag{4.16}$$

The coefficients K_1 and K_2 have to be determined from calibration studies using laboratory-generated test aerosol particles. The function $f(T)$ that corrects for multiple scattering from the filter matrix has also to be determined by calibration studies. An example of such an extensive study is given by Sheridan *et al.* (2005) for the Reno aerosol optics study (RAOS) in 2002. Current literature reports correction

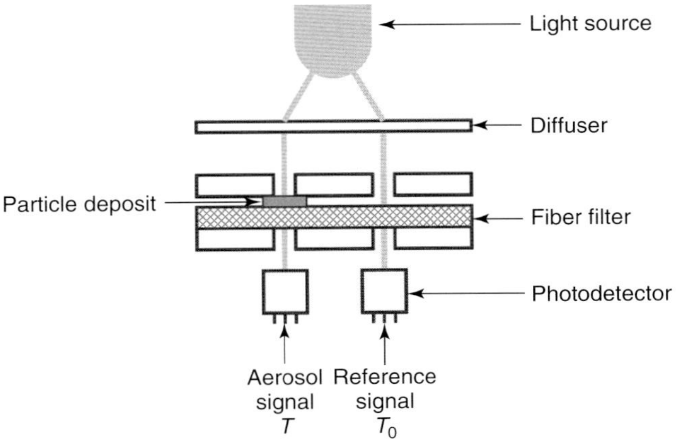

Figure 4.14 Schematic of filter PSAP setup. (Source: Adapted from Bond, Anderson, and Campbell (1999). Copyright 1999 Mount Laurel, N.J. Reprinted with permission.)

functions for the PSAP (Bond, Anderson, and Campbell, 1999; Virkkula, 2010; Virkkula *et al.* 2005) and for the Aethalometer (Arnott *et al.* 2005; Collaud Coen *et al.* 2010). Concerning the use of the PSAP, the comment by Ogren (2010) on the adequate use of correction functions has to be considered.

In contrast to the transmittance approach, MAAP uses a radiative transfer scheme for data interpretation that includes the treatment of light scattering (Petzold and Schönlinner, 2004). Intensive method intercomparison has demonstrated that the MAAP reduces the cross-sensitivity of the absorption measurement to light-scattering and filter effects (Petzold *et al.* 2005a). However, all filter-based methods may be affected by substantial biases in absorption data in the case of high organic aerosol loading (Cappa *et al.* 2008).

4.5.2.2 *In Situ* Methods

Absorption of visible radiation by particulate matter can also be quantified by measuring the transformation of electromagnetic energy into heat energy by the absorption process. The detection of the transferred energy is performed by measuring the resulting acoustic wave by photoacoustic spectrometry or the resulting refractive index change of the air by interferometry (Moosmüller, Chakrabarty, and Arnott, 2009). In contrast to filter-based methods, these *in situ* methods are not affected by particle-filter matrix effects.

Among *in situ* methods, photoacoustic spectrometry is the most mature technique for airborne applications. Moosmüller, Chakrabarty, and Arnott (2009) and references therein present a detailed description of photoacoustic spectrometry. Briefly, light-absorbing particles contained in a cylindrical cavity are illuminated by laser radiation of power P_{las} at wavelength λ_{las}. Part of the incoming radiance I_0 is scattered or transmitted (I_{sca}, I_{tra}), and the remaining fraction I_{abs} is absorbed either by the particles and/or by the embedding gas. The absorbed radiation generates a temperature increase ΔT in the surrounding gas. Since the aerosol is kept at

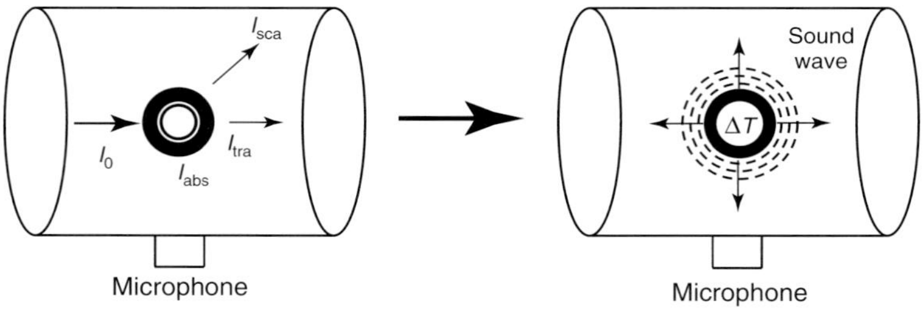

Figure 4.15 Schematic of photoacoustic spectrometry.

a constant volume, ΔT results in a pressure increase Δp. Figure 4.15 shows a schematic of the photoacoustic spectrometry approach.

The incoming laser radiation is modulated at a frequency ν that matches one of the resonance frequencies of the cylindrical cavity containing the aerosol. Because of this resonance effect, the acoustic wave generated by the light absorbed by particles and/or the gas is amplified by the quality factor Q of the acoustic resonator. The resonance frequency of the resonator depends not only on the resonator geometry but also on the adiabatic compressibility γ of the gas, which is a function of pressure and temperature. Thus, constant tuning of the laser modulation frequency is required for airborne use. Airborne installations must also successfully eliminate vibrational and flow noise from affecting the signal.

Summarizing, the combined absorption coefficient of the gaseous compounds $b_{abs,p,gas}$ and the particulate compounds $b_{abs,p}$ depends on the acoustic resonator constant C_{res} and on the laser power p_{las} according to

$$b_{abs,p}(\lambda_{las}) + b_{abs,p,gas}(\lambda_{las}) = \frac{C_{res}(\nu, Q, \gamma)}{p_{las}} \tag{4.17}$$

In contrast to filter-based methods, the photoacoustic signal is not affected by aerosol scattering. However, if the surrounding gas also absorbs radiation at the laser operation wavelength, the signal has to be corrected for this contribution. Working at a wavelength of ~500 nm, NO_2 is a potential gaseous absorber. In the near IR, H_2O absorption bands have to be considered; see Moosmüller, Chakrabarty, and Arnott (2009). Therefore, the sensitivity of photoacoustic spectrometry to light absorption by gases can also be used for the determination of the resonator constant C_{res}.

As was demonstrated during the intercomparison study RAOS (Petzold *et al.* 2005a; Sheridan *et al.* 2005), all available methods agree well for the studied aerosol particles with respect to the measurement of the aerosol absorption coefficient $b_{abs,p}$, as long as the required correction and/or calibration of methods and the adjustment to a common wavelength for comparison are carefully considered. Figure 4.16 demonstrates the equivalency of the different methods.

4.5.2.3 **Airborne Application**

All described methods have different limitations and benefits. As a basic prerequisite for the installation of an instrument in the pressurized cabin of a research

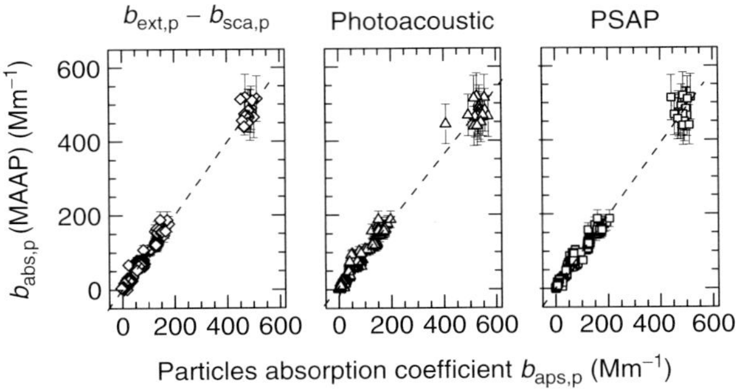

Figure 4.16 Equivalency of aerosol absorption methods (Petzold *et al.* 2005a). (Source: Copyright 1999 Mount Laurel, N.J. Reprinted with permission.)

aircraft is its leak-tightness, because the instrument is at ambient pressure while the cabin is at a pressure > 800 hPa. So far, this criterion is met only by *in situ* instruments such as photoacoustic spectrometers or by single-filter instruments such as the PSAP. Filter-type instruments such as the Aethalometer and MAAP are not leak-tight and can therefore be operated only on nonpressurized aircraft for boundary layer studies.

Furthermore, filter-based methods are sensitive to variations in inlet pressure that modify filter optical properties during ascent, descent, or turns of the aircraft. Figure 4.17 shows an example from absorption measurements inside a strong dust layer during the approach into Dakar airport in Senegal (Petzold *et al.* 2011). Between 15:25 and 15:30 UTC, the aircraft ascended above the dust layer and descended again. During the altitude change, the PSAP raw signal dropped below zero, reached a constant value at the upper level, and showed the similar behavior

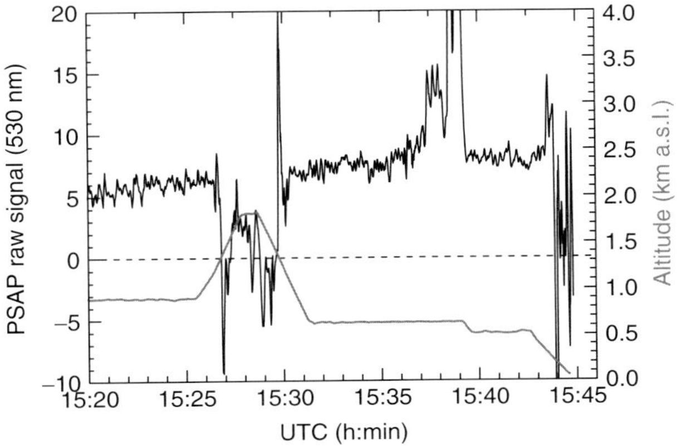

Figure 4.17 Time series of PSAP raw signal (530 nm; black line) and flight altitude (gray line); data originate from a descent into Dakar airport inside a strong dust layer (Petzold *et al.* 2011).

during the descent into the dust layer. During the small altitude change at 15:39 UTC, no such behavior is visible. The strong enhancement in the PSAP raw signal between 15:37 and 15:39 UTC is connected to the urban pollution plume of Dakar.

It turned out that for the analysis of PSAP data flight levels at constant altitude avoid measurement artifacts from changing pressure and enhances instrument sensitivity because absorption coefficient measurements can be averaged over the constant level sequences. A detailed discussion of flights strategies for measuring absorption coefficients is given, for example by Petzold, Kramer, and Schönlinner (2002), Weinzierl *et al.* (2009), and Andrews, Sheridan, and Ogren (2011).

Despite the disadvantage of data correction for aerosol scattering and filter loading by particles that requires the simultaneous measurement of aerosol absorption and aerosol scattering (PSAP, Aethalometer) or of the limited applicability to airborne platforms (Aethalometer, MAAP) due to pressure integrity issues, the filter-based methods offer a simple and robust instrument setup. They have the further advantage of accumulating light-absorbing particulate matter on a filter matrix so that reducing the temporal resolution enhances the sensitivity of the method, which allows to measure low-absorption coefficients even under background conditions. PSAP instruments were used to measure light absorption in biomass-burning plumes after intercontinental transport (Petzold *et al.* 2007) or recently in the Arctic troposphere (Brock *et al.* 2011).

The first deployment of a photoacoustic instrument on board of an aircraft is reported by Arnott *et al.* (2006). The effects of changing ambient pressure and temperature on the generation of an acoustic signal, and in particular on the acoustic resonator constant C_{res} (Eq. (4.17)) are carefully evaluated by Arnott *et al.* (2006) to characterize the acoustical performance of the resonant photoacoustic instrument for accurate measurements of the aerosol light absorption coefficient aloft. This publication serves as a key reference for the application of photoacoustic spectroscopy onboard of an aircraft. Recently, a five-channel photoacoustic instrument was successfully operated onboard of an aircraft (Lack *et al.* 2012).

4.5.3
Extinction Due to Aerosol Particles

The aerosol particle extinction coefficient $b_{ext,p}$ at ambient conditions can be measured directly by cavity ring-down spectroscopy (CRDS). Strawa *et al.* (2003) describe the first airborne prototype of a CRDS instrument. This method uses a pulsed laser beam that travels a number of times between two highly reflective mirrors (better than 99.96%) on both sides of a sensing volume. The path length can be made long enough to accurately determine the extinction, and scattering is determined simultaneously. The laser pulse bounces between the two mirrors inside the ring-down cavity like a ping-pong ball. The amount of light lost at each bounce is collected and detected with a photomultiplier or similar detector. The intensity of the light decreases exponentially depending on the mirror reflectivity and the extinction of the material inside the cavity. The extinction coefficient is then obtained by the difference between measurements made when the cell contains filtered air and when the cell contains a particle-laden flow as

$$b_{\mathrm{ext,p}} = \frac{1}{c} \cdot \left(\frac{1}{t_{\mathrm{p}}} - \frac{1}{t_0} \right) \tag{4.18}$$

where c is the speed of light, and k_{p} and k_0 are the ring-down times of the aerosol laden flow and filtered air, respectively.

CRDS is an emerging method for measuring aerosol light extinction by compact and robust instruments. Successful airborne applications are reported by Schmid *et al.* (2006); Brock *et al.* (2011); Langridge *et al.* (2011). While first airborne applications of CRDS instruments were limited to the boundary layer (Schmid *et al.* 2006), improvements in the lower detection limits permit now an applications to measurements in the free troposphere (Brock *et al.* 2011). There are no particular issues with operating CRDS instruments on airborne platforms. However, careful considerations of potential cross-sensitivities to light-absorbing gases and relative humidity are required (Langridge *et al.* 2011).

4.5.4 Inversion Methods

Data inversion methods offer another approach for inferring aerosol optical properties from measured microphysical aerosol data, in particular, for those applications where one key optical parameter cannot be measured directly but has to be inverted from other data and model assumptions. Inversion methods are particularly valuable approaches for airborne applications because available sensitive methods for microphysical parameters such as size distributions or PSAP absorption can be combined with aerosol optical properties based, for example, on Mie theory, to determine the complete set of aerosol optical properties, including the complex refractive index.

One approach is described by Petzold *et al.* (2009) who use size distribution measurements and combine them with PSAP-based absorption measurements. The information on the scattering coefficient that is required from the correction of the PSAP signal (Eq. (4.15)) is calculated from measured size distributions, so that

$$b_{\mathrm{PSAP}}(\lambda) + \int_{-\infty}^{\infty} Q_{\mathrm{sca,p}}(D_{\mathrm{p}}, \lambda, m) \cdot \frac{\pi \cdot D_{\mathrm{p}}^2}{4} \cdot \frac{\mathrm{d}N}{\mathrm{d}\log D_{\mathrm{p}}} \mathrm{d}\log D_{\mathrm{p}} \rightarrow b_{\mathrm{abs,p}}, b_{\mathrm{sca,p}}, \widetilde{\omega}_{\mathrm{p}} \tag{4.19}$$

An iterative approach is applied for achieving consistent results that match the PSAP absorption data at three wavelengths and the size distribution data. Having calculated the scattering coefficient permits the correction of the PSAP signal b_{PSAP} and the resulting determination of $b_{\mathrm{abs,p}}$, $b_{\mathrm{sca,p}}$, the single-scattering albedo $\widetilde{\omega}_{\mathrm{p}}$, and the complex refractive index m. Relative uncertainties in the determination of the imaginary part of the refractive index, k, are 10–16% for values $> 10^{-3}$ and 25% for values $< 10^{-3}$. The respective relative uncertainty for the absorption Ångström exponent, $å_{abs,p}$ is 20% (Petzold *et al.* 2009). The concepts of multi-instrument data inversion are described in more detail in Section 4.3.4.

4.6 CCN and IN

Measurements of the nuclei of cloud particle formation are important for defining cloud properties such as droplet or ice crystal number concentrations and size spectra. Droplets nucleate on CCN when water supersaturation occurs due to updrafts or cooling. IN are a smaller aerosol subset, which produce ice by several heterogeneous mechanisms, depending on aerosol properties, temperature, supersaturation with respect to ice and liquid water, and time. This section explores the scientific motivations for studying CCN and IN, emphasizes the current state of the art in measurement methods and calibration procedures, and summarizes the challenges and emerging techniques related to airborne observations.

The largest climate change uncertainties are the effects of aerosol particles on clouds, for example, through subsequent indirect radiative forcing and precipitation alterations (e.g., Lohmann and Feichter, 2005). Hence, measurements of cloud-forming particles (CCN and IN) are fundamental.

CCN ability to activate into cloud droplets is determined by aerosol physical and chemical properties. The equilibrium saturation ratio (S_v^{eq}) at the surface of a droplet of a given diameter (D_p) and temperature (T) is described by a modified Köhler equation (Köhler, 1936), (1998) as

$$S_v^{eq} = \exp\left(\frac{4\,b' \cdot M_w}{k_B \cdot T \cdot \rho_w \cdot D_p} - \frac{\Phi \cdot M_w}{\pi/6 \cdot \rho' \cdot D_p^3 - \sum_i m_i} \cdot \left|\sum_i \frac{v_i \cdot m_i}{M_i}\right|\right) \tag{4.20}$$

where $k_B = 1.3806 \times 10^{-23}\,\mathrm{J\,K^{-1}}$ represents the Boltzmann constant; b' is the surface tension of the solution; M_w and M_i are the respective molecular weights of water and solute, respectively; m_i is the dissolved solute mass; ρ_w and ρ' are the densities of water and aqueous solution, respectively; ν_i is the number of ions into which a solute molecule dissociates (van't Hoff factor), and Φ is the osmotic coefficient of the aqueous solution. It is seen that S_v^{eq} is reduced by dissolved ions (2nd or Raoult term in Eq. (4.20) exponential), which thus overcome surface tension effects (1st or Kelvin term in Eq. (4.20) exponential) that otherwise prevent the formation of small water droplets.

The peak or critical supersaturation (S_c) at a wet critical diameter (D_c) for a given particle is determined by the balance between Raoult and Kelvin effects. Solution droplets smaller than D_c remain at stable sizes based on ambient humidity. Droplets larger than D_c are activated and continue to grow as long as the ambient supersaturation is larger than the equilibrium of the droplet surface. In general, S_c of each particle depends inversely on the number of soluble ions that it contains and its physical size (i.e., a higher S is required to activate smaller particles or those with fewer soluble ions). Smaller molecular weights and higher densities of the solution ions increase CCN efficiency. Insoluble components add mass and size without affecting S_c. CCN measurements are necessary because aerosol chemistry and mixing state are usually insufficiently known to accurately determine S_c.

Only CCN with S_c less than maximum S_v^{eq} in an air parcel, or S_{max}, form activated (super-micrometer) droplets, which can interact with solar radiation (determine cloud albedo) and initiate precipitation. The updraft velocity (w) and the CCN spectrum near cloud base determine S_{max}, which then determines maximum droplet concentrations (N_{Dmax}). The variability of w and CCN concentrations creates a distribution of S among various cloud parcels. Atmospheric S cannot be measured directly, but can be inferred by comparisons between CCN spectra and measured cloud droplet concentrations (N_D) (Snider *et al.* 2003; Hudson, Noble, and Jha, 2010). Observed N_D are usually less than N_{Dmax} mainly due to evaporation and dilution by clear air (entrainment).

Single variable parameterizations simplify estimates of CCN concentrations in climate models (Gunthe *et al.* 2009; Petters and Kreidenweis, 2007) and have been applied to a range of aerosol types (Pringle *et al.* 2010). It has long been known that estimating CCN concentrations based on the measured size distribution and an assumed pure salt composition often leads to an overprediction of CCN concentrations (Bigg, 1986; Snider *et al.* 2003). Shulman *et al.* (1996) and Roberts *et al.* (2002) showed that CCN activity can be approximated by simple component mixtures of soluble and insoluble species. Recent studies (Jimenez *et al.* 2009; Sullivan *et al.* 2009; George and Abbatt, 2010; Roberts *et al.* 2010) suggest that aerosol particles from various sources undergo atmospheric processes that tend to homogenize their chemical and physical properties with respect to CCN activity. Understanding the evolution of CCN in the atmosphere will allow more accurate representation of CCN in climate models.

Homogeneous ice nucleation occurs when droplets formed on CCN freeze at temperatures below −38 °C. Heterogeneous IN activate ice formation either directly onto a surface from the vapor phase (deposition nucleation; see Figure 4.18) or initiate freezing within a droplet (Vali, 1985) as part of the cloud formation process at sufficiently low temperatures (condensation freezing; Figure 4.18) or after cloud droplets or haze particles have cooled to sufficiently low temperatures (immersion freezing; Figure 4.18). A fourth heterogeneous process (contact freezing) occurs when IN collide with the droplet surface from outside or from within the droplet (contact-freezing inside-out); see Durant and Shaw (2005). IN measurements are essential because classical nucleation theory has not proven to be adequate alone for quantifying heterogeneous ice nucleation. Nevertheless, no single device can reproduce all possible IN mechanisms.

The predictive utility of IN measurements can be tested versus cloud ice concentrations, but processes such as secondary ice formation can complicate such efforts. IN measurements should at least attempt to quantify IN sources and IN temporal and spatial variability in relation to other aerosol properties. IN measurements are also useful, in principle, for constraining the roles of heterogeneous, homogeneous, and secondary ice formation mechanisms. There has been some progress made in defining the specific contributions of IN to ice

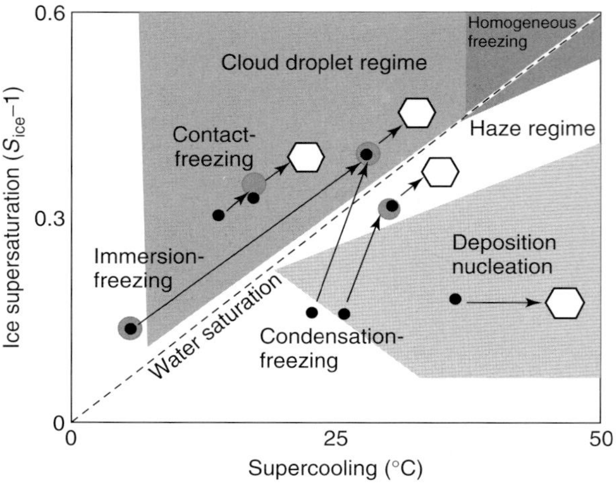

Figure 4.18 Ice nucleation mode regimes that IN instrumentation seek to define. (Courtesy of P. deMott.)

formation in clouds where secondary processes are minimal (Eidhammer *et al.* 2010).

CCN and IN measurements are most often done in diffusion chambers with quite different technical requirements. Both measurement approaches seek to carefully reproduce the appropriate cloud conditions, which are more complex for IN.

4.6.1 CCN Measurements Methods

Single-S CCN counters have existed for decades (Twomey, 1959). Although natural clouds usually form by expansion, this is not a useful technique for CCN measurements. Chemical diffusion chambers (Twomey, 1959) soon gave way to static thermal-gradient diffusion chambers (Squires and Twomey, 1966; Twomey and Wojciechowski, 1969) that exploit the nonlinearity of the saturation vapor density as a function of temperature between two parallel plates. Solving Fick's second law of diffusion (truncating the higher order terms), vapor pressure e at position x between two water-coated surfaces is determined as a function of the saturation vapor pressures at warm, $e_{s,w}(T_w)$, and cold, $e_{s,w}(T_c)$, walls and the plate separation as

$$e = e_{s,w}(T_C) + \frac{e_{s,w}(T_W) - e_{s,w}(T_C)}{\Delta x} \cdot x \tag{4.21}$$

Similarly, the temperature profile is found by solving Fourier's law of heat conduction:

$$T = T_C + \frac{T_W - T_C}{\Delta x} \cdot x \tag{4.22}$$

Because the saturation vapor pressure at any position is an exponential function of temperature following the Clausius–Clapeyron equation, water vapor supersaturation is generated between the two surfaces.

Static instruments use batch processing with an analog light-scattering signal from the droplets (Lala and Juisto, 1977; Snider and Brenguier, 2000) or droplet photographs (Delene *et al.* 1998). CCN spectra (concentration versus S) are obtained by changing the plate temperatures, but this requires more time than is available for aircraft. The static chamber of Snider *et al.* (2003) has since been optimized for airborne measurements allowing four-point spectra to be obtained in 140 s.

To achieve higher temporal resolution for aircraft, the next generation of CCN instruments used OPCs to count and even size the droplets from a continuous flow cloud chamber (Sinnerwalla and Alofs, 1973; Hudson and Squires, 1976; Fukuta and Saxena, 1979). Hudson (1989) extended the continuous flow technique to develop a CCN spectrometer with an S range of 0.01 to 1% by using a series of increasing S zones and using the consequent dispersed sizes of the droplets to deduce the S_c of the particles. Roberts and Nenes (2005) exploited differences in mass and heat diffusion with a cylindrical continuous flow thermal-gradient CCN instrument chamber that generates a constant S along the streamwise axis (Figure 4.19). The Droplet Measurement Technologies (DMT) CCN instrument incorporates this architecture. CCN spectra in this device can be obtained in airborne applications by changing the flowrate (Moore, Nenes, and Medina, 2010). Since changes in pressure affect heat and mass transfer, a pressure-controlled inlet is virtually required for airborne operation (Moore *et al.* 2011). Size-resolved CCN measurements complement number size distributions to determine aerosol hygroscopicity and mixing states (Kuwata and Kondo, 2009). A thorough review and assessment of the performance of various CCN instruments is discussed in much greater detail by Nenes *et al.* (2001a) and Rose *et al.* (2008). Nonetheless, direct measurements at S below 0.1% remains a challenge as most CCN instruments are limited to larger S due to short growth times (Saxena and Carstens, 1971; Nenes *et al.* 2001a). CCN spectra can be achieved with sufficient temporal resolution for airborne studies by both the Desert Research Institute (DRI) instantaneous CCN spectrometer (Hudson, 1989) and the streamwise CCN instrument (Roberts and Nenes, 2005; Moore, Nenes, and Medina, 2010).

4.6.2 IN Measurement Methods

Aircraft IN measurement methods have been largely guided by what is reasonable and possible for measuring the extremely small numbers of these particles (frequently one in one million total particles), whose low numbers belie their vital role in initiating precipitation in cold clouds. Most desirable is to determine IN concentrations by all mechanisms at the temperatures, relative humidities, time, and histories of the sample air. In place of flying several instruments, simulation of exposure to specific temperatures and S is thought to represent the maximum IN concentrations. Aircraft IN devices are as follows: (i) cloud chambers, (ii) mixing

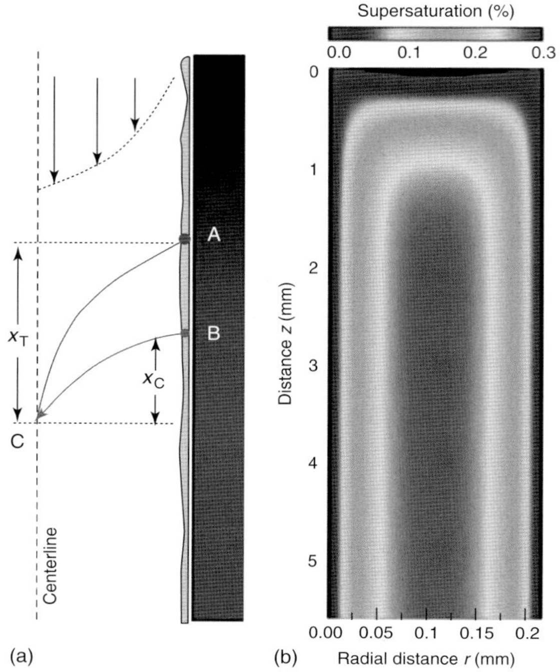

Figure 4.19 The linear thermal-gradient technique for CCN measurements shown in cylindrical symmetry. (a) This illustrates the development of the supersaturation profile at the centerline, exploiting the difference in diffusion of water vapor (B) and heat (A). (b) This shows the resultant supersaturation profile in the column. The supersaturation is always positive throughout the instrument and is nearly constant at the centerline where particles are focused. (Source: Adapted from Roberts and Nenes (2005). Reprinted with permission.)

devices, (iii) continuous flow diffusion chambers (CFDCs), (iv) contact-freezing methods, or (v) capture for offline analysis or processing. Capture and offline analysis involves the use of collected volumes of air on filters or other hydrophobic templates for processing in devices that produce S using thermal and vapor gradients and observe the surfaces via microscopy (Langer and Rogers, 1975; Klein *et al.* 2010; Zimmermann *et al.* 2007). Most modern designs of such devices focus a flow of relative humidity-conditioned air over the collection surface. Much research has documented limitations of porous filters, namely, no a priori knowledge of the needed sample volume, that part of the in surface may be hidden within the filter body or by each other particles, that particles may be affected by heat transfer material sometimes used to mate the filter to cold plates, and the possibility of vapor competition effects on ice nucleation when the first crystals form at the filter surface. Some of these limitations may be mitigated; the elimination of particle exposure to heat transfer and chemical oils or vapor contacting particles appear to be critical (Klein *et al.* 2010b). This method should readily assess IN activation by deposition nucleation, freezing of haze droplets, and condensation freezing above water saturation. Particles identified as IN can be subsequently probed for composition or subjected to repeated condensation/freezing cycles to look for memory effects. Other advantages of this method are relatively large sample volumes for assessing low IN number concentrations and the ease of collection from an ambient aerosol inlet. Collections of this type may also be released into water for droplet immersion freezing studies (Ardon-Dreyer, Levin, and Lawson, 2011).

Portable IN cloud chambers characteristically form a liquid cloud on the total ingested aerosol or, sometimes on added CCN, that is, a mixed-phase cloud is formed. This was the principle used for the "NCAR counter" (Langer, 1973; Super *et al.* 2010). This involves mixing aerosol into a manufactured cloud of drops or generation of cloud by mixing air streams of different humidity. Cloud S is typically not controlled. IN may activate by deposition, condensation and immersion freezing, and perhaps even contact freezing if additional CCN are added to limit IN participation in droplet formation. These methods offer assessment of maximum IN activation in mixed-phase clouds and potentially large sample volumes when time resolution is not a concern. Limitations are that deposition nucleation cannot be separately assessed, the thermodynamic path of particles (relative humidity and temperature) is not well controlled during cloud formation, and the continuous presence of cold cloud in contact with chamber walls can lead to frost control issues.

CFDCs are designed to allow cloud and ice formation to ensue under controlled temperature and humidity conditions during continuous sampling. The theory of operation is the same as for static diffusion chambers, see Eqs. (4.20) and (4.21), regarding creating ice and water supersaturations within the space between two walls held at different temperatures and coated with ice in this case (Rogers, 1988). For CFDCs with a vertically downward flow orientation, steady-state flow may be calculated as a superposition of a buoyant circulation and a viscous (Poiseuille) flow, which skews maximum downward velocity and an initially center-positioned aerosol lamina toward the colder wall (Rogers, 1988). An example of predicted temperature and supersaturation profiles are shown in Figure 4.20(a).

Continuous flow largely eliminates vapor competition effects that can exist in static instruments and offers the possibility of continuous and near real-time measurements of a wide dynamic range of IN number concentrations, and ease of interfacing with methods for collecting activated IN for chemical analysis via electron microscopy (Kreidenweis *et al.* 1998) or single-particle mass spectrometry (Cziczo *et al.* 2003). Designs differ only in the applied geometry/orientation and the means for detecting ice formation. A cylindrical design of nested warm and cold walls (to eliminate horizontal boundary conditions) with a sheathed flow of sample air between the walls (Figure 4.20(b)) was the first to be applied toward aircraft measurements (Rogers *et al.* 2001), but a vertical, parallel flat-plate design has recently also been developed toward aircraft use, which may offer ease of construction, ice detection, and maintenance (Stetzer *et al.* 2008). The version described by Chou *et al.* (2011) is the basis for the DMT spectrometer for ice muclei (SPIN) instrument.

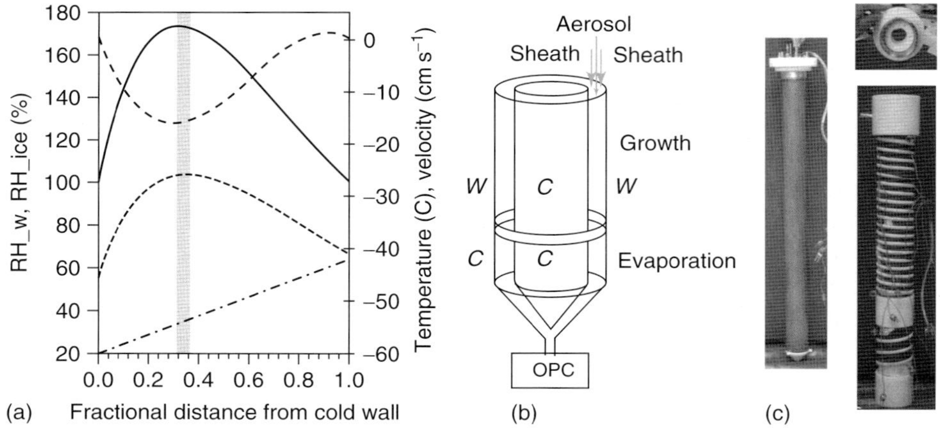

Figure 4.20 Example of CFDC characteristics and construction. (a) Calculations of two-dimensional profile of flow velocity (dashed), temperature (dash-dotted), and relative humidity with respect to water (dotted, RH_w) and ice (solid, RH_{ice}). (b) A cylindrical geometric configuration of warm (light shading) and cold (darker shading) walls in a current CFDC (Richardson, 2009). C for cold and W for warm walls. (c) Images (clockwise) of the inner cylinder (interior coils), capping knife-edge inlet surrounded by sheath flows, and outer cylinder (exterior coils) of this device.

Operation of CFDC instruments for specific processing conditions (controlled RH, *T*) may permit assessment of the individual or joint contributions of different ice nucleation mechanisms. For example, the conditions for the onset of homogeneous freezing have been isolated at temperatures below −38 °C (DeMott *et al.* 2009; Richardson *et al.* 2010). The mechanisms contributing above water saturation are less clear since particles must undergo increasing RH and decreasing temperature as they transit to the steady conditions in the chamber, and uncertainty in RH is large enough across the aerosol lamina that cloud activation conditions are not precisely defined. Since most particles may be activated as CCN above water saturation, and residence times are limited to less than 10 s, it is expected that contact freezing is not well assessed by such measurements. Detection of nucleated ice crystals is performed using systems to detect the optical size growth of ice crystals in excess of droplets or to detect the ice phase via scattering or depolarization properties. Detection by size alone requires restriction on the size of aerosol particles sampled, while phase discrimination requires ice growth to certain sizes and special calibration. As for cloud chambers, CFDC instruments must take special precautions to document and reduce formation of artifact counts from frost growing at sites of opportunity in these devices.

A new device using mixing to generate controlled *S* at supercooled temperatures for IN activation has been described by Bundke *et al.* (2008). This device may be described as employing the continuous flow mixing chamber (CFMC) method. Supersaturation is produced from the turbulent mixing of cold dry air with warm humidified air in a closed-loop system. The sample flow is controlled by mass conserving principles using a separate mass flow controller at the exhaust of the system. Ice particles grown to sizes above 4 μm during about 10 s residence time are discriminated from droplets by their individual circular depolarization properties and counted in an optical detector. The use of an additional fluorescence channel for detecting biological IN has also been demonstrated (Bundke *et al.* 2010). Potential advantages of CFMC instruments are as follows: (i) a high sample flow (up to 10 LPM) for improved counting statistics at low IN concentrations, (ii) demonstrated use of active monitoring of RH and T in the ice activation region, and (iii) capability for rapid variation of S and T through varying the flow rates of the different particle-free gas flows. These benefits remain to be fully demonstrated, as does the application of this method in aircraft operations. Possible limitations of the present CFMC methods relate to the complex depolarization and frost point measurements that require careful adjustment and calibration. In addition, since activated IN are dependent on the initial water vapor content for their growth as ice crystals, there must be an upper concentration limit for which not all particles can grow to a detectable size when in competition for water vapor. This limitation may be solved, when recognized, by diluting the IN sample flow with particle-free air, albeit at the expense of lower sample volume.

Deshler and Vali (1992) review the few methods for contact-freezing measurements potentially applicable for aircraft measurement, which are available only in conference/workshop publications: Cooper's method of placing manufactured drops on a cold filter surface to initiate contact with collected aerosol particles, and Vali's drop freezing counter that uses electrostatic precipitation to scavenge particles onto large supercooled drops.

4.6.3 Calibration

4.6.3.1 CCN Instrument Calibration

Since water saturation cannot be directly measured at the low saturation ratios found in clouds, instruments must be calibrated based on known hygroscopic properties of the calibration aerosol. The calibration of CCN instruments is done by generating classified aerosol particles of known dry sizes and composition, which thus specifies S_c, and relating the number of cloud chamber activated droplets to the total classified particle concentration. The resulting activation curve determines the critical dry diameter at which the CCN instrument grows the particles into droplets; this diameter is related to the instrument S based on the Köhler theory. For the DRI instruments, a calibration curve relates S_c to cloud chamber droplet sizes in terms of channel numbers. For many years, the calibrations of static CCN instruments were determined based on theoretical calculations of temperature gradient (Katz and Mirabel, 1975). However, recent studies have indicated discrepancies between observed and calculated S due to the difficulty of properly measuring temperature at the wall surfaces and the impact of thermal resistance between walls and the wetted surface material on the measured temperatures (Snider *et al.* 2006; Lance *et al.* 2006).

Recent studies (Kuwata and Kondo, 2009; Rose *et al.* 2008) indicate that shape factors play a role in the calibration of CCN instruments. For example, cubic salts, such as NaCl, introduce up to 18% errors in instrument calibrations, especially at high *S*. Furthermore, shape factors are influenced by the rate of drying the salt before classification. Ammonium sulfate has been recommended due to its sphericity (Rose *et al.* 2008), although the value of its van't Hoff factor (number of ions that dissolve) is questionable and may vary with size (Chýlek and Wong, 1998; Low, 1969). The residence time in the growth section of the CCN instrument becomes the limiting factor in directly measuring CCN (approx. 10 s are needed to obtain measurements at 0.1% *S*). In addition, for calibration at supersaturations less than 0.1%, considerations must be given to the threshold size to distinguish activated CCN droplets from large deliquesced interstitial aerosol (Roberts and Nenes, 2005).

Continuous flow CCN chambers also require calibrations at a particular internal instrument pressure. The DRI instruments (Hudson, 1989) were the first to employ such calibrations, which were absolutely necessary for the continuous flow instruments. Calibrations on the continuous flow streamwise CCN instrument have been mapped for a given pressure and temperature gradient (Roberts *et al.* 2010); however, data during ascents and descents must be removed. Hence, pressure controllers are required on the continuous flow systems to maintain supersaturation calibrations for continuous flow airborne CCN measurements.

4.6.3.2 **IN Instrument Calibration**

Calibration and validation of ice nucleation instruments is not straightforward and remains undemonstrated. The requisite needs for IN instrument calibration include validation of calculated thermodynamic conditions and generation of IN with at least specific activation RH properties that can be delivered over a range of number concentrations from very low to very high. That different ice nucleating materials display a range of behaviors that are not well documented and sometimes are not reproducible is the chief complication.

Many investigators have tested IN instruments for their ability to activate silver iodide (AgI) particles. However, AgI particles can possess an extremely wide range of IN properties depending on generation method, size, and trace chemistry (Finnegan and Chai, 2003). Consequently, comparison to historical data may not be very meaningful. Other widely available INs are certain clay minerals such as kaolinite and montmorillonite. A reliable method for producing different sizes of these types of particles with reproducible IN characteristics has not been demonstrated. The biological IN Snomax™, a manufactured version of the bacteria pseudomonas syringe, holds some promise as a useful IN calibrant. These particles are observed to activate at relatively precise relative humidity as a function of temperature, at least by some methods (Möhler, DeMott, and Stetzer, 2008).

An additional calibration need is for the definition of artifact counts in each method. For CFDC instruments, this is achieved using filtered sample air. For real-time sampling, application of procedures for assessing background counts from frost needs to be done at regular intervals and to occur for the full range of processing conditions (temperature and RH). Finally, the processing of IN data into meaningful experimental units for reporting may itself be considered a calibration issue for each device, as governed by Poisson sampling statistics.

4.7 Challenges and Emerging Techniques

4.7.1 Particle Number

Overall, CPCs for use on aircraft have matured into robust, reliable, compact, and easy-to-use instruments. Nevertheless, several challenges remain. One important requirement is the need to detect particles less than 3 nm, as freshly nucleated particles may need several hours to grow to sizes larger than 3 nm at which point they can be detected by current CPC technology. Especially under clean ambient conditions where only small amounts of precursor condensable material limit the growth rate of these small particles, it would be desirable to detect the particles just after they have grown beyond the size of the critical cluster. Two instruments, the pulse-height CPC (PH-CPC) and the particle size magnifier (PSM) have recently been developed and applied to ground-based ambient and laboratory measurements (Gamero-Castano and Mora, 2000; Sipila *et al.* 2008; Sipila *et al.* 2009; Sipila *et al.* 2010). These techniques are capable of detecting particles less than 3 nm and application to aircraft-based measurements seems feasible in future.

Another issue concerns the heating of the particles (i) when their velocity relative to the aircraft is being slowed down during the sampling process or when the sample flow is allowed to adapt to the temperatures of the sampling lines inside the warm aircraft cabin and (ii) when changing from ambient conditions (e.g., $-55\,^{\circ}$C at tropopause altitudes) to the region of fixed temperature in the condenser unit (e.g., $+10\,^{\circ}$C). It is conceivable that a fraction of the particles may shrink in size during this process due to evaporation of semivolatile substances when being heated by several tens of degrees. If the particles shrink below the lower size cut-off of the CPC, they are not counted anymore and only a lower limit of the actual particle concentration is measured. This is particularly critical for the differential measurements provided by CPC batteries and could influence the derived results considerably. To our knowledge, this issue has not been addressed so far. One recommendation important for future aircraft-borne measurements of nucleation mode particles between 2 and 20 nm would be to modify CPCs and their sampling inlet lines in such a way that the aerosol is maintained at ambient temperature conditions to reduce such measurement artifact.

4.7.2 Particle Size

Given the high speed of aircraft, airborne measurements with high time resolution are often necessary to resolve spatial variations of aerosol properties. The widely used scanning mobility particles sizer (SMPS) often takes about 1 min or more to measure a size distribution. This may be too slow to capture variations over spatial scales less than a few kilometers. A number of fast instruments based on electrical mobility were recently developed. By simultaneously detecting particles of a wide range of electrical mobility, these instruments measure sub-micrometer size distribution within 1 s. One of them is fast integrated mobility spectrometer (FIMS), which combines a separator and a condenser in a continuous rectangular channel (Kulkarni and Wang, 2006a; Kulkarni and Wang, 2006b; Wang, 2009). Inside the separator, under the influence of an electric field, charged particles are separated into different flow streams based on their electrical mobility. The separated particles then grow into super-micrometer droplets in the condenser. The grown droplets are illuminated at the exit of the condenser and their images are captured by a high-speed CCD camera. The images provide both the particle concentration and the droplet position, which directly relates to the particle electrical mobility. By simultaneously measuring particles of different sizes/mobilities using single-particle counting, the FIMS provides sub-micrometer aerosol size distribution within 1 s (Olfert and Wang, 2009; Olfert, Kulkarni, and Wang, 2008; Wang, 2009).

Other fast instruments include electrical aerosol spectrometer (EAS) (Horrak *et al.* (1998); Mirme *et al.* (2010); Tammet, Mirme, and Tamm (2002)), differential mobility spectrometer (DMS) (Biskos, Reavell, and Collings (2005)), and engine exhaust particle sizer (EEPS) (Johnson, Rice, and Brown (2004)). All three instruments have geometry of concentric cylinders. Charged particles are introduced near the central rod and migrate radially outward from the rod under influence of electric field. Particles of different mobilities are concurrently detected by a column of electrically isolated electrode rings that are coaxial with the central rod. The particle number concentration is determined by measurement of the electrical current collected on the various electrodes using electrometers. Compared to single-particle optical counting in FIMS or CPC, the low sensitivity of electrometer often limits these instruments to measurements of high-concentration combustion aerosol particles. Future improvements of electrometer technology may allow the fast measurement speed of these instruments to be fully taking advantage of for airborne measurements of ambient aerosol particles.

Airborne measurements also subject instruments to rapid variation of sampling conditions, including temperature, pressure, and relative humidity. The maximum voltage at which a DMA can operate is limited by electrostatic breakdown, which occurs when electric field exceeds approximately $10^4\,\mathrm{V\,cm^{-1}}$ at atmospheric pressure. The limit of electric field strength, and thereby the maximum classifying voltage, decreases with decreasing sampling pressure (i.e., increasing sampling altitude). A recently developed opposed migration aerosol classifier (OMAC) classifies aerosol particles with minimal diffusion effects at substantially lower voltage than in traditional DMA (Flagan, 2004). This unique feature helps maintain the dynamic size range at reduced maximum classifying voltage, which is necessary to prevent arcing at high sampling altitudes. OMAC also allows miniaturization of the SMPS system, which is important for measurements onboard research aircraft especially unmanned aerial vehicles.

4.7.3
Aerosol Optical Properties

Although measurement methodologies are available now a days for the airborne measurement of principal aerosol optical properties, there are significant challenges for *in situ* instrument development. The measurement of aerosol light absorption still requires substantial improvement. Filter-based methods, such as PSAP when operated on an aircraft, can measure absorption coefficients as low $0.1\,\text{Mm}^{-1}$ (Petzold, Kramer, and Schönlinner, 2002), but suffer from measurement artifacts during pressure changes that limit the data analysis to constant-pressure flight legs (Andrews, Sheridan, and Ogren, 2011; Petzold *et al.* 2007). One potential solution to this problem may arise from the development of a single-filter leak-tight MAAP version, which shows reduced cross-sensitivity to filter effects (Petzold *et al.* 2005a). Another limitation of filter-based absorption measurement methods is the reduced time resolution of $\geq 30\,\text{s}$ at low-to-moderate absorption coefficient levels of $\leq 1.0\,\text{Mm}^{-1}$. Concerning time resolution, photoacoustic spectroscopy may report 1 Hz data. However, applications to airborne research require an instrument sensitivity of $\leq 0.1\,\text{Mm}^{-1}$, while current photoacoustic instruments used for airborne research are limited to an uncertainty range of $\pm 1.0\,\text{Mm}^{-1}$ (Arnott *et al.* 2006).

The measurement of light extinction due to particles may be further improved by a modification of CRDS, the so-called cavity-attenuated phase shift particle extinction monitor (Massoli *et al.* 2010). This new instrument has not yet been operated on board of an aircraft but shows a robust and compact design that makes it a promising technique for the development of an airborne *in situ* extinction instrument. Combining improved techniques for measuring light extinction and light absorption by particles would in turn provide more precise *in situ* measurements of the single-scattering albedo. Currently a combination of CRDS and PSAP is the most advanced approach for airborne applications (Brock *et al.* 2011).

One pressing key topic in atmospheric aerosol research is the observation of aerosol optical properties at ambient RH. Actually, light extinction can be measured at ambient RH by means of high spectral resolution LIDAR instruments, which are described in Chapter 9, whereas most extractive instruments are operated at dry conditions. Except for the scattering coefficient, when humidified integrating nephelometers can be used, no extractive methods are available so far for measuring at ambient humidity levels.

Particle asphericity or the particle asymmetry parameter, respectively, is another key parameter for which no adequate measurement technique exists. For super-micrometer-sized crystals, the polar nephelometer provides excellent results; see (Crépel *et al.* 1997) and Chapter 5. This instrument can also be applied for the measurement of super-micrometer-sized mineral dust particles of volcanic ash. Furthermore, the cloud-aerosol spectrometer (CAS; see Chapter 5) will be equipped with a depolarization channel, which then may permit the separation of spherical and nonspherical super-micrometer-sized particles. Comparable approaches for sub-micrometer-sized aerosol particles, however, are not yet available.

4.7.4
Chemical Composition of Aerosol Particles

Many challenges remain in the quantification of the aerosol chemistry onboard aircraft. Majors are related to the collection of coarse particles, largely related not only to the performances of particle inlets but also to the time resolution of the measurements. Advances via mass spectrometry and ablation/incandescence techniques have been made in achieving time resolution of the order of minutes for particles in the sub-micrometer range, yet for broad classes of particle compounds. However, the quantification of the aerosol bulk chemistry, for example, of the organic fraction speciation at the functional or molecular scale, still relies on filter sampling and postfield analysis, which require exposure times often incompatible with airborne sampling strategy. The development of particle concentrators, large particle inlets, and online chromatographic analysis techniques should be pursued.

4.7.5
CCN Measurements

Some key technical measurement challenges for CCN measurements include measurement of large CCN and discerning surface chemical effects on CCN activation. Giant CCN, those having sizes larger than 1 μm, represent a measurement challenge due to their low S and low number concentrations. Their existence is scientifically important because they may modulate large droplet number concentrations and the formation of drizzle drops (Woodcock, 1952; Feingold *et al.* 1999; Blyth *et al.* 2003; Reiche and Lasher–Trapp, 2010; Hudson, Jha, and Noble, 2011). The condensation coefficient, the ratio of molecules that actually stick to a condensing droplet upon collision, is usually considered to be a constant, although its value is still uncertain by two orders of magnitude. If the condensation coefficient varies, this would alter the relation between CCN number concentration and W (Nenes *et al.* 2001a). This may be the case if CCN are partially soluble, form films around droplets (Feingold and Chuang, 2002), or if droplet surface tension varies (Facchini *et al.* 1999). Hegg *et al.* (2001) and Ruehl, Chuang, and Nenes (2009) present evidence of this possibility. A final technical challenge is the fact that no universal CCN calibration standard exists.

4.7.6
IN Measurements

Numerous technical and measurement challenges remain for IN measurements. Technical issues surround efforts at long period operation and detection of IN acting by different mechanisms. First, all continuous-measuring IN counting devices typically have influences from frost formation on surfaces within the flow that affect measuring low IN number concentrations at device detection limits (Prenni *et al.* 2009). No autonomous instruments for aircraft or climatological sampling yet exist, and frost development is a critical issue for such a development. Possible mitigation may evolve through research on improved means for coating and adhering ice to walls in chambers. Novel ideas are also needed to limit the size and power consumption of any autonomous IN device. Simpler and more reliable detection methods for nucleated ice versus droplets are also needed for continuous flow instruments. Strong consideration of liquid and ice growth kinetics may be needed as well, before applying measurement devices to very low temperature studies, as there may be limitations for detection of ice by size or by phase at very small crystal sizes (Richardson, 2009). There are no present instruments for measuring contact-freezing nucleation from an aircraft. Another important measurement issue not being addressed is sampling of particles through an inlet that might preserve particle phase state (constant RH and T) at ambient conditions so as to measure possible "preactivated" IN (Roberts and Hallett, 1968). Any solution would seem to involve a special inlet that overcomes the dynamic heating effect with active or passive forced cooling. Finally, there are no present IN instruments with demonstrated capability for rapid scanning of IN spectra (number concentration dependence on RH and T). CFMC devices are hypothetically capable of doing this online, and Rogers (1994) describes the potential conversion of a CFDC for similar online scanning. One could also imagine construction of multichannel continuous flow devices. Spectral dependence on T and RH is presently possible only using offline processing of captured particles.

Emerging research issues for IN measurements include isolating time-dependent impacts on ice nucleation, performance of additional closure evaluations in relation to ice formation in clouds, and developing a database for discerning aerosol chemical impacts on IN populations. Methods for real-time detection of composition and size of IN have been demonstrated (Cziczo *et al.* 2003), but it remains to be seen if this or other novel methods will find application in aircraft studies due to the extremely difficult sampling statistics. Future effort would seem to first require methods for preconcentrating particles for IN instruments.

No ideal calibration standards yet exist for IN instruments. An issue is both controlling ice nucleation activity of particles to be consistent, as well as fixing ice nucleation mechanism. Workshop activities using IN instruments and cloud parcel simulation chambers may facilitate improvements in this area, as well as providing insights to solving other technical and research issues (Möhler, DeMott, and Stetzer, 2008; DeMott *et al.* 2011).

5
In Situ Measurements of Cloud and Precipitation Particles

Jean-Louis Brenguier, William D. Bachalo, Patrick Y. Chuang, Biagio M. Esposito, Jacob Fugal, Timothy Garrett, Jean-Francois Gayet, Hermann Gerber, Andy Heymsfield, Alexander Kokhanovsky, Alexei Korolev, R. Paul Lawson, David C. Rogers, Raymond A. Shaw, Walter Strapp, and Manfred Wendisch

5.1
Introduction

5.1.1
Rationale

Clouds play a leading role in the Earth system, particularly in the global energy balance and hydrological cycle. Indeed, clouds contribute to about half of the Earth's albedo, a key parameter that determines the fraction of solar incident energy reflected by the Earth. Clouds also regulate the greenhouse effect that controls the temperature at the Earth surface.

Clouds form and evolve from near the surface (fog) to the stratosphere. At a first approximation, cloud thickness determines cloud albedo regardless of the altitude; however, the top altitude of the cloud is crucial for regulating the greenhouse effect. Indeed, low-level clouds that radiate at almost the same temperature as the surface do not contribute noticeably to the greenhouse effect. Deep convective and stratospheric clouds absorb terrestrial infrared radiation and radiate at a much cooler temperature down to 200 K, hence actively contributing to the greenhouse effect.

Clouds play a key role in the redistribution of energy and water from the equator to the poles and vertically in the atmosphere. More than half of this energy is stored as latent heat ($\approx 2.5 \times 10^6$ J kg^{-1}) when liquid water evaporates at the surface. When the air is cooled, either by vertical ascent or emission of long-wave radiation, down to the point when its relative humidity (RH) exceeds 100%, the water vapor condenses and releases the latent heat. The huge amount of energy released by this mechanism is clearly visible in the rapid development of thunderstorms and tropical cyclones.

In today's climate and forecast models, only cloud bulk properties, that is, water mixing ratios for vapor, liquid, and ice phases, are derived from conservation of heat

Airborne Measurements for Environmental Research: Methods and Instruments, First Edition.
Edited by Manfred Wendisch and Jean-Louis Brenguier.

and total water. How water is exchanged between these categories, how particles grow in each category, how they interact with radiation, atmospheric chemical compounds, and ultimately with dynamics also depend on additional properties of the cloud particles such as their number concentration, size, phase, and shape (for ice crystals this is referred to as *habit*). The study of these processes is commonly referred to as *cloud microphysics*. The most recent model developments today thus aim at implementing a more realistic description of cloud microphysics to account for the spatial and temporal variability of the hydrological cycle and cloud radiative forcing and their susceptibilities to anthropogenic pollution.

When the RH exceeds 100% in a volume of air, water vapor condenses on particle embryos, also referred to as *cloud condensation nuclei* (CCN) for liquid droplets and *ice nuclei* (IN) for ice crystals. The droplet concentration thus varies with the CCN properties of the air mass, from a few particles per cubic centimeter (cm^{-3}) in a very pristine environment to more than a thousand in polluted air masses. The ice crystal concentration varies from a few per liter at temperatures slightly colder than 0 °C to a few thousands at temperatures below −30 °C. Once cloud particles are formed and start depleting excess water vapor, the supersaturation decreases to a very small value. Particles then grow by vapor diffusion.

Once particles reach a size large enough for their fall velocity to become significant, they begin to collide and coalesce (stick together) to form larger drops or ice crystals and the largest particles fall out of the cloud (precipitate). Depending on the temperature (altitude), liquid particles can freeze and release additional latent heat, thus enhancing the convection. Overall, the cloud particle sizes vary from less than a micrometer (μm) for freshly activated CCNs to millimeters for the largest rain drops and to centimeters for hailstones. The ice crystal shape (habit) is also highly variable: the Magono–Lee classification contains almost 80 different particle types, but in real clouds, an almost infinite number of shapes can be observed. More detailed information on these complex processes can be found in Pruppacher and Klett (1997).

This chapter is dedicated to airborne measurements of cloud particles, from their bulk properties such as liquid and ice water content (IWC) to the full characterization of the particle size distribution (PSD) and shapes.

5.1.2
Characterization of Cloud Microphysical Properties

In numerical cloud models, heat and the total water mixing ratio are the conservative variables from which the RH, supersaturation, and partitioning between the water vapor, liquid, and ice phases can be derived. These mixing ratios are precisely defined, although difficult to measure because of the large range of sizes, shapes, and densities of the cloud particles that carry the condensed water. A full characterization of cloud microphysics is challenging, and it is common in numerical models to define particle categories based on size, shape, and density, for example, droplets, rain drops, small ice crystals, snow, graupel, and hailstones. Each category is then characterized by its size distribution $f(D)$ ($cm^{-3}\ \mu m^{-1}$), which represents the number concentration density of particles with diameter D. It then becomes feasible to parameterize microphysical processes using various moments M_p of the distributions:

$$M_p = \int_0^\infty D^p \cdot f(D)\,dD \qquad (5.1)$$

such as particle number concentration or zeroth moment, the integral radius or first moment (water vapor depletion by condensation is driven by this moment), the integrated particle cross section or second moment (light extinction is derived from this moment), the water mass concentration, also referred to as the *liquid water content* (LWC) proportional to the third moment, the precipitation mass flux that is proportional to the fifth moment for small particles and to the fourth moment for bigger ones, and the Radio Detection and Ranging (RADAR) reflectivity that is proportional to the sixth moment.

Such relationships are rigorous for small droplets of pure water that are spherical with a density of $1.0\,g\,cm^{-3}$ and a refractive index of 1. For rain drops, the concept of particle diameter becomes questionable because large drops become more oblate as they fall and the size depends on the orientation. The situation becomes more complicated for ice crystals with anisotropic shapes and ice aggregates whose density varies from 1 to less than $0.2\,g\,cm^{-3}$.

When performing *in situ* measurements of cloud microphysics, it is thus crucial to concomitantly document various attributes of the cloud particle ensemble that are also useful in numerical models. Instrumented aircraft are, therefore, equipped with a suite of instruments to measure the size-differentiated number and mass concentration (over specific size ranges), the bulk liquid and IWCs, and the optical properties such as the extinction coefficient and asymmetry factor.

The definition of hydrometeor size is not unique and depends on the measurement technique as well as the cloud particle type. Throughout the remainder of the chapter, we use the abbreviation, D, to refer to the diameter of a spherical hydrometeor. There are, however, different types of diameter, for example, geometric, aerodynamic, and optical, whereby the geometric diameter refers to the physical size, the aerodynamic diameter is related to the terminal velocity and response to changes in air flow, and the optical diameter is in reference to how the cloud particle interacts with the light and depends on the wavelength of the light and the refractive index of the particle. The optical diameter is also what is measured by those instruments that use light scattering to measure the particle size, as detailed below. The definition of "size" becomes even more nebulous for ice crystals since these are mostly asymmetric and to define their size requires the use of more than a single dimension or, at the least, some commonly accepted dimension, as detailed in Section 5.4 when we discuss data processing and interpretation.

5.1.3 Chapter Outline

There is a large variety of airborne instruments that measure cloud microphysical properties. The earliest airborne measurements in clouds were made by capturing impressions of cloud droplets and ice crystals on surfaces exposed to the passing airstream. This principle is still in use for some specific applications and is described in Section 5.2.

Impactors and replicators evolved into electro-optical spectrometers to measure particle size, hot-wire sensors to measure liquid and IWC, and other techniques for deriving the optical properties of cloud particles. The new generation of airborne microphysical instruments can be classified into two categories: single-particle counters, which detect and characterize cloud particles individually and then describe cloud properties by cumulating a large number of such single-particle measurements, and ensemble integrators, which directly measure some integral properties of a particle ensemble.

Sections 5.2–5.4 describe impactors, single-particle counters, and ensemble integrators, respectively. Each of these sections begin with a brief history of how the technique was first conceived, followed by sufficient detailed information to describe the fundamental principles, then diagrams and photos that show how the technique is implemented, ending up with a discussion of the measurement issues that have to be understood and taken into account when interpreting the measurements.

Details of the more theoretical aspects, including relevant equations and figures, can be found in Section A.4, given in the Supplementary Online Material provided on the publisher's web site. The interested reader can also find much more complete information on all of the techniques and the instruments that implement them by reading the many papers that have been published and are listed in the references.

Very few scientific questions concerning the evolution of clouds and their environmental impact can be addressed with a single instrument, as has been already emphasized, and the compilation, analysis, and interpretation of the measurements from these sensors requires a basic understanding how they operate and complement each other, what limits their implementation, and what are the sources of uncertainties. Section 5.5 discusses how diverse instruments can be combined to document the whole range of particle sizes and shapes or independently measure some integral properties of the particle ensemble for improving the accuracy of airborne measurements.

Airborne instrument development continues as new techniques are needed to address outstanding problems in cloud microphysics. Section 5.6 describes some emerging technologies such as the interferometric laser imaging for droplet sizing (ILIDS) that relies on a simplified implementation of holography, although limited to liquid particles, and the backscatter cloud probe (BCP), a miniaturized optical particle counter (OPC) for integration on in-service (commercial) aircraft and unmanned aerial vehicles (UAVs).

Table 5.1 provides a quick guide to the instruments that are discussed in this chapter, the techniques that they implement, and other useful information that gives the reader a broad perspective of the many types of sensors available and their specifications. The sampling rate (5th column) refers to maximum value, using the full exposed length of laser beam. The depth of field (DOF) may be smaller below a threshold particle size. In addition, volumetric rate may be reduced if probe

Table 5.1 Overview of airborne microphysical instruments for cloud measurements

Parameter measured	Measurement technique	Instrument	Range	Sampling rate	Calibration method	Section	Primary references
particle size	impaction with imaging	VIPS	>5 μm	$0.5\,l\,s^{-1}$	n/a	5.2	Heymsfield and McFarquhar (1996b)
		Cloudscope	>5 μm	$0.5\,l\,s^{-1}$	n/a	5.2	Arnott *et al.* (1995)
	light scattering and interference	FSSP–100	2–50 μm	$40\,cm^3\,s^{-1}$	Beads or drop generator for sizing, positioned scatterer for DOF and beam width	5.3.2	Knollenberg (1976), Knollenberg (1981)
		FFSSP	1–50 μm	$40\,cm^3\,s^{-1}$	Same as FSSP		Brenguier *et al.* (1998)
		CDP	2–50 μm	$25\,cm^3\,s^{-1}$	Beads or drop generator for sizing, analytical sample area	5.3.2	Lance *et al.* (2010)
		FCDP	1–50 μm	$25\,cm^3\,s^{-1}$	same as CDP		
		CAS	0.5–50 μm	$25\,cm^3\,s^{-1}$	Beads for sizing, analytical sample area	5.3.2	Baumgardner *et al.* (2001)
		CAS-DPOL	0.5–50 μm	$25\,cm^3\,s^{-1}$	same as CAS	5.3.2	Baumgardner *et al.* (2012)
		BCP	5–75 μm	$25\,cm^3\,s^{-1}$		5.3.2	Baumgardner *et al.* (2012)
		CPSD	0.5–50 μm	$25\,cm^3\,s^{-1}$		5.3.2	Baumgardner *et al.* (2012)

(continued overleaf)

Table 5.1 *(continued)*

Parameter measured	Measurement technique	Instrument	Range	Sampling rate	Calibration method	Section	Primary references
		SID-2	2–70 μm	$100\ cm^3\ s^{-1}$		5.3.2	Hirst *et al.* (2001)
		SID-3	2–140 μm	$100\ cm^3\ s^{-1}$		5.3.2	Cotton *et al.* (2009)
		PDI	1–2000 μm	Flexible	Drop generator	5.3.2	Bachalo (1980)
		HOLODEC	5–2000 μm	$cm^3\ s^{-1}$	USAF Resolution target	5.3.4	Fugal *et al.* (2004)
		ILIDS	20–200 μm	$10\ cm^3$ per frame, $1\ l\ s^{-1}$ for 100 Hz frame rate	n/a	5.6.1	
particle size	optical array imaging	2D–C/2DG	25–800 μm 25–1600 μm	$5\ l\ s^{-1}$ $10\ l\ s^{-1}$	Beads, Reticle	5.3.3	Knollenberg (1970), Knollenberg (1976), Knollenberg (1981)
		2D-P	20–6400 μm	$170\ l\ s^{-1}$	Beads, reticle	5.3.3	Knollenberg (1970), Knollenberg (1976), Knollenberg (1981)
		1D-260-X	10–620 μm	$3\ l\ s^{-1}$	Beads, reticle	5.3.3	Knollenberg (1970), Knollenberg (1976), Knollenberg (1981)
		CIP	25–1550 μm	$16\ l\ s^{-1}$	Beads, reticle	5.3.3	Baumgardner *et al.* (2001)
		CIP-GS	15–900 μm	$10\ l\ s^{-1}$	Beads, reticle	5.3.3	Baumgardner *et al.* (2001)
		PIP	100–6400 μm	$160\ l\ s^{-1}$	Beads, reticle	5.3.3	Baumgardner *et al.* (2001)
		CPI	>3 μm	$400\ cm^3\ s^{-1}$	Beads	5.3.3	Lawson *et al.* (2001)
		2D-S	10–1280 μm	$8\ l\ s^{-1}$	Beads	5.3.3	Lawson *et al.* (2006)
		HVPS-3	150–19200 μm	$312\ l\ s^{-1}$	Beads, reticle	5.3.3	Lawson, Stewart, and Angus (1998)
		3V-CPI	4.6–1280 μm	$400\ cm^3\ s^{-1}$	Beads	5.3.3	www.specinc.com

LWC/TWC	Heated wire	Johnson Williams LWC		1.2 l s^{-1}	Tunnel icing blade/cylinder	5.4.1	Neel (1955)
		King LWC probe and LWC–100	0.01–3 g m^{-3} at 100 m s^{-1}	4 l s^{-1}	Tunnel icing blade/cylinder	5.4.1	King, Parkin, and Handsworth (1978b)
		Nevzorov LWC and TWC	0.005–3 g m^{-3} at 100 m s^{-1}	4–5 l s^{-1}	Tunnel icing blade/cylinder	5.4.1	Korolev *et al.* (1998a)
		Multiwire WCM-2000 LWC and TWC	0.005–7 g m^{-3} at 200 m s^{-1}	1.2–5 l s^{-1}	Tunnel icing blade/cylinder	5.4.1	Lilie *et al.* (2004)
		T-probe LWC, TWC	n/a	12–70 l s^{-1}	Tunnel icing blade/cylinder	5.4.1	Vidaurre and Hallett (2009b)
particle size	inlet–based evaporators	Ruskin TWC probe	n/a	8 l s^{-1}	n/a	6.4.1	Ruskin (1965)
		UK-MRF	0.005–20 g kg^{-1}	8 l s^{-1}	Tunnel icing blade/cylinder	6.4.1	Nicholls, Leighton, and Barker (1990)
		HTW isokinetic evaporator	5–2500 μmol mol^{-1}	9.5 l s^{-1}	Humidity measurement/calibration	6.4.1	Weinstock *et al.* (2006)
		CLH	0.005–1 g m^{-3}	3 l s^{-1}	Humidity measurement/calibration	6.4.1	Davis *et al.* (2007)
		FISH	0.5–1000 μmol mol^{-1}	l s^{-1}	Humidity measurement	6.4.1	Schiller *et al.* (2008)
		IKP	0.05–10 g m^{-3} at 200 m s^{-1}	8 l s^{-1}	Tunnel, icing blade/cylinder	6.4.1	Davison, MacLeod, and Strapp (2009)

(continued overleaf)

Table 5.1 (*continued*)

Parameter measured	Measurement technique	Instrument	Range	Sampling rate	Calibration method	Section	Primary references
		CVI	0.03–2 $g\,m^{-3}$	$3\,l\,s^{-1}$	Tunnel, icing blade/cylinder	6.4.1	Noone *et al.* (1988)
	Mass accretion	Icing rate detector		$16\,l\,s^{-1}$	Tunnel, icing blade/cylinder	5.4.1	
	Light scattering	PVM-100A	>0.002 $g\,m^{-3}$	$11\,l\,s^{-1}$	Reference Scattering medium	5.3.2	Gerber, Arends, and Ackerman (1994)
extinction coefficient	Light attenuation or scattering	CEP	0.2 to >200 km^{-1}	$6000\,l\,s^{-1}$		5.4.2	
		CIN		$300\,l\,s^{-1}$		5.4.2	Gerber *et al.* (2000)
		PN	1–1000 μm	$5\,l\,s^{-1}$		5.4.2	Gayet *et al.* (1997)
Asymmetry factor	Light scattering	CIN		$300\,l\,s^{-1}$		5.4.2	Gerber *et al.* (2000)
		PN	1–1000 μm	$5\,l\,s^{-1}$		5.4.2	Gayet *et al.* (1997)
effective radius	Light scattering	PVM–100A	3–35 μm $g\,m^{-3}$	$11\,l\,s^{-1}$	Reference scattering medium	5.3.2	Gerber, Arends, and Ackerman (1994)

2D-C: two-dimensional cloud probe; 2D-P: two-dimensional precipitation probe; 2D-S: two-dimensional stereo probe; BCP: backscatter cloud probe; CAS: cloud and aerosol spectrometer; CAS-DPOL: cloud and aerosol spectrometer with depolarization; CDP: cloud droplet probe; CIN: cloud integrating nephelometer; CIP: cloud-imaging probe; CLH: closed path tunable diode laser hygrometer; CPI: cloud particle imager; CPSD: cloud particle spectrometer with depolarization; CVI: counterflow virtual impactor; FCDP: fast CDP; FISH: fast *in situ* stratospheric hygrometer; FSSP: forward-scattering spectrometer probe; FFSSP: fast FSSP; HOLODEC: holographic detector for clouds; HTW: Harvard total water hygrometer; HVPS: high-volume precipitation spectrometer; IKP: isokinetic TWC probe; ILIDS: interferometric laser imaging for droplet sizing; PDI: phase Doppler interferometer; PIP: precipitation imaging probe; PVM: particle volume monitor; PN: polar nephelometer; SID: small ice detector; UK-MRF: United Kingdom Meteorological Research Flight; and VIPS: video ice-particle sampler.

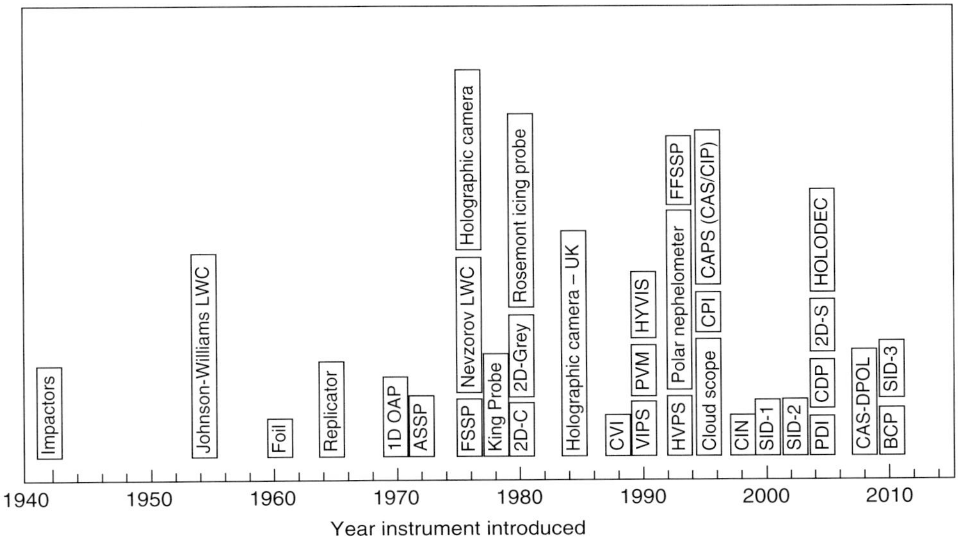

Figure 5.1 The time evolution of instrument development can be seen in this chart that summarizes the introduction of instruments beginning in 1940.

temporarily suspends collection in high particle concentrations ("overload"). For particle-imaging probes, estimates are based on the full width of the shadowing diode array, assuming the center of the image is within the array, and without any image reconstruction of partial images. The PMS 1D-260X automatically rejects particles that exposed over diode array end elements, and its sample volume is accordingly adjusted.

Figure 5.1 gives the reader a historical reference for when the different instruments were first introduced. Figure 5.2 illustrates the range of sizes, expressed here as maximum geometric diameter, of all the probes that measure single-particle properties.

5.1.4
Statistical Limitations of Airborne Cloud Microphysical Measurements

Clouds are turbulent phenomena with high spatial and temporal variability, that is, cloud particle properties such as concentration, size, and shape vary noticeably on scales of millimeters and seconds. Even if it was feasible, the full characterization of each particle properties would not be useful in the framework of existing models. Airborne measurements rather aim at characterizing the statistical properties of a specified cloud volume in terms of particle size and shape distributions. In that perspective, the main limitation of airborne measurements is that the volume swept out by cloud probes have highly elongated "spaghettilike" shapes aligned along the flight track of the aircraft, that is, airborne *in situ* cloud sampling can be considered as quasi-1D measurements. Depending on the concentration of

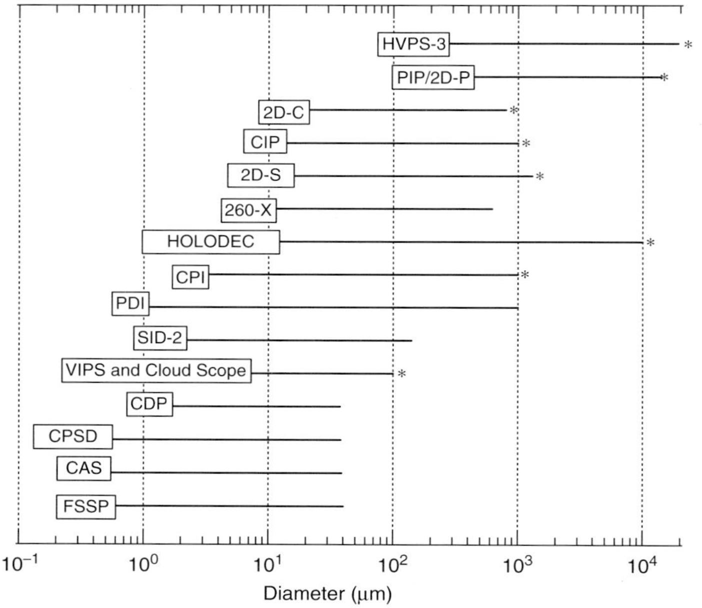

Figure 5.2 The names of all the instruments that measure the size of individual cloud particles are positioned on this chart according to their lower size threshold and size range, where the lower size threshold begins at the right-hand side of the instrument label box and the length of the line is the nominal size range. Upper size ranges of imaging probes with an asterisk at the right may be extended by sizing partial or high axis ratio images along the airspeed direction of the image.

cloud particles, the length of the sampled cloud volume necessary for a statistically significant characterization of the ensemble of particles varies from a few meters to several kilometers. The cloud microstructure may have significant variations at such spatial scales, which may not be reflected by the cloud sampling from highly elongated volumes. Another limitation is related to the fact that cloud sampling by aircraft is extended over some period of time, which may be comparable with the characteristic lifetime of the cloud. This hinders the analysis of the spatial correlations of the microstructure, since the measurements are associated with different stages of the cloud evolution. These limitations should be seriously taken into consideration during analysis and interpretation of airborne *in situ* microphysical measurements. Cloud particle number concentrations can be as low as a few per cubic meter to as high as several thousand per cubic centimeter. The very low concentrations require instruments with large sample volumes, hence large sample area, to avoid poor sampling statistics. Figure 5.3 shows the sample volumes for 13 of the instruments that measure PSDs for an airspeed of 100 ms^{-1} and a sample time of 1 s. The sample volume is the product of the sensitive sample area of the laser beam, airspeed, and sample time. The sample area is the effective beam width times the DOF. The impactors (VIPS and Cloudscope) and

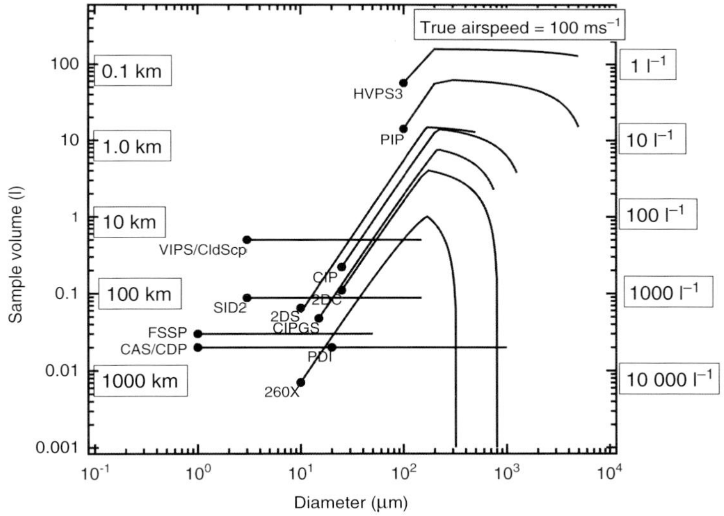

Figure 5.3 The sample volumes swept out by the cloud spectrometers are shown as a function of particle size at an airspeed of 100 m s^{-1} and sample time of 1 s. The numbers in boxes inside on the left-hand side show how many kilometers of cloud would have to be sampled to get at least 100 particles if the concentration is 1.0 l^{-1}. The numbers in the boxes to the right-hand side of the figure show what concentration of particles would be needed to sample 100 particles over a 100 m cloud path at this sample volume. The cloud spectrometers are described in Chapter 5. HVPS3: high-volume precipitation spectrometer–version 3; PIP: precipitation imaging probe; VIPS: video ice-particle sampler; CldScp: cloud scope; CIP: cloud-imaging probe; CIP-GS: cloud-imaging probe-grey Scale; 2DC: 2D cloud probe; 2DS: 2D stereo probe; Sid2: small ice detector-version 2; FSSP: forward-scattering spectrometer probe; CAS/CDP: cloud aerosol spectrometer/cloud droplet probe; PDI: phase Doppler interferometer; and 260X: 1D cloud probe (OAP-260X).

light-scattering probes (FSSP, CAS, CDP, PDI, and Sid2) have fixed effective beam widths and DOFs that do not depend on the particle size such as the imaging probes (2D-C, CIP, 2D-S, HVPS), whose DOF increases with the square of the particle diameter and whose effective beam width decreases with the diameter. In addition to the instrument sample volumes, the vertical scale has also been labeled to show how these sample volumes correspond to getting a statistical sample of 100 particles, the number of particles needed to reduce the counting uncertainty to less than 10%, assuming that the cloud particles are distributed randomly along the flight path. To the right of the left vertical scale are the lengths of cloud that would have to be sampled if the concentration is l^{-1}. For example, all of the light-scattering instruments would require 200–800 km of cloud samples to get a statistically representative sample. The imaging probes also need 50–200 km of cloud in the size range from 10 to 25 μm. To the right of the right vertical axis, it is shown what concentration of particles would be required to sample 100 particles over a path length of 100 m (1 s of sampling at 100 ms^{-1}).

While it may seem to be a good idea to increase the sample area of single-particle spectrometers as large as possible to accommodate low concentrations, there is a trade-off if the same instrument is to be used for measuring medium or high concentrations because the probability of particle coincidence in the sensitive volume of the probe would also increase (Section 5.3.1). To mitigate possible artifacts due particle coincidences, it is thus judicious to associate single-particle spectrometers with integrating probes that are not sensitive to particle coincidence.

5.2
Impaction and Replication

5.2.1
Historical

The earliest measurements in cloud from an airborne platform can be attributed to those made by Wigand in the early twentieth century (Pruppacher and Klett, 1997) who described the shapes of ice crystals and graupel particles in 1903. The measurement method is not clearly documented but presumably particles were captured on cooled material with a dark background. The measurements of ice crystals from a fixed wing aircraft were made by Weickman (1945) who collected a comprehensive set of ice crystals with many habits over a range of temperatures. He exposed glass slides covered with viscous shellac to the air stream, and then examined and photographed them after the flight. The pace of airborne measurements accelerated in the late forties as improved impaction and replication techniques were used to capture impressions of cloud droplets and crystals on surfaces exposed to the passing airstream. These were simple frames holding a single slide or more sophisticated, multislide "cloud guns" (Golitzine, 1950; Clague, 1965; Spyers–Duran, 1976). Glass slides were prepared with various coatings, for example, carbon black (soot), MgO powder, viscous oil, or formvar. The slide was then exposed to the airstream for a short, timed interval. More repeatable timing of the exposure was attained using a spring-loaded or compressed gas to propel the slide past the aperture where the droplets impinged on the slide (Hindman, 1987). The cloud droplets leave size-proportional pits in the soot or powder, or are captured as water bubbles in the oil. The ice crystals are replicated in formvar-coated slides. Oil-coated slides have also been used to capture intact ice crystals, whereby the slides are exposed to the airstream then placed in a cold box for examination later.

5.2.2
Measurement Principles and Implementation

Impactor-type probes are especially useful for complementing the electro-optical spectrometers with more detailed information about crystal sizes smaller than about 100 μm. A particularly desirable feature of impactor probes is that they have

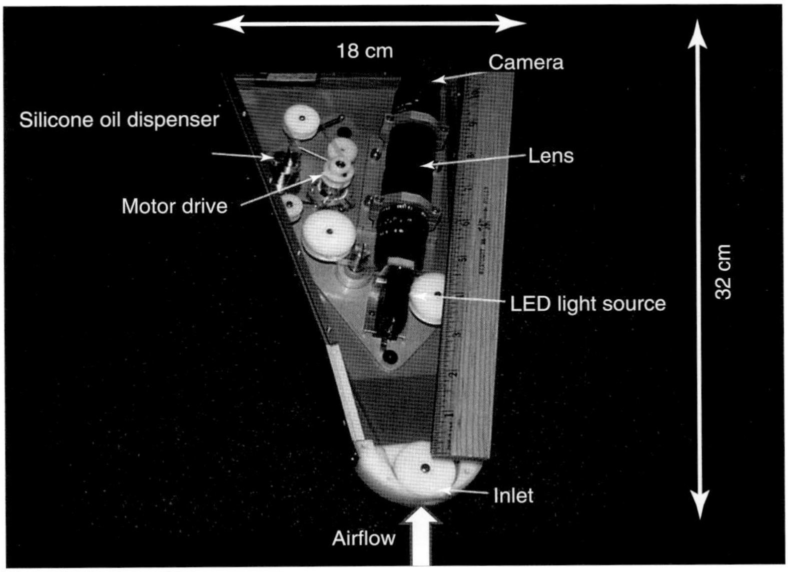

Figure 5.4 NCAR VIPS as viewed from above. The inlet is the small rectangular area at the leading surface of the probe. (Courtesy of A. Heymsfield.)

larger sample areas than the light-scattering probes and imaging probes for sizes smaller than 50 μm. They have potentially better resolution for discerning particle habit and the ice-particle cross-sectional areas can be measured directly from which the mass of individual ice particles can be estimated for some crystal habits.

The video ice-particle sampler (VIPS), developed at NCAR, can obtain images of particles larger than about 5 μm (Heymsfield and McFarquhar, 1996a; Schmitt and Heymsfield, 2009) by collecting particles on an 8 mm wide film, coated with silicone oil, then imaging them with video microscopes at two magnifications. Ice crystals are captured when they impact a transparent moving belt exposed to the airstream. A video microscope (Figure 5.4) records the crystal images that are subsequently analyzed with image analysis software. Derived properties for each particle include their projected area and maximum dimension, from which number concentrations as a function of size, that is, PSDs, are generated. The area of the VIPS inlet, 0.05 cm^2, and the aircraft speed determine the sample volume of the instrument (Table 5.1). The collection efficiency is reduced for particles smaller than 10 μm, although calculations show that the efficiency is 94% or greater for larger particles (Ranz and Wong, 1952). A second video microscope, coupled with a heater positioned beneath the transparent moving belt, is used to melt, photograph, and derive masses for individual ice particles from the image of the droplet formed from the melted crystal.

The Cloud Scope, developed at the Desert Research Institute (Hallett *et al.* 1998), detects particles between five and up to a few hundred micrometers that impact on a sapphire window (Arnott *et al.* 1995). As shown in Figure 5.5, it is

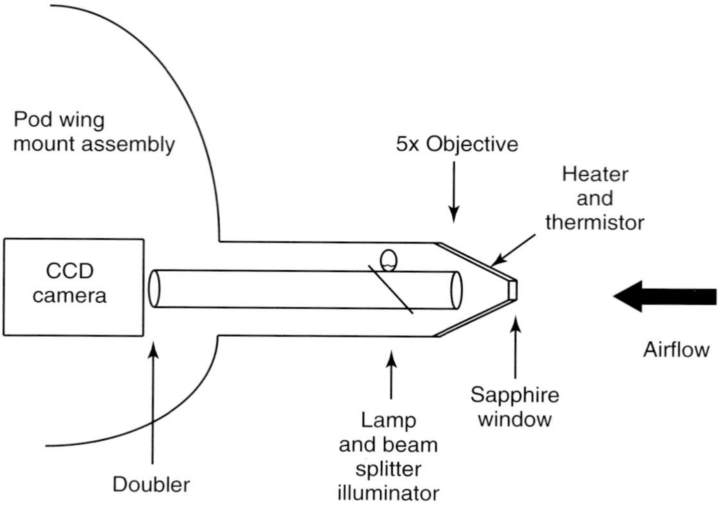

Figure 5.5 Schematic depiction of the Desert Research Institute (DRI) Cloud Scope.

a charge-coupled detector (CCD) video camera attached to an optical microscope. Data are recorded with an analog video camera but can be retrofitted with a digital recording capability. The window is at a stagnation point of the flow so that adiabatic compression heats the window to sublimate the ice crystals; hence, subsequent, impinging crystals are imaged without superimposition on earlier impacted crystals. The rate of sublimation of an ice particle is proportional to its mass and from this the IWC can be estimated. A beam splitter-miniature bulb combination provides illumination for night operation. Several shutter speeds are available to accommodate daytime lighting conditions by direct sunlight. Images are digitized and analyzed with software packages such as the National Institute of Health's IMAGE analysis software to determine particle sizes, shapes. and concentrations. Although the sample volume of the Cloud Scope is influenced by the rate of sublimation of the ice particles, it is nominally about the same as the VIPS (Table 5.1).

5.2.3 Measurement Issues

In the VIPS, ice crystals larger than a couple of hundred micrometers will break into a number of pieces, depending on the air speed, density, and crystal habit. Particle breakup on the edge of the probe inlet is thought to be insignificant when there are low concentrations of particles larger than a few hundred microns. Similar breakup on the Cloud Scope has been used to document ice crystal shattering (Vidaurre and Hallett, 2009a).

Sizing and sample volume uncertainties, in both the VIPS and Cloud Scope, lead to a distortion, most often a broadening, of the derived PSD. The sources of the uncertainties are diverse, but the main source is generally related to the collection efficiencies of the two instruments as a function of ice crystal size and shape. The issue of air flow distortion and size sorting as a result of how particles follow the streamlines around the instrument inlet is discussed in much greater detail in Chapter 6. In brief, particles will follow a trajectory that is dependent on the free stream air velocity, particles mass, cross-sectional area, and shape factor. Detailed, numerical calculations have been carried out to predict these trajectories for known crystal habits, but given that these habits are not known *a priori*, and often the crystals are complex in shape, the results from these modeling studies serve as a first approximation to the expected uncertainties. The reader is referred to Chapter 6 for more information on this measurement issue. Another source of uncertainty is that oil sometimes contains contamination (bubbles or dust) and adds background noise to the image strip.

5.3
Single-Particle Size and Morphology Measurements

Electro-optical spectrometers, for measuring the size of individual cloud particles, were introduced by Robert Knollenberg who developed the 1D optical array probe (1D-OAP) in 1970 (Knollenberg, 1970), the axial scattering spectrometer probe (ASSP) and forward-scattering spectrometer probe (FSSP) in 1972 (Knollenberg, 1976; Knollenberg, 1981), and the two-dimensional (2D) OAP (2D-OAP) in 1976 (Knollenberg, 1981). The majority of the instruments that were developed after 1980, still in use for measuring the size of individual cloud particles, are those whose designs are based on the "Knollenberg probes" (also referred as *PMS probes* because they were originally built and sold by Particle Measuring Systems). Many studies have been published on the performance of the light-scattering probes (Baumgardner, 1983; Cerni, 1983; Dye and Baumgardner, 1984; Baumgardner, Strapp, and Dye, 1985; Baumgardner and Spowart, 1990; Brenguier and Amodei, 1989a; Brenguier and Amodei, 1989b; Brenguier *et al.* 1993; Wendisch, Keil, and Korolev, 1996a; Brenguier *et al.* 1998; Schmidt, Lehmann, and Wendisch, 2004) and OAPs (Curry and Schemenauer, 1979; Joe and List, 1987; Korolev *et al.* 1991; Gayet, Brown, and Albers, 1993; Gayet, Febvre, and Larsen, 1996; Baumgardner and Korolev, 1997; Korolev, Strapp, and Isaac, 1998b; Strapp *et al.* 2001; Jensen and Granek, 2002). In fact, not only have the PMS probes received far more scrutiny than any of their successors discussed below, but also very little has changed from the original design. Only the optical and electronic components have been improved as a result of the general advances in technology over the past 40 years. At the same time that Knollenberg was developing his instrument, work was being undertaken at NCAR to implement a technique to photograph cloud particles in flight (Cannon, 1970). The "Cannon" camera, using high-speed photography, was successfully employed on a sailplane to capture images of ice crystals larger than 10 μm. Although the data processing was tedious, the images were of sufficient resolution to identify size and habit. Airborne holography was also being developed during that same decade (Trolinger, 1975; Trolinger, 1976) but was not routinely used on aircraft until the mid-1980s when it was employed on the aircraft of the British Meteorological Office (Brown, 1989). Further advances were made 15 years later, once again taking advantage of technological advances in optics and electronics (Fugal *et al.* 2004; Fugal, Schulz, and Shaw, 2009; Fugal and Shaw, 2009).

This section describes the principles and implementation of single-particle spectrometers, instruments that detect particles one after the other and measure their individual size and shape. Particles are classified in categories (liquid, different types of ice crystals) if necessary and grouped in size classes. When a large number of particles have been classified, the size distribution is approximated as a size spectrum in discrete bins. There is, however, an intrinsic limitation to all single-particle counters, namely, their sampled volume. Indeed, particles are detected when they enter a sensitive volume v, within some fraction of a light beam, which should be small enough so that the probability will be small in a way that more than a single particle will be present in this volume at the same time. Typically, $v \ll 1/N$ where N is the particle concentration. Over a time period Δt, a volume of air is sampled along the flight path. To build a statistically representative estimate of the ambient size distribution, many particles need to be counted. Given that the detection volume is small, the sample period may be relatively long, depending on the magnitude of N. Consequently, single-particle counters document a cylinder through the cloud with a small cross section (typically less than a square millimeter, but with a length that depends on the sample period and velocity of the airplane). This means that the small features of the cloud, that is, fluctuations in the particle concentration are smoothed out. This intrinsic limitation and possible mitigations are discussed in Section 5.3.1.

The next issue is to determine particle size and shape. For very small particles with size below the diffraction limit ($<100\,\mu m$), the light scattered by an illuminated particle provides indirect information about its size and shape. Single-particle light-scattering spectrometers are described in Section 5.3.2. For larger particles, images can be directly recorded and processed to classify particle types and sizes. Imaging probes are described in Section 5.3.3.

The holographic technique described in Section 5.3.4 establishes a bridge between single-particle counters and integrators. Indeed, it consists in recording the interference of a large number of particles with an illuminating coherent beam. It is thus intrinsically not sensitive to particle coincidence and a large sensitive volume can be used, hence providing robust statistics for retrieval of the PSD. Data processing then provides a comprehensive description of each particle as with a single-particle spectrometer. The limitation though is the amount of data to be recorded and processed that still prevents continuous measurements along the flight path and only shows snapshots of the particle spatial distribution in clouds.

5.3.1
Retrieval of the PSD

Single-particle spectrometers provide a discretized approximation of the PSD $f(D)\,\mathrm{d}D$ function with the particle number concentration in a series of size classes. When n_i, the number of particles counted in each size class $[D_i, D_{i+1}]$, is large, $N_i = n_i/V$, the measured number concentration of particles with sizes between D_i and D_{i+1}, provides a good approximation of

$$F_i = \int_{D_i}^{D_{i+1}} f(D)\,\mathrm{d}D \tag{5.2}$$

where V is the volume of air sampled during the sampling period Δt. Optical particle spectrometers are currently located outside the aircraft and a light beam is exposed to the airflow. Each time a cloud particle crosses the sensitive volume of the beam v, it is detected, sized from scattered light for the small ones or the image for larger ones, and classified. If S is the section of the sensitive volume perpendicular to the airflow, the sampled volume of air is equal to $V = S \cdot V_\mathrm{a} \cdot \Delta t$, where V_a is the air speed in the probe sampling section. To improve the statistical significance of the sample, that is, for N_i to provide a good approximation of F_i, the number of counted particles $n_i = N_i/V$ shall be large (the uncertainty scales with $n^{-1/2}$). However, with an aircraft speed of 100–200 m s^{-1}, Δt shall be small enough to document small-scale spatial fluctuations of the droplet size distribution. An efficient way to improve statistics is, thus, to increase the cross section of the beam sensitive volume. This option is constrained by the occurrence of particle coincidences in the sensitive volume of the light beam that seriously affect counting and sizing. Assuming particles are randomly distributed in the volume of air, the Poisson theory tells us that the probability of having more than one particle in the sensitive volume scales like

$$P(>1) = 1 - X\,, \quad \text{with} \quad X = \exp(-N \cdot v) \tag{5.3}$$

Ideally, the optical setup should thus be designed with the largest possible cross section S and the smallest possible sensitive volume v, hence the smallest possible width parallel to the flight path, which is a real challenge for optical systems. Because, the particle number concentration decreases when the particle size increases, each particle spectrometer is designed to fit a specific particle size range, for instance, a sensitive volume of the order of less than 0.2 mm^3 for the FSSP (cloud droplets) and up to 100 cm^3 for precipitating drops with the 2D-C. A short sampling period is suited to characterize the spectrum mode diameter and width at small spatial scales, but the wings of the spectrum with much lower relative droplet concentrations are poorly characterized statistically, resulting in random fluctuations from one class to the next. For these rare particles, counts can be cumulated over a longer time period, hence loosing information on their small-scale structures. Also note that in real instruments with nonuniform distribution of laser beam light intensity, the geometry of the sensitive volume depends on the particle size. In the above formulae, v and S, hence the sampled volume V, are therefore dependent on the particle size, an additional complication when processing the collected data.

5.3.1.1 Correction of Coincidence Effects

Coincidences of particles in the detection beam result in count loses and a distortion of the spectrum toward larger sizes. When looking at the total particle concentration $N(D^-, D^+)$, where D^- and D^+ are the lower and upper boundaries of the instrument size range, coincidence losses can be statistically corrected assuming droplets are randomly crossing the detection beam. The actual number of counts during the sampling period or the actual rate n_a can thus be derived from the counted rate n_m, as

$$n_a = \frac{n_m}{X} \tag{5.4}$$

An alternative approach consists in measuring the probe activity A, that is, the fraction of time the probe detects particles, that is, $A = n_m \cdot \tau/T$, where τ is the particle transit time through the detection beam and T is the sampling period. The actual rate can then be derived as

$$n_a = n_m \cdot (1 - A) \tag{5.5}$$

Recording the interarrival times between detections (Baumgardner, Strapp, and Dye, 1985) provides a third and independent statistical estimation of the actual rate since the cumulated frequency distribution of the interarrival times δt scales like

$$P(\delta t > 1) = \exp(-n_a \cdot t) \tag{5.6}$$

Using these three redundant but independent approaches provides a very robust estimation of the coincidence losses (Brenguier, Baumgardner, and Baker, 1994). The redundancy also provides a way to check the quality of the measurements and detect failures during the flight such as condensation on the optics (Brenguier *et al.* 1993). This theory though is only valid with retriggerable probes, that is, probes that are able to detect a new particle as soon as the previous one has left the detection beam. Some instruments, such as the FSSP, however, have an electronic dead time during which particle detection is not possible. A modified and slightly more complicated set of equations shall then be used for the first two techniques based on counted rate and activity (Brenguier and Amodei, 1989b), while the third one based on interarrival times is not affected by the probe dead time (Baumgardner, Strapp, and Dye, 1985). In the FSSP, coincidence effects become significant when the concentration exceeds $\sim$500 cm^3.

Particle sizing is also affected by coincidences that lead to a distortion of the size distribution toward larger sizes (Cooper, 1988) and an overestimation of the derived LWC (Burnet and Brenguier, 1999; Burnet and Brenguier, 2002). Correction of the coincidence effects for each size class, however, and retrieval of the actual spectrum is hardly feasible because coincidences of particles are counted as a single count with a larger size. A straightforward solution starting with the measured spectrum to derive the actual one will thus require an enormous amount of matrix calculations. Inverse calculation is, however, feasible starting with a guess to derive a measured spectrum and iteratively converging to an accurate solution (Brenguier *et al.* 1998; Schmidt, Lehmann, and Wendisch, 2004). Such corrections are tedious and have rarely been applied in the processing of the very large data sets collected during airborne experiments. For bigger particles, however, postprocessing of the images recorded by single-particle imaging probes provides opportunities to identify and reject such events.

5.3.1.2 Optimal Estimation of the Particle Concentration

When the number of counted particles is too small for a robust statistical estimation of the spectrum (fine-scale measurements), it is still feasible to analyze the data following a Bayesian approach. Optimal nonlinear estimators suited for Poisson processes are capable of calculating the probability density function of the possible values of the particle concentration with a spatial resolution smaller than the particle interarrival time (Pawlowska, Brenguier, and Salut, 1997). The resulting PDF can then be examined to derive the most probable concentration value and its confidence level. Such algorithms are extremely powerful at detecting microscale structures in clouds, but they also require a significant amount of computer time and, hence, are limited to very specific studies.

5.3.2 Single-Particle Light Scattering

The airborne instruments that are presently used to measure *in situ* the size distributions of cloud droplets and ice crystals in the small particle range, here defined as 1–50 μm, are the FSSP, cloud droplet probe (CDP), fast FSSP (FFSSP) and fast CDP (FCDP), cloud and aerosol spectrometer (CAS), cloud and aerosol spectrometer with depolarization (CAS-DPOL), small ice detector (SID), and the phase Doppler interferometer (PDI). The FSSP, CAS, SID, PDI, and CDP are shown (not to scale) in Figure 5.6, and the FFSSP and FCDP are depicted in Figure 5.7. All these instruments are based on the scattering of light by single particles in a focused laser beam.

5.3.2.1 Measurement Principles and Implementation

The operating principle of the FSSP (Dye and Baumgardner, 1984), CDP (Lance *et al.* 2010), CAS (Baumgardner *et al.* 2001), CAS-DPOL (Baumgardner *et al.* 2011), and SID (Cotton *et al.* 2009; Kaye *et al.* 2008) is based on the concept that the intensity of scattered light in some specific scattering angles can be theoretically related to the particle size if the shape and refractive index of a particle is known, as well as the wavelength of the incident light. The theoretical basis for this interaction is described in detail in Section A.4, given in the Supplementary Online Material provided on the publisher's web site. This theory, known by its originator, Mie (1908), is applied in OPCs by collecting scattered light, over some range of angles that depend on the design of the instrument, from particles that pass through a light beam of controlled intensity and wavelength. This collected light is converted to an electrical signal whose amplitude is then associated with the size of the

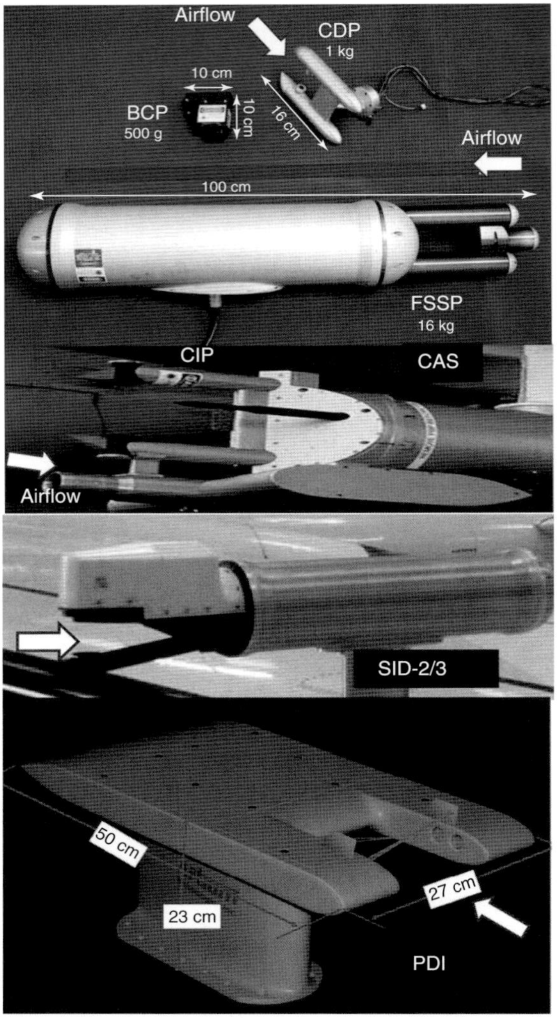

Figure 5.6 The FSSP, CAS, CDP, SID-1 and -2, and PDI shown here represent the range of single-particle spectrometers currently flown on atmospheric research aircraft.

particle. We refer to this size as an optical diameter due the manner in which it is derived. On the basis of the particle scattering phase function, the scattering cross section is the property that defines how a particle interacts with the incident light and describes the relative intensity of scattered light within the solid angle defined by the collection optics of a particular instrument. The scattering cross section is a function of the particle's physical cross-sectional area, shape, refractive index, and wavelength and polarization of the incident light. The FSSP, CAS, CDP, and SID differ primarily in their optical configurations with respect to the angles over which

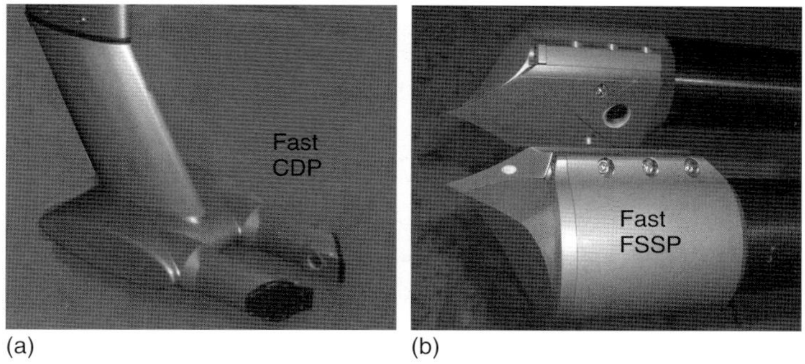

Figure 5.7 Photos of (a) fast CDP (FCDP) and (b) fast FSSP (FFSSP). (Reprinted with permission from SPEC.)

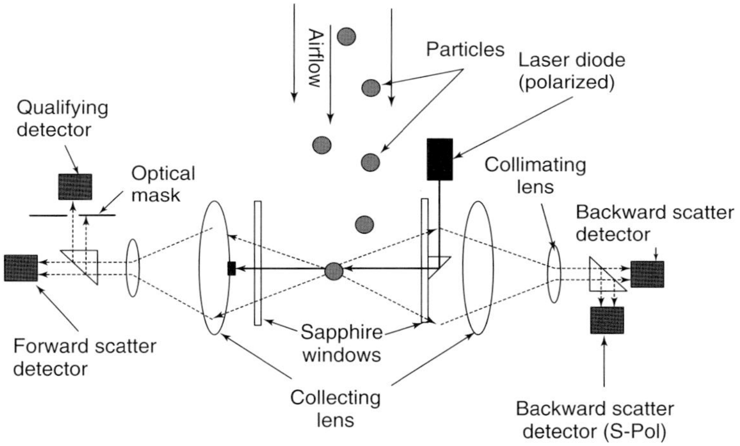

Figure 5.8 The optical path of the CAS-DPOL is shown here, illustrating the basic, optical components of a typical single-particle spectrometer.

the scattered light is collected (collection angles) and in their sampling sections that have evolved over time to reduce problems of ice crystal shattering. The optical configurations are described in greater detail below and the issue of ice crystal shattering is discussed in Chapter 6.

The collection angles used in the FSSP, CDP, CAS, and CAS-DPOL are between the half-angles of approximately 3.5° and 12°. The CAS and CAS-DPOL also implement an additional set of optics to collect backscattered light from 168° to 176°. The near forward angles are used because the largest fraction of light scattered from a particle is in the forward direction when the particle diameter is larger than the wavelength of the incident light. Figure 5.8 is a diagram of the light collection system that is used in the CAS-DPOL. The FSSP and CDP optics are essentially the same, but exclude the backscattering collection optics. The information from the backscattering signal is used for particle shape discrimination as described below.

The polarized diode laser provides the source of monochromatic light (the FSSP uses a nonpolarized, multimode TM00 He–Ne gas laser). Lenses focus the beam to a diameter of about 0.2 mm in the FSSP and about a millimeter in the CAS, CAS-DPOL, and CDP. The larger beam is needed in the latter instruments because the diode laser produces a beam with Gaussian intensity cross section and only approximately 15% of the most intense region is used for particle sizing. The gas laser produces a beam with many modes, approximating a top-hat intensity profile, so that approximately 60% of the beam is used in the FSSP.

The beam falls on a light absorbing region, a "dump" spot, which blocks the incident laser light from reaching the detectors. The forward scattered light is collected by the optics and focused onto two detectors via a beam splitter (in the CAS and CAS-DPOL, part of the light backscattered is also collected). The acceptance angles are determined by the diameter of dump spot and the aperture of the primary collection lens.

The location of the particle in the beam is a critical piece of information needed to interpret the measurement since this defines one part of the sample volume, designated the DOF, used to calculate the sampled volume and, hence, the number concentration. For accurate sizing, the OPC must accept and size only particles that pass through a uniformly intense region of the laser beam where they are in focus, creating the need to "qualify or DOF accept" the particle. A comparison of voltage pulses from the sizing and qualifying detectors is used to determine whether the particle was in the DOF. The qualifying detector has an optical mask that restricts scattered light from reaching it, depending on where the particle is with respect to the center of focus. If the particle is outside the region defined as the DOF, a comparison of the signals from the sizing and qualifying detectors will produce a decision to reject the particle.

Once the scattered light is converted by the photodetector to an electrical signal, the peak analog voltage is digitized and processed by comparing its magnitude with a table of values that correspond to nominal optical diameters under the assumption of a known refractive index for the particle and spherical shape. A frequency histogram is updated corresponding to what size category is selected from the table and the histogram has the associated size bin incremented by one. This frequency histogram, that is, PSD, is transmitted to the data system at a fixed download rate. Some of the OPCs also send each qualified particle's peak voltage and its time of detection to the data system to provide much higher resolution information.

The fast-FSSP is an improved version of the standard probe in which the peak voltage is converted into 256 size classes instead of 16 in the standard probe. Moreover, pulse width and particle arrival time are recorded for each detected particle with a time resolution of 85 ns (16 MHz digital sampling). With these two additional parameters, it is possible to derive independent and more accurate estimations of the number concentration based on counting rate, activity, and PDF of interarrival times (Section 5.3.1). The continuous series of counts can also be analyzed to perform optimal estimation of the concentration at the microscale and the high peak voltage resolution can be used for a stand-alone absolute calibration

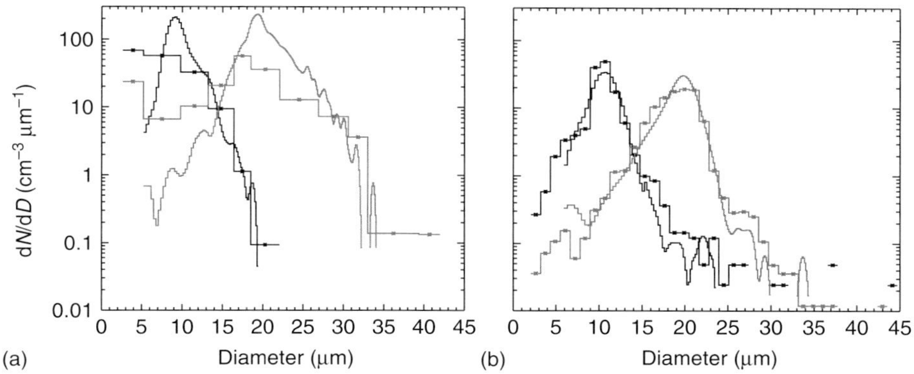

Figure 5.9 Droplet size distributions as measured with (a) the FSSP-100 (lines with symbols) and the fast-FSSP (lines with no symbols) during SCMS (black and gray curves indicate two different times (0.3 s accumulation time) and (b) the SPP-100 (lines with symbols) and the fast-FSSP (lines without symbols) during DYCOMS-II (again, black and gray curves indicate two different times, with 1 s (black lines) and 2 s (gray lines) accumulation time). (From Brenguier *et al.* (2011).)

of the probe (Brenguier *et al.* 1998). Figure 5.9 shows droplet spectra measured with three single-particle scattering probes. In Figure 5.9(a), measurements performed in cumulus clouds with the standard FSSP and fast-FSSP are compared, illustrating the impact of the coarse FSSP size resolution on the characterization of the smallest droplet concentration density at small sizes. In Figure 5.9(b), measurements performed with the CDP (SPP) and fast-FSSP in stratocumulus clouds are compared, showing a very good agreement between the two probes. The fast-FSSP has been flown on a number of different aircraft participating in field projects for the study of warm convective clouds and marine stratocumulus (Burnet and Brenguier, 2007; Burnet and Brenguier, 2010; Brenguier, 1993; Brenguier and Chaumat, 2001b; Brenguier *et al.* 2000b; Brenguier *et al.* 2000a; Brenguier, Pawlowska, and Schüller, 2003; Brenguier, Burnet, and Geoffroy, 2011; Chaumat and Brenguier, 2001b; Gerber *et al.* 2005; Malinowski *et al.* 2007; Pawlowska and Brenguier, 2003; Schüller, Brenguier, and Pawlowska, 2003).

A commercially available version of the fast-FSSP is the fast CDP (FCDP), which time-stamps each individual particle to within 25 ns and records individual arrival time, transit time, signal, and qualifier pulse heights. In addition, the electronics digitizes both the qualifier and signal waveforms for postanalysis.

The relationship between peak voltage and particle diameter is established by calibration using monodispersed particles of known size and refractive index, typically crown glass beads, to determine the relationship between the photon intensity reaching the photodetector and the output of the detector. The relationship between the photon intensity and particle diameter is determined from Mie theory as described previously. A number of different approaches have been taken to calibrate these instruments with varying degrees of complexity. More information on calibrating the FSSP, CAS, CDP, and SID can be found in (Dye and Baumgardner

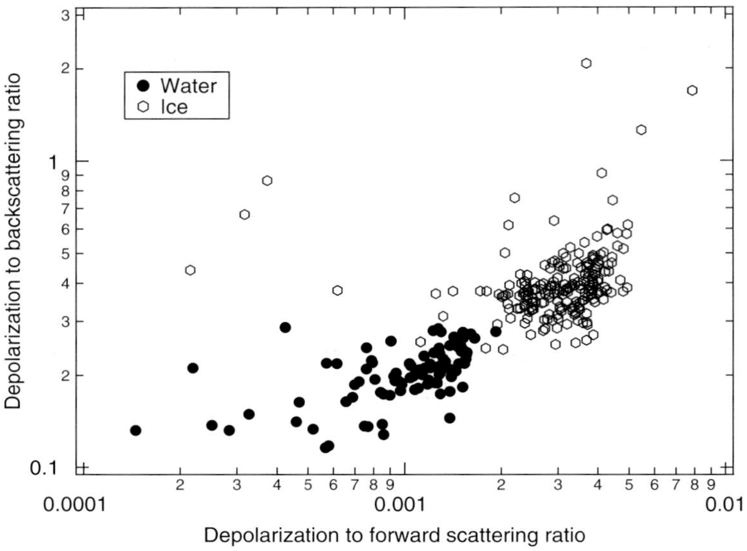

Figure 5.10 The CAS-DPOL is able to separate water droplets from ice crystals by measuring the amount of depolarization of light scattered by the individual particles, as shown in this correlation plot. The data for this figure are from measurements made in clouds over the North Atlantic.

(1984); Wendisch, Keil, and Korolev (1996a); Brenguier *et al.* (1998); Baumgardner *et al.* (2001); Nagel *et al.* (2007); Lance *et al.* (2010); Rosenberg *et al.* (2012); Cotton *et al.* (2009)).

The shape of ice crystals cannot be determined with the FSSP or CDP from the measurement over one solid collection angle and nonspherical ice crystals will be undersized, depending on their deviation from spherical shape (Borrmann, Luot, and Mishchenko, 2000). The CAS was designed to collect scattered light from two directions to discriminate water from ice based on deviation from sphericity. As described in Baumgardner *et al.* (2005), comparison of light scattered in the forward and backward directions provides an indication, at selected sizes, of the sphericity of a particle.

Adding measurements of the depolarization ratio provides a complementary information on particle types, as shown in Figure 5.10, with the comparison of the depolarization to the forward and backward scattering ratios.

The SID was specifically designed to discriminate ice from water based on the angular distribution of light scattered in the forward direction. There are currently three SIDs that have been used for airborne measurements, the SID-1, described by Hirst *et al.* (2001), the SID-2 discussed by Cotton *et al.* (2009), and the SID-3, the latest version of the SID (Kaye *et al.* 2008). The SID-1 has six optical detectors arranged to measure the azimuthal distribution of light scattered over a forward-scattering angle by individual cloud particles passing through a laser beam. Each detector is centered on a scattering angle of 30° with a lens half-angle of 10°.

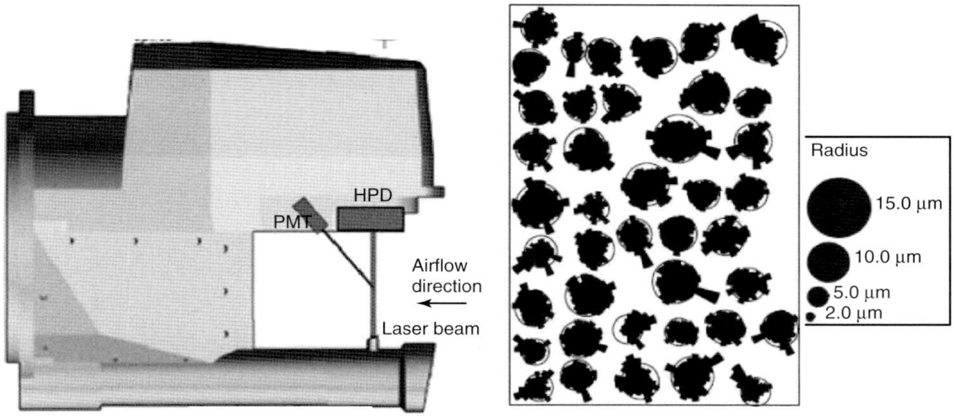

Figure 5.11 This schematic shows a side view of the SID-2 and the optical path with detectors; the diagram with the detector geometry shows the locations of the detectors and images are examples of the scattering patterns produced from ice crystals measured in a supercooled cumulus clouds.

The SID-1 is able to count and size cloud particles down to 2 μm optical diameter. The upper size limit is approximately 70 μm because of the electronic dynamic range. The SID-2 is designed to discriminate the phase of each cloud particle rather than just the dominant phase in a small time interval as with SID-1. The scattered light in the SID-2 is detected by a custom manufactured hybrid photodiode (HPD). The HPD is a segmented silicon photodiode that contains 27 independently sensed photodiode elements, with 3 central and 24 outer ones arranged azimuthally. The outer elements cover a forward-scattering angle of 9–20°. Particles trigger the HPD measurement when scattering is detected coincidentally by two photomultiplier tubes (PMTs). The overlap of the PMT sensing volumes within the laser beam defines the area of the triggering zone. The HPD of the SID-2 was replaced with a CCD in the SID-3, providing a much higher resolution image of the forward scattering and more information about shape of the ice crystal. The SID-2 and -3 can discriminate individual supercooled liquid drops from small ice particles, based on their scattering pattern, and estimate the size of the ice particles up to approximately 140 μm. However, coincident cloud drops can produce "false-irregulars" that may be interpreted as ice particles. A schematic view of the instrument head, which protrudes from the canister, and an example of some of the scattering patterns measured in a supercooled cumulus, are shown in Figure 5.11. The SID-2 and -3 have incorporated various improvements over the SID-1; the most important are as follows: an increased azimuthal resolution of the scattered light detection, the probe geometry is designed to reduce any possible shattering of large cloud particles and the peak rather than the integrated signal from each particle is recorded. The latter improvement decreases the air speed dependency of the measured signal.

Instruments known as phase Doppler interferometers (PDIs), phase Doppler particle analyzers (PDPAs), or phase Doppler analyzers (PDA) (all names are equivalent; here "PDI" is used for consistency) are based on light-scattering interferometry (Bachalo, 1980; Bachalo, 1983; Bachalo, 1994; Bachalo and Houser, 1984; Bachalo and Houser, 1985; Sankar and Bachalo, 1992; Jackson, 1990; Sankar *et al.* 1990; Albrecht *et al.* 2003), an extension of the well-known laser Doppler velocimeter (LDV) technique (Durst, Melling, and Whitelaw, 1981). The scattering by spheres much larger than the instrument wavelength is approximated by geometrical optics. Size and velocity are determined by measuring sinusoidal scattering signals on adjacent detectors as particles move through an interference fringe pattern formed in the intersection of two laser beams. As such, unlike other common airborne droplet sizing techniques, the method is independent of the laser beam and light-scattering intensities except for the need to produce adequate signal-to-noise ratio (SNR) and maintain precise alignment of the optical components.

PDI instruments use a laser operating in single longitudinal mode to produce a coherent polarized beam that is split into two equal intensity beams. As shown in Figures 5.12 and 5.13, the two beams are focused to an intersection region to form the sample volume. An interference fringe pattern is formed with the fringe spacing determined by the laser wavelength and beam intersection angle. Spherical particles passing through the beam intersection scatter light that is detected by the receiver. The receiver, located at a suitable off-axis angle, is composed of a lens to collect the light and image it to a small selectable aperture. The collected light is partitioned into three segments and directed to three photodetectors. The signals have a sinusoidal character that are detected and processed using digital sampling and the Fourier transform. Each detector forms a nearly identical signal that is phase shifted relative to the signals from the other detectors. The frequency of each signal is proportional to the particle velocity. Using geometrical optics theory, the phase shift between each of the three possible signal pairs from the three detectors (i.e., 1–2, 1 –3, and 2–3) is linearly proportional to the droplet diameter. With three pairs of detectors, phase ambiguity (phase shifts greater than 360°) is resolved and redundant measurements of each droplet are obtained for additional signal validation and averaging.

The PDI measures drop velocity and size. The velocity measurement is a useful diagnostic that, when compared to the aircraft velocity, confirms the accuracy of the drop sizing. The PDI technique is capable of measuring drop diameters in the size range of approximately 1 μm to 1 mm diameter. A single instrument is limited to a size dynamic range of approximately 50:1 due to photodetector performance. Minimum drop size is determined by the smallest necessary SNR, and for instruments designed for cloud drop measurements, it is estimated to be between 1 and 3 μm. The maximum size is limited by the assumption of drop sphericity. Automatic adjustment of the detector gain is used to move the 50:1 dynamic range over the drop size range of interest. For realistic operations, the upper cloud drop size range is limited primarily by counting statistics and not instrument capabilities. In stratocumulus, a reasonable upper-bound has been found to be ~100 μm diameter (Chuang *et al.* 2008).

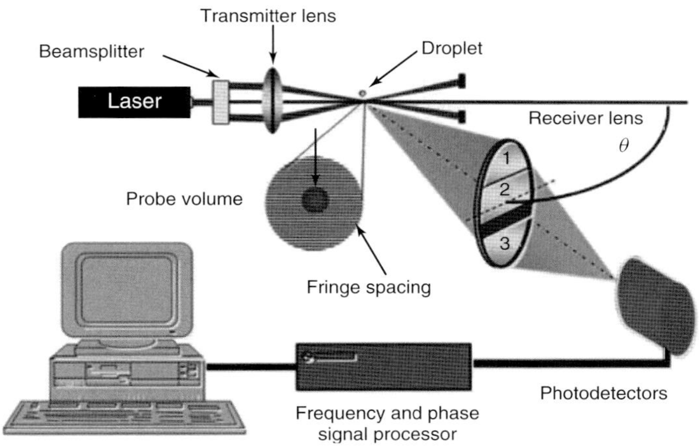

Figure 5.12 Principle of drop sizing for a PDI system. Two coherent laser beams of unequal frequency intersect to form a sample volume with an interference fringe pattern. The scattering from a droplet passing through this volume is collected by receiver optics, where the fringe spacing Λ is now proportional to drop diameter, and the fringes move at the particle velocity. Multiple sensors on the collection side collect sinusoidal signals that are processed to provide particle size and velocity (courtesy of Artium Technologies Inc.).

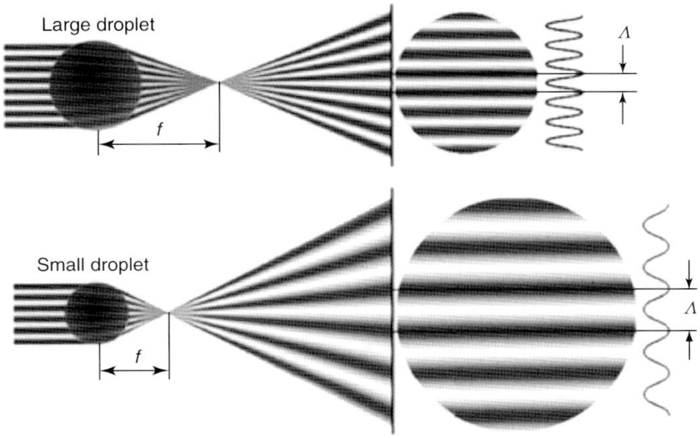

Figure 5.13 Sizing of droplets using the PDI (courtesy of Artium Technologies Inc.).

PDI instruments have a flexible, well-defined sample volume that can be adjusted to suit expected drop concentration and diameter ranges. The sample area length is defined by an optical slit width in the receiver optics as well as the off-axis light detection angle and optical focal lengths. The laser beam has a Gaussian intensity profile, leading to a sensitive width that varies with particle diameter. The effective beam diameter is determined in real time for each diameter class using light beam transit time measurements (Esposito and Marrazzo, 2007). For airborne applications, the volume sampled is the sample area multiplied by the air speed. The resulting concentration uncertainty is estimated to be ~5% although it also depends on the counting statistics.

The optical size calibration is relatively insensitive to optical contamination and does not vary provided that the instrument geometry remains fixed. Calibration is performed using monodispersed droplets at the factory, with a fundamental uncertainty of 1% or 0.5 μm, whichever is larger (Bachalo, 1994). Routine monitoring of the receiver and transmitter optics alignment assures continuing sizing accuracy. With appropriately chosen receiver optics, Mie resonances add only a relatively small uncertainty of about 0.3 μm at the low end of the size range.

Since its inception in 1982, the PDI method has been applied widely to a large range of spray characterization applications; however, it has not been used on aircraft except for a very limited number of applications and there are only a few publications where measurements from the PDI have been benchmarked against the more traditional "legacy" instruments such as the FSSP (Chuang *et al.* 2008). Hence, unlike the instruments that have been in use for more than 30 years and have been critically scrutinized in hundreds of papers, the PDI remains promising but largely untested as an airborne instrument.

5.3.2.2 **Measurement Issues**

The four principle sources of uncertainty in the retrieval of the PSD using single-particle light-scattering spectrometers are associated with (i) counting errors, (ii) the uncertainty on the sampled volume used to translate counts into number concentration, (iii) the estimation of particle size, and (iv) artifacts caused by the shattering of ice crystals on surfaces near the instrument sample volume.

(i) *Counting Errors* The uncertainty in counting stems primarily from coincidence errors and poor sampling statistics. Coincidence errors and correction techniques to mitigate them have been discussed in Section 5.3.1. When additional information about the probe activity and/or the PDF of particle interarrival times are recorded, these techniques can noticeably reduce counting errors even at very high particle counting rates. Sampling statistics is intrinsic to particle concentration measurements and it limits the ability of these instruments to explore small scales at which the number of counts is not sufficient. Bayesian techniques can be used to overcome this limitation, but their implementation requires a significant amount of calculation (Section 5.3.1). Assuming state-of-the-art correction techniques are used for coincidence correction and the sampling period is long enough to strengthen the statistics, particle counting uncertainty can thus be reduced to less than 10%.

(ii) *Uncertainty of Sample Volume* The definition of the sampled volume is more challenging. The air velocity through the probe inlet can be accurately measured, but the sampling section is difficult to characterize. Indeed, the beam light intensity of particle spectrometers is not perfectly uniform. Within the size range of a particle spectrometer, the smallest particles are only detected in a limited fraction of the beam where the intensity is maximum.

This effect is illustrated in Figure 5.8(a) with a number concentration density measured in the first size class of the FSSP one order of magnitude greater than the one measured with the fast-FSSP. In the middle of the size range, the sampling section is nominal, but it drops down again at the upper limit because particle sizes are close to the beam size and the probability of particles crossing the beam edge increases. Laboratory calibration is, therefore, necessary to precisely characterize the sensitive beam fraction as a function of the particle size. In summary, if most of the particle sizes are in the middle of the instrument size range, the uncertainty can be reduced to about 20%, but it increases considerably if the particles are at the lower or upper limit of the instrument range. The uncertainty on the particle number concentration (the counting rate divided by the sampled volume) comes therefore primarily from the uncertainty on the sampled volume.

(iii) *Estimation of Particle Size* The uncertainty in particle sizing with all the light-scattering probes, except the PDI, is a result of the nonmonotonic relationship between particle size and scattering cross section (Mie, 1908; Baumgardner, 1983; Dye and Baumgardner, 1984; Brenguier *et al.* 1998), nonuniformity of the lasers (Baumgardner and Spowart, 1990; Wendisch, Keil, and Korolev, 1996a; Lance *et al.* 2010), and particle coincidence (Cooper, 1988; Brenguier *et al.* 1998). As illustrated in Figure 5.14, where the scattering cross section for the CDP is shown, particles of different optical diameters can have the same cross section. This introduces an ambiguity when converting the scattered light to a size. At the present time, correction algorithms are being developed to decrease the uncertainty due to this problem, but these algorithms are still in the exploratory stage. The uncertainty in sizing caused by the nonuniformity in laser intensity and from coincident particles can also be decreased using the same algorithms that are being developed to correct for the Mie ambiguities. These algorithms are similar to the one developed by Cooper (1988) and Brenguier *et al.* (1998) to correct for the oversizing of coincident particles, whereby an inversion technique is applied to the size distribution using a model of the instrument response.

The fundamental assumption with the approach used to derive size from light scattering is that particles are spheres. Significant deviation from sphericity in theory leads to measurement uncertainty in the real-time processing algorithm. Some theoretical studies have been done and correction algorithms developed to predict the response of the FSSP to spheroids with different aspect ratios (Borrmann, Luot, and Mishchenko, 2000), but users of these instruments and their data should use caution when interpreting the measurements in mixed phase or all ice clouds (Gayet, Febvre, and Larsen, 1996).

(iv) *Splashing and Shattering* The issue of measurements of water droplet or ice crystals generated by large droplet splashing or crystal shattering is not discussed here, other than to say that it can be a serious problem that the data analyst should be aware of and use caution when evaluating measurements from clouds with large drops or ice crystals larger than approximately 100 μm. This issue is discussed in much more detail in Chapter 6.

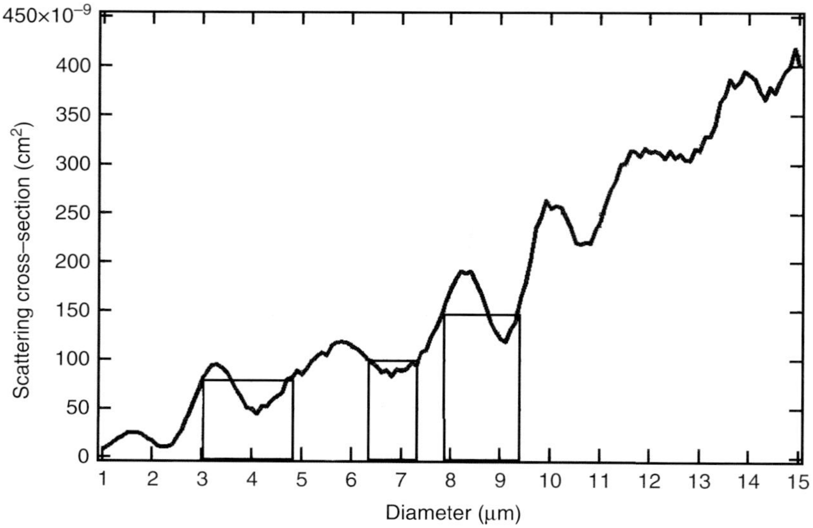

Figure 5.14 The scattering cross section as a function of optical diameter is shown here for one size interval where the nonmonotonic behavior leads to multiple diameters having the same cross section. In this example, the boxes illustrate three size intervals where multiple sizes have similar cross sections.

5.3.2.3 **Summary**

In summary, measurements using the single-particle light-scattering techniques:

(i) cover a range of optical diameters from 1 to 50 μm (forward-scattering probes) to as large as 1000 μm (the PDI, using alternating size ranges);
(ii) discriminate water droplets from ice crystals with pattern analysis of SID measurements or from the amount of depolarization detected with the CAS-Depol;
(iii) have concentration uncertainties due to small sample volumes and coincident particles;
(iv) can mis-size particles due to the nonmonotonic relationship between optical diameter and light scattering and due to particle coincidence in the beam; and
(v) cannot accurately measure the size of aspherical ice crystals.

5.3.3 Single-Particle Imaging

An alternative to scattering for determining the size of a particle is to capture its image. This has the advantage of extracting information about the shape as well as the size. All of the instruments currently in use implement optical arrays to capture the image and hence their general classification as OAPs. The OAPs that are most widely used are the 2D cloud and precipitation probes (2D-C and 2D-P),

Figure 5.15 The six instruments shown here are those that use imaging to capture the information on the size and shape of cloud particles. The CIP, PIP, 2D-S, HVPS1 (Version 1, developed in 1993), and CPI do single-particle imaging, while the HOLODEC records ensembles of particles.

the cloud-imaging probe (CIP), the precipitation imaging probe (PIP), the cloud particle imager (CPI), the 2D stereo probe (2D-S), the high-volume precipitation spectrometer (HVPS), and the 3V-CPI, which combines the CPI and 2D-S into a single instrument. Figure 5.15 shows some of these instruments (the 2D-C and 2D-P look almost the same as the CIP and PIP). Shown also in this figure is the holographic detector of clouds (HOLODEC) that captures images of particle ensembles.

A detailed look at the HVPS is given in Figure 5.16.

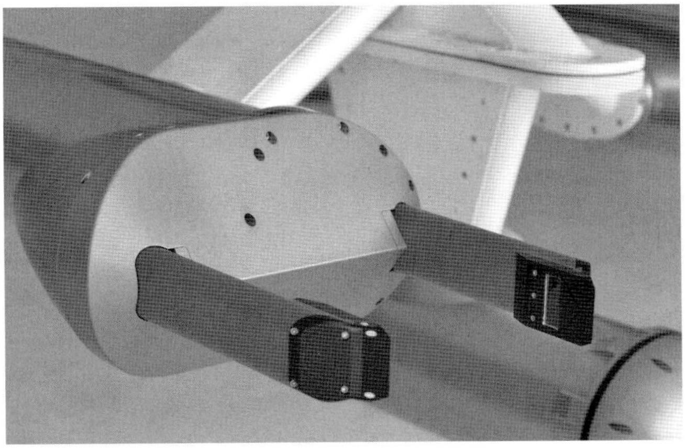

Figure 5.16 A detailed look at the HVPS3 (Version 3) developed in 2009, mounted on King Air aircraft. (Reprinted with permission from SPEC.)

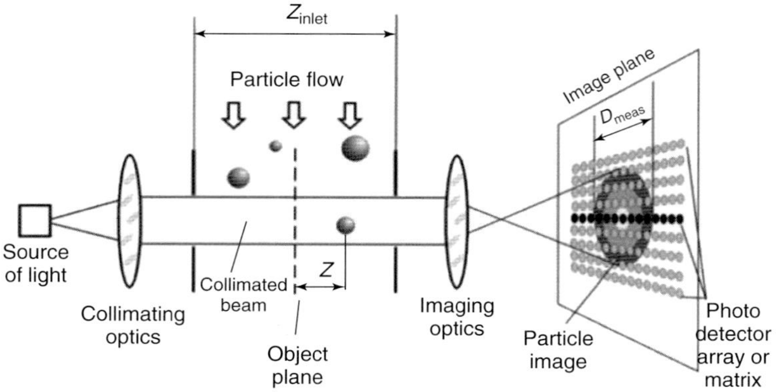

Figure 5.17 The basic principle of particle imaging is shown in this diagram. (Courtesy of A. Korolev.)

5.3.3.1 Measurement Principles and Implementation

An important property of the imaging technique is its insensitivity to the material and refractive index of the particle. This is a significant advantage over the light-scattering methods (Section 5.3.2), where the shape and refractive index of a particle play an essential role in the determination of its size. The basic design used in particle-imaging instruments, shown in Figure 5.17, consists of a light source, collimating /imaging optics, and an image recorder. The image of the particle passing through the collimated beam is projected by the optics onto the image plane where it is recorded by a linear or rectangular array of detectors. All existing imaging probes employ the white field method, that is, the particle image appears as a dark shadow on a white background on the detector array.

The distribution of light intensity over the particle image is a result of interference between the incident, reflected, refracted, and diffracted light; however, forward scattered, reflected light is usually so minor that it can be neglected. The refracted light, after a few internal reflections, scatters at large angles. The refracted light that passes through the droplet diverges at angles approaching 90°. The effect of the refracted component in the droplet's image is limited mainly by a small area around the central bright peak, only when the droplet is close to the object plane. Furthermore, for an ice crystal with a rough surface, the refracted component has little effect on the forward scattered light and its shadow images are mainly defined by diffraction. Thus, the diffracted wave is all that is required for the shadow reconstruction. Diffraction depends only on the shape of the particle. The result of diffraction for coherent and noncoherent light is appreciably different; however, given that OAPs only use laser light sources, our discussion only concerns coherent illumination.

The detailed equations and theoretical discussion that describe diffraction of light by an opaque, circular object are presented in Section A.4, given in the Supplementary Online Material provided on the publisher's web site. The important points are the following (Korolev *et al.* 1991): 1. The diffraction image can be presented as a function of only one dimensionless variable, $Z_d = \lambda \cdot |Z|/R^2$, where λ is the wavelength of the incident light, Z is the distance of the object from the image plane, and R is the radius of the object. 2. Two droplets with different radii, R_1 and R_2, with distances Z_1 and Z_2 from the object plane, give the same diffraction image if $|Z_1|/|Z_2| = R_1^2/R_2^2$. The images for such droplets are different only by the scale factor R_1/R_2. 3. The diffraction image does not depend on the sign of Z, that is, the image of the droplet will be the same size at equal distances on either side of the object plane.

Korolev *et al.* (1991) experimentally proved that for coherent illumination the diffraction images of glass beads and water droplets can be well approximated by the Fresnel diffraction theory for an opaque disc. Properties of the diffraction images simplify the analysis of water droplet images in the OAPs with coherent illumination. Although the properties were obtained for circular discs, there are valid physical reasons to consider that they remain valid for particles with arbitrary shapes.

Figure 5.18 shows diffraction images, consisting of a series of concentric rings, of an opaque disc at different dimensionless distances Z_d. A distinct feature of

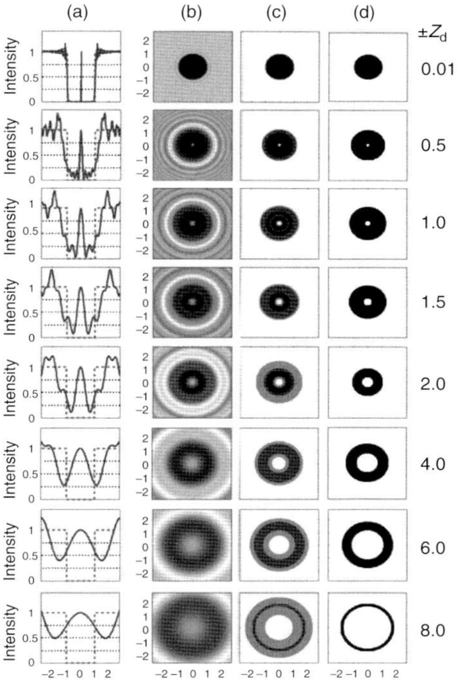

Figure 5.18 Diffraction images of an opaque disc calculated for different Z_d. (a) Changes of the intensity along the radial direction. (b)–(d) 2D distribution of light across the diffraction images at 25 %, 50 %, and 75 % intensity levels. (Korolev *et al.* 1998a. Copyright 1998 American Meteorological Society (AMS). Reprinted with permission by AMS.)

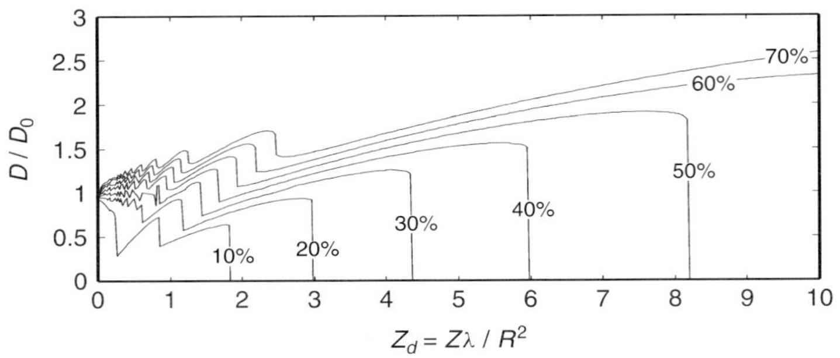

Figure 5.19 The relationship between the dimensionless diameter of the diffraction image D/D_0, measured at different intensity levels, and dimensionless distance Z_d from the object plane is shown in this figure. (From Korolev, Strapp, and Isaac, 1998b. Copyright 1998 American Meteorological Society (AMS). Reprinted with permission by AMS.)

the diffraction images is the bright spot in the center of the image, known as the *Poisson spot* after the scientist who had posited it. The Poisson spot should not be confused with the single or multiple gap in the images of ice crystals that are related to the transmitted light through plane-parallel facets, which are typical for ice crystals (e.g., hexagonal plated or columns).

The three right columns in Figure 5.18 simulate ideal digital images at intensity thresholds levels of 25% (b), 50% (c), and 75% (d). Such intensity level thresholds are used in "gray" probes (e.g., OAP-2DG, CIP-GS) or "mono" (black-and-white imagery) probes (e.g., OAP-2DC, CIP, 2D-S). As seen from Figure 5.18(b)–(d) the apparent diameter of the image D changes with Z_d. Figure 5.19 shows the dependence of $D(Z_d)$ at different intensity levels (Korolev, Strapp, and Isaac, 1998b).

The dependence of the measured size versus distance from the object plane in the imaging probes was studied by Joe and List (1987); Hovenac and Hirleman (1991) and Reuter and Bakan (1998). The research by Korolev *et al.* (1991), Strapp *et al.* (2001) and more recently Jensen and Granek (2002) showed that the theoretically predicted image sizes (Figure 5.19) are in good agreement with those measured in laboratory calibrations, although time-response effects must also be considered particularly for the older technology OAPs. A wide range of particle sizes can be measured with the imaging technique. There are no theoretical limitations for the maximum measured size. In practice, it is limited by the optical aperture or the dimension of the probe's inlet Z_{inlet}. The minimum size is set by the diffraction limit, $d_{min} = \lambda/(2\,\mathrm{NA})$, where λ is the wavelength and *NA* is the numerical aperture of the imaging optics. For visible wavelengths, the theoretical limit of the image resolution is approximately equal to 0.1 µm. The image resolution for the OAPs is significantly coarser than the theoretical limit and it is determined by the numerical aperture ($0.12 > \mathrm{NA} > 0.03$). Realistically, the best image resolution for probes with the above characteristics will be limited to approximately 2 µm.

The imaging optics in the OAPs are a microscope with a long working distance. For the existing OAPs, the working distance ranges from approximately 4 to 15 cm depending on the magnification. There are two methods used for image recording.

The first method scans a linear photodiode array and the second provides instant particle imaging with a 2D photodetector array (e.g., CCD).

In the first method, the linear array is scanned at a rate proportional to the particle velocity such that at each time interval that the particle moves a distance across the array that is equal to the resolution along the array, the on/off state of all the elements, a data "slice," is recorded. The sequence of individual slices forms a digital image of the entire particle. The sample rate of the array depends on the air speed and size resolution and varies from 0.5 to 20 MHz for the different OAPs. The advantage of the optical array method is that it allows for a continuous recording of the sequence of particles passing through the sample area, if processing, downloading, and recording time permits. However, it is sensitive to the adjustment of the scanning rate and may result in the distortion of the aspect ratio of the image if adjusted incorrectly. The OAPs that use the linear array technique are the 2D-C and 2D-P (Knollenberg, 1970; Knollenberg, 1981); the CIP, PIP, and CIP-GS (Baumgardner *et al.* 2001); and the HVPS (Lawson, Stewart, and Angus, 1998) and 2D-S (Lawson *et al.* 2006).

Table 5.1 summarizes the features of each of these instruments that, with the exception of the CIP-GS and 2D-GS, CPI and 3V-CPI are all monoscale (on/off) that use a threshold level of 50% to identify a valid image of a particle. This means that the light level on each of the diode elements is monitored and when any one of these levels decreases by 50% or more, the state (greater than or less than 50%) of all of the elements in the array is recorded. The CIP-GS is a grayscale probe whose light level thresholds are programmable, and any three of 256 levels can be used; however, the instrument is typically operated with three levels of 25%, 50%, and 75%. The advantage of the grayscale is that it gives some additional details on the features of ice crystals and is more accurate in identifying particles in and out of focus as the examples shown in Figure 5.20.

The 2D-S, where the "S" designates "stereo", is slightly different from the other OAPs because of the two sets of arms orthogonal to one another (Figure 5.15). The original design was meant to decrease the uncertainty in measuring the smallest particles (less than 100 μm) whose effective DOFs are less than the width of the arm separation. The idea was to image some of the particles simultaneously where the two beams intersect (stereo), restricting the sample volume and avoiding out-of-focus particles. The DOF of a 50 μm particle is only about 2 mm and small changes in alignment under normal flight conditions make it difficult to maintain a coincident sample volume. The principal advantage of the 2D-S is the 10 μm size resolution, 256 elements, and high-speed electronics that provide images with good detail (Figure 5.21).

Another method for image recording uses a square CCD or CMOS photodiode array (i.e., a digital camera). An entire image is recorded on the photodetector matrix as particles pass through the sample volume. This method requires pulse illumination and a triggering system for the identification of the presence of the particle in the sample volume. The CPI and 3V-CPI are currently the only instruments in airborne operation that use this technique. Figure 5.22 shows schematics of the CPI and 3V-CPI electro-optics. A particle detection system (PDS)

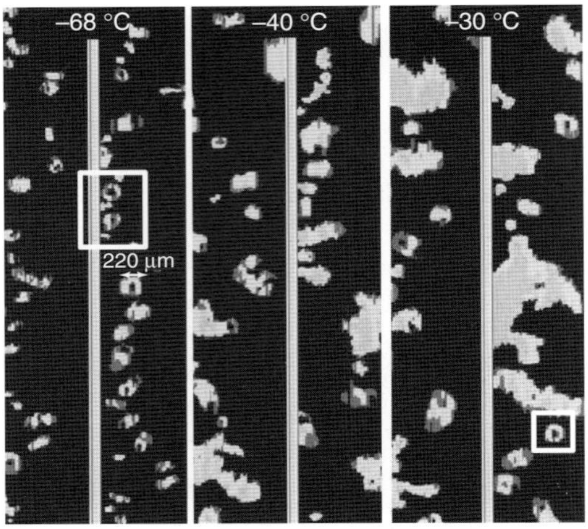

Figure 5.20 These images from the CIP-GS show the shapes of ice crystals in three temperature regimes where we see pristine crystals at −68 °C in a cirrus cloud, and more amorphous aggregates of crystals at −40 °C and −30 °C in a cumulus cloud. The images within the white squares are out of focus, as discussed more detail below under measurement issues.

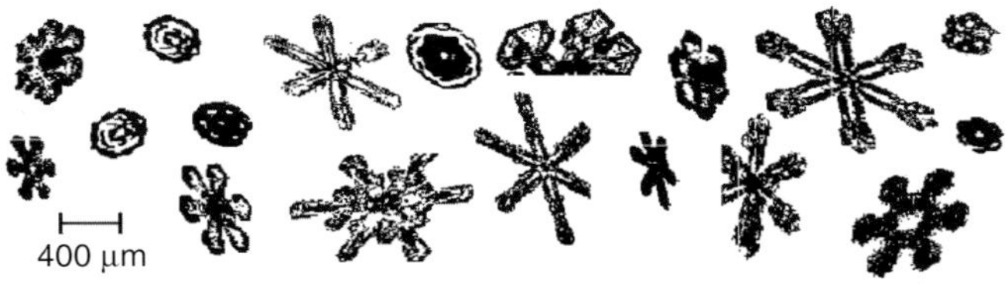

Figure 5.21 These images are samples of crystals imaged with the 2D-S.

triggers the pulsed imaging laser when a particle is exactly at the object plane of the imaging optics. The images measured by the CPI are insensitive to airspeed as long as the duration of the imaging laser is short enough. The CPI uses single photodiode detectors in the PDS, which leads to uncertainties in counting of smaller (<100 μm) particles, due to inhomogeneities in the PDS laser beams and irregularities in particle geometry. The PDS system in the CPI is replaced with two 128-photodiode arrays in the 3V-CPI and the same (fast) electronics as the 2D-S, so that the 3V-CPI is effectively a combination of CPI and 2D-S probes. The 2D-S probe runs continuously and provides individual size distributions and images with 10 μm resolution from each 128-photodiode array. In addition, particles that pass through the overlap region of the PDS laser beams trigger the CPI and image particles at a rate of 400 frames per second with 2.3 μm pixel resolution, producing three independent views of these particles. The 8-bit digital camera in the CPI produces a full 256 gray levels, providing digital images that can be used to determine ice crystal habit and to distinguish ice crystals from water drops based

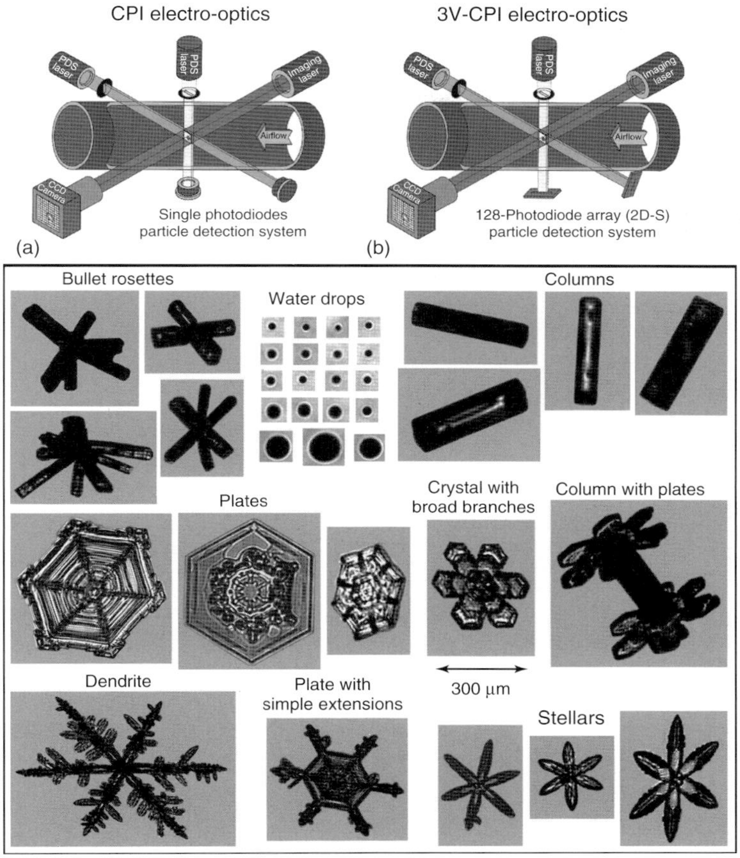

Figure 5.22 CPI optics shown schematically in (a), and 3V-CPI optics in (b), both use two PDS continuous lasers to trigger a high-power, pulsed imaging laser that illuminates the particle whose image is recorded on the CCD camera. The fundamental difference between the CPI and 3V-CPI is that the 3V-CPI uses 128-photodiode arrays for the PDS, which also doubles as a fully functional 2D-S probe. The images in (c) illustrate the detail that is possible. (Reproduced in modified form from King *et al.* (2007), with permission from Cambridge University Press.)

on sphericity comparison (Korolev *et al.* 1999; Lawson *et al.* 2001; Stith *et al.* 2002; Lawson *et al.* 2010).

5.3.3.2 **Measurement Issues**

Uncertainties in the particle size and concentration are both linked to the relationship between the diffraction pattern and distance from the center of focus. In the original design of the OAP, Knollenberg (1970) empirically determined that the diameter of the image of a particle will be within 10% of the geometric diameter if the particle in the beam causes a decrease in intensity on the linear array >50%. He also determined that the region, often designated the DOF, within which a particle will produce this decrease in the laser intensity is equal to DOF = $A \cdot R^2/\lambda$, where $A = 6$ and R is the radius of the particle. It was shown by Korolev (2007), however, that particles can be much farther away from the center of focus than defined by the calculated DOF and still create an image that shadows at greater than 50%. In these cases, when particles are out of focus, they can be recognized as "donuts", that is, images that are missing shadowed pixels in their center portion, and they will also be larger than the size of the actual particle. In Figure 5.20 the white squares show examples of these types of out-of-focus particles. Korolev (2007) derived a method to calculate the actual size of these images and their distance from the center of focus so that the correct sample volume can be computed to calculate the number concentration. The definition of particle size becomes ambiguous when the images are nonspherical and remains an open discussion among the users of OAP data. This discussion is explored in greater detail in Section 5.5 with examples of how particle size can be defined in various ways and how it affects PSDs.

As with the single-particle light-scattering probes, the splashing/shattering of hydrometeors on the arms of the OAPs produces fragments that have to be detected and removed from the analysis. Various analysis techniques have been developed to do this and are discussed in greater detail in Chapter 6. The optimum solution suggested by both Korolev *et al.* (2011) and Lawson (2011) is to minimize the problem by an improvement in the design of the probe arms and remove remaining shattered particles using postprocessing software algorithms, solutions that are also discussed in Chapter 6.

5.3.3.3 Summary

In summary, optical imaging encompasses the following:

(i) a wide range of measured sizes about $<5\,\mu m$;
(ii) sizing independent of the particle shape, composition, or refractive index but dependent on the particle orientation and position in the sample volume of the probe;
(iii) a dependence of image size on the distance from the object plane and errors in sizing as much as a factor of 1.8 requiring correction algorithms;
(iv) DOF and sample area are functions of particle size;
(v) sizing of particles with decreasing numbers of pixels subject to increasing sizing errors is subject to large errors related to image digitization and smaller images also have larger uncertainty in the DOF definition, which may result in large errors in concentration; and
(vi) in general, shadow image formation depends on the position of the particle in reference to the object plane, chromatic properties of light, degree of coherence, beam divergence, numerical apertures of optics, and optical aberrations.

5.3.4 Imaging of Particle Ensembles – Holography

Holography provides the particle position in three-dimensional (3D) space, and its shape and size for each particle in a dilute collection of cloud droplets and ice particles inside a localized, 3D sample volume. In-line holography is particularly useful in atmospheric applications because of its relatively simple and robust optical setup. The holographic method experienced some of its earliest development as a result of its application to the measurement of atmospheric particles (Thompson, 1974; Trolinger, 1975). Since then in-line holography has been applied to various problems of atmospheric interest using photographic emulsions (Trolinger, 1975; Kozikowska, Haman, and Supronowicz, 1984; Brown, 1989; Borrmann and Jaenicke, 1993; Uhlig, Borrmann, and Jaenicke, 1998) and digital cameras (Lawson and Cormack, 1995; Fugal *et al.* 2004; Raupach *et al.* 2006; Fugal and Shaw, 2009).

Holography has several advantages relative to other cloud microphysical measurement techniques: (i) it provides a well-defined sample volume independent of particle size, given an appropriate reconstruction method, (Fugal, Schulz, and Shaw 2009) and air speed; (ii) it measures over a wide range of particle sizes; (iii) the relative spatial locations of particles is measured, thereby enabling studies of particle clustering or ice crystal shattering (Fugal and Shaw, 2009); and (iv) it allows relatively large, contiguous volumes of cloud to be sampled instantaneously, in a localized volume, without requiring assumptions regarding statistical homogeneity. The primary, relative disadvantages of digital holography are the large data rates, which constrain the frequency at which holograms can be recorded. This places demands on the data storage logistics and increases the complexity and time required in data processing, with the latter including digital reconstruction and particle detection and characterization.

In simple terms, an in-line hologram is an interference pattern resulting from the superposition of an incident plane wave E_R (the reference wave) and light scattered by the dilute suspension of illuminated particles, E_S (Figure 5.25a,b). Thus, the electric field at the measurement plane is $E_H = E_R + E_S$. In practice, the intensity or the modulus squared of the superimposed waves is measured, yielding

$$\begin{aligned} I_H(x, y) &= E_H \cdot E_H^* \\ &= E_R \cdot E_R^* + E_R \cdot E_S^* + E_R^* \cdot E_S + E_S^* \cdot E_S \end{aligned} \tag{5.7}$$

The first term on the second line is the constant background, which can be filtered out before the numeric reconstruction (Pan and Meng, 2003). The second and third terms represent the interference of the reference and object waves and are called the *real* and *virtual images*, respectively. The overlap of the real and virtual images is a disadvantage of the in-line geometry, but especially for small particles, the virtual image results in relatively minor changes to the background. The last term represents the diffraction pattern and is typically negligible for small cloud particles. An approximate analytical expression for $I_H(x, y)$ for a spherical particle is discussed in Section A.4.4 given in the Supplementary

Online Material provided on the publisher's web site. Suffice it to say, that the detailed fringe pattern depends on both the particle diameter and its spatial position along the optical axis, allowing a size and 3D position to be obtained through reconstruction. When the real image is reconstructed, the virtual image reconstructs as a blurry background appearing around the reconstructed "focused" real image. The effect of the virtual image on the quality of the real image can be reduced (Raupach, 2009), and this method can also be used to probe the internal structure of ice particles in holograms. The "focus" position for the reconstructed real image of a given particle yields the 3D position of the particle, and the image at this position gives the shape and size of that particle (Figure 5.23c).

In practice, images of sampled particles are obtained by numerical reconstruction of the recorded digital holograms. There are a wide range of possible approaches to the reconstruction problem, each with limitations and advantages. The most straightforward reconstruction methods, based on standard Fourier optics (Goodman, 1996), are suitable for detecting cloud droplets and ice crystals, and in determining particle size, shape, and location. The basic method is to estimate particle size and location using the intensity of digitally reconstructed images stepped uniformly along the optical axis. Optimal reconstruction methods and hologram filtering might differ for various particle shapes, sizes, and experimental conditions. Under typical aircraft flight conditions, the sampling rate of an in-line hologram instrument is not dependent on air speed, but on the individual sample volume size and the frame rate of the camera. The minimum particle detection size is, in general, determined by the greater of two criteria: either $\gtrsim 2$ pixels wide to resolve a particle or diffraction-limited resolution as found by $Dp = 2.44\,\lambda \cdot z_{\max}/D_{\text{ap}}$, where D_p is the diffraction-limited scale diameter of the particle, λ is the wavelength, $z_{\max}$ is the furthest distance in the sample volume from the camera, and D_{ap} is the scale diameter of the aperture or camera (Fugal and Shaw, 2009). The maximum detection size is some fraction of the detector size as determined typically by the automated hologram processing code's ability to reliably bring such a large particle in focus. Particle sizes can be calculated by counting the number of pixels contained in each focused particle and calculating an equivalent diameter for a circle of equal area. For spherical particles, this pixel counting method gives a precision approximately equal to the square root of the pixel size (Lu *et al.* 2008). It is possible to avoid dependence of the sample volume on particle size by using a low-pass filter as described by Fugal, Schulz, and Shaw (2009), which enforces a uniform minimum detectable particle size throughout the sample volume. Effectively it makes particles of the same size and shape appear the same throughout reconstructed images, independent of the reconstructed distance. This is done at the expense of overlooking some smaller particles that would be visible nearer the camera, but not at the far end of the sample volume.

Figure 5.23(d) shows a sample of ice crystals reconstructed from holograms taken with HOLODEC 2 during a research flight conducted over Wyoming, USA, in 2011. Figure 5.23(e) shows a 3D view of cloud droplet positions in $1\,\text{cm}^3$ from a single hologram from the same flight, with droplets smaller and greater than

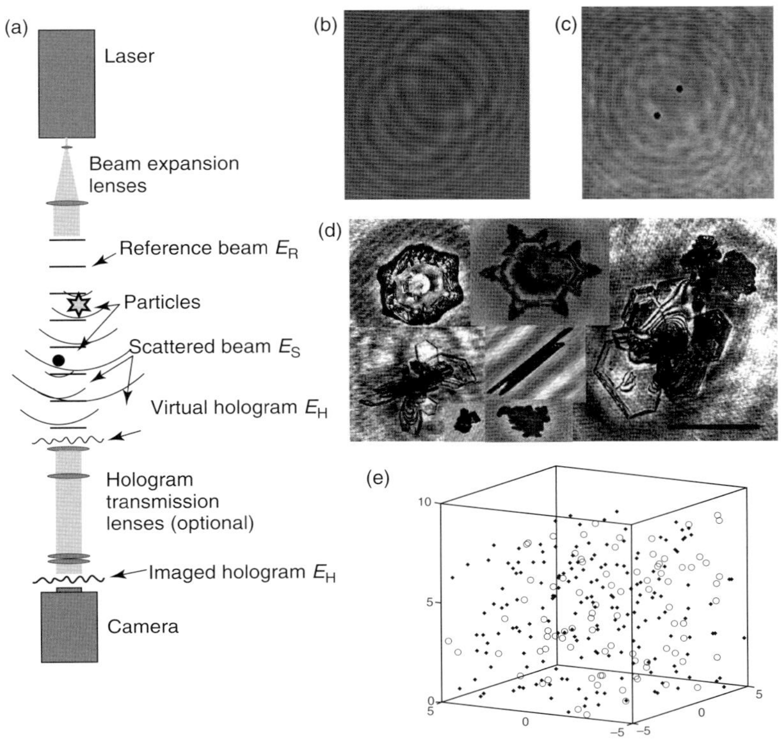

Figure 5.23 (a) A typical configuration for in-line holography (with an optional set of hologram transmission lenses for adjustment of effective pixel size), illustrating the reference and scattered waves, and the resulting interference pattern. (b) A hologram of two 35 μm droplets located 21.4 mm from the camera. The hologram was taken with a 1024 × 768, 4.65-μm square pixel camera with a 532 nm wavelength laser beam. (c) The reconstruction of those droplets, with the real image being the dark, disc-shaped images of the droplets, and the virtual image the wavy or blurry rings surrounding each droplet. (d) Seven ice crystals reconstructed from holograms taken with HOLODEC 2 during a research flight. The black bar in the lower right of the image is 100 μm wide. (e) A volumetric view of cloud droplet positions measured in 1 cm^3 from a single hologram from HOLODEC 2. For illustration, droplets with diameter less than 10 μm are shown as dots and droplets with diameter greater than 10 μm are shown as open circles (Beals *et al.* 2012).

10 μm shown by different symbols (Beals *et al.* 2012). These panels illustrate the utility of holography in characterizing populations of liquid and solid particles in clouds. With a sufficiently large sample volume, the holographic method allows for a statistically significant size distribution estimate from a single hologram, meaning from a spatially localized sample volume. This minimizes the need for averaging over scales that are not microphysically connected, as well as the associated assumptions of statistical homogeneity involved with such large-scale averaging.

5.4 Integral Properties of an Ensemble of Particles

In the previous section, we discussed methods for measuring cloud particle properties on an individual particle basis. From these measured properties of individual droplets and crystals, integral properties of the particle ensemble can then be derived, such as LWC, phase function, and extinction. Unless the clouds are all water, however, the wide variety of crystal shapes and effective densities will lead to large uncertainties that can be avoided by using techniques that measure the integral properties of particle ensembles directly. Such instruments, referred to as *integrators*, are intrinsically not affected by particle coincidence; hence, they can be designed with a larger sample volume than single-particle counters and thus provide better statistics about the small-scale spatial distribution of cloud microphysics. Integrators are designed to measure properties directly proportional to moments of the size distribution. For example, hot wires described in Section 5.4.1 are configured to measure latent heat release proportional to the liquid and/or IWC. Optical integrators such as the cloud extinction probe (CEP) and the particle volume monitor model 100 (PVM-100A), described in Section 5.4.2, measure transmitted or scattered light proportional to light extinction and LWC (second and third moments of the size distribution). Nephelometers, such as the cloud integrating nephelometer (CIN; Section 5.4.3), extend these measurements to forward and backward scattered light or measure the full scattering phase function as in the polar nephelometer (PN).

5.4.1 Thermal Techniques for Cloud LWC and IWC

Early measurements of cloud LWC and total condensed water content (TWC) were accomplished by taking bulk samples (e.g., rime ice samples or ice-particle captures), and in fact, such measurements still provide the basis for important engineering standards such as the certification of aircraft for flight into icing conditions. Since then various techniques have been developed to continuously measure or infer LWC and TWC: hot-wire devices for melting or evaporating water, inlet-based evaporators, optical devices, and even riming devices. In the following section, hot-wires and icing sensors are discussed. Devices that take a large air sample, evaporate the hydrometeors, and measure the resulting water vapor also fall into the category of thermal techniques. Because they also have issues related to bringing particles into the sampling section of a sensor, the details of this type of instrument are included in Chapter 6, Section 6.4.2. Their specifications, however, are included in Table 5.1 since they measure an important property of clouds.

5.4.1.1 Hot-Wire Techniques

In the early 1950s, the first commercial instrument was developed to measure cloud LWC using a hot-wire exposed to cloud (Neel, 1955), providing the possibility of a continuous record of the structure of LWC by rapid sampling. The device

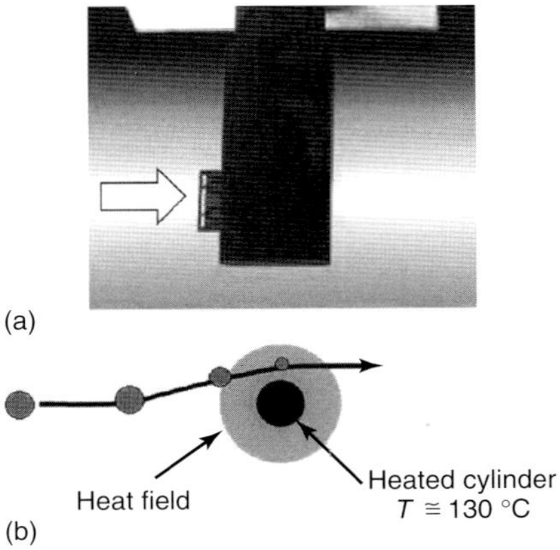

Figure 5.24 (a) This shows a commercial hot-wire sensor and (b) the schematic representation of a cloud particle evaporating as it passes the heated element. T ranges between 100 and 130 °C.

operated on the principle that the LWC could be deduced from the amount of power delivered to the wire to evaporate cloud droplets. In the mid-1970s and early 1980s, hot-wire instruments were developed that incorporated an important change: the devices were run with a constant wire temperature and therefore their response characteristics were more readily derived from "first principles." In addition, these sensors, in theory, would not require a calibration (Merceret and Schricker, 1975; King, Parkin, and Handsworth, 1978b). The "King" device was developed for commercial sale and has since been widely used in the community. Since the introduction of this instrument, a variety of other hot-wire devices have been developed that incorporate the constant-wire-temperature technique (Nevzorov, 1980; Lilie *et al.* 2004; Vidaurre and Hallett, 2009b). Figure 5.24 shows a typical hot-wire sensor and a schematic that shows conceptually how hydrometeors approaching the heated element begin to evaporate as they enter the heat field around the element.

These instruments measure the amount of power delivered to the wire to keep it at a constant temperature in cloud. Cooling of the wire results from convective heat loss to the airstream and the amount of heat delivered to evaporate water droplets:

$$P_t = P_{dry} + P_{LWC} \tag{5.8}$$

where P_t is the total power measured across the wire, P_{dry} is the dry-air convective heat loss across the wire, and P_{LWC} is the power expended to evaporate droplets (King, Parkin, and Handsworth, 1978b).

The convective heat loss (P_{dry}) for a cylindrical wire can be expressed as a function of the dimensionless Nusselt (Nu) and Reynolds (Re) numbers and has been parameterized by calibration flights in cloud-free air and wind tunnel studies (Zukauskas and Ziugzda, 1985). The calculated LWC is derived from the following equation (King, Parkin, and Handsworth, 1978b):

$$W_{\mathrm{L}} = \frac{P_{\mathrm{t}} - P_{\mathrm{dry}}}{\left[L_{\mathrm{v}} + (T_{\mathrm{e}} - T_{\mathrm{a}})\right] \cdot V_{\mathrm{a}} \cdot L \cdot D \cdot \epsilon_1} \tag{5.9}$$

where l_{v} is the latent heat of evaporation, T_{e} is the evaporative temperature of water, T_{a} is the air temperature, V_{a} is the air velocity, $S = L \cdot D$ is the sensor cross-sectional area, and ϵ_1 is the overall collection efficiency of the sensor for droplets. Nevzorov (1980) pointed out that T_{e} should be more correctly the "equilibrium" temperature, accounting for diffusional vapor transfer from the warm wire to its immediate environment.

Several types of hot-wire devices incorporate a reference hot-wire not exposed to direct impact with water droplets to improve the estimation of the dry-air convective heat losses. The overall baseline uncertainty due to the removal of the dry term is typically of the order of $0.03\,\mathrm{g\,m^{-3}}$ (King, Parkin, and Handsworth, 1978b), which varies according to flight maneuvers, and can be improved slightly using the reference wire. With a careful analysis and manual setting of the noise baseline to zero, the sensitivity of some devices can be as low as several milligram per cubic meter for specific situations (Korolev *et al.* 1998a). Variations in these hot-wire systems are primarily in the type (solid or wound) and geometries of the wires, which affect their capture efficiencies for water droplets and ice particles. Relative to other LWC and TWC estimation devices, hot-wires remain an attractive option because of their relatively small size and low power consumption.

Wire geometries with a capture volume have also been used to estimate TWC in mixed-phase conditions. With an independent measurement of LWC, the IWC can then also be estimated as long as the LWC does not dominate the TWC. King and Turvey (1986) proposed a wire-wound half-cylinder TWC-heated element rather than a cylindrical wire, operating as an ice-particle melting sensor where TWC could be calculated in a similar manner as that of LWC. In the mid-1980s, the Nevzorov LWC/TWC device was introduced into airborne measurements (Nevzorov, 1980; Korolev *et al.* 1998a) and used an evaporating rather than a melting TWC hot-wire. Several newer hot-wire TWC devices have since been developed based on similar principles (Lilie *et al.* 2004; Vidaurre and Hallett, 2009a). Figure 5.25 illustrates the physical configuration of the Nevzorov hot-wires, as well as the concept of the phase-separation capability of this and similar devices. A cylindrical "LWC" hot-wire (Figure 5.25a) and a noncylindrical wire-wound "TWC" collector with a conical capture volume pointed into the airflow (Figure 5.25b) are each attached to the leading edge of a flow-correcting vane.

In this design, ice particles are assumed to impact and bounce off the surface of the cylindrical LWC sensor, resulting in very little heat loss, while liquid droplets spread out into a film and quickly evaporate, resulting in a relatively large heat loss. The TWC collector is assumed to capture both ice particles and water droplets efficiently, and the heat loss to the wire will reflect both the liquid and ice-particle mass of the cloud. In reality, both wires react to liquid and ice particles and so care is necessary when interpreting the measurements. Considering all the contributions to both the LWC and IWC wire responses, the following equations adapted from

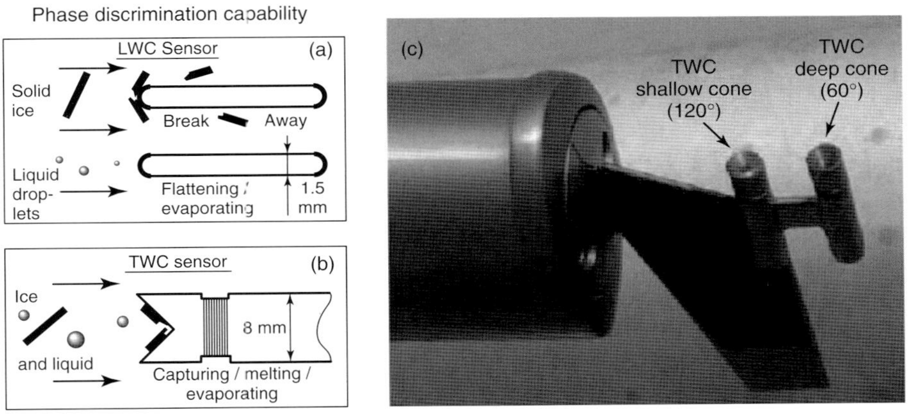

Figure 5.25 The standard Nevzorov probe has two sensors: (a) a cylindrical sensor as the one shown in Figure 5.24 that is in concept sensitive to only liquid water droplets (LWC sensor) and (b) a conical sensor that can temporarily capture water drops and/or ice crystals long enough for them to melt and evaporate (TWC sensor). (c) The photo shows pedestal-mounted Nevzorov rotating sensor vane that keeps the sensors oriented into the airflow. This particular version of the vane is equipped with two TWC sensors of the different cone geometry for special performance testing. Copyright American Meteorological Society (AMS). Reprinted with permission by AMS.

Korolev *et al.* (1998a) are used to estimate IWC and LWC from the measured responses of each wire:

$$W_{\mathrm{I}} = \frac{\epsilon_1 \cdot W_{\mathrm{Tm}} - \epsilon_3 \cdot W_{\mathrm{Lm}}}{C \cdot (\epsilon_1 \cdot \epsilon_4 - \epsilon_2 \cdot \epsilon_3)} \tag{5.10}$$

$$W_{\mathrm{L}} = \frac{W_{\mathrm{Tm}} - \epsilon_4 \cdot C \cdot W_{\mathrm{I}}}{\epsilon_3} \tag{5.11}$$

$$W_{\mathrm{T}} = W_{\mathrm{L}} + W_{\mathrm{I}} \tag{5.12}$$

where W_{L}, W_{T}, and W_{I} are the true LWC, TWC, and IWC values, respectively. W_{Lm} and W_{Tm} are the measured LWC and TWC, respectively. The collection efficiencies of the LWC and TWC sensors for water droplets and ice crystals are given by ϵ_1, ϵ_3, ϵ_2, and ϵ_4, respectively. C is a latent heat enhancement factor for ice particles to include both melting and evaporation. At aircraft speeds of $100\,\mathrm{m\,s^{-1}}$, the LWC sensor measures a false signal from ice-particle impacts that is about 10–20% of the measured IWC (Korolev *et al.* 1998a Cober *et al.* 2001b).

5.4.1.2 **Mass-Sensitive Devices**

Another instrument that uses phase change to measure hydrometeor mass is the Rosemount icing detector (RID) that is sensitive only to supercooled water droplets. Shown schematically in Figure 5.26, this technique is based on changes in the natural frequency of a vibrating cylinder due to accretion of ice on its surface. The

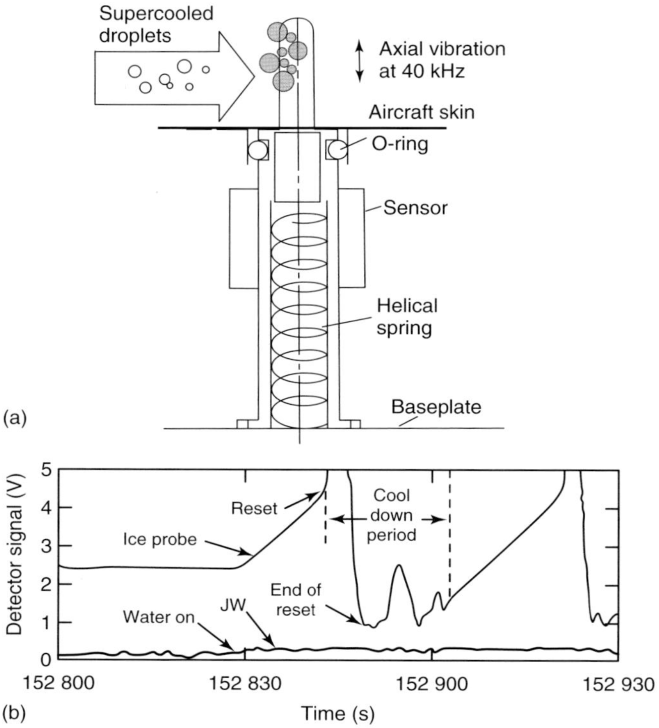

Figure 5.26 (a) The Rosemount Icing probe consists of a steel cylinder vibrated longitudinally with a resonant frequency of 40 kHz. (b) The signal from the sensor is a steadily increasing voltage that is proportional to the ice mass. When a maximum voltage of 5 V is reached, a heater melts the ice. (Copyright American Meteorological Society (AMS). Reprinted with permission by AMS.)

cylinder is vibrated using magnetostrictional excitation that causes axial oscillations of the cylinder at a natural frequency of 40 kHz. As supercooled water droplets impact the cylinder, they freeze and their added mass decreases the oscillation frequency. This change in frequency is proportional to the ice mass and through calibration, it is converted to LWC. Ice mass is allowed to build up on the sensor until the signal reaches a threshold that turns on a heater and removes the ice. No further accretion of ice will occur during this heating cycle or a variable period after the heater is turned off. The actual amount of time when no useful data are taken depends on the temperature, airspeed, and LWC. More details can be found in Baumgardner and Rodi (1989); Heymsfield and Miloshevich (1989); Cober *et al.* (2001b); Cober, Isaac, and Korolev (2001a) and Mazin *et al.* (2001) (Figure 5.26).

5.4.1.3 **Measurement Issues**

In principle, cloud droplets that impact a hot-wire will spread out and form a thin film over the wire so that evaporation should be rapid and complete if the wire temperature remains above the equilibrium temperature. The diameters of the wires have typically been chosen between 0.5 and 2 mm, at least a factor of 10 larger than typical cloud droplet diameters. Calculated collision efficiencies have been commonly incorporated into estimates of LWC and tend to exceed 0.8 for droplets larger than about 10 μm at typical aircraft speeds, except for the largest wire diameters. The accuracy of hot-wire LWC estimates has been established by comparison to reference measurements provided by icing cylinder or blade rime ice measurements in wet wind tunnels (Stallabrass, 1978; Ide and Oldenburg, 2001). In these types of tests, using a commonly selected droplet median volume diameter (MVD) of 20 μm, hot-wires have been found to agree with such reference methods to within about 15% (King *et al.* 1985; Strapp *et al.* 2002), indicating that the hot-wire accuracy approaches the estimated accuracy of the reference method; however, the fraction of the LWC measured by cylindrical wires has been found to decrease with increasing MVD (Biter *et al.* 1987; Strapp *et al.* 2002), dropping to as low as 50% (200 μm MVD, 100 m s^{-1}, 1.85 mm cylindrical wire) at high MVD. This decrease in sensitivity with droplet MVD had been recognized since the early use of hot-wires (Owens, 1957; Neel, 1973) and has been attributed to droplet splashing and re-entrainment of water after impact with the wire. There has been no theoretical treatment of this effect to date, and LWC corrections to higher MVD droplet clouds must rely on the empirical response measurements from wind tunnel studies. In addition, hot-wires of different geometries have been found to have varying sensitivity with respect to MVD. For example, LWC measured with larger diameter cylindrical hot-wires is less sensitive to increasing MVD than smaller diameter hot-wires. Noncylindrical hot-wires with "capture" volumes have been found to have the highest efficiencies at high MVD (Ide, 1999; Strapp *et al.* 2002; Schwarzenboeck *et al.* 2009), but tend to have larger inertial collision efficiency corrections at small MVD due to their typically larger wire diameter. The difference in the amount of water lost from the wires of different geometry at high MVD has, in fact, been proposed as a possible method for the direct estimation of droplet MVD (Strapp *et al.* 2002), and one recent hot-wire device incorporates three impact wires and a reference wire to estimate both LWC and droplet MVD (Lilie *et al.* 2004). Such a method for MVD estimation relies on an empirical calibration of the device in a wind tunnel and the accuracy of the tunnel MVD estimate itself. The sensitivity to the droplet distribution with this device has not been determined. While noncylindrical TWC sensors measure most of the typical cloud LWC (Ide, 1999; Strapp *et al.* 2002; Lilie *et al.* 2004), significant fractions of ice particles can be ejected from the capture volume as has been demonstrated using high-speed video photography (Emery *et al.* 2004; Strapp *et al.* 2005; Korolev *et al.* 2008). Studies of the response of different hot-wire geometries to ice are currently incomplete; however, variations of up to a factor of three in IWC measurements have been observed in ice-particle wind tunnels and glaciated natural clouds (Strapp *et al.* 2005; Korolev *et al.* 2008) when comparing probes with deeper capture volumes to those more shallow. The determination of absolute IWC efficiencies of hot-wire sensors has been impeded by the lack of a suitable reference standard for ice-particle testing. Although facilities now exist with a reference IWC estimate, for example, Strapp, MacLeod, and Lilie (2008), artificially produced ice particles in these facilities do not accurately represent natural ice crystals and complementary aircraft-based testing is needed in natural clouds (Isaac *et al.* 2006; Korolev *et al.* 2008).

5.4.2
Optical Techniques for the Measurement of Cloud Water

5.4.2.1 The PVM

The PVM model 100A is closely related to a class of instruments termed *laser-diffraction particle-sizing instruments* (Azzopardi, 1979; Hirleman, Oechsle, and Chigier, 1984). The basic features of the PVM are shown in Figure 5.27. A collimated laser irradiates an ensemble of particles that scatter light onto a narrow circular region of a sensor after passing through a transform lens. The distance for the center of focus of the laser to the sensor defines the scattering angles. The PVM has a large-area photodiode in front of which is placed a fixed spatial filter with varying radial transmissions that converts the scattered light to measures of the integrated particle volume concentration, C_v (volume of the aerosol particles per volume of the suspending medium). The PVM has been used primarily to measure C_v in warm water clouds where $C_v = \mathrm{LWC} \cdot \rho^{-1}$ (ρ = water density). As shown

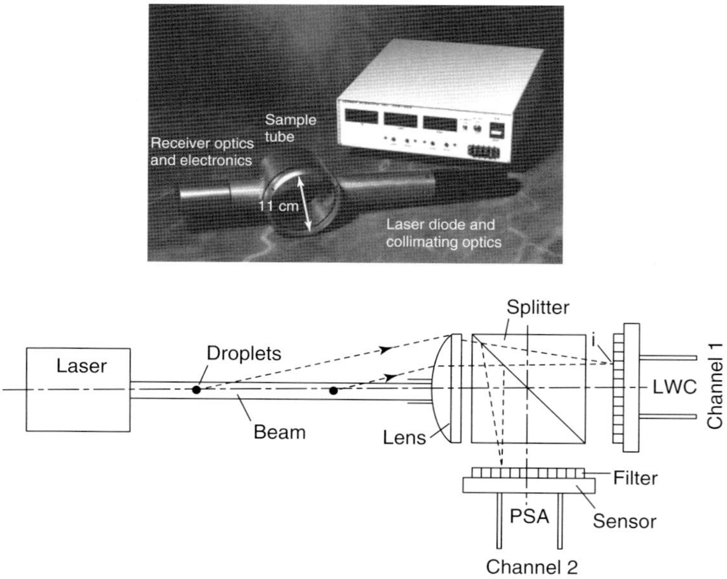

Figure 5.27 This illustration shows the basic optical configuration for the PVM. There are two channels of information. The first channel is from the LWC detector and the second one is from the particle surface area (PSA) detector. The component labeled "lens" is the transform lens referred to in the text. (Copyright Gerber Scientific, Inc. (GSI). Reprinted with permission by GSI. Courtesy of H. Gerber.)

in Figure 5.27, the second channel of the PVM converts the scattered light to the integrated surface area (PSA) of drops, which is proportional to the light extinction by the particles ($\sigma_{ext} = 2\pi \cdot Q_{ext} \cdot \text{PSA}$). The ratio of LWC/PSA is, therefore, proportional to the effective diameter (D_{eff}) of the droplet size distribution.

The conversion of the scattered light to LWC and PSA depends on establishing the proper radial transmissions of the spatial filters that cause the scattered light impinging on the photodiodes to be proportional to D^3 (D is the droplet diameter) for LWC and D^2 for PSA. The means for specifying such filter transmissions are described in Gerber (1991a) and Gerber, Arends, and Ackerman (1994) and are briefly summarized here using diffraction theory. The light diffracted by the droplets and incident on the photodiodes is given by

$$F(D) = \sum_{i=1}^{N} f_i(D) \cdot T_i \tag{5.13}$$

where $f_i(D)$ is the flux of light falling onto a circular annulus i of the spatial filter, T_i is the fraction of light transmitted through the filter at i, and N is the number of adjacent annuli, here $N = 25$. To derive the radial transmission values of the two filters, the assumptions are a one–dimensional light beam, a uniform distribution of the droplets in the beam, and an optimal linear response with LWC for droplets sizes as large as $D \sim 45$ m. The values of T_i are shown in Gerber (1991a), for the filter used in the PVM-100A. A similar procedure is used to establish a second spatial filter for the PSA channel.

The PVM-100A probe, shown in Figure 5.27, consists of a 11.4 cm diameter sampling tube that faces into the direction of flight, attached to two smaller support cylinders. The longer of the cylinders contains the laser-diode light source (modulated at 100 kHz) and collimating optics. The shorter cylinder holds the optics and detectors for the LWC and PSA channels. The laser beam converges at an angle of 1.7° to focus onto the light trap located in front of the transform lens located in the short end. The beam illuminates a volume of 1.25 cm^3 in the inlet of the probe. The projected, sensitive area of the beam in this section is 1.14 cm^2. The forward scattered light by the cloud particles in the center section is collected over the range of half-angles from ~0.2° to 6°.

The outputs of the detectors in the probe are processed by additional electronics located within the aircraft. A low-pass filter controls the time response and resolution of the measurements. For example, with the filter set for at a 1000 Hz roll off, the PVM-100A can resolve features in the LWC at a length scale of 10 cm at an aircraft speed of 100 m s^{-1}. In this case, the swept out volume of the cloud is ~10 cm^3.

There are two principal methods for calibrating the PVM-100A to establish the scaling constants that relate detector voltages to LWC and PSA and to establish the droplet size range over which those parameters are valid. The first method compares the measurement of LWC by the PVM-100A to a measurement of the infrared optical extinction, b_{ext}, (Chylek, 1978). The relationship between LWC and b_{ext} is

$$\text{LWC} = \frac{2\rho \int D^3 \cdot f(D)\,\mathrm{d}D}{3\int Q_{ext} \cdot D^2 \cdot f(D)\,\mathrm{d}D} \cdot b_{ext} \tag{5.14}$$

where $f(D)$ is the size distribution of the droplets and Q_{ext} is the Mie efficiency factor for extinction. Chylek (1978) suggested that an approximate linear relationship existed between Q_{ext} and D for an infrared wavelength of 11 μm and droplets with diameters <28 μm so that

$$\mathrm{LWC}(\mathrm{g\,m^{-3}}) \approx \frac{b_{ext}(\mathrm{km^{-1}})}{128} \tag{5.15}$$

The PVM was compared to an infrared transmissometer in the CALSPAN Corp. cloud chamber to establish the scaling constants and validate the linearity of the PVM output for different LWCs (Gerber, 1998).

The second calibration method placed the PVM-100A into the CHIEF fog chamber at ECN (Netherlands Energy Research Foundation). This chamber has a carefully controlled, continuous flow environment near 100% RH, the capability of generating LWC with different MVD, and a reliable filter collection method by which the liquid water in the chamber is collected, weighed, and converted to an LWC. The PVM-100A was compared to the ECN filter method for fogs with MVD ranging from 10 to 41 μm (Gerber, Arends, and Ackerman, 1994). For MVD smaller than ~30 μm, a comparison was made between the infrared (CALSPAN) and filter (ECN) approaches for establishing the PVM-100A scaling constants; good agreement was found suggesting that the PVM-100A provides a good absolute measure of LWC. The LWC channel of the PVM-100A has an accuracy of ~10% over a droplet size range of about 4–30 μm diameter, while the PSA channel measures with similar accuracy to about 70 μm diameter. The precision of the probe is ~0.002 V, which is equivalent to 2 mg $\mathrm{m^{-3}}$ of LWC. The strengths of the PVM-100A are its fast response time, accuracy, and precision over the indicated droplet size range and stability of the measurements. Numerous publications have included PVM-100A measurements. For example, the probe's fast response has permitted new looks at fine-scale behavior in clouds including the finding of a "scale-break" in the LWC power spectra of clouds (Davis *et al.* 1999; Gerber *et al.* 2001) and at the entrainment effect in stratocumulus (Gerber *et al.* 2005) and cumulus (Gerber *et al.* 2008). Examples of the type of detail that can be measured with the PVM-100A is shown in Section 5.5.

The PVM-100A begins to lose sensitivity to droplets larger than 30 μm (Gerber, Arends, and Ackerman, 1994; Wendisch, Garrett, and Strapp, 2002). The design calls for an optimum upper droplet size response of ~45 μm diameter for LWC measurements, close to the response of the FSSP droplet spectrometer, but the loss of sensitivity is a result of the convergence of the laser beam. By using a parallel beam, the size response improves, but at the cost of reducing the sample volume and increasing statistical sampling error from the random spatial distribution of droplets. The roll-off in LWC affects the derivation of D_{eff}, which can be measured accurately from D_{eff} ~8 μm to ~25 μm. The PVM-100A cannot distinguish water droplets from ice crystals and the weighting functions are based on spherical droplets with the density of liquid water. Hence, caution should be exercised when interpreting measurements in mixed phase or ice clouds.

5.4.2.2 Angular Optical Cloud Properties

The importance of clouds for modulating the amount of solar radiation that reaches the earth and terrestrial radiation that is absorbed has already been underscored in the introduction. The optical properties that are most commonly used in climate models to describe the interactions between clouds and the fluxes of short- and long-wave radiation are the extinction coefficient, single-scattering albedo, and asymmetry factor. These three parameters, which are dependent on the wavelength of light, are also discussed in Chapters 4 and 7 hence, here we only touch briefly on their definition before describing how they are derived or measured directly in clouds with *in situ* instruments. Assuming randomly oriented particles, the extinction coefficient is derived from the following relationship (Gayet *et al.* 2002):

$$b_{\text{ext}}(\lambda) = \sum_i Q_{\text{ext}}(\lambda_i) \cdot N_i \cdot S_i \tag{5.16}$$

where $Q_{\text{ext}}(\lambda_i)$ is the extinction efficiency at the wavelength λ_i (assumed to have a value equal to 2.0 in the range of measured sizes, that is, the large particle approximation) and N_i the concentration of particles having a geometrical cross section S_i. The asymmetry factor g is equal to the mean value of μ (the cosine of the scattering angle), weighted by the angular scattering phase function $P(\mu)$:

$$g = \frac{1}{2} \int_{-1}^{1} \mu \, P(\mu) \mathrm{d}\mu \tag{5.17}$$

The phase function is defined as the energy scattered per unit solid angle in a given direction to the average energy in all directions. The asymmetry factor is a measure of the relative magnitude of light that is scattered in the forward and backward directions.

The asymmetry parameter is calculated from (Gerber *et al.* 2000)

$$g = f + (1 - f) \cdot \left(\frac{\overline{R}_1 - \overline{R}_2}{R_1 + R_2} \right) \tag{5.18}$$

where

$$\overline{R}_1 = \frac{1}{2} \int_{\theta_1}^{\pi/2} \beta(\theta) \cdot \cos 2\theta \, \mathrm{d}\theta \tag{5.19}$$

$$\overline{R}_2 = \frac{1}{2} \int_{\pi/2}^{\theta_2} \beta(\theta) \cdot |\cos 2\theta| \, \mathrm{d}\theta \tag{5.20}$$

and

$$R_1 = \frac{1}{2} \int_{\theta_1}^{\pi/2} \beta(\theta) \cdot \sin\theta \, \mathrm{d}\theta \tag{5.21}$$

$$R_2 = \frac{1}{2} \int_{\pi/2}^{\theta_2} \beta(\theta) \cdot \sin\theta \, \mathrm{d}\theta \tag{5.22}$$

The single-scattering albedo is defined as the ratio of the scattering coefficient to the extinction coefficient and is equal to one when the absorption is negligible. While accurate estimates of these parameters can be obtained in clouds with only water droplets, through application of Mie theory to measured distributions of spherical droplets, the same accuracy is not possible for clouds composed of ensembles of ice crystals. Estimates can be made through the theoretical application of ray-tracing of light beams to idealized crystal shapes (Iaquinta, Isaka, and Personne, 1995; Macke, Mueller, and Raschke, 1996; Yang and Liou, 1998) or by using a combination of radiative transfer models, measured microphysical properties, and solar radiative flux profiles (Wielicki *et al.* 1990; Stackhouse and Stephens, 1991; Francis *et al.* 1994; Spinhirne, Hart, and Hlavka, 1996). Values of b_{ext} have also been derived indirectly through the integration of size distributions of cloud particle projected area (Heymsfield and McFarquhar, 1996a). Hence, the development of instruments that can directly measure these optical properties was an important step toward improving our understanding of cloud and radiation interactions. Not only are better parameterizations needed in climate models, but *in situ* measurements are also needed for comparisons with microphysical properties derived from ground-based and space-borne remote sensors, many of which derive the cloud properties from measurements of light scattering and extinction.

The PVM-100A, described in the earlier section, measures the extinction coefficient (Gerber, 1991a). The PN, introduced by Gayet *et al.* (1997), was the first airborne instrument to measure the phase function of cloud particle ensembles, from which the extinction coefficient and asymmetry factor are derived. The PN was closely followed by the CIN in 1998 (Gerber *et al.* 2000) that is designed to measure the asymmetry factor and extinction coefficient. Most recently, the CEP was developed to measure the extinction coefficient with good sensitivity in clouds with low concentrations of ice. The next three sections discuss the PN, CIN, and CEP measurement principles and implementation in greater detail.

5.4.2.3 **The PN**

The PN measures light scattered from suspended particles in a liquid or gas colloid by employing a light beam and a detector set to several sides (usually 90°) of the source beam. The measured intensity of scattered light depends on the PSD and the wavelength of the incident light. Hence, the PSD, as well as the shape of particles, can be estimated from measurements made at a sufficient number of angles (Huffman and Thursby, 1969; Sassen and Liou, 1979; Barkey and Liou, 2001).

The PN, constructed at LaMP (Laboratoire de Météorologie Physique, Clermont-Ferrand, France) is the first instrument to perform direct *in situ* measurements of the scattering phase function of cloud particles over a broad range of sizes (from a few micrometers to about 1 mm diameter). The term *polar* refers to a polar coordinate system where in the PN the instrument's laser beam is the polar axis and the measurements are made at a number of angles around this axis. The prototype version of the instrument is described by Gayet *et al.* (1997), where the preliminary results were from measurements made in cloudy

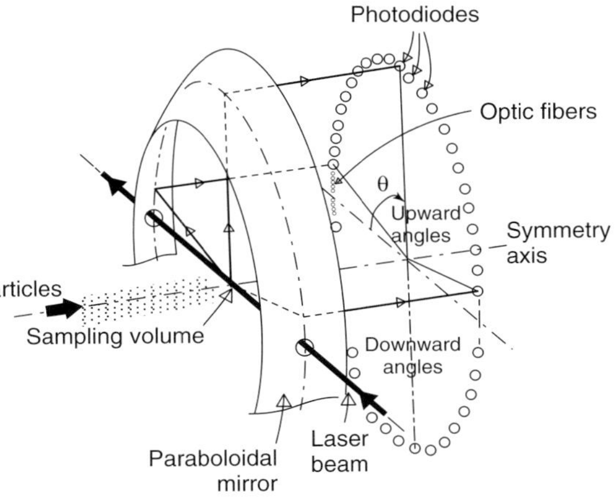

(a)

(b)

Figure 5.28 (a) This schematic diagram of the PN shows the optical configuration and general layout of components. (b) The photo shows the PN mounted on the DLR Falcon-20, next to a 2D-C OAP during a field campaign to measure cirrus.

conditions with a mountain top wind tunnel (Crépel *et al.* 1997). The airborne version was developed at LaMP with similar concepts to those of the prototype and designed to fit inside a standard PMS canister. The instrument, shown in Figure 5.28, has been used during many airborne research projects (Gayet *et al.* 1998; Auriol *et al.* 2001; Gayet *et al.* 2002; Gayet *et al.* 2007).

The probe measures the scattering phase function of an ensemble of cloud particles that intersect a collimated laser beam near the focal point of a paraboloidal mirror (Figure 5.28). The laser beam is provided by a high-power (1.0 W) multimode laser diode operating at $\lambda = 804$ nm. The light scattered at polar angles from $\pm 3.5^{\circ}$ to $\pm 169^{\circ}$ is reflected onto a circular array of 41 photodiodes. At small angles ($3.5^{\circ} < \theta < 10^{\circ}$), the angular resolution is 1.58° and 7° for larger angles ($10^{\circ} < \theta < 165^{\circ}$). The sensitive volume ($0.2\,\text{cm}^3$) is defined by a 10 mm long and 5 mm diameter laser beam. The data acquisition system provides a continuous sampling volume by integrating the signals from each of the detectors over a selected period. For example, the sampling volume (V), determined by the sampling area and the aircraft cruise speed of $\sim 200\,\text{m s}^{-1}$, with jet aircraft, is $\sim 1000\,\text{cm}^{-3}$ for an acquisition frequency of 10 Hz. This means that the detection threshold is close to one particle per liter at this frequency.

Direct measurement of the scattering phase function allows particle types (water droplets or ice crystals) to be distinguished and optical parameters to be derived, that is, scattering coefficient and asymmetry parameter.

The calibration procedure (Gayet *et al.* 1997) consists of the experimental determination of the relationship between the measured voltage output and the light power reaching the photodiode of each of the channels associated with the polar scattering angle. The volume scattering cross section is calculated according to the incident laser power, the length of the sampling volume, and the optical aperture of the detector. The incident laser power is provided by a calibrated LED.

The PN measurements are generally validated by comparing the microphysical parameters retrieved from the scattering phase function (Dubovik, 2004) with those derived from direct size distribution measurements by a single-particle optical spectrometer such as the FSSP or 2D-OAP. A numerical inversion method enables the retrieval of the volume particle size distribution and additional microphysical and optical parameters such as LWC, N, scattering coefficient B_{sca}, and asymmetry parameter, g. Mie theory is used to describe light-scattering characteristics for water droplets and for small ice particles that have a shape close to spherical. For large ice crystals, the geometrical optics approximation is used, that is, the maximum dimension of the ice crystal is much larger than the wavelength of incident light. Closure measurements are also systematically performed on bulk extinction measurements (extinction) to validate the calibration.

The PN measurements at the near forward and backward directions ($\theta < 15^\circ$ and $\theta > 162^\circ$) are generally not reliable because of the contamination by diffracted light caused by the edges of holes that are drilled on the paraboloidal mirror (Gayet *et al.* 1997). The magnitude of the light pollution (and therefore of the signal offsets on the corresponding channels) strongly depends on the laser diode alignment settings.

The derivation of scattering coefficient and asymmetry parameter requires the integration of the scattered light over the scattering angle from 0° to 180°.

To mitigate the restricted range of available scattering angles, techniques have been developed that improve the accuracy of the extinction coefficient measurements to about 25% (Gayet *et al.* 2002).

Equation (5.18) assumes that the diffractive and refractive components of the scattered light at small scattering angles $\theta_1 < 15^\circ$ and $\theta_2 > 162^\circ$ can be separated (Gerber *et al.* 2000). The ratio f of the diffracted portion of the scattered light to the total scattered light has been determined for nonabsorbing large spheres to be f = 0.56 from Mie calculations for typical measured water droplet spectra. Having no *a priori* information of the thermodynamic phase of cloud particles, this value of f is also assumed for ice particles; the value is close to the average value that can be estimated from the theoretical results of Takano and Liou (1995) for a variety of ice crystal shapes. Nevertheless, because of the limitations of theoretical approaches, particularly for irregular shaped ice crystals, significant uncertainties may remain in the measured asymmetry parameter; however, the relative fluctuations of this parameter probably reflect the actual cloud optical properties. The accuracy of measurements of the asymmetry parameter is about 4% as estimated from Mie theory for corresponding mean droplet size spectra for water particles measured by an FSSP.

The iterative inversion method developed by Jourdan *et al.* (2003) is based on a bicomponent (water droplet and ice crystal) representation of cloud composition and uses the nonlinear least-square fitting of the scattering phase function with smoothness constraints on the desired PSDs. The inversion method yields two volume particle equivalent size distributions simultaneously, one for (spherical) water droplets and another for ice crystals that are assumed to be hexagonal shaped with different aspect ratios and randomly oriented in 3D space. Figure 5.29 displays results for two typical cloud types, mixed phase and all ice. The average measured

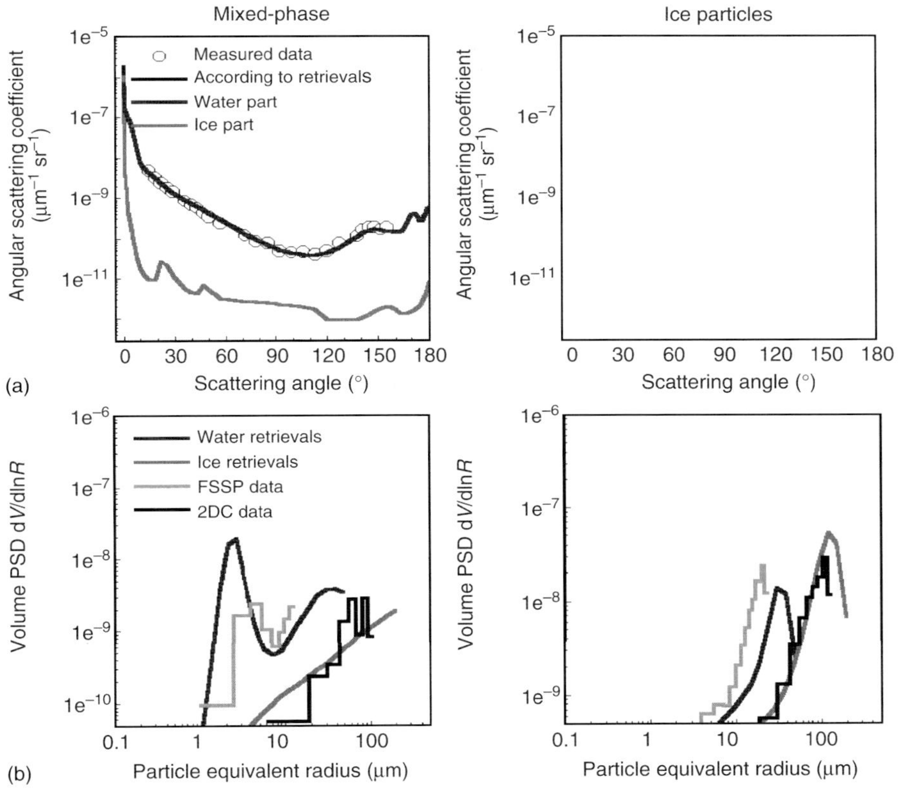

Figure 5.29 Inversion of the averaged scattering phase functions for two cloud types: (i) mixed-phase cloud and (ii) ice cloud. (a) Measured, retrieved, and extrapolated scattering phase functions. The contributions on scattering properties are displayed for both particle compositions (water and ice). (b) Direct (FSSP and 2D-C probe) and retrieved PSDs. From Jourdan *et al.* (2003). (Copyright 2003 American Geophysical Union (AGU). Reprinted with permission by AGU.)

scattering phase functions are plotted (a) along with theoretical scattering phase functions obtained according to the retrieval results. Comparison with the retrieved water and ice PSDs and direct measurements with an FSSP-100 and 2D-C probe are shown in (b). The results show that the retrieved PSDs agree well with the direct measurements. Although the water droplet contribution dominates the scattering features of the cloud, the contribution of ice particles is large enough to be detected with the inversion technique, particularly at the side scattering angles. In ice clouds, the presence of small spherical ice particles is detected (Figure 5.29b) as are the large hexagonal crystals.

The PN measurements contain a considerable amount of information on cloud composition that can be traced to the systematic differences in scattering patterns for water droplets and ice crystals. In addition to providing the phase function for an ensemble of particles, the ability of the inversion technique to discriminate the contribution of the components leads to the retrieval of representative PSDs. In Section 5.5, another example is shown from the PN, highlighting the value of this type of measurement for better understanding of how clouds impact climate.

5.4.2.4 **The CIN**

The CIN, the "g-meter," was developed to address the need for direct *in situ* measurements of the asymmetry parameter, g, and the optical extinction coefficient b_{ext} at visible wavelengths in clouds (Gerber, 1998; Gerber *et al.* 2000). The CIN operates in almost the same way as integrating nephelometers used for aerosol measurements. A collimated laser-diode beam illuminates cloud particles that pass through the probe's aperture. These particles scatter light that is collected by a set of Lambertian sensors (sensors whose response does not vary with the direction from which light is coming). The absolute and relative magnitudes of the signals received by each of the sensors are used to infer the light-scattering properties of the cloud particles. The CIN is designed to measure the asymmetry parameter and extinction coefficient of cloud particle ensembles whose diameters range from about 4 to 2500 μm diameter at the wavelength of the instrument laser (635 nm).

The four Lambertian sensors integrate light scattered into two angular domains, separated by a 90° scattering angle. A schematic of the CIN design is shown in Figure 5.30. The instrument has two forward (F and cF) and two back (B and cB) channels. Baffles exclude scattering from the forward 10° and the rear 5°. For measurement of cF and cB, the scattered light is passively cosine-weighted using a quarter-circle mask.

In an idealized nephelometer, the asymmetry parameter is calculated from the angular scattering coefficient $\beta(\theta)$. As a practical matter, sensing the full angular distribution of scattering requires an infinitely long sensor. This is not possible and the fraction of energy diffracted by 10 μm spheres into the forward direction is large. For example, between 0° and 1°, integrated scattering already approaches 50%.

It is known that for large particles, half of the energy is concentrated in the diffraction cone and half comes from the geometrical optics scattering regime. Therefore, if the truncation angle θ is carefully selected, the integrated forward

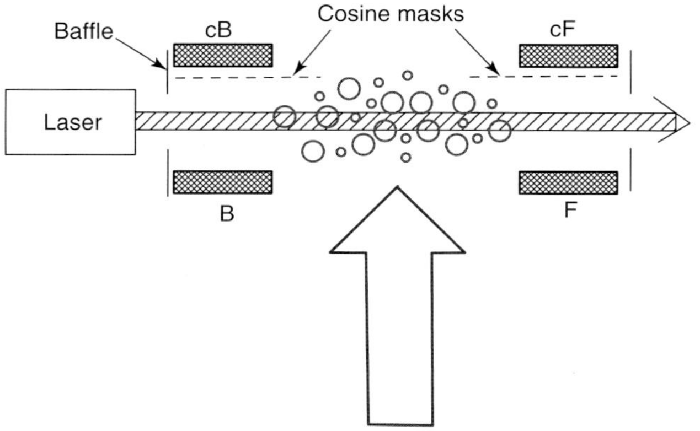

Figure 5.30 Schematic of the CIN. Laser light passes through a cloud of particles and is scattered to four sensors, two that measure forward-scattering (cF and F) and two that measure backscattering (cB and B). Cosine masks weight the scattered light by the cosine of the scattering angle (Gerber *et al.* 2000).

scattered energy becomes less sensitive to the precise choice of θ and can be estimated based on theoretical considerations. Truncation occurs even in the backward scattering regime, but it was found that values of g and b_{ext} are minimally sensitive to this truncation.

Provided the forward scattered fraction f is known, then Eq. (5.18) can be modified to (Gerber *et al.* 2000)

$$g = f + (1 - f) \cdot \left(\frac{\text{cF} - \text{cB}}{\text{F} + \text{B}} \right) \tag{5.23}$$

and the extinction coefficient is given by

$$b_{ext} = \frac{\text{F} + \text{B}}{1 - f} \tag{5.24}$$

where absorption processes are neglected. The parameters cF, cB, F, and B are given by integrals $\overline{R}_1, \overline{R}_2, R_1, and R_2$, respectively; see Eqs. (5.20–5.22). The difference is that the CIN performs the integration during measurements themselves and not in the data processing chain as it takes place, as in the case of the PN described in the previous section.

Gerber *et al.* (2000) showed that a suitable truncation angle is $\theta' = 10^\circ$. In this case, over a wide range of ice crystal habits and particle sizes, the fraction of energy f scattered into $\theta < \theta'$ is bounded by 0.52 and 0.54 for a broad range of possible droplet distributions and 0.57 ± 0.02 for a range of idealized ice crystals (Figure 5.31).

The CIN sampling cross section is about $30\,\text{cm}^2$, and so the sampling volume is $300\,\text{l}\,\text{s}^{-1}$ at $100\,\text{m}\,\text{s}^{-1}$ aircraft speeds. Since the probe's measurement technique is purely optical, and the sample volume relatively large, measurements can, in principle, be obtained at very high sample rates. However, in practice, the highest useful rate is set by the optical density of the clouds and the magnitude of intrinsic

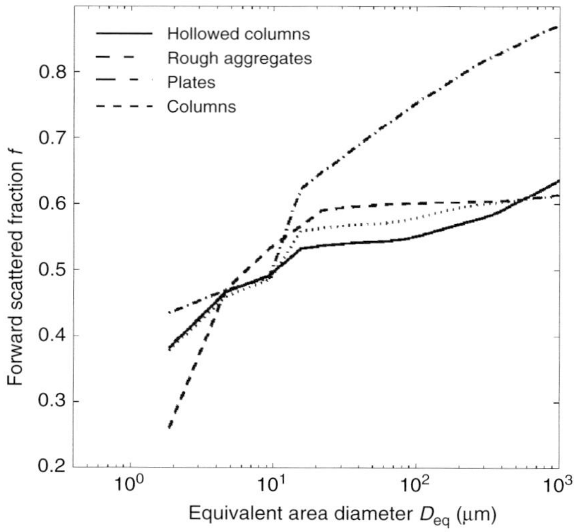

Figure 5.31 Value of f assuming that $\theta' = 10^\circ$ for various ice crystal habits and equivalent area diameters D_{eq}. (Adapted from Garrett (2007). Copyright 2007 American Meteorological Society (AMS). Reprinted with permission by AMS.)

instrument noise. More information on the probe calibration and operation can be found in Gerber *et al.* (2000); Garrett, Hobbs, and Gerber (2001); Garrett (2007); Gerber (2007).

The CIN has been flown on a number of different aircraft participating in field projects for the study of Arctic clouds (Gerber *et al.* 2000; Garrett, Hobbs, and Gerber, 2001), cirrus formation and evolution (Garrett *et al.* 2003, 2005; Garrett, 2007, 2008), and mid-latitude hurricanes and Atlantic stratocumulus.

CIN measurements of the asymmetry parameter in ice clouds show values that tend to be much lower than those determined from ray-tracing calculations for idealized ice crystals. Characteristic values of g measured by the CIN in ice clouds are 0.75±0.03, with values as low as 0.68 in clouds composed of complex dendritic and stellar snowflakes (Garrett, Hobbs, and Gerber, 2001). By comparison, ray-tracing calculations yield values of g ranging from 0.77 to 0.92, depending on the idealized crystal habit and size. Measurements also show a dependence of g on particle effective diameter that is remarkably weak compared to theoretical estimates. For example, Fu (1996) developed parameterizations for the single-scattering properties of cirrus clouds for use in climate models. In the visible, nonabsorbing range of wavelengths, the theoretically calculated value for g for hexagonal prisms increases by 0.06 as the effective diameter increases from 10 to 140 μm. By comparison, measured values of g increase by just 0.01–0.02 over the same size range (Garrett *et al.* 2003; Garrett, 2008). Measurements and models only agree for the smallest sizes.

Thus, a fundamental inconsistency exists between values of g that are theoretically derived from idealized representations of ice crystal shapes and values measured

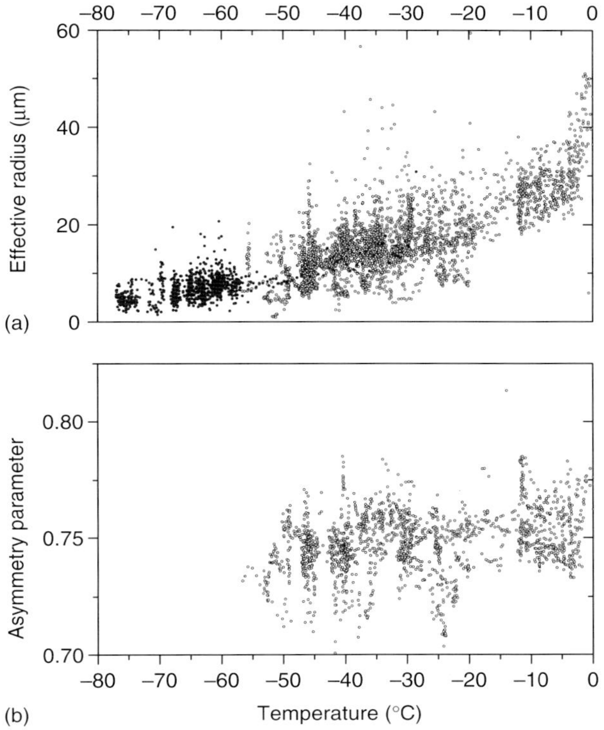

Figure 5.32 Effective radius r_e (a) and asymmetry parameter g (b) versus temperature T within cirrus anvils sampled over southern Florida. Each point represents the average for 10 s of flight time (Garrett *et al.* 2003). (Copyright 2003 American Geophysical Union (AGU). Reprinted with permission by AGU. Courtesy of T. Garrett.)

within cirrus cloud. The differences are important, for they imply up to 50% greater backscattered solar energy for a given cirrus optical depth. A discussion of possible reasons for discrepancies between models and measurements is given by Garrett (2008). Note that some models in fact agree well with measurements. Koch fractals and stochastically distorted spheres (Macke, Mueller, and Raschke, 1996) provide values of $g^G \sim 1/2$ and, therefore, $g \sim 3/4$ (although particles of such shapes do not exist in ice clouds).

The effective radius and asymmetry factor, as a function of temperature, are shown in Figure 5.32 derived from measurements made with the CIN in cirrus clouds over South Florida. The effective radius was derived from the measurements of the extinction coefficient by the CIN and IWC measured with an evaporator (Garrett *et al.* 2003).

5.4.2.5 **The CEP**

The CEP measures the extinction coefficient, b_{ext}, in clouds and precipitation using transmissiometric method for the measurement of the attenuation of light between the receiver and transmitter. The advantages of the transmissiometric technique are as follows: (i) calculations of b_{ext} from first principles based on the Beer–Bouguer law, (ii) large sample volume, which enables statistically significant measurements, and (iii) minimal effect of ice shattering on measurements due to large spatial separation (up to 10 m or higher) between the transmitter and receiver.

Early attempts to use airborne transmissiometers for measurements of visibility in clouds go back to the works of Kampe (1950) and Weickmann and Kampe (1953) who used an incandescent lamp, a collimator, and a photocell for measuring the light intensity. The light source was separated by a few meters from the photocell and both were mounted on the wing. Zabrodsky (1957) built an airborne double pass transmissiometer where light traveled to a retro-reflector and back to a photodetector. Nevzorov and Shugaev (1972); Nevzorov and Shugaev (1974) built an advanced version of this transmissiometer with improved stability and high sensitivity, a design that allowed for the collection of a large data set of the extinction coefficient in different types of clouds (Ruskin, 1965; Kosarev *et al.* 1976; Korolev *et al.* 2001). King and Handsworth (1979) built a single-pass transmissiometer with an ultraviolet source of light generated by a germicidal lamp and Zmarzly and Lawson (2000) designed a multipass and multiwavelength extinctiometer.

The CEP consists of an optical unit that combines a transmitter and receiver as well as a retro-reflector. Figure 5.33 shows a general schematic of the optical unit and a photo of the CEP mounted on a Convair 580. A collimated light beam is generated by an optical system consisting of a (1) superbright LED with a wavelength $\lambda = 635$ nm, (2) diffuser, (3) condenser, (4) pinhole, and (5) an objective lens. The beam travels from the optical unit to the retro-reflector (6) and then returns to the optical unit. After passing though the objective and beam splitter (7) its intensity is measured by a photodetector (8). The optical chopper (9) modulates the light beam and controls turning on and off the LED with the help of the optical coupler (10). The optical chopper consists of a sequence of holes, dark areas, and mirrors.

The measurement follows a sequence with four stages modulated by a chopper to quantify the intensity of the LED background and attenuated light with the same photodetector. Using the above scheme minimizes the effect of changes in the photodetector sensitivity during flight (e.g., caused by temperature drift) on the measurements of the extinction coefficient.

As shown in Figure 5.33(a), the CEP has been installed on a Convair-580 aircraft with the optical unit mounted inside the wing tip canister and the retro-reflector inside a hemispherical cap at the rear side of a PMS probe canister. The distance between the optical unit and the retro-reflector was $L = 2.35$ m. The separation between the optical unit and the retro-reflector may vary from 1 to 10 m, depending on the type of aircraft.

The sample area of the probe is defined by the length of the beam (L) and the diameter of the reflector ($d = 25$ mm) and is calculated as $S = L \cdot d$. For the installation on the Convair-580, the sample area is $S \approx 0.06\,\text{m}^2$. At a typical airspeed of $100\,\text{m}\,\text{s}^{-1}$, the corresponding cloud volume sampling rate is approximately $6\,\text{m}^3\,\text{s}^{-1}$. This allows measurements of a statistically significant extinction coefficient of cloud particles with concentrations of few particles per cubic meter.

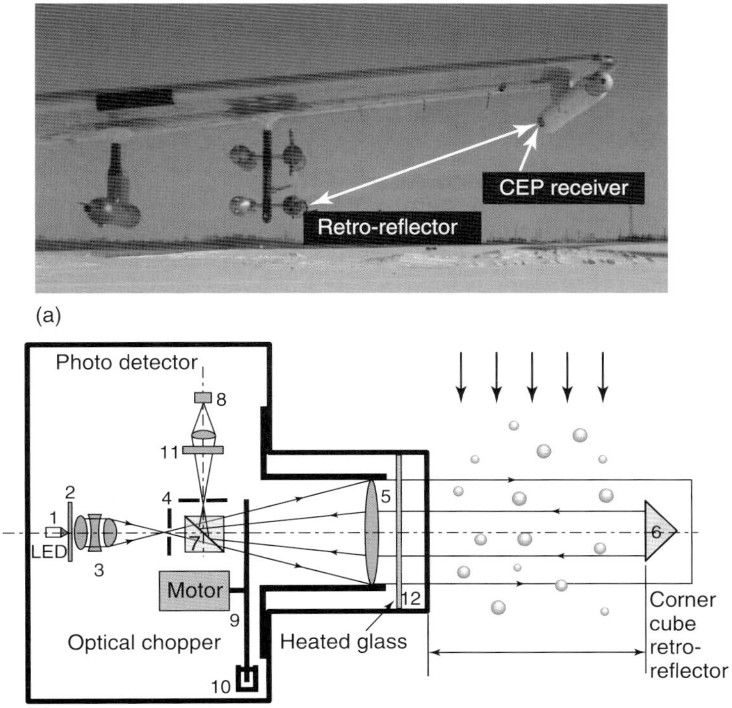

(a) (b)

Figure 5.33 (a) The photo illustrates the installation of the CEP on a Convair 580. (b) This figure shows the schematic diagram of the CEP: (1) LED 635 nm; (2) diffuser; (3) condenser; (4) pinhole; (5) objective; (6) cone cube retro-reflector; (7) beamsplitter; (8) photodetector; (9) optical chopper; (10) optical coupler; (11) filter; and (12) front heated glass. (Courtesy of A. Korolev.)

The extinction coefficient measured by the CEP is calculated based on the Beer–Bouguer law:

$$b_{\text{ext}} = -\frac{1}{2L} \cdot \ln\left(\frac{I}{I_0}\right) \tag{5.25}$$

Here I and I_0 are the output signals that characterize the radiant fluxes transmitted in clouds and in clear sky, respectively. On the basis of flight tests, the threshold sensitivity of the CEP was found to be approximately 0.2 km^{-1}. The upper limit of the measured extinction coefficient is estimated to be no less than 200 km^{-1}.

5.4.2.6 **Measurement Issues**

Contamination of the measurements from spurious particles generated by the shattering of ice crystals on the inlets is a potential issue with the PN and CIN. Hence, the interpretation of measurements in clouds with even moderate concentrations of large ice crystals larger than about 100 μm should be done cautiously, keeping in mind that these types of artifacts may lead to faulty conclusions. Comparisons between these instruments and others on the aircraft from which similar parameters can help identify regions of cloud where shattering might occur.

Because the CIN uses PMTs to amplify the scattered signal, it is subject to "shot" noise due to leaking of ambient light into the instrument. Shot noise limits the instrument SNR and produces a baseline offset to the channel signals. In the past, shot noise has limited instrument sensitivity to measurements of b_{ext} greater than about 0.5 km^{-1} when ambient light is high. The sensitivity can be improved if airborne measurements of ambient light intensity are also made, for example, with a solar pyranometer. Since the magnitude of the zero offset is linearly correlated with ambient radiant flux, a suitable offset correction can be calculated based on flight in clear air (Garrett, Hobbs, and Gerber, 2001). In addition, because the noise is "white", the SNR can be increased by averaging sequential data. Because g is sensitive to the strength of the signal in the back channels, and in general the back channel signal is relatively low, more averaging may be required for measurements of g than for b_{ext}.

A final consideration is uncertainty in the value of f for the amount of light scattered in the forward 10°. In general, uncertainty in f is ±0.01 for liquid droplets and ±0.02 for ice crystals. While this uncertainty has an almost negligible effect on estimates of b_{ext}, for g it introduces uncertainties of approximately ±0.003 for liquid clouds and ±0.01 for ice clouds (Gerber *et al.* 2000). In total, estimated uncertainties based on combined errors from calibration, noise, and uncertainty in f are approximately 15% for b_{ext} and ±0.02 for g.

5.5
Data Analysis

The preceding review of existing airborne instruments for cloud microphysical measurements shows a large variety of measurement principles and technologies and it also reveals that none of these instruments covers the whole range of sizes and shapes of the cloud particles. Because of coincidence artifacts, each single-particle counter is thus designed for a specific size range; the technology of ensemble integrators also limits their usage to some specific particle categories; spherical liquid particles can be characterized using simpler techniques than ice crystals that require full characterization of crystal shapes.

The performance of these instruments is not uniform over their specified size range. Usually, maximal performance is achievable in the middle of the size range and, as a band-width, it declines on both limits of the range. For instance, in single-particle optical spectrometers, the sample volume decreases at the lowest size limit because detection is only feasible in the most intense fraction of the laser beam and the sizing accuracy is degraded when the SNR approaches its detectability threshold. At the largest size limit, the sample volume is no longer suited to the typically very low number concentration of large particles and the counting statistics is poor. For these reasons, instrumented aircraft shall carry a comprehensive suite of complementary instruments, using different measurement principles and providing overlapping size ranges, from the smallest (micrometer) to the largest (centimeter) possible particle sizes.

Wind tunnels with cloud particle generation and airspeed similar to the aircraft one (100–200 $m\,s^{-1}$) are rare. They are commonly used to calibrate LWC, IWC, and icing probes (Spyers–Duran, 1968; Strapp and Schemenauer, 1982; King *et al.* 1985; Ide and Oldenburg, 2001; Twohy, Strapp, and Wendisch, 2003; Wendisch, Garrett, and Strapp, 2002). Instrument calibration is hardly feasible in flight since there are very few natural calibration standards for cloud microphysics. The adiabatic reference is an exception. Indeed, it is feasible to derive from thermodynamics the adiabatic LWC (LWC produced in a closed adiabatic convective cell from the cloud base to the observation level). This value is an absolute maximum in nonprecipitating clouds. Adiabatic samples are not expected to occur frequently because of entrainment-mixing processes, but short samples can reach this maximum within a few hundreds of meters above cloud base. Detection of LWC peak values measured with the fastest instrument (finest spatial resolution) can then serve as an absolute standard to calibrate this instrument with respect to the adiabatic maximum. This instrument can then be used to calibrate additional slower instruments after spatial averaging. To mitigate these limitations, it is common to perform in-flight intercalibration (Heymsfield and Miloshevich, 1989; Gayet, Brown, and Albers, 1993; Burnet and Brenguier, 1999), during which, measurements from different instruments are compared over a large and diverse data set to assess their relative performance.

This section aims at illustrating the diverse methodologies that are currently applied to take the best from a suite of airborne microphysical instruments.

5.5.1.1 Adjustment to Adiabaticity

In a nonprecipitating convective cell, conservation of total water implies that the liquid water mixing ratio at any level z is equal to the saturation mixing ratio at cloud base minus the saturation mixing ratio at that level:

$$\mathrm{LWC_{ad}} = \left[r_\mathrm{s}(z_\mathrm{b}) - r_\mathrm{s}(z)\right] \cdot \rho_\mathrm{a} \tag{5.26}$$

where r_s is the saturation mixing ratio, z_b is the cloud base altitude, and ρ_a is the air density. The LWC is further reduced by entrainment-mixing with dry environmental air from the side in isolated cumuli and from the top in stratocumulus clouds. The occurrence and size of adiabatic samples thus decreases with increasing altitude. A few hundreds of meters above cloud base, however, such samples can be detected and used as a standard for absolute calibration of LWC measurements. Figure 5.34 shows an example of LWC measurements performed on the NCAR C-130 during the RICO campaign in fair weather cumuli. The 1000 Hz measurements from the PVM probe are validated against the adiabatic reference. The samples are then averaged at 1 Hz and compared to the King-probe hot-wire measurements (blue) and LWC values derived by integration of the droplet size distributions measured with a FSSP probe (red). This analysis demonstrates that the three probes agree to within 20%, at least in the range of low LWC values below 1 $g\,m^{-3}$.

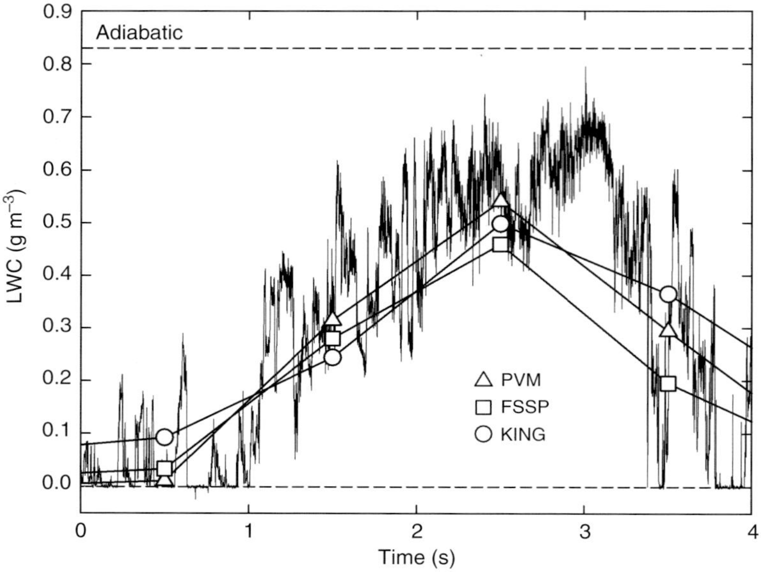

Figure 5.34 1000-Hz measurements of LWC made on the NCAR C-130 during the RICO campaign in fair weather cumuli. After averaging at 1 Hz (open triangles), they are compared to measurements made with the King hot wire (open circles) and LWC derived from integrating droplet size distributions measured with the FSSP (open squares). (Courtesy of H. Gerber.)

This methodology has been extensively discussed in Cooper (1978); Heymsfield and Parrish (1978); Paluch (1979); Boatman and Auer (1983); Jensen *et al.* (1985); Pontikis, Rigaud, and Hicks (1987); Blyth *et al.* 2003; Vaillancourt and Yau (2000); Brenguier and Chaumat (2001b); Chaumat and Brenguier (2001a); Kollias *et al.* (2001).

5.5.1.2 **Instrument Intercalibration**

Basically, in-flight intercalibration consists in comparing measurements performed by different types of instruments over a large data set. Ideally, the data set shall include very diverse cloud microphysical conditions, as well as diverse operating conditions such as varying aircraft speeds, pitch angles, etc. Examination of how each instrument responds to these diverse conditions provides useful information about their performance at the limits of their size range (samples with diverse particle mean sizes), about their susceptibility to particle coincidences (samples with low and high particle concentrations), about their response time (samples at varying aircraft speeds), and about sensitivity to air flow alignment (samples with varying aircraft pitch angles).

The comparison shall also be applied to diverse cloud microphysical parameters measured using different principles. For instance, the LWC derived from integration of the size distribution measured with a single-droplet spectrometer can be compared to the one directly measured with an integrator. On a subset of data samples with MVD in the range between 10 and 20 μm (nominal range for the PVM-100) and increasing droplet concentrations, the PVM-100 can be used as a reference to qualify the coincidence effects on the single-particle counter measurements, and reciprocally, on a second subset with droplet number concentrations in the range from 100 to 200 cm^{-3} (negligible coincidence artifacts) and increasing MVD, the single-particle spectrometer will be used as a reference to qualify the response of the PVM-100 to very small and large droplet MVDs at the limits of its size range.

The following example shown in Figure 5.35 illustrates the methodology with data collected during the VAMOS ocean–cloud–atmosphere–land study (VOCALS) experiment on board the NCAR C-130 using three single-particle spectrometers (FSSP, CDP, and fast-FSSP) and two integrators (PVM-100 and King-probe). The first set of figures (a) shows comparisons of the droplet concentration values measured with the spectrometers. They suggest good agreement between the CDP and fast-FSSP, while the FSSP concentration measurements are slightly overestimated. In fact, the intercalibration revealed disfunctioning of the FSSP probe in the first four size classes that have therefore been excluded to reduce the bias. The second set of figures (b) compares LWC derived from the measured droplet spectra with the values directly measured with the PVM-100 and the King-probe. Overall, the agreement is good except for the FSSP that underestimates LWC, hence suggesting that the droplet sizes are underestimated. In summary, the intercalibration validates the CDP, fast-FSSP, PVM, and King measurements, and it reveals that the removal of the first FSSP size classes does not fully resolve its biases.

5.5.1.3 **Instrument Spatial Resolution**

The issue of spatial resolution has been introduced in the introduction, showing that single-particle spectrometers provide comprehensive information about the PSD and particle shapes, as long as counts are cumulated over a time period long enough for the measured spectrum to accurately represent the actual size distribution. This limitation can be partly mitigated using optimal estimation (Section 5.3.1.2), although at the expense of a significant computing cost. In contrast, ensemble integrators are designed with a sampling volume large enough for providing statistically significant measurements even at very high sampling rates. In particular, the PVM-100A, with its sampling volume of 1.25 cm^3, can be operated at 1000 Hz, that is, a spatial resolution of 10 cm with an aircraft flying at 100 $m\,s^{-1}$.

Figure 5.36 shows the comparison between 1000 Hz LWC measurements with the PVM-100A and the LWC derived from fast-FSSP droplet size distribution measurements at 100 Hz. At 100 Hz, the fast-FSSP LWC shows large fluctuations that can be attributed to the randomness of the counting. At a higher rate, the LWC values would no longer be practicable. Optimal estimation efficiently reduces the counting noise and it is useful for detecting the main features of the LWC small-scale fluctuations (Pawlowska, Brenguier, and Salut, 1997).

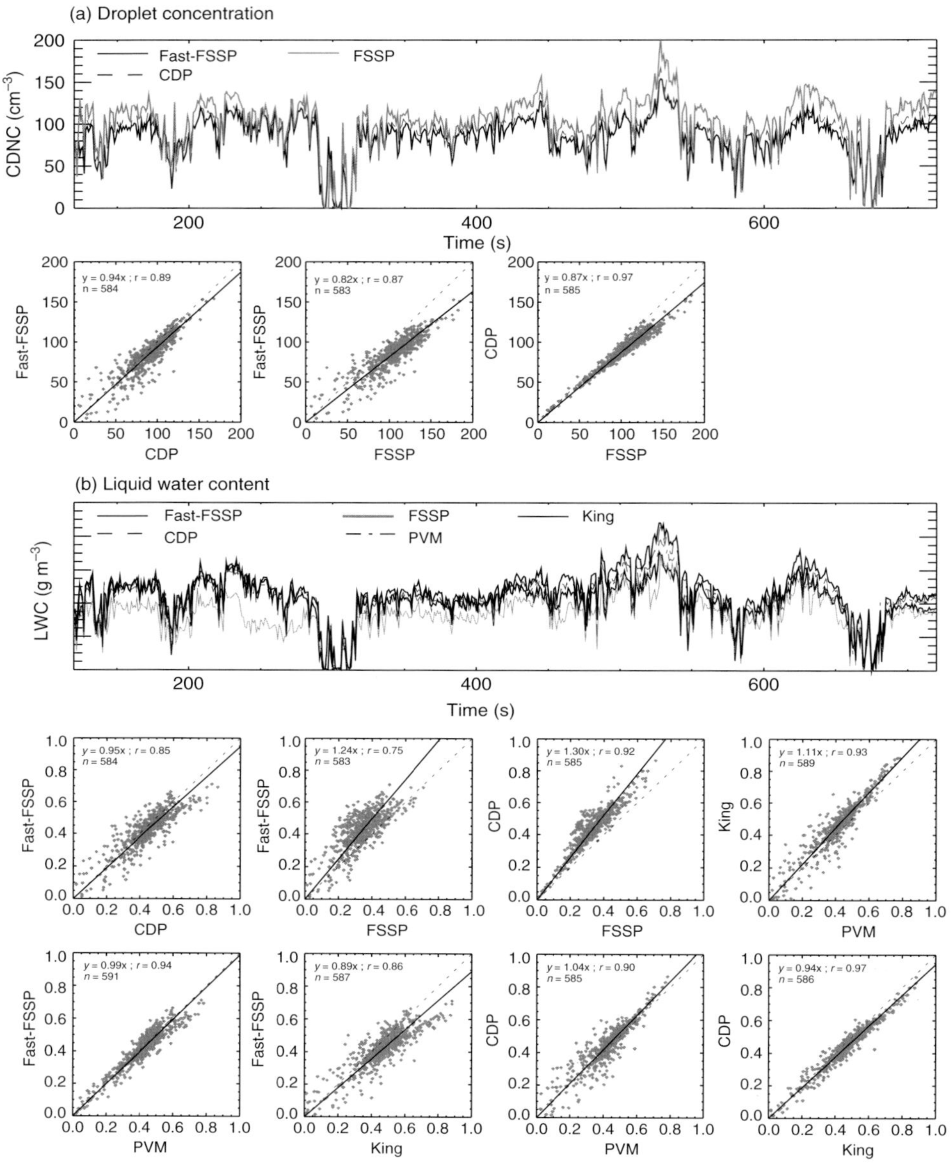

Figure 5.35 Intercalibration of single-particle spectrometers and integrators for cloud droplet measurements. From NCAR C130 RF07 flight during the VOCALS–2008 campaign. Courtesy of F. Burnet, Météo France.

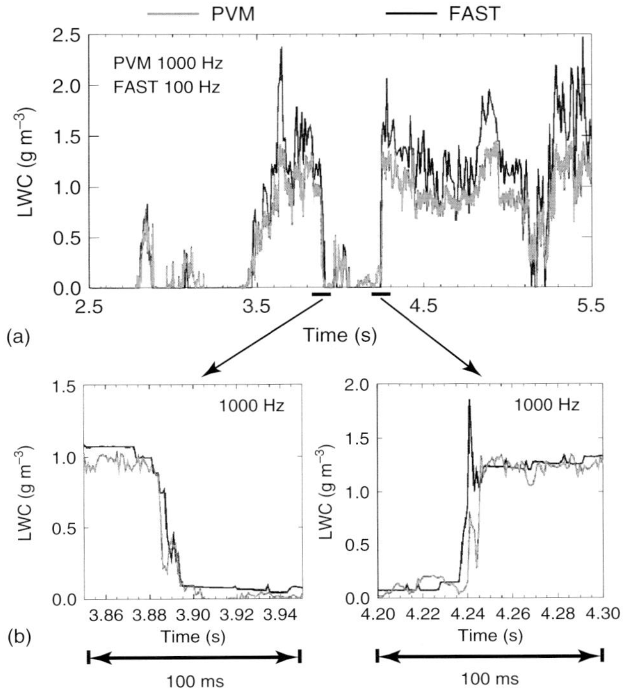

Figure 5.36 High-resolution LWC measurements. (a) 1000 Hz LWC measurements with the PVM-100A (gray) compared to LWC derived from integration of the droplet size distribution measured with the fast-FSSP at 100 Hz (black). (b) 1000 Hz PVM-100A measurements (gray) compared to LWC derived from optimal estimation at 1000 Hz of the fast-FSSP droplet size distribution. (Courtesy of F. Burnet, Météo France.)

5.5.1.4 Integrating Measurements from Scattering and Imaging Probes

The overlap of PSDs using measurements from the various cloud particle probes, for example, a scattering probe, cloud particle probe, and PIP, has been a focus of the scientific community since the 1970s. Recent progress in the development, calibration, and characterization of optical cloud particle probes has led to some progress in this area. For example, Figure 5.37 shows measurements from two scattering probes, an optical imaging cloud probe with two channels, and a PIP. Also shown are representative 3V-CPI images taken in the 12-s penetration at $-17\,^{\circ}$C of a (mixed-phase) tropical cumulus cloud during the 2011 ice in clouds experiment- tropical (ICE-T). The overlap in the size distributions is reasonably good and a composite size distribution is generated using transition points where the individual size distributions overlap. The region where the ice and water size distributions overlap, from about 20 to 200 μm, are separated using 3V-CPI images and size distribution measurements from the 2D-S portion of the 3V-CPI probe. Bulk properties, such as effective particle radius, equivalent RADAR reflectivity, liquid, and IWC, are then determined from the ice and water size distributions.

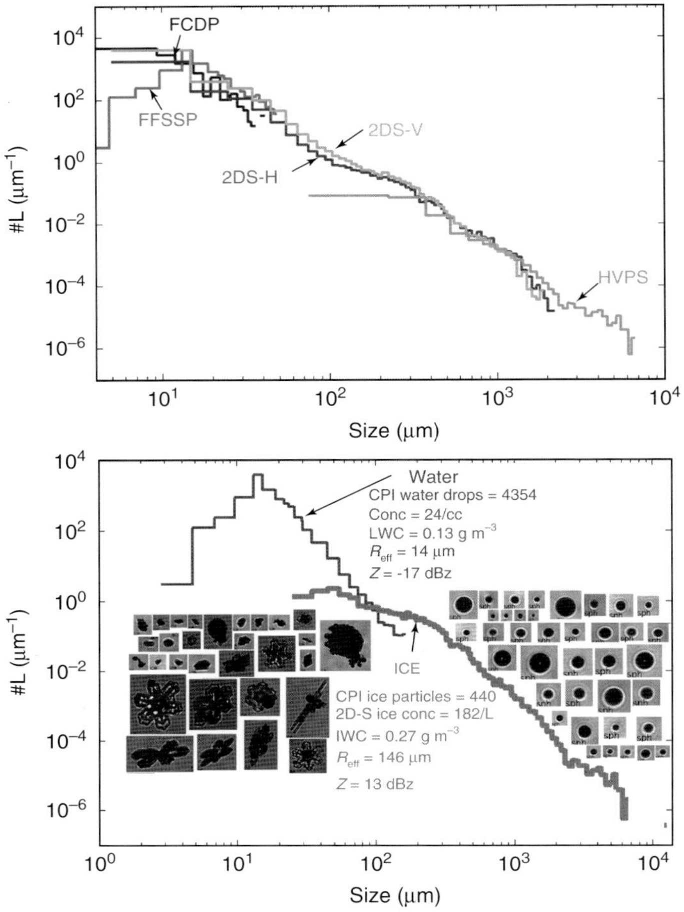

Figure 5.37 Example of overlap of PSDs using measurements from the various cloud particle probes. (Reprinted with permission from SPEC.)

5.5.1.5 Integrating Cloud Microphysical and Optical Properties

Figure 5.38 illustrates the results from a young contrail (2.5 min aged) sampled near −60 °C (Febvre *et al.* 2009). Figure 5.38 (a) represents the average scattering phase function (without normalization) measured by the PN and the theoretical phase function calculated from the FSSP-300 size distribution assuming spherical ice particles. The results show that the calculated phase function agrees quite well with the observations. At the same time, the model slightly overestimates the scattered energy at forward angles (6–10°) and around 50°. The differences may result from a slight deviation from the spherical shape (Febvre *et al.* 2009). Figure 5.38(b) displays the FSSP PSD along with the particle size spectrum retrieved on the basis of the scattering phase function measured by the PN (Dubovik, 2004). Compared to the direct observations, the retrieved size spectrum shows roughly the same mode

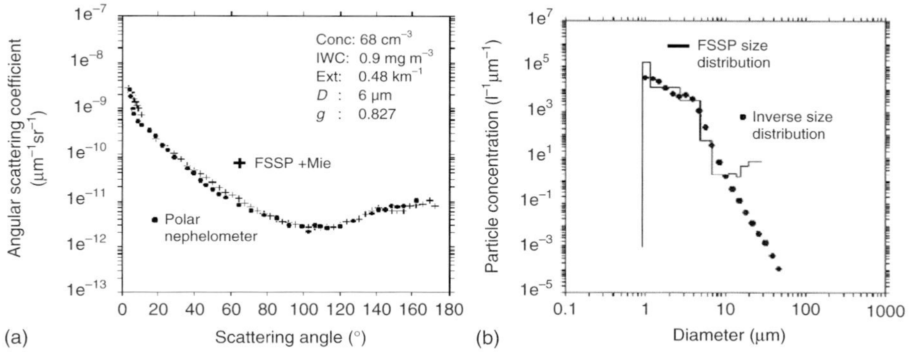

Figure 5.38 Properties of a young contrail (2.5 min aged) sampled at −60 °C level. (a) Measured (PN, red-circle symbols) and theoretical (black dots) scattering phase functions, the latter being calculated from the FSSP size distribution and assuming ice spheres. (b) Direct PSD (FSSP) and retrieved particle spectrum (black dots) from the measured scattering phase function. Microphysical and optical parameters are also reported. (From Febvre *et al.* (2009). Copyright 2009 American Geophysical Union (AGU). Reproduced/modified with permission by AGU.)

size and spreading for particle diameters smaller than 10 µm. The few particles larger than 10 µm measured by the FSSP contribute only slightly to the optical properties even when assumed to be irregular shaped ice crystals. The results show that quasispherical ice particles with diameter smaller than 5 µm control the optical properties of the plume shortly after formation.

5.5.1.6 Evaluation of OAP Images

The analysis of ice crystal images from OAPs is complicated by two factors: (i) the complexity of particle shapes and (ii) the dependence of the sample volume on particle size. As has been previously noted, there can be many different definitions of size, when referring to ice crystals. The scientific community has not reached a consensus with respect to this definition, especially when constructing PSD. As shown in Figure 5.39, there are a number of characteristic dimensions that can define the shape and size of an ice crystal image, that is, the maximum width, maximum length, projected length, perimeter, and area. For images produced by OAPs, the width and length are given by the dimensions across the linear array of diodes and the number of image "slices", respectively. A frequently used dimension by those who process the data from the OAPs is the "area equivalent diameter" (Baker and Lawson, 2006). This is the diameter of a circle that has the same area as the crystal area.

The effect that using different size definitions has on constructing PSD can be observed in Figure 5.40, which shows the number concentrations (Figure 5.40a) and mass concentrations (Figure 5.40b) as a function of size, where size is defined as area equivalent diameter, projected length, maximum width, maximum size, and average size, respectively. The projected length is the square root of the sum of the squares of the width and length. The maximum size is the longest of the width

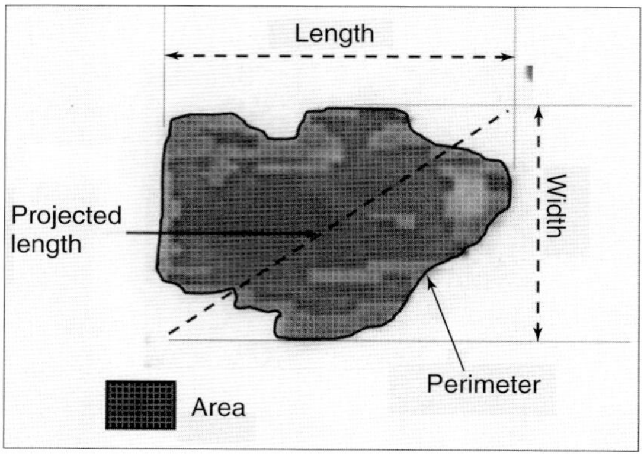

Figure 5.39 This diagram shows an ice crystal image measured with the CIP-GS with characteristic dimensions that are used in the determination of particle size.

and length and the average size is the average of the length and width. Clearly the choice of size can have a major impact on the shape of the size distribution, which then propagates in the derivation of mass, extinction, or any other quantity calculated from the moments of the PSD.

As was discussed earlier in this chapter, the sample volume of OAPs is a function of the particle size, where DOF = $6\ R^2/\lambda$; however, this definition assumes a water droplet where R is the radius. Hence, when calculating DOF for ice crystal images, there is uncertainty due to the definition of R. Given that the area equivalent diameter is normally much smaller than any of the other definitions of size, the DOF will also be much smaller because it is a function of the square of the size. This is a major contribution to the differences seen in the shapes of the size distributions in Figure 5.40. A smaller DOF means a smaller total sample volume and hence a larger concentration for a particular size category.

The other issue when calculating sample volume is the dimensions of the effective width of the sample area. This is the dimension across the linear array (Heymsfield and Parrish, 1978). The sample volume is the product of the DOF, effective width, airspeed, and time. The width is referred to as *effective* since it also varies with the size of the particle and can commonly be calculated two ways: "center in" (no end reject) or "all in". There is always a finite probability that the image of a particle will not fall completely within the width of the diode array. In these cases of a partial image, the derived size will be an underestimate of the actual size. Techniques have been developed (Heymsfield and Parrish, 1978; Korolev *et al.* 2001) to reconstruct partial images and thus increase the useful sample volume if it can be determined that at least the center of the particle is within the array. In this case, the effective width = (RES)(Ndiodes)+2D, where RES is the resolution of the OAP, Ndiodes is the number of diodes in the array, and D is the particle size. If particles are only accepted if the image falls completely within the array (all in), then the effective width is (RES)(Ndiodes$-D-1$). The effective width decreases with particle size because the probability increases that the particle will cover one

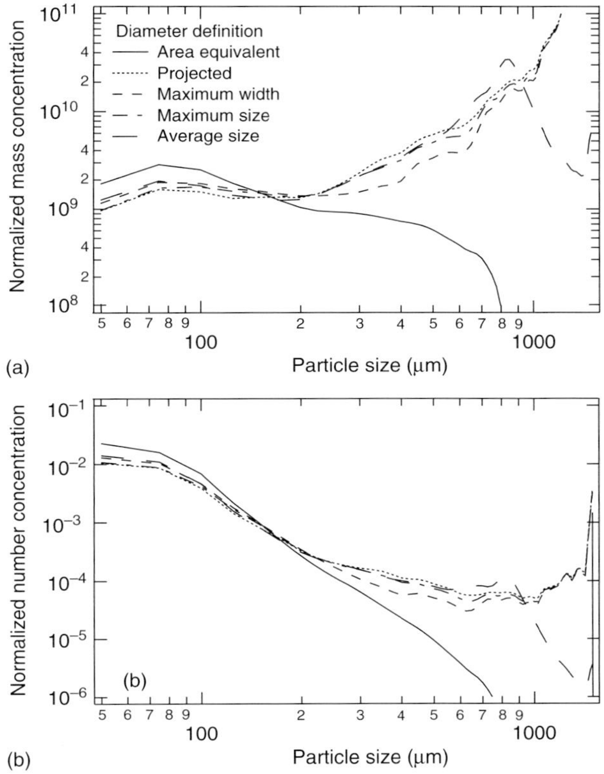

Figure 5.40 The PSDs displayed in this figure show the number (a) and volume (b) concentrations as a function size, defined as area equivalent diameter, projected length, maximum width, maximum size, and average size (see text for size definitions).

of the diodes on either end of the array and be rejected. Figure 5.41 illustrates the impact on the size distribution of using the two different effective width definitions and for two different size definitions. Just as the DOF depends on how the size is calculated, so does the effective width.

5.6 Emerging Technologies

There are a number of evolving technologies for measuring cloud properties but have yet to be widely deployed. The ILIDS, BCP and the cloud particle spectrometer with depolarization (CPSD) all address one or more measurement deficiency in the current suite of cloud measurement technologies. The ILIDS provides large sample volume capabilities, the BCP is an instrument to be used to acquire statistics on clouds from commercial aircraft, and the CPSD addresses the issue of ice shattering.

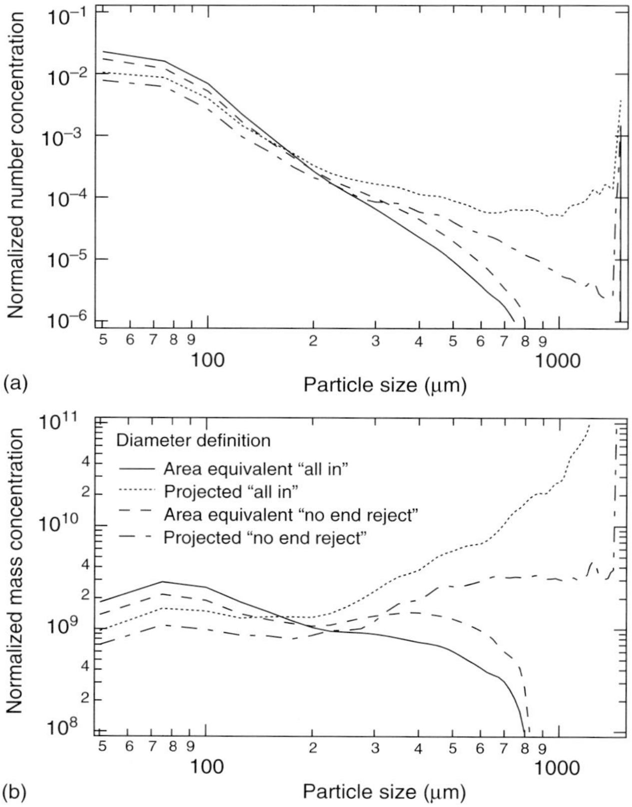

Figure 5.41 The PSDs displayed in this figure show the number (a) and volume (b) concentrations as a function size, defined two different ways and for two different methods of calculating the sample volume.

5.6.1
Interferometric Laser Imaging for Droplet Sizing

The ILIDS is a technique that provides the absolute instantaneous size and spatial distribution of transparent and spherical particles (droplets and bubbles) in a section of a flow. It was first introduced by König, Anders, and Frohn (1986), and further improved by Glover, Skippon, and Boyle (1995) in image acquisition and data processing. Figure 5.42 shows the typical experimental set up.

A laser sheet is projected through a group of droplets and the scattered light is collected by the receiving optics. Different set ups can be found in the literature, around an off-axis angle of 66° (Glover, Skippon, and Boyle, 1995; Kobayashi, Kawaguchi, and Maeda, 2000) or 90° (Mounaïm-Rousselle and Pajot, 1999). The glare points associated with reflected and refracted rays can be observed in the focal plane (Figure 5.42). The size of the particle is determined by measuring the distance

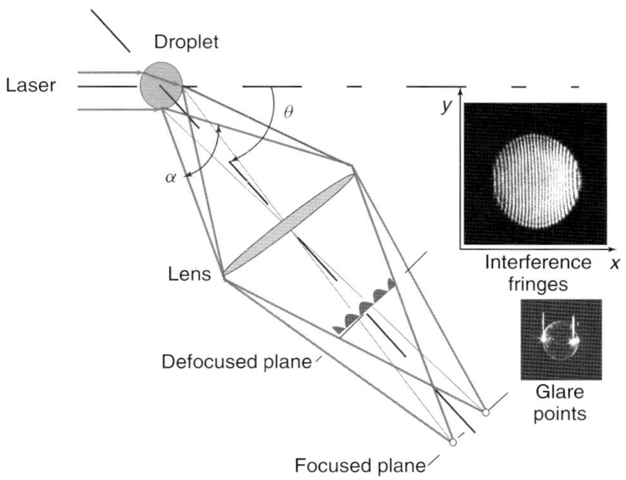

Figure 5.42 ILIDS typical set up showing the optical components and their layout. (Adapted from Percheron *et al.* (2007). Copyright 2007 Elsevier. Reprinted with permission by Elsevier.)

between these two points (van de Hulst, 1981) and requires a very high resolution to determine the droplet's size with good accuracy. Another approach is to observe the droplets outside of the focal plane. For the ILIDS diagnostic, the images are captured by a CCD camera positioned on a nonfocal plane (out-of-focus imaging). The camera observes the interferometric pattern of the laser light scattered by the particles. Each droplet is thus associated with a fringe pattern in a circle. Using a geometrical optics approach, König, Anders, and Frohn (1986) calculated the optical path difference between reflected and refracted rays and deduced the relation between the interfringe spacing, the droplet diameter, and the index of refraction. For each droplet, the fringe frequency (N) is linked to the droplet diameter (D), by a factor which depends on the aperture angle of the collecting system (α) (α is linked to the receiving optics parameters such as the numerical aperture, the focal length, the out-of-focus distance, etc.), on the scattering angle (Φ), the refractive index of the droplet (m), and the wavelength of the laser sheet (λ).

$$d = \frac{2\,\lambda \cdot N}{\alpha} \cdot \left[\cos(\theta/2) + \frac{m \cdot \sin(\theta/2)}{\sqrt{m^2 - 2\,m \cdot \cos(\theta/2) + 1}}\right]^{-1} \tag{5.27}$$

Up to now, the ILIDS technique was commonly used in laboratories for fluid mechanics research on spray, but progress in optics technology such as the miniaturization of high energy lasers with ultrashort pulses and high-resolution CCD cameras allow the ILIDS technique to be transferred to aircraft for performing airborne measurements. A new airborne spectrometer based on ILIDS principle is being developed under the auspices of EUFAR (European Facility for Airborne Research) project to measure the size of droplets in clouds.

The ILIDS optical setup is simple, but unlike the PDI (Section 5.3.2) its optical alignment is not critical. The coherent light originates from a pulsed Nd:YAG laser ($\lambda = 532$ nm) with a high repetition rate and polarized perpendicularly to the scattering plane in order to increase the fringe visibility. The laser beam is then extended using a cylindrical lens to create a laser sheet. The laser pulse duration is less than 10 ns such that the ILIDS technique is able to measure a frozen droplet field at aircraft velocities up to 200 m s^{-1}. Images are acquired by a high-speed CCD camera that is synchronized with the laser frequency. The design parameters of the optical setup are specified in order to reach a range of droplets sizes from 20 μm to 200 μm with a large measurement volume (100 mm x 100 mm x 1 mm). The range can be extended to smaller droplets with a reduced sample volume. The cloud volume sampled by the ILIDS is then dependent on the repetition rate of the laser and the camera that can easily sample from 30 Hz, with a very compact, air-cooled laser, up to more than 100 Hz with a water-cooled laser. Image processing must be fast enough to perform real-time processing during a flight through clouds.

Image processing is fast enough to perform real-time preprocessing during the flight. The common approach for the ILIDS interferogram analysis is to detect the fringe pattern on each droplet of the image and to determine its frequency to derive the associated droplet size and concentration. The drawback of this method, for airborne application, is the computation time due to fringe patterns detection. Therefore, image analysis is not performed on each droplet individually but globally over a whole image. The high-speed analysis of the ILIDS interferogram is based on fast-Fourier transforms developed by Quérel *et al.* (2010). The validation of the ILIDS technique was achieved in two steps. A comparison was made with backscattering measurements performed on a drop by drop jet and then against PDI measurements in full cone sprays (Quérel *et al.* 2010; Lemaitre, Porcheron, and Nuboer, 2007). Results show a good agreement between ILIDS and both techniques with a relative difference of 3% for the Sauter mean diameter. Besides, the ILIDS technique offers the potential to discriminate water droplets from ice for which no interference fringes appear.

5.6.2
The Backscatter Cloud Probe

The BCP is a very small, light weight, and low-power OPC that has been developed for the In-Service Aircraft for a Global Observing System (IAGOS) project. As shown in Figure 5.43, the BCP uses a single detector measuring light backscattered over a solid angle of 144° to 156°. As there is no qualifying detector to reject particles that pass through the less intense portions of the beam, an inversion is required to retrieve the size distributions; however, since the measurement is made off-axis from the illuminating laser, the sample volume is fairly small and the retrieval is simple and fast to implement.

The BCP is installed on the cargo hatch of commercial aircraft (Figure 5.43b,d), and its small size and weight, less than 1 kg (Figure 5.43c), make it an ideal instrument for unattended operation and acquisition of cloud statistics during

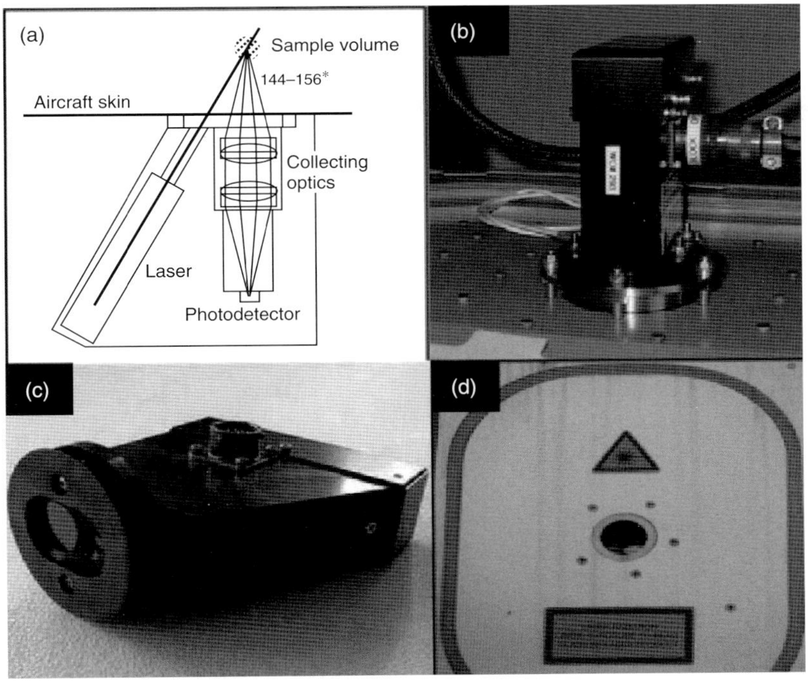

Figure 5.43 The BCP, whose optical layout is shown in (a), is mounted on the cargo hatch of commercial aircraft (b) and (d) and is quite small as shown in photo (c).

the frequent flights of commercial aircraft. The BCP is being flown, along with a number of other gas, aerosol and meteorological instruments on cargo aircraft as a part of the IAGOS mission of accumulating a larger data base of atmospheric information using commercial aircraft.

5.6.3
The Cloud Particle Spectrometer with Depolarization

The CAS-DPOL was discussed in Section 5.3.2 with respect to its capabilities for differentiating liquid cloud particles from ice. One of the limitations of the CAS-DPOL, however, is the design of its geometry that incorporates a tube to direct the particles through the sample volume (Figure 5.6). As discussed in the next chapter, the leading edge of the inlet presents a surface where ice crystals can shatter and whose fragments subsequently contaminate the measurement in a difficult to correct way.

This deficiency has been corrected with the development of the cloud particle spectrometer with sepolarization (CPSD) whose measurements are over the same size range as the CAS-DPOL and differentiates liquid and ice using depolarization. As shown in Figure 5.44, the CPSD has no inlet as it allows free-flowing air past

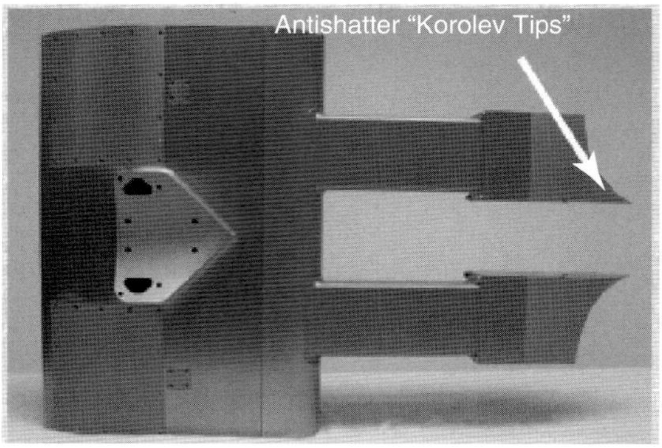

Figure 5.44 The CPSD incorporates an aerodynamic design that prevents the contamination of measurements from ice crystals that shatter on the inlet. In this design, the anti-shattering "Korolev tips" direct shattered fragments away from the sample volume located between the arms.

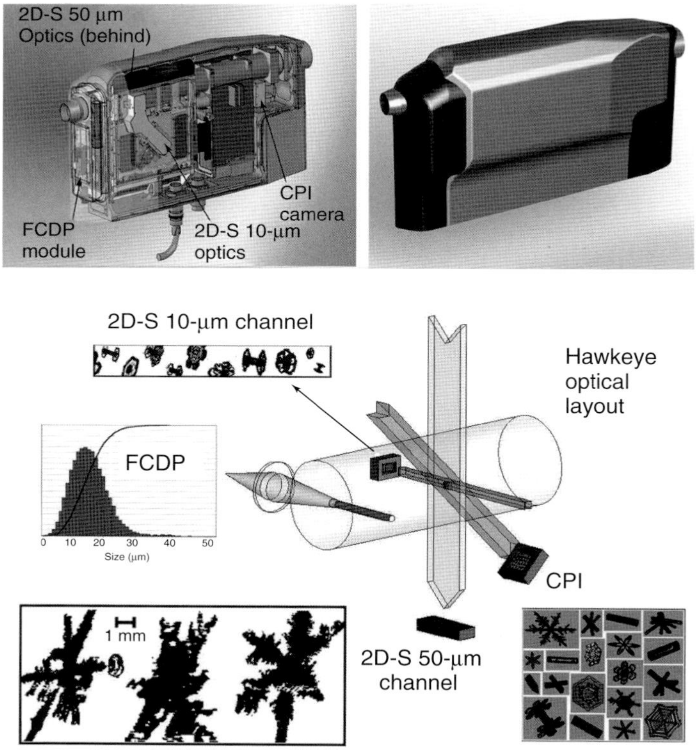

Figure 5.45 Schematic of the Hawkeye that combines the FCDP, 2D-S, and CPI. (Reprinted with permission by SPEC.)

the arms that direct the illuminating laser beam and collects the scattered light. In addition, these arms have tips that incorporate the design of Alexei Korolev that is also discussed in the next chapter.

5.6.4
Hawkeye Composite Cloud Particle Probe

The Hawkeye, shown in Figure 5.45, combines the FCDP, 2D-S, and CPI. In this case, one channel of the 2D-S has been converted from 10 to 50 μm resolution, so that it images particles from 50 μm to 6.4 mm. The Hawkeye was originally developed for autonomous operation on the Global Hawk UAV and will be installed and flown on the Global Hawk in October 2012. The Hawkeye is currently (April 2012) being test flown on the SPEC Learjet.

Acknowledgments

We thank Dr. Darrel Baumgardner for his significant and very useful contributions to this chapter.

6
Aerosol and Cloud Particle Sampling

Martina Krämer, Cynthia Twohy, Markus Hermann, Armin Afchine, Suresh Dhaniyala, and Alexei Korolev

6.1
Introduction

Airborne aerosol and cloud particle measurements are important to extend our knowledge of their distribution, properties, and interactions over the vertical extent of the atmosphere. In principle, a broad range of platforms can be used for atmospheric measurements. These range from fixed tethered balloons (Siebert *et al.* 2003, for example) slow-speed platforms such as kites (Baslet, Jensen, and Frehlich 1998; Muschinski *et al.* 2001) and balloons (Deshlet *et al.* 2003), to high-speed aircraft or rockets (Rapp *et al.* 2010). While the higher speed aircraft are capable of sampling large areas, the slower speed platforms have their advantages in lower costs, higher spatial sampling resolutions, and less airflow distortion by the platform. The effect of air compressibility and associated changes in the air temperature (evaporation and condensation) must be considered for air flows with Mach numbers larger than 0.3. The particle stopping distance, which is important for determining the impaction of particles onto a surface or the deviation of the aspiration efficiency of an inlet system from unity, increases linearly with the air velocity. Hence, aerosol and cloud particle measurements from high-speed aircraft are most susceptible to sampling artifacts. Most of the limited literature dealing with airborne aerosol and cloud sampling refer to aircraft platforms; consequently, we also focus here on aircraft. However, most of the described effects are of relevance for other airborne platforms too.

Atmospheric research aircraft deploy a great variety of sensors to measure atmospheric state parameters (temperature, pressure, and altitude), gas-phase constituents, and particles. For all those measurements, especially for aerosol and cloud particles, it is fundamental to have an air sample representative of the free stream atmospheric environment. However, appropriate air sampling from aircraft is a challenge, because air entering a sampling inlet or sensor can be influenced by the aircraft itself or by the sampling device (Wendisch *et al.* 2004). Movement of air around the aircraft fuselage can affect not only the airflow speed and direction but also the concentrations of various atmospheric components, especially particles.

Airborne Measurements for Environmental Research: Methods and Instruments, First Edition.
Edited by Manfred Wendisch and Jean-Louis Brenguier.

Once particles enter a sampling inlet, additional biases can occur. These potential changes are a function of sensor location and shape, particle size and composition, and the type of measurement being made. In some cases, errors due to sampling location and inlet type may be larger than errors inherent in the measurement itself. Errors can be minimized if flow speed, flow direction, and particle behavior is known for sampling locations and inlets, so the best environment can be chosen for a specific measurement. Once a location and inlet is chosen, remaining biases must be understood and quantified for a complete analysis of the measurement uncertainty.

When the Wright Flyer flew in 1903, its speed was 13 m/s, its weight was 275 kg, and it had a tiny open-air fuselage. The aerodynamic heating on stagnation surfaces under such conditions was much <1 K. There were only three instruments onboard: a stopwatch, an anemometer, and a tachometer. In contrast, modern jet aircraft may have speeds exceeding 200 m/s, weigh tens of thousands of kilograms, and have fuselages several meters wide. Atmospheric measurements have been made from various aircraft for many years; in fact, airflow-induced changes in pressure are utilized for standard cockpit instrumentation such as airspeed indicators. However, early efforts to measure aerosol and cloud particles from aircraft focused on technical details of the instruments themselves, rather than how these instruments might be influenced by air flowing around the aircraft surface. For example, optical array probes measuring large drops and ice crystals were often mounted near the fuselage surface, a region where some of these large hydrometeors never reach. One of the first attempts to observe ice clouds was made by Weickmann (1949), who mounted a microscope in the blister of a reconnaissance plane and sampled ice crystals at various temperatures (Figure 6.1). Though the concentrations measured by this method were likely biased by the sampling method, information on ice crystal sizes and habits could be obtained.

Wind tunnels and in-flight testing have provided valuable information on airflow effects on these measurements, but they are labor-intensive. Computational fluid dynamics (CFD) methods to predict airflow around aircraft have progressed in recent years from relatively simple potential flow codes to faster and more accurate representations that include boundary layer and compressibility effects. Improved meshing capabilities allow not only the aircraft shape to be modeled but also the shape of complex inlet and instrument geometries used for particle sampling. In addition, the trajectories of particles with all interesting sizes can now be simulated with little additional effort. The combination of computational methods and empirical testing is now used to provide ever better measurement capability. Furthermore, empirical equations to calculate the sampling efficiencies of standard instruments are straightforward to derive from both modeling and measurements.

This chapter presents an overview on the influence of the aircraft itself on the sampling of aerosol and cloud particles (Section 6.2). Furthermore, the methods and the problems of aerosol and cloud particle sampling are outlined in Sections 6.3 and 6.4, respectively. Finally, a summary is given and some guidelines are provided in Section 6.5.

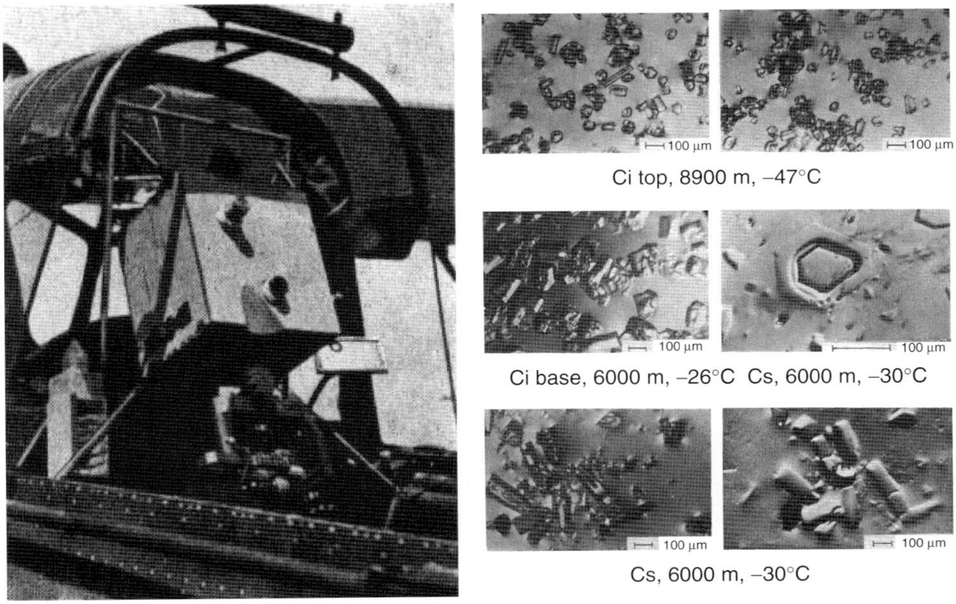

Figure 6.1 Left: Microscope mounted in the blister of the short-distance reconnaissance plane Hs 126 by Weickmann (1949). Right: Ice crystal pictures at different temperatures from the microscope. (Copyright 1949 Deutscher Wetterdienst (DWD). Reprinted with permission by DWD.)

The aim of this chapter is to introduce the field of aircraft-borne aerosol and cloud particle sampling to experienced researchers as well as students. The chapter gives a compact overview to the field, including further literature for more detailed information. To simplify matters, in the following, no distinction will be made between different aerosol chemical or physical properties, other than size.

6.2
Aircraft Influence

In this section, the influence of the measurement platform on aerosol and cloud particle measurements is discussed. Aircraft provide unique sampling characteristics for atmospheric research, such as long-range or high-altitude capability, which allow the whole troposphere or even the stratosphere to be explored. These advantages are gained by high aircraft speed in the range of 50–250 m/s, which in turn represents the biggest disadvantage of using aircraft for aerosol and cloud measurements. The body of the aircraft disturbs both the ambient air (Section 6.2.1) and the particles contained within it (Section 6.2.2). As gas molecules, to first approximation, can be considered to have zero mass (i.e., no inertia), and most of the particles do, trajectories of the disturbed airflow and of the particles are not necessarily the same. This difference and other aircraft-related issues can lead to large measurement biases of up to 100% (or more) in particle concentrations, and also to measurement artifacts of different types (Section 6.2.3).

6.2.1
Flow Perturbation

From a fluid dynamics point of view, there is no difference between the aircraft moving with a speed of 150 m/s through the more or less stationary atmosphere and a stationary aircraft in a wind field with a velocity of 150 m/s. For convenience, the latter coordinate system is usually chosen. As for any body in a flow field, a boundary layer develops around the aircraft hull in flight. Inside this boundary layer, turbulence leads to particle deposition and thus measurements inside this boundary layer are no longer representative of the ambient, undisturbed air. Hence, any measurement instrument or inlet system should be mounted well outside this aircraft boundary layer. Ideally, the aircraft boundary layer thickness at the point of measurement is known, for instance, when provided by the aircraft manufacturer or determined by CFD studies. However, if these ways of gaining detailed information are not feasible, as an approximate guideline one can use 1/60th of the distance from the aircraft nose to the point of measurement to estimate the boundary layer thickness (Bohl, 1998). Thus, an inlet system mounted 6 m behind the aircraft nose should extend at least 10 cm to sample outside the boundary layer. As is explained in Section 6.2.2, there are other reasons why an inlet might have to be even farther from the fuselage. As an example, Figure 6.2 shows the boundary layer thickness of an artificial aircraft-like body at the lower fuselage, 7 m downstream of the aircraft nose, as determined by CFD modeling. The boundary layer thickness is usually defined by the ratio of the local velocity to the free stream velocity, often using a threshold of 99% for the upper limit. For technical reasons (e.g., aerodynamic load, bird strike, or icing), it is desirable for an inlet system to be small (small dynamic drag and surface area). Hence, there is always a compromise to make between the analytical requirements (particle population similar to undisturbed conditions) and the technical and security requirements (surface mounting as small as possible).

When choosing the best position for mounting an instrument or inlet system on an aircraft, not only the boundary layer thickness is important. The curvature of the aircraft fuselage upstream of this position influences the speed and the direction of the local wind vector. The flow speed at *in situ* sampling points is usually required to calculate the actual volume of air sampled and can affect even gas species concentration and fluxes (Cooper and Rogers, 1991).

Also, as misalignment between the flow vector and an inlet system can lead to particle losses (Section 6.3), it is important to know the local wind direction, which unfortunately changes with the flight conditions (true air speed (TAS), altitude, and pitch angle; see Section 2.3.2 and Fig. 2.1). Nevertheless, this effect can and should be partly accounted for by building a mean correction angle into the inlet

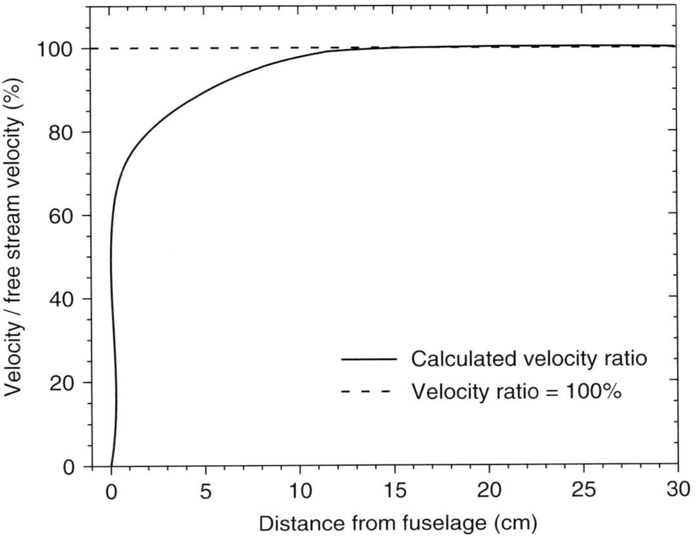

Figure 6.2 Boundary layer thickness of an artificial aircraft-like body at the lower fuselage, 7 m downstream the aircraft nose. Fuselage diameter is 2.5 m. Data were obtained by CFD modeling. for a flight altitude of 6 km and a Mach number of 0.55. (Courtesy of M. Hermann.)

geometry or by using a shroud to align the flow stream (Section 6.3.3). Again, CFD studies are a very useful tool in this context.

As is discussed in Section 6.2.2, regions of strong curvature also lead to a separation of particles and hence to either positive or negative measurement biases. Considering the two major regions of influences, the boundary layer and the aircraft hull curvature, leads to the following preferable measurement positions for particle instruments or inlets on aircraft: well ahead of the aircraft nose, on the lower region of the aircraft fuselage, below the center of each wing, and (with some restrictions) also on the upper region of the fuselage, well behind the cockpit, but upstream of the region influenced by the wings and/or engines.

Finally, it should be noted that not only the aircraft but also the aerosol instrument or inlet body influences the airflow upstream of the sampling point. The distance upstream to which this influence is noticeable can be estimated as two to three times (depending on the cross-sectional area) the diameter of the respective element in the plane perpendicular to the wind vector. For instance, a wing pod with a 20-cm diameter on a jet aircraft will influence the air in a region from the tip to ~50 cm upstream of the tip. The effect of the reduced flow velocity and the associated increased temperature and pressure on the particle measurements must be determined individually for each instrument. For the forward scattering spectrometer probe (FSSP) mounted under the wing of a DHC06 Twin Otter aircraft, for instance, Norment (1988) found a low bias in the measured droplet concentration due to airflow distortion of about 24%. Half of this distortion was caused by the instrument itself, and the other half by the aircraft, although the airspeed was relatively low (~50 m/s). Similar results were found in the studies by Drummond and MacPherson (1985) and MacPherson and Baumgardner (1988) for aircraft flying at speeds up to about 100 m/s. Unfortunately, for high-speed jet aircraft, there is only little information on the measurement influence of the instrument or inlet body available.

6.2.2
Particle Trajectories

In the context of aircraft-borne aerosol measurements, the mass of gas molecules in the sampling air can be considered to be zero, that is, they do not have inertia. However, aerosol and cloud particles do. Therefore, in regions where the gas streamlines have strong curvature, for instance, at the aircraft nose, particle trajectories may cross the gas streamlines, leading to different particle concentrations downstream compared with those in the ambient air upstream. A useful parameter in this context is the modified particle Stokes number, as defined by King (1984):

$$\mathrm{Stk}_{\mathrm{mod}} = \frac{2\,a^2 \cdot U_0 \cdot \rho}{9\,\eta \cdot b} \tag{6.1}$$

where a is the particle radius, U_0 the air velocity, ρ the particle density, η the air viscosity, and b the fuselage radius. With respect to particle or droplet size range, two extreme behaviors result. Very small particles or droplets, with modified Stokes numbers $\mathrm{Stk}_{\mathrm{mod}}$ of the order of 0.1 or below, follow the streamlines and are not subject to any inertial separation process. Hence, the concentration of these particles or droplets at the measurement point is the same as in the undisturbed ambient air. Very large droplets (modified Stokes number of the order of 200 or above) are virtually unaffected by the change in the flow vector; they have so much inertia that they follow a straight line and impact on the aircraft fuselage (see lower panel of Figure 6.3). Particles or droplets in an intermediate size range only partly follow the curved streamlines, and particles with a modified Stokes number of the order of 10 are most affected by the change in the flow vector (King, 1984). The behavior of these particles or droplets is most difficult to predict, but can be quantified using a CFD analysis that includes particle trajectory calculations. Enhancement factors can be calculated as the ratio of the flux of particles in the sampling region to that in the free stream (King, 1984).

Some results of one of the earliest, but still instructive analysis are shown in Figure 6.3. King (1984) used a two-dimensional (2D) potential flow model to calculate gas flow streamlines and particle trajectories around a Fokker F-27 aircraft. In the upper panel of Figure 6.3, the predicted gas flow streamlines at 90 m/s are displayed. Originally, they are equally spaced perpendicular to the flow (left side), indicating the same flow velocity. However, due to the aircraft body, the streamlines are compressed over the cockpit, indicating regions of higher airspeed (right side) compared with the free stream. In the lower panel of Figure 6.3, trajectories

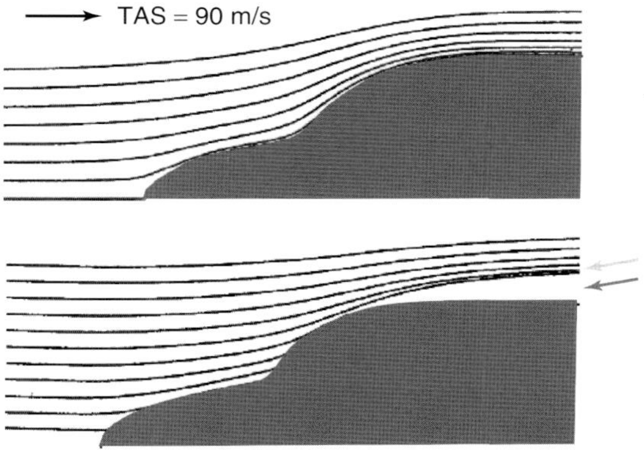

Figure 6.3 Air flow streamlines (upper figure) and trajectories of 100-μm particles (lower figure) around a Fokker F-27 aircraft fuselage at 90 m/s TAS (King, 1984). Arrows in the lower figure indicate regions of particle enhancement (light gray) and shadow zones (dark gray). (Copyright 1984 American Meteorological Society (AMS). Reprinted with permission by AMS.)

for 100-μm water droplets are displayed for the same flight conditions. As these droplets have high inertia, most of the trajectories end at the aircraft fuselage, that is, the droplets impact on the aircraft. However, some of the trajectories were partially deflected leading to regions devoid of droplets (shadow zone, dark gray arrow) or with increased droplet concentration (enhancement zone, light gray arrow). Although the droplets in this case are relatively large, this effect also occurs for micrometer-sized particles on jet aircraft with much higher TASs. Hence, when measuring such particles, cloud droplets, or ice crystals, it is important to know where shadow and enhancement zones on the aircraft platform are located. Note that the same fuselage station (distance from nose of aircraft) may be subject to both, the shadow zone, if a probe is positioned close to the aircraft fuselage, and the enhancement zone, in case the probe is farther away from the fuselage.

While for measurements of atmospheric gases, sampling outside the boundary layer may be sufficient, this is generally not adequate for particle sampling where shadow and enhancement zones are wider than the aircraft boundary layer. As a simple approximation based on potential flow analysis, King (1984) showed that the maximum shadow zone depth is 15–20% of the fuselage radius. This maximum occurs at a particle radius, $a_{\max}$, that has a modified Stokes number, $\mathrm{Stk}_{\mathrm{mod,max}}$, of 6:

$$a_{\max} = \sqrt{\frac{27\,\eta \cdot b}{U_0 \cdot \rho}} \tag{6.2}$$

Detailed potential flow calculations were also performed by Norment (1985), Geller, Rader, and Kempka (1993), and Twohy and Rogers (1993) to determine aircraft shadow and enhancement zones. Although measurements have validated the general characteristics of potential flow results for aircraft shapes (King *et al.* 1984; Twohy and Rogers, 1993), the influence of the aircraft boundary layer or high-speed flight cannot be fully simulated using potential flow models, which assume inviscid, incompressible ($\lesssim 70\,\mathrm{m/s}$) flow. Fortunately, in the past decade, computer capabilities have improved and CFD programs without these restrictions have become more user-friendly; hence, aircraft boundary layers and particle trajectories can in many cases be determined accurately for specific aircraft and conditions.

6.2.3 Measurement Artifacts

Shadow and enhancement zones for particles (Section 6.2.2) are the most prominent particle measurement bias caused by an aircraft body. However, there are other issues to be considered. Owing to the aircraft body, air may decelerate, which according to the laws of thermodynamics leads to an increase in air temperature (ram heating). Depending on the aircraft airspeed and the magnitude of deceleration at a given location, this warming can be up to 30 K. Particles passing through these zones of increased temperature might partly evaporate and thus change size as well as chemical composition. On the other hand, in zones of airflow acceleration, the temperature decreases, which might lead to condensation of vapor molecules and even formation of ice crystals in supercooled clouds (Heymsfield *et al.* 2010).

Measurements well outside the aircraft boundary layer are important not only to prevent particle losses in this turbulent flow region but also to prevent memory effects caused by resuspension of particles deposited at the fuselage before. For the NASA DC-8 research aircraft, Vay *et al.* (2003) showed that air from a cabin air vent stayed within the aircraft boundary layer for at least 20 m downstream (Figure 6.4). Hence, it is likely that resuspended particles starting at the fuselage will also stay within the boundary layer. The investigations by Vay *et al.* (2003) emphasize the importance of knowing the streamlines along the aircraft to prevent artifacts caused by air vents. In Figure 6.4, it is shown that inlets mounted at the right-hand side of the DC-8 at positions 1 and 3 (first and third dark gray ellipses) are partly sampling cabin air emitted at an upstream air vent (medium gray ellipse), thus leading to measurement artifacts. When mounting particle-measuring instrument or an inlet system on an aircraft, one should not only check for upstream air vents, but ideally there also would be no obstructions upstream of the point of measurement. This is not trivial, because, if, for instance, an inlet is mounted 15 m behind the aircraft nose at the lower aircraft fuselage, it is likely that another equipment, such as antennas or other inlets, is located further upstream. These obstacles may not only influence the measurements by causing turbulence and hence particle losses in a wake downstream, but also can capture particles due to impaction, in this way reducing the particle concentration downstream. Surface mountings might also generate artifact particles when cloud droplets or ice particles hit the obstacle with high speed (Cohen, 1991; Yarin, 2006; Vidaurre and Hallett, 2009a); also see Section 6.4.1 and Figure 6.21. This may cause a large number of smaller broken droplets or crystals to be generated; in addition, water films flowing along the obstacle surface in combination with the usually sharp rear edges can act as spray

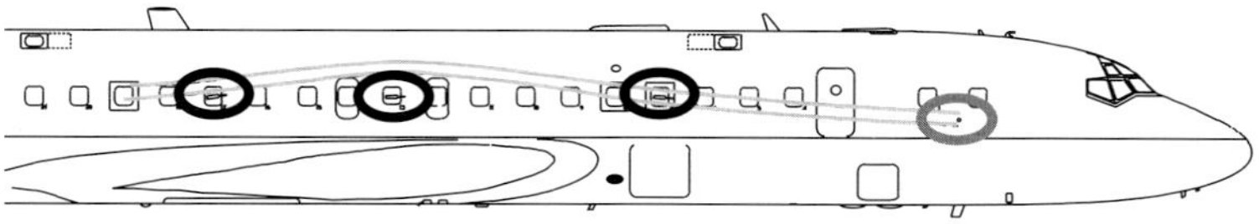

Figure 6.4 Inlet positions (dark gray ellipses) and vent (middle gray circle) at the NASA DC-8 research aircraft during the AFWEX campaign in 2000 (Vay *et al.* 2003). Air flow streamlines are highlighted in light gray. (Copyright 2003 American Meteorological Society (AMS). Reprinted with permission by AMS.)

generator creating additional particles. Obstructions upstream may also induce electrical effects such as corona discharge or field charging, which can influence the particle sampling efficiency of the inlets (Romay *et al.* 1996).

Unfortunately, there is, to our knowledge, no literature available, which allows quantifying all the above effects. Hence, the best way, if possible, is to avoid mounting positions with upstream surface mountings.

6.3 Aerosol Particle Sampling

Aircraft-borne inlet systems are often needed to transport particles from the outside to the measurement devices in the aircraft when an outside mounting is not feasible (e.g., due to weight, size, or temperature range limits of the measurement device). In this section, all issues related to aircraft-borne aerosol inlet systems and the inlet particle sampling efficiency are discussed. We focus here on aircraft-specific topics, while more general information on aerosol particle measurements is provided in the excellent textbooks by Hinds (1999), Baron and Willeke (2005), and Vincent (2007). Generally, there are many physical processes that can lead to particle losses during sampling and transport, but for aircraft-borne measurements, the number of dominant processes can be reduced to half a dozen (Section 6.3.1). Owing to these loss processes, not all particles are sampled representatively into the inlet and transported efficiently to the measurement device (Section 6.3.2). Since the turn of the new millennium, there have been several innovative approaches for constructing aircraft-borne inlet systems with high and characterized particle sampling efficiency (Section 6.3.3). This efficiency is strongly dependent on particle size, and thus inlets can even be used to prevent particles or droplets from entering the measurement system (Section 6.3.4). But even for state-of-the-art inlets, sampling biases or artifacts may occur, and every inlet user should be aware of these problems (Section 6.3.5).

6.3.1 Particle Loss Processes

The major goal of aircraft-borne aerosol measurements is to carry out representative measurements; that is, the investigator needs to know the physical and chemical state of the aerosol particles in the undisturbed atmosphere. Unfortunately, because of the extreme measurement conditions (low temperature and pressure and high airspeed), there are always processes that lead to particle losses or particle modification. Hence, it is important to quantify these processes. In principle, there are several ways to accomplish this task: the use of empirical equations from the literature (Baron and Willeke, 2005; von der Weiden, Drewnick, and Borrman, 2009), CFD modeling studies (Ram, Cain, and Taulbee, 1995; Cain and Ram, 1998; Twohy, 1998; Dhaniyala *et al.* 2003; Krämer and Afchine, 2004; Eddy, Natarajan, and Dhaniyala, 2006), experimental quantification of particle losses by wind tunnel studies (Chandra and McFarland, 1997; Cain, Ram, and Woodward, 1998; Murphy and Schein, 1998; Twohy, 1998; Hegg *et al.* 2005; Hermann *et al.* 2001; Irshad *et al.* 2004), or in-flight testing and comparison (Huebert, 1990; Porter *et al.* 1992; Weber *et al.* 1999; Blomquist *et al.* 2001; Huebert *et al.* 2004, Moore *et al.* 2004; Hegg *et al.* 2005; McNaughton *et al.* 2007). The above order is intentional; it represents an increase in effort from the first approach to the last, and at the same time an improved accuracy in the derived sampling efficiency. For example, to carry out a wind tunnel or flight study to determine an inlet sampling efficiency is preferable, because it yields the sampling efficiency under realistic conditions, and it is also the most complex way to do it.

Aerosol particles' size modes are usually defined as nucleation mode: $D_p = 0.001\text{–}0.010\,\mu\text{m}$; Aitken mode: $D_p = 0.010\text{–}0.100\,\mu\text{m}$; accumulation mode: $D_p = 0.100\text{–}1.000\,\mu\text{m}$; and coarse mode: $D_p > 1.000\,\mu\text{m}$; with D_p the particle diameter (also see Chapter 4). For particles in the larger end of the coarse-mode size range, sampling losses due to gravitational settling may be important. These losses can be calculated reliably using empirical equations provided in the literature. Coarse-mode particles, and also sometimes accumulation mode particles, are subject to inertial deposition in the inlet system. These losses occur mainly in bends of the sampling line. In addition, free stream turbulence and turbulence in the sample flow can enhance loss of particles of all sizes in the sampling line. Although sampling losses due to some of these processes (e.g., turbulence in tubes and bend deposition) can be estimated from empirical equations (Hinds, 1999; Baron and Willeke, 2005; Vincent, 2007), the influence of other factors, such as turbulence in diffusing inlets and in the free stream, are less well known. Because of the difficulties of modeling turbulence and its effect on particle trajectories, CFD modeling can only be used to determine particle losses in inlet systems when the applicability of the turbulence model to the problem is first established. For laminar flow conditions, however, the use of CFD modeling to determine sampling characteristics is a valid approach. Finally, small submicrometer particles, mainly in the Aitken and nucleation mode, are subject to diffusive losses under both laminar and turbulent flow conditions. Again, in principle, there are empirical equations available to quantify these losses realistically. However, most of these equations were determined for simple geometries, such as straight tubes. The net transport efficiency is often determined as the product of the transport efficiencies for several independent sections, consisting of individual straight sections or tube bends. In the field, and particularly onboard aircraft, the sampling line geometries are often more complex. Moreover, effects such as tube orientation and upstream flow history are also important, as demonstrated by Wang, Flagan, and Seinfeld (2002a). These effects are not accounted for in the empirical relations. Hence, for complex sampling lines, an experimental determination of the transport efficiency in the laboratory is the best choice.

6.3.2 Sampling Efficiency

An aircraft-borne aerosol inlet system consists of an inlet, located outside the aircraft, and the sampling line that transports the particles from the inlet to the measurement device inside the aircraft. Similarly, the overall sampling efficiency of an inlet system has two components: (a) the inlet efficiency and (b) the transport efficiency through the sampling line. The inlet efficiency is derived as the product of the fraction of particles entering the inlet (aspiration efficiency) and the fraction of these particles being transmitted through the inlet (transmission efficiency). Individual efficiency values usually range from 0.0 to 1.0 (but can sometimes even be larger) and are multiplied to obtain the overall inlet sampling efficiency. Possible particle modifications are not accounted for in these efficiencies and must be treated separately. Fortunately, for many processes, the above efficiencies can be determined from the available empirical equations (Hinds, 1999; Baron and Willeke, 2005; Vincent, 2007). However, empirical equations for efficiencies are not easily established for some inlet configurations, for example, for a turbulent conical diffuser (Section 6.3.3), and for such cases, the efficiencies must be determined differently (e.g., by wind tunnel studies).

6.3.2.1 Inlet Efficiency

6.3.2.1.1 Aspiration Efficiency The aspiration efficiency of an aerosol inlet is a measure of how well aerosol particles can follow the streamlines at the tip of the inlet. Generally, it is most desirable to operate an inlet with an aspiration efficiency of 1.0 for all particle sizes. This is achieved by orienting the inlet parallel to the streamlines of the free stream flow (isoaxial sampling) and providing a sampling flow velocity equal to that of the free stream (isokinetic sampling, upper left panel of Figure 6.5). The use of a shroud upstream of the inlet (Section 6.3.3) can aid isoaxial sampling and minimize the effect of free stream turbulence on particle sampling. Note that the particle Stokes number Stk is as traditionally defined in Chapter 4, and not the modified Stokes number $\mathrm{Stk}_{\mathrm{mod}}$ discussed in Section 6.2.2, Eq. (6.1).

Aspiration efficiencies versus particle Stokes number are shown in Figure 6.6, calculated as a function of inlet velocity U and free stream velocity U_0 by using the empirical functions derived by Belyaev and Levin (1974) and those from CFD modeling by Krämer and Afchine (2004). Curve 4, a straight line with aspiration efficiency 1.0, represents the isokinetic sampling case ($U = U_0$). For all other cases, the aspiration efficiency will deviate from 1.0 for particles with Stokes

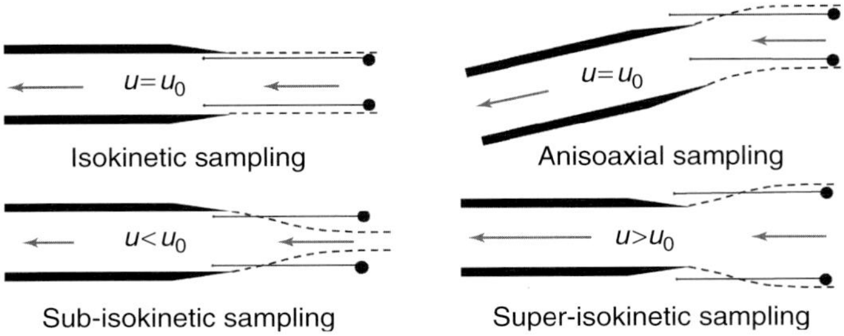

Figure 6.5 Sampling regimes for aerosol particle inlets with different velocity ratios and angles to the free stream. The dashed lines denote the region of the free stream from where the gas streamlines enter the inlet, the black dots illustrate large particles that do not follow the gas streamlines, and particle tracks are indicated by thin solid lines. (Adapted from Baron and Willeke (2005). Reproduced with permission of John Wiley and Sons, Inc.)

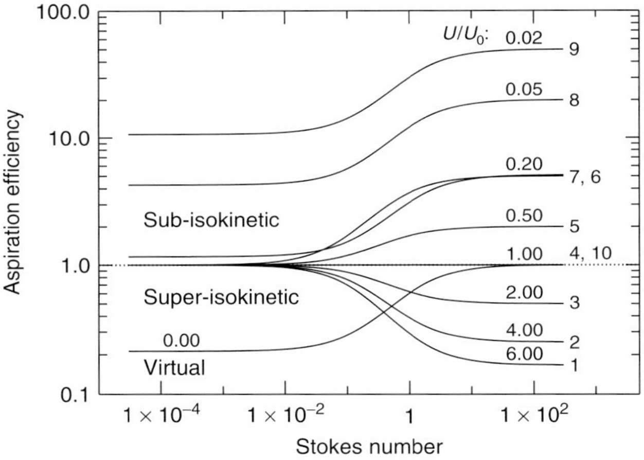

Curves 1–3:	1.0	≤	u/u_0	≤	6.0	B&L	(super-isokinetic)
Curve 4:			u/u_0	=	1.0	B&L	(isokinetic)
Curves 5–6:	0.2	<	u/u_0	<	1.0	B&L	(sub-isokinetic)
Curves 7–9:			u/u_0	≤	0.2	K&A	(sub-isokinetic, nearly virtual)
Curve 10:			u	=	0.0	K&A	(virtual)

Figure 6.6 Aspiration efficiencies calculated using the formula by Belyaev and Levin (1974), referred to as B&L, and Krämer and Afchine (2004), referred to as K&A. Aspiration efficiency of anisoaxial sampling (not shown here) is comparable to that for super-isokinetic sampling and depends on U/U_0 and the angle between inlet and free flow. U is the flow velocity inside the inlet and U_0 the velocity of the free flow.

numbers larger than about 0.1. For super-isokinetic inlets ($U > U_0$, Figure 6.5, lower right panel and Figure 6.6, curves 1–3), large particles are undersampled, while sub-isokinetic sampling ($U < U_0$, Figure 6.5, lower left panel and Figure 6.6, curves 5–9) leads to an enhancement of the large particles. The characteristics of anisoaxial sampling (Figure 6.5, upper right panel) are comparable to the super-isokinetic case. For inlets on high-speed aircraft, however, the difficulty of maintaining high sampling velocities makes it challenging to sample isokinetically, so their sampling is typically sub-isokinetic ($U < U_0$). For an aircraft-borne inlet with a 5-mm entrance diameter, flown at 500-hPa altitude with 150-m/s airspeed and a particle density of 1.5 g/cm^3, a Stokes number of 0.1 corresponds to a particle diameter of about 0.6 μm. Hence, anisokinetic sampling from aircraft can result in concentration changes even for submicrometer particles.

6.3.2.1.2 **Transmission Efficiency of the Inlet** The transmission efficiency accounts for all particle loss processes after the particles have entered the inlet. Anisokinetic sampling can enhance turbulence inside the inlet, and thus increase particle losses due to turbulent inertial deposition and turbulent diffusion. To minimize the risk of flow separation and turbulence generation at large angles of attack, rounded inlet lips are desirable; because for these lips, the flow attaches to the inlet surface and follows the curvature of the lips. Unfortunately, an inlet with rounded lips has a comparatively large cross-sectional surface area, which can serve as an impactor for coarse-mode and cloud particles. Hence, rounded lips tend to cause measurement artifacts inside clouds (Section 6.3.5). Moreover, rounded inlet lips force a larger volume flow through the inlet entrance compared with sharp-edged inlet lips. On jet aircraft already flying with high Mach number (0.70–0.85), this increased volume flow can generate shock waves inside the inlet and thus cause measurement artifacts (Section 6.3.5).

6.3.2.2 Transport Efficiency Inside the Sampling Line

The transport efficiency through the sample line accounts for all particle loss processes inside the sampling line (Section 6.3.1). Concerning these loss processes, fortunately reliable empirical equations from the literature are available (Hinds, 1999; Baron and Willeke, 2005; Vincent, 2007). However, with respect to particle modification (e.g., evaporation or condensation) and hence changes in chemical composition, one has to rely on CFD studies or experimental approaches.

6.3.3 Inlet Types

In the following two subsections, we present state-of-the-art aircraft-borne aerosol inlets and discuss their advantages and disadvantages. We focus on inlets capable of transporting the aerosol particles from the outside to the measurement devices inside the aircraft. There are also particle sampling systems that collect the particles outside the aircraft, for instance, on filters, which are analyzed after the flight, typically for particle chemical composition. Examples of such sampling systems are provided by Huebert, Lee, and Warren, (1990); Ram, Cain, and Taulbee, (1995); Cain, Ram, and Woodward, (1998); and Huebert *et al.* (2004).

6.3.3.1 Solid Diffuser-Type Inlet

To sample particles in the Aitken or accumulation mode, a solid diffuser-type inlet (SDI) is a good choice and is frequently used onboard aircraft, for example, Jonsson *et al.* (1995); Huebert *et al.* (2004); Hermann *et al.* (2005b); Schwarz *et al.* (2006); and McNaughton *et al.* (2007). An SDI consists of a pylon, the inlet tube, and optionally a shroud (Figure 6.7). The pylon is needed to place the point of sampling well outside the aircraft boundary layer (Section 6.2) and usually has an elliptic footprint to reduce aerodynamic drag. To sample isokinetically, aircraft-borne aerosol inlets sample a high-velocity volume flow, which is not consistent with the requirements of the airborne-aerosol instruments. Therefore, the sample flow must be slowed down, which is usually achieved by using a diffuser as the leading part of the inlet tube. In the diffuser, the tube diameter is expanded and thus the flow velocity is reduced (from 100–250 to 1–10 m/s). Practical considerations require the diffuser section to be short. However, a short diffuser will have a large expansion angle and this might result in flow separation, increased turbulence, and, hence, particle losses. Consequently, a compromise between the diffuser length and the diffuser opening angle must be made. As a guideline, in Figure 6.8, the maximum diffuser half-angle, which assures nonseparating flow in conical diffusers is displayed versus the Reynolds number at the diffuser entrance (Bohl, 1998). But even following this guideline, flow separation can still occur, if the diffuser gets too long. In the middle section of the SDI, a sampling line that extracts the desired sample flow may be mounted parallel to the flow. In the displayed type of SDI, a nozzle at the rear end accelerates the flow again to about the free stream velocity. The optional shroud is an aerodynamically formed tube that primarily serves the purpose to align the gas flow vectors with the centerline of the inlet tube (Torgeson and Stern, 1966), thus reducing particle losses associated with high angles of attack (Section 6.3.2). The shroud geometry can also be designed to reduce the sample flow velocity upstream of the inlet diffuser, whereby a shorter diffuser or a smaller diffuser angle can be used.

The advantages of an SDI are that it is comparatively easy to design, works passively (i.e., it needs no pump for the inlet, only for the relatively small sampling line volume flow), and provides a high particle sampling efficiency for Aitken and accumulation mode particles. For the nucleation and coarse-mode particles, however, particle losses are much higher in the SDI, of the order of several tens of percent (Hermann *et al.* 2001), which makes it unsuitable for these kinds of particles.

6.3.3.2 Isokinetic Diffuser-Type Inlet

An advanced version of the SDI is the isokinetic diffuser-type inlet (IDI) from Météo France/Comat, which was flown aboard the French ATR 42 aircraft (Figure 6.9). In principle, the IDI is a classical SDI. However, while the SDI has a fixed orientation with respect to the aircraft fuselage, the IDI has a mechanical control unit just

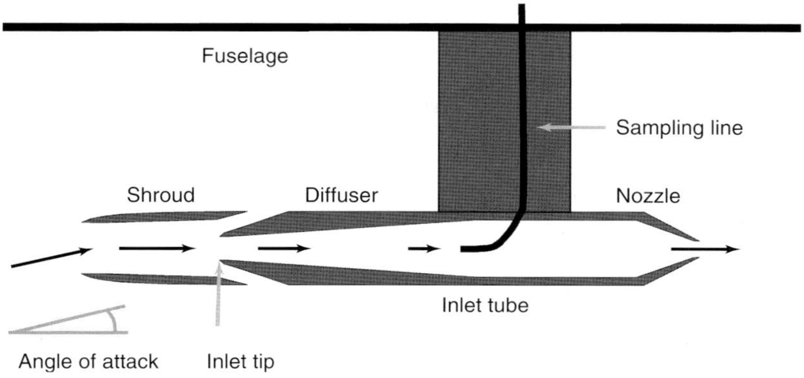

Figure 6.7 Schematic diagram of an SDI. (Courtesy of M. Hermann.)

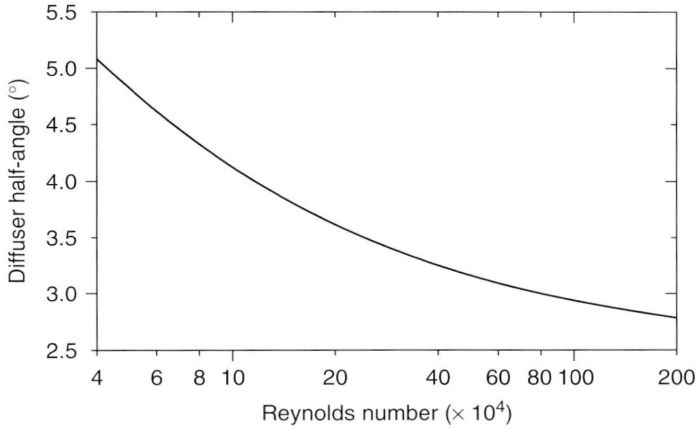

Figure 6.8 Maximum diffuser half-angle for nonseparating flow in conical diffusers as a function of the Reynolds number at the diffuser entrance, from the textbook by Bohl (1998). (Reprinted with permission by Vogel Buchverlag.)

below the aircraft fuselage to align the inlet centerline to the free stream flow vector. The alignment of the inlet is checked by four opposed pressure sensors in the inlet tube. This helps reduce aspiration losses not compensated for by the shroud, but is mechanically demanding.

6.3.3.3 **Low-Turbulence Inlet**

A substantial advancement in the design of aircraft-borne aerosol inlets was the invention of the low-turbulence inlet (LTI) (Wilson *et al.* 2004; Huebert *et al.* 2004). This inlet was particularly designed to sample coarse-mode particles with high sampling efficiency, which is not accomplished by an SDI. In an SDI, these particles get lost mainly due to turbulence in the inlet diffuser and the straight middle section of the inlet tube. Hence, the idea of the LTI is to reduce the turbulence by applying boundary layer suction, that is, the sampled flow near the walls is removed by

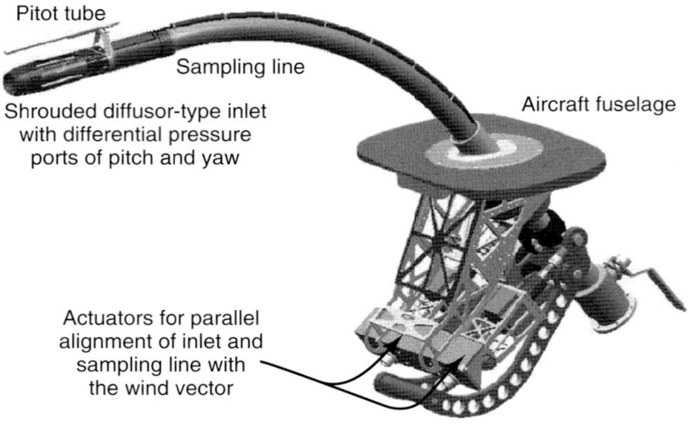

Figure 6.9 Sketch of the IDI. By courtesy of Météo France/Comat, Toulouse, France.

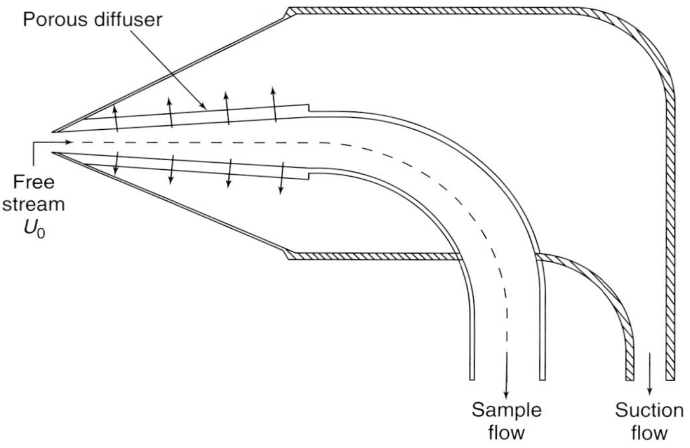

Figure 6.10 The LTI of the University of Denver. (Adapted from Baron and Willeke (2005). Reproduced with permission by John Wiley and Sons, Inc.).

suction through the porous diffuser wall (Figure 6.10). Only a small flow around the centerline is transmitted to the measurement devices inside the aircraft. As the Reynolds numbers at the entrance of aircraft-borne inlets are of the order of 10^4–10^5, a large fraction of the incoming flow must be removed before nonturbulent conditions (turbulence intensity <1%) are reached (Figure 6.11). The advantage of the LTI is its high sampling efficiency for super-micrometer particles. However, because of the increasing inertia with increasing particle diameter, particles larger than 1 μm are enhanced in the sample flow, as they cannot follow the rapidly changing streamlines in the suction region. These enhancement factors must be known for data interpretation, for which CFD studies are useful. A disadvantage of the LTI is the need for a big pump (several hundred liters per minute), which either consumes a lot of power or, in the case of a venturi pump, represents a large

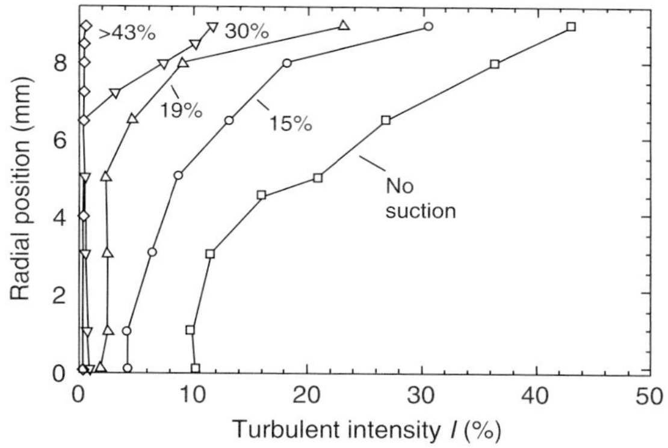

Figure 6.11 Turbulence intensity in the LTI as a function of the applied suction flow and the radial position (Wilson *et al.* 2004). (Copyright 2004 Mount Laurel, N.J. Reprinted/modified with permission. Courtesy of J.C. Wilson.)

flow resistance, potentially reducing aircraft performance. Consequently, the LTI is usually flown on larger aircraft, such as the National Center for Atmospheric Research (NCAR) C-130, the National Oceanic and Atmospheric Administration (NOAA) WP-3D, and the British BAe-146, where power is less limited.

6.3.3.4 **Nested Diffuser-Type Inlet**

A new approach for reducing the turbulence inside an aerosol inlet is the nested diffuser-type inlet (NDI), which is currently developed for the German research aircraft High Altitude and Long Range Research Aircraft (HALO). The NDI, like the SDI, consists of a pylon, a diffuser inlet tube, and a shroud. However, inside the diffuser inlet tube, a second diffuser inlet tube is nested, and inside this second tube, a third diffuser is located, followed by the sampling line (Figure 6.12). The nesting of the diffusers results in reduced turbulence inside the inlet and thus higher sampling efficiency for super-micrometer particles. Compared with the LTI, the major advantage of the NDI is that it does not need a large pump and hence in principle can also be applied on small aircraft. Its disadvantage is that the turbulence is not as strongly reduced as in the LTI, leading to final CFD-derived turbulence intensities of 2–12%, depending on the turbulence model used.

6.3.4
Size Segregated Aerosol Sampling

Particles sampled by inlets on high-speed aircraft may experience changes in size because of the airflow-induced temperature changes (Yang and Jaenicke, 1999) or differences in temperature and vapor pressure in the inlet compared with that in the atmosphere (Section 6.3.5). The change in particle size within the inlet system complicates analysis of particle size–sensitive properties. Often, a characterization of a subset of atmospheric particles, for example, interstitial particles in tropospheric liquid or ice clouds or in polar stratospheric clouds,

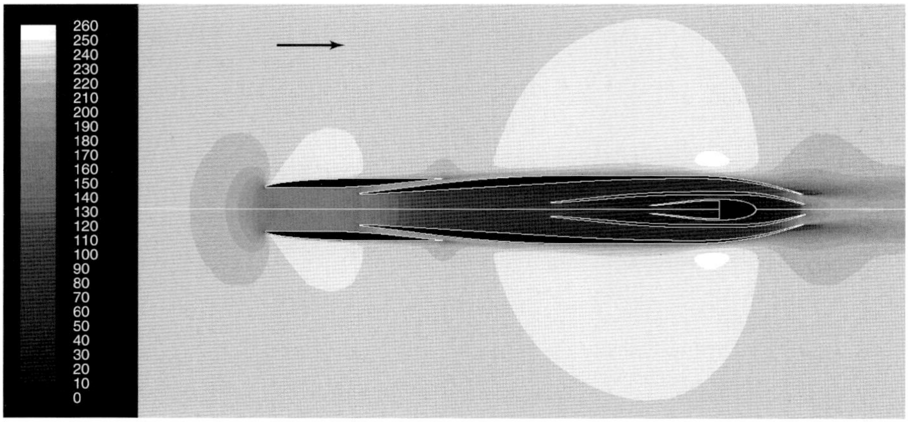

Figure 6.12 2D velocity contour plot (in m/s) for the axial symmetric NDI as currently (2012) under construction for HALO. The CFD modeling results were obtained for 50 000" flight altitude and 235-m/s airspeed (Hermann, 2010). (Courtesy of M. Hermann.)

is desired. For such applications, systems that can size-classify particles before sampling are designed (particle/cloud separating inlets are described either here or in Section 6.4, according to whether the focus is aerosol particle or cloud analysis).

A simple approach for separating cloud droplets from aerosol particles is the use of a shrouded inlet oriented at a large ($> \sim 8^{\circ}$) angle of attack (Eisele *et al.* 1997). While aerosol particles will still follow the curved streamlines into the inlet, droplets have too much inertia to reach the point of sampling. A similar approach was chosen for the "football inlet," where the inlet body shields one of the two inlets from cloud droplets or ice crystals (Section 6.4 and Figure 6.26).

Although formally a sampling device for cloud droplets, the counterflow virtual impactor (CVI) is used onboard aircraft to analyze the subset of atmospheric particles involved in cloud formation and growth. Conversely to the football inlet, a CVI samples only the cloud droplets and prevents interstitial aerosol particles (and gases) from entering the sampling line by applying an inert gas, directed counter to the free stream flow direction (Ogren, Heintzenberg, and Charlson (1985) and Figure 6.13). Owing to this flow resistance, only larger droplets with sufficient inertia pass this virtual barrier and are sampled. As droplets are impacted into a warm, dry sample line, water and other volatile particle material evaporates and residual aerosol particles and volatile gases can then be analyzed downstream of the CVI using a whole range of aerosol and gas-phase instrumentation. The droplet cut-size is determined by the amount of counterflow and the length of the internal porous tube. The minimum cut diameter (50% efficiency) for airborne CVIs typically can be varied over a range of approximately 5–20 μm, depending on the geometry and ambient conditions. The variable minimum cut diameter can even be used to derive the cloud droplet size distribution by using an inversion method (Noone and Heintzenberg, 1991).

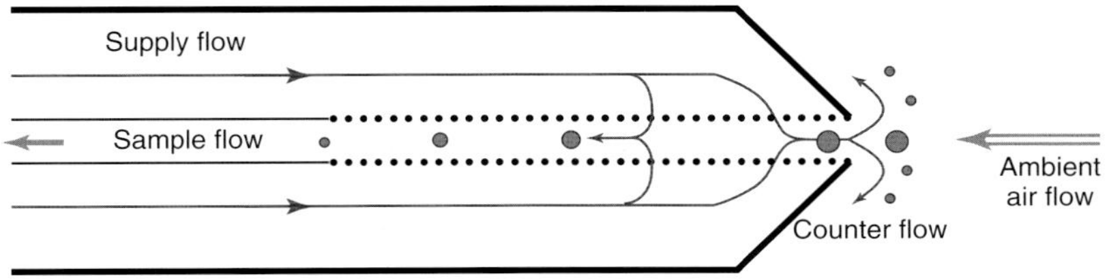

Figure 6.13 Schematic of a CVI inlet. The warm, dry carrier gas flows in opposing directions after passing through the porous center tube. (Courtesy of C. Twohy.)

Some CVIs also employ upstream shrouds to straighten airflow ahead of the counterflow region. Flow characterization and calibrations of aircraft-borne and ground-based CVIs were carried out in several studies (Anderson, Charlson, and Covert, 1993; Laucks and Twohy, 1998; Twohy, 1998; Chen *et al.* 2005; Boulter *et al.* 2006; Mertes *et al.* 2007; Lin and Heintzenberg, 1995). Composition of individual residual particles from ice clouds have been measured using the CVI in the studies by Heintzenberg, Okada, and Ström (1996); Ström and Ohlsson (1988); Petzold *et al.* (1998); Twohy and Poellot (2005), and others. For more details also see Section 6.4.2, where the use of the CVI specifically for measurement of cloud-condensed water content is discussed.

The CVI technique was extended for selective measurements of aerosol- and gas-phase compositions by Dhaniyala *et al.* (2003) (Figure 6.14). The primary goal of the inlet was to alternately sample and exclude particles larger than $\sim$0.1 μm when measuring aboard a high-speed aircraft ($\sim$200 m/s TAS). The cut-size of 0.1 μm was chosen because the mass of atmospheric particles smaller than this size was assumed to be negligible and thus, measurement of gas-phase semivolatiles with this inlet will not be affected by the presence of particles. That inlet, called the chemical ionization mass spectrometer (CIMS) inlet, is operated to detect volatile species (HNO_3) in the gas phase and in particles. It has a linear geometry (i.e., 2D rather than axisymmetric) with an external shroud to eliminate any dependence of inlet performance on the angle of attack. An inner shroud is used to direct the flow at high-speed toward the sample port over a narrow channel width. Close to the sample port, there are two small airfoil-shaped "blades" that provide the appropriate flow conditions required for aerosol and gas sampling. For aerosol sampling, these blades are positioned such that the inlet operates as a CVI (Figure 6.14, top right panel). The design of the CIMS inlet enables particle sampling with a smaller cut-size than that typically possible with previous CVI designs (Figure 6.14, bottom panel). For gas sampling, a stepper motor moves one of the blades and occludes the inlet opening, resulting in a configuration where the sample flow is extracted perpendicular to the bulk flow (Figure 6.14, top left panel). Aircraft-based measurements during the Stratospheric Aerosol and Gas Experiment III Ozone Loss and Validation Experiment 2000 campaign suggested that the inlet performed largely as established with CFD simulations (Dhaniyala *et al.* 2003).

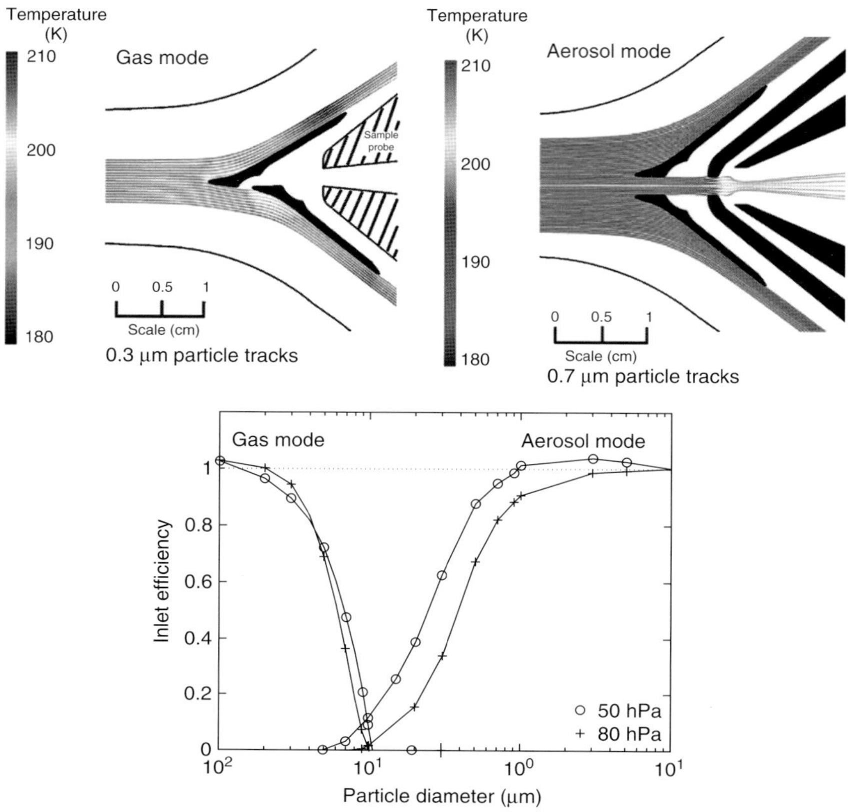

Figure 6.14 Top panels: Sketch of the CIMS, or modified CVI inlet, from the study by Dhaniyala *et al.* (2003). A moving blade system can be moved to operate the inlet as a particle-excluding gas inlet (top left panel) or as an aerosol CVI inlet (top right panel). Lower panel: sampling efficiency curves of both inlet modes at two different ambient pressures.

6.3.5 Sampling Artifacts

Sampling artifacts are any perceived distortion in the sampled particle concentrations relative to that in the atmosphere or other data errors caused by the instrument of observation. In aircraft-borne aerosol research, the high airspeed of the measurement platform and the temperature increase from outside the aircraft to the cabin is primarily responsible for potential measurement artifacts. The influence of the aircraft or the instrument housing on particle measurements is discussed in Section 6.2.3. In this section, we focus on measurement artifacts that might occur in connection with the inlet.

Particle evaporation occurs in almost all aircraft-borne inlet systems. This is because, typically, the sampled air must be slowed down from the aircraft speed

(~100–250 m/s) to a moderate sampling velocity (~1–10 m/s) consistent with the flow requirements of the analysis instrument (Section 6.3.3). From the first law of thermodynamics, it is known that a slowing flow will experience heating (~20 K for flow slowed to stagnation from 200-m/s airspeed). Additional heating occurs when the air sample is brought into the cabin, where temperatures are normally ~303 K. This overall heating of 20–90 K (depending on the flight conditions) results in the evaporation of volatile material, thus changing the size distribution and chemical composition of the particles. For example, even ambient relative humidities of 100% are reduced to relative humidities of <0.3% after a 70-K heating process in an inlet system. The evaporation can be very fast, <0.1 s Eidhammer and Deshler (2005), and can even lead to the total evaporation of sulfuric acid/water particles in the nucleation mode (Hermann *et al.* 2001). Attempts to actively cool the air down to counter this effect are difficult and have not generally been successful. Hence, particle measurements downstream of an inlet system on a jet aircraft are always a measurement of a heated and at least partially dried aerosol.

Particles deposited on the inner walls of an inlet tube by particle loss processes are not necessarily permanently removed from the gas stream. Although adhesive forces are large, particularly for submicrometer particles (Hinds, 1999), collisions between deposited particles and subsequently sampled particles might dislodge the deposited particles and resuspend them (Ziskind, 2006). An illustrative example of such a process was provided by Murphy *et al.* (2004), who used an aerosol mass spectrometer to characterize aerosol particles in the upper troposphere. During the Atmospheric Composition Change campaign in September 2000, they had the opportunity to measure the exhaust plume of a space shuttle, yielding a very characteristic chemical particle signature. On the next day, when flying through cirrus clouds, again this plume fingerprint was measured, although no space shuttle was there. The authors came to the only conclusion that ice crystals must have resuspended space shuttle plume particles from the inlet wall, where they were deposited the previous day. One possibility to partly avoid contamination of the inlet system particularly while on ground, in the boundary layer, or near the airport (with high aerosol loading) is to flush the inlet system backward with particle-free air, as is done within the Civil Aircraft for Regular Investigation of the Atmosphere Based on an Instrument Container system (*www.caribic-atmospheric.com*). In their study Murphy *et al.* (2004), and also Kojima *et al.* (2004), show that ice crystals can potentially abrade material from the inner wall of a stainless steel inlet and generate artifact particles that can contaminate the sample.

Beside abrasion of the inlet wall by ice crystals and release of deposited particles, flying through clouds, in fact, is a prominent source of artifacts in aerosol measurements. Similar to particle impaction and breakup on the aircraft body, inlet tips also represent surfaces where cloud droplets or ice crystals might impact and breakup, for example, Cohen (1991); Yarin (2006); and Vidaurre and Hallett (2009a). For the NCAR C-130 aircraft, Weber *et al.* (1998) compared aerosol measurements made downstream of two inlet systems; one with a relatively large inlet surface area compared with the inlet sampling cross-sectional area (large area ratio) and one with a relatively small inlet surface area compared with the inlet sampling cross-sectional area (small area ratio). Although the C-130 aircraft speed (100 m/s) is not very high, the authors found that water droplets impacting the lips of the large area ratio inlet led to the generation of many small particles downstream. Sampled aerosol particle concentrations were increased by at least an order of magnitude compared with the measurements downstream of the small area ratio inlet. For ice particles, this effect was not seen, but should not be generally excluded. Artifacts due to droplet breakup might not only occur due to particle impaction on surfaces but also as they pass through a shock wave (e.g. Chang and Kailasanatha, 2003). Although inlets should be designed (with the help of CFD models) to prevent shock waves occurring in the inlet tube, certain flight maneuvers might lead to increased flow velocities and hence to conditions where shock waves occur. In this case, particle measurements downstream of the shock wave may be strongly biased.

Finally, aerosol measurements might also be affected when flying through clouds, if the aircraft collects electrical charges (Illingworth and Marsh, 1986; Jones, 1990). This charging can cause strong electrical fields and even possibly corona discharge at pointed surface mountings, such as aerosol inlet tips. A corona discharge at the inlet tip can charge the sampled aerosol particles and, thus affect the inlet aspiration and transmission efficiencies (Romay *et al.* 1996). Moreover, discharge effects at the aircraft fuselage or the inlet can occasionally lead to artifact particles (Zahn *et al.* 2002). There is, unfortunately, almost no literature to deal with this potential source of measurement artifacts.

Sampling lines onboard aircraft are mostly made of stainless steel, because it is conductive and hence does not lead to particle losses due to electrostatic charging. Furthermore, it resists compression even under vacuum conditions. However, the logistics of aircraft-based measurements often result in the usage of at least one short piece of flexible tubing in the sampling line. This flexible tubing is needed because aircraft "breathe," that is, because of the pressure difference between outside and inside the pressurized cabin, the aircraft expands during ascent and shrinks during descent. For a large jet aircraft, this mechanical dilatation amounts to several millimeters. Hence, any totally rigid sampling line might rupture after take-off. The most commonly used flexible tubing element in any aerosol measurement system is conductive silicon tubing. Recent studies have shown that these tubings emit siloxanes, which can lead to artifacts in chemical composition and optical and hygroscopicity measurements downstream the tubing (Timko *et al.* 2009; Yu *et al.* 2009). Therefore, precaution must be used in selecting flexible tubing for aerosol measurement systems.

6.4
Cloud Particle Sampling

Cloud sampling from aircraft is a particular challenge, because some cloud drops and ice crystals are big enough to have substantial inertia at the high aircraft cruising speeds. Therefore, they tend to depart from the gas streamlines (Section 6.2.2) and impact (and possibly break up) on any obstacle or bend. Section 6.4.1 is dedicated to such sampling issues. Airborne instruments for in-cloud measurements are described in Chapter 5, while those sampling cloud particle ensembles (bulk) are found in Section 6.4.2.

6.4.1
Cloud Sampling Issues

6.4.1.1 Effect of Mounting Location

Mounting locations of cloud microphysical instrumentation on the aircraft play a crucial role in the accuracy of the measurements. Even if a cloud probe itself is capable of highly accurate measurements, a poor mounting location may be a cause of significant biases of the measurements. As discussed in Section 6.2.2, an aircraft fuselage causes a significant disturbance of the airflow, which results in regions of enhanced and depleted concentration in its vicinity. Ice particles bouncing from the nose of the fuselage at low pressure (i.e., in the upper troposphere) may travel away from the fuselage tens of centimeters and contaminate measurements of the particle probes mounted on the fuselage.

There are therefore two major requirements for the mounting locations of the cloud microphysical instrumentation: minimal disturbance of the airflow and minimal contamination by ice particles bounced from the airframe. Of all possible aircraft mounting locations, the under-wing pylons are considered most suitable for mounting of cloud probes and provide the best quality measurements. An example of such arrangement is shown in Figure 6.15 (upper photo). Miniature probes with relatively small dimensions and weight, such as hot-wire probes (Chapter 5), can be mounted on booms, as shown in Figure 6.15 (lower photo).

6.4.1.2 Effect of Probe Housings

The housings of cloud particle probes disturb the airflow (i.e., change its speed and direction) before its passing through the sample volume. Cloud particles suspended in the air will respond to this disturbance in accordance with their aerodynamic dimensions, with varying degrees of deviation from streamlines and spatial sorting. Larger particles, due to higher inertia, are less susceptible to the changes in airflow and may cross streamlines, whereas smaller particles follow the streamlines better. The spatial sorting and redistribution of particles having different sizes may eventually affect the size distribution measured in the sample volume. The effect of the particle sorting is more pronounced at low altitudes, when the air has higher density and the drag forces related to the changes in the airflow will have a stronger effect on the particle trajectories. Another consequence caused by the air disturbance is related to changes in the local flow within the sample volume, which may affect the calculation of particle concentration. Deceleration or acceleration of the airflow may also cause heating or cooling of the air, potentially changing the size of the particles.

Estimation of the airflow disturbance can be performed based on CFD calculations. Figure 6.16 shows examples of such calculations for the airflow around

Figure 6.15 Example of mounting location of cloud spectrometer and impactor and cloud particle probes OAP-2D-C, OAP-2D-P, High Volume Precipitation Spectrometer (HVPS), and 2D-S on the under-wing pylons (upper photo) and King, Nevzorov, and SEA hot-wire probes on the under-wing boom on the NRC Convair 580 (lower photo). (Courtesy of A. Korolev.)

the OAP-2DC, Cloud Imaging Probe (CIP), and FSSP probes. It can be seen from Figure 6.16d that for the case of FSSP, the airflow may undergo a sequence of deceleration and acceleration before passing trough the sample volume. For the case of OAP-2DC and CIP, the deviation of the airspeed across the sample volume from its background value is relatively small and does not exceed a few percent (Figure 6.16e, f, h, and i).

Deceleration in front of probe canisters may also cause changes in orientation of ice crystals and deformation of droplets when they pass through the sample volume. A few examples of alignment of columns and flipping of stellar crystals in the sample volume of the CIP are shown in Figure 6.17. The orientation of the ice particles may also be affected by proximity of the probe to the airframe, which may interfere with airflow changes around the probe's housing. The drag forces experienced by ice particles may also result in their break up into smaller fragments; see the following section.

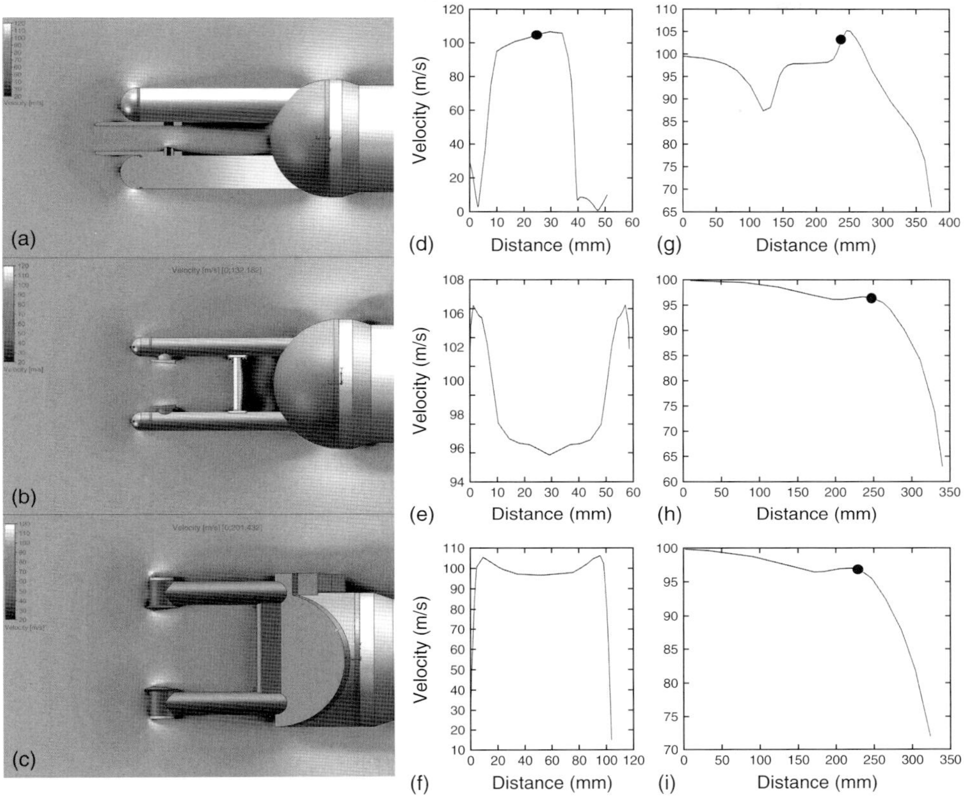

Figure 6.16 Calculation of the airflow around (a) FSSP, (b) OAP-2DC, and (c) CIP. The upper left panel shows a cross section of the FSSP inlet tube and the lower arm. Calculations were carried out for $U = 100\,\text{m/s}$; $p = 800\,\text{hPa}$; $T = -10\,^{\circ}\text{C}$. The CIP flow calculations were performed without the pitot tube and hot-wire sensor installed on a standard version. (d–f) Changes in the air velocity along the axis of the sample volume (laser beam); (g–i) air velocity changes along the line parallel to the airflow going through the center of the arms or the inlet tube. Black dots indicate the location of the sample area. (Courtesy of A. Korolev.)

6.4.1.3 Droplet Splashing and Breakup

Large drops have been shown to break up and produce enhanced concentrations of small droplets and aerosol particles during high-speed sampling (Weber *et al.* (1998); Twohy, Strapp, and Wendisch, (2003)). The probability of drop breakup is determined by the Weber number, which is the ratio of destabilizing hydrodynamic force to stabilizing surface tension force. Drops will fragment when a critical value of the Weber number is exceeded. Experimental studies indicate that the critical Weber number for water drops is approximately 10–12 (Pilch and Erdman, 1987; Tarnogrodzki, 1993). The critical diameter of droplet expected to break up (or "shatter") is approximately 100-μm diameter for speeds of order of 100 m/s.

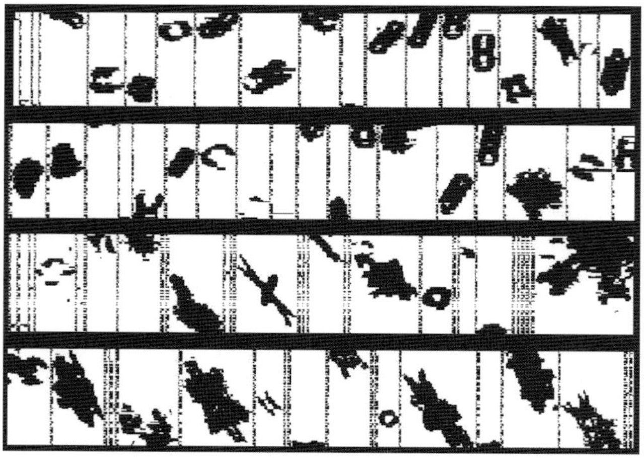

Figure 6.17 Examples of changes in particle orientation in front of the CIP in the case of columns (two top charts) and stellar ice crystals (two bottom charts). The changing orientation of ice crystals in these specific cases is related to the proximity of the probe's canister relative to the airframe. (Courtesy of A. Korolev.)

Therefore, virtually all drizzle and precipitation sized drops may break up if they contact surfaces upstream of instrument sampling points. In addition, strong shear in airflow along a drop trajectory may also induce breakup.

Drops slightly exceeding the critical size may produce only a few droplet fragments, but the number of fragments will increase with the parent drop size (Komabayasi, Gonda, and Isono, 1964). Breakup and splashing of droplets may result in underestimation of liquid water content (LWC) in hot-wire probe measurements (Biter *et al.* 1987); Strapp *et al.* 2002) or enhance the concentration of small droplets measured by particle spectrometers (Baker *et al.* 2009a,b). As the impact of liquid droplets with the solid surface is essentially inelastic, the splashed droplets do not travel across the airflow as far as ice particles experiencing bouncing (discussed in the following section). The fate of the drops striking airborne surfaces may also depend on the orientation of the obstacle and the sampling conditions; for example, in supercooled clouds, they may flatten over a surface and freeze, contributing an icing hazard to aircraft flight control surfaces.

6.4.1.4 **Ice Particle Bouncing and Shattering**

6.4.1.4.1 **General Features** Bouncing and shattering of ice particles may be a source of significant errors in microphysical measurements in ice and mixed-phase clouds. The effect of shattering has been studied in a number of papers, for example, Gardiner and Hallett (1985); Gayet, Febvre, and Larsen, (1996); Field *et al.* (2003), Field, Heymsfield, and Bansemer (2006); Korolev and Isaac (2005), McFarquhar *et al.* (2007); Heymsfield (2007); Vidaurre and Hallett, (2009a); Jensen *et al.* (2009); Korolev *et al.* (2011); and Lawson (2011), see also Figure 6.21.

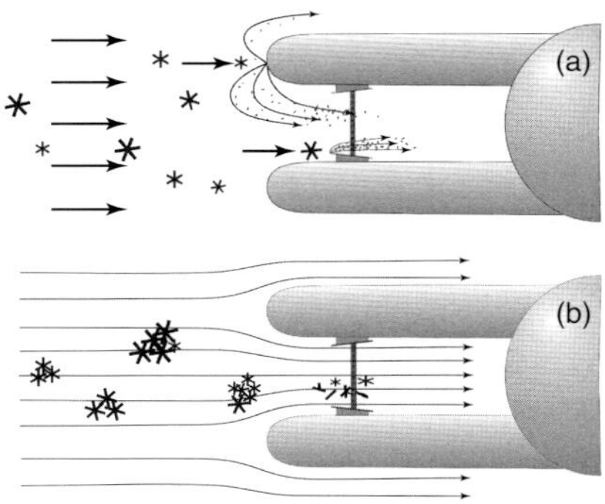

Figure 6.18 Conceptual diagram of two types of cloud particle fragmentation during sampling: (a) resulting from the impact of ice particles with the probes housing and (b) related to the break up of aggregates due to stresses related to the wind shear and vorticity caused by the probes housing, from the study by Korolev *et al.* (2011). (Copyright 2011 American Meteorological Society (AMS). Reprinted with permission by AMS.)

During sampling, ice particles may shatter and break up for two reasons. First, before entering the instrument sample volume, ice particles may impact on the probe's upstream tips or inlet and shatter into small fragments, which may cause multiple artificial counts of small ice (Figure 6.18a). Second, ice particles may be fragmented due to stresses experienced in the regions with high vorticity generated by the probes housing (Figure 6.18b). The first type of shattering may affect nearly all types of ice particles, whereas the second type mainly affects aggregates with weak bonding between ice particles (e.g., aggregates of dendrites or needles) or some types of naturally fragile ice particles (e.g., bullet rosettes and dendrites).

The appearance of the shattered fragments of the first and second types measured by imaging probes is shown in Figure 6.19. The shattered particles of the second type appear as a cluster of relatively large fragments with characteristic sizes of hundreds of micrometers and millimeters (Figure 6.19c, f, g, and h). The fragments resulting from the shattering of the first type have much smaller sizes (Figure 6.19a, b, d, e, i, and j) and the number of the fragments in the view field of imaging probes may exceed a hundred (Figure 6.19a, b, and e). The analysis of 2D imagery registered by the imaging probes suggests that the first type of shattering is more common than the second one.

High-speed video recording conducted in wind tunnels showed that ice particles bouncing from the surface of the probes may travel up to 10 cm across the airflow at 100 m/s and 1013 hPa before passing through the sample area (Emery *et al.* 2004; Korolev *et al.* 2011). For the same conditions, ice particles may bounce forward of the airflow up to 1 cm at 1013 hPa and an airspeed of 100 m/s. Figure 6.20a–c shows the snapshots from high-speed videos with the trajectories of particles bouncing from the OAP-2DC and CIP arm tips and from an FSSP inlet tube.

The effect of shattering on the microphysical measurements depends on a large number of different parameters. These parameters can be split into two categories. The first category is related to external environmental conditions such as airspeed,

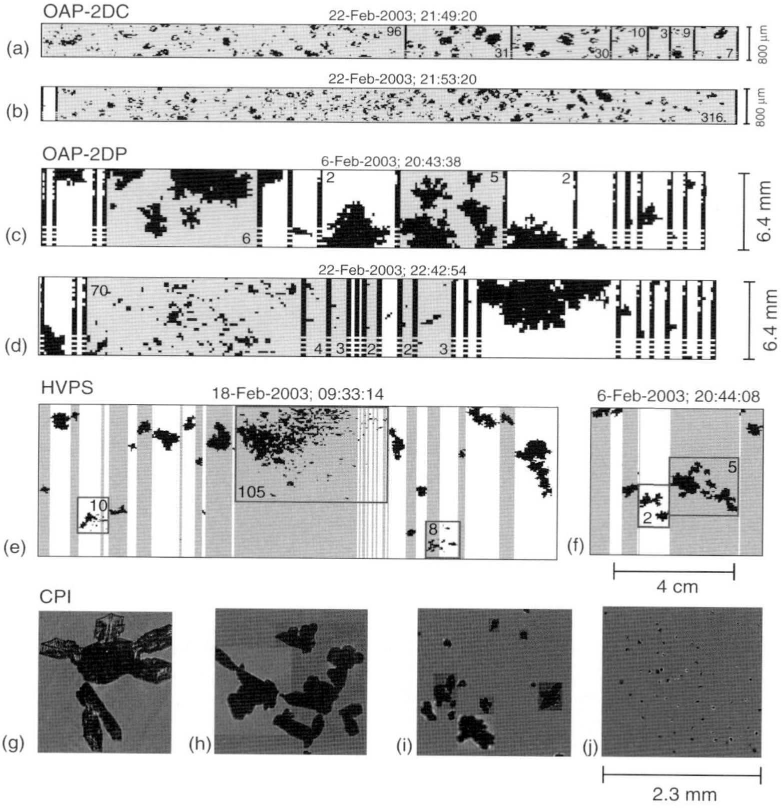

Figure 6.19 Appearance of the images of shattered particles registered by imaging probes with different viewing aperture and pixel resolution: OAP-2DC at 25-μm pixel resolution (a and b), OAP-2DP at 200-μm pixel resolution (c and d), HVPS at 200-μm pixel resolution (e and f), and CPI at 2.3-μm pixel resolution (g–f). The numbers inside the image frames indicate the number of isolated images associated with the image of the shattered particle. The 2D images were collected by the Environment Canada from the NRC Convair 580, from the study by Korolev and Isaac (2005). (Copyright 2005 American Meteorological Society (AMS). Reprinted with permission by AMS.)

pressure, temperature, ice particle size, habit, density, and orientation. The second category of the parameters is related to the property of the probe's housing and electronics: configuration of arm tips or inlets and their temperature, angle of attack, surface roughness of the tips, proximity of the edge of the tips or inlets to the sample volume, size of the sample volume, particle size threshold sensitivity, and response time of the electronics.

The electronic response time and particle size threshold sensitivity (e.g., pixel resolution and minimum detectable size) define the capability of the probe to measure small shattered fragments. If the response time is too large or the pixel resolution is too coarse, then small fragments may not be detected by the probe's electronics.

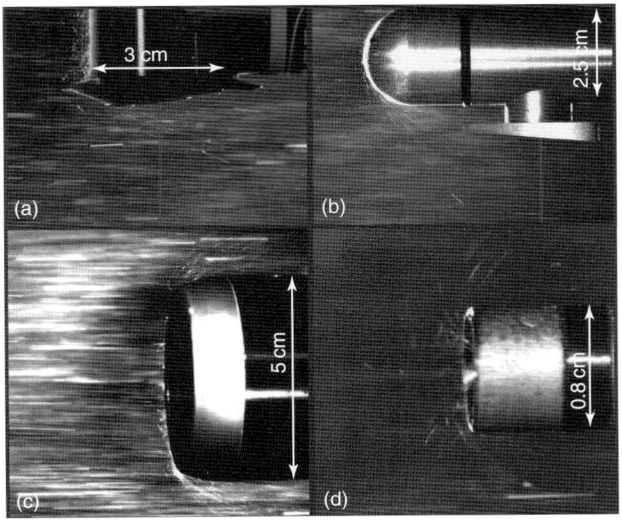

Figure 6.20 High-speed video snapshots of the trajectories of ice particles bouncing from the arm tips of CIP (a), OAP-2DC (b), FSSP inlet tube (c), and Nevzorov LWC/IWC hot-wire sensor (120° cone) (d). Frames from high-speed videos were taken in ice sprays in the Cox wind tunnel at an airspeed of 80 m/s, from the study by Korolev *et al.* (2011). The high-speed videos associated with these snapshots can be downloaded from ftp://depot.cmc.ec.gc.ca/upload/hsvideo/. (Copyright 2011 American Meteorological Society (AMS). Reprinted with permission by AMS.)

Ice particle bouncing may also result in an underestimation of ice water content (IWC) measured by hot-wire probes and bulk cloud samplers (Section 6.4.2), which are used to measure LWC/IWC, such as the Nevzorov probe and Science Engineering Associates (SEA) LWC/IWC probe. High-speed videos recorded in wind tunnels and natural clouds have shown that ice crystals can bounce away from the surface of LWC/IWC hot-wire sensors, resulting in an underestimation of IWC (Emery *et al.* 2004; Isaac *et al.* 2006). Figure 6.20d shows a snapshot from a high-speed video demonstrating that the trajectories of ice particles bounced out of the hot-wire LWC/IWC Nevzorov shallow cone. Korolev *et al.* (2008) showed that bouncing may result in a underestimation of IWC measured by the Nevzorov LWC/IWC shallow cone sensor. A modified deep cone sensor with a 60° angle captures nearly all ice particles, thus minimizing the effect of bouncing.

6.4.1.4.2 **Techniques for Identifying Shattering Events** The shattering effect has a large impact on the small part of the size distribution and it appears to be quite subtle for particle diameters larger than 500 μm. Shattering may result in overestimation of the concentration by two orders of magnitude for the FSSP, OAP-2DC, and CIP measurements. As the extinction coefficient and mass of the typical size distributions are normally dominated by the larger particles, these parameters estimated from 2D particle images are significantly less affected by

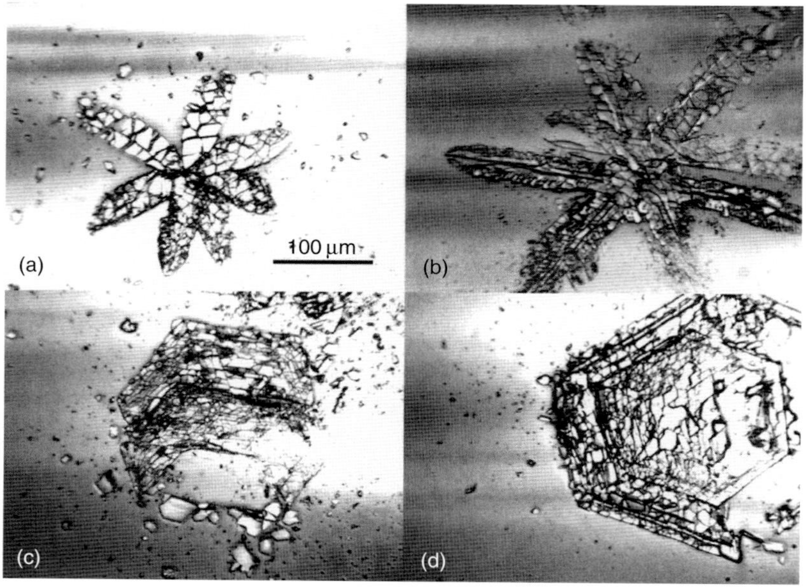

Figure 6.21 Images of broken ice crystals collected by a replicator at 130 m/s. The size of the smallest fragment goes down to 5 μm, from the study by Vidaurre and Hallett (2009a). (Copyright 2009 American Meteorological Society (AMS). Reprinted with permission by AMS.)

shattering than the number concentration is Korolev *et al.* (2011); Lawson, (2011); Jensen, *et al.* (2009). In imaging probes, the shattering events can be identified and filtered out based on one of the two following algorithms.

Korolev and Isaac (2005) developed an algorithm based on the analysis of the image frames with fragmented particles. Examples of such images are shown in Figure 6.19. This algorithm has a limited efficiency for identifying shattering events due to a relatively low occurrence of the image frames with fragmented images in the particle imaging probes.

Another algorithm is based on the fact that after impact, a shattered particle forms a cluster of closely spaced fragments. Some of these fragments pass through the sample volume with interarrival times (t_{int}) much shorter than that of intact particles. If the frequency distributions of the interarrival times associated with the shattered and intact particles have a bimodal distribution with well-separated modes, then the interarrival times can be used for identification of shattering events and filtering them out (Cooper, 1978; Field and Heymsfield, 2003; Field, Heymsfield, and Bansemer, 2006; Baker *et al.* 2009a,b, Lawson, 2011).

An example of such a frequency distribution deduced from the measurements of OAP-2DC is shown in Figure 6.22a. Particles with the interarrival time interval $t_{int} < t_c$ are considered to be a result of shattering, whereas particles with $t_{int} > t_c$ are considered to be intact ones. If the short and long interarrival modes are overlapped with each other, then the interarrival time algorithm has a reduced performance

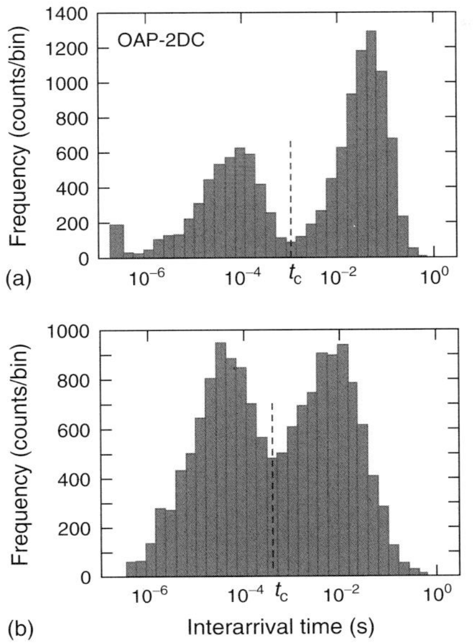

Figure 6.22 Frequency distributions of the interarrival time for cloud particles measured by the OAP-2DC when (a) short and long interarrival time modes are well separated and (b) when short and long interarrival time modes overlap. The mode associated with the short interarrival time is assumed to result from shattered particles, whereas the natural particles are assumed to form the longer interarrival mode. The dotted lines show the cut-off time (t_c) used during data processing for filtering the shattered events. (Courtesy of A. Korolev.)

(Figure 6.22b). In this case, this algorithm rejects some fraction of intact particles along with the shattered ones, and some fraction of the shattered particles will be accepted. Such a situation is typical for mixed-phase clouds and for some ice clouds.

The importance of shattering events is influenced by the natural concentration of small particles (Jensen *et al.* 2009; Lawson, 2011). In regions where natural concentrations are low, the contribution of shattered particles to the total particle concentration can be large, while in cases where concentrations are high, the contribution of shattered particles is minimal.

The interarrival time algorithm is based on the assumption that all particles with short interarrival times are associated with shattered events only, whereas particles with long interarrival times result from the measurements of intact particles. Korolev *et al.* (2011) showed that some events from the short interarrival time may be associated with intact particles, whereas the long interarrival mode may contain shattered particles. The first situation occurs if the arrival of the intact particle coincides with the shattering event. The second case occurs if only one fragment from a shatter group ends up intersecting the sample volume, while the other group-fragments pass outside. The interarrival times of these single-fragment particles will be indistinguishable from the natural particles.

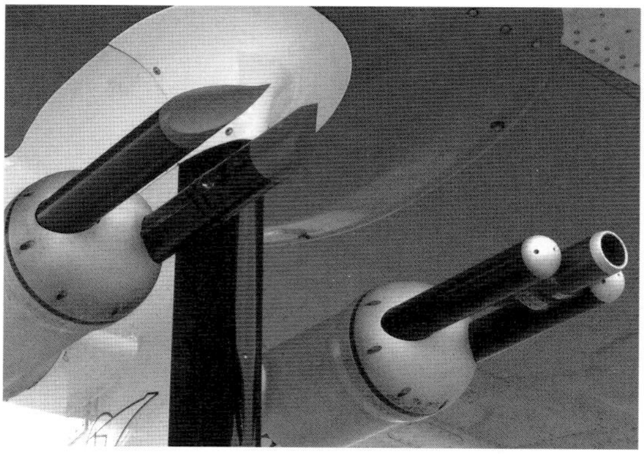

Figure 6.23 Modification of the FSSP housing to mitigate the shattering effect. The standard FSSP is mounted on the right side (NRC Convair 580). (Courtesy of A. Korolev.)

6.4.1.4.3 **Methods of Mitigating Shattering** The shape of arm tips and inlets plays a critical role in the effect of shattering on ice particle measurements. The shape of the tips defines the area that deflects particles toward the probe's sample volume, the angle of impact with particles, and the direction in which the particles are scattered after bouncing from the probe's surface. The angle of impact also determines the exchange of kinetic energy between the ice particle and the tip and thus affects the number of fragments generated by the collision.

One of the possible ways to mitigate shattering is to minimize the surface area of tips or inlets that may bounce particles into the sample area. The modified tips should deflect bouncing particles and shed water away from the sample volume and optical field apertures (Korolev, 2011).

An example of one such modification for the FSSP is shown in Figure 6.23, where the inlet tube of the modified FSSP was removed and the original hemispherical tips were replaced with new designs.

Comparisons of the simultaneous measurements of cloud particles by the probes with the standard and modified housing configurations showed that the modified tips can effectively mitigate the shattering effect (Korolev *et al.* 2011).

Figure 6.24 shows comparisons of simultaneous measurements between standard and modified OAP-2DCs, Figure 6.25 shows similar plots for the 2D-S probes. It is evident that the probes with the standard tips in this case show a larger number of small particles not observed in probes with modified tips. A comparison of the performances of the modified tips and the antishattering algorithms by Korolev *et al.* (2011) suggests that the modified tips are much better at mitigating shattering than postprocessing algorithms for the OAP-2DC and CIP. However, for the 2D-S probe with a superior pixel resolution (10 μm) and faster response time, the efficiency of the arrival time algorithm becomes comparable or better than that of the antishattering tips (Lawson, 2011). Altogether, both the antishattering tips and the postprocessing algorithms are important steps toward mitigating shattering effects in the final data set, and ideally, they should be used together if possible.

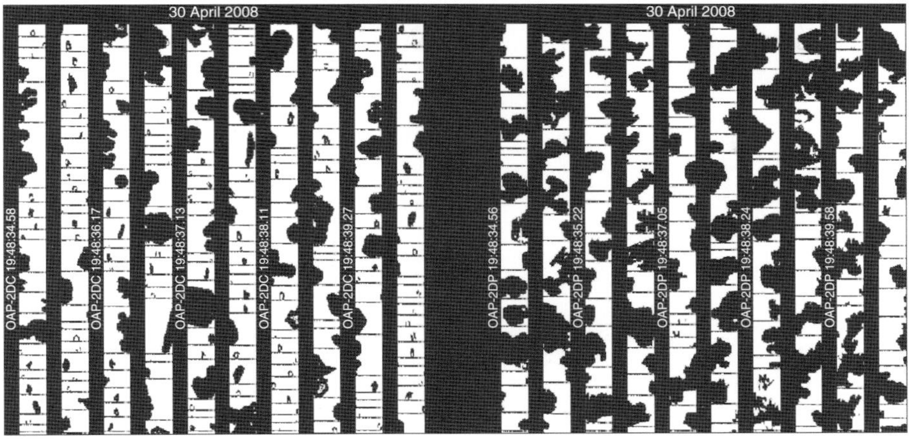

Figure 6.24 Comparisons of the ice particles images measured by two OAP-2DCs with the standard (left) and modified tips (right). The images on the left have many more small particles relative to those on the right. The majority of the small particles on the left result from the probe tip shattering. The pixel resolution of both OAP-2DCs is 25 μm. The image sampling was conducted in a frontal Cs-Ns cloud system in the vicinity of Fairbanks, Alaska, on 30 April 2008. (Courtesy of A. Korolev.)

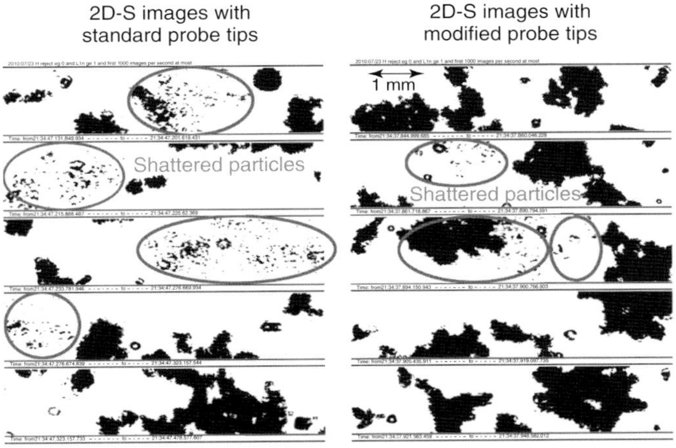

Figure 6.25 Examples of 2D-S images from two 2D-S probes flown side by side on a Learjet underneath a precipitation anvil on 23 July 2010. One probe had standard 2D-S probe tips (left) and (right) the other with tips modified to reduce the effects of shattering. (Adapted from Lawson (2011).)

6.4.2
Bulk Cloud Sampling

Two types of instruments are described here for bulk measurements of (a) the LWC and IWC and (b) the chemical composition of the cloud water.

6.4.2.1 Cloud Water Content – Inlet-Based Evaporating Systems

The water content of clouds can be measured by various techniques, which are described in Section 5.4. Here, we cover those systems that sample cloud water via inlets (the sampling characteristics of inlets are described in Section 6.3.2), evaporate the condensed water, and determine subsequently the amount by means of a hygrometer. Two types of inlets are used for that purpose: the first are heated forward-facing inlets as described in Section 6.3.2 and the second is a CVI (also see Section 6.3.4). The difference between these inlets is that the CVI measures solely the condensed-phase cloud particles and excludes the water vapor from the gas phase, while a forward-facing inlet samples both, the condensed and the gas-phase water (= total water).

To determine the LWC/IWC (i.e., LWC and/or IWC; LWC/IWC is used throughout the text) from a total water measurement, it is necessary to measure the gas-phase water independently. This can be carried out by using the same instrument but with a backward- facing inlet, sampling only the gas-phase water, or an open-path laser absorption hygrometer May (1998). LWC/IWC is then calculated by taking the difference between the instrument water vapor content after hydrometeor evaporation and the independently measured free stream water vapor content.

When measuring the IWC of cirrus clouds, the forward-facing inlet has to be designed taking into account that the water signal from the ice is often small compared with the signal of the gas phase. Thus, the inlet should be optimized to sample strongly sub-isokinetically (see Section 6.3.2.1) so that the concentration of large ice crystals is intensely enhanced. This can be done by reducing the flow velocity in the inlet (U) to very low values in comparison to the aircraft cruising speed (U_0) as shown in Figures 6.5 and 6.6. But, as the particles are not sampled isokinetically, adjustments must be made to account for the difference in particle collection efficiencies with size. To this end, either the size distribution of the cloud particles has to be known or, because the collection efficiencies reach a constant value for particles larger than a certain size (Figure 6.6), the inlet should be designed to sample only near the particle size range of this constant value, such as that for a CVI.

To avoid a falsification of the measurement of LWC/IWC as shown in Section 6.2.2 (Figure 6.3), forward-facing inlets should ideally have a strut length exceeding the particle shadow and enhancement zones. The walls of the inlet tip should be as thin as possible to avoid a contamination of the IWC from ice crystals impinging on the inlet wall, where they may possibly shatter and bounce into the inlet (for more detail on ice particle bouncing and shattering, see Section 6.4.1).

For all inlet-based evaporating systems, hydrometeors enter an inlet exposed directly to the air stream. The inlet acts as a deep capture volume, so that the bouncing and reentrainment observed in hot-wire devices (Emery *et al.* 2004; Strapp *et al.* 2005; Korolev *et al.* 2008) is minimized, providing a measurement advantage for large water drops (Twohy, Strapp, and Wendisch, 2003) and ausstreichen also for large ice particles. As mentioned above, LWC/IWC is deduced from enhanced humidity due to hydrometeor evaporation, usually measured by an hygrometers such as a Lyman–α or TDL (Tunable Diode Laser) absorption hygrometer. Ruskin (1965) described an isokinetic flow-through device with a 1-cm inlet, in which captured hydrometeors would melt and evaporate due to warming by heating elements and impaction with fine screens. The pressure inside the inlet was kept at free stream static air pressure, so that all particles should be captured with near 100% efficiency, an unachievable performance for hot-wires. Similar devices were described by Kyle (1975) and Nicholls, Leighton, and Barker, (1990), where the former was used for measurements of heavy rain in convective clouds and the latter primarily for measurements of ice clouds. Residence times for isokinetic devices can be quite short, of the order of milliseconds, so the design must be such that hydrometeors melt and evaporate efficiently before reaching the exhaust and without accumulating within the system. Design details vary according to the application.

The fast *in situ* stratospheric hygrometer (FISH) is based on the Lyman photofragment fluorescence technique (Zöger *et al.* 1999; Schiller *et al.* 2008). The FISH is sensitive for H_2O mixing ratios from approximately 0.5–1000 ppmv (bitte!) and thus well suited for investigations in the upper troposphere and lower stratosphere over a large dynamical range. The time resolution is 1 s, determined by the exchange time of air through the measurement cell. The air enters by a heated forward-facing inlet tube mounted outside the research aircraft sampling total water, that is, the sum of both gas-phase molecules and ice particles. The large ice particles are sampled with enhanced efficiency compared with the gas molecules; see Schiller *et al.* (2008) and also Figure 6.6. The aspiration coefficient (or enhancement factor) of the inlet increases from approximately one for particles with diameter smaller than 0.6 μm to its maximum value E_{max}, which is typically achieved for particle radii larger than 2–4 μm. The value of E_{max} depends on the air density, temperature, and velocity and can vary between 3 and 10. Davis *et al.* (2007) described the closed-path TDL hygrometer (CLH), another sub-isokinetic sampler. CLH measures H_2O mixing ratios larger than 50 ppmv. The inlet of CLH is designed to have an Emax of about 50 to enhance the small water concentrations of ice clouds well above the lower detection limit.

Weinstock *et al.* (2006) describe the Harvard total water hygrometer, which is an isokinetic evaporator specifically designed and calibrated for low-IWC cirrus measurements down to several milligrams per cubic meter. The isokinetic device of Davison, MacLeod, and Strapp (2009) and Davison and MacLeod (2009) was designed for high-IWC environments of up to 10 g/m^3 at 200 m/s and incorporates a helical long-path evaporation coil where particles are centrifugally forced to the heated wall for deceleration, melting, and evaporation; also see Figure 6.13. The flow system avoids blockage surfaces such as screens where partially melted particles can accumulate.

The CVI, developed to measure evaporated hydrometeor residuals (Section 6.3.4), has also been used to measure LWC/IWC (Ogren, Heintzenberg, and Charlson 1985; Noone *et al.* 1988; Noone *et al.* 1993; Twohy, Schanot, and Cooper 1997).

Hydrometeors of sufficient inertia ($\propto$ size) enter the inlet against a counterflow and travel into an evaporation chamber with an initially dry carrier gas. The humidity subsequently measured in the device is solely due to hydrometeor evaporation and independent of the free stream humidity. The CVI has a detection limit as low as $1\,mg/m^3$ but suffers from hysteresis at relatively high water contents. Owing to the counterflow, CVI instruments typically have a cut-size of the order of $5-10\,\mu m$, below which particles are not detected efficiently. This is not a major limitation in most clouds, except for polluted clouds/fogs with many small droplets. The cloud spectrometer and impactor is a CVI inlet coupled with electronics and TDL detector in a large canister that can fly externally to the aircraft.

6.4.2.2 **Chemical Composition of Cloud Water – Bulk Sampling Systems**

6.4.2.2.1 **Ice Clouds** In recent years, measuring nitric acid (HNO_3) in ice has been one focus of research in the upper troposphere and lower stratosphere. As at the low temperatures prevailing at these altitudes HNO_3 occurs not only in the gas phase, and also in the aerosol particles and ice crystals, separating HNO_3 in the ice needs a special sampling strategy (comparable to the above-mentioned combination of a forward/backward inlet system to measure the IWC in cirrus clouds). To accomplish this, two inlets are used simultaneously: one samples the gas phase plus the larger ice crystals, while the other samples the gas phase plus the smaller aerosol particles. The difference between the two yields the ice-bound HNO_3.

NOAA's "football" inlet (Kelly *et al.* 1989; Dhaniyala *et al.* 2004) is designed for that purpose, but this sampler has two forward-facing inlets housed on a prolate spheroid-shaped blunt body (which resembles an American football, Figure 6.26a). One of the inlets is placed at the front end of the body, while the other inlet is located near the aft end. Both the inlets sample strongly sub-isokinetically to enhance the low amount of HNO_3 in the particles, as already suggested above for IWC detection. The front inlet samples particles of all sizes with varying enhancement factors depending on the ambient pressure (Dhaniyala *et al.* 2004), while the sampling characteristics of the rear inlet is influenced by the shape of the housing. Appropriate shaping of the blunt-body housing can result in elimination of particles larger than a certain cut-size from the rear inlet. CFD calculation results of the enhancement factor of the two inlets on the football geometry are shown in Figure 6.26b. One should be aware, however, that for large angles of attack, flow separation upstream of the rear inlet might occur. In this case, the rear inlet might sample air from the football's boundary layer. In addition, the impaction and shatter of larger particles/droplets on the front end of such a sampler might contaminate the rear inlet. Both the problems of flow separation and shatter-particle contamination might be minimized or eliminated by the appropriate use of boundary-layer suction on the blunt-body housing.

A forward/backward inlet system to detect HNO_3 in ice is described in the study by Voigt *et al.* (2005). Here, the aerosol particles plus gas phase are sampled with a backward-facing inlet by taking advantage of the fact that the larger ice

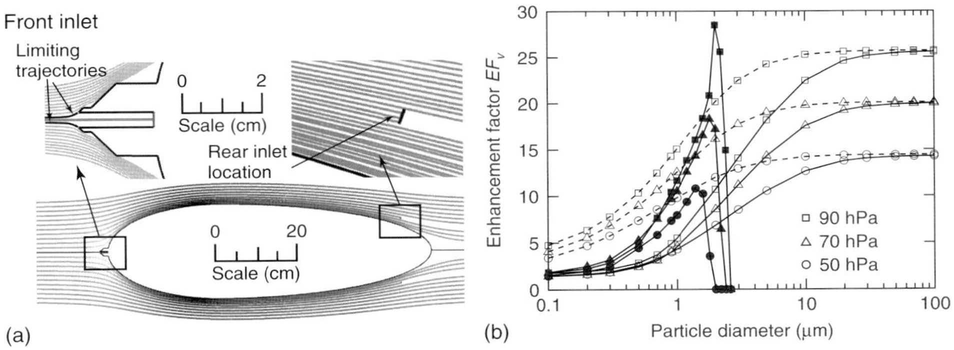

Figure 6.26 (a) Geometry of the football inlet showing the location of the front and rear inlets. (b) Calculated enhancement factor EF (= aspiration efficiency) as a function of particle size for the rear (solid lines with solid symbols) and front (solid lines with open symbols) inlets for varying ambient pressures. In comparison, corresponding empirical enhancement values for a thin-walled inlet (dashed lines) are shown: the presence of the blunt body results in lower particle enhancements for sizes smaller 10 μm (Dhaniyala *et al.* 2004). (Copyright 2004 Mount Laurel, N.J. Reprinted with permission.)

crystals are not able to follow a 360° turn of the streamlines. The forward-looking inlet is thin-walled. Thus, the problems of flow separation and shatter-particle contamination discussed for the football inlet system might be reduced.

6.4.2.2.2 **Liquid Clouds** To evaluate the chemical composition of liquid cloud water, the cloud particles may be sampled and stored in a closed vessel for subsequent off-line chemical analysis, such as ion chromatography of the soluble substances or analysis of the insoluble particles (size distribution and/or elemental composition by electron microscopy).

This has been realized by using, among others, the slotted rod sampler (Mohnen 1980; Huebert, Vanbramer, and Tschudy, 1988), the isokinetic cloud probe system (ICPS) (Maser *et al.* 1994), or the wide-stream impaction cloud water collector (WICC) (Brinkmann *et al.* 2001). As an example, the WICC, which is mounted on top of the aircraft's fuselage, is shown in Figure 6.27 (bottom plate: cross section of the whole sampler). WICC is a virtual impactor like the CVI, but with a gutter-shaped collection surface facing the air flow (left top plate: vertical projection of the collection area). The cloud droplets are sampled by inertial impaction (the surface determining the impaction characteristics are the virtual plate closing the gutter) into a quasi stagnant layer and then run down into the sampling vessel by gravitational settling. The vessel is mounted inside the aircraft cabin so that several samples could be taken during a flight by manual replacement of the vials. The outer design of the WICC is wing shaped to prevent generation of turbulence by the collector. To prevent sample contamination while flying in cloud-free air, the gutter can be revolved out of the stream during the flight from inside the cabin (see right top plate).

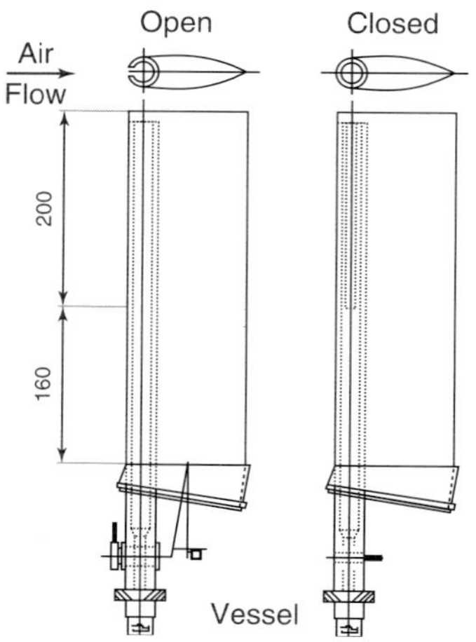

Figure 6.27 Wide-stream impaction cloud water collector (WICC). Bottom: cross section of the whole sampler; top: vertical projection of the collection area; left: collection area in sampling position (open gutter); right: collection area in nonsampling position (gutter closed); free stream direction of the air as indicated. Dimensions in millimeters (Brinkmann *et al.* 2001). (Copyright 2001 Wiley-VCH Verlag GmbH&Co. KGaA. Reprinted with permission.)

6.5 Summary and Guidelines

When intending to observe aerosol or cloud particles from aircraft, the first step is to select the appropriate instrument (Chapters 4 and 5).

Once the instrument is chosen, the new researcher will quickly notice that it is not sufficient to know the capability of the measuring device, but one must also know how to obtain a representative particle measurement on the high-speed aircraft platform.

Chapter 6 summarizes the art and the issues of aerosol or cloud particle sampling and we recommend in the following approach to attain the best possible measurement. First, choose carefully the position of your instrument or its inlet by taking into account the flow field around the aircraft and the trajectories of the particles in the desired size range (Section 6.1). If this information is not available and you do not have the capacity to perform CFD calculations, a compromise could be to follow guidelines derived for an aircraft comparable in shape and air speed. Next, the sampling characteristics of the instrument or its inlet (Section 6.3) should be determined, either by using empirical equations, CFD modeling, wind tunnel experiments, or in-flight intercomparisons. Note that in case an inlet is needed for air supply to the instrument, its sampling characteristics are composed of the aspiration and transmission efficiency, as well as the transport efficiency through the sampling line. Furthermore, in case large cloud particles are to be sampled (Section 6.4), the instrument should be designed in a way that shattering of the particles while sampling should be avoided or at least minimized. Last, during the analysis of particle measurements, all sources of possible uncertainties, also including errors in the measurement itself (Chapters 4 and 5), should be taken into account, applied corrections should be named, and error bars should always be given.

In conclusion, without a good position of an instrument on the aircraft and without the knowledge of its sampling characteristics and subsequent thorough data analysis, the measured particle concentrations could diverge from the true concentrations by up to several hundred percent. In most cases, the larger the particles being measured, the higher is the probability for large errors. However, it should be emphasized that overcoming the challenge of performing reliable airborne measurements makes the actual sampling even more exciting!

7
Atmospheric Radiation Measurements

Manfred Wendisch, Peter Pilewskie, Birger Bohn, Anthony Bucholtz, Susanne Crewell, Chawn Harlow, Evelyn Jäkel, K. Sebastian Schmidt, Rick Shetter, Jonathan Taylor, David D. Turner, and Martin Zöger

7.1
Motivation

Most of the energy exchange within the universe is realized by electromagnetic radiation. For our home planet Earth, the electromagnetic radiation incident at the top of atmosphere (TOA) is the only significant source of energy for almost all processes, exceeding the next largest source, radioactive decay from the Earth's interior, by approximately four orders of magnitude (Sellers, 1965). Solar radiation powers the complex and tightly coupled circulation dynamics, chemistry, and interactions among the atmosphere, oceans, cryosphere, and land that maintain the terrestrial environment. As a consequence, all processes that modify the distribution and deposition of electromagnetic radiation within the atmosphere and at the Earth's surface are important for climate (the consequence of long-term energy balance), weather (driven by short-term energy imbalance), and the regulation of the biosphere.

Solar radiation is scattered and absorbed within the atmosphere and at the Earth's surface. Scattering is a directional redistribution of the radiative energy. Radiation that is absorbed at the surface and within the atmosphere is converted into thermal energy, which determines the rate and spectral composition of infrared (IR) emission. Thus, solar radiation is the source of almost all the IR radiation emitted by the surface and atmosphere. Without the Sun, the Earth's surface temperature would be approximately 30 K, determined primarily by the heat flux from its interior (Davies and Davies, 2010), and Earth's emission almost 10 000 times less, derived from the Stefan–Boltzmann law of emission (Eq. (7.35)). Within the atmosphere, IR-active gases – those which absorb and emit IR radiation, sometimes called greenhouse gases – are responsible for warming the surface by approximately 33 K. A detailed discussion on Earth's radiative energy budget, including updated estimates of all the components, can be found in the study by Trenberth and Fasullo (2011).

In addition to fundamental energy balance, solar radiation is important for remote sensing – the measurement of one physical variable to derive another – of

Airborne Measurements for Environmental Research: Methods and Instruments, First Edition.
Edited by Manfred Wendisch and Jean-Louis Brenguier.

the atmosphere. Many active and passive ground-based, airborne, and space-borne methods of remote sensing of physical and chemical atmospheric and surface properties are based on the propagation of solar and terrestrial radiation through the atmosphere and its interaction with the surface. The measurement of electromagnetic radiation is the most common method of atmospheric remote sensing.

7.2 Fundamentals

In this section, we introduce basic radiative quantities. More detailed introductions of fundamentals of atmospheric radiative transfer can be found in specialized textbooks (Wendisch and Yang (2012) and references therein).

7.2.1 Spectrum of Atmospheric Radiation

A regularly oscillating charge produces a harmonic electromagnetic wave, which is characterized by the frequency of oscillation ν in units of $\text{Hz} = \text{s}^{-1}$ or the wavelength λ in units of m. ν is defined by the number of cycles of oscillations per second, and λ is the distance between the maximum amplitudes of the harmonic electromagnetic wave during one cycle of oscillation. Both quantities are linked by the following relation:

$$\nu = \frac{c}{\lambda} \tag{7.1}$$

In Eq. (7.1), the oscillating harmonic electromagnetic wave propagates through a vacuum with the speed of light of $c = 2.997925 \times 10^8\ \text{m s}^{-1}$. Alternatively to ν, the electromagnetic wave can be described by the wavenumber $\tilde{\nu}$ in units of m^{-1} or more commonly in cm^{-1}:

$$\tilde{\nu} = \frac{1}{\lambda} \tag{7.2}$$

In the atmosphere, there is an abundance of both periodically oscillating and randomly moving electric charges (e.g., oscillating molecules of atmospheric gases and moving electrons, atoms, and ions of liquid and/or solid atmospheric particles) generating electromagnetic waves with a wide range of frequencies. The frequencies of electromagnetic waves in the atmosphere range from a few cycles per second to more than 10^{26} Hz. The wavelengths may vary from hundreds of thousands of kilometers to the diameter of an atomic nucleus. These electromagnetic waves are ubiquitous; their superposition in the atmosphere comprises atmospheric radiation.

Radiation at one wavelength is called monochromatic, while electromagnetic radiation dispersed over narrow wavelength intervals, or bands, is called spectral radiation. Electromagnetic radiation at several noncontiguous wavelength bands

Table 7.1 Wavelength, frequency, and wavenumber for certain ranges of electromagnetic radiation.

Name of range	Wavelength λ	Frequency $\nu = c/\lambda$	Wavenumber $\tilde{\nu} = 1/\lambda$ (cm^{-1})
Ultraviolet (UV)	10–370 nm	30 000–810 THz	10^6–27 027
UV-C	10–280 nm		
UV-B	280–315 nm		
UV-A	315–370 nm		
Visible (VIS)	370–750 nm	810–400 THz	27 027–13 333
Near-infrared (NIR)	750–2 000 nm	400–150 THz	13 333–5000
Thermal infrared (TIR)	3–50 μm	100–6 THz	3333–200
IR	0.002–1 mm	150 THz–300 GHz	5000–10
Microwave (MW)	0.3 mm–30 cm	1 000–1 GHz	33–0.033
Solar	0.2–5 μm	1 500–60 THz	50 000–2000
Terrestrial	>5 μm	<60 THz	<2000

comprise a discrete spectrum. A continuous spectrum of radiation spans an infinite number of wavelengths. Atmospheric radiation is considered to be nearly continuous. Solid and liquid substances consist of atoms and ions, which move randomly, at a broad spectrum of frequencies, and emit an almost continuous spectrum of radiation. Atmospheric gases, however, mainly consist of periodically oscillating and moving molecules (at fixed frequencies) that emit radiation in discrete spectral lines or a conglomerate of discrete spectral lines that form a spectral band. In addition, emission of electromagnetic radiation from the Sun is modulated by absorption and emission within the Sun's photosphere and chromosphere and by absorption and scattering by the terrestrial atmosphere but is otherwise a continuous spectrum of radiation.

The full range of wavelengths, frequencies, and wavenumbers of atmospheric electromagnetic radiation is classified into several categories; those which are relevant to the topics in this book are summarized in Table 7.1.

For atmospheric, electromagnetic radiation, we make a primary distinction between the two major spectral ranges: solar ($0.2\,\mu m < \lambda \leq 5\,\mu m$) and terrestrial ($\lambda > 5\,\mu m$). The thermal infrared (TIR) spectral range ($3\,\mu m < \lambda < 50\,\mu m$) is also used. Regardless of the nomenclature, it is best to identify the spectral range for any application to avoid confusion that often occurs among different disciplines.

7.2.2 Geometric Definitions

In atmospheric radiative transfer, it is most convenient to use spherical polar coordinates to define direction (Figure 7.1). The zenith angle θ is the angle between an arbitrary direction and the vertical base vector $\widehat{\mathbf{e}}_3$ of the Cartesian coordinate

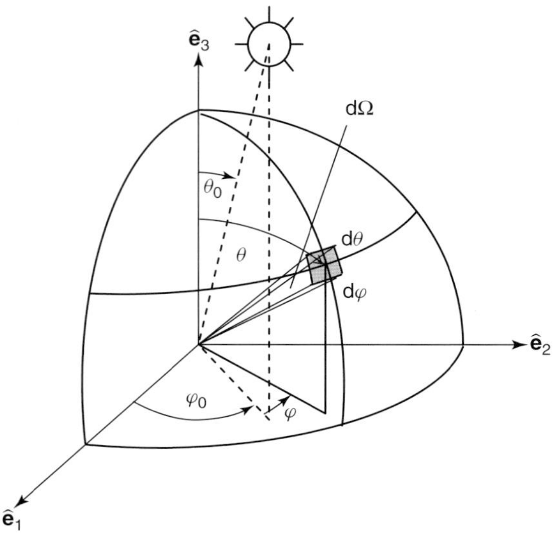

Figure 7.1 Illustration of definition of spherical polar coordinates and solid angle. The vectors $\widehat{\mathbf{e}}_1$, $\widehat{\mathbf{e}}_2$, and $\widehat{\mathbf{e}}_3$ represent the base vectors of the Cartesian coordinate system.

system given by the basis vectors $\widehat{\mathbf{e}}_1$, $\widehat{\mathbf{e}}_2$, and $\widehat{\mathbf{e}}_3$. In the zenith (overhead) direction, $\theta = 0^\circ$; in the direction toward the horizon, $\theta = 90^\circ$. The nadir direction is defined by $\theta = 180^\circ$. It is often more convenient to use zenith angle distance $\mu = \cos\theta$. The azimuth angle φ is the angular distance from a reference in the plane perpendicular to zenith. It is defined counterclockwise from a reference such that $0^\circ \le \varphi < 360^\circ$. In this book, we use the projection of the solar direction on the $(\widehat{\mathbf{e}}_1, \widehat{\mathbf{e}}_2)$ plane as the reference. The solar direction (θ_0, φ_0) is indicated by subscript "0."

In addition, we introduce the solid angle Ω, with the unit of "steradian" (abbreviation sr), to define a set of directions encompassed by an area projected onto the unit-sphere. Solid angle is a two-dimensional analogue to the planar angle. The differential solid angle element $\mathrm{d}\Omega$ is given by the following equation (Figure 7.1):

$$\mathrm{d}\Omega = \sin\theta \, \mathrm{d}\theta \, \mathrm{d}\varphi \tag{7.3}$$

7.2.3
Vertical Coordinate: Optical Depth

Instead of the geometrical altitude z, we use the vertical optical depth $\tau(\lambda, z)$ at the given wavelength λ as a transformed vertical coordinate. τ is a measure of the opacity of the atmosphere and is defined as

$$\tau(\lambda, z) = \int_z^{z_{\mathrm{TOA}}} b_{\mathrm{ext}}(\lambda, z') \, \mathrm{d}z' \tag{7.4}$$

Optical depth is the vertical integral over altitude from z to the TOA of the volume extinction coefficient, $b_{\mathrm{ext}}(\lambda, z)$, a quantity that gives the attenuation due to absorption and scattering by atmospheric gases and particles. For more details on the determination of $b_{\mathrm{ext}}(\lambda, z)$, see the textbook by Wendisch and Yang (2012). For the case of $z = 0$ in Eq. (7.4), the optical depth is called the optical thickness. The optical depth can be expanded into the individual components that contribute to the attenuation solar radiation:

$$\tau(\lambda, z) = \tau_{\mathrm{p}}(\lambda, z) + \tau_{\mathrm{Rayl}}(\lambda, z) + \sum_i \tau_{\mathrm{abs},i}(\lambda, z) \tag{7.5}$$

where τ_{p} represents the optical depth of particles (scattering and absorption), τ_{Rayl} is the optical depth due to molecular scattering (also called Rayleigh scattering optical depth). $\tau_{\mathrm{abs},i}$ represents the absorption optical depth of gases (primarily water vapor, oxygen, ozone, carbon dioxide, methane, and nitrogen dioxide) with subscript i indicating the individual gas component.

7.2.4 Quantitative Description of Atmospheric Radiation Field

The following fundamental quantities are introduced to quantify the radiation field in the atmosphere. We begin with the radiant energy, E_{rad}, in units of Joules (J). This is a physical quantity that can be measured applying first principles and using established methods. E_{rad} provides the basis for derived radiation quantities. We define the radiant energy flux, or radiant power, Φ, in units of Watts ($1\,\mathrm{W} = 1\,\mathrm{J\,s^{-1}}$) to be

$$\Phi(\tau) = \frac{\mathrm{d}E_{\mathrm{rad}}}{\mathrm{d}t} \tag{7.6}$$

where t represents time. The radiant energy flux density, or irradiance, F (often incorrectly called flux), with units of $\mathrm{W\,m^{-2}}$ is

$$F(\tau) = \frac{\mathrm{d}\Phi}{\mathrm{d}A} = \frac{\mathrm{d}E_{\mathrm{rad}}}{\mathrm{d}t\ \mathrm{d}A} \tag{7.7}$$

F represents the radiant energy per unit time (radiant flux) and per unit area transported by electromagnetic radiation through a plane or deposited on a surface. A differential element of area in this plane is given by $\mathrm{d}A$, and its orientation with respect to the radiation field is arbitrary. Thus, F represents the rate of radiant energy transport (radiant flux) per unit area. Radiance I, with units of $\mathrm{W\,m^{-2}\,sr^{-1}}$, is defined as

$$I(\tau, \theta, \varphi) = \frac{\mathrm{d}\Phi}{\mathrm{d}A_{\perp}\ \mathrm{d}\Omega} = \frac{\mathrm{d}\Phi}{\cos\theta\ \mathrm{d}A\ \mathrm{d}\Omega} = \frac{1}{\cos\theta}\frac{\mathrm{d}F}{\mathrm{d}\Omega} \tag{7.8}$$

with $\mathrm{d}A_{\perp}$ representing the differential area perpendicular to the direction of propagation. Integration of Eq. (7.8) over the hemisphere yields the relationship between irradiance and radiance:

$$F(\tau) = \int_{2\pi} I(\tau, \theta, \varphi) \cdot \cos\theta\ \mathrm{d}\Omega \tag{7.9}$$

In atmospheric applications, the reference unit area is usually defined as horizontal. Therefore, upward $F^{\uparrow}$ and downward $F^{\downarrow}$ irradiances are obtained from radiance I by applying Eqs. (7.3) and (7.9):

$$F^{\downarrow}(\tau) = \int_0^{2\pi} \int_{\pi}^{\pi/2} I(\tau, \theta, \varphi) \cdot \cos\theta \cdot \sin\theta \, d\theta \, d\varphi \tag{7.10}$$

and

$$F^{\uparrow}(\tau) = \int_0^{2\pi} \int_0^{\pi/2} I(\tau, \theta, \varphi) \cdot \cos\theta \cdot \sin\theta \, d\theta \, d\varphi \tag{7.11}$$

Net irradiance, F_{net}, is the difference between upward and downward irradiance:

$$F_{\text{net}} = F^{\downarrow} - F^{\uparrow} \tag{7.12}$$

The dimensionless albedo, α, is the ratio of upward to downward irradiance:

$$\alpha = \frac{F^{\uparrow}}{F^{\downarrow}} \tag{7.13}$$

For the special case of isotropic radiation (radiance independent of direction), $I(\theta, \varphi) = I_{\text{iso}} = \text{constant}$, we obtain

$$F_{\text{iso}}^{\downarrow} = \pi \cdot I_{\text{iso}} = F_{\text{iso}}^{\uparrow} \tag{7.14}$$

Actinic flux density F_{act}, sometimes called average intensity, is the integral of radiance over solid angle:

$$F_{\text{act}}(\tau) = \int_0^{2\pi} \int_0^{\pi} I(\tau, \theta, \varphi) \cdot \sin\theta \, d\theta \, d\varphi \tag{7.15}$$

Like irradiance, actinic flux density has units of W m^{-2}. It represents the energy flux on a unit sphere, normalized by the cross section of the sphere and therefore is related to flux divergence.

Irradiance, F, radiance, I, and actinic flux density, F_{act}, can be either spectral or band-integrated (broadband) quantities. For example, we define the spectral flux density F_{λ} by

$$F_{\lambda} = \frac{dF}{d\lambda} \tag{7.16}$$

The spectral flux density F_{λ} is the irradiance per unit wavelength interval. As a result, the units of a spectral radiant energy quantity contains an additional term nm^{-1}. The band-integrated flux density $F(\lambda_1, \lambda_2)$ includes radiant energy contributions from wavelengths within a wavelength interval (λ_1, λ_2). Thus, the band-integrated flux density $F(\lambda_1, \lambda_2)$ is obtained by

$$F(\lambda_1, \lambda_2) = \int_{\lambda_1}^{\lambda_2} F_{\lambda} \, d\lambda \tag{7.17}$$

For example, the solar irradiance is defined by $\lambda_1 = 0.2\,\mu m$, $\lambda_2 = 5\,\mu m$ (Table 7.1). In a similar way, spectral (index λ) and band-integrated radiance and actinic flux density are defined.

Irradiance and radiance may be divided into the contributions from scattering (diffuse) and direct transmission indicated by the subscripts "dif" and "dir", respectively. The sum of both corresponds to the total irradiance or radiance, with the subscript "tot." Occasionally, the total radiation component is called global; however, we omit this term here. For example, for the total irradiance, we obtain

$$F_{tot} = F_{dir} + F_{dif} \tag{7.18}$$

Directly transmitted solar radiation does not interact with the atmospheric components on the path from the Sun to the receiver. It is the solar radiation that has been neither scattered nor absorbed. Diffuse solar radiation arises from photons that have been scattered; terrestrial emission can be neglected in the solar spectral wavelength range. It should be noted that radiation scattered in the forward direction is indistinguishable from directly transmitted radiation. Measurements of F_{dir} (see Section 7.3.4.2.1 on Sun photometry) may need to account for errors due to forward scattering. For the terrestrial radiation, the direct portion of radiation (emission from the Sun) and scattering are usually neglected and it is assumed that all radiation results from terrestrial emission. However, there is no fixed wavelength where solar radiation ends and terrestrial radiation begins.

7.2.5 Basic Radiation Laws

7.2.5.1 Lambert–Bouguer Law

The Lambert–Bouguer law (sometimes known as Beer's law) describes the exponential attenuation of directly transmitted solar spectral radiation, as it penetrates into the atmosphere. This is the simplest form of the equation of radiative transfer, with no sources (scattering and emission; e.g., Wendisch and Yang (2012)). If applied to the directly transmitted solar spectral irradiance, $F_{dir,\lambda}$, we obtain

$$F_{dir,\lambda}(\tau) = F_{dir,\lambda,TOA} \cdot \exp\left[-\frac{\tau(\lambda, z)}{\mu_0}\right] \tag{7.19}$$

$F_{dir,\lambda,TOA}$ represents the extraterrestrial direct solar spectral irradiance incident at the TOA. To account for the changing Earth–Sun distance, we include a quadratic distance factor and rewrite Eq. (7.19) as a function of z instead of τ:

$$F_{dir,\lambda}(z) = \left(\frac{\overline{R}_{S-E}}{R_{S-E}}\right)^2 \cdot \overline{F}_{dir,\lambda,TOA} \cdot \exp[-\tau(\lambda, z) \cdot M(\theta_0, z)] \tag{7.20}$$

$\overline{R}_{S-E} = 149.6 \times 10^9$ m is the average Earth–Sun distance, or 1 AU (astronomical unit), and R_{S-E} is the actual Earth–Sun distance, which can be calculated from astronomical ephemeris information. $M(\theta_0, z)$ represents the relative optical air mass at the given solar zenith angle, θ_0, and altitude, z. $\overline{F}_{\text{dir},\lambda,\text{TOA}}$ is the direct solar irradiance at the TOA at the mean Sun–Earth distance.

The relative optical airmass, $M(\theta_0, z)$, is defined as the optical path length at altitude z for a given solar zenith angle, θ_0, relative to the optical path length in the vertical direction. When the Sun is directly overhead, the relative optical airmass is equal to 1, and for most solar zenith angles ($\theta_0 < 60^\circ$), the relative optical airmass can be approximated to a high degree of accuracy by

$$M(\theta_0, z) = \frac{1}{\cos\theta_0} = \frac{1}{\mu_0} \tag{7.21}$$

However, for increasingly large solar zenith angles ($\theta_0 \geq 60^\circ$), as the Sun approaches the horizon, the optical path length diverges from (becomes less than) Eq. (7.21) due to the sphericity of the Earth's atmosphere. Other correction must be made for altitude-dependent effects of refraction. Various formulae and approximations have been developed to account for these effects and can be found in the literature (Kasten and Young, 1989). It has also been shown (Thomason, Herman, and Reagan, 1983) that the airmass is dependent on the vertical distribution of the attenuators in the atmosphere and that for highest accuracies, the airmass for each attenuating agent (aerosol particles and various gas species) should be determined individually.

7.2.5.2 **Planck Law**

An object with temperature T will generally emit radiation at all wavelengths (or frequencies or wavenumbers). However, there is an upper bound of that emitted radiation, which is quantified by the Planck law, or Planck function, given here in units of spectral radiance:

$$B_\lambda(T) = \frac{2\,h\cdot c^2}{\lambda^5}\,\frac{1}{\exp[h\cdot c/(k_\text{B}\cdot\lambda\cdot T)] - 1} \tag{7.22}$$

where $h = 6.6262 \times 10^{-34}$ J s is the Planck constant and $k_\text{B} = 1.3806 \times 10^{-23}$ J K^{-1} represents the Boltzmann constant. Planck's law as a function of frequency is

$$B_\nu(T) = \frac{2\,h\cdot\nu^3}{c^2}\cdot\frac{1}{\exp[h\cdot\nu/(k_\text{B}\cdot T)] - 1} \tag{7.23}$$

A blackbody is a hypothetical body that absorbs all the incident radiation (hence the term black), and in all wavelength bands and directions, the maximum possible emission is realized. Blackbody radiation is isotropic; the radiation emitted by a blackbody depends only on its temperature. Therefore, any radiance can be inverted to get a temperature of an equivalent blackbody, the brightness temperature T_B, defined later in Eq. (7.30).

The Planck function for a specific temperature peaks at a wavelength that is inversely proportional to temperature. The wavelength for which the Planck function attains its maximum can be determined by finding where its first derivative is zero. As a result, Wien's displacement law relates temperature to the wavelength of maximum radiance for the wavelength form of the Planck function:

$$\lambda_{max} = \frac{k_W}{T} \tag{7.24}$$

where $k_W = 2897\,\mu m\,K$ is Wien's constant. Solar radiation (Sun with $T_S = 6000\,K$) peaks at a wavelength of $\lambda_{max} = 0.48\,\mu m$. For typical terrestrial temperatures of 200–300 K, the wavelength Planck function peaks between 9.6 and 14.4 µm.

The Rayleigh–Jeans approximation, which is important in microwave remote sensing ($\lambda > 0.3\,mm$), is the limiting form of the Planck distribution at long wavelength ($\lambda \to \infty$) or low frequency ($\nu \to 0$). Actually, it is the product of λ and T (or the ratio of ν and T) that matters. However, we do not assume extreme temperatures here. For $\lambda \to \infty$ ($\nu \to 0$), Planck's law can be approximated by

$$B_\lambda(T) \approx \frac{2\,c \cdot k_B}{\lambda^4} \cdot T \tag{7.25}$$

or in the frequency domain by

$$B_\nu(T) \approx \frac{2\,k_B \cdot \nu^2}{c^2} \cdot T \tag{7.26}$$

At the other extreme, for $\lambda \to 0$ ($\nu \to \infty$), the Planck's function reduces to Wien's approximation

$$B_\lambda(T) \approx \frac{2\,h \cdot c^2}{\lambda^5} \cdot \exp[-h \cdot c/(\lambda \cdot k_B \cdot T)] \tag{7.27}$$

or as a function of frequency:

$$B_\nu(T) \approx \frac{2\,h \cdot \nu^3}{c^2} \cdot \exp[-h \cdot \nu/(k_B \cdot T)] \tag{7.28}$$

7.2.5.3 **Kirchhoff's Law**

The Planck function quantifies the maximum possible radiance emitted by a hypothetical blackbody, at temperature T. In reality, the emitted radiance by a real body is less than that emitted by an ideal blackbody, by a factor equal to the absorptivity of the real body. This is known as Kirchhoff's law. Applying another form of Kirchhoff's law that equates absorptivity to emissivity ε, the relationship between the emitted radiance I_λ of the real body to that of a blackbody is

$$I_\lambda = \varepsilon(\lambda) \cdot B_\lambda(T) \tag{7.29}$$

7.2.5.4 **Brightness Temperature**

T_B is defined as the temperature identical to the real temperature T if the emitting body was a blackbody. In the frequency domain, we have the implicit definition of the brightness temperature T_B by

$$B_\nu(T_B) = \varepsilon(\nu) \cdot B_\nu(T) \tag{7.30}$$

We apply the definition of the brightness temperature (Eq. (7.30)) to the Planck function in the frequency domain (Eq. (7.23)) to obtain

$$\frac{2\,h \cdot \nu^3}{c^2} \cdot \frac{1}{\exp[h \cdot \nu/(k_\mathrm{B} \cdot T_\mathrm{B})] - 1} = \varepsilon(\nu) \cdot \frac{2\,h \cdot \nu^3}{c^2} \cdot \frac{1}{\exp[h \cdot \nu/(k_\mathrm{B} \cdot T)] - 1} \tag{7.31}$$

Simple algebra leads to a general relation between the brightness temperature T_B and the temperature T:

$$T_\mathrm{B}^{-1} = \frac{k_\mathrm{B}}{h \cdot \nu} \cdot \ln\left\{1 + \frac{\exp[h \cdot \nu/(k_\mathrm{B} \cdot T)] - 1}{\varepsilon}\right\} \tag{7.32}$$

In the domain of validity of the Rayleigh–Jeans approximation ($\lambda \to \infty$ or $\nu \to 0$), the brightness temperature can be approximated more easily if we apply the Rayleigh–Jeans approximation in the frequency domain (Eq. (7.26)) to Eq. (7.30):

$$\frac{2\,k_\mathrm{B} \cdot \nu^2}{c^2} \cdot T_\mathrm{B} = \varepsilon(\nu) \cdot \frac{2\,k_\mathrm{B} \cdot \nu^2}{c^2} \cdot T \tag{7.33}$$

This finally yields

$$T_\mathrm{B} = \varepsilon \cdot T \tag{7.34}$$

Eq. (7.34) yields in both the wavelength and frequency domain, if the assumptions of the Rayleigh–Jeans approximation are met.

7.2.5.5 Stefan–Boltzmann Law

The Stefan–Boltzmann law defines the relationship between spectrally integrated blackbody irradiance and temperature. It gives the maximum possible emitted broadband irradiance at a given temperature. The Stefan–Boltzmann law is obtained by integrating the Planck function (Eq. (7.22)) over wavelength and over the hemisphere. Because blackbody radiation is isotropic, the blackbody broadband irradiance F_BB is

$$F_\mathrm{BB}(T) = \pi \int_0^\infty B_\lambda(T)\,\mathrm{d}\lambda = \sigma \cdot T^4 \tag{7.35}$$

The Stefan–Boltzmann constant σ is

$$\sigma = \frac{2\,\pi^5 \cdot k_\mathrm{B}^4}{15\,c^2 \cdot h^3} \approx 5.671 \times 10^{-8}\ \mathrm{W\,m^{-2}\,K^{-4}} \tag{7.36}$$

7.3 Airborne Instruments for Solar Radiation

The role of airborne solar radiation measurements in atmospheric sciences is many-fold. Perhaps the most fundamental application is in determining vertical heating rates and the earliest airborne measurements were devoted to these studies. As measurement capabilities improved, applications expanded, but most can be classified into two major categories: atmospheric energetics and remote sensing.

One important exception is atmospheric chemistry. These broad categorizations are distinguished by their measurement variables (Section 7.2), mainly irradiance and radiance.

Energetics, which includes radiative energy balance and budgets, perturbation to balance (forcing), and deposition (divergence), require the measurement of irradiance or, alternatively, the transformation of measured radiance to irradiance through angular distribution models. Upward and downward hemispheric irradiance are necessary to derive all the energetic parameters of interest: net irradiance, albedo, and transmittance, the ratio of transmitted to incident irradiance, vertical irradiance divergence, and the difference between net irradiance at the horizontal planar boundaries of a layer.

Remote sensing of cloud and aerosol properties is most commonly applied to measurements of radiance but hemispheric irradiance is also useful, particularly for horizontally uniform layers where assumptions of homogeneity over the hemisphere are valid. However, in measurement applications, the distinction between radiance and irradiance is really one of field-of-view (FOV) rather than the fundamental definitions in Section 7.2. Radiant energy is the measured variable for all radiometers, defined by the instrument solid angle, detector area, spectral bandpass, and integration time. The output from sensors with a hemispheric FOV is typically converted to an irradiance scale with the assumption that the radiation field is uniform over the detector area and integration time. No assumption of uniformity over bandpass is made; instead, for modeling studies, the instrument bandpass function is convolved with a higher resolution model spectrum. Narrow-FOV measurements are typically converted to a radiance scale with the additional assumption that the field is uniform in the set of directions defined by the instrument solid angle.

In the first four subsections of Section 7.3, instruments to measure broadband (Section 7.3.1) and spectral (Section 7.3.2) irradiance and radiance, spectral actinic flux density (Section 7.3.3), and directly transmitted solar spectral irradiance (Section 7.3.4) are introduced. Table 7.2 gives an overview of the commonly used techniques to measure solar radiation from aircraft, as well as their advantages and disadvantages. In Section 7.3.5, one special problem common in airborne solar irradiance measurements is addressed, which is the correction for horizontal misalignments of irradiance sensors due to aircraft attitude.

7.3.1
Broadband Solar Irradiance Radiometers

7.3.1.1 Background

The interaction of solar radiation with the atmosphere and Earth's surface is complex, as discussed in Section 7.1. Solar radiation can be absorbed or reflected at the surface. It can be scattered by air molecules (Rayleigh's scattering law), aerosol particles, and clouds, and absorbed by gas molecules (H_2O, O_2, O_3, NO_2, CO_2, etc.), condensed water in clouds, and aerosol particles. Because these processes are wavelength dependent, spectrally resolved measurements of solar radiation can

Table 7.2 Overview of measurement techniques commonly applied in solar radiation measurements.

Measurement	Spectral characterization	Instrument	Advantages		Disadvantages		Applications
Irradiance or radiance	Broadband, see Section 7.3.1	Thermopile pyranometers	Commercially available. No moving parts.	Broader spectral range than most spectrometers. Does not require numerical integration.	Requires modications for aircraft use; Large time constant	Integrated quantity does not provide insight into constituent contributions to measured signal. Spectral response and its variability may be difcult to detect.	Cloud absorption; cloud and aerosol radiative forcing; Clear-sky radiative transfer; Flux divergence and heating rates; Broadband Albedo, transmittance and reflectance.
		Pyroelectric solar radiometers	Fast; Phase-sensitive detection/ reduced noise; Insensitive to thermal drift.		Chopping required; Moving parts.		
	Continuous spectral, see Section 7.3.2	Spectrometers, spectroradiometers	Ideally suited for revealing underlying causes of measured irradiance signal. Necessary for remote sensing. Easier to characterize spectral response. Solid-state detectors less sensitive to harsh thermal environment than bolometers.		Some spectral regions not sampled. Numerical integrations required to get broadband equivalent.		Spectral cloud absorption; Spectral aerosol forcing; cloud and aerosol remote sensing; Aerosol–Cloud Interactions; Spectral albedo, spectral transmittance and spectral reflectance.
Actinic flux density, see Section 7.3.3	Discrete bands	Filter radiometer	No moving parts. Single diode detectors.		Insufficient spectral coverage for deriving photolysis frequencies.		Photolysis frequencies, aerosol particle single scattering albedo (when combined with spectral flux divergence).
	Continuous spectral	Monochromatic	Covers full spectrum. Excellent stray light rejection.		Slow.		
		Spectroradiometer	Covers full spectrum. Fast sampling rate is best suited for aircraft measurements.		Poor stray light rejection		
Directly transmitted irradiance, see Section 7.3.4	Discrete bands	Pointing sun photometer	Pointing eliminates any error induced by non–ideal angular response from global irradiance measurement. Rapid time sampling.		Sun-Pointing on a moving platform is challenging. Proneto forward scattering error, especially in cases of high turbidity.		Spectral aerosol optical thickness, aerosol radiative forcing, column water vapor, column ozone.
		Shadow-band radiometer	Does not require pointing; May be simpler to implement in aircraft applications.		Added uncertainty due to aircraft attitude (for unstabilized flight), angular response, and ambiguity in extracting direct component of signal. Slow time sampling		

provide a wealth of information related to atmospheric composition and particle microphysical properties. However, for many applications, it is the integral over the complete solar spectrum that is important because that determines the total amount of solar radiative energy available to drive circulation dynamics or the hydrological cycle. The measured broadband solar irradiance is therefore a fundamental quantity used to characterize the radiative energy budget of the Earth–atmosphere system.

7.3.1.2 Instruments

Many of the airborne instruments that are used to measure the broadband solar irradiance are thermal detectors or bolometers. There are various types of thermal detectors but the basic principle behind their operation is the conversion of radiative energy to heat energy. In thermal detectors, the incident radiant flux is absorbed by some type of blackened receiver, inducing a temperature gradient that either generates a voltage that is measured (such as for thermocouples) or is balanced by a known electrical power (such as for active cavity type bolometers). The receiver is coated with a high-emissivity flat black paint to maximize the absorption of the radiation. Ideally, these black paints should absorb well at all wavelengths across the solar through IR spectrum. To limit the detected radiation to the solar part of the spectrum, the incident radiant flux energy needs to be filtered. This is usually done by the proper choice of an overlying glass dome, which also protects the sensor from the environment. In some instruments, interference filters are also used for band selection.

The thermal detectors typically used to measure the broadband solar irradiance from an aircraft fall into two basic categories: (i) thermopile solar radiometers called pyranometers, and (ii) pyroelectric solar radiometers.

7.3.1.2.1 **Thermopile Solar Radiometers (Pyranometers)** Pyranometers are the most common type of instruments used to measure the broadband solar irradiance from an aircraft. These instruments use thermopile detectors that consist of multiple thermocouples connected in series. Thermocouples are precise temperature sensors made of two different metals connected together at each end. A temperature difference between the junction of the metals at one end (the hot junction) and the junction of the metals at the other end (the reference junction or cold junction) produces an electric potential, or voltage, across the junctions, which is proportional to the temperature difference (Coulson, 1975). As thermocouples generate very small voltages, multiple thermocouples are connected in series into a thermopile to generate measurable voltages. In pyranometers, the hot-junction end of the thermopile is coated with a high-emissivity flat black paint and serves as the receiver end of the sensor. The reference junction end of the thermopile is connected to the body of the instrument, which has high thermal inertia. Two glass domes are placed over the detector (Figure 7.2) to serve as spectral filters and provide environmental protection. The spectral transmissivity of the domes varies with the type of glass used, but typical bandpasses are from 0.2 to 3.6 μm, effectively filtering out terrestrial radiation. The hemispheric shape of the domes ensures that incident light from any angle is equally transmitted. A desiccator

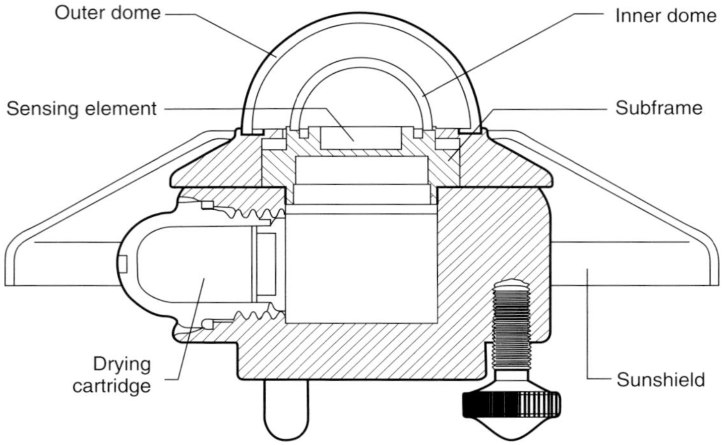

Figure 7.2 Schematic of Kipp & Zonen pyranometer, from Kipp & Zonen (2010). (Figure reproduced with permission from the manufacturer.)

inside the body of the pyranometer prevents the formation of condensate on the inside of the domes. The purpose of the outer glass dome is to protect the sensor from the elements because the heating of the thermopile can be affected by wind and moisture. However, the temperature of the outer glass dome may vary due to the ambient temperature, wind, rain, or thermal emission to the sky. It will therefore emit a varying amount of IR radiation downward onto the thermopile, inducing a voltage offset unrelated to the incident solar irradiance. The purpose of the inner glass dome, which is less exposed to the elements, is to minimize the effects of the outer glass dome on the signal by blocking the downward IR emission from the outer glass dome.

As a pyranometer measures the temperature difference between the hot and cold junctions of the thermopile, the signal can be affected by changes in the body temperature of the instrument, which serves as the reference junction. Offsets in the signal can also be introduced due to temperature differences between the inner glass dome and the body of the instrument. Modern pyranometers employ various combinations of thermistors and temperature compensation circuitry to account for these offsets. However, minimizing and characterizing the effects of heat exchange between the glass domes, the sensor, and the body of the instrument is an ongoing challenge for pyranometers.

Commercially available pyranometers for surface-based measurements have been manufactured for many years and can be obtained from a number of companies. Currently, there are no commercially available pyranometers that are customized for use on an aircraft.

Simply mounting off-the-shelf commercial pyranometers on an aircraft can be problematic because instruments designed for surface-based measurements have housings that are not practical for mounting on an aircraft, and the sensors typically generate low-voltage output signals (in the millivolt range) that can be susceptible to electronic noise in a research aircraft environment. For best performance, commercial pyronometers are modified for aircraft use. The modifications include amplification of the signal at the sensor and customizing the instrument housing (which is the thermopile reference junction) for easier mounting on an aircraft and to allow room for an amplifier to be mounted directly below the sensor (Laursen, 2003; Ramana *et al.* 2007; Bucholtz *et al.* 2008; Guan *et al.* 2010).

7.3.1.2.2 **Pyroelectric Solar Radiometers** Radiometers that utilize pyroelectric detectors have also been used to measure the broadband solar irradiance from aircraft (Valero, Gore, and Giver, 1982). Pyroelectric detectors utilize the temperature dependency of electric polarization in a pyroelectric crystal (Geist and Blevin, 1973). When a pyroelectric crystal is heated, the lattice spacings in the crystal are changed. This causes a change in the electric polarization of the crystal and leads to the charging of the crystal surfaces. Connecting an electric circuit to the surface of the crystal will generate a current that is proportional to the rate of change in the temperature in the crystal (Coulson, 1975; Hengstberger, 1989). That is, pyroelectric detectors require a change in the temperature of the crystal to generate a signal, as opposed to thermopile detectors that generate a voltage simply due to the temperature difference between the hot and cold junctions. Pyroelectric solar radiometers must therefore continually chop the incoming solar radiation, periodically blocking the detector from the incident solar radiation to modulate the signal.

Pyroelectric based radiometers have distinct advantages over pyranometers for aircraft based measurements. First, because pyroelectric crystals have a very fast response time (in the nanosecond range) so the frequency of the chopping can be very fast (100 Hz or faster). Therefore, pyroelectric detectors have much faster time constants than pyranometers, a plus when making measurements from a fast moving aircraft. Also, because the chopping generates an alternating current (AC) signal, synchronous amplification techniques can be used, which increase the signal-to-noise ratio (SNR). In addition, the chopping eliminates the effect of any drift in the sensor, which will just be seen as a direct current signal underlying the AC signal generated by the chopping. Chopping also enables implementation of very accurate electrical calibration. This is achieved by balancing optical power when the detector is open to incident solar radiation with an equivalent amount of electrical power when the chopper blade covers the detector aperture. The advantage of this type of electrical substitution is that the measurement is unaffected by drifts in responsivity, amplifier gain, and temperature (Valero, Gore, and Giver, 1982), making it particularly suitable for airborne operations where rapid and large temperature changes are quite common. The main disadvantages of pyroelectric detector radiometers are their increased complexity with a moving mechanism (chopper) and their more complicated electronics for phase-sensitive detection and closed-loop control.

The first, and only, pyroelectric solar radiometers specifically designed for aircraft measurements of the broadband solar irradiance were developed by Valero, Gore, and Giver (1982). Figure 7.3 shows a schematic of an early version of the instrument.

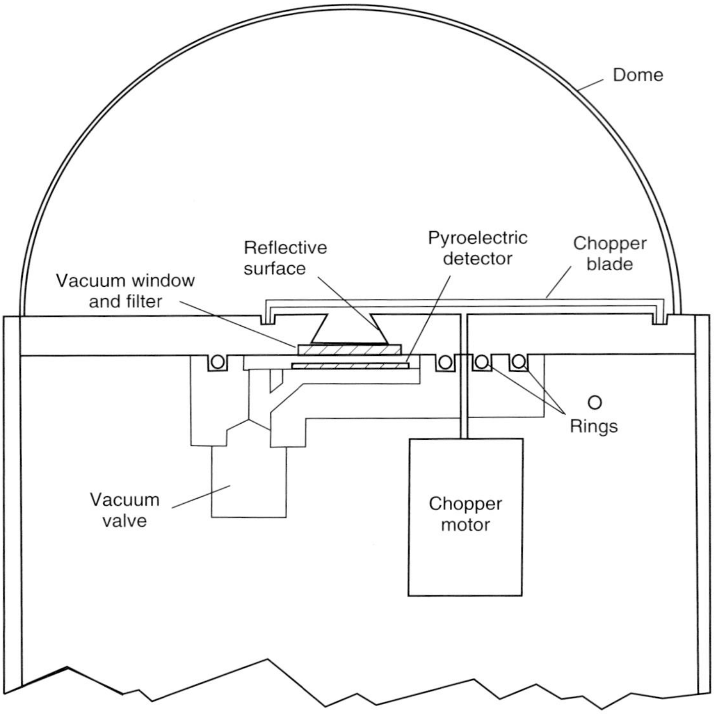

Figure 7.3 Schematic of a pyroelectric solar radiometer, from the study by Valero, Gore, and Giver (1982). (Copyright 1982 Optical Society of America (OSA). Reprinted in modified form with permission by OSA.)

In this instrument, the pyroelectric sensor was coated with a high-emissivity flat black paint to absorb the incident radiant flux and to heat the pyroelectric crystal.

7.3.1.3 **Calibration**

The accurate conversion of the measured signal from a thermopile pyranometer or a pyroelectric solar radiometer to SI units requires the calibration of (i) radiometric power response, (ii) angular response, (iii) spectral response, and (iv) the relative comparison of instruments. The calibration methods described here apply to both thermopile pyranometers and pyroelectric solar radiometers. For a well-calibrated instrument, an absolute accuracy of 2–5% and a precision of 1–3% can be expected.

7.3.1.3.1 **Radiometric (Power) Calibration** The purpose of this calibration is to determine the calibration constant, the factor that converts the measured voltage, in units of volts, into solar irradiance, in units of $\mathrm{W\,m^{-2}}$. This is the primary calibration for these instruments and various techniques have been developed. All these techniques require comparison of the research radiometer with an absolute or primary standard reference radiometer.

The most straightforward method is a side-by-side comparison of the research radiometer with a standard instrument of the same type. The standard radiometer is typically a primary or secondary standard radiometer traceable to the World Radiation Reference standard in Davos, Switzerland. The research radiometer is placed directly alongside the standard radiometer with both viewing the same light source. The light source can be a solar simulator lamp in the laboratory, an integrating sphere, or the Sun in an outdoor side-by-side comparison. As both radiometers view the same light source under the same conditions, it is straightforward to transfer the calibration from the standard radiometer to the research radiometer.

A common method for determining the calibration of a broadband solar irradiance radiometer is called the direct-diffuse shadow method or the alternating shade method (ASTM, 2005). In this method, the research radiometer is placed outside on a clear day next to a self-calibrating, absolute cavity pyrheliometer (e.g., the Eppley Hickey–Frieden cavity radiometer). Absolute cavity pyrheliometers use electrical substitution principles. The temperature rise in a blackened cavity caused by the incident solar radiation (i.e., the optical power) is matched by electrical heating in a reference cavity when the Sun is shuttered from the detector. When electrical power applied to the reference matches the optical power, the incident solar irradiance can be determined. The absolute cavity pyrheliometer has a narrow FOV (approximately 5°) and is placed on a solar tracker to continually measure the directly transmitted solar irradiance. The research radiometer, which has a hemispheric FOV, receives direct and diffuse or global solar irradiance. It must be alternately shaded and unshaded to periodically measure only the diffuse component of the solar irradiance. Subtracting the measured diffuse signal from the measured total signal gives the measured direct signal. The ratio of the measured direct signal (in volts) from the radiometer to the direct solar irradiance component (in $W\,m^{-2}$) measured by the absolute pyrheliometer, multiplied by the cosine of the solar zenith angle, gives the responsivity of the radiometer, in units of $V/(W\,m^{-2})$.

A variation of the alternating shade method is called the component method. In this method, instead of alternately shading the research radiometer, it remains unshaded and continually measures the total solar irradiance. An identical radiometer is placed nearby and is continuously shaded to give the diffuse component. The difference between the total solar from the research radiometer and the diffuse solar from the second radiometer gives the direct measured signal, which is then compared with the direct component measured by the absolute cavity pyrheliometer to determine the responsivity. This method can work because the diffuse component is typically much smaller than the direct component, so any error in the measurement of the diffuse component has a minimal effect on the calculated responsivity.

Another approach is to directly measure the direct component of the solar irradiance with the research radiometer by placing a tube over the radiometer that limits its FOV to match that of the absolute pyrheliometer. The radiometer is placed on a solar tracker to continually measure the direct component of the solar irradiance, which can then be compared with the direct component measured by the absolute pyrheliometer to determine the calibration constant (Valero *et al.* 1997; Uchiyama *et al.* 2007).

7.3.1.3.2 **Angular Response Calibration** From Eq. (7.9), it is evident that instruments used to measure irradiance must have an angular response that is proportional to the cosine of the zenith angle of incident radiation. In practice, an ideal cosine response is rarely achieved. Most instruments have responses that are within a few percent of a perfect cosine for incident zenith angles less than about 60°–70° or $\cos\theta > 0.5$ (it should be noted that cosine, rather than an angle, is the appropriate independent variable), but deviations become significantly greater at higher incident zenith angles ($\geq 70^\circ$) (Michalsky, Harrison, and Berkheiser, 1995; Valero *et al.* 1997). Azimuthal deviations can also occur.

The angular response of thermopile pyranometers is mainly a function of the glass domes and the spectral and spatial uniformity of the black coated thermopile detector. The angular response of the pyroelectric solar radiometers is a function of the glass dome, the diffuser, the light trap, the interference filter, and the uniformity of the black coated pyroelectric detector. The diffusers used in the radiometers by Valero, Gore, and Giver (1982) were custom-designed to optimize the cosine response, especially at higher incident zenith angles.

To characterize the angular response, the radiometer views a collimated stable light source as the instrument is rotated in steps through a range of zenith and azimuth angles across the complete hemisphere. Maintaining the stability of the light source throughout the duration of the angular calibration and alignment of the radiometer in the collimated beam is necessary to avoid misinterpreting variations due to the output of the lamp or in misalignment with variations in the angular response of the instrument.

7.3.1.3.3 **Spectral Response Calibration** The purpose of this calibration is to determine the spectral response of the instrument. The spectral response of a solar radiometer is the product of the spectral transmissivity, absorptivity, and reflectivity of each of its individual optical components (Schreder *et al.* 2004). Some manufacturers provide this information for each component. Ideally, however, the spectral response of the complete instrument should be determined, and this should be done periodically to account for any changes or degradation in the optical components, for example, the degradation spectral response of the sensor black paint or the degradation in the dome transmission.

Determining the relative spectral response requires a light source that can be tuned to select different monochromatic wavelengths across the solar spectrum. A typical setup for this type of calibration consists of a broadband light source (e.g., a tungsten or xenon lamp) that serves as the input to a scanning monochromator. The output from the monochromator is either collimated or diffuse (e.g., in an integrating sphere) to fill the FOV of the radiometer. A reference photodiode also views the output from the monochromator to account for variations in the output of the broadband light source. The radiometer views the collimated or diffuse monochromatic light as the monochromator scans across the wavelengths of the solar spectrum. The relative spectral response is given by the ratio at each wavelength of the signal from the radiometer divided by the signal from the reference photodiode, normalized to the ratio at the wavelength of maximum response (Schott, 2007; Schreder *et al.* 2004; Martinez, Andujar, and Enrique, 2010a).

7.3.1.3.4 **Relative Comparison of Instruments** In addition to the absolute, angular, spectral calibrations described earlier, it can be informative to perform relative comparisons between a set of calibrated solar radiometers. These tests are used to check the consistency of the calibrations, to determine any offsets between instruments, and to discover outliers that may indicate instrument or calibration problems.

These comparisons involve placing the solar radiometers outside, side by side, and comparing measurements over a large dynamic range in signal, under a variety of sky conditions (clear to cloudy), and over a range of solar zenith angles. In field studies where similar solar radiometers will be mounted on the top and bottom of multiple aircraft and on the surface, flight legs where the aircraft fly wingtip to wingtip, and low flights over surface sites can be made to compare instruments under actual flight conditions.

Radiometer comparisons at the surface and in-flight side by side are especially useful for applications where the relative instrument responses are more important than their absolute calibrations. This includes quantities that are derived from the difference or ratio of direct measurements, such as net irradiance (from which layer absorption or heating rates can be derived) or albedo (see Schmidt *et al.* (2010) for more detail). Any offset between instruments determined from relative, side-by-side comparisons can be used to correct the derived quantities.

7.3.1.4 Application

Measurements of the downward broadband solar irradiance have been made routinely for many years at surface-based sites all around the world, such as in the Baseline Surface Radiation Network (BSRN) and the Atmospheric Radiation Measurements (ARM) program. The potential of aircraft to extend these measurements throughout the vertical atmospheric column has long been recognized and measurements of the broadband solar irradiance have been made on a variety of research aircraft using various instrumentation throughout the years (Fritz, 1949; Kuhn and Suomi, 1958; Roach, 1961; Cox and Griffith, 1979; Ackerman and Cox, 1981; Rawlins, 1989; Saunders *et al.* 1992; Hayasaka, Kikuchi, and Tanaka, 1995; Wendisch *et al.* 1996b; Valero *et al.* 1997; Hignett *et al.* 1999; Wendisch and Keil, 1999; Valero *et al.* 2003; Guan *et al.* 2010; Long *et al.* 2010; Osborne *et al.* 2001; Ramana *et al.* 2007). The usual practice is to mount an identical pair of solar radiometers to the top and bottom of a research aircraft to simultaneously measure the downward and upward broadband solar irradiance. In this configuration, many useful radiative properties can be measured. For example, low-altitude flight patterns can be used to map out the broadband albedo of the surface (Eq. (7.13)). By flying level flight legs at multiple altitudes, the net irradiance can be measured at each altitude. The difference between net irradiance at the top and bottom of a layer, the vertical irradiance divergence, is equivalent to layer absorption in the absence of horizontal irradiance divergence and can be used to derive the instantaneous layer heating rate.

Aircraft measurements of broadband solar radiation have been the focal point of a number of field missions directed at the study of fundamental clear-sky radiative transfer, cloud-anomalous absorption, and radiative forcing from dust and anthropogenic pollution, to name but a few. For example, Haywood *et al.* (2003, 2011) describe the application of these measurements to the perturbation of the radiative energy budget by Saharan dust. The United States (US) Department of Energy sponsored two airborne field campaigns implementing broadband solar radiometers Atmospheric Radiation Enhanced Shortwave Experiment (ARESE I and II) to study the absorption of solar radiation by clouds (Valero *et al.* 1997, 2003; Ackerman, Flynn, and Marchand, 2003; Li *et al.* 2003).

7.3.1.5 **Challenges**

Despite its potential to address many fundamental issues regarding the solar radiative balance of the atmosphere, the measurement of broadband solar irradiance from an aircraft can be challenging. The thermal-based sensors that have been used to date must deal with many temperature and heat exchange issues that can adversely affect the measurements. These effects can be challenging enough when making measurements from the surface. They are only exacerbated when making measurements from an aircraft. For example, offsets can be introduced into pyranometer measurements due to temperature differences between the glass domes and the case of the instrument or by a change in the temperature of the body of the instrument as the aircraft climbs or descends. Additional offsets may occur from differential heating of the glass dome when it is exposed to the airstream (Saunders *et al.* 1992).

Because of these temperature effects, along with a number of other factors, instrumentation to measure the solar irradiance from an aircraft has transitioned to spectral or hyperspectral type instruments (Section 7.3.2). These are typically spectrometer-based sensors using temperature-stabilized solid-state detectors that are less sensitive to environmental temperature variability than bolometers. However, to obtain the broadband solar irradiance, spectra from these instruments must be integrated over all wavelengths, which can introduce errors. Also, spectrometers may not cover as wide a wavelength range as thermal-based sensors and may miss some of the solar radiation incident at shorter or longer wavelengths.

Another challenge in making solar irradiance measurements from an aircraft is provided by the natural spatial and temporal variation of the atmosphere. For example, when attempting to measure the amount of absorption of solar radiation within an atmospheric layer by measuring the net irradiance at multiple altitudes, care must be taken to account for any change in the atmosphere during the time it took to fly each altitude leg. This is usually done by averaging measurements over flight legs to account for spatial inhomogeneity, minimizing the time between altitude legs as much as possible, and accounting for the change in solar zenith angle over the time frame of the measurements.

Other problems arise from aircraft irradiance measurements, which require corrections. For example, the instruments described in this section require unobstructed hemispheric views of the zenith and nadir scenes. However, because the instruments are usually mounted on the fuselage, the tail of the aircraft blocks some of the scattered radiation. By estimating its subtended solid angle, the obstruction from the tail and other objects (e.g., other instruments and aircraft antennae) can be used to correct the measured irradiance.

Another challenge occurs when flying in heavy aerosol layers. The leading edge of the radiometer dome may become coated with particles, or even pitted and scoured, in cases with dust particles. Haywood *et al.* (2011) have shown that by flying special patterns, the impact on measured irradiances can be minimized.

The biggest challenge, however, in making aircraft measurements of the broadband (and spectral) solar irradiance is dealing with offsets introduced into the measurements due to the constantly changing attitude of the sensor as the aircraft pitches and rolls in flight. These offsets can be extremely large and difficult to correct, as discussed in Section 7.3.5.

7.3.2 Solar Spectral Radiometers for Irradiance and Radiance

7.3.2.1 Instruments

This section describes spectral irradiance sensors similar to those employed by Wendisch *et al.* (2001) and Pilewskie *et al.* (2003). It is not a complete summary of all spectral radiometers (also called spectroradiometers) deployed on aircraft. However, other than differences in dispersive elements, detectors, and FOV, almost all instruments share the common elements discussed in this section. The basic elements of a typical spectral radiometer deployed on aircraft are shown in the diagram in Figure 7.4. Light enters through the fore-optic. For irradiance measurements, a hemispheric light collector is used; for radiance measurements, a telescope is applied. Because the device in Figure 7.4 measures irradiance, the fore-optic (also called the light collector or optical inlet) is a diffusive element, which transmits or reflects optical power proportional to the cosine of the angle that the incident light makes with respect to the aperture normal. For radiance measurement devices, the fore-optic is a telescope.

The most common types of diffusers are transmitting devices composed of, for example, flashed opal, Teflon, Delrin, or any variety of diffusely scattering materials. One drawback for any transmitting diffuser is the limited spectral range over which it scatters diffusely. Flashed opal, for example, shows a remarkable fidelity to cosine response in the VIS and very near-infrared (NIR), but becomes almost completely transparent at wavelengths just beyond 1000 nm. Another limitation is the absorption properties of the material. Some plastic materials that work well in the VIS have appreciable absorption in the NIR. This must be taken into account in the radiometric calibration. Finally, all materials exhibit departure from ideal

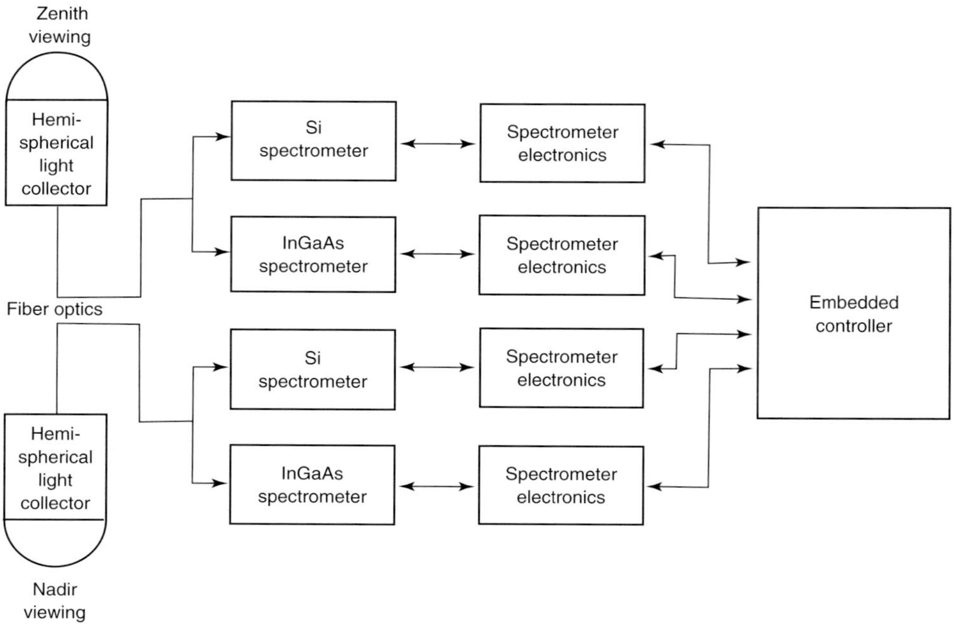

Figure 7.4 Basic elements of an airborne spectral radiometer to measure spectral upward and downward irradiance in the VIS and NIR.

cosine response to light incident near the horizon up to elevation angles as high as 30° and sometimes even 45°. This can be mitigated to some extent by the geometrical design of the diffuser, replacing a flat surface with a "top-hat" design that increases the transmission of light arriving at glancing incidence.

An integrating sphere made out of highly reflective Spectralon is an alternative option to a transmitting diffuser, one which reduces all three of the nonideal behaviors discussed earlier. Spectralon is highly reflective (very low absorption) throughout the NIR; the integrating sphere typically has very good response even at glancing incidence. Aircraft spectral radiometers deployed by Pilewskie *et al.* (2003) and Wendisch *et al.* (2001) implemented integrating spheres modeled after a design by Crowther (1997), which had very good cosine angular response at low elevation angles and near uniformity throughout the spectrum, but not completely without undesirable features. There was a region centered at $\cos\theta$ of 0.85, where angular response exceeded ideal cosine by 10–15%, requiring a postprocessing correction. A newer design, replacing the conical baffle by Crowther (1997) with a flat cylindrical design has reduced the bias to <5% (Kindel, 2010).

Light from the diffuser may be transmitted directly onto the entrance slit of a spectrometer where it is dispersed into its component wavelengths. However, for ease of aircraft integration and field calibration, the instruments designed by Pilewskie *et al.* (2003), Wendisch *et al.* (2001), and Shetter *et al.* (2003) implemented optical fibers to transmit light from the aircraft skin-mounted light collector to the rack-mounted spectrometers located in the aircraft cabin. The light collectors are small lightweight devices that greatly simplify aircraft integration, improve the adaptability to a broad collection of platforms, and make field calibration easier and more accessible. There is an additional benefit from fiber optics: they highly depolarize incident light, the consequence of which is to reduce the impact of polarizing optical elements on radiometric accuracy.

Perhaps the most important elements of the spectral radiometer are the spectrometer or monochromator and the detector. A description of the full range of dispersive elements, including gratings, prisms, and circular variable filters, is beyond the scope of this treatise, as is the range of photoelectric detectors, which includes photovoltaic and photoconductive devices, photomultiplier tubes, and array detectors such as charge-coupled devices (CCDs), complementary metal oxide semiconductors, and diode arrays. At the core of the instruments deployed by Pilewskie *et al.* (2003), Wendisch *et al.* (2001), and Shetter *et al.* (2003) were monolithic spectrometers from Zeiss, which included VIS and NIR holographic and ruled gratings and linear silicon (Si) and indium-gallium-arsenide (InGaAs) diode arrays all enclosed in a small lightweight package. Four of these units (two covering the spectral range from 300 to 1000 nm and two covering from 1000 to 2200 nm) were integrated into a package no larger than a desktop computer for the instrument deployed by Pilewskie *et al.* (2003) and Wendisch *et al.* (2001) to simultaneously monitor zenith and nadir spectral irradiance.

The final component of the spectral radiometer is the instrument controller and data system, which commands the instrument, potentially for automated operations (required, e.g., on the NASA high-altitude ER-2 aircraft and in unmanned aerial vehicles), amplifies, conditions, and digitizes the detector signal output voltage, writes the digitized output to a flight-ruggedized storage device, and in some instances, applies calibrations, displays to a monitor, and serves a substream of data to a central aircraft computer for telemetry to the ground. The extent and sophistication of these operations depend on the requirements of the particular experiment and the potential need for real-time monitoring either on the aircraft or on the ground.

7.3.2.2 **Calibration**

The measured radiant flux (power) Φ over a spectral bandpass is given by

$$\Phi = R \cdot \overline{F}_\lambda \cdot \Delta\lambda \cdot \Delta A \tag{7.37}$$

where $\overline{F}_\lambda$ is the average irradiance over the bandpass spectral width, $\Delta\lambda$, and detector area ΔA. R is called the responsivity relating the measured power to incident irradiance (Kostkowski, 1997). Note that a similar equation applies for radiance by replacing $\overline{F}_\lambda$ by $\overline{I}_\lambda \cdot \Delta\Omega$.

This is an idealized representation of the measurement equation. In practice, R combines a number of calibrations and assumptions to account for nonideal instrument response. These include nonlinearity, nonideal angular response, polarization response, and other effects such as spectral aberrations and order-sorting, for example.

7.3.2.2.1 **Radiometric (Power) Calibration** The most common method of calibrating spectral irradiance instruments is through the use of standards laboratory (e.g., the National Institute of Standards and Technology (NIST) in the US) traceable FEL 1000-W lamps. Note that FEL is the lamp-type designation, not an acronym, of NIST. These are quartz-halogen lamps that can be approximated as point sources at the prescribed calibration distance of 50 cm. Primary standards are calibrated directly at the national standards laboratory. Because their calibrations certificate is limited by total usage time (due to geometric distortion of the filament and contamination in the surface of the quartz envelope, among other characteristic degradations), primary lamps are often used only to calibrate secondary lamps that then bear the greatest operation time in transferring a calibration to the field spectrometer. Secondary lamps can also be purchased from vendors at a fraction of the cost of a primary standard. Neglecting the nonideal behaviors listed earlier, the measured spectrum from the standard source is used to derive the responsivity, R, in Eq. (7.37). Typical output from a spectrometer measurement, after signal conditioning and digitization, is in units of digital number. Eq. (7.37) is used to derive R using the laboratory-supplied values for $\overline{F}_{\lambda}$. It is unlikely that the wavelength scale of the calibration source spectrum will match that of the instrument being calibrated, requiring an interpolation of the calibration spectrum onto the instrument wavelength scale. Note that near-identical procedures are applied for the calibration of radiance sensors, with the exception that the calibration source is either a reflectance panel such as Spectralon, illuminated with a calibrated FEL lamp, or a calibrated integrating sphere with lamps mounted internally and appropriately baffled.

Field instruments require frequent calibration due to the drift in R over time. It is most often impractical to use the same primary or secondary laboratory standard in the field, particularly on aircraft where mounting spectral radiometers and their optical light collectors may be tedious and time consuming. A practical approach for field calibration is to use portable field calibrators to monitor drift in R over the duration of the experiment. These are typically lower power lamps, 200–300 W, and custom-designed for ease of mounting on the aircraft-integrated spectral radiometers. This fundamental approach has been employed independently by Pilewskie *et al.* (2003), Wendisch *et al.* (2001), and Shetter *et al.* (2003) with similar but unique field calibration devices. Each of the instruments in these studies used optical fibers for connecting the throughput from the aircraft skin-mounted light collectors to the rack-mounted spectrometers. This greatly simplified the field calibration procedure, requiring only the dismounting of the relatively lightweight optical head rather than the entire instrument.

7.3.2.2.2 **Angular Response Calibration** Any spectral radiometric measurement may be sensitive to the direction of incident radiation. We describe here the most common requirement for calibrating spectral radiometers: hemispheric irradiance measurements that, by the definition in Section 7.2, must be proportional to the cosine of the angle that a pencil of incident radiation makes with the normal to the entrance aperture. Because irradiance is power density, it is the projected area of the aperture in the field of incident radiation that determines its magnitude. The standard procedure for calibrating irradiance sensors, described in Section 7.3.1 for broadband sensors, is to use a collimated source and rotate the instrument through 90° in zenith angle θ (in equal intervals of cosine zenith, μ) and 180° in azimuth angle φ (in equal intervals of azimuth). Measurements at every μ, φ pair are related to the measurement at normal incidence. The inverse of that ratio may be used to correct the measured signal when the angular distribution of the source radiation field is known, as it would be, for example, for the directly transmitted solar irradiance.

7.3.2.2.3 **Spectral Response Calibration** The wavelength scale for spectral radiometers must be calibrated. Drift in the wavelength scale distorts the spectral signal and propagates into power uncertainty, especially when a highly wavelength-dependent source such as an FEL lamp is used for absolute radiometric calibration. Line sources such as lasers and mercury lamps are the most common calibration standards. A conversion from instrument "channel" number to wavelength is made by numerically fitting measurements of multiple and distinct spectral lines to their known line centers. Any line source may be used for wavelength calibration as long as the line center is known to a high degree of accuracy and the line widths are much narrower than the spectral resolution of the instrument. When the latter is realized, the resulting measured broadened spectral shape represents the instrument slit function. The width of this function at a magnitude of one half of the maximum amplitude represents the spectral resolution of the instrument at the source line center wavelength. Ideal slit functions are triangular, Gaussian, or quasi-Gaussian symmetric functions. The exact form of the slit function can be used either to deconvolve the measured signal to retrieve the true spectral distribution of the measured source or, more commonly, to convolve a high- resolution computed spectrum to compare with measured spectra.

7.3.2.3 Application

The application of airborne spectral radiometers in atmospheric sciences has increased over the past decade after a historical reliance on its broadband counterpart for examining problems such as cloud absorption, aerosol radiative forcing, and the fundamental measurement of surface and atmospheric albedo. Despite its fundamental importance in deciphering the underlying causal mechanisms in atmospheric radiative processes, continuous spectral radiation measurements were slow in their implementation. Perhaps the best example of the intractability of broadband solar irradiance measurements and subsequent success of spectrally resolved irradiance was in the resolution of anomalous cloud absorption. The initial report by Cess *et al.* (1995) on a bias between modeled and measured broadband cloud absorption was subsequently debated in the literature for more than a decade and was one motivating factor for the development of the solar spectral flux radiometer (SSFR) instrument (Pilewskie *et al.* 2003) and, independently, the actively stabilized spectral modular airborne radiation system (SMART) albedometer (Wendisch *et al.* 2001; Wendisch and Mayer, 2003). Although discrepancies remain to this day between spectral measurements and models, it was determined that for previous broadband studies, both observations and theory had underestimated uncertainties. The new spectral measurements determined unequivocally that spectral absorption features in clouds could be explained by water and that no "unexplained absorber" was required to reconcile observations and theory within experimental and computational uncertainties (Pilewskie, Rottman, and Richard, 2005). Newer studies have now revealed the spectral signature of a three-dimensional (3D) bias in "apparent absorption," the absorption derived from vertical flux divergence and commonly applied to aircraft measurements (Schmidt *et al.* 2010). Even sampling corrections based on the assumption that cloud absorption vanishes in the visible were shown to have spectral biases that can only be fully understood by spectrally resolved measurements (Kindel *et al.* 2011).

One of the earliest utilities of spectral radiometry was the identification of cloud thermodynamic phase (Pilewskie and Twomey, 1987a,b). This has been updated in recent studies to quantify the relative contribution from liquid water and ice to mixed-phase cloud spectral reflectance (Ehrlich *et al.* 2008, 2009). With spectrally resolved solar irradiance measurements, aerosol radiative effects can be determined without the interference from water vapor absorption and the spectral absorption of aerosol particles was derived (Pilewskie *et al.* 2003; Pilewskie, Rottman, and Richard, 2005; Redemann *et al.* 2006; Bergstrom *et al.* 2007; Schmidt *et al.* 2009; Russell *et al.* 2010).

The spectral signature of mineral dust (Haywood *et al.* 2011) and volcanic ash from the 2010 Eyjafjallajökull eruption (Newman *et al.* 2012) were derived from measurements from an instrument nearly identical to that of Wendisch *et al.* (2001) and Pilewskie *et al.* (2003). However, the studies by Haywood *et al.* (2011) and Newman *et al.* (2012) did not rely on absolute calibration. Instead, radiative transfer models in aerosol-free conditions were used for calibration and applied to aerosol conditions.

Measurements of spectral surface albedo have been performed, for example, by Webb *et al.* (2004) and by Wendisch, (2004). The impact of ice crystal shape on spectral irradiances was extensively studied by Wendisch *et al.* (2005) and Wendisch, Yang, and Pilewskie (2007). A new method of deriving aerosol layer absorption by utilizing the divergence form of the equation of radiative transfer applies spectrally resolved measurements of irradiance and actinic flux density from two cases and offers promise for the improved quantification of aerosol absorption (Bierwirth *et al.* 2010).

Other studies have utilized spectrally resolved airborne measurements for improving cloud and simultaneous cloud and aerosol remote sensing and for quantifying the influence of aerosol particles on retrieved cloud properties. Schmidt *et al.* (2009) analyzed spectral irradiance in scenes with heterogeneous clouds imbedded in pollution aerosol particles, requiring 3D radiative transfer to reconcile observations with results from large eddy simulations. Coddington *et al.* (2010) explored the effect of absorbing aerosol layers above clouds by comparing simultaneous satellite and airborne spectrometer retrievals. Kindel *et al.* (2010) examined the consistency between radiance- and irradiance-derived cloud retrievals from the Moderate Resolution Imaging Spectroradiometer (MODIS) airborne simulator and SSFR. Finally, the general utility of solar spectral observations for remote sensing applications has been provided in the studies by Coddington, Pilewskie, and Vukicevic (2012) and Schmidt and Pilewskie (2011).

7.3.3 Spectral Actinic Flux Density Measurements

7.3.3.1 Background

Airborne measurements of spectral actinic flux density are made to determine photolysis frequencies (or photolysis rate constants) of atmospheric photolysis processes. The photolysis frequency $J(\mathrm{A} \rightarrow \mathrm{B})$ of an atmospheric gas molecule A constitutes the first-order rate coefficient of the dissociation process:

$$\mathrm{A} + h \cdot \nu \rightarrow \mathrm{B} \tag{7.38}$$

which depends on the available actinic radiation. J in units of s^{-1} is given by

$$J(\mathrm{A} \rightarrow \mathrm{B}) = \int_0^{\infty} \sigma_{\mathrm{A}}(\lambda, T, p) \cdot \phi_{\mathrm{B}}(\lambda, T, p) \cdot F'_{\mathrm{act},\lambda}\, \mathrm{d}\lambda \tag{7.39}$$

σ_{A} is the absorption cross section of the molecular species A in units of cm^2, ϕ_{B} represents the quantum yield of the photoproduct B (dimensionless), both parameters are accessible by laboratory work. T is the air temperature and p the atmospheric pressure. Note that in Eq. (7.39), the quantity $F'_{\mathrm{act},\lambda}$ (spectral photon actinic flux density) is introduced instead of the spectral actinic flux density, $F_{\mathrm{act},\lambda}$, defined in Eq. (7.15). $F_{\mathrm{act},\lambda}$ must be converted to the corresponding spectral photon actinic flux density $F'_{\mathrm{act},\lambda}$ applying the following relation:

$$F'_{\mathrm{act},\lambda} = \frac{\lambda}{h \cdot c} \times 10^{-13}\, F_{\mathrm{act},\lambda} \tag{7.40}$$

In Eq. (7.40), λ is given in units of nm, $F_{\mathrm{act},\lambda}$ in units of $\mathrm{W/m^2/nm}$, and $F'_{\mathrm{act},\lambda}$ in units of $\mathrm{cm^{-2}/nm/s}$.

7.3.3.2 Instruments

There are two radiometric types of instruments to derive atmospheric photolysis frequencies of photodissociation processes: (i) filter radiometers and (ii) spectroradiometers (Hofzumahaus, 2006).

7.3.3.2.1 Filter Radiometers

Filter radiometry is used to measure the spectrally integrated solar actinic flux density in discrete spectral intervals. The front-end optical elements are typically frosted quartz domes that provide isotropic response (Volz-Thomas *et al.* 1996). Light is transmitted through bandpass filters whose wavelengths have been selected to optimally sample the molecule-specific products of absorption cross sections ϕ_{A} and quantum yields σ_{A} of species A (Eq. (7.39)). A filter radiometer, therefore, covers only one photolysis process and needs calibration with a reference instrument. Furthermore, instrument-specific nonlinearities have to be considered (Bohn *et al.* 2004). Nevertheless, because of their fast time response (about 1 s) and compact design, filter radiometers are well suited for airborne operations. A single optical receiver can be coupled to several radiometers.

7.3.3.2.2 **Spectroradiometers** Spectroradiometry is used to measure spectral actinic flux density over the continuous UV–VIS spectral range. This is the most versatile method to determine photolysis frequencies because following Eq. (7.39), J can be derived for any molecule if the photolysis process is represented within the instrumental wavelength range and the molecular parameters are known. Spectral resolutions of 1–2 nm are sufficient for this purpose. Similar to filter radiometers, optical receivers typically consist of frosted quartz domes to obtain an isotropic angular response. Light is transmitted to a monochromator or spectrometer where the radiation is detected as a function of wavelength. Detectors used with scanning monochromators are typically photomultiplier tubes; multichannel spectrometer (MCS) utilizes photodiode arrays (PdAs) or CCDs. Scanning monochrometers used by, for example, Shetter and Müller (1999), Hofzumahaus *et al.* (2002), and Shetter *et al.* (2003) typically complete a spectral scan within 10–90 s. During aircraft operations, the time-consuming scanning procedure can lead to a distortion of the spectrum, especially under scenes with inhomogeneous clouds or inhomogeneous surface albedo. Furthermore, aircraft motion may induce jitter on moving mechanisms. On the other hand, scanning double-monochromators provide stray light rejection, which is critical for accurate UV-B measurements in the atmosphere. Stray light may be a problem for MCS whose advantages are high time resolution (about 1 s) and mechanical stability because of fixed gratings. CCD detectors have the advantage of a higher sensitivity (of the order of one magnitude) compared with PdA detectors and a time resolution of <0.5 s. Airborne measurements using MCS techniques were presented by, for example, Jäkel *et al.* (2005) and Stark *et al.* (2007).

7.3.3.3 Calibrations

Spectroradiometers used to measure actinic radiation can be calibrated radiometrically using spectral irradiance sources that can be traced to international standards laboratories (Section 7.3.2). The angular calibration is performed in a manner similar to that for irradiance sensors (Section 7.3.1). The receiver optics is rotated in zenith angle and in azimuth angle. For the isotropic angular response required for actinic radiation measurements, the signal at all zenith angles must be equivalent to the signal measured at normal incidence. In reality, the angular response of the receivers deviates from complete isotropy.

Two considerations must be made for actinic flux density measurements: a deviation of the optical receiver from ideal isotropy and stray light MCS systems in the UV-B spectral region.

The actinic flux density is over a 4π sr FOV. In practice, F_{act} is measured on aircraft with two receivers covering the upper and the lower hemisphere. However, measurements may be biased by radiation from the opposite hemisphere because of the unavoidable extension of the receiver (Figure 7.5). This effect is amplified for measurements of the downward component $F_{act}^{\downarrow}$ when flying above highly reflective

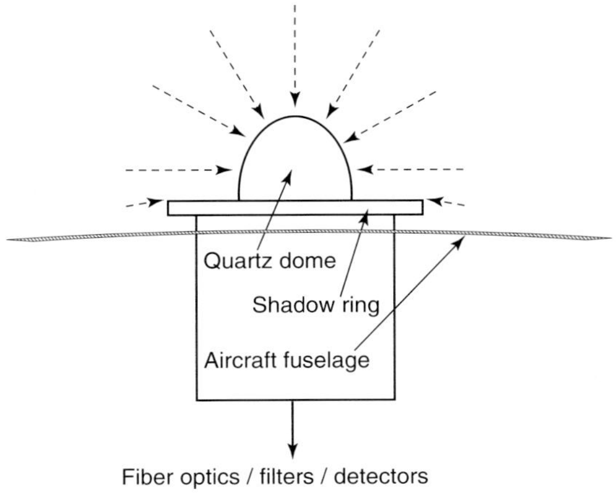

Figure 7.5 Schematic sketch of an optical receiver for actinic radiation. Dashed arrows illustrate the direction and the sensitivity toward incoming radiation. The shadow ring is built into the aircraft shell.

surfaces or clouds. In particular, for the upward component $F_{act}^{\uparrow}$, the cross talk to the upper hemisphere can cause significant uncertainties. On the ground, this problem can be minimized by using large horizontal shadow rings, but this is not an option on aircraft for aerodynamic reasons. Therefore, corrections were introduced by Hofzumahaus, Kraus, and Müller (1999) and Jäkel *et al.* (2005). Radiative transfer simulations were applied to estimate a spectral correction function that depends on several parameters, such as surface albedo, cloud albedo, aerosol load, and solar zenith angle. To limit the magnitude of these corrections, it is useful to adjust the angular response of the two receivers for an optimum 4π sr response even if this may result in imperfections at larger zenith angles if only one hemisphere is considered. Thus, the adjustment strategy is slightly different from that for ground-based measurements.

MCS systems with PdA detectors exhibit errors in the UV-B spectral range because of low signal, low detector sensitivity, and a large stray light contribution. The uncertainty generally increases with decreasing wavelength. It also depends on solar zenith angle (θ_0) and total ozone column. Although CCDs have higher sensitivities, than many PdAs, stray light is still a problem. Accurate measurements of F_{act} in the UV-B range (280–315 nm) are required for the determination of the ozone photolysis frequency $J(O^1D)$. Currently, there are two strategies to deal with the problem. The first is to adjust the procedure of background subtraction, depending on external conditions. This approach can be tested and optimized by comparison with double-monochromator reference instruments (Bohn *et al.* 2008). The second method is to limit the measurements to a wavelength range considered reliable, for example, $\lambda \geq 305$ nm, and to fill the gap using parameterizations based on radiative transfer calculations (Jäkel, Wendisch, and Lefer, 2006). For example, Figure 7.6 shows simulated spectral $J(O^1D)$ for two values of solar zenith angle θ_0. It can be seen that for small θ_0, the contribution below 305 nm is significant.

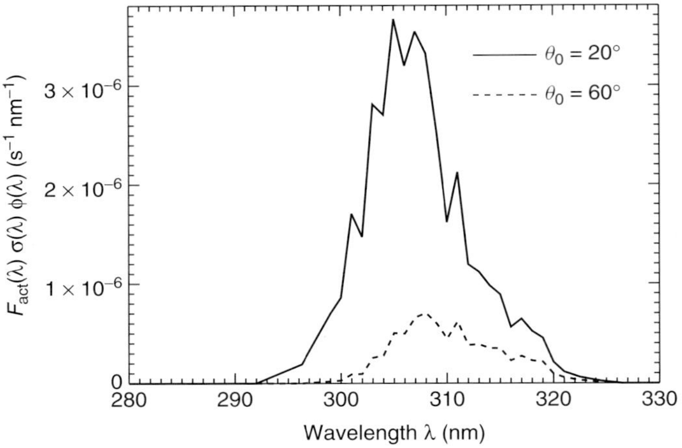

Figure 7.6 Clear-sky spectral photolysis frequency of ozone calculated for solid zenith angles of $\theta_0 = 20°$ (solid line) and $\theta_0 = 60°$ (dashed line) at ground level for a total ozone column of 250 DU. From the study by Jäkel, Wendisch, and Lefer (2006). (Copyright 2006 Springer. Reprinted in modified form with permission from Springer Science + Business Media B.V.)

For larger θ_0, $J(O^1D)$ is shifted toward higher wavelengths. The position of the short-wavelength limit depends on the position of the atmospheric cutoff, which is a function of the total ozone column in Dobson units (DU) and θ_0.

7.3.3.4 **Application**

Accurate determination of actinic flux density requires direct measurements because an alternative way, the conversion from irradiance to actinic flux density, would require knowledge of angular dependence of scattered light. The first airborne spectral actinic radiation measurements were performed in the 1990s (Shetter and Müller, 1999; Crawford *et al.* 1999) with the scanning actinic flux spectroradiometer. During the past decade, MCS systems with higher temporal resolution than scanning instruments have been used for airborne applications. In particular, studies of the cloud effect on the actinic radiation field require fast spectral measurements. This knowledge is needed for regional and global chemistry models, which often rely on simple parameterizations for cloud effects on actinic flux density. Aerosol may modify actinic radiation by ±20%, while clouds can amplify actinic radiation by as much as 200% in the most upper part of the cloud and decrease it by 90% below the cloud (Lefer *et al.* 2003). The cloud effect is wavelength dependent (Jäkel *et al.* 2005). Therefore, the photolysis frequencies of diverse species show different cloud responses.

Within the European Influence of Clouds on the Spectral Actinic Flux in the Lower Troposphere project, the 3D actinic radiation field was characterized using measurements from several aircraft (Thiel *et al.* 2008). One of the instruments (Wendisch *et al.* 2001) was equipped with a horizontal stabilization system to assure a clear separation between hemispheres, improving the angular response correction and facilitating comparisons with radiative transfer simulations of upward and downward actinic radiation (Kylling *et al.* 2005).

Stark *et al.* (2007) used airborne measurements of the photolysis of the nitrate radical (NO_3), which is an essential oxidant in the troposphere. They reported the importance of photolysis to NO_3 loss during daytime and for the catalytic destruction of tropospheric ozone. Airborne charge-coupled device actinic flux spectroradiometer (CAFS) data were analyzed by Petropavlovskikh *et al.* (2008) to validate collocated daytime Aura Microwave Limb Sounder (MLS) partial ozone columns. The method used measured downward actinic flux densities in conjunction with radiative transfer simulations. The results of three validation campaigns have shown an overall agreement of better than 5%, within the combined uncertainties of CAFS and MLS.

7.3.4
Directly Transmitted Solar Spectral Irradiance

7.3.4.1 Background

Aircraft measurements of the directly transmitted solar spectral irradiance are made primarily to obtain the spectral optical depth or thickness of the atmosphere τ (Eq. (7.4)), and based on that the aerosol optical depth of aerosol particles τ_p (Eq. (7.5)) in a cloud-free atmosphere. The optical depth of aerosol particles is one of the fundamental parameters needed to characterize the radiative effects of atmospheric aerosol particles. These measurements can also be used to infer the size distribution of aerosol particles (King *et al.* 1978a) and to estimate the column abundance of ozone (Livingston *et al.* 2005) and water vapor (Bruegge *et al.* 1992; Schmid *et al.* 1996; Halthore, 1997).

Instruments that measure directly transmitted solar irradiance do not require radiometric calibration by reference standard sources or detectors to convert to SI units. For linear detectors, measured voltage, $V_{\text{dir},\lambda}(\theta_0,\ z)$, is directly proportional to the directly transmitted solar irradiance:

$$V_{\text{dir},\lambda}(\theta_0, z) = C \cdot F_{\text{dir},\lambda}(\theta_0, z) \tag{7.41}$$

where C is a calibration constant unique to each detector. Using Eqs. (7.5) and (7.41), and incorporating the airmass factors for each attenuator, Eq. (7.20) can be rewritten:

$$V_{\text{dir},\lambda}(\theta_0, z) = \left(\frac{\overline{R}_{\text{S-E}}}{R_{\text{S-E}}}\right)^2 \cdot \overline{V}_{\lambda,\text{TOA}} \cdot \exp\left[-\tau_{\text{p}}(\lambda, z) \cdot M_{\text{p}}(\theta_0, z) - \tau_{\text{Rayl}}(\lambda, z) \cdot M_{\text{Rayl}}(\theta_0, z) - \sum_i \tau_{\text{abs},i}(\lambda, z) \cdot M_{\text{abs},i}(\theta_0, z)\right] \tag{7.42}$$

where $\overline{V}_{\lambda,\ \text{TOA}}$ is the voltage that the instrument would read at the TOA at the mean Sun–Earth distance. Note that the calibration constant C has been eliminated. The spectral optical depth of aerosol particles can then be obtained by rearranging Eq. (7.42):

$$\tau_{\mathrm{p}}(\lambda, z) = \frac{1}{M_{\mathrm{p}}(\theta_0, z)} \cdot \left\{ \ln\left[\left(\frac{\overline{R}_{\mathrm{S-E}}}{R_{\mathrm{S-E}}}\right)^2 \cdot \overline{V}_{\lambda,\mathrm{TOA}}\right] - \ln V_{\mathrm{dir},\lambda}(\theta_0, z) \right.$$
$$\left. - \tau_{\mathrm{Rayl}}(\lambda, z) \cdot M_{\mathrm{Rayl}}(\theta_0, z) - \sum_i \tau_{\mathrm{abs},i}(\lambda, z) \cdot M_{\mathrm{abs},i}(\theta_0, z) \right\} \quad (7.43)$$

Eqs. (7.42) and (7.43) assume only single scattering, which is a good approximation for many applications. However, for higher aerosol optical depths, multiple scattering into the FOV of the sun photometer can increase the measured signal, causing an underestimation in the derived aerosol optical depth. Techniques have been developed to account for biases due to multiple scattering (Shiobara and Asano, 1994; Kokhanovsky, 2007).

For retrieving the optical depth of aerosol particles τ_{p}, the filter band-center wavelengths are selected to avoid strong gas absorption bands. Most of the terms on the right-hand side of Eq. (7.43) can, therefore, be directly calculated to a high degree of accuracy. $V_{\mathrm{dir},\lambda}(\theta_0,\ z)$ is the measured signal. The molecular scattering optical depth τ_{Rayl} can be accurately calculated (Bucholtz, 1995), and the absorption optical depths for the various trace gases (ozone and NO_2) can be sufficiently estimated using satellite or climatology data combined with model radiative transfer calculations (Russell *et al.* 1993). $\overline{V}_{\lambda,\ \mathrm{TOA}}$, the voltage the instrument would read at the TOA for an average Sun–Earth distance, is the only component on the right-hand side of Eq. (7.43) that needs to be determined through a calibration technique called the Langley plot method (described in the following sections).

7.3.4.2 **Instruments**

Two types of instruments have been used to measure the directly transmitted solar irradiance from an aircraft. The first type of instruments, (i) Sun photometers, actively track the Sun. The second type, (ii) shadow-band solar radiometer, utilizes a periodically oscillating armature to isolate the direct component of solar radiation.

7.3.4.2.1 **Sun Photometers** One of the primary instruments used to measure the directly transmitted solar spectral irradiance from an aircraft is called a Sun photometer. They typically consist of the following components: multiple, narrow bandpass (typically 0.5- to 1.0-μm wide) detectors spanning the solar spectrum; a cylindrical tube, sometimes called a Gershun tube, that limits the FOV of the instrument to greater than the 0.5° angular width of the Sun to allow for some pointing offsets (typical FOVs are 1–1.5°); a two-axis, azimuth-elevation automated drive mechanism for finding and tracking the Sun; and a specialized detector (typically a four-quadrant detector) for accurate solar pointing.

While Sun photometers have been used extensively on the surface to make measurements of the direct solar spectral radiance (Holben *et al.* 1998), their use on aircraft has been much more limited. This is mainly because aircraft Sun photometers, while following the same basic design as described earlier, must deal with the added challenge of finding and tracking the Sun from a continuously moving platform. Some attempts have been made to use handheld Sun photometers, manually pointing them at the Sun from the interior of the aircraft (Shaw, 1975; Dutton, DeLuisi, and Herbert, 1989; Maring *et al.* 2003; Porter *et al.* 2007). While this approach is inexpensive, it can be difficult for the operator to steadily point the Sun photometer accurately enough to obtain high-quality data (Porter *et al.* 2007). It also requires pointing through aircraft window material, the transmittance through which must be factored into the calibration.

A second approach has been to use an automated sun-tracking system located inside the aircraft, pointing the Sun photometer through a window at the Sun (Stone *et al.* 2010). This approach provides more accurate sun-tracking and minimizes the need to harden the instrument to withstand the harsher environment on the outside of an aircraft, but the measurements require specific flight-tracks to keep the Sun in view. Again, window transmittance must be derived.

The best approach to obtain highly accurate Sun photometer measurements from an aircraft is to mount the instrument outside on top of the fuselage. Two such aircraft Sun photometers have been developed at NASA. The six-channel NASA Ames Airborne Tracking Sunphotometer (AATS-6; Figure 7.7) was developed in the late 1980s (Matsumoto *et al.* 1987; Russell *et al.* 1993) and a 14-channel version (AATS-14) was later developed in the mid-1990s.

Data analysis for Sun photometers is relatively straightforward. The instruments are calibrated using the Langley plot method to provide exo-atmospheric voltages, discussed later. Highly accurate aerosol optical depth data can be retrieved. In addition, during a research flight, estimates of the aerosol optical depths may be used for real-time flight planning. The main drawback to these instruments is that they are complex and expensive to develop, primarily because of the need to actively track the Sun from a moving aircraft. This has limited the use of Sun photometers on research aircraft.

7.3.4.2.2 **Shadow-Band Solar Radiometers** The second type of instrument that has been used to measure the directly transmitted solar spectral irradiance from an aircraft is known as a shadow-band radiometer. This type of instrument does not require any active tracking of the Sun. Instead, it utilizes a continuously rotating shadow arm to isolate the directly transmitted solar irradiance. Shadow-band radiometers have been developed for surface-based measurements (Harrison, Michalsky, and Berndt, 1994) and for airborne atmospheric research, for example, the total direct-diffuse radiometer (TDDR) (Valero, Ackerman, and Gore, 1989).

The TDDR utilized a Spectralon diffuser, shaped to optimize the cosine angular response required for irradiance measurements. The diffuser was covered by a quartz glass dome for protection. Underneath the diffuser, an inverted, mirrored, parabolic cylinder collimated the light transmitted through the diffuser to the detector plane below. The TDDR had seven wavelength channels spanning the VIS solar spectrum from the UV to the NIR. Each channel consisted of a silicon photodiode with an overlying interference filter. The filter spectral bandpasses were 10 nm full width at half maximum (FWHM), with center wavelengths chosen to avoid gas absorption bands (Valero *et al.* 1997).

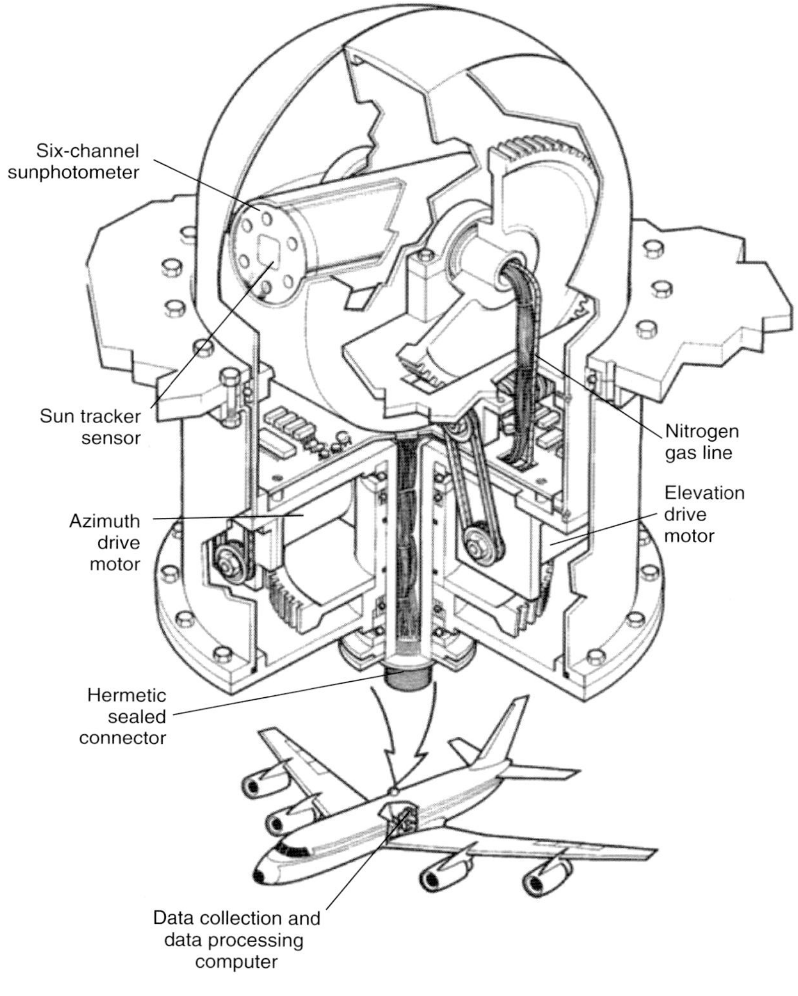

Figure 7.7 The six-channel NASA Ames Airborne Tracking Sunphotometer (AATS-6), from Matsumoto *et al.* (1987). (Copyright 1987 American Meteorological Society (AMS). Reprinted in modified form with permission by AMS.)

The advantage of a shadowband radiometer is that it not only provides a measurement of the direct component of the downward solar spectral irradiance but also the total and diffuse components of the sunlight (Valero, Ackerman, and Gore, 1989; Pilewskie and Valero, 1993). For example, the shadow arm of the TDDR was a hemispheric metal ring that continuously rotated over the top of the instrument with a period of approximately 1 min. Regardless of the location of the Sun in the sky, the arm eventually shadowed the aperture of the instrument blocking the direct component of the solar radiation. This shadowing could be seen

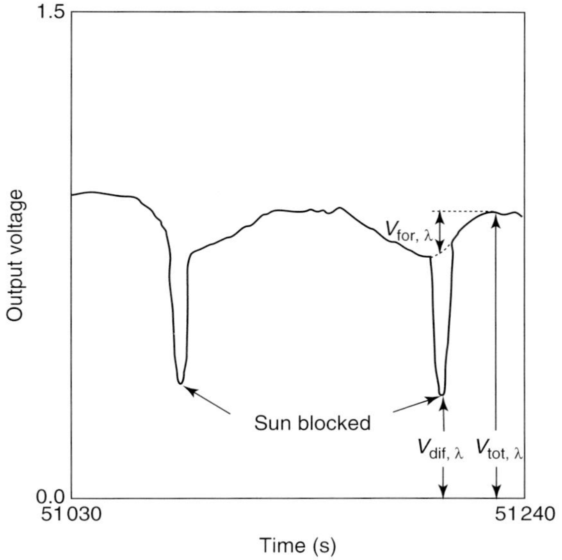

Figure 7.8 Example of a time series of measured TDDR voltages, from Pilewskie and Valero (1993). (Copyright 1993 Elsevier. Reprinted in modified form with permission by Elsevier.)

as a dip in the raw voltage signal from the TDDR (Figure 7.8). For shadowband radiometers when the arm is completely out of the FOV, the total, $V_{tot,\lambda}$, or global solar radiation (i.e., the direct + diffuse) is measured (Figure 7.8). Eventually, the arm blocks the direct sunlight, and the diffuse radiation, $V_{dif,\lambda}$, is measured. By carefully choosing the point in the dip where the shadow arm just begins to block the solar disk, $V_{for,\lambda}$, both the forward scattered radiation (i.e., the solar aureole) and the small portion of the diffuse sky blocked by the shadow arm can be taken into account in the diffuse term. The difference between the measured total and diffuse radiation gives the direct component (Eq. (7.18)):

$$V_{dir,\lambda} = V_{tot,\lambda} - V_{dif,\lambda} - V_{for,\lambda} \tag{7.44}$$

Once the directly transmitted component of the solar irradiance is determined, the spectral aerosol optical depth can be retrieved utilizing the Lambert–Bouguer attenuation law (Eq. (7.43)). Shadowband radiometers are also calibrated using the Langley plot method.

7.3.4.3 **Calibration**

The Langley method is the primary technique used to determine the exo-atmospheric calibration voltage constant, $\overline{V}_{\lambda,\, TOA}$, for a Sun photometer or shadow-band radiometer. This method, originally developed in the early 1900s to obtain estimates of the TOA solar irradiance (at the time called "solar constant") (Langley, 1903), is based on the Lambert–Bouguer attenuation law (Eq. (7.20)), applied to voltages and not irradiances (see also Russell *et al.* (1993)):

$$V_{\mathrm{dir},\lambda}(\theta_0, z) = \left(\frac{\overline{R}_{\mathrm{S-E}}}{R_{\mathrm{S-E}}}\right)^2 \cdot \overline{V}_{\lambda,\mathrm{TOA}} \cdot \exp[-\tau(\lambda, z) \cdot M(\theta_0, z)]$$
$$= V_{\lambda,\mathrm{TOA}} \cdot \exp[-\tau(\lambda, z) \cdot M(\theta_0, z)] \quad (7.45)$$

where

$$V_{\lambda,\mathrm{TOA}} = \left(\frac{\overline{R}_{\mathrm{S-E}}}{R_{\mathrm{S-E}}}\right)^2 \cdot \overline{V}_{\lambda,\mathrm{TOA}} \quad (7.46)$$

$V_{\lambda,\,\mathrm{TOA}}$ is the voltage the instrument would read at the TOA at the time of the calibration. Taking the natural logarithm of both sides of Eq. (7.45) yields the linear equation:

$$\ln[V_{\mathrm{dir},\lambda}(\theta_0, z)] = \ln(V_{\lambda,\mathrm{TOA}}) - \tau(\lambda, z) \cdot M(\theta_0, z) \quad (7.47)$$

Linear regression of the natural log of $V_{\mathrm{dir},\lambda}(\theta_0,\ z)$ against airmass, M, yields a line with y-intercept equal to $\ln(V_{\lambda,\,\mathrm{TOA}})$ and a slope equal to the atmospheric optical depth, $\tau(\lambda,\ z)$. The Langley method takes advantage of this linear relationship and consists of measuring $V_{\mathrm{dir},\lambda}(\theta_0,\ z)$ as the Sun rises or sets (i.e., over a range of airmass values), plotting $\ln[V_{\mathrm{dir},\lambda}(\theta_0,\ z)]$ versus M, and then extrapolating the data to zero airmass (i.e., the intercept) to determine $\ln(V_{\lambda,\,\mathrm{TOA}})$, from which the exo-atmospheric calibration voltage constant, $\overline{V}_{\lambda,\,\mathrm{TOA}}$, can be obtained from Eq. (7.46). The optical thickness of aerosol particles is eventually obtained applying Eq. (7.43).

Care must be taken to minimize error from Langley calibrations. The primary assumption in this technique is that the aerosol optical depth remains constant over the time span of the measurements. For this reason, calibrations are usually carried out at high-altitude locations, to get above the surface mixing layer and into an area where turbidity is lower and more stable than at lower altitudes. A discussion of various modifications and refinements to the Langley method have been suggested to better account for a variable atmosphere and to improve on the accuracy of these calibrations (Michalsky *et al.* 2001). A description of these techniques can be found in the studies by Shaw (1983), Tanaka, Nakajima, and Shiobara (1986), Forgan (1994), Shiobara *et al.* (1991), and Schmid and Wehrli (1995).

7.3.4.4 **Application**

The AATS-6 and AATS-14 have flown on a variety of aircraft, providing highly accurate measurements of the directly transmitted solar spectral irradiance during numerous field studies (Russell *et al.* 1993, 1996; Schmid *et al.* 2003; Redemann *et al.* 2005). Both AATS instruments employ photodiode detectors with preamplifiers and interference filters with a typical bandpass of 5 nm. Each channel is temperature controlled and has a baffled tube that limits the FOV to 2°. AATS-6 has six channels spanning 380–1021 nm, while AATS-14 extends this range with 14 channels spanning 354–1558 nm. The center wavelengths for most of the channels for both instruments are located in regions of the spectrum dominated by the effects of aerosol particles, but additional channels are included in spectral regions affected by ozone, nitrogen dioxide, and water vapor. To initially locate the Sun, both instruments scan the sky to find the brightest spot. They then use differential Sun sensors to control azimuth-elevation motors to lock on to the Sun. As long as the Sun remains visible (i.e., above the aircraft and not blocked by clouds), the instruments can continuously track the Sun. This provides great flexibility, allowing measurements to be made during a variety of aircraft maneuvers such as spiral climbs and descents.

The TDDR has been used extensively on various aircraft studies providing high-quality aerosol optical depth data. The strengths of shadow-band radiometers for aircraft measurements lie in their small, compact size and relatively straightforward mechanical design. However, because they have a hemispheric FOV, only measurements when the instrument is at level, or near-level, provide useful data, and all near-level data must be corrected for the attitude of the instrument. In addition, because the TDDR shadow arm has a period of about 1 min, direct solar radiation measurements are sparser than that for Sun photometers. On fast moving aircraft, this can be problematic. Finally, the analysis of shadow-band radiometer measurements can be challenging because it can be difficult to fully automate the process of finding the dips in the data and choosing the proper inflection point for each dip that corresponds to the time when the shadow arm just begins to block the solar disk. Manual analysis of the data is almost always needed. Despite these limitations, with careful analysis, TDDR-type shadow-band radiometers can provide high-quality aerosol optical depth measurements.

7.3.5 Solar Radiometer Attitude Issues

7.3.5.1 Background

Aircraft measurements of broadband solar (Section 7.3.1) or solar spectral (Section 7.3.2) irradiance are carried out in many climate- and weather-related field studies to characterize the radiative energy budget within the atmospheric column. Typically, this is done by mounting identical radiometers directly to the top and bottom of a research aircraft to simultaneously measure the downward and upward solar irradiance. However, radiometers that are rigidly affixed to an aircraft's fuselage will tilt in pitch and roll along with the aircraft during flight. Because of the cosine response of irradiance sensors, attitude variations from a horizontally level position introduce biases in the measured signal that could, if left uncorrected, be mistaken for true solar irradiance variability.

As mentioned in Section 7.2.4, and explained in the study by Bucholtz *et al.* (2008), Eq. (7.9) is actually a general definition of irradiance that could be applied to any unit surface area independent of the orientation of that unit area. The zenith angle is always defined with respect to that unit area (Figure 7.1). For the broadband and solar spectral radiometers described earlier, the unit area and zenith angle are specified by the entrance aperture of the radiometer. The measured irradiance, F, will therefore vary with the attitude of the radiometer because the zenith angle with respect to the radiometer, for any point in the sky, will vary with the attitude of the radiometer, causing the normal component of the measured radiance to vary.

For atmospheric radiative budget studies, it is the solar irradiance at a given altitude with respect to a horizontally level unit surface area that is required. This is challenging to measure from an aircraft because aircraft are rarely level in flight. The irradiance measured by a radiometer rigidly mounted to an aircraft will therefore be different from the irradiance measured by a horizontally level radiometer. The difference will depend on the tilt of the radiometers and the angular distribution of the sky radiation. The farther the radiometer is from level, and the farther the radiation field is from isotropic, the larger the difference will be between the measured irradiance and the desired horizontally level irradiance.

This effect is most pronounced in solar irradiance measurements because in the absence of high-order scattering from clouds or aerosol particles, downward solar radiation is dominated by the directly transmitted solar irradiance, with a smaller contribution coming from the radiation scattered by the atmosphere, called diffuse radiation. Directly transmitted solar radiation is proportional to the cosine of the solar zenith angle. However, for radiometers fix-mounted on aircraft, the cosine of the solar zenith angle with respect to the radiometer is not only a function of the position of the Sun in the sky but also of the pitch, roll, and heading of the aircraft as illustrated in Figure 7.9.

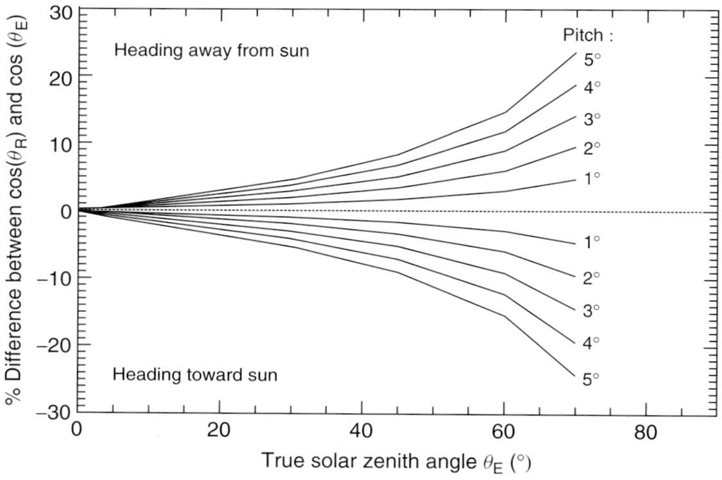

Figure 7.9 Illustration of the difference between the cosine of the solar zenith angle with respect to a radiometer rigidly mounted to the aircraft (θ_R) and the cosine of the true solar zenith angle with respect to Earth-centered coordinates (θ_E) for a horizontally level radiometer. The difference varies with the true solar zenith angle and the aircraft's pitch and heading. Aircraft typically fly in a slightly pitched up attitude. For this case, the measured solar irradiance while heading toward the Sun will be less than that for a heading away from the sun, redrawn from the study by Bucholtz *et al.* (2008). (Copyright 2008 American Meteorological Society (AMS). Reprinted in modified form with permission by AMS.)

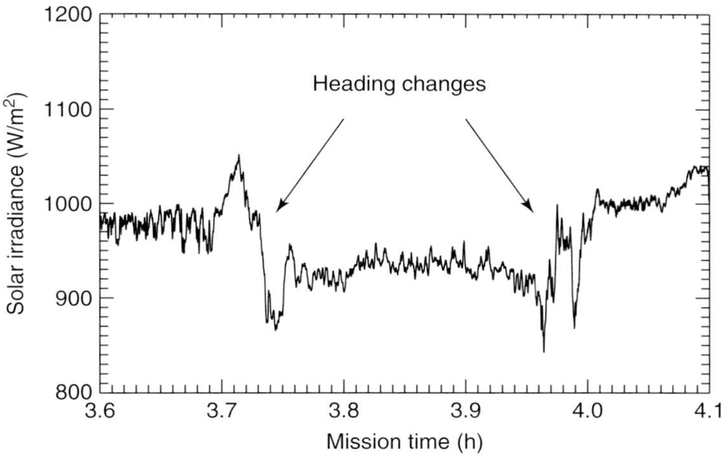

Figure 7.10 Typical example of the offset introduced into the downward solar irradiance measured by a radiometer rigidly mounted to the aircraft due to nothing more than changes in the heading of the aircraft. The aircraft maintained a constant altitude for this entire flight segment, redrawn from the study by Bucholtz *et al.* (2008). (Copyright 2008 American Meteorological Society (AMS). Reprinted in modified form with permission by AMS.)

Changes in solar zenith angle with respect to the radiometer will change the measured signal, even when the Sun's position in the sky and the direct component of irradiance is fixed. Because the direct solar radiation is usually much larger than scattered solar radiation, it is typical to see large changes in the measured signal from fix-mounted radiometers that are generated by nothing more than aircraft heading changes (Figure 7.10).

7.3.5.2 After-Flight Software Corrections for Fixed Instruments

Various methods have been developed to minimize and correct for the offsets introduced into the solar irradiances measured by aircraft-mounted radiometers. Simple methods include both keeping the aircraft as level as possible when making measurements and ignoring any data for further analysis where the aircraft pitch and roll angles exceed a certain value ($<1^\circ$ are acceptable at best). More involved methods include various techniques to correct the measured irradiances back to a level orientation in postprocessing (Hammer, Valero, and Kinne, 1991; Bannehr and Glover, 1991; Saunders *et al.* 1992; Bannehr and Schwiesow, 1993; Valero *et al.* 1997; Boers, Mitchell, and Krummel, 1998).

The following technique applies to fix-mounted instruments and is applicable to the direct portion of solar radiation only. Although not as effective as active stabilization (see Section 7.3.5.3 below), postflight correction procedures are widely used because active stabilization is technically challenging and expensive. To correct the data for horizontal attitude deviations, the following correction factor is applied (Bannehr and Schwiesow, 1993):

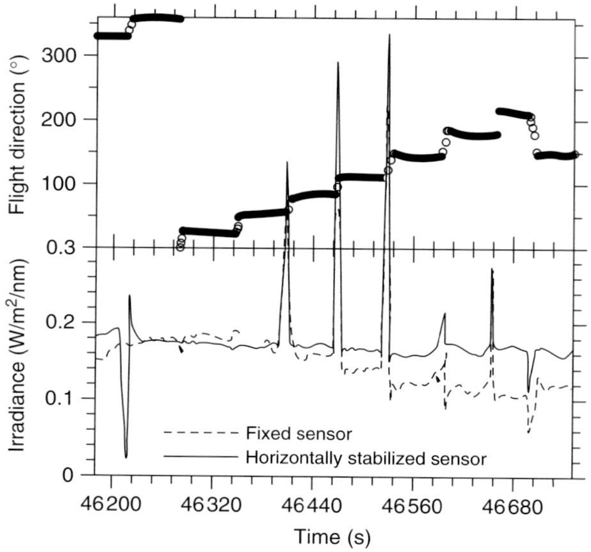

Figure 7.11 Spectral irradiance data recorded with the SMART albedometer at 550 nm wavelength. This figure is adopted from Wendisch *et al.* (2001). (Copyright 2001 American Meteorological Society (AMS). Reprinted in modified form with permission by AMS.)

$$k = \frac{\sin h_0}{\cos h_0 \sin \Phi \sin(\varphi_0 - \psi) - \cos h_0 \sin \Theta \cos \Phi \cos(\varphi_0 - \psi) + \sin h_0 \cos \Theta \cos \Phi} \tag{7.48}$$

with the solar angle coordinates (indicated by subscript "0") of $h_0 = \pi - \theta_0$, the solar altitude angle (θ_0 is the solar zenith angle), and φ_0, the solar azimuth angle (see also Figure 7.1). The aircraft angular coordinates (no index) are given as ψ, the true heading of the aircraft with respect to the Earth-fixed coordinate system, Θ being the aircraft pitch angle, and Φ representing the aircraft roll angle. There is one additional problem with this formula. If the radiometer is not level with the aircraft fixed coordinate system, then there is an additional bias that needs to be determined.

An example of solar irradiance measurements not corrected for attitude biases is shown in Figure 7.11. Eq. (7.48) is applied in an iterative fashion adopted until the jumps in the dashed line in Figure 7.11 are completely removed.

At any time and for any geophysical position, the direction of the Sun in Eq. (7.48) is calculated using the common formulae given, for example, by Iqbal (1983).

Triangular (or box) flight patterns at constant high altitudes under cloud-free conditions above the flight altitude are useful for determining sensor misalignments (Figure 7.11). Haywood *et al.* (2011) have shown that pirouette maneuvers when the aircraft is on the ground are even more effective at eliminating measurement biases induced by alignment offsets.

These postflight software correction methods are difficult and time consuming, and they do not always completely remove the attitude offsets in the measurements.

They are especially problematic for inhomogeneous sky conditions, such as partly cloudy skies, because it can be very difficult to distinguish changes in the measured signal that are due to changes in the attitude of the aircraft from changes due to the varying sky conditions.

7.3.5.3 **Stabilized Platforms**

The best method for improving the accuracy of solar and IR irradiance measurements from aircraft, and greatly simplifying the data analysis, is to keep the radiometers horizontally level in flight. This has become possible within the past 12 years because of the availability of airborne stabilization technology such as fiber optic gyros (FOGs) and fast, accurate, magnetic torque motors that can be exploited for airborne atmospheric science research.

Currently, only a handful of stabilized platforms have been developed for aircraft solar irradiance measurements (Wendisch *et al.* 2001; Bucholtz *et al.* 2008). While differing in specifications and components, each system shares some fundamental design features. Specifically, the current stabilized platforms work by sensing changes in the attitude of the platform with respect to Earth- centered coordinates and the angular offsets between the platform and the aircraft. These changes are countered through a series of actuators and motors. A real-time inertial navigation system (INS) is fundamental to the operation of the platforms. The INS senses the inertial motion of the platform to obtain its attitude. The system consists of a highly accurate inertial measurement unit (IMU) for rate and acceleration data, a Global Positioning System (GPS) for positional information, and software for real-time analysis. It employs FOGs to measure the payload plate angular rate and silicon accelerometers to measure linear acceleration. From this information, the attitude (pitch and roll) of the payload plate assembly with respect to Earth-centered coordinates is determined. A major source of error to positional stability originates from the inherent drift of the IMU. To compensate for this drift, a GPS is used and integrated with the IMU to improve the performance.

The first horizontally stabilized system developed specifically for aircraft radiometry is described by Wendisch *et al.* (2001) and Jäkel *et al.* (2005). This system actively keeps an airborne spectral albedometer and actinic flux radiometer level to better than 0.2° for pitch and roll angles within the range of $\pm 6^\circ$. A next-generation version of this platform, named the stabilized platform for airborne solar radiation measurements (SPARM) is installed on the National Center for Atmospheric Research (NCAR) HIAPER research aircraft.

Bucholtz *et al.* (2008) describe an independently developed, actively stabilized, horizontally level platform for aircraft solar and IR radiometer measurements named the stabilized radiometer platform (STRAP). STRAP is able to stabilize as many as three radiometers to better than $\pm 0.02^\circ$ for aircraft pitch and roll angles of up to approximately $\pm 10^\circ$. The system update rate of 10 Hz compensates for most pitch and roll changes experienced in normal flight and in turbulence.

The STRAP design differs from the stabilized system described by Wendisch *et al.* (2001). One of the main differences is that STRAP has the IMU directly attached to the bottom of the payload plate that holds the radiometers, while Wendisch *et al.*

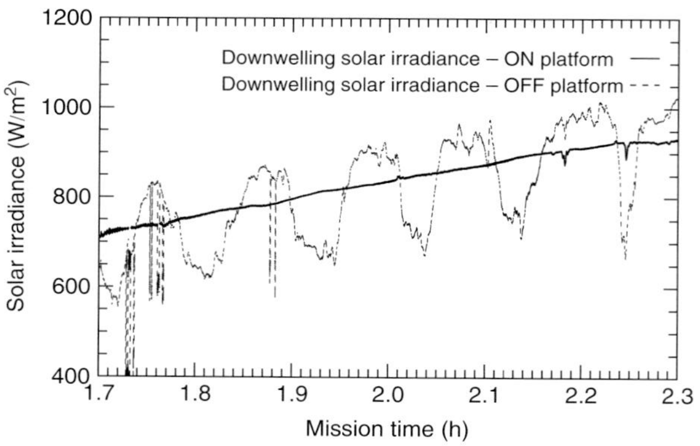

Figure 7.12 Comparison of the downwelling solar irradiance measured during a spiral ascent by a radiometer that is horizontally stabilized (solid line) and one that is rigidly mounted to the aircraft (dashed line), redrawn from the study by Bucholtz *et al.* (2008). (Copyright 2008 American Meteorological Society (AMS). Reprinted in modified form with permission by AMS.)

(2001) have the IMU fixed to the aircraft and not to the sensor plate. There are advantages and disadvantages to both designs. The Wendisch *et al.* (2001) design requires only one IMU to stabilize both an upward and downward looking sensor. This reduces costs because only one IMU is needed, but because the attitude of the sensors is not directly measured by the IMU, the leveling accuracy that can be obtained is more limited. The STRAP has the IMU directly attached to the payload plate so the attitude of the radiometers can be measured to very high accuracy. However, to stabilize both upward and downward sensors, a second IMU and STRAP system would be required adding to the cost.

To illustrate the performance capability and utility of stabilized platforms for airborne radiometry, Figure 7.12 shows the downwelling solar irradiance measured by solar radiometers on top of an aircraft as it did a spiral ascent under clear skies from approximately 60–5300 m in altitude. The pitch of the aircraft during the spiral was approximately 7°–8° (i.e., nose-up). The roll varied between approximately -2° to -10° (negative roll means the left or port wing down), but the average roll was approximately -5°. The measurements from an identical pair of solar pyranometers – one mounted on STRAP and the other rigidly mounted nearby directly to the aircraft – are compared to illustrate the improved accuracy from active stabilization (Bucholtz *et al.* 2008). As shown in Figure 7.12, the irradiance measured by the radiometer rigidly mounted to the aircraft has huge, periodic variations in magnitude ($\pm 100\,\mathrm{W/m^2}$) as the aircraft circles and climbs in the spiral, while the irradiance measured by the radiometer on the stabilized platform steadily and smoothly increases. This segment of the flight occurred in the morning and the steady increase in the solar irradiance is due to the rising Sun. Solar radiometer measurements during a spiral were never used in the past because the pitch and roll angles were too large to correct for in the data. With the stabilized platforms, solar irradiance data can now be obtained in a spiral. This new capability means that accurate measurements of the solar irradiances, the flux divergence, and the heating rates can now be obtained for the vertical atmospheric column without the need for multiple altitude straight and level legs. This allows characterization of the radiative budget profile within a shorter amount of time (approximately 20–40 min) and within a closer proximity to any surface-based measurements. This capability expands the utility of aircraft radiometer measurements and increases the data collection efficiency.

7.3.5.4 **Challenges**

The advantages of utilizing stabilized platforms for aircraft measurements of the solar broadband and solar spectral irradiance are obvious. Stabilized platforms reduce errors from nonlevel flight and simplify postprocessing by eliminating correction procedures. In addition, because stabilized platforms keep the solar radiometers level for pitch and roll angles of anywhere from $\pm 6^\circ$ to $\pm 10^\circ$, the quantity of useful data from any given flight is greatly increased and the restrictions on flight patterns are reduced. As shown in Figure 7.12, useful solar irradiance measurements can be obtained even in spiral ascents or descents.

There are two main reasons why stabilized platforms are not more widely used. The primary reason is the cost. To develop current systems required hundreds of thousands of dollars to over half-a-million dollars. These high costs were due to the equipment, especially the IMUs and specialized motors, and also due to the limited number of companies that have the expertise to develop such platforms and the small market for stabilized platforms for atmospheric science measurements. The latter means that these platforms are currently not off-the-shelf items but must be custom-made.

The other reason for the limited use of stabilized platforms has been the lack of expertise with these systems in the atmospheric science community. The current systems were built by private companies and the science community has had to rely on these companies for maintenance and repair, adding to the cost of having a stabilized platform.

Fortunately, platform component costs are coming down and the expertise within the atmospheric engineering and scientific community is increasing as we gain more experience with the current systems. This will lead to stabilized platforms that are more affordable and reliable, making them more attractive for national research aircraft institutions to acquire.

7.4 Terrestrial Radiation Measurements from Aircraft

In the following sections, instruments to measure terrestrial atmospheric radiation are introduced. Section 7.4.1 describes how to measure broadband TIR irradiance with pyrgeometers with no spectral resolution. The following two sections include descriptions of low-resolution (Section 7.4.2) and highly resolved (Section 7.4.3) spectral TIR radiance measurements. Last but not the least, Section 7.4.4 deals with airborne microwave spectral radiance measurements.

7.4.1 Broadband TIR Irradiance Measurement with Pyrgeometers

7.4.1.1 Instruments

The measurement of the TIR irradiance is made with a pyrgeometer. A pyrgeometer provides a voltage that is proportional to the radiation exchange between the instrument and the atmosphere (or ground) in its hemispherical FOV. The detector signal output can be positive or negative. If the sky is colder than the pyrgeometer, the net radiative energy flow will be from instrument to the sky and the output is negative. If the scene is warmer than the pyrgeometer, the net flow is from scene to instrument so the instrument will give a positive output. To calculate the net TIR radiation, it is necessary to know the temperature of the instrument housing close to the detector with high precision and in a time frame synchronized with the detector signal.

A pyrgeometer consists of a thermopile and a silicon dome. Ideally, the dome transmits all terrestrial radiation without any attenuation and reflects all solar radiation. There is, therefore, no thermal emission from the dome. In an ideal instrument, the voltage, V, from the thermopile is linearly related to the net gain or loss of radiant power. The thermopile absorbs and emits as a blackbody at a measured temperature T. The incoming irradiance, F_{inc}, is given by

$$F_{\mathrm{inc}} = \frac{V}{\eta} + \sigma \cdot T^4 \qquad (7.49)$$

where η is the sensitivity of the instrument and σ the Stefan–Boltzmann constant given by Eq. (7.36). In reality, silicon domes typically have a transmissivity that varies between 0.2 and 0.4 over the spectral range of interest (4–50 μm). This low transmissivity causes temperature differences to arise between the thermopile and the dome that need to be determined.

Commercial instruments are available from numerous sources, of which Kipp & Zonen and Eppley are two of the main providers. The World Meteorological Organization (WMO) operates the BSRN and specifies the operating conditions for these instruments on the ground. They recommend screening from the direct solar beam and ventilation to remove heating effects of the dome. The sensitivity of this instrument to dome absorption and emission causes the largest errors associated with airborne operations. Aircraft installations do not allow for solar shielding, and ventilation across the dome is nonuniform. Furthermore, ventilation on the ground is aimed at thermal stabilization between the dome and the instrument body, but on an aircraft, ventilation of the instrument body is often not possible or complete. Airborne installations also suffer from dynamic heating of the dome's leading edge caused by deceleration of the air which, once again, is nonuniform.

Philipona, Fröhlich, and Betz (1995) showed that the incoming irradiance is better represented as

$$F_{\text{inc}} = \frac{V}{\eta} \cdot (1 + k_1 \cdot \sigma \cdot T_{\text{B}}^3) + k_2 \cdot \sigma \cdot T_{\text{B}}^4 - k_3 \cdot \sigma \cdot (T_{\text{D}}^4 - T_{\text{B}}^4) \tag{7.50}$$

where k_1, k_2, and k_3 are constants that are a function of the dome transmissivity and reflectivity, emissivity of the dome, and emissivity of the detector. T_{B} and T_{D} are the body and dome temperatures, respectively.

To put things in context, it is useful to consider the typical thermal environment of a pyrgeometer. Foot (1986) described the installation on the C-130 Hercules that the UK Met Office operated from the early 1970s until 2001. He argued that the pyrgeometer temperatures were typically of the order of 5 °C to 10 °C warmer than the ambient air and that T_{B} and T_{d} can differ by 0.5 °C–1.0 °C. Foot (1986) also highlighted the fact that the thermal mass of the dome is significantly lower than that of the detector assembly. Therefore, the sensitivity of the instrument to changes in air temperature (often seen in airborne test flights) is nonuniform; thermal equilibration is required before data can be interpreted with confidence. Foot (1986) concluded that the dome temperature cannot be measured accurately enough at a single point. This is in agreement with Philipona, Fröhlich, and Betz (1995), who measured their ground-based instrument's dome temperature at three separate points. Foot (1986) also concluded that internal temperature gradients were present in typical airborne pyrgeometer installations. Rather than applying a correction for internal temperature gradients, Foot developed a new thermopile, known as the Foot thermopile (Foot, 1985) (Figure 7.13).

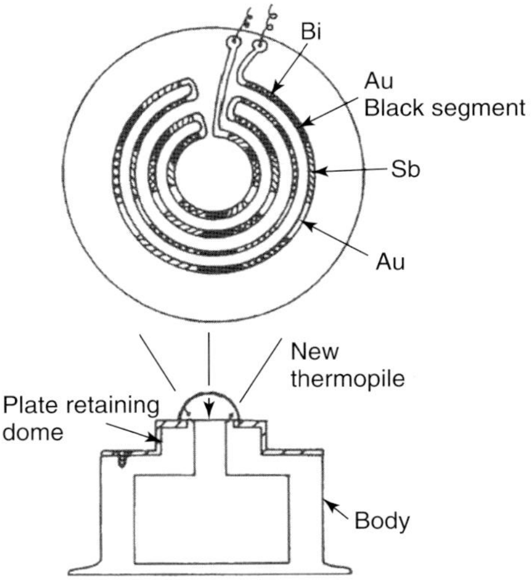

Figure 7.13 A schematic of the Foot Thermopile for airborne pyrgeometers. Figure modified from the original in Foot (1986). (Copyright 1986 American Meteorological Society (AMS). Reprinted in modified form with permission by AMS.)

The Foot thermopile has been used ever since on both the Met Office research aircraft: the C-130 and the new Met Office and Natural Environment Research Council jointly funded BAe146. The Foot thermopile consists of 12 pairs of antimony and bismuth junctions arranged in a circular pattern. Gold is used to connect between these metals and alternate gold stripes are painted with Nexel optical black paint. The differential effect is generated between the different emissivities of the paint (>0.99) and the gold (<0.01). Foot (1986) reported airborne tests that showed this new thermopile structure to be significantly less sensitive to temperature gradients introduced by the aircraft changes in airspeed and temperature.

7.4.1.2 Calibration

Calibration of pyrgeometers is conducted by mounting the pyrgeometer to view the interior of a thermally stabilized blackbody target (Philipona, Fröhlich, and Betz, 1995). Care needs to be taken to achieve thermal equilibration of the instrument by ventilating clean, dry air. Calibration over a range of blackbody temperatures allows the various coefficients to be defined, provided that detailed measurements are made of the dome and instrument temperatures. Following this type of calibration, a ground-based pyrgeometer can achieve an accuracy of approximately $\pm 2\,W/m^2$. Analysis of data gathered on airborne pyrgeometers suggests that the highest achievable accuracy is $\pm 10\,W\,m^{-2}$ due to significant uncertainties in the instrument body and dome temperatures. The measurement of broadband-terrestrial irradiance continues to be a significant challenge for airborne platforms.

7.4.2 TIR Spectral Radiance

7.4.2.1 Instruments

In general, a system for detecting IR radiation consists of the optical system, a detector, and a signal processing system. In developing an airborne system, all these components need to be considered on the basis of the spectral range of interest. The combination of specific window materials and sensitive detectors tuned to selected wavelengths facilitates fast response instrumentation. Conversely, interferometry has slower response but wider spectral coverage.

There is a wide range of materials that transmit energy in the TIR, each with specific spectrally dependent transmittance properties. Care must be taken in the choice of material used in airborne instrumentation due to the potentially harsh environment that optics can experience. A tropospheric research aircraft may operate over a typical temperature range from $+40^\circ C$ to $-60^\circ C$ and a relative humidity range from 0% to 100%. It may also suffer from abrasion by sea salt, dust, and ice.

An TIR radiometer usually consists of the following elements:

(i) *Calibration targets:* Typically a hot and cold blackbody target that ideally is observed through the same optical chain as the atmosphere.
(ii) *Window:* A transmissive material that is the interface between the atmosphere and the detector assembly. Zinc selenide (ZnSe) is a common material that has

Table 7.3 Types of TIR detectors: PZT = lead zirconate titonate, TGS = triglycine sulfate, $LiTaO_3$ = lithium tantalate, PbS = lead sulfide, PbSe = lead selenide, InSb = indium antimonide, HgCdTe = mercury cadmium telluride (MCT).

	Type	Detector	Spectral response (μm)	Operating temperature (K)
Thermal	Thermocouple/pile	Golay cell	Depends on window material	300
	Bolometer	PZT, TGS		300
	Pneumatic cell	$LiTaO_3$		300
	Pyroelectric detector			300
Quantum	Photoconductive	PbS	1–3.2	300
		PbSe	1.5–5.2	300
		InSb	1–6.7	213
		HgCdTe (MCT)	0.8–25	77
	Photovoltaic	InAs	1–3.8	77
		InSb	1–5.5	77
		HgCdTe	2–16	77

a fairly flat spectral response and transmits radiation across a wide spectral range. It often requires an anti-reflective coating and has a transmittance of around 65%. There are other TIR window materials that have higher transmittance, but they are often less applicable to the airborne environment.

(iii) *Filter:* A narrow bandpass highly transmissive material that allows selection of specific wavelengths to the detector. Some instruments have single filters in front of a detector, while others utilize a filter wheel to rotate a selection of filters in front of the detector.

(iv) *Detector:* They are classified as thermal or quantum types. Thermal detectors convert absorbed TIR energy into thermal energy and their photo-sensitivity is independent of wavelength. Most thermal detectors do not require cooling but have slow response times and poor detection capability. Quantum detectors usually offer higher performance and faster response but often require cooling and exhibit photo-sensitivity that is wavelength dependent. Table 7.3 shows the types of TIR detectors.

7.4.2.2 **Calibration**

Photovoltaic (quantum) detectors consist of a p–n junctions. Incoming photons generate the transfer electrons from one material to another, which generates a change in the potential difference across the detector. These are fast response detectors. In photoconductive (quantum) detectors, the absorbed photons change the number of free electrons and electron holes in the detector material, causing a change in the conductivity of the detector. Photodiode detectors can be utilized in a photovoltaic or photoconductive mode. In many detectors, the current detected with no incident radiation (the dark current) is temperature dependent. Therefore, in many radiometers, the incoming observed radiance at the detector is alternated between the atmospheric scene and a black target to allow compensation for the change in dark current during calibration.

7.4.2.3 Application

Simple single channel instruments are utilized widely by atmospheric research aircraft for routine measurement of the radiance leaving the surface. One such commonly used commercially available instrument is the Heitronics KT19 series. For laboratory use, the manufacturer's calibration is usually sufficient for this type of instrument. However, for airborne operations where the environment is more variable, most users utilize external calibration targets periodically inserted into the FOV of the instrument to ensure accurate calibration. Other widely used instruments include the Cimel Electronique Airborne Radiometer CE-332, which measures at three TIR wavelengths.

With a knowledge of the surface emissivity and the downward radiance, filter radiometers sensitive in the atmospheric window region (9–12 μm) allow fast and accurate measurement of sea surface temperature (SST). Single channel radiometers can also be used to measure the air temperature at the aircraft flight level by measuring the radiance at a wavelength in the middle of the strong carbon dioxide absorption feature at 4.3 μm. At this wavelength, the absorption in the atmosphere is so strong that the radiometer is sensitive only to radiation over a path length of around 10 m. Mounting a radiometer normal to the aircraft horizontally allows a direct measure of air temperature at flight altitude. Because the radiance at 4.3 μm is low, the 15-μm carbon dioxide absorption band may be better suited for colder air temperature measurements. The key benefit of this radiometric air temperature measurement over conventional platinum resistance thermometer techniques is that it does not suffer from wetting and hence allows in-cloud measurement of air temperature. An in-cloud temperature probe of this type was operated for many years on the Met Office C-130 aircraft and an Ophir Air Temperature Radiometer is currently operated on the NCAR C-130 (Beaton, 2006).

7.4.3 TIR Interferometry

7.4.3.1 Background

At TIR wavelengths, the absorption and emission by gases in the atmosphere becomes an important element of the radiative transfer through the atmosphere. In the TIR, the dominant atmospheric absorbing gaseous species are water vapor, carbon dioxide, and ozone. However, other gases such as methane, carbon monoxide, and CFCs also show characteristic absorption features in this spectral region. Furthermore, in the atmospheric window regions, the IR radiance is determined by the temperature and emissivity properties of the land or sea surface. Transfer of TIR radiation is also significantly influenced by the presence of clouds and coarse-mode aerosol particles such as sea salt, mineral dust, and volcanic ash.

The gas absorption coefficient changes rapidly with frequency ν. Individual gas molecules possess discrete energy levels associated with different vibrational and rotational states. The structure of the absorption features is often complex and is further complicated because each line in a band is broadened by collision or Doppler broadening; mechanisms that vary with atmospheric temperature and pressure.

This ability to resolve complex spectral details leads to the potential to measure the concentration of species in the atmosphere and to differentiate between gaseous and particulate absorption features.

7.4.3.2 **Instruments**

Michelson interferometers can be used in the laboratory to measure the detailed atmospheric absorption structure and are flown on a few research aircraft across the world. Radiation incident on a Michelson interferometer is partially reflected and partially transmitted by a beam splitter. A variable optical path difference is introduced between the reflected and transmitted paths before recombination of the beams. The resulting interference pattern is monitored as a function of the optical path difference. This interferogram contains all the spectral information about the source, which can be recovered using the technique of Fourier transform analysis (Griffiths and Haseth, 1986). The optical path difference sampling interval is determined by the fringe spacing of a reference interference spectrum generated by a known laser source that passes through the same optical path (e.g., a He–Ne laser). The maximum optical path difference ($\mathrm{OPD}_{\mathrm{max}}$) defines the maximum spectral resolution of the instrument, which is given by the wavenumber sampling interval $\delta\tilde{\nu} = 1/(2 \times \mathrm{OPD}_{\mathrm{max}})$. This is the spectral-domain Nyquist sampling interval: the continuous spectrum limited by the maximum Optical Path Difference (OPD) of the interferometer can be completely reconstructed from any set of discrete samples with regular spacing, $\delta\tilde{\nu}$ (Griffiths and Haseth, 1986).

Laboratory-based interferometers may have $\mathrm{OPD}_{\mathrm{max}}$ of a few meters allowing extremely fine detail of absorption spectra to be sampled. The $\mathrm{OPD}_{\mathrm{max}}$ of airborne interferometers needs to be chosen with care. The separation of the beams is achieved by physically moving one mirror in the optical path relative to the other. The longer the path length, the more physical space is required and the more prone the system is to vibrations in this movement. Long mirror scans also take longer and the balance between resolution and spatial sampling along the flight track needs to be considered in the trade off. The random noise in an interferogram can be reduced by coaddition of the interferograms. The noise reduces by a factor inversely proportional to the square root of the number of scans (e.g., coaddition of 100 interferometer scans will reduce the random noise component to 0.1 of that present in 1 interferometer scan). When specifying the performance of an airborne interferometer, the noise sensitivity of the target (atmosphere, surface, or clouds) needs to be considered. Factors such as vibration, aircraft speed, and scan time result in typical airborne interferometers having $\mathrm{OPD}_{\mathrm{max}}$ of around 1 or 2 cm.

For illustration, the Airborne Research Interferometer Evaluation System (ARIES) flown on the Met Office C-130 and later the FAAM BAe146-301

Atmospheric Research Aircraft (Wilson, Atkinson, and Smith, 1999) is discussed. The core of ARIES is a rugged commercial interferometer (MR200) developed by ABB Bomem of Canada. A front-end optical head was built by the Met Office to allow the interferometer to view two calibration targets and the atmosphere in the zenith or nadir direction. On the FAAM BAe146, the ARIES interferometer is mounted in a large blister on the port side of the aircraft, providing a clear view of the zenith and the nadir (including across track nadir views out to 60°) scenes. The interferometer is suspended from a frame with antivibration mounts designed to remove the impact of aircraft vibrations. A schematic diagram of the ARIES interferometer optics is given in Figure 7.14.

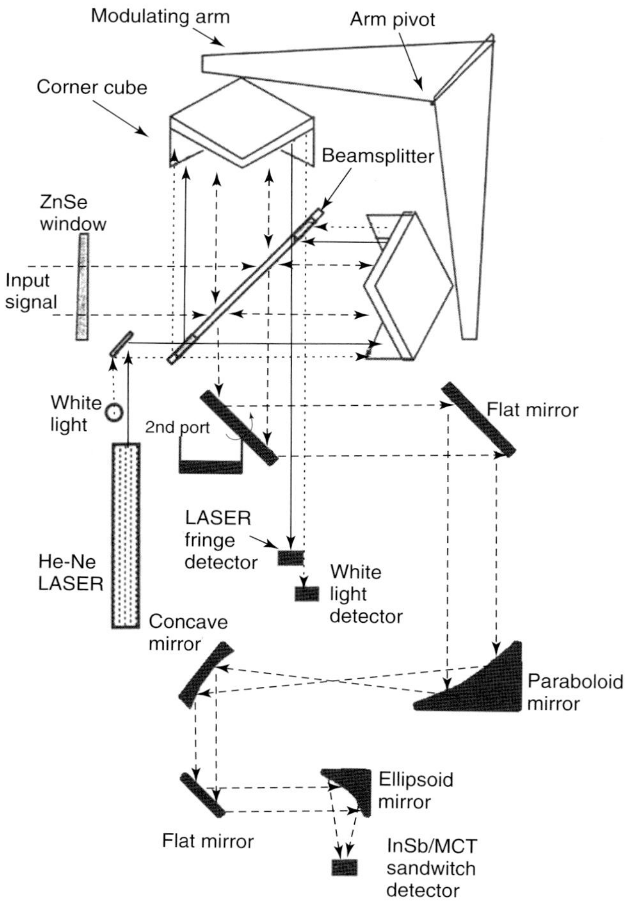

Figure 7.14 A schematic of the ARIES optical layout. Figure modified from the original in Wilson, Atkinson, and Smith (1999). (Copyright 1999 American Meteorological Society (AMS). Reprinted in modified form with permission by AMS.)

After exiting the pointing optics, radiation passes into the main optics of the interferometer through a zinc selenide window and onto a potassium bromide beam splitter. The main optics of ARIES are contained in a sealed aluminum box, which is charged to a positive pressure with dry nitrogen gas to avoid absorption or emission by the gases within the interferometer. The radiation is directed on to a pair of corner-cube mirrors, which are mounted on a pivot arm. The pivot arm can conduct two full scans in each direction in 1 s. The motion of the pivot arms moves the mirrors in opposite directions with respect to the beam splitter. This differential motion introduces a OPD_{max} of 1.037 cm, which equates to a wavenumber sampling interval of 0.482 cm^{-1}. After reflection by the corner cube mirrors, the radiation interferes at the beamsplitter before being directed to two IR detectors arranged as a sandwich cooled by a Stirling cooler to around 80 K. The two detectors are mercury cadmium telluride (MCT; 5.5–18 μm, 550–1800 cm^{-1}) and indium antimonide (InSb; 3.3–5.9 μm, 1700–3000 cm^{-1}). MCT detectors normally suffer from a nonlinearity effect, where their sensitivity varies with the magnitude of the incoming radiation and must be corrected (Fiedler, Newman, and Bakan, 2005). ARIES is a dual-port interferometer. The measured signal is proportional to the difference between an internal second-port blackbody and the scene. The optical path difference sampling interval is set by the fringe spacing of an internal He–Ne laser. When switched on, the ARIES instrument uses a broad spectra white light source to determine the mirror positions related to the zero path difference fringe. Once this position of zero path difference has been found, the white light source is turned off.

7.4.3.3 Calibration

The calibration of an interferometer, like other thermal radiometers, requires views of well-characterized blackbody targets. The ARIES instrument has two calibration targets, which can be temperature controlled. One of these targets is usually maintained at a temperature of 333 K (a hot target) and the other is set to a few degrees above the ambient temperature (cold target). The temperature is measured using a network of platinum resistance thermometers embedded within the targets. The targets consist of an array of pyramids coated with an IR black Nextel paint. The emissivity of the targets is measured by direct comparison of calibrations using these targets with those of a national standard calibration target. The interferogram recorded while viewing a calibration target is transformed to a complex radiance spectrum using a Fourier transform:

$$S_k = \sum_{j=0}^{N-1} V_j \cdot \exp(2\pi \cdot \mathrm{i} \cdot j \cdot k/N) \tag{7.51}$$

where j and k are integers, N is the number of points on the interferogram, $\mathrm{i} = \sqrt{-1}$, V_j is the real interferogram at point j, and S_k is the complex raw spectrum at point k. The complex raw spectra S are related to the views of the two blackbody targets that result in the following equations:

$$S_\mathrm{n} = G \cdot (I_\mathrm{n} - I_\mathrm{pn} - C) \tag{7.52}$$

where the subscript n refers to either the cold or hot calibration target, I_n is the scene radiance spectrum from that calibration target, I_{pn} is the radiance spectrum from the internal second-port blackbody during the measurement of calibration target n, and G and C are the complex gain and radiance offset, respectively. The radiance spectra I are computed from the Planck functions for the two calibration scenes, allowing the derivation of the complex gain and radiance offset. Following calibration, an atmospheric scene radiance spectrum I can be found. Although the resulting scene radiance will be a complex array, it is normally assumed that the complex part is pure noise and the radiance spectrum is the real part only.

It should be noted that because of different phase effects, data associated with each direction of movement of the pendulum mirror scan should be treated separately. That is, separate calibrations should be applied. The resulting radiance spectra can be averaged to reduce noise, but in the calibration phase, each direction of spectrum should be dealt with independently. Michelson interferometers, such as ARIES, can have excellent noise performance due in part to the Jacquinot advantage, which allows for a large energy throughput due to the ability to use large apertures. The noise of interferometers is expressed as the noise-equivalent temperature difference. For ARIES, this varies across the spectrum but is <0.1 K in the center of the atmospheric window region.

The application of Fourier transform spectroscopy to atmospheric science is beneficial because of the high spectral resolution and the high accuracy achievable. However, to obtain meaningful retrievals of physical quantities such as temperature, water vapor, and cloud properties, one requires a precise knowledge of the effect of the instrument characteristics on the acquired data. The instrument line shape (ILS) of a Fourier transform spectrometer (FTS) is governed by the angular distribution of light in the interferometer. This influence is sometimes referred to as the off-axis effect. The ILS is defined as the instrument response to a monochromatic scene. It is mainly governed by the windowing effect of the Fourier transform, and it is also affected by other effects such as FOV configuration, diffraction, optical aberration, and misalignment. The main contribution to the ILS of an FTS is the truncation of the interferogram arising from the finite optical path difference measurement scan. Its effect is to convolve the spectrum with a sinc function corresponding to the optical path difference measurement window.

Another effect to consider is the finite source size. Light rays propagating through the interferometer at slightly different angles undergo slightly different paths. The resultant ILS is a distorted sinc function shifted toward lower wavenumbers. In analysis of airborne FTS data, it is imperative that the ILS be understood for the instrument. This is usually achieved by observing a known monochromatic source that completely fills the instrument FOV. Comparison of observed spectra with calculated spectra needs to be done by applying the ILS to the computed spectra to transform the spectra in to that which would be observed by the instrument. This is normally achieved by performing a Fourier transform analysis on the computed spectra, convolving the resulting interferogram with the known ILS, and then transforming the spectra back to radiance space.

7.4.3.4 Principal Component Noise Filtering

Principal component analysis (PCA) is a multivariate analysis technique commonly used to reduce the dimensionality of a data set that has a large number of interdependent variables, like a set of IR radiance spectra from an interferometer. PCA essentially performs a singular value decomposition on the covariance matrix of the observations, mapping the spectrally correlated observations into decorrelated quantities expressed as principal component coefficients. The PCA generates an array of Eigenvectors of principal component scores, which are ranked according to magnitude. The highest order Eigenvectors relate to the most spectral information (i.e., correlated structures that are apparent in the largest number of most of spectra), while the lower order Eigenvectors relate to uncorrelated spectral features or noise. If one can determine beyond which point in the Eigenvector sequence the content is only related to instrument noise, it is theoretically possible to reconstruct the original spectra using only the leading Eigenvectors and hence reduce the noise in the observation. This is of particular relevance to hyperspectral interferometer data where the positions of water vapor lines or lines from other trace gases are coherent from spectrum to spectrum, and they would be expected to be resolved in the high-order Eigenvectors.

The upper panel in Figure 7.15 shows a section of a radiance spectrum measured by ARIES in the middle of the 4-μm band of carbon dioxide. The spectrum is one of several hundred gathered during a nadir-viewing sequence over the Gulf of Mexico. The center panel shows the same spectrum having been passed through a PCA-based noise filter where the spectrum has been reconstructed using only the first 14 Eigenvectors. The power of this technique is self-evident from this figure that clearly shows that the key spectroscopic detail has been maintained, while the unwelcome instrument noise has been removed. The lowest panel in Figure 7.15 shows the residual of the original spectrum and the noise-filtered spectrum, which can be seen to be random in nature, that is, pure noise.

To conduct successful noise filtering, a large set of spectra needs to be gathered and normalized by the initial noise estimate of the instrument. PCA is preformed on this noise-normalized matrix of radiances, after which the spectra are reconstructed from a subset of the leading Eigenvectors and the normalization is removed by multiplication of the original noise estimate. Selection of the cutoff beyond which Eigenvectors are ignored in the reconstruction is critical to ensure useful coherent information is not being discarded. If one generates a purely random synthetic array of data of the same magnitude as the measurement matrix and conducts PCA on these synthetic data, then by analyzing at which point the Eigenvalue of the real data is the same as the Eigenvalue of the synthetic data one can determine the component beyond which the information content is noise.

Turner *et al.* (2006) provide an alternate way to determine the optimal number of Eigenvectors to use in the reconstruction to remove the maximum amount of random error. An example of the noise reduction is shown in Figure 7.16 where the Eigenvalue of 1 of the 2422 individual observation spectra gathered over the Gulf of Mexico is shown as a function of its component number and a similar line for the synthetic matrix is plotted. The two lines cross at component 19, indicating

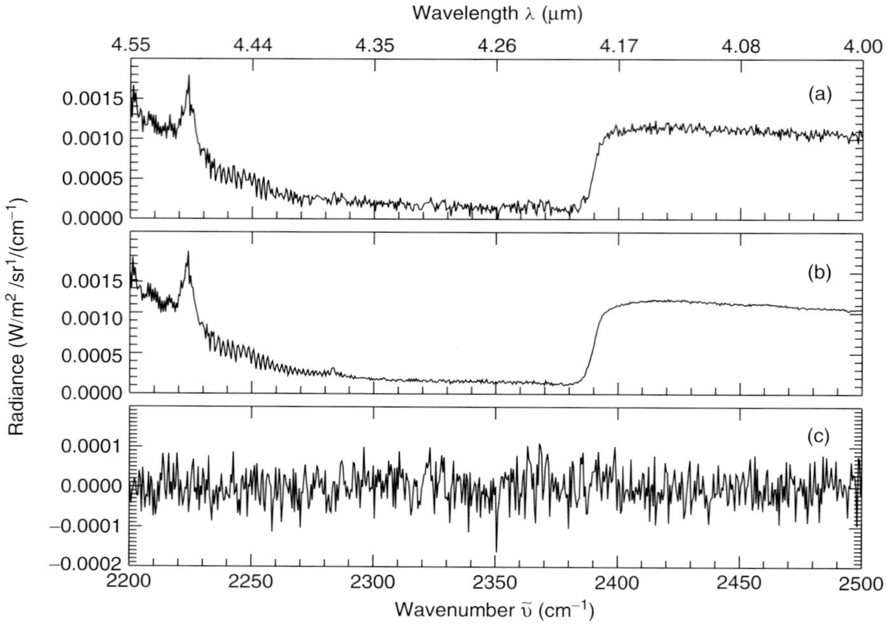

Figure 7.15 The use of PCA in noise filtering interferometer data. The upper panel shows the measured spectrum, the middle panel shows the same spectrum reconstructed using only the first 14 Eigenvectors. The lower panel shows the residual of these two spectrum which has the structure of random noise. (a) Unfiltered radiance, (b) PCA-filtered radiance, (c) Residual noise.

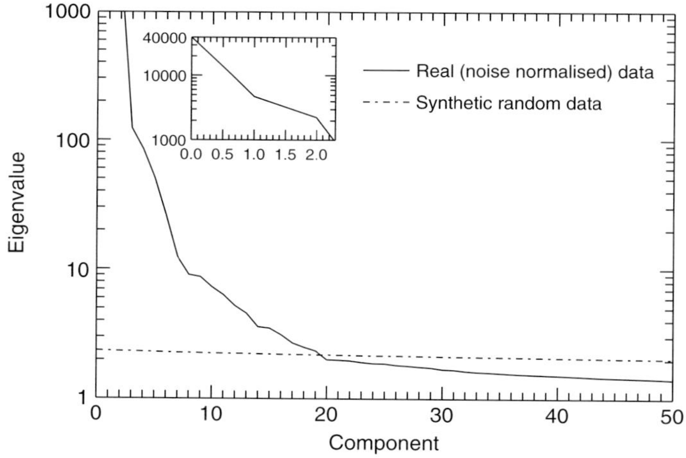

Figure 7.16 Eigenvalue of one of the 2422 individual observation spectra gathered over the Gulf of Mexico is shown as a function of its component number (solid line) and a similar line for the synthetic matrix is plotted (dotted line).

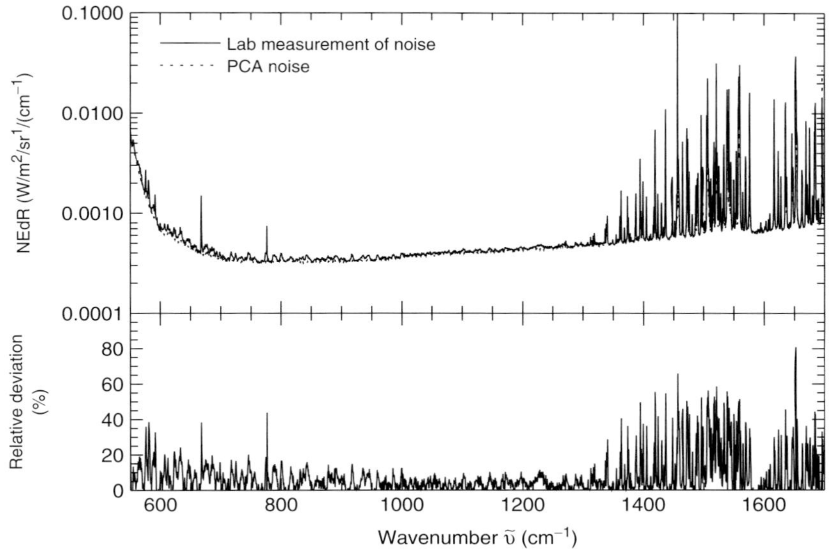

Figure 7.17 The instrument noise characteristics of the ARIES instrument as measured in the laboratory presented as the noise-equivalent radiance observed whilst looking at a laboratory calibration target held at a constant 199 K (solid line). The PCA noise estimate is the average of the residuals of the observed minus reconstructed radiance spectrum from 4577 nadir spectra re-constructed using the leading 30 Eigenvectors.

that all the information in the observed spectra can be reconstructed using only the leading 19 elements of the Eigenvector matrix. The number of Eigenvectors required to represent all the information content of the original spectra is scene dependent. If the matrix of observations contains spectra from different surface types, with and without clouds, aerosol particles, and so on, then the number of Eigenvectors needed in the reconstruction would increase.

This technique can also be used to compress data for storage. The process of PCA is also particularly useful as a means of monitoring the performance of the instrument because the noise characteristics should remain constant. Figure 7.17 shows the instrument noise characteristics of the ARIES instrument measured in the laboratory and presented as the noise-equivalent radiance observed while looking at a laboratory calibration target held at a constant temperature of 199 K (solid line).

The dotted line shows the PCA noise estimate, which is the average of the residuals of the observed minus reconstructed radiance spectrum from 4577 nadir spectra reconstructed using the leading 30 Eigenvectors. This figure clearly demonstrates that the noise performance of the ARIES instrument on the aircraft is comparable to that achieved in the laboratory, evidence of the measures taken to reduce vibration artifacts induced by a four-engine jet aircraft. The continual monitoring of this type of data allows the user of the instrument to have confidence in the quality of the data. The instrument noise of ARIES can be seen to vary significantly across the spectrum. Some of this is due to the inherent sensitivity of the detectors that have poorer performance for lower wavenumbers. Other features, such as noise spikes around 1500 cm^{-1}, are strong water vapor absorption lines, which arise from water vapor in the optical path between the ZnSe window and the external calibration targets.

7.4.3.5 Application

Figure 7.18 shows a TIR spectrum measured by the ARIES instrument viewing in the nadir from an altitude of around 10 km over open ocean. The spectrum has been noise filtered using the PCA technique described earlier, reconstructing the radiance spectrum using the leading Eigenvectors. Instead of spectral radiances, the brightness temperature T_B, as defined in Section 7.2.5, paragraph (b), is plotted as a function of wavenumber or wavelength.

The TIR spectrum is dominated by spectrally resolved line features associated with absorption and emission by gases in the atmosphere. The spectrum has been coded by letters A to F to identify the regions where the signal is dominated by the key gaseous species: carbon dioxide (B), ozone (C), water vapor (D), methane (E), and nitrogen dioxide (F). Atmospheric window regions are indicated by A. Here, the signal is dominated by the underlying surface that is modulated by the water vapor

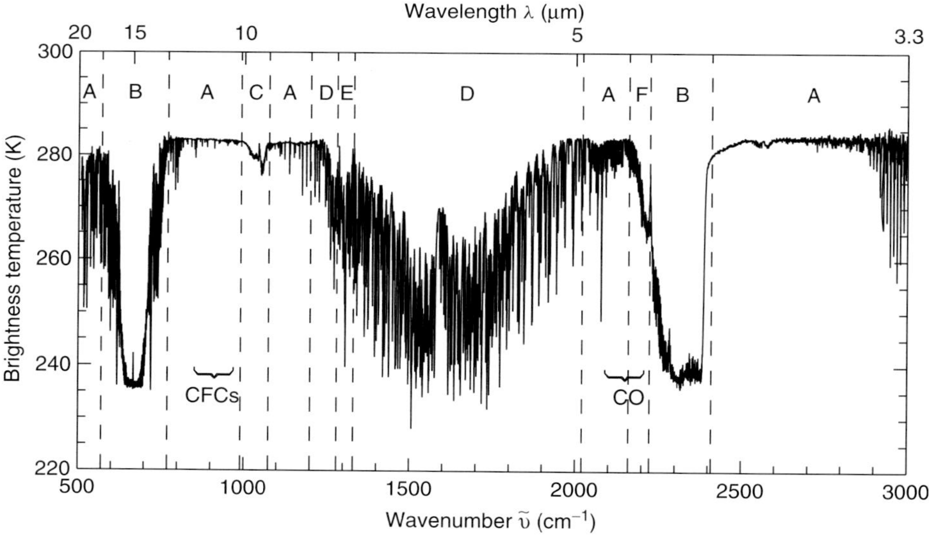

Figure 7.18 A TIR PCA noise-filtered spectrum measured by the ARIES instrument viewing in the nadir direction from an altitude of around 10 km over open ocean. The spectrum has been coded to identify the regions where the signal is dominated by the key gaseous species: Atmospheric window regions (A), carbon dioxide (B), ozone (C), water vapor (D), methane (E) and nitrogen dioxide (F). Also indicated are spectral regions where other atmospheric trace gases exhibit clearly distinguishable spectral features (CFCs, CO etc).

continuum. The figure also indicates spectral regions where other atmospheric trace gases exhibit clearly distinguishable spectral features (CFCs, CO, etc.) It is evident from this figure that there is a wealth of data available from this instrument. For example, it is possible to retrieve profiles of temperature, water vapor, and ozone and total column amounts of carbon monoxide. The window region is sensitive to the temperature and emissivity of the surface, facilitating retrieval of these two properties for land surface studies or remote sensing purposes.

Figure 7.19 shows the brightness temperature spectrum across the window region for a range of different scenes to illustrate the information content. The upper panel (a) is a clear-sky spectrum over the ocean; between 800 and 1000 cm^{-1}, where there is little atmospheric absorption. The high brightness temperatures are representative of the surface temperature. Below 770 cm^{-1}, the brightness temperature drops, because increased carbon dioxide absorption results in emission from higher (colder) layers in the atmosphere. Sampling across the wings of the

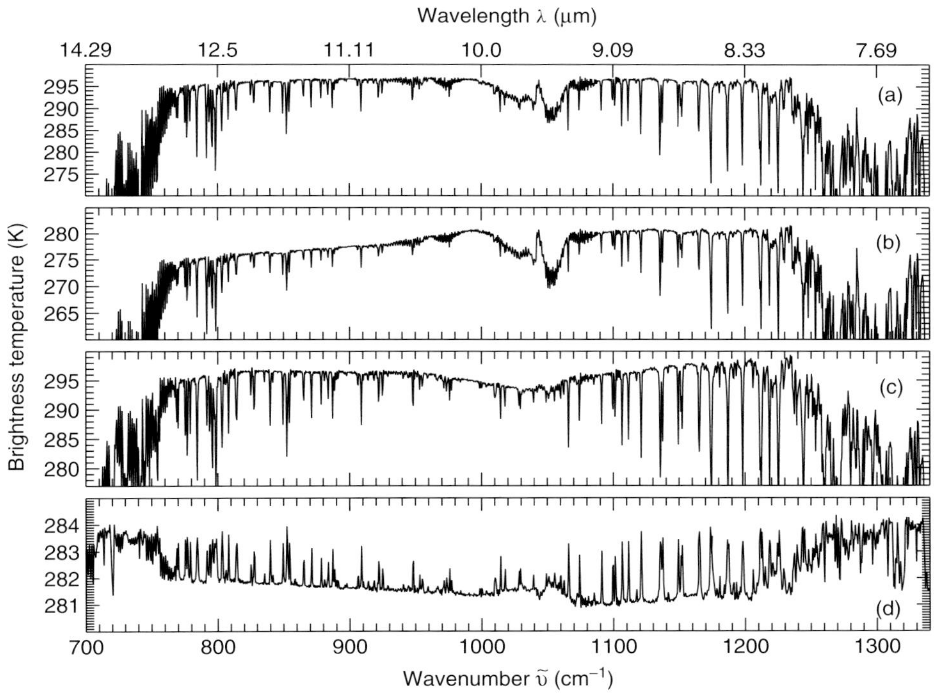

Figure 7.19 The brightness temperature spectrum across the window region for a range of different scenes as an indication of the information content. The upper panel is a clear-sky spectrum over ocean. The second panel is from a view down through cirrus clouds. The third panel shows the upward spectrum observed through a dense desert dust episode off the coast of Africa. The fourth panel shows an upward spectrum gathered whilst flying low over the surface near Oklahoma at night time. (a) Clear sky over ocean, (b) Cirrus, (c) Saharan dust, (d) Low altitude over land.

carbon dioxide band to the atmospheric window region enables retrieval of the atmospheric temperature profile.

The second panel (b) of Figure 7.19 is from a view through cirrus clouds. Here, the scattering and absorption of the ice crystals create a gradient in the window region, which is dependent on ice crystal size and amount. The third panel (c) shows the upward spectrum observed through a dense desert dust episode off the coast of Africa. The characteristic "V" shape across the window region in this spectrum is caused by desert dust absorption, directly traceable spectral behavior of dust refractive indices, and suitable for deriving the total column dust abundance. The fourth panel (d) of Figure 7.19 shows an upward spectrum gathered, while flying low over the surface near Oklahoma at night time. The spectral structure of the surface emissivity is evident in the broad curve across the window region. In this spectrum, the surface is cooler than the atmosphere above, resulting in emission lines from the water vapor in contrast to the absorption lines seen in the other panels. It also indicates the benefit of sampling the spectrum at high spectral resolution where the impact of gases, hydrometeors, and surfaces can be independently identified. The analysis of IR interferometer data is enhanced with additional observations of the temperature and water vapor structure of the atmosphere from dropsondes.

7.4.4 Microwave Radiometers

7.4.4.1 Background

The microwave (MW) spectral range covers the longest wavelengths (0.3 mm to 30 cm) of terrestrial radiation (Table 7.1). For historical reasons mainly based on technological development, the spectral notation is typically in frequency (1000–1 GHz) instead of wavelength. The MW spectral range offers some distinct advantages for remotely sensing geophysical parameters. First, the high transmissivity of the atmosphere (>80% below 40 GHz) permits surface properties to be observed. Second, because wavelengths are relatively large compared with atmospheric particles, scattering can be ignored for most applications.

If scattering is neglected, the vertical ($\mu = 1$) upward spectral radiance $I_\nu^\uparrow$ observed by a nadir viewing airborne MW radiometer at a frequency ν flying at altitude z is given by

$$\begin{aligned} I_\nu^\uparrow(z) &= \varepsilon(\nu, z=0)\cdot B_\nu[T(z=0)]\cdot \exp\left[-\int_0^z b_{\text{abs}}(\nu,\zeta)\,\mathrm{d}\zeta\right] \\ &+\int_0^z b_{\text{abs}}(\nu,z')\cdot B_\nu[T(z')]\cdot \exp\left[-\int_{z'}^z b_{\text{abs}}(\nu,\zeta)\,\mathrm{d}\zeta\right]\mathrm{d}z' \end{aligned} \tag{7.53}$$

where the surface emissivity is $\varepsilon(\nu,\ z=0)$, surface temperature is $T(z=0)$, the Planck function is $B_\nu(T)$ (Eq. (7.23)), and the volumetric absorption coefficient is $b_{\text{abs}}(\nu,\ z)$. The derivation of Eq. (7.53) can be found in textbooks on radiative transfer (e.g., Wendisch and Yang, (2012)). The upward radiance sensed by a downward looking MW radiometer is composed of two terms. (i) The surface term is determined by surface emission (emissivity times the emission of a blackbody given by the Planck function), which is reduced on its way through the atmosphere by absorption processes. (ii) The second term quantifies atmospheric emission and absorption between the surface and the sensor altitude z.

MW upward radiances can be converted to corresponding temperatures T using the Rayleigh–Jeans approximation of the Planck function (Eq. (7.26)), valid for long wavelengths or small frequencies, or more precisely for $h \cdot \nu/(k_B \cdot T) \ll 1$:

$$T \approx \frac{c^2}{2\,k_B \cdot \nu^2} \cdot B_\nu = \frac{c^2}{2\,k_B \cdot \nu^2} \cdot \frac{I_\nu^\uparrow}{\varepsilon} \tag{7.54}$$

Please note that the radiance described by the Planck function is isotropic, that is, $B_\nu = B_\nu^\uparrow$. For deriving Eq. (7.54), we have applied Kirchoff's law (Eq. (7.29)), in the form of

$$B_\nu = \frac{I_\nu^\uparrow}{\varepsilon} \tag{7.55}$$

As the Rayleigh–Jeans approximation is not strictly valid for all relevant temperatures in the MW range, a Planck equivalent should be considered (Lipton, Moncet, and Uymin, 2009). Furthermore, we introduce the atmospheric transmissivity $tr(\nu,\ z,\ z')$ of the layer between flight altitude z and a height z', where z' is an altitude between the ground ($z = 0$) and z, for nadir-viewing geometry ($\mu = 1$) by the following expression:

$$tr(\nu, z, z') = \exp\left[-\int_{z'}^{z} b_{abs}(\nu, \zeta)\,d\zeta\right] \tag{7.56}$$

Using the Rayleigh–Jeans approximation in the form of Eq. (7.54) and the definition of transmissivity by Eq. (7.56), we obtain from Eq. (7.53):

$$\varepsilon(\nu, z) \cdot T(z) \cdot \frac{2\,k_B \cdot \nu^2}{c^2} = \varepsilon(\nu, z = 0) \cdot B_\nu[T(z = 0)] \cdot tr(\nu, z, z' = 0) + \int_0^z b_{abs}(\nu, z') \cdot B_\nu[T(z')] \cdot tr(\nu, z, z')\,dz' \tag{7.57}$$

We transform to

$$\varepsilon(\nu, z) \cdot T(z) = \varepsilon(\nu, z = 0) \cdot \frac{c^2}{2\,k_B \cdot \nu^2} \cdot B_\nu[T(z = 0)] \cdot tr(\nu, z, z' = 0) + \int_0^z b_{abs}(\nu, z') \cdot \frac{c^2}{2\,k_B \cdot \nu^2} \cdot B_\nu[T(z')] \cdot tr(\nu, z, z')\,dz' \tag{7.58}$$

We apply Eq. (7.54)

$$\varepsilon(\nu, z) \cdot T(z) = \varepsilon(\nu, z = 0) \cdot T(z = 0) \cdot tr(\nu, z, z' = 0) + \int_0^z b_{abs}(\nu, z') \cdot T(z') \cdot tr(\nu, z, z')\,dz' \tag{7.59}$$

We utilize the equation relating the brightness temperature and the temperature $[T_B(\nu,\ z) = \varepsilon(\nu,\ z) \cdot T(z)]$ valid for the Rayleigh–Jeans domain (Eq. (7.34)). This yields

$$T_B(\nu, z) = T_B(\nu, z = 0) \cdot tr(\nu, z, z' = 0) + \int_0^z b_{abs}(\nu, z') \cdot T(z') \cdot tr(\nu, z, z')\, dz' \quad (7.60)$$

From Eq. (7.59), it is clear that for high atmospheric transmissivity (typical at low frequencies), the measurement is dominated by the surface term. Because surface emissivity $\varepsilon(\nu,\ z = 0)$ depends on several parameters (frequency, polarization, surface material, wind speed, etc.), multifrequency MW measurements allow surface properties such as snow, sea ice, soil moisture, SST, and ocean salinity to be derived (see Section 7.4.4.3.1 Applications). The strongest surface emissivity contrast occurs between ocean (around 0.5) and land (around 0.9) surfaces. Therefore, the low ocean emissivity provides a cold background to observe thermal emission of atmospheric constituents. This is illustrated in Figure 7.20, top panel, by the increased brightness temperatures over ocean due to liquid water emission below 200 GHz. With increasing frequency, the water vapor continuum absorption increases, making the atmosphere opaque. Therefore, even for upper atmospheric (Müller *et al.* 2008; Mees *et al.* 1995) and astronomical observations above 1 THz, it is necessary to fly above the tropopause.

Several atmospheric gases show rotational absorption lines in the MW spectral range. The strong oxygen absorption complex around 60 GHz has been used since

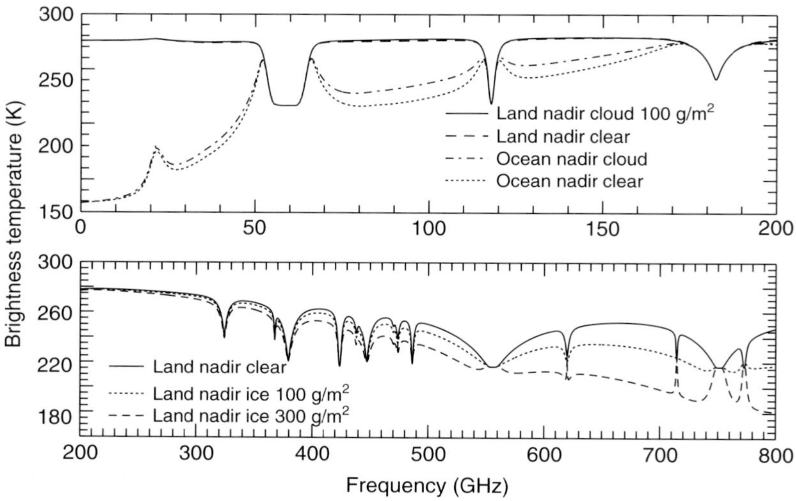

Figure 7.20 Brightness temperatures in the MW spectral range simulated for the US standard atmosphere for nadir viewing over ocean and land, respectively. Top panel: A cloud with liquid water path of 100 g/m^2 has been added. Lower panel: An ice cloud with two different amounts of ice water path.

1978 (by TIROS-N) in microwave sounding units on board polar orbiting satellites for temperature sounding. Within the band center, the atmosphere is highly opaque, allowing for the retrieval of the stratospheric temperature. Moving away from the center, observed brightness temperatures increase as the emission systematically arises from lower (warmer) atmospheric layers (Figure 7.20). Introducing the weighting function W for a nadir-viewing geometry in units of inverse meters (m^{-1})

$$W(\nu, z, z') = \frac{\mathrm{d}tr(\nu, z, z')}{\mathrm{d}z} = b_{\mathrm{abs}}(\nu, z) \cdot tr(\nu, z, z') \tag{7.61}$$

the radiative transfer equation in the MW spectral range (Eq. (7.60)) can be written as

$$\begin{aligned} T_{\mathrm{B}}(\nu, z) = {} & T_{\mathrm{B}}(\nu, z = 0) \cdot tr(\nu, z, z' = 0) \\ & + \int_0^z W(\nu, z, z') \cdot T(z')\,\mathrm{d}z' \end{aligned} \tag{7.62}$$

At frequencies near the center of absorption lines where optical thickness is large and atmospheric transmissivity is low, the first term of Eq. (7.62) disappears and the weighting functions give a clear description of the vertical resolution of the temperature measurement (Figure 7.21). In addition to the two oxygen absorption features at 60 and 118 GHz, two pressure-broadened water vapor lines occur below 200 GHz (at 22 and 183 GHz). While the weaker 22-GHz line is commonly used

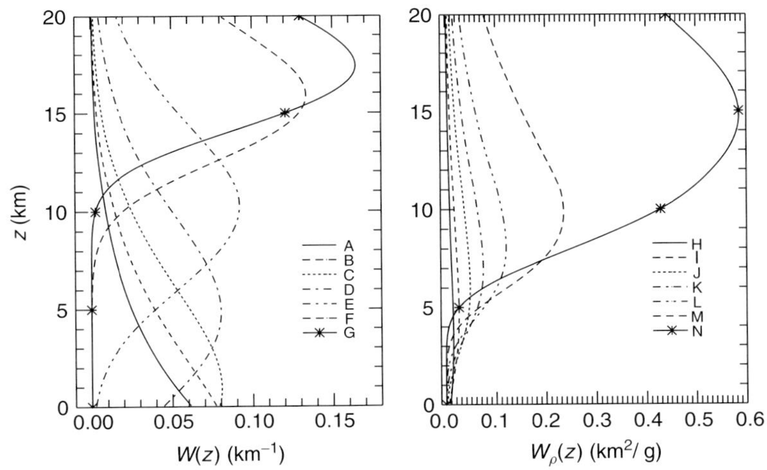

Figure 7.21 Nadir-viewing weighting functions for US 1976 Standard Atmosphere over black surface. Channels of the Hamburg Microwave Package (HAMP) instrument (Table 7.4), along the 60-GHz oxygen band (left panel) and the 183-GHz water vapor line (right panel) are shown. Curve notation as follows: A = 50.3 GHz, B = 51.76 GHz, C = 52.8 GHz, D = 53.75 GHz, E = 54.94 GHz, F = 56.66 GHz, G = 58.0 GHz, H = 183.31 ± 12.5 GHz, I = 183.31 ± 7.5 GHz, J = 183.31 ± 5.0 GHz, K = 183.31 ± 3.5 GHz, L = 183.31 ± 2.5 GHz, M = 183.31 ± 1.5 GHz, N = 183.31 ± 0.6 GHz.

from the ground for water vapor profiling (Westwater *et al.* 2005), the stronger line at 183 GHz is used on the advanced microwave sounding unit (AMSU) onboard of the NOAA and MetOp (*http://www.esa.int/esaME/index.html*) satellites, as well as on several airborne instruments. With a pressure broadening of about 3 MHz/hPa, tropospheric profiling can be realized with several channels spaced by a few GHz along the line, as is done with AMSU. High spectral resolution observations with channel bandwidths of 1 MHz or lower can be exploited for stratospheric and mesospheric profiling. For this purpose, however, airborne instruments employ an upward geometry. In this respect, trace gas profiling for stratospheric chemistry is of high interest as several species such as ozone, ClO, and HCl show significant emission lines above 200 GHz.

At lower MW frequencies, the emission by cloud liquid water is simply proportional to the liquid water content and does not depend on drop size; this is a characteristic of a molecular scattering regime where the atmospheric particles are much smaller than the wavelength. By combining measurements in window regions (i.e., 35 and 90 GHz) with information from water vapor absorption channels (i.e., 23.8 GHz), the liquid water path (LWP) can be derived over oceans. LWP has also been derived from measurements along the wings of the 183-GHz water vapor line (Zuidema *et al.* 2012). To overcome the problem of increasing water vapor emission and surface contribution, the aircraft flew below the cloud with the radiometer pointing upward. Most of the water vapor information comes from the 23.8-GHz channel, which has a nearly constant response with height and therefore simultaneously allows the determination of the integrated water vapor (IWV). With higher frequencies and particles of larger diameter D (rain drops), the size parameter $\pi \cdot D/\lambda$ increases, Mie theory needs to be considered and assumptions on the drop size distribution needs to be made.

Below roughly 60 GHz, ice clouds are transparent to MW radiation. At higher frequencies, scattering effects by frozen particles influence the upward MW signal. Therefore, a brightness temperature depression can be observed as the upward MW signal is reduced due to scattering by frozen particles (Figure 7.20). The strength of this scattering effect depends strongly on particle properties such as size, shape, and orientation. Therefore, precipitation rate is often derived from a scattering index (Grody, 1991). It has to be noted that this assumes a relation between the frozen particles aloft and the liquid precipitation at the ground. The advantage of scattering techniques is that they can be applied above ocean and land. At shorter wavelengths, for example, submillimeter, scattering by smaller cloud ice particles becomes important (Figure 7.20) and facilitates the determination of ice water path. It should also be noted that hydrometeor scattering may lead to polarization effects.

To retrieve the geophysical parameters, a set of representative atmospheric states is often used as input for radiative transfer simulations. Subsequently, a statistical retrieval algorithm is derived as the relation between the given atmospheric state, for example, IWV and LWP, and the corresponding set of calculated brightness temperatures. The atmospheric state is typically taken from a large set of radiosonde profiles with diagnosed cloud properties or high-resolution numerical weather prediction models.

7.4.4.2 **Instruments**

Several different types of MW radiometers have been flown over the past two decades on various aircraft for diverse applications ranging from land surface imaging, thermodynamic profiling, and trace gas observations to observations of cloud and precipitation properties (Table 7.4). In the simplest configuration, an MW radiometer is mounted on an aircraft with the antenna pointing at a fixed angle either toward the surface or the sky for upper atmosphere or astronomical applications. If the aircraft is flying at high altitudes, cross-track or conical scanning is desirable to acquire spatial information. An example of cross-track scanning is the National Polar-Orbiting Operational Environmental Satellite System (NPOESS) aircraft sounder test bed – MW spectrometer (NAST-M) (Blackwell *et al.* 2001) with 17 channels along the two oxygen bands at 60 and 118 GHz. Cross-track scanning is technically simple, but this approach cannot be used if polarization measurements are desired. Fortunately, polarization effects are not of interest for applications such as temperature profiling.

Because polarization differences become important for observations of the Earth's surface, conical scanning radiometers such as the conical scanning millimeter-wave imaging radiometer (CoSSIR) (Wang *et al.* 2007) have a fixed angle of incidence and a vertical axis of rotation. The polarimetric scanning radiometer (PSR) (Stankov *et al.* 2008) concept makes use of a scan-head, which allows both cross-track and conical scanning. The scan-head can be integrated into various aircraft platforms and contains several fully polarimetric or dual-orthogonal linearly polarized radiometers. Similarly, the microwave airborne radiometer scanning system (MARSS) (McGrath and Hewison, 2001) is operated in an external pod that has the advantage that no additional window materials are used. These windows can lead to undesired attenuation effects that must be determined in the calibration. Such effects become problematic with increasing frequencies where stable materials such as high-density polyethylene show significant attenuation and contribute significantly to the observed signal (Mees *et al.* 1995).

To treat the observation as a pencil beam (a narrow FOV), a highly directional antenna with a beam width, commonly defined as FWHM, in the order of a few degrees is needed. Therefore, the diffraction limit causes the antenna size to become relatively large (up to 0.5 m for $<$10 GHz), which often proves to be a limiting factor for the operation on a smaller aircraft. Further antenna requirements are low spill-over losses and good side-lobe suppression. Symmetric Gaussian-shaped beams are achieved with corrugated feed horns that are sometimes illuminated with a scanning system. Flight altitude and antenna beam width determine the spatial resolution at the Earth surface. For example, for a flight altitude of 6 km and a beam width of 8°, the spatial resolution at the surface of the Earth is about 2 km.

Because of the low radiances emitted in this spectral range, MW radiometers need to amplify the signal received at the antenna by about 80 dB. Over the past decade, low-noise amplifiers (LNAs) have become available for frequencies up to

Table 7.4 Synthesis of existing airborne instruments.

Sensor	Name	Reference	Aircraft/campaigns	Frequencies (GHz)
AMMR	Airborne multichannel microwave radiometer	Lobl *et al.* (2007)	NASA P-3/TOGA-COARE, Wakasa Bay	21 37
AMSOS	Airborne microwave stratospheric observing system	Müller *et al.* (2008)	Learjet 35A T-781/ THESEO, Lautlos, Scout-O3	High spectral resolution 183.3 ± 0.5 (H_2O) 175.9 ± 0.5 (O_3)
ASUR	Airborne submillimeter radiometer	Mees *et al.* (1995)	FALCON, DC-8/THESEO, SOLVE, PAVE	High spectral resolution Lines between 600 and 670
CoSSIR	Conical scanning submillimeter-wave imaging radiometer	Evans *et al.* (2005)	ER-2/ CRYSTAL-FACE TC4	3 × 183.3, 220 3 × 380.2 3 × 487.25, 640, 874
CoSMIR	Conical scanning millimeter-wave imaging radiometer	Wang *et al.* (2007)	ER-2/SSMI/S validation	50.3, 52.8, 53.6 91.665 V, H, 150 183.3 ±1, ±3, ±6.6
GVR	G-band vapor radiometer	Zuidema *et al.* (2012)	NCAR C-130/ VOCALS-REx	183.31 (±1, ±3, ±7, ±14)
HAMP	Halo microwave package		HALO/NARVAL	7 × (22–31) 7 × (51–58), 90 4 × 118 (±1.4, ±2.3, ±4.2, ±8.5) 7 × 183 (±0.6, ±1.5, ±2.5, ±3.5, ±5.0, ±7.5, ±12.5)
MAR-SCHALS	Millimeter-wave airborne receiver for spectroscopic characterization of atmospheric limb sounding	Del Bianco *et al.* (2007)	M55 Geophysica Scout-O3	High spectral resolution in Band B 294–305.5 Band C 316.5–325.5 Band D 342.2–348.8
MARSS	Microwave airborne radiometer scanning system	McGrath and Hewison (2001)	BAe-146 BBC BAe-146/BBC	90 150 3 × 183.3 90 (±1, ±3, ±7)
MIR	Millimeter-wave imaging radiometer	Racette *et al.* (1996)	ER-2 CAMEX, Wakasa Bay, SUCCESS	85, 150, 220, 3 × 183.3 (±1, ±3, ±7) 3 × 325 ±1, ±3, ±8
MTP	Microwave temperature profiler	Denning *et al.* (1989)	ER-2, DC-8, L188C/PAVE, TC-4/TEXAS 2000P	57.3 58.8
NAST-M	NPOESS aircraft sounder testbed-microwave	Blackwell *et al.* (2001)	ER-2 CAMEX	8 × 60 9 × 118
PSR	Polarimetric scanning radiometer	Stankov *et al.* (2008)	P-3, DC-8, Wakasa Bay, SMEX ER-2, WP 57F/ Wakasa Bay, SMEX	8.7, 21.5, 37, 7 × 55, 89, 7 × 118, 7 × 183.3, 3 × 325, 340, 5 × 380, 5 × 425 & 500

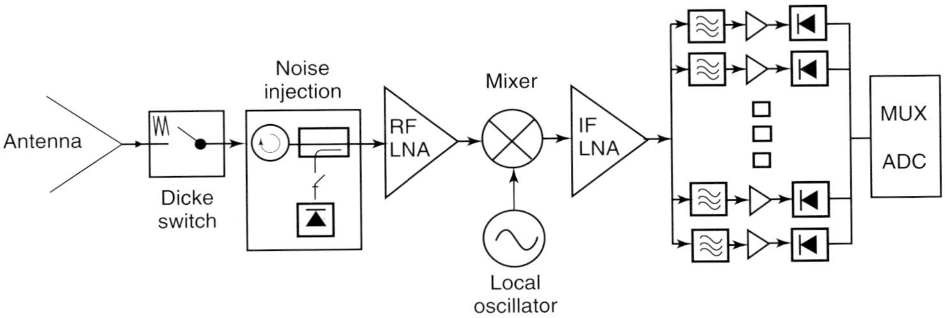

Figure 7.22 Schematic view of a multifrequency heterodyne MW receiver. LNAs are implemented in the RF and IF. After the signal splitting bandpass filtering, amplification and detection takes place in the individual channels. The analogue signals are combined by a multiplexer and converted by analogue digital converters to digital counts.

about 100 GHz. Therefore, only a few low-frequency radiometers are based on direct detection techniques and heterodyne receivers are more frequently implemented (Figure 7.22). In a heterodyne receiver, the radio frequency (RF) is down-converted to an intermediate frequency (IF) by a mixer using a frequency stable local oscillator (LO). In principle, two sidebands with

$$\nu_{RF} = |\upsilon_{LO} \pm \upsilon_{IF}| \tag{7.63}$$

are converted to the IF. This is often exploited when observations along a single symmetric absorption line are performed. In this case, the LO is set to the line center frequency, and contributions from equally distant frequencies on both sides of the line are combined into a single channel. Thus, a single LO can be used for several frequency channels along the line and respective profiling. This double sideband approach increases the SNR by a factor of two. If this technique is used for window channels, low IF center frequencies have to be used to realize typical channel bandwidths of a few GHz. If only one frequency band is received, a single sideband filter, for example, of Martin–Puplett or Fabry Perot type, has to be implemented.

Multichannel MW radiometers typically split the IF signal into the different channels by bandpass filters (filter bank; see right of IF LNA in Figure 7.22). Each channel's signal is further amplified, detected, and converted to digital counts. For profiling of upper atmospheric gases, the pressure-broadened line shape needs to be observed with a frequency resolution of about 1 MHz. In this case, instead of a filter bank, a back-end spectrometer such as the acousto optical spectrometer or digital fast Fourier transform spectrometer with high spectral resolution over about 1 GHz can be used (Müller *et al.* 2008).

For accurate observations of brightness temperatures, the signal calibration is of utmost importance. Classically, two blackbody calibration targets of well-known temperature and emissivity are used to derive the gain and system noise temperature (T_{sys}), which are subsequently used to calculate brightness temperatures. To take system nonlinearities into account, an additional reference such as a noise diode is necessary. Even if a system is thermally well stabilized, gain fluctuations typically occur on shorter time scales than variations in T_{sys}. Therefore, Dicke-type radiometers use magnetically controlled Dicke switches to direct the receiver input alternately to the antenna or an internal reference load. This reduces the integration time by a factor of two, but with a cycling of several Hz, gain fluctuations can be effectively minimized. For absolute calibration, the full system, including the antenna and the scanning unit, needs to be characterized. For this purpose, scan patterns often include the viewing of external targets. Because of the relatively large apertures, the characterization of the external load in terms of mean radiating temperature and emissivity is challenging. Nevertheless, through careful consideration of the different error sources, absolute calibrations uncertainties of about half a Kelvin and with a much better radiometric resolution are achieved.

7.4.4.3 **Application**

7.4.4.3.1 **Land Surface Properties: Snow Emissivity** To illustrate the use of MW remote sensing of land, this section details recent work that has been carried out to investigate the MW emissivity of snow-covered surfaces in the millimeter wavelength range. Such emissivities are important for the retrieval of atmospheric variables using MW sounding instruments such as the AMSU and the Special Sensor Microwave Imager/Sounder (SSMI/S) (English, 1999, 2008). Resent research has also shown that snow emissivities at these frequencies can be related to the snow water equivalent (SWE) and, in particular, the partitioning of SWE between coarse and fine layer fractions in an Arctic tundra snow pack (Harlow, 2010).

Recent studies have investigated the interaction between the surface and the downward MW radiation near the surface. In most work, this interaction has been approximated by specular reflection that would occur for a smooth planar interface. Much of the early literature on retrieval of snow emissivities assumed specular reflection (Hewison and English, 1999; Hewison, 2001; Haggerty and Curry, 2001; Prigent *et al.* 2005; Karbou, 2005; English, 2008). In the study by Mätzler (2005), the importance of diffuse scattering effects for the determination of emissivities with satellite data was outlined. In the study by Mätzler (2007), several bistatic scattering functions to describe the angular dependence of scattering are investigated, and snow-covered surfaces are found to be Lambertian through analysis of AMSU data at 50 GHz.

The surface interaction term was investigated for 89, 157, and 183 GHz using MARSS airborne measurements in the study by Harlow (2009). With data acquired at 150 m above the surface, the atmospheric absorption and attenuation of the overlying atmosphere may be ignored and the surface reflection and emission effects more accurately determined. Attenuation of the MW signals in the thin atmospheric layer under the aircraft is taken into account in radiative transfer with the use of dropsonde measurements of temperature and humidity and RADAR measurements of aircraft height above the surface. Using the three 183-GHz channels (183±1, 183±3, and 183±7 GHz), two approximations to the emissivity and effective temperature may be calculated using the upward and downward

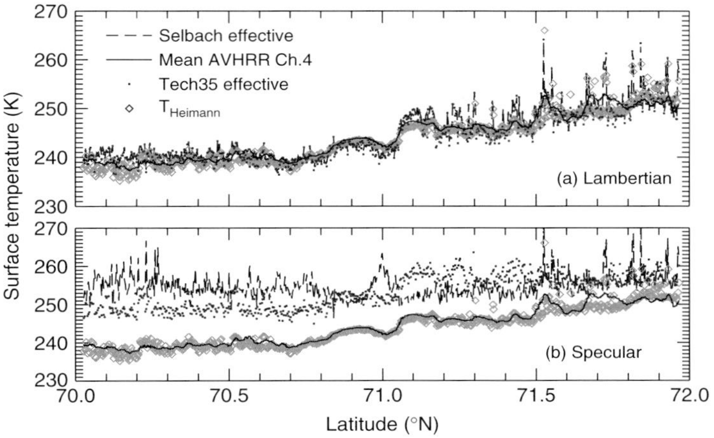

Figure 7.23 Retrieved surface temperatures under (a) specular surface reflection and (b) Lambertian surface scattering for a transect flight from snow-covered wetlands (70–70.8 °N) to land fast ice (70.8–71.4 °N) to broken pack ice (north of 71.4 °N). (Figure adapted from Harlow (2011). Copyright 2011 IEEE. Reprinted in modified form with permission by IEEE.)

measurements of brightness temperature. These two techniques are called the Selbach and the Technote-35 methods; both have been calculated under specular and Lambertian surface interaction assumptions using MARSS' scan over multiple sky-viewing look angles (Harlow, 2009).

In Figure 7.23, the remotely sensed 183-GHz effective temperatures of the surface along a flight track are depicted for an assumed Lambertian surface in the top panel and a specular surface in the bottom panel. When the surface interaction is assumed to be a specular reflection, the retrieved Selbach and Technote-35 effective temperatures disagree within a root mean square difference of 4.4 K and are 5–15 K warmer than the IR surface temperatures retrieved with Advanced Very High Resolution Radiometer (AVHRR) and the Heimann radiometers. In contrast, the two estimates of 183-GHz effective temperature agree to within 0.5 K when the surface interaction is assumed to be Lambertian scattering, and they agree closely with the IR estimates of the surface temperature. This agreement is seen as strong evidence that the surface interacts with the radiation in a diffuse manner.

Figure 7.24 depicts the emissivities along the same flight track as Figure 7.23, in this case for the Lambertian assumption only. In this case, the Selbach and Technote-35 183-GHz emissivities agree to within 0.002, so only the Selbach emissivity is depicted. Both Figures 7.23 and 7.24 depict data from a transect run from frozen wetlands (70–70.8 °N) to various types of land fast ice (LFI; 70.8–71.4 °N) to broken first-year pack ice (north of 71.4 °N) along an South South East (SSE) to North North West (NNW) line passing through Smith Bay on the North Slope of Alaska. The effective temperature and emissivity traces mirror the changing snow and ice conditions along the flight track. A detailed analysis of these emissivities in relation to the surface snow conditions may be found in the study by Harlow (2011).

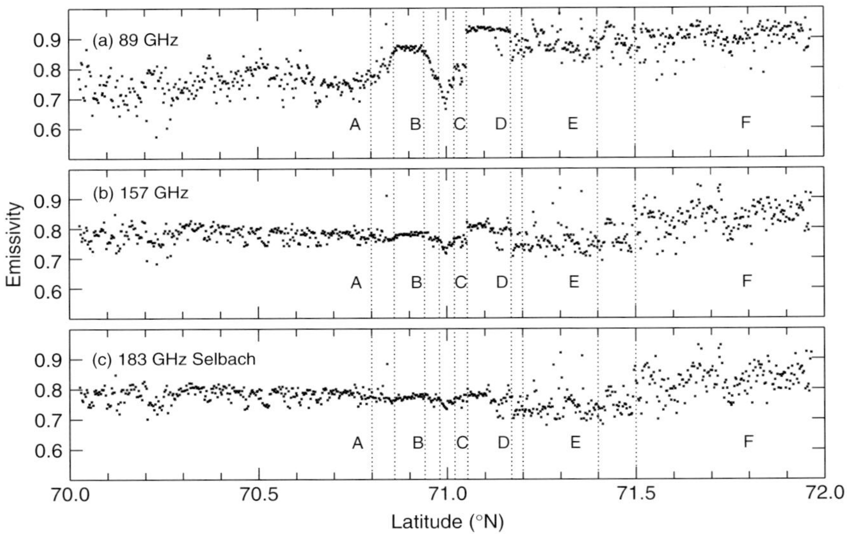

Figure 7.24 Retrieved Lambertian sea ice emissivities associated with the surface. The labels are explained as follows: A, land; B, LFI-SB; C, Ridge Rubble; D, Yng LFI; E, Hetero LFI; F, Broken FYI. (Figure adapted from Harlow (2011). Copyright 2011 IEEE. Reprinted in modified form with permission by IEEE.)

7.4.4.3.2 **Atmospheric Profiling** Before atmospheric profiles can be retrieved from multispectral MW observations, one needs to be able to solve the forward problem, that is, calculate the MW signal from a given atmospheric profile. Unfortunately, there are still gaps in our knowledge about the correct description of gaseous absorption, and in particular, absorption by the water vapor continuum. Attempts to validate gas absorption models have been addressed by airborne profiling using *in situ* and remotely sensed temperatures (English *et al.* 1994; Hewison, 2006), although ground-based observations have also been used to evaluate the accuracy of MW absorption models (Turner *et al.* 2009).

Temperature profiling is carried out primarily by using multiple channels along the 60-GHz absorption complex with frequencies similar to those of AMSU (e.g., NAST-M and MARSS), although the 118-GHz oxygen absorption line is also used. Higher vertical resolution profiles around a confined height interval can be derived by the microwave temperature profiler (Denning *et al.* 1989) that employs only two frequency channels (Table 7.4) but performs limb-like elevation scanning. With this approach, potential temperature surfaces (isentropes) and more recently, mixing layer heights have been derived in the lower stratosphere (Nielson–Gammon *et al.* 2008).

Several airborne MW radiometers employ channels along the 183-GHz line similar to those of AMSU for profiling upper and mid tropospheric water vapor (CoSSIR, CoSMIR, MARSS, and PSR). Some also include further window channels that can show brightness temperature depressions caused by scattering from frozen hydrometeors. High spectral resolution observations together with an upward looking geometry provide information on stratospheric and mesospheric constituents (Müller *et al.* 2008). In addition, profiles of several other gaseous species active in stratospheric chemistry from ozone to chlorine monoxide can be observed by this technique with instruments covering frequencies up to 670 GHz (Mees *et al.* 1995).

7.4.4.4 **Challenges**

Airborne MW remote sensing for satellite validation studies (Blackwell *et al.* 2001; Crewell *et al.* 1994; Lobl *et al.* 2007) has played an important role in the past and will certainly continue to do so. In addition, airborne demonstrator instruments provide an opportunity to test new prototype techniques before being applied to space. An example of this is the application of submillimeter observations to derive global ice water path distributions (Buehler *et al.* 2007) for which the manufacturing of a prototype is currently underway.

A particular advantage of airborne observations is their ability to bridge the scale between point-wise *in situ* or ground-based observations and large-scale satellite observations. Because of the relatively low spatial resolution of MW satellite pixel beam, filling problems are particularly evident for highly variable parameters such as clouds and precipitation. Airborne observations can help address scale-related issues in both satellite retrieval algorithms and in larger scale atmospheric modeling. In this respect, sensor synergy with auxiliary instrumentation such as cloud RADAR, Light Detection and Ranging (LIDAR), and solar and IR radiation observations is highly valuable for deriving atmospheric parameters with improved accuracy. A classical example is the determination of liquid water profiles over ocean. Here, the drop size-dependent RADAR signal can be adjusted by the integral amount provide by an MW radiometer with high accuracy. Sensor synergy allows the most complete view on the atmospheric state and, therefore, provides valuable input for process studies. The recent development of research aircraft with relatively large payloads and long distance coverage, for example, HIAPER in the United States and HALO in Germany, will certainly accelerate this development.

8
Hyperspectral Remote Sensing

Eyal Ben-Dor, Daniel Schläpfer, Antonio J. Plaza, and Tim Malthus

8.1
Introduction

Hyperspectral remote sensing (HRS) and imaging spectroscopy (IS) are the same technologies that provide detailed spectral information for individual pixels of an image. While HRS refers mostly to remote sensing (from a distance), the emerging IS technology covers wide spatial–spectral domains, from microscopic to macroscopic HRS/IS. IS is an innovative development of the charge-coupled device (CCD), which was invented in 1969 by the two 2009 Nobel prize in Physics winners Willard Boyle and George Smith. In 1972, Goetz applied the CCD technology for spectral applications, and after developing the first field portable spectrometer, a combined spatial and spectral capability was designed and successfully operated from orbit (LANDSAT program).

HRS/IS is a technology that provides spatial and spectral information simultaneously. It enables the identification of targets and other phenomena as the spectral information is presented on a spatial rather than point (pixel) basis. HRS/IS are tools with many applications, such as geology, ecology, geomorphology, limnology, pedology, atmospheric science, and forensic science. As such HRS/IS technology is applied by decision makers, farmers, environmental watchers in both the private and government sectors, city planners, stock holders, and others. The use of HRS/IS sensors is still relatively costly and requires professional manpower to operate the instrument and process the data.

Today, in addition to the growing number of scientific papers and conferences focusing on this technology, the HRS/IS discipline is very active: commercial sensors are being built, orbital sensors are in advanced planning stages, national and international funds are being directed toward using this technology, and interest from the private sector increases. The aim of this chapter is to provide the reader with a comprehensive overview of this promising technology from historical to operational perspectives.

Airborne Measurements for Environmental Research: Methods and Instruments, First Edition.
Edited by Manfred Wendisch and Jean-Louis Brenguier.

8.2 Definition

HRS/IS (HRS from now on) provides high spatial/spectral resolution radiance data from a distance. This information enables the identification of targets based on their spectral properties (mainly absorption features of chromophores). This approach has been found to be very useful in many terrestrial, atmospheric, and marine applications (Clark and Roush, 1984; Goetz and Wellman, 1984; Gao and Goetz, 1990; Dekker *et al.* 2001; Asner and Vitousek, 2005). The classical definition for HRS is given by Goetz *et al.* (1985): "The acquisition of images in hundreds of contiguous registered spectral bands such that for each pixel a radiant spectrum can be derived." This definition covers the spectral regions of VIS (visible), NIR (near infrared), SWIR (shortwave infrared), MWIR (midwave infrared), LWIR (longwave infrared), and recently also the UV (ultraviolet). For the specific wavelengths, refer to Table 7.1 in Chapter 7 and Table 8.1. It includes all spatial domains and platforms (microscopic to macroscopic; ground, air, and space platforms) and all targets (solid, liquid, and gas). For this technology, Not only "high number of bands" but also high spectral resolution, that is, a narrow bandwidth (FWHM, full width at half maximum), and a large sampling interval across the spectrum are needed. The accepted FWHM for HRS technology is 10 nm (Goetz, 1987). New applications, such as assessing vegetation fluorescence, require bandwidths of less than 1 nm (Guanter, Estellés, and Moreno, 2006; Grace *et al.* 2007).

HRS can also be defined as "spatial spectrometry from afar" that adopts spectral routines, models, and methodology and merges them with spatial information. While conditions are constant and well controlled in the laboratory, in the acquisition of high-quality spectral data in airborne/spaceborne cases, significant interference is encountered, such a lower signal-to-noise ratio (SNR) induced by the short dwell time of data acquisition over a given pixel, atmospheric attenuation of gases and aerosol particles and the uncontrolled illumination conditions of the source and objects. This makes HRS a challenging technology that involves many disciplines, including atmospheric science, electro-optical engineering, aviation, computer science, statistics and applied mathematics, and more. The major aim of

Table 8.1 Wavelength ranges applied in HRS.

Name of Range	Abbreviation	Wavelength, λ (μm)
Ultraviolet	UV	0.28–0.35
Visible	VIS	0.35–0.7
Near infrared	NIR	0.7–1
Shortwave infrared	SWIR	1–2.5
Midwave infrared	MWIR	3–5
Longwave infrared	LWIR	8–12
Thermal infrared	TIR	3–50
Infrared	IR	1–1000

HRS is to extract physical information across the spectrum (radiance) to describe inherent properties of the targets, such as reflectance and emissivity. Under laboratory conditions, the spectral information across the UV-VIS-NIR-SWIR-MWIR-LWIR spectral regions can be quantitatively analyzed for a wealth of materials, natural and artificial, such as vegetation, water, gases, artificial material, soils, and rocks, with many already available in spectral libraries. If a HRS sensor with high SNR is used, an analytical spectral approach yields new products (Clark, Gallagher, and Swayze, 1990; Krüger, Erzinger, and Kaufmann, 1998). The high spectral resolution of HRS technology combined with temporal coverage enables better recognition of targets and an improved quantitative analysis of phenomena, especially for land use cover application.

Allocating spectral information temporally in a spatial domain provides a new dimension that neither the traditional point spectroscopy nor air photography can provide separately. HRS can thus be described as an "expert" geographic information system (GIS) in which surface layers are built on a pixel-by-pixel basis rather than a selected group of points with direct and indirect chemical and physical information. Spatial recognition of the phenomenon in question is better performed in the HRS domain than by traditional GIS techniques. HRS consists of many points (actually the number of pixels in the image) that are used to generate thematic layers, whereas in GIS, only a few points are used to describe an area of interest (raster vs vector).

Figure 8.1 shows the concept of the HRS technology. Each individual pixel is characterized by a complete spectrum of ground targets (and their mixtures) that can be quantitatively analyzed within the spatial view. The capability of acquiring quantitative information from many points on the ground at almost the same time provides another innovative aspect of HRS technology. It freezes time for all spatial pixels at almost the same point, subsequently permitting adequate temporal

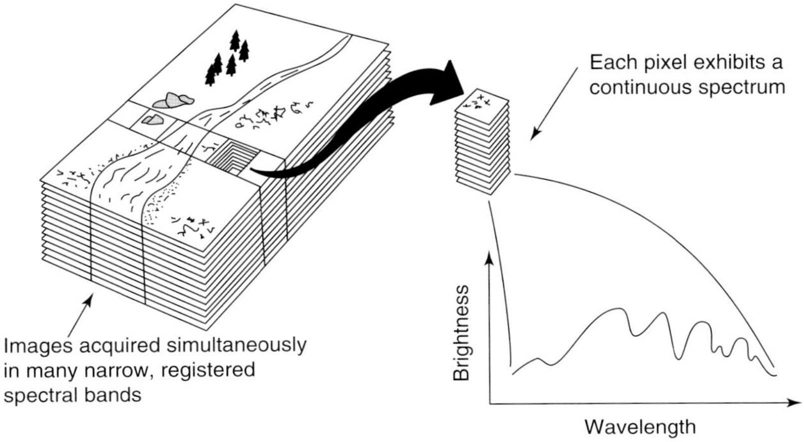

Figure 8.1 The concept of HRS/IS: each pixel element has a continuous spectrum that is used to analyze the surface and atmosphere, after Goetz (1987).

analysis. HRS technology is thus a promising tool that adds many new aspects to the existing mapping technology and improves our capability to remote-sense materials from far distances.

8.3 History

Alex Goetz is considered a mentor and pioneer scientist in HRS technology together with his colleague Gregg Vane. Goetz (2009) and MacDonald, Ustin, and Schaepman (2009) reviewed the history of HRS development since 1970. HRS technology was driven by geologists and geophysicists who realized that the Earth's surface mineralogy consists of significant and unique spectral fingerprints across the SWIR, MWR, and LWIR spectral regions (later the VIS–NIR spectral region was also explored). This knowledge was gained from comprehensive work with laboratory spectrometers and was followed by a physical explanation of the reflectance spectral response of minerals in rocks and soil. Hunt and Salisbury (1970, 1971); Hunt, Salisbury, and Lenhoff (1971a,b); Stoner and Baumgardner (1981); Clark (1999) created the first collection of available soil and rock spectral libraries.

HRS capability leans heavily on the invention of the CCD assembly in 1969 (Smith, 2001), which provided the first step toward digital imaging. These achievements acted as a precursor to establishing a real image spectrometer that would rely on the commercial hybrid focal plane array that was available at that time (in 1979). The first sensor of this kind was used in the shuttle mission SMIRR (shuttle multispectral infrared radiometer) in 1981. In 1983, Goetz and Vane started to build an airborne HRS sensor (airborne imaging spectrometer, AIS), which was sensitive in the SWIR region (Goetz, 2009).

The 2D detector arrays (32×32 elements) consisted of HgCdTe detectors generated images at wavelength greater than 1.1 μm. The array detector did not need a scan and provided sufficient improvement in the SNR to suit airborne applications. The AIS was a rather large instrument and was flown onboard a C-130 aircraft. It had two versions, with two modes being used in each: the "tree mode" from 0.9 to 2.1 μm and the "rock mode" from 1.2 to 2.4 μm.

The instantaneous field of view (IFOV) of the AIS-1 was 1.91 mrad and of the AIS-2 2.05 mrad; the ground instantaneous field of view (GIFOV) (from 6 km) was 11.4 and 12.3 m, and the FOV was 3.7° and 7.3°, respectively. The image swath was 365 m for AIS-1 and 787 m for AIS-2, with a spectral sampling interval of 9.3 and 10.6 nm, respectively. The AIS-1 was flown from 1982 to 1985 and the AIS-2, a later version with spectral coverage of 0.8–2.4 μm and 64-pixel width (Vane and Goetz, 1988), was operated in 1986. In those days, methods to account for atmospheric attenuation were not available; nonetheless, by simple approximation, the sensor and the HRS concept were able to show that minerals can be identified and spatially mapped over an arid-environment terrain. The proceedings of a conference that summarized the activity and first results of the AIS missions were published by the National Aeronautics and Space Administration (NASA) in 1985 and 1986. At that time, spectral libraries of mineral and rock material had not yet been developed. Rowan proved that the HRS technology was able to detect the mineral buddingtonite from afar and solved the mystery of unrecognized spectral feature at that time.

In 1984, Vane started to build AVIRIS (Airborne Visible and Infrared Imaging Spectrometer). The first developed AVIRIS lasted three years (1984–1987), with its first flight taking place in 1987. Although being a relatively low-quality SNR instrument, the first AVIRIS demonstrated excellent performance relative to the AIS. The sensor covered the entire VIS-NIR-SWIR region with 224 bands (around 10 nm FWHM), with 20 m GIFOV and around 10×10 km swath. It was a whiskb-room sensor with a SNR of around 100 carried onboard an ER-2 aircraft from 20 km altitude. Since then, the AVIRIS sensor has undergone upgrades. The major differences are its SNR (100 in 1987 relative to >1000 today), spectral coverage (400–2500 nm vs 350–2500 nm today) and spatial resolution (20 m vs 2 m today).

The instrument can fly on different platforms at lower altitudes and has opened up new capabilities for potential users in many applications. Even today, with many new HRS sensors having become available worldwide, the AVIRIS sensor is still considered the best HSR sensor (Goetz, 2009). This is due in large part to careful maintenance and upgrade of the sensor over the years and to the growing interest of the HRS community in using the data. The AVIRIS program has established an active HRS community in the United States and then in Europe that has rapidly matured. On the basis of this capability and success, other sensors have been developed and built over the past two decades worldwide.

8.4 Sensor Principles

Imaging spectrometers typically use a two-dimensional (2D) matrix array (e.g., a CCD or focal plane array (FPA) that produces a 3D data cube (spatial dimensions and a third spectral axis). These data cubes are built in a progressive manner by (i) sequentially recording one full spatial image after another, each at a different wavelength, or (ii) sequentially recording one narrow image (1 pixel wide, multiple pixels long) swath after another with the corresponding spectral signature for each pixel in the swath. Some common techniques used in airborne or spaceborne applications are depicted in Figure 8.2. The first two approaches shown are basic ones, used to generate images such as those used in LANDSAT (Figure 8.2a) and SPOT (Figure 8.2b). They show the concept of measuring reflected radiation in a discrete detector or in a line array.

Multichannel sensors such as LANDSAT TM are optical mechanical system in which discrete, fixed detector elements are scanned across the target perpendicular to the flight path by a mirror and these detectors convert the reflected solar photons from each pixel in the scene into an electronic signal. The detector elements are placed behind filters that pass broad portions of the spectrum. One approach to

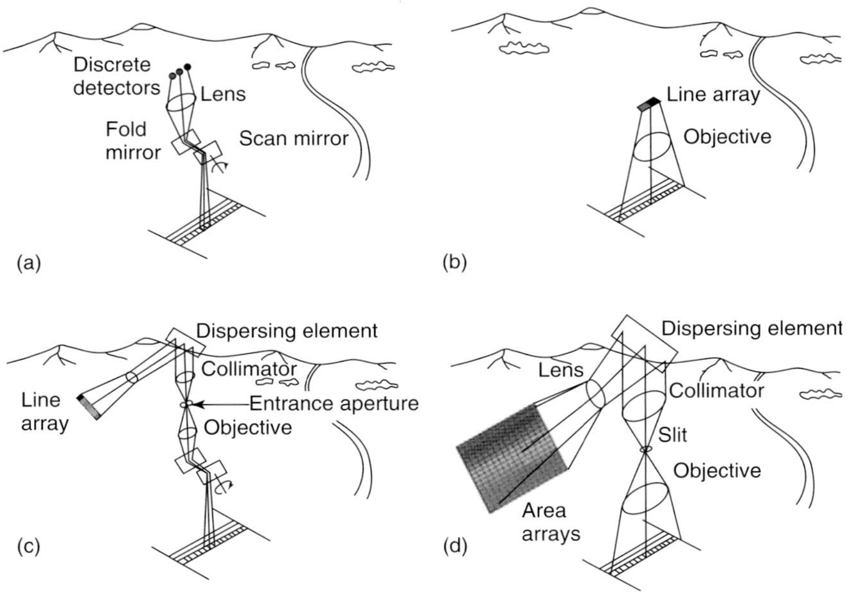

Figure 8.2 Four approaches to sensors for multispectral imaging. (a) multispectral imaging with discrete detectors (LANDSAT type); (b) multispectral imaging with line arrays (SPOT type); (c) imaging spectroscopy with line arrays (AVIRIS type, whiskbroom); and (d) imaging spectroscopy with area array (AISA type, pushbroom). (Source: Taken from Goetz (1987).)

increase the residence time of a detector in the IFOV is to use line arrays of detector elements (Figure 8.2b). This type of sensor is presented by the French satellite sensor SPOT.

There are limitations and trade-offs associated with the use of multiple line arrays, each with its own spectral band-pass filter. If all the arrays are placed in the focal plane of the telescope then the same ground locations are not imaged simultaneously in each spectral band. If a beam splitter is used to facilitate simultaneous data acquisition, the signal is reduced by 50% or more for each additional spectral band acquired in a given spectral region. Furthermore, instrument complexity increases substantially if more than 6–10 spectral bands are desired. Two other approaches to HRS are shown in Figure 8.2c,d. The line array approach is also known as *whiskbroom configuration* and the area array approach as *pushbroom configuration*. The line array approach is analogous to the scanner approach (Figure 8.2b), except that the light from a pixel is passed into a spectrometer where it is dispersed and focused onto a line array. Thus, each pixel is simultaneously sensed in as many spectral bands as there are detector elements in the line array. For high spatial resolution imaging of ground IFOVs of 10to 30 m, this concept is suitable only for an airborne sensor that flies slowly enough so that the integration time of the detector array is a small fraction of the integration time. Because of the high velocities of spacecraft, an imaging spectrometer designed for the Earth's orbit requires the use of two distinguished area arrays of the detector in the focal plane of the spectrometer (Figure 8.2d), thereby obviating the need for an optical scanning mechanism (pushbroom configuration).

The key to HRS is the detector array. Line arrays of silicon, sensitive to radiation at wavelengths of up to 1.1 μm, are available commercially in dimensions of up to 5000 elements in length. Area arrays of up to 800 × 800 elements of silicon were developed for wide-field and planetary camera. However, the state of infrared array development for wavelength beyond 1.1 μm is not so advanced. Line arrays are available in several materials up to few hindered detector elements in length. Two of the most attractive materials are mercury-cadmium-telluride (HgCdTe) and indium antimonite (InSb). InSb arrays of 512 elements with very high quantum efficiency and detectors with similar element-to-element responsivities have developed. The InSb arrays respond to wavelengths from 0.7 to 5.2 μm. A comprehensive description of both pushbroom and whiskbroom technologies with advantages and disadvantages can be found in Sellar and Boreman (2005).

8.5 HRS Sensors

8.5.1 General

The growing number of researchers in the HRS community can be seen by their attendance at the yearly proceedings of the AVIRIS Workshop Series, organized by JPL since 1985 (starting with AIS, and today with HyspIRI) and other workshops organized by international groups such as WHISPERS (Work group on Hyperspectral Image and Signal Processing: Evaluation in Remote Sensing) and EARSEL SIG IS (European Remote Sensing Laboratory Special Interest Group on Imaging Spectroscopy). In 1993, a special issue of *Remote Sensing of Environment* was published, dedicated to HRS technology in general and to AVIRIS in particular (Vane, 1993). This broadened the horizon for many potential users who still had not heard about HRS technology, ensuring that the activity would continue. Today, new HRS programs are up and running at NASA, such as the M^3 (Moon Mineralogy Mapper) project in collaboration with the Indian Space Agency to study the moon's surface (Pieters *et al.* 2009b), along with preparations to place a combined optical and thermal hyperspectral sensor in orbit, the HyspIRI (Hyperspectral Infrared Imager) project (Knox *et al.* 2010). In addition to the AIS and AVIRIS missions, NASA also successfully operated a thermal hyperspectral mission known as TIMS (thermal infrared multispectral scanner) in circa 1980–1983 (Kahle and Goetz, 1983) and also collaborated on other HRS initiatives in North America. The TIMS and, then later, the ASTER spacecraft sensors showed the thermal region's promising capability for obtaining mineral-based information. Apparently, the thermal infrared (TIR) HRS capability because of its costs and performance was set aside, and it has only recently begun to garner new attention, in new space initiatives (HyspIRI) and in new airborne sensors (e.g., TASI-600 (Thermal Airborne Spectrography Imager) and MASI-600 (Midwave Infrared Airborne Spectrographic Imager) from ITRES (Innovation Technology Research Excellence and Science), HyperCam from TELOPS, SEBASS (Spatially Enhanced Broadband Array Spectrograph System) from Aerospace Corporation, and Owl from SpecIm). In parallel to the US national HRS activity, a commercial HRS sensor was developed in circa 1980. The Geophysical & Environmental Research Corporation (GER) of Millbrook, New York, developed the first commercial HRS system which acquired 576 channels across 0.4–2.5 μm in 1981, first described by Chiu and Collins (1978). After the GER HRS, came a 63–channel sensor (GERIS-63) that was operated from around 1986 to 1990: this was a whiskbroom sensor that consisted of 63 bands (15–45 nm bandwidth) across the VIS-NIR-SWIR region with a 90° FOV (Ben-Dor, Goetz, and Shapiro, 1994). The sensor was flown over several areas worldwide and demonstrated the significant potential of the HRS concept. Although premature at that time, GER then began to offer commercial HRS services. However, it appears that the market was not yet educated enough and the very few scientists who were exposed to this technology at that time could not support the GER activity. Thus, the GER initiative was ahead of its time by about two decades, and it reestablished its commercial activity in 2000. The GER sensor was brought to Europe in May and June 1989 for demonstration purposes, and a campaign organized by several European users (known as EISAC-89, European Imaging Spectroscopy Airborne Campaign) was conducted. The results of this mission were impressive and pushed the European community to learn more about this technology (Itten, 2007). At around the time of the first AIS mission (1981), the Canadians had also developed an imaging device known as FLI (fluorescence line imager).

In the mid-1980s, Canada Moniteq Ltd. developed and used a limited pushbroom scanner, the FLI/PMI, with 228 channels across 430–805 nm (Borstad *et al.* 1985; Gower and Borstad, 1989). This sensor was also brought to the EISAC-89, and in 1991, the first *EARSeL Advances in Remote Sensing* issue (Volume 1, Number 1, February 1991), which was dedicated to HRS, provided the outcomes of this campaign, demonstrating that atmospheric attenuation, calibration, and validation were the major issues that needed to be tackled. It is interesting to note that most of the authors were satisfied with the results but their demand for more data was blocked by an inability to access data and sensors until DLR (Deutsches Zentrum für Luft- und Raumfahrt, German Aerospace Center) entered the scene. The interest of DLR in HRS began in around 1986 when they announced plans for ROSIS, Reflective Optics System Imaging Spectrometer (a pushbroom instrument offering 115 bands between 430 and 850 nm), which only became operational in 1992 and was continuously upgraded until 2003 (Holzwarth *et al.* 2003; Doerffer *et al.* 1989; Kunkel *et al.* 1991). In 1996, DLR owned and operated the DAIS 7915 (Digital Airborne Imaging Spectrometer) sensor (see further on) and then operated the HyMAP (Hyperspectral MAPping) in several campaigns in Europe and Africa. They recently own the HySpeX sensor, together with GFZ in Germany (2012) that will enable freedom and comfort to operate HSR sensor without leaning on a third party. Both bodies (DLR and GFZ) together with other German groups initiated, in 2007, a new and ambiguous initiative to place high-quality HRS sensor in orbit, termed Environmental Mapping and Analysis Program (EnMAP; see further on).

On the basis of the growing interest of the European Union (EU) scientific community in HRS technology, especially after the successful EISAC-89 campaign, it was obvious that AVIRIS, the most advanced sensor at that time, would be brought to Europe for a large campaign. AVIRIS was deployed in the Mac-Europe campaign in 1991 (Clevers, 1999) on-board the NASA ER-2 aircraft, and covered test sites in Germany, The Netherlands, France, Iceland, Italy, England, Spain, and Austria (Itten *et al.* 1992). The success of the campaigns on the one hand and the complexity and cost involved in bringing AVIRIS (or any other HRS sensor) on the other were the driving forces for a new initiative in Europe to be independent in term of sensors, data availability, research capacity, and experience. This led to the purchase of HRS sensors by several bodies in Europe: in Germany (Compact Airborne Spectrographic Imager (CASI), by the Free University of Berlin and DAIS 7915 by DLR) and Italy (multispectral infrared and visible imaging spectrometer (MIVIS), by the Italian National Research Council, CNR). In addition, plans were made for the development of more general sensors for the benefit of all European Community (EC) members and were established via the European Space Agency (ESA) Program for the Development of Scientific Experiments (PRODEX) project APEX, Airborne Prism Experiment (Itten *et al.* 2008), and by some limited commercial activities.

The DAIS 7915 was a GER whiskbroom instrument characterized by 72 channels across the VIS-NIR-SWIR region and 7 bands in the TIR region (3.0–12.6 μm). It had a 26° FOV and GIFOV between 5 and 20 m. This instrument was offered in 1996 as a large-scale facility instrument to European researchers and served as a test bed in a large number of international flight campaigns. Although it was not the ideal sensor in terms of SNR and operational capabilities, the DAIS 7915 was operated by DLR until 2002 when it could no longer satisfy the higher SNRs being requested by the community. The experience gained from the DAIS 7915 campaigns was very valuable in terms of opening up the HRS field to more users, developing independent operational and maintenance capabilities, educating the younger generation and opening fruitful discussions among emerging HRS community members in Europe.

Italy's activity in HRS technology began in 1994 with the purchase and operation of the MIVIS system, a Daedalus whiskbroom sensor, by the CNR. The MIVIS is a passive scanning and imaging instrument that is composed of four spectrometers that simultaneously record reflected and emitted radiation. It has 102 spectral bands from the VIS-NIR-SWIR to the TIR spectral range and the wavelength ranges between 0.43 and 12.7 μm, with an IFOV of 2 mrad and a digitized FOV of 71.1°. The band position was selected to meet research needs that were already known at that time for environmental remote sensing, such as agronomy, archaeology, botany, geology, hydrology, oceanography, pedology, urban planning, atmospheric sciences, and more. Under the Laboratio Aero Per Ricerche Ambientali (LARA) project, the CNR has flown the instrument very intensively since 1994 on-board a

CASA 212 aircraft, acquiring data mostly over Italy and also in cooperation with other nations, such as Germany, France, and the United States (Bianci *et al.* 1996).

In Canada, a new airborne VIS–NIR sensor was developed in 1989 by ITRES (Alberta, Canada), known as CASI. The sensor was a pushbroom programmed sensor aimed at monitoring vegetation and water bodies. ITRES provided data acquisition as well as processing services and also sold a few instruments to individuals who operated the system and then developed measurement protocols for a limited market (e.g., the Free University of Berlin in 1996). In 1996, ITRES developed a research instrument for the Canadian Center for Remote Sensing (CCRS) known as SFSI (Shortwave Infrared Full Spectrum Imager), and recently (2010), they developed an instrument for the LWIR region (8–11.5 μm) named TASI-600 and an instrument for the MWIR region (3–5 μm) named MASI-600 with 64 channels (55 nm bandwidth). The CASI offers several modes, between 512 bands (spectral modes) and 20 preselected bands (spatial modes), with intermediate numbers of spectral bands and pixels being programmable. The spectral range is between 0.4 and 1 μm with a FOV of 29.6° and a IFOV of 2.1 mrad.

The SFSI provides 120 bands (115 used in practice) across the 1.219–2.445 μm spectral region. The FOV is 9.4° and across-track pixels' IFOV is 0.33 mrad. The TASI-600 is a pushbroom thermal imager with 64/32 spectral channels ranging from 8 to 11.5 μm with 600 pixels across track. The FOV is 38°, and the IFOV is 0.49 mrad. The MASI-600 has 64 bands across 3–5 μm with 32 μm bandwidth and a FOV of 40° and an IFOV of 1.2 mrad. ITRES provides to the community also the Shortwave Infrared Airborne Spectral Imager (SASI) sensor operates across the SWIR region (0.950–2.450 μm) with 100 spectral bands at 15 nm sampling interval and 40° FOV. The National Research Council of Canada funded modifications to the SASI sensor to have 160 spectral channels covering the 0.85–2.50 μm spectral range and 38° FOV.

8.5.2 Current HRS Sensors in Europe

Another HRS company, the Finnish Spectral Imaging Ltd. (Specim), has gone quite a long way and can be considered an important benchmark in the HRS arena. From 1995, when the company was founded, they were able to significantly reduce the cost of HRS sensors, making them available to many more users. Two airborne sensors, AISA (airborne image spectrometer for different application)-Eagle and -Hawk for the VIS–NIR and SWIR regions, respectively, were developed, using the PGP (prism-grating-prism) concept invented by Specim in the 1990s. The PGP design enables the construction of a small low-cost spectrometer that is suitable for industrial and research purposes in the wavelength range of 0.32–2.70 μm. Its small size and ease of maintenance and operation, along with the ability to mount the sensor on-board small platforms, have made the Specim sensor accessible to many users who could not otherwise afford to enter the expensive HRS field.

According to Specim, in 2010 more than 70 instruments had been sold worldwide, reflecting the growing interest in this technology in general and in low-cost capability in particular. This revolution has enabled user independence in terms of data acquisition and operation while providing a breakthrough in HRS strategy in Europe: no longer does one need to count on joint campaigns; the user can plan the mission and the flight, and process the data for his/her particular needs at a relatively low cost. Although the SNR and data performance of the new sensors was not at the level of AVIRIS or HyMAP, the Specim products enabled enlarging HRS capabilities in mission planning, simulation, flight operation, data acquisition, archiving, corrections, calibration, and education. Riding on their success, Specim announced, in 2009, that contracts for a total value of € 1.4 million had been signed with government institutions and private remote sensing companies in Germany, Malaysia, Brazil, and China.

Recent achievements in HRS technology are due, to a certain extent, on the fact that more companies are building and manufacturing small-size HRS sensors for ground and air applications (e.g., HeadWall Photonics: *http://www.headwallphotonics.com/*). While the VIS–NIR sensor is much easier to build, as it is based on available and reliable detectors, the SWIR region is still more problematic.

Two more activities in Europe can be considered important in HRS technology: the first is Instituto Nacional de Tecnica Aeroespacil (INTA) Spain's activity in HRS and the second is the Norwegian company Norsk Elektro Optikk (NEO), which manufactured a new HRS sensor. In 2001, INTA entered the HRS era by first exploring the field and then running a joint venture with Argon ST (a company resulting from a merger between Daedalus Enterprises and S.T. Research Corporation) in 1998, conducting their first campaign in circa 2003 in Southern Spain. The follow-up campaigns demonstrated the HRS concept's promise and, in 2005, the Airborne Hyperspectral Scanner (AHS) was purchased by INTA: it was first operated in Spain and then in other European countries as well. The AHS consisted of 63 bands across the VIS-NIR-SWIR region and 7 bands in the TIR region with a FOV of 90° and IFOV of 2.5 mrad, corresponding to a ground sampling distance (GSD) of 2–7 m. This sensor was flown on-board a CASA 212 aircraft and operated by personnel from INTA. The sensor has been operational in Spain and Europe (via ESA and VITO (Vlaams Instituut Voor Technologisch Onderzoek) since 2005 and remains in good condition until today (2012). The system is well maintained and undergoes a yearly checkup at Argon ST laboratories. Experience gained over the years, along with mechanical upgrading (both electronic and optical), ensures that the sensor will stay operational for a long time.

In about 1995, NEO developed a small HRS satellite sensor (HISS, Hyperspectral Imager for Small Satellites) for ESA, covering the spectral range from 0.4 to 2.5 μm. As ESA did not have any immediate plans for launching such an instrument at the time, the experience gained from the HISS was used to develop a hyperspectral camera for airborne applications – the Applied Spectral Imaging (ASI). The first prototype was built in 1998–1999. In 2001, a collaboration with the Norwegian Defense Research Establishment (FFI) was initiated, which is still continuing today. In the framework of this cooperation, the ASI camera participated in a multinational military measurement campaign in France in 2002. An upgraded version of the instrument was flown in 2003 and 2004 in different multinational military field trials. In 2004, airborne HRS data were also acquired for several local civilian research institutions. The cooperation with these institutions was continued in 2005 when a further upgraded version of the instrument was flown successfully, including a HRS camera module covering the part of SWIR region (0.9–1.7 μm), in addition to the VIS and NIR regions (0.4–1.0 μm).

All these research activities led to the development of a line of hyperspectral cameras (HySpex) that are well suited for a wide variety of applications in both the civilian and military domains. Main characteristics of the sensor are coverage of the entire range (0.4–2.5 μm) with more than 400 bands with 3.7 and 6.25 nm band width two different sensors (the VNIR 640 and SWIR 320). The sensor underwent several experiments in Europe with proven success but has not yet aggressively entered the commercial remote sensing arena.

Beside the AVIRIS sensor, today the HyMAP sensor has become available: this is a commercially designed and operated system that was based on the Probe-1 sensor (operated in circa 1998 by Applied Signal and Image Technology (ASIT), USA). Several campaigns in the United States demonstrated the promising commercial capability of HRS technology (Kruse *et al.* 2000). Integrated Spectronics, Australia, designed and operated the HyMAP sensor for rapid and efficient wide-area imaging for mineral mapping and environmental monitoring. The sensor can be defined as a high SNR instrument with high spectral resolution, ease of use, a modular design concept, calibrated spectroradiometry, proven in-field operation, and heavy load capacity. It is a whiskbroom sensor with 100–200 bands (usually 126) across the 0.45–2.45 μm spectral region with bandwidths ranging from 10 to 20 nm.

The SNR is in the range of 500 : 1 with 2–10 m spatial resolution. It is characterized by a 60°–70° swath width and furnished with an on-board radiometric and spectral calibration assembly. In 1999, a group shoot using the HyMAP sensor was conducted in the United States. A report by Kruse *et al.* (2000) declared the sensor to be the best available at the time. Since then, the HyMAP sensor has been operated worldwide, providing high-quality HRS data to its end users and opening up a new era in HRS data quality. It has been operated in Europe, Australia, the United States, Asia and Africa in specific campaigns and through Hy Vista activity, which provides end-to-end solutions for the potential customer. HyMAP can thus also be considered a benchmark in HRS technology, which was reached in circa 1999 by Probe-1 and then afterwards by HyMAP sensors. The problem with HyMAP is that the sensor is limited and is operated only by HyVista, and hence, its use is strongly dependent on their schedule and availability. Moreover, the cost of the data is still prohibitive for the daily use capability that is desired from HRS technology. It can be concluded that there is still a significant gap between high SNR and low cost/easy operation in sensors: ideally, this gap might be bridged by fusing the AISA and HyMAP characteristics that are based on two different technologies: pushbroom and whiskbroom, respectively. As more and more companies undertake moving HRS technology forward, we believe that in the near future such a fusion will be possible and we will see more low-cost, high-quality data and more applications emerging from this capability.

In 2011, the APEX has become available to the European research community after a long prototyping and development (Itten *et al.* 2008). It has been built in ESA's Prodex program by Swiss–Belgium collaborative efforts and is operated by VITO, Belgium. This system may be considered a new breakthrough in HRS technology, as it is the first airborne pushbroom system offering a complete coverage of the spectral range between 0.4 and 2.5 μm in one integrated system. APEX provides the same spatial resolution of 1–5 m at 1000 across-track pixels for both the VIS–NIR and SWIR spectral range. The prism design optics allows for very high spectral resolution in the visible part down to 1 nm, whereas the SWIR is resolved with 7 nm. Its data is currently evaluated for various IS applications and the system is to be used for cross-calibration purposes for ESA satellites and alike.

The above provides only the milestone stages in HRS technology over the years. Several of the sensors and activities may not have been mentioned. The reader is therefore directed to a comprehensive description of all HRS sensors until 2008 made by Prof. Gomez from George Mason University in the United States and to a summary of all remote sensing organizations worldwide and all institutes, private sectors, and abbreviations commonly used with this technology at *http://www.tau.ac.il/~rslweb/pdf/HSR.pdf.* A historical list of HRS sensors compiled by Michael Schaepman is available at *http://www.geo.unizh.ch/~schaep/research/apex/is_list.html.*

8.5.3 Satellite HRS Sensors

Among the airborne HRS benchmarks mentioned earlier, orbital HRS activity has contributed greatly to the blossoming HRS activity. The first initiative to place an HRS sensor in orbit took place in the early 1990s when a group of scientists chaired by Goetz started work on the NASA HRS mission HIRIS (High Resolution Imaging Spectrometer). This was part of NASA's High-Resolution Imaging Spectrometer Earth Observation System program. The idea was to place an AVIRIS-like sensor in orbit with a full range between 0.4 and 2.5 μm and a spatial resolution of 30 m. A report that provides the capacity of this sensor, including its technical and application characteristics, was issued in several copies (Goetz, 1987). This report was the first document that showed the intention to go into space with HRS. The HIRIS mission was terminated, apparently because of the Challenger space shuttle disaster, which significantly changed the space programs at NASA.

The scientists, however, agreed that using HRS in orbit is an important task that needs to be addressed (Nieke *et al.* 1997). A report by Hlao and Wong (2000) submitted to the US Air Force in 2000 assessed the technology as premature and still lagging behind other remote sensing technologies such as air photography. The next benchmark in orbital HRS was Hyperion, part of the NASA New Millennium Program (NMP). The Hyperion instrument was built by TRW Inc.

(Thompson Ramo Woddbridge) using focal planes and associated electronics remaining from the Lewis spacecraft, a product of the NASA Small Satellite Technology Initiative (SSTI) mission that fell in 1997. The integration of Hyperion took less than 12 months from Lewis's spare parts and was sent into orbit on-board the EO-1 spacecraft. The mission, planned for 3 years, is still operational today (2012) with a healthy sensor and data, although the SNR is poor. The instrument covers the VIS-NIR-SWIR region from 0.422 to 2.395 μm with two detectors and 244 bands of 10 nm bandwidth. The ground coverage FOV provided a 7.5 km swath and 30 m GSD. The first data sets cost around US$2500 and had a lower SNR than originally planned. Nonetheless, over the years, and despite its low quality, the instrument has brought new capability to sensing the globe by temporal HRS coverage, justifying the effort to place a better HRS sensor in space. As of the summer of 2009, Hyperion data are free of charge, which has opened up a new era for potential users. In circa 2001, the CHRIS (Compact High Resolution Imaging Spectrometer) sensor was launched into orbit on-board the PROBA (Project on Board Autonomy) bus. It was developed by the Sira Electro Optic group and supported by the ESA. The CHRIS sensor is a high spatial resolution hyperspectral spectrometer (18 m at nadir) with a FOV resulting in 14 km swath.

One of its most important characteristics is the possibility of observing every ground pixel at the same time, in five different viewing geometry sets (nadir, $\pm 55^\circ$ and $\pm 36^\circ$). It is sensitive to the VIS–NIR region (0.41–1.059 μm), and the number of bands is programmable, with up to 63 spectral bands. Although limited in its spectral region, the instrument provides a first view of the bidirectional reflectance distribution function (BRDF) effects for vegetation and water applications, and it is robust, as it is still operating today. The "early" spaceborne planning missions in both the United States and Europe comprised, among others, the following projects: IRIS (Interface Region Imagery Spectrograph), HIRIS (NASA), GEROS (German Earth Resources Observer System, USA), HERO (Hyperspectral Environmental and Resource Observer, CSA), PRISM (Process Research by an Imaging Space Mission), SPECTRA (Surface Processes and Ecosystem Changes Through Response Analysis, all ESA), SIMSA (Spectral Imaging Mission for Science) and SAND (Spectral Analysis of Dryland). Although most of these initiatives were not further funded and are not active today, they demonstrated government agencies' interest in investing in this technology, albeit with a fearful and cautious attitude. Other orbital sensors, such as MODIS (Moderate Resolution Imaging Spectrometer), MERIS (Medium Resolution Imaging Spectrometer), and ASTER (Advanced Spaceborne Thermal Emission and Reflection), can also be considered part of the HRS activities in space, but in terms of both spatial (MODIS and MERIS) and spectral (ASTER) resolution, these sensors and projects still lag behind the ideal HRS sensor that we would like to see in orbit with high spectral (more than 100 narrow bands) and spatial (less than 30 m) resolutions. It is important to mention, however, that a new initiative to study the moon and Mars using HRS technology took place by a collaboration between NASA and ISA (India), within which the M^3 mission to the moon has recently provided remarkable results by mapping a thin layer of water on the moon's surface (Pieters *et al.* 2009b,a). In addition, missions to Mars, such as CRISM (Compact Reconnaissance Imaging Spectrometer for Mars) show that it is now understood that HRS technology can provide remarkable information about materials and objects remotely.

EnMAP is a German hyperspectral satellite mission providing high-quality hyperspectral image data on a timely and frequent basis. Its main objective is to investigate a wide range of ecosystem parameters encompassing agriculture, forestry, soil and geological environments, coastal zones, and inland waters. This will significantly increase our understanding of coupled biospheric and exospheric processes, thereby enabling the management and guaranteed sustainability of our vital resources. Launch of the EnMAP satellite is envisaged for 2015 (updated in 2012). The basic working principle is that of a pushbroom sensor, which covers a swath (across-track) width of 30 km, with a GSD of 30 × 30 m. The second dimension is given by the along-track movement and corresponds to about 4.4 ms exposure time. This leads to a detector frame rate of 230 Hz, which is a performance-driving parameter for the detectors, as well as for the instrument control unit and the mass memory. HyspIRI is a new NASA initiative to place a HRS sensor in orbit and is aimed at complementing EnMAP, as its data acquisition covers the globe periodically.

It is important to mention that other national agencies are aiming to place HRS sensor in orbit as well. A good example is PRISMA of the Italy's space agency. PRISMA is a pushbroom sensor with swath of 30–60 km, GSD of 20–30 m (2.5–5 m peroxyacetylnitrate (PAN)) with a spectral range of 0.4–2.5 μm. The satellite launch was foreseen by the end of 2013, but it seems that some delay is encountered and the new lunch date is unknown.

To keep everyone up to date and oriented on the efforts being made in HRS pace activities, a volunteer group was founded in November 2007 by Dr Held and Dr Staenz named ISIS (International Satellite Imaging Spectrometry) (Staenz, 2009). The ISIS group provides a forum for technical and programming discussions and consultation among national space agencies, research institutions, and other spaceborne HRS data providers. The main goals of the group are to share information on current and future hyperspectral spaceborne missions and to seek opportunities for new international partnerships to the benefit of the global user community. The initial "ISIS Working Group" was established following the realization that there were a large number of countries planning HRS ("hyperspectral") satellite missions with little mutual understanding or coordination. Meetings of the working group have been held in Hawaii (IGARSS 2007), Boston (IGARSS 2008), Tel Aviv (EARSeL 2009), Hawaii (IGARSS 2010), Vancouver (IGARSS 2011) and Munich (IGARSS 2012). The technical presentations by the ISIS group have garnered interest from space agencies and governmental and industrial sectors in this promising technology. An excellent review on current and planned civilian space hyperspectral sensor for Earth observation (EO) is given by Buckingham and Staenz (2008).

8.6
Potential and Applications

Merging of spectral and spatial information, as is done within HRS technology, provides an innovative way of studying many spatial phenomena at various resolutions. If the data are of high quality, they allow near-laboratory level spectral sensing of targets from afar. Thus, the information and knowledge gathered in the laboratory domain can be used to process the HRS data on a pixel-by-pixel basis. The "spheres" that can feasibly be assessed by HRS technology are atmosphere, pedosphere, lithosphere, biosphere, hydrosphere, and cryosphere. Different methods of analyzing the spectral information in the HRS data are known, the basic one consisting of comparing the pixel spectrum with a set of spectra taken from a well-known spectral library. This allows the user to identify specific substances, such as minerals, chlorophyll, dissolved organics, atmospheric constituents, and specific environmental contaminants, before moving ahead with other more sophisticated approaches (Section 8.8.4). The emergence of hyperspectral imaging moved general remote sensing applications from the area of basic landscape classification into the realm of full spectral quantification and analysis. The same type of spectroscopy applications that have been utilized for decades by chemists and astronomers are now accessible through both nadir and oblique viewing applications. The spectral information enables the detection of indirect processes, such as contaminant release, based on changes in spectral reflectance of the vegetation or leaves. The potential thus lies in the spectral recognition of targets using their spectral signature as a footprint and on the spectral analysis of specific absorption features that enable a quantitative assessment of the matter in question. Although many applications remain to be developed, within the past decade, significant advances have been made in the development of applications using hyperspectral data, mainly because of the extensive availability of today's airborne sensors. While, a decade ago, only a few sensors were available and used in the occasional campaign, today, many small and user-friendly HRS sensors that can operate on any light aircraft are available.

Hydrology, disaster management, urban mapping, atmospheric study, geology, forestry, snow and ice, soil, environment, ecology, agriculture, fisheries, and oceans and national security are only a few of the applications for HRS technology today. In 2001, van der Meer and Jong (2001) published a book with several innovative applications for that time. Since then, new applications have emerged and the potential of HRS has been discussed and analyzed by many authors at conferences, in proceedings papers and full-length publications. In a recent paper, Staenz (2009) provides his present and future notes on HRS, which very accurately summarize the technology up to today. In the following, we paraphrase and sharpen Staenz's points. It is clear from the numerous studies that have been carried out that HRS technology has significantly advanced the use of remote sensing in different applications (e.g., AVIRIS 2007). In particular, the ability to extract quantitative information has made HRS a unique remote sensing tool. For example, this technology has been used by the mining industry for exploration of natural resources, such as the identification and mapping of the abundance of specific minerals. HRS is also recognized as a tool to successfully carry out ecosystem monitoring, especially the mapping of changes because of human activity and climate variability. This technology also plays an important role in the monitoring of coastal and inland waters. Other capabilities include the forecasting of natural hazards, such as mapping the variability of soil properties that can be linked to landslide events, and monitoring environmental disturbances, such as resource exploitation, forest fires, insect damage, and slope instability in combination with heavy rainfall. As already mentioned, HRS can be used to assess quantitative information about the atmosphere such as water vapor content; aerosol load; and methane, carbon dioxide, and oxygen content. HRS can also be used to map snow parameters, which are important in characterizing a snow pack and its effect on water runoff. Moreover, the technology has shown potential for use in national security, for example, in surveillance and target identification, verification of treaty compliance (e.g., Kyoto Accord on Greenhouse Gas Emission), and disaster preparedness and monitoring (Staenz, 2009). Some recent examples show both the quantitative and exclusive power of HRS technology in detection of soil contamination (Kemper and Sommer, 2003), soil salinity (Ben-Dor *et al.* 2002), species of vegetation (Ustin *et al.* 2008), atmospheric electromagnetic emissions of methane (Noomem, Meer, and Skidmore, 2005), detection of ammonium (Gersman *et al.* 2008), asphalt condition (Herold *et al.* 2008), water quality (Dekker *et al.* 2001), and urban mapping (Ben-Dor, 2001). Many other applications can be found in the literature and still others are in the R&D phase in the emerging HRS community. Nonetheless, although promising, one should remember that HRS technology still suffers from some difficulties and limitations. For example, the large amount of data produced by the HRS sensors hinders this technology's usefulness for geometry analysis or visual cognition (e.g., building structures and roads) and one has to weight the added value promised by the technology for one's applications. There are other remote sensing tools and the user should consult with an expert before using HRS technology. Since the emergence of HRS, many technical difficulties have been overcome in areas such as sensor development, data handling, aviation and positioning, and data processing and mining. However, there are several main issues that require solutions to move this technology toward more frequent operational use today. These include a lack of reliable data sources with a high SNR are required to retrieve the desired information and temporal coverage of the region of interest; although analytical tools are now readily available, there is a lack of robust automated procedures to process data quickly with a minimum of user intervention; the lack of operational products is obviously due to the fact that most efforts to date have been devoted to the scientific development of HRS; interactions with other HRS communities have not yet developed – there are many applications, methods, and know-how in the laboratory-based HRS disciplines but no valid connection between the communities; systems that can archive and handle large amounts of data and openly share the information with the public are still lacking; only a thin layer of the surface can be sensed; there is no standardization for data quality or quality indicators (QI); not much valid experience exists in merging HRS data with that of other sensors (e.g., LIDAR, SAR (Synthetic Aperture RADAR)); many sensors have emerged in the market but their exact operational mechanism is unknown, biasing an accurate assessment; thermal HRS sensors are just starting to emerge, whereas point thermal spectrometers are existing (Christensen *et al.* 2000); oblique view and ground-based HRS measurements have not yet been frequently used; the cost of deriving the information product is too high, since the analysis of HRS data is currently too labor intensive (not yet automated); it is not yet recognized by potential users as a routine vehicle as, for example, is air photography; not too many experts in this technology are available. Several authors have summarized this technology's potential to learn from history, such as Itten (2007); Schaepman *et al.* (2009) and Staenz (2009).

It is anticipated that HRS technology will catch up when new high-quality sensors are placed in orbit and the data become available to all (preferably in reflectance values), when the air photography industry uses the HRS data commercially, and when new sensors that are inexpensive and easy to use are developed along with inexpensive aviation (such as unmanned aerial vehicle, UAV).

8.7 Planning of an HRS Mission

In this section, we describe major issues for the planning of a mission for an airborne campaign: we do not cover the possible activities involved for a spaceborne mission. Planning a mission is a task that requires significant preparation and knowledge of the advantages and disadvantages of the technology. The idea behind using HRS is to get an advanced thematic map as the final product which no other technology can provide. In the planner's mind, the major step toward achieving the main prerequisite of a thematic map is to generate a reflectance or emission image from the raw data.

First, a scientific (or applicable) question has to be asked, such as Where can saline soil spots be found over a large area? For such a mission, the user has to determine whether there exists spectral information on the topic which is being covered by the current HRS sensor. This investigation might consist of self-examination or a literature search of both the area in question and the advantageous of using HRS (many times, HRS is an overkill technology for answering simple thematic questions). Once this investigation is carried out, the question is What are the exact spectral regions that are important for the phenomenon in question and what pixel size is needed? In addition, the question of what SNR values will enable such detection should be raised. Having this information in hand, the next step is to search for the instrument. Sometimes, a particular instrument is available, and there is no other choice. In this case, the first spectral investigation stage should focus on the available HRS sensor and its spectral performances (configuration, resolution, SNR, etc.) infrastructure. It is recommended that the spectral information on the thematic question be checked at the sensor-configuration stage. In some sensors, especially pushbroom ones, it is possible to program the spectral configuration using a new arrangement of the CCD assembly.

In this respect, it is important that the flight altitude be taken into consideration (for both pixel size and integration time) along with aircraft speed. Most sensors have tables listing these components and the user can use them to plan the mission frame. As within this issue, the user can configure the bands with different FWHM and positions; it should be remembered that combined with spatial resolution, this might affect the SNR. When selecting the sensor, it is important to obtain (if this is the first use) a sample cube to learn about the sensor's performance. It is also good to consult with other people who have used this equipment. Getting information on when and where the last radiometric calibration was performed as well as obtaining information about the sensor stability and uncertainties is very important. It is better if the calibration file of the sensor is provided but if not, the HRS owner should be asked for the last calibration date and its temporal performances.

Quality assurance (QA) of the sensor's radiance must be performed in order to assure a smooth step to the next stage, namely, atmospheric correction. Methods and tools to inspect these parameters were developed under EUFAR JRA2 initiative and recently also by Brook and Ben-Dor (2011).

The area in question is generally covered by 30% overlap between the lines. This has to be carefully planned in advance taking into consideration the swath of the sensor and other aircraft information (e.g., stability, time in the air, speed and altitude preferences, navigation systems). A preference for flying toward or against the direction of the Sun's azimuth needs to be decided on, and it is recommended that the Google Earth interface be used to allocate the flight lines and to provide a table for each line with starting and ending points for all flight lines. One also needs to check if the GPS is available and configure the system to be able to ultimately allocate this information in a readable and synchronized form.

A list of go/no go items should be established. For instance, a forecast for the weather should be on hand 24 h in advance, with updates every 3 h. If possible, a representative should be sent to the area in question to report on cloud coverage close to acquisition time. In our experience, one should be aware of the fact that a 1/2 cover over the area in question will turn into almost 100% coverage of the flight lines that appeared to be free of clouds. Moreover, problems that may emerge at the airport need to be taken into consideration, such as the GPS is not functioning or the altitude obtained from air control is different from that which was planned. The go/no go checklist should be used for these issues as needed. Each go/no go list is individual, and one should be established for every mission.

The aircrew members (operators, navigator, and pilot) must be briefed before and debriefed after the mission. A logbook document should be prepared for the aircrew members (pilot and operator) with every flight line reported by them. It is important to plan a dark current acquisition before and after each line acquisition. Acquisition of a vicarious calibration site (in the area of interest or on the way to this area) in question should also be planned for, that is well prepared and documented in advance. If possible, radio contact with the aircrew should be obtained at a working frequency before, during, and after the overpass. A ground team should be prepared and sent to the area in question for the following issues: (i) calibrating the sensor's radiance and examining its performance (Brook and Ben-Dor, 2011), (ii) validating the atmospheric correction procedure, and (iii) collecting spectral information that will be useful further on for thematic mapping (e.g., chlorophyll concentration in the leaves). The ground team should be prepared according to a standard protocol, and it should be assured that they are furnished with the necessary equipment (such as video and still cameras, field spectrometer, maps, Sun photometer, and GPS). After data acquisition both from air and ground, the data should be immediately backed up and quality control checks run to determine data reliability. Afterward, the pilot logbook, ground documentation, and any other material that evolved during the mission should be collected.

In general and to sum up the above, a mission has to be lead by a senior person who is responsible to coalesce the end user needs, the ground team work, the airborne crew activity, and the processing stages performed by experts. He/she is responsible to interview the end user and understand the question at hand and is responsible to allocate a sensor for the mission and meet with the sensor owner and operator ahead of the mission and arrange a field campaign by a ground team. Other responsibilities such as arranging logistics and briefing of all teams as well as backing up the information just after the mission end, that is, at the airport are also part of their duties and are very important. A checklist and documents on every stage are available in many bodies (e.g., DLR, TAU, the Tel Aviv University), but in general, it can be developed by any group by gathering information from main HSR leading bodies (DLR, NASA, INTA).

8.8 Spectrally Based Information

A remotely sensed object interacts with electromagnetic radiation where photons are absorbed or emitted via several processes. On the Earth's surface (solid and liquid) and in its atmosphere (gasses and aerosol particles), the interaction across the UV-VIS-NIR-SWIR-TIR regions is sensed by HRS means to give additional spectral information relative to the common multiband sensors. The spectral response of the electromagnetic interaction with matter can be displayed as radiance, reflectance, transmittance, or emittance, depending on the measurement technique and the illumination source used. Where interactions occur, a spectrum shape can be used as a footprint to assess and identify the matter in question. Variations in the position of local minima (or maxima, termed "peaks") and baseline slope and shape are the main indicators used to derive quantitative information on the sensed material. The substance (chemical or physical) that significantly affects the shape and nature of the target's spectrum is termed "chromophore." A chromophore that is active in energy absorptance (e.g., chlorophyll molecule in vegetation) or emission (e.g., fluorescence) at a discrete wavelength is termed a "chemical chromophore." A chromophore that governs the spectrum's shape (such as the slope and albedo

Table 8.2 A summary of possible chromophores in all spheres of interest for our planet by remote sensing using the spectral information available on the Earth surface.

Sphere	Pedosphere	Lithosphere	Biosphere	Hydrosphere	Cryosphere	Atmosphere
0.35–1 μm VIS–NIR						
Abs-electronic	Fe, Ni+	Fe, Ni+	Chlorophyll+	Chlorophyll+		
Scattering particles	Particle size	Particle size	Leaf structure	Particle size	Particle size	Mie, Rayleigh
Emission - electronic	—	—	Fluorescence	—	—	
Abs-overtones	—	OH- 3 ν Overtone	H_2O	H_2O	—	O_2, H_2O, O_3, NO_2
1–2.5 μm SWIR						
Abs-electronic						
Scattering particles	Albedo-particle size	Albedo-particle size	Leaf structure	—	Particle	Mie
Emission–electronic	—	—	—	—	—	—
Abs-Overtones combination modes	OH, C–H, N–H+	+	+	H_2O	H_2O	H_2O, CO_2, O_2, CH_4
3–12.5 μm MWIR–LWIR						
Abs-electronic	—	—	—	—	—	—
Scattering particles	—	—	—	—	—	Mie
Emission–electronics	Temperature	Temperature	Temperature	Temperature	Temperature	—
Abs Fundamentals	Emissivity, SI–O, Al–O, Fe–O	Emissivity, SI–O, Al–O, Fe–O	Emissivity C=O	Emissivity H_2O, OM	Emissivity	SO_4 NH_3 CO_2

+, some other causes for the spectral mechanism visualization.

(e.g., particle size, refraction index)) is termed "physical chromophore." Often, the spectral signals related to a given chromophore overlap with the signals of other chromophores, thereby hindering the assessment of a specific chromophore. The spectrum of a given sample, which is the result of all chromophore interactions, can be used to analyze and identify the matter if a spectral-based method for that end spectrum is used. Fourier and other spectral tools (e.g., wavelet transforms, principle component analysis) that are usually applied to laboratory spectra can be excellent tools for application to HRS data provided the data are of good quality. A comprehensive review of chemical and physical chromophores in soils and rocks, as an example, can be found in Irons, Weismiller, and Petersen (1989); Ben-Dor, Irons, and Epema (1999); Clark (1999); Malley, Martin, and Ben-Dor (2004); McBratney and Rossel (2006). A compilation table that provides the chromophores of known Earth targets in all spheres is given in Table 8.2. The table, which covers all spectral regions (VIS, NIR, SWIR, MWIR, and LWIR), may be of interest for HRS technology from field, air, and space levels.

The chemical chromophores in the VIS-NIR-SWIR regions refer to two basic chemical mechanisms: (i) overtones and combination modes in the NIR–SWIR region that emerge from the fundamental vibrations in the TIR regions and (ii) electron processes in the VIS region that are in most cases crystal-field and charge-transfer effects. The physical chromophores in this region refer mostly to particle size distribution and to refraction indices of the matter in question. The electronic processes are typically affected by the presence of transition metals, such as iron, and although smeared, they can be used as a diagnostic feature for iron minerals (around 0.80–0.90 μm crystal field and around 0.60–0.70 μm charge transfer).

Accordingly, all features in the UV-VIS-NIR-SWIR-TIR spectral regions have a clearly identifiable physical basis. In solid–fluid Earth materials, three major chemical chromophores can be roughly categorized as follows: (i) minerals (mostly clay, iron oxide, primary minerals-feldspar, Si, insoluble salt, and hard-to-dissolve substances such as carbonates, and phosphates), (ii) organic matter (living and decomposing), and (iii) water (solid, liquid, and gas phases). In gaseous Earth materials, the two main chemical chromophores are (i) gas molecules and (ii) aerosol particles of minerals, organic matter, and ice.

Figure 8.3 presents a summary of possible chromophores in soils and rocks (Ben-Dor, Irons, and Epema, 1999). Basically, the (passive) electromagnetic sources for HRS are the radiation of Sun and Earth (terrestrial) (Sun: VIS-NIR-SWIR, Sun and Earth: TIR). Assuming that in a photon pack emitted from a given source (F; F_0 for Sun, F_e for Earth), some photons may be absorbed (F_α), reflected (F_ρ), or transmitted (F_τ) at a given wavelength and incident angle. The energy balance (in terms of flux densities) on a given target for every boundary (atmosphere, geosphere, and hydrosphere) can be written as follows:

$$F = F_\tau + F_\alpha + F_\rho \tag{8.1}$$

where $F = F_0 + F_e$ or any other incident light hitting the target (e.g., F_g, F_{wf}, and F_{ws} for ground, water floor, and water surface, respectively; Figure 8.4a,b). If we assume that we know the source energy (e.g., F_0), dividing Eq. (8.1) by F_0 gives

$$1 = \tau + \alpha + \rho \tag{8.2}$$

where τ is the transmittance, α the absorptance, and ρ the reflectance coefficients, respectively. These coefficients describe the proportion of F_τ, F_α, and F_ρ, respectively and range from 0 to 1. In an ideal cases, the Sun emits photons (F_0) that pass through the atmosphere and hit the ground (F_g) and then are reflected back

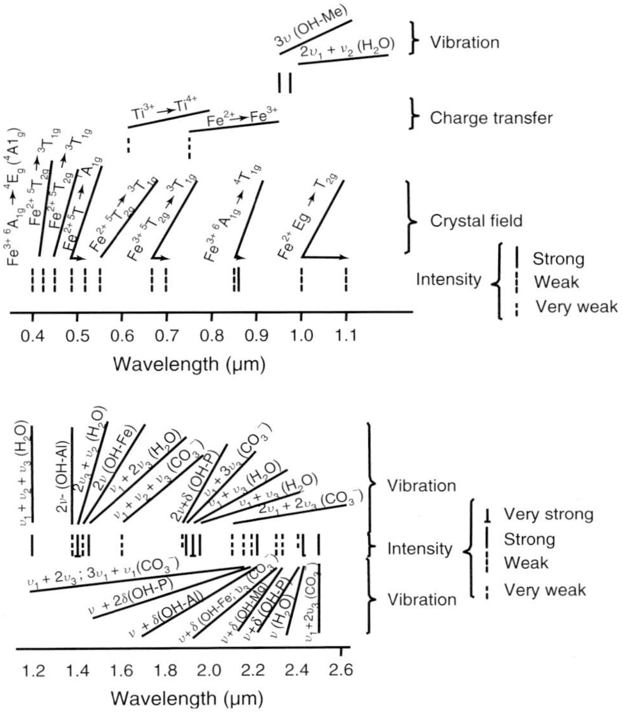

Figure 8.3 Compilation of chromophores in soil and rocks: VIS–NIR electronic processes and overtones, SWIR overtones and combination modes. (Source: Taken from Ben-Dor, Irons, and Epema (1999).)

to the sensor. Figure 8.4 provides a schematic view of two targets (solid and water) and three mediums (atmosphere, geosphere, and hydrosphere). This schematic illustration shows an ideal case demonstrating the role of each coefficient at every sphere. Across the spectral range where the atmosphere is (semi) transparent to the photons (known as *atmospheric window*) or where the atmospheric attenuation is modeled, the atmospheric transmittance (τ_a) and reflectance (ρ_a, term scattering) change based on the atmospheric condition. Accordingly, it is possible to estimate the atmospheric absorptance (α_a) if both τ_a and ρ_a are known; Eq. (8.2) can be written as

$$1 = \tau_a + \alpha_a + \rho_a \tag{8.3}$$

where τ_a and ρ_a are measured or estimated by radiative transfer models (or local measurements), enabling extracting of the atmosphere absorptance (α_a) as well as to provide the new energy flux densities that hit the ground surface (F_g, irradiance).

F_g serves afterwards as an energy source for the geosphere and hydrosphere interactions with the electromagnetic radiation. Figure 8.4a illustrates the interaction of F_g with the Earth's solid surface (geosphere) and Figure 8.4b presents the

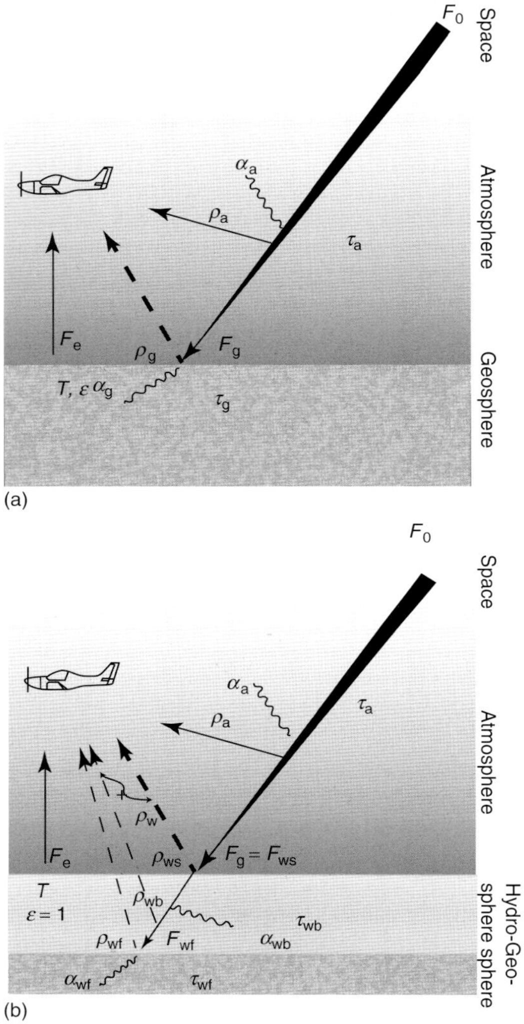

Figure 8.4 Schematic views of two and three mediums (atmosphere and geosphere (a) and atmosphere, hydrosphere, and geosphere (water floor surface) (b), respectively). The following symbols are used: F_0, emission by the Sun (Planck); F_e, Earth emission (Planck); T, Earth temperature; ϵ, emissivity; τ_a, transmittance of atmosphere; α_a, absorptance of atmosphere; ρ_a, reflectance of atmosphere; τ_g, transmittance of solid Earth surface (ground); α_g, absorptance of solid Earth surface (ground); ρ_g, reflectance of the solid Earth surface (ground); τ_{wb}, transmittance of water body; α_{wb}, absorptance of water body; ρ_{wb}, reflectance of water body; ρ_{ws}, reflectance of water surface; ρ_{wf}, reflectance of water floor surface; τ_{wf}, transmittance of water floor surface; α_{wf}, absorptance of water floor surface; F_{ws}, irradiance at surface level; and F_{wf}, irradiance at sea floor level.

interaction with a water body (hydrosphere). In the first example (Figure 8.4a – geosphere), the surface is considered to be opaque and thus the transmittance (τ_g) is set to zero. Accordingly, Eq. (8.2) for the geosphere becomes

$$1 = \alpha_g + \rho_g \tag{8.4}$$

As α_g reflects the material's chemical and physical properties, its extraction can be made by measuring the surface reflectance ρ_g only. Accordingly, this procedure (to extract the surface reflectance) is a key issue in the HSR arena. In the second example (Figure 8.4b, hydrosphere), the water surface is not opaque and thus F_g (also F_{ws}) is divided between direct and indirect portions of photons: those which are reflected from the water to the atmosphere (expressed as reflectance coefficient of the water surface (ρ_{ws}) and those which are scattered from the water body back (expressed as the water body reflectance ρ_{wb}), respectively:

$$\rho_w = \rho_{ws} + \rho_{wb} \tag{8.5}$$

Also, transmittance photons penetrate into the water medium (expressed by the coefficient of transmittance (τ_{wb}), whereas some photons are absorbed by the water body (expressed by the coefficient of absorptance; α_{wb}). As all of the coefficients in this cases are greater than zero, Eq. (8.3) can be written as

$$1 = \tau_{wb} + \alpha_{wb} + (\rho_{ws} + \rho_{wb}) \tag{8.6}$$

As ρ_w depends on the water condition, it is likely that in shallow clean water, where $\rho_{ws} + \rho_{wb} \to 0$, there will be enough photons to penetrate the water body and hit the water floor surface (F_{wf}). This energy (F_{wf}) is then reflected back to the water volume depending on the water floor surface characteristics. In this case, the water floor can be considered opaque (part of the geosphere), leading to $\tau_{wf} = 0$. Accordingly, Eq. (8.2) for the water floor becomes

$$1 = \alpha_{wf} + \rho_{wf} \tag{8.7}$$

where α_{wf} is the water floor absorptance and ρ_{wf} the water floor reflectance. The radiance acquired on-board an HSR sensor is a product of the Sun irradiation (F_0) that is attenuated by all the above mentioned processes (reflectance, absorptance, and transmittance) and their corresponding coefficients. As previously discussed, the absorptance is the most important coefficient for the diagnosis of each sphere since it relates to the chemical and physical properties of the sensed matter. Doing so spectrally can discriminate between the chemical compound being in the atmosphere, geosphere, and hydrosphere. In each sphere, different ways to extract α are taken. To estimate α for the geosphere case, only one degree of freedom is valid in Eq. (8.4) and thus acquiring the reflectance (ρ_g) is an easy task. In the case of atmosphere and hydrosphere, the estimation of the absorptance α is more complicated as two degrees of freedom in Eqs. (8.5) and (8.6) exist (ρ_a, ρ_{wb}, τ_a, and τ_{wb}). This makes the analysis in this case more complicated. In general, all coefficients are playing an important role in order to recover the at-sensor radiance; a full solution to extract ρ_g, $\rho_a + \tau_a$, and ρ_{wf} from the at-sensor radiance is further discussed in the atmospheric removal section (Section 8.9.2).

It is important to mention that all the previous discussion is schematic in order to illustrate how energy decays from the sun to the sensor while interacting with several materials in each sphere. This also highlights how some of the coefficients are important for the HSR concept (e.g., extracting reflectances for the geosphere). No consideration to BRDF, topography, and adjacency effects were taken in this discussion.

The original source of energy (F_0, F_e) can be calculated (or measured) according to Planck's displacement law of a black body entity (depending on its temperature). This calculation shows that the radiant frequencies are different using the Sun (VIS-NIR-SWIR) or Earth (TIR) and thus demonstrates separate HRS approaches using the Sun (mostly performed) and the Earth (just emerging) as radiation sources. When the Sun serves as the radiant source, the reflectance of the surface ρ_g is used as a diagnostic parameter to map the environment. When the Earth serves as the radiant source, the emissivity (ϵ) and the temperature (T) are used as diagnostic parameters. These parameters can be derived from the acquired radiances using several methods to remove atmospheric attenuation (mostly τ_a, and then after separating between T and ϵ (in the TIR region) or extracting ρ_g (in the VIS-NIR-SWIR region)). The reflectance and emissivity are inherent properties of the sensed matter that do not change with external conditions (e.g., illumination or environmental conditions) and hence are used as diagnostic parameters. They both provide, if high spectral resolution is used, spectral information about the chromophores within the matter being studied.

According to Kirchhoff's law, the absorptivity of a perfect black body material is equal to its emissivity (in equilibrium) and thus reflectance has a strong relation to emissivity across the spectral region studied, that is, $\epsilon = 1 - \rho_g$. In atmospheric windows where $\tau \neq 0$ across the VIS-NIR-SWIR-MWIR and LWIR region, HRS can be performed using atmospheric correction techniques as shown in Figure 8.5. While the LWIR (8–12 μm) is sufficient for remote sensing of the Earth (if the temperature is known), as is the VIS-NIR-SWIR region, the MWIR (3–5 μm) region is more problematic for HRS remote sensing of the Earth, as both Sun and Earth Planck functions provide low radiation in their natural position (Sun 6000 K, Earth 300 K) and overlap across this region. Hence, the MWIR region across 3–5 μm is usable for hot (Earth) targets that enable the dominant photons to be above the Sun's background across this region. It should be noted that ρ_g and α_g are important parameters for assessing the Earth's surface composition, but if they are known in advance (e.g., ground targets with known ρ_g), τ_a and α_a can be extracted at specific wavelengths and hence can provide information about the atmospheric constituents (gases and aerosol particles). In other words, HRS can also be a tool to quantitatively study the atmosphere.

While in the VIS region, only limited information on terrestrial systems is available, important information about many of the Earth's materials can be extracted from the NIR–SWIR region. This is because in the VIS region, the electronic processes responsible for broad spectral features are dominant, whereas in the NIR–SWIR region, overtone and combination modes of fundamental vibrations responsible for noticeable spectral features are dominant. Many of

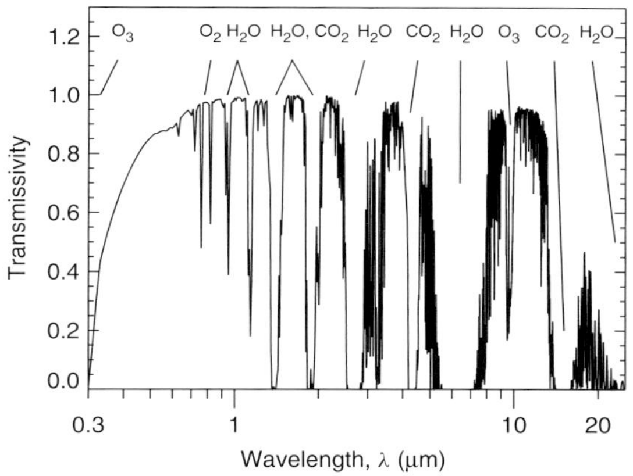

Figure 8.5 Spectral distribution of simulated solar atmospheric transmissivity. Absorption bands correspond to O_3: $< 0.3\,\mu m$; $9.6\,\mu m$; O_2: $0.76\,\mu m$; H_2O: 0.72, 0.82, 0.94, 1.1, 1.38, 1.87, and $2.7\,\mu m$; $6.3\,\mu m$; CO_2: 1.4, 2.0, and $2.7\,\mu m$; $4.3, 15\,\mu m$. The simulations were performed with MODTRAN4 (Version 3 Revision 1, MOD4v3r1) with the Sun in the zenith, the spectral results were averaged over a FWHM of 10 nm. (Source: Adapted from Wendisch and Yang (2012).)

the Earth's materials show significant spectral absorption in the NIR–SWIR region, which serves as a unique fingerprint for mineral identification (Hunt and Salisbury, 1970,1971; Hunt, Salisbury, and Lenhoff, 1971a,b). In addition, atmospheric gases, such as oxygen, water vapor, carbon dioxide, and methane, produce specific absorption features in the VIS-NIR-SWIR regions (Goetz, 1991). Lying in the narrow band's width (usually more than 10 nm) that HRS is capable of generating are spatial qualitative and quantitative indicators for ecologists, land managers, pedologists, geologists, limnologists, atmospheric scientists, and engineers, for which the selection of appropriate methods is dependent on the particular management objectives and the characteristics of the indicators.

In general, the above mentioned spectral information is part of the radiance at sensor, among other factors (such as sun angle, viewing angle, terrain relief, and atmosphere attenuation). To extract the spectral information that are considered inherent properties of the sensed matter, a special data analysis stage must be applied.

8.9 Data Analysis

8.9.1 General

Data processing is performed following a chain procedure, an example of which is given in Figure 8.6. It starts with quality assessment (and assurance, QA) of the raw data and data preprocessing to obtain reliable radiance information and later,

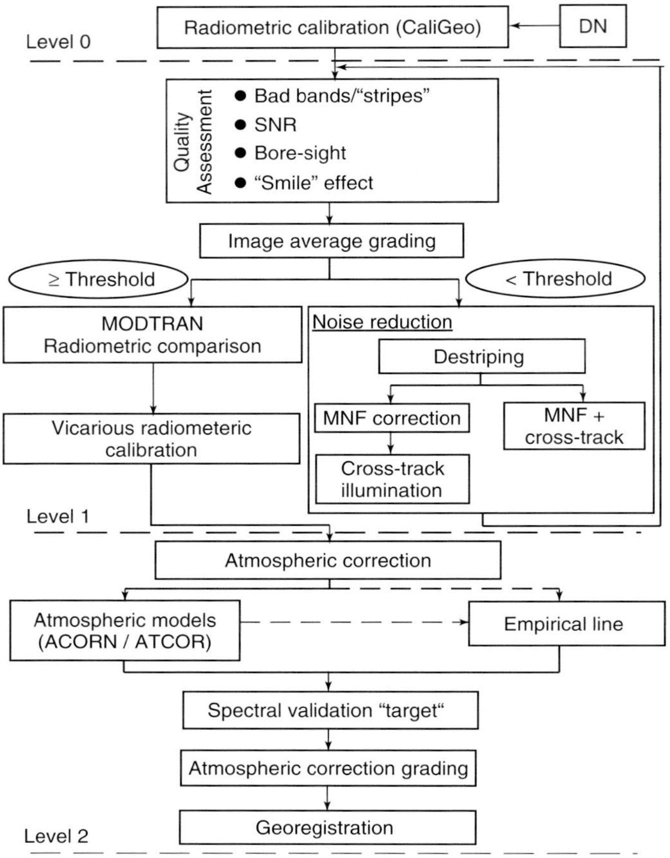

Figure 8.6 A data-processing chain, as used at RSL-TAU (Remote Sensing Laboratory at the Tel Aviv University) with the AISA-Dual sensor. Note that at three stages, quality assurance is crucial: the raw data (including radiance), the atmospheric correction stage (reflectance and emittance), and the thematic mapping stage.

a final product (thematic maps). For each stage, QI are used (denotes "grading" in Figure 8.6). Then, the data should undergo atmospheric correction to yield reliable reflectance (or emittance) data. The cube is then transferred to the "thematic processing" stage in which background knowledge (supervised classification) or the absence of information (unsupervised classification) are used.

8.9.2
Atmospheric Correction

As most of the HRS sensors are operating across the VIS-NIR-SWIR region, the current section deals with atmosphere correction in this wavelength region. The results are either directional surface reflectance quantities or the spectral albedo values (Nicodemus *et al.* 1977). Note, we use the term "reflectance" hereafter as a generic term of a relation between reflected and incoming radiative flux (Eq. (8.2)). This radiometric conversion of the measured radiance to reflectance is referred to as "atmospheric correction" already in early remote sensing literature (Dozier and Frew, 1981). Note that the term "correction" is appropriate as long as data are adjusted to match a given ground reference by empirical methods. However, it may be misleading for methods relying on physical radiative transfer models. The term "atmospheric compensation" would be a more appropriate description in this case, as the atmospheric effects are compensated from correctly calibrated imagery; however, this term has not yet been widely established.

In this section, some empirical normalization methods are summarized first, which allow for fast and efficient atmospheric correction. The basics of methods based on radiative transfer models are given in the subsequent subsections.

8.9.2.1 Empirical Reflectance Normalization

All empirical atmospheric correction methods have in common that a priori knowledge about the surface spectral albedo is put in relation to the imagery in order to find factors for a normalization of the atmospheric effect (Smith *et al.* 1987). Hereafter, a collection of methods that are suited for systems of unsecured calibration state and if fast results are required are compiled. These methods may be applied on uncalibrated image data, that is, directly on the digital number $\mathrm{DN_s}$.

The flat-field approach (Dobbins and Boone, 1998) uses a spectrally flat spectrum from within the image for normalization to calculate a flat-field (quasi-) reflectance ρ_{ff} such that

$$\rho_{\mathrm{ff}} = \frac{\mathrm{DN_s}}{\mathrm{DN_{ff}}} \tag{8.8}$$

where $\mathrm{DN_s}$ is the (uncalibrated) digital number signal at the sensor and $\mathrm{DN_{ff}}$ is the signal of a selected spectrum. This normalization may result in reflectance values above 100% as the selected flat-field reflectance is usually below 100 %.

The known/bright target approach uses the known (or assumed) reflectance ρ_{b} of one typically bright target in the image such that the whole (calibrated) image data may be normalized by the at-sensor measurement $\mathrm{DN_b}$ at the target by the transformation:

$$\rho = \frac{\mathrm{DN_s}}{\mathrm{DN_b}} \cdot \rho_{\mathrm{b}} \tag{8.9}$$

A variation of the bright target approach is the "quick atmospheric correction" (QUAC) method (Bernstein *et al.* 2008). Instead of taking one pixel as a reference, the relation of a generic expected average reflectance to the average signal in the image is taken as reference for correction of the full image, also known as *IARR method* (Kruse, 1988).

The empirical line correction uses a combination between dark and bright targets in a scene. If two or more objects are known, a linear function is derived for each spectral band between measured signal and reflectance. The linear fit is performed between the known reflectances ρ_i and the respective measurements DN_i, such that a slope $\Delta\rho/\Delta\mathrm{DN}$ of the function $\rho(\mathrm{DN})$ with a typical offset for dark objects $\mathrm{DN_{dark}}$ can be found. This function is then used for normalization of all spectra of the image using the equation

$$\rho = (\mathrm{DN_s} - \mathrm{DN_{dark}}) \cdot \frac{\Delta\rho}{\Delta\mathrm{DN}} \tag{8.10}$$

The empirical line works satisfactorily well for flat terrain and small FOV imagery but is at its limit in mountainous areas and if repeatable corrections are required for an image series.

8.9.2.2 At-Sensor Radiance Description

Other than the empirical correction methods, the physical atmospheric correction of HRS data relies on an appropriate description of the at-sensor radiance from known parameters. In HRS, the at-sensor radiance is composed of three major components comprising the direct reflected and the backscattered radiance from the surface and the atmosphere. The thermal emission may be neglected for the wavelength range up to 2.50 μm as long as the temperature of the surface is below 350 K. Thus, the at-sensor radiance I_s may be in a good approximation described as a sum of the direct ground-reflected radiance $I_{g,dir}$, the so-called adjacency radiance $I_{g,adj}$, and the atmospheric radiance I_{atm}:

$$I_s = I_{g,dir} + I_{g,adj} + I_{atm} \tag{8.11}$$

We use ρ as the in-field hemispherical–directional reflectance factor (also denoted as $\mathrm{HDRF_{meas}}$), ρ_{adj} as the large-scale spectral albedo of the surface, and s as the spherical albedo of the atmosphere. The terms in Eq. (8.11) may then be written in a good approximation for the direct component

$$I_{g,dir} = \frac{1}{\pi} \cdot F_g \cdot \rho_g \cdot \tau_u \tag{8.12}$$

the adjacency radiance

$$I_{g,adj} = \frac{1}{\pi} \cdot F_g \cdot \rho_{adj} \cdot \tau_{u,adj} \tag{8.13}$$

and the atmospheric radiance

$$I_{atm} = \frac{1}{\pi} \cdot F_0 \cdot s \tag{8.14}$$

The term F_0 is the top-of-atmosphere irradiance and F_g is the total irradiance (solar flux) on a ground surface element, which may be written as

$$F_g = F_0 \cdot \tau_d \cdot \cos\varphi + F_{dif} \cdot V_{sky} + F_{ter} \tag{8.15}$$

The latter depends on the local solar incidence angle φ and includes the total diffuse irradiance F_{dif}, scaled by the fraction of the visible sky (skyview factor V_{sky}) and the terrain irradiance F_{ter}. The parameter τ_d is the downward atmospheric transmittance; τ_u and $\tau_{u,adj}$ are the upward transmittances of the atmosphere for the direct and the adjacency radiative paths, respectively.

A different formulation of the at-sensor radiance is derived, if the adjacency term is written using the back-reflected radiance from the ground coupled by the single scattering albedo s of the atmosphere, compare (Tanré *et al.* 1979). Here, all ground-reflected radiance is summarized in the term $I_{g,tot}$. This results in the following relation:

$$I_s = I_{g,tot} + I_{atm} = \tau_u \cdot F_g \cdot \rho_g \cdot \frac{1}{\pi \cdot (1 - \rho_{adj} \cdot s)} + I_{atm} \tag{8.16}$$

where the parameters are as described for Eq. (8.11). Such formulations of the at-sensor radiance are the basis for the atmospheric correction task.

8.9.2.3 Radiative-Transfer-Based Atmospheric Correction

Radiative transfer codes (RTCs) such as MODTRAN®-5 (Berk *et al.* 2005) or 6 S-V (Vermote *et al.* 2006) are well suited for forward simulation of the at-sensor signal from given boundary conditions. However, they are not built for the task of inversion for surface reflectance properties from radiometric images. For this purpose, atmospheric correction software is required. Examples of such software are TAFKAA (Gao *et al.* 2000), ACORN (Atmospheric Correction Now) (Green, 2001), HAATCH (High Accuracy Atmospheric Correction for HRS Data) (Qu, Goetz, and Heidbrecht, 2001), ATCOR (Atmospheric Topographic Correction) (Richter and Schläpfer, 2002), or FLAASH (Fast Line-of-Sight Atmospheric Analysis of Spectral Hypercubes) (Cooley *et al.* 2002). Such software allows an efficient inversion of the calibrated imagery on the basis of set of equations bellow. By inversion and reformulation of Eq. (8.11), the surface reflectance may be retrieved by the following equation:

$$\rho_g = \frac{\pi \cdot d^2 \cdot \left(I_s - I_{g,adj} - I_{atm}\right)}{\tau_u \cdot \left(\tau_d \cdot F_0 \cos\varphi + F_{dif} \cdot V_{sky} + F_{ter}\right)} \tag{8.17}$$

The components of this equation are to be derived from (i) physical model of a radiative transfer code: I_{atm}, F_{dif}, τ_d, and τ_u; (ii) boundary conditions of terrain: incidence angle φ and sky view factor V_{sky}; (iii) astronomical data: the average extraterrestrial solar constant F_0 and the dependency on the relative Earth–Sun distance described by parameter d; and (iv) iteration of atmospheric correction: $I_{g,adj}$ and terrain irradiance F_{ter}.

As all the parameters except F_0 and d vary per pixel, it is not feasible to calculate the radiative transfer directly for each pixel. Precalculated look-up tables (LUTs) are normally employed. These LUTs are interpolated with the pixel properties to find the applicable parameters for the correction.

A different approach is to perform the atmospheric compensation in the "apparent reflectance" domain ρ_s after dividing the at-sensor radiance by the ground solar irradiance, propagated to the at-sensor level $F_{0,s}$:

$$\rho_s = \frac{\pi \cdot d^2 \cdot I_s}{F_{0,s}} \tag{8.18}$$

$$\rho_{atm} = \frac{\pi \cdot d^2 \cdot I_{atm}}{F_{0,s}} \tag{8.19}$$

$$\rho_{adj} = \frac{\pi \cdot d^2 \cdot I_{adj}}{F_{0,s}} \tag{8.20}$$

These terms are typically used over flat ground, introducing a total transmittance term $\tau_{tot} = \tau_d \cdot \tau_u$, which relates the at-sensor reflectance to the ground reflectance. The inversion of Eq. (8.16) for reflectance results in

$$\rho_g = \frac{(\rho_s - \rho_{atm}) \cdot (1 - \rho_{adj} \cdot s)}{\tau_{tot}} \tag{8.21}$$

If the adjacency reflectance is further assumed to be the same as the pixel reflectance (i.e., $\rho_{adj} = \rho_g$), the equation is reduced to

$$\rho_g = \frac{(\rho_s - \rho_{atm})}{\tau_{tot} + (\rho_s - \rho_{atm}) \cdot s} \tag{8.22}$$

This is a basic atmospheric correction equation, which may be used in simple atmospheric correction programs or for fast inversion of a radiative transfer code. Note that working in the reflectance domain is critical for airborne instruments, as this approximation relies on accurate knowledge of the radiance at sensor level. An additional modeling step is required to infer the at-sensor radiance level $F_{0,s}$ as a reference value.

8.9.3
Process of Complete Atmospheric Correction

A complete atmospheric correction as implemented in atmospheric correction routines follows these steps: (i) create a LUT, containing the parameters of the above equations in relation to the parameters at a fixed number of data points (covering the expected range); (ii) calculate skyview factor, height, and incidence angle from DEM (digital elevation model), using the solar zenith and azimuth angles; (iii) derive atmospheric parameters from imagery (i.e., water vapor and aerosol load of the atmosphere); (iv) make fixed preselections (e.g., flight altitude and aerosol model); (v) invert the LUT, that is, derive the parameters by multilinear interpolation for each pixel; (vi) use Eq. (8.10) or (8.22) to perform the atmospheric correction; and (vii) perform the last two iteration of steps for adjacency correction and for the calculation of the terrain irradiance.

Some variations of this procedure exist, as the parameterization of the problem may differ and the LUT may be precalculated or calculated for each scene directly while correcting the data. An ideal high level standard procedure combines geometric and atmospheric processing (Schläpfer and Richter, 2002). Linked parameters are the viewing angle per pixel, the absolute distance from the aircraft to each pixel location, or the relative air mass between sensor and pixel. Furthermore, other DEM related parameters, such as height, slope, or aspect, are required for radiometric correction algorithms and can only be used if the image is brought to the same geometry as the DEM. The dependencies within the atmospheric correction part lead to iterative reflectance retrieval steps, specifically for adjacency correction purposes. The final step of the processing should be a correction of the reflectance anisotropy (i.e., a BRDF correction). Some details regarding crucial correction steps are given hereafter.

8.9.3.1 Atmospheric Parameter Retrieval

Airborne HRS sensors offer the inherent capability for automatic retrieval of the radiometrically critical parameters atmospheric water vapor content and aerosol load (optical thickness). For the atmospheric water vapor, the 0.94/1.13 μm water vapor absorption bands are typically used for the retrieval of columnar water vapor over land on a per-pixel basis (Schläpfer *et al.* 1998). The aerosol optical thickness is normally calculated using the dark dense vegetation (DDV) approach (Kaufman and Tanré, 1996) interpolating the aerosol load to areas not covered by vegetation. These two methods allow for a mostly autonomous atmospheric correction of HRS data.

8.9.3.2 Adjacency Correction

The correction of the atmospheric adjacency effect is of high relevance, especially for limnological applications (Tanré *et al.* 1987). The effect is significant in a horizontal range from 100 m up to 1.5 km starting at flight altitudes of 1000 m above ground and higher. Thus, each pixel has to be corrected with respect to the average reflectance of the adjacent areas. This can be carried out in an efficient way by the definition of a spatial convolution function, which takes a distance-weighted average of the adjacent area in the image to calculate an adjacency weighting factor. The corresponding radiance has to be simulated in the radiative transfer code as indirect ground-reflected radiance according to the aforementioned parameterization.

8.9.3.3 Shadow Correction

Cast shadows, cloud shadows, and shadows from building are often present in HRS data. They receive mostly diffuse irradiance that is sufficient to provide enough signal for data analysis with state of the art sensor systems. Correction approaches try to classify the shadowed areas first and then apply a separate correction model to these parts of the image such that shadows are removed in optimal situations (Adler-Golden *et al.* 2002; Richter and Muller, 2005). The correction model takes into account the diffuse nature of the irradiance in the cast shadow areas and needs to consider the skyview factors correctly for an accurate correction.

8.9.3.4 BRDF Correction

The derivation of spectral albedo (i.e., the bihemispherical reflectance, BHR) from directional reflectance values is the task of BRDF correction. The operational correction of BRDF effects in images is not yet solved satisfactorily, but progress has been made on this issue (Feingersh, Ben-Dor, and Filin, 2010). The correction of the BRDF effects may be performed if the BRDF properties of the observed target(s) and the (diffuse) irradiance distribution is known. For operational use, an anisotropy factor needs to be calculated and applied for each pixel. This factor accounts for the relation between measured hemispherical–directional reflectance ($HDRF_{meas}$) and the spectral albedo (BHR; also known as *white sky albedo*). The anisotropy factor has to be inferred from an appropriate BRDF model or from measurements. The finally calculated spectral albedo product is a quantity that may be easily compared in multitemporal analysis and may be used for unbiased object classification.

8.9.4
Retrieval of Atmospheric Parameters

On the basis of the relatively high spectral resolution obtain by the HRS sensors, one can use the specific absorption features of atmospheric gases (natural or contaminated) and evaluate their column content on a pixel-by-pixel basis. This may provide an innovative way of mapping the gases' spatial distribution and of spotting new quantitative information on the atmospheric conditions at very high spatial resolution. The gases that are active across the UV-VIS-NIR-SWIR-TIR spectra are divided into two sectors: (i) a major sector in which the spectral response of the gases is well detected (high fraction and strong absorption) and (ii) a minor sector in which the spectral response is low and difficult to assess because of the low fraction of the gases and relatively weak absorption features. The major gas group is composed of O_2, H_2O, and CO_2, whereas the minor gas group consists of O_3, N_2O, CO, and CH_3 in the UV-VIS-NIR-SWIR and SO_2 and NO_2 in the TIR. Figure 8.7 provides the absorptance features of the above gas components across the UV-VIS-NIR-SWIR-TIR spectral region along with the atmospheric windows. The advantage of assessing the above gases on a pixel-by-pixel basis is significant. It can help accurately extract surface reflectance by estimating the gases' absorption (and hence their atmospheric transmission) on a pixel-by-pixel column basis. Consequently, calculating water vapor directly from the image, first demonstrated by Gao and Goetz (1995), is now a very common way of achieving high performance of atmospheric correction methods (Section 8.8). While H_2O is considered to be a nonuniformly spatially distributed gas, other major gases, that is, CO_2 and O_2, are known to be well mixed – hence, their use as indicators to assess atmospheric phenomena that might affect the spatial distribution of the gas in question. For example, over rough terrain, if spatial changes are encountered using a well-mixed gas, this might indicate different elevations, as the column pixel volume over high terrain consists of less molecules than that over low terrains for a particular gas. On the basis of this, Ben-Dor and Kruse (1996) and Green (2001) showed that it is possible to construct a DEM structure of the studied area solely from the HRS radiance information and the CO_2 peak. Furthermore, as O_2 is also a well-mixed gas, it can be used to estimate, on a pixel-by-pixel basis, the Mie scattering effect across the VIS–NIR region and hence can be used to better extract the surface reflectance, assuming that the scattering is not a spatially homogeneous phenomenon. Using one absorption peak of the H_2O at 1.38 μm, Gao, Goetz, and Wiscombe (1993) showed that a nonvisible cirrus cloud can be detected and mapped based on the high scattering properties of the ice particles within the cloud volume. Ben-Dor (1994) suggested taking precautions in using these absorption peaks over high terrain and bright targets, and in another paper, (Ben-Dor, Goetz, and Shapiro, 1994) suggested that the O_2 peak be used to map the cirrus cloud distribution in the VIS–NIR region (0.760 μm). On the basis of this idea, Schläpfer *et al.* (2006) was able to quantitatively assess a smoke plume over a fire area using the scattering effect on the O_2 absorption peak. Alakian, Marion, and Briottet (2008) developed a method to retrieve the microphysical and optical properties in aerosol plumes (L-APOM) in the VIS region as well. Recently, Chudnovsky *et al.* (2009) mapped a dust plume over the Bodele depression in northern Chad using Hyperion data and the SWIR region. Another innovative study that shows the applicability of HRS in the atmosphere was performed by Roberts *et al.* (2010). They showed that if high SNR data are available, it is also possible to detect the distribution of minor gases. Using AVIRIS 2006 data over a marine (dark) environment, they were able to detect, on a pixel-by-pixel basis, the emission of methane over the Coal Oil Point (COP) marine seep fields, offshore of Santa Barbara, California, and the La Brea Tar Pits in Los Angeles, California. In the TIR region, there are several examples of the detection of plumes of toxic gases based on the fundamental vibration peak across the atmospheric windows between 2.5 and 16 μm. Using SO_2 emission in the LWIR region at 8.58 and 8.82 μm, Shimoni *et al.* (2007) were able to spot shade on a plume emitted over an industry refinery zone with additional information extracted from the VIS region. Figure 8.7 provides a summary for the absorption positions of some of the above mentioned gases across the VIS-NIR-SWIR-TIR spectral region. In summary, it can be concluded that HRS technology is not only capable of deriving surface information but also has the proven capability to extract quantitative information on atmosphere constituents in an innovative way that none of the current remote sensing means can provide.

8.9.5 Mapping Methods and Approaches

Over the past few years, many techniques for mapping and processing of HRS data have been developed (Schaepman *et al.* 2009). The special characteristics of hyperspectral data sets pose different processing problems, which must be tackled under specific mathematical formulations, such as classification (Landgrebe, 2003; Richards and Jia, 2006) or spectral unmixing (Adams, Smith, and Johnson, 1986). These problems also require specific dedicated processing software and hardware platforms (Plaza and Chang, 2007).

In previous studies (Plaza *et al.* 2009), available techniques were divided into full-pixel and mixed-pixel techniques, where each pixel vector defines a spectral signature or fingerprint that uniquely characterizes the underlying materials at each site in a scene. Mostly based on previous efforts in multispectral imaging,

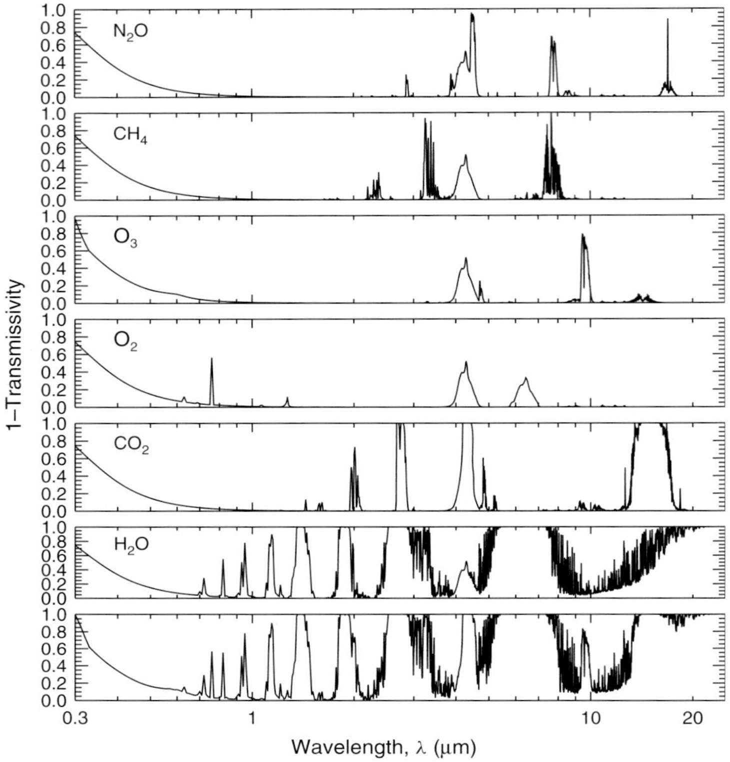

Figure 8.7 Extinction spectra of atmospheric gases across the VIS-NIR-SWIR-TIR region. The simulations were performed with MODTRAN4 (version 3 Revision 1, MOD4v3r1), with the Sun in the zenith; the spectral results were averaged over a FWHM of 10 nm. (Source: Adapted from Wendisch and Yang (2012).)

full-pixel techniques assume that each pixel vector measures the response of one single underlying material (Schaepman *et al.* 2009). Often, however, this is not a realistic assumption. If the spatial resolution of the sensor is not good enough to separate different pure signature classes at the macroscopic level, these can jointly occupy a single pixel, and the resulting spectral signature will be a composite of the individual pure spectra, often called *endmembers* in hyperspectral terminology (Boardman, Kruse, and Green, 1995). Mixed pixels can also result when distinct materials are combined into a homogeneous or intimate mixture, which occurs independently of the spatial resolution of the sensor. To address these issues, many spectral unmixing approaches have been developed under the assumption that each pixel vector measures the response of multiple underlying materials (Kruse, 1988; Keshava and Mustard, 2002).

Spectral unmixing has been an alluring goal for exploitation, from the earliest days of hyperspectral imaging (Goetz *et al.* 1985) until today. Regardless of the spatial resolution, the spectral signatures collected in natural environments are invariably a mixture of the signatures of the various materials found within the spatial extent of the ground instantaneous field view of the imaging instrument (Adams, Smith, and Johnson, 1986). In this case, the measured spectrum may be decomposed into a combination of pure spectral signatures of soil and vegetation, weighted by areal coefficients that indicate the proportion of each macroscopically pure signature in the mixed pixel (Keshava and Mustard, 2002). The availability of hyperspectral imagers with a number of spectral bands exceeding the number of spectral mixture components (Green *et al.* 1998) has allowed casting the unmixing problem in terms of an over determined system of equations in which, given a set of pure spectral signatures (called endmembers), the actual unmixing to determine apparent pixel abundance fractions can be defined in terms of a numerical inversion process (Harsanyi and Chang, 1994; Bateson and Curtiss, 1996; Plaza *et al.* 2004; Berman *et al.* 2004; Chang *et al.* 2006; Rogge *et al.* 2006; Wang and Chang, 2006; Zaer and Gader, 2008).

A standard technique for spectral mixture analysis is linear spectral unmixing (Heinz and Chang, 2001; Plaza *et al.* 2004), which assumes that the spectra collected by the spectrometer can be expressed in the form of a linear combination of endmembers weighted by their corresponding abundances. It should be noted that the linear mixture model assumes minimal secondary reflections and multiple scattering effects in the data-collection procedure, and hence, the measured spectra can be expressed as a linear combination of the spectral signatures of materials present in the mixed pixel (Figure 8.8a). Although the linear model has practical advantages, such as ease of implementation and flexibility in different applications (Chang, 2003), nonlinear spectral unmixing may best characterize the resultant mixed spectra for certain endmember distributions, such as those in which the endmember components are randomly distributed throughout the instrument's FOV (Guilfoyle, Althouse, and Chang, 2001). In those cases, the mixed spectra collected by the imaging instrument are better described by assuming that part of the source radiation is multiply scattered before being collected at the sensor (Figure 8.8b). In addition, several machine-learning techniques have been applied to extract relevant information from hyperspectral data during the past decade. Taxonomies of remote sensing data-processing algorithms (including hyperspectral analysis methods) have been developed in the literature (Richards and Jia, 2006 Schowengerdt, 1997). It should be noted, however, that most available hyperspectral data-processing techniques focus on analyzing the data without incorporating information on the spatially adjacent data, that is, hyperspectral data are usually not treated as images but as unordered listings of spectral measurements with no particular spatial arrangement (Rogge *et al.* 2006). The importance of analyzing spatial and spectral patterns simultaneously has been identified as a desired goal by many scientists devoted to multidimensional data analysis.

In certain applications, however, the integration of high spatial and spectral resolution is mandatory to achieve sufficiently accurate mapping and detection results. For instance, urban area mapping requires sufficient spatial resolution to distinguish small spectral classes, such as trees in a park or cars on a street (Bruzzone and Marconcini, 2006). Owing to the small number of training samples

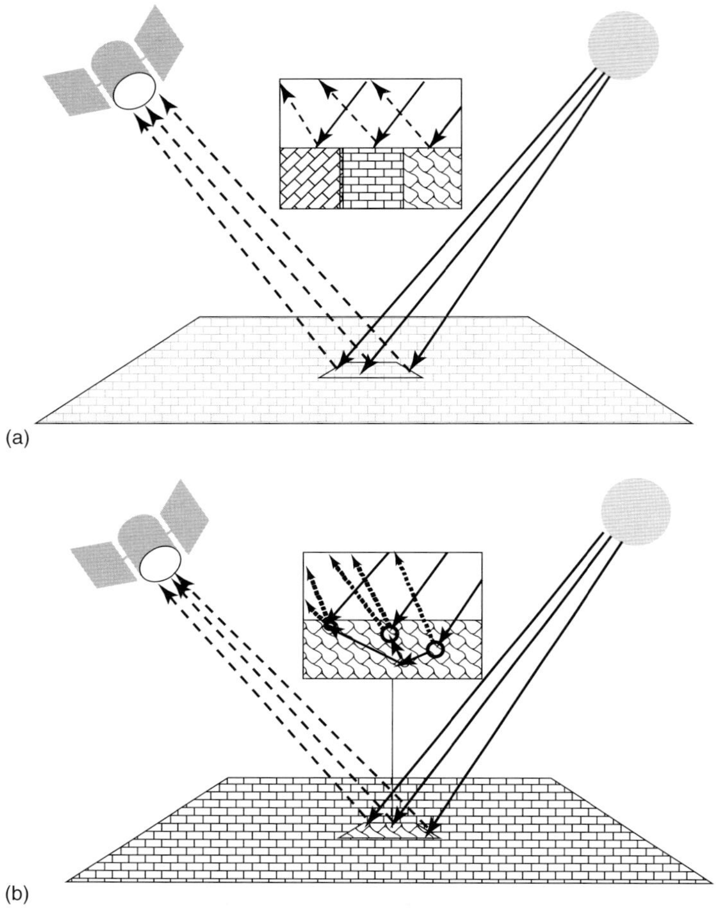

Figure 8.8 Linear (a) versus nonlinear (b) mixture models: single versus multiple scattering.

and the high number of features available in remote sensing applications, reliable estimation of statistical class parameters is another challenging goal (Foody, 1999). As a result, with a limited training set, classification accuracy tends to decrease as the number of features increases. This is known as the *Hughes effect* (Landgrebe, 2003). High-dimensional spaces have been demonstrated to be mostly empty, thus making density estimation even more difficult. One possible approach to handling the high-dimensional nature of hyperspectral data sets is to consider the geometrical properties rather than the statistical properties of the classes. The good classification performance demonstrated by support vector machines (SVMs) using spectral signatures as input features (Prasad, 2008) can be further increased by taking advantage of semisupervised learning and contextual information. The latter is performed through a combination of kernels dedicated to spectral and contextual information, while in the former, the learning is provided with some supervised information in addition to the wealth of unlabeled data. Among the great many methods proposed in the literature for such approaches, we focus on the transductive SVM for semisupervised learning (Bruzzone and Marconcini, 2006) or a composite kernel-based methodology for contextual information integration at the kernel level (Camps-Valls *et al.* 2006) have shown great success in practice.

As most of the methods reviewed here deal with endmember extraction and data mining from the reflectance or emittance cubes (an unsupervised approach), there are methods in which the endmembers are known in advance or the spectral model to map the pixels has already been developed (supervised approach). One of the first and most usable endmember-based approaches in HRS is the Spectral Angle Mapper (SAM), developed by Kruse *et al.* (1993), which is based on the angle calculated between two spectral vectors: the pixel and the selected endmember. Since then, many other spectral-based techniques have been developed, where most recently, spectral-based models that are generated in a spectral domain (e.g., PLS or neural network) are implemented on a pixel-by-pixel basis to the image cube in question. This method enables quantitative mapping of selected properties on the Earth's surface such as infiltration rate (Ben-Dor *et al.* 2004) organic matter content (Stevens *et al.* 2008), salinity (Ben-Dor *et al.* 2002), and more.

Finally, although the mapping and classification techniques described above hold great promise for hyperspectral data processing, they also introduce new computational challenges. With the recent explosion in the amount and complexity of hyperspectral data, parallel processing and high-performance computing (HPC) practices have necessarily become requirements in many remote sensing missions, especially with the advent of low-cost systems such as commodity clusters (Plaza and Chang, 2007). On the other hand, although hyperspectral analysis algorithms map nicely to clusters and networks of workstations, these systems are generally expensive and difficult to adapt to on-board data-processing requirements introduced by several applications, such as wildland fire tracking, biological threat detection, monitoring of oil spills, and other types of chemical contamination. In those cases, low-weight and low-power integrated components are essential to reducing the mission's payload and obtaining analyzed results quickly enough for practical use. In this regard, the emergence of specialized hardware devices such as field-programmable gate arrays (FPGAs) has helped in bridging the gap toward real-time analysis of remotely sensed hyperspectral data.

8.10 Sensor Calibration

8.10.1 General

In combination, calibration and validation (cal/val) can be regarded as a single process that encompasses the entire remote sensing system, from sensor to data product. The objective of both is to develop a quantitative understanding and characterization of the measurement system and its biases in both space and time (National Research Council, 2007). Calibration of hyperspectral sensor data is a critical activity for a number of reasons. First, we need to have confidence in the reliability of data delivered by such sensors. Second, as many of the products that we are deriving from hyperspectral data are quantitative, we need to know that the data from which they are derived are accurate (this holds for qualitative data as well). We often test the accuracy of remote sensing data products by performing validation of the subsequent data sets; thus, the raw data delivered by sensors must be well calibrated and the products derived from them are also well validated. cal/val are, therefore, activities that form an integral component of the efficient use of any form of EO data and in the maintenance of the scientific value of EO data archives. As HRS data are acquired in DN values, but for most applications, we need radiometric information as an input to extract reflectance or emissivity values, accurate transfer from one stage to another is crucial. In this respect, radiometric and spectroscopic assurance is required. Radiometric calibration refers to the process of extracting physical units from the original raw spectroscopic data and assigning the channels in the sensors to a meaningful wavelength.

Cal/val is, therefore, a fundamentally important scientific activity and should be a continuous component in any remote sensing program, providing an independent check on the performance of both space-based airborne and ground-based hyperspectral sensors and processing algorithms.

In general, one can say that the calibration of EO data is critical if we are to reliably attribute detected changes observed in data to real environmental changes occurring at ground level. Without calibration, we are unable to rule out the influence of other factors, such as instrument error or influences of the atmosphere. This problem is exacerbated by the myriad of sensors operated by multiple countries and organizations. Calibration allows the traceability of sensor data to the same physical standards and is routinely required as sensors decay throughout their lifetime. Calibration is thus critical if we want to reliably extract information from measured radiance, compare information acquired from different regions and different times, compare and analyze HRS observations with measurements provided by other instruments, and extract information from spectral image measurements using physically based computer models.

Validation refers to the independent verification of the physical measurements made by a sensor as well as the derived geophysical variables. Validation allows for the verification and improvement of the algorithms used (e.g., for atmospheric correction and vegetation state). To achieve this, conventional, ground-based observations are required using calibrated and traceable field instrumentation and associated methods. To this end, several indicators are valid and developed to check the accuracy of the calibration stage and provide the user with a reliable feeling about his data set.

The definition of all the common terms used here for cal/val are taken from the Committee of Earth Observation Satellites (CEOS) as follows:

(i) *Calibration*. The process of quantitatively defining the responses of a system to known, controlled signal inputs;
(ii) *Validation*. The process of assessing, by independent means, the quality of the data products derived from the system outputs;
(iii) *Traceability*. Property of a measurement result relating the result to a stated metrological reference (free definition and not necessarily SI) through an unbroken chain of calibrations of a measuring system or comparisons, each contributing to the stated measurement uncertainty;
(iv) *Uncertainty*. Parameter that characterizes the dispersion of the quantity values that are being attributed to a measured mean, based on the information used;
(v) *Vicarious Calibration*. Vicarious calibration refers to techniques that make use of natural or artificial sites on the surface of the Earth for post calibration of airborne or spaceborne sensors.

8.10.2
Calibration for HSR Sensor

Calibration translates electrical output DN values (voltages or counts) to reliable physical-based units (radiometric information) by determining the transfer functions and coefficients necessary to convert a sensor reading. The coefficients are extracted throughout a careful measurement stage in the laboratory using well-calibrated facilities and traceable standards. There are a number of components ensuring a thorough calibration approach. Radiometric and spectral responses need to be accurately monitored through the lifetime of a sensor to monitor changes in response, as it ages over time. In the case of spaceborne hyperspectral sensors, both prelaunch and post (on-orbit) launch calibrations are undertaken, either directly or using vicarious targets-on-orbit and vicarious calibration enable taking into account changes in calibration over time using the moon's surface (Kieffera *et al.* 2003). Airborne hyperspectral sensors have the advantage over spaceborne sensors that they can be removed from the aircraft and resubjected to rigorous laboratory calibration tests similar to those performed for prelaunch calibration of spaceborne sensors. This is often performed before and after a flying "season." The calibration coefficients from each season can also be used to track the sensor's deterioration over its years of operation.

8.10.2.1 **Preflight Calibration**

The three key components to prelaunch calibration are radiometric, spectral, and spatial. Achieving radiometric calibration involves the use of a calibrated integrating sphere whose ideal output is homogeneous and large enough to illuminate all elements in a sensor array with the same radiance. An absolute radiometric calibration determines the relationship between sensor signals and radiance for all spectral channels. Varying the output of the integrating sphere also allows for the study of the linearity between sensor response and radiance and the assessment of the SNR at radiance levels similar to those encountered when sensing the Earth's surface (Gege *et al.* 2009).

Spectral calibration typically uses a monochromator to produce a collimated narrow beam of light that is blocked by transmission filters and is thus tunable to different wavelengths. Measurements performed here allow for determination of spectral response function, center wavelength, spectral smile, spectral sampling distance, the spectral range of pixels, and spectral resolution and to perform a wavelength calibration (Oppelt and Mauser, 2007).

Spatial calibration (geometric) can most accurately be achieved with the movement of a point light source across the sensor array whose beam is controlled by a slit (Gege *et al.* 2009). This allows for along-track and across-track calibration of the sensor array. Measurements performed here allow for the derivation of line spread function across track; center coordinates for each CCD in the array; across-track sampling distance; pixel instantaneous FOV; total sensor FOV; and the modulation transfer function (the reparability of adjacent targets as a function of distance and contrast (Oppelt and Mauser, 2007).

8.10.2.2 **In-Flight/In-Orbit Calibration**

This involves the use of in-built calibration sources and vicarious calibration or cross-calibration to other satellite sensors. The critical issue at this stage is to be able to monitor changes in sensor performance over time (Pearlman *et al.* 2003). For example, Hyperion, the first fully spaceborne hyperspectral sensor, relied on the diffuse reflectance of an in-built Spectralon ™reflectance surface illuminated by the Sun or a lamp, in calibrations performed once every two weeks. The moon and other opportunistic Earth surface targets were also used to monitor sensor performance over time (Jarecke and Yokoyama, 2000; Pearlman *et al.* 2003; Ungar *et al.* 2009). Cross-calibration to data from the LANDSAT 7 ETM+ sensor was also frequently performed.

EnMAP, the new German-built hyperspectral sensor scheduled for launch in 2015, will carry for calibration a full aperture diffuser, coupled with an integrating sphere with various calibration lamps. A shutter mechanism also allows for dark measurements to be performed. APEX, a joint Belgian–Swiss airborne sensor development, carries an in-flight characterization facility using a stabilized lamp coupled with vicarious and cross-calibration (Nieke *et al.* 2008), (Itten *et al.* 2008).

8.10.2.3 **Vicarious Calibration**

Vicarious calibration is also used as an in-flight check on sensor performance (Green and Shimada, 1997; Green and Pavri, 2000; Secker *et al.* 2001). The approach can use homogeneous targets on the land surface (e.g., dry lake beds, desert sands, ice sheets, water bodies, and so on) or artificial targets of varying brightness if the sensor has sufficient spatial resolution (Brook and Ben-Dor, 2011). The sites or targets must be well characterized, and ideally, reflectance and, if possible, radiance should be measured at the ground surface using calibrated spectroradiometers simultaneously with sensor overflight. Increasingly sophisticated ground-based instrumentation is being used to provide autonomous and near-continuous measurement of the characteristics at many of these sites (Brando *et al.* 2010). Correction involves either top-down (correction of "top-of-atmosphere" sensor data to ground-leaving reflectance using an atmospheric correction model) or bottom-up (correction of ground target reflectance to top-of-atmosphere radiance using a radiative transfer model taking into account atmospheric transmission and absorption, e.g., MODTRAN). Increasingly, a combination of measurements obtained at varying scales and resolutions (e.g., *in situ*, airborne, and satellite) are being used to provide the basis for assessment of the on-orbit radiometric and spectral calibration characteristics of spaceborne optical sensors (Green, Pavri, and Chrien, 2003a).

The smaller pixel sizes of airborne imagery compared to typical image satellite resolutions, along with targeted deployment, means that artificial vicarious calibration targets can be rapidly deployed in advance of specific airborne campaigns. Such targets can also help overcome the difficulties of finding sufficient natural homogeneous targets of varying brightness. Supervised vicarious calibration (SVC) (Brook and Ben-Dor, 2011) uses artificial agricultural black polyethylene nets of various densities as calibration targets, set up along the aircraft's trajectory. The different density nets, when combined with other natural bright targets, can provide full coverage of a sensor's dynamic range. The key to the use of any form of vicarious calibration target is the use of simultaneous field-based measurement of their reflectance properties and positions; uncertainties are reduced if a number of calibration targets are used, a large number of reflectance measurements are made of each target, and their positions are accurately located (Secker *et al.* 2001).

Vicarious calibration, therefore, provides an indirect means of QA of remotely sensed data and sensor performance that is independent of direct calibration methods (use of on-board radiance sources or panels). This is important as on-board illumination sources may themselves degrade over time.

In all calibration efforts, traceability, the process of ensuring measurements, is related through an unbroken chain of comparisons to standards held by National Metrology Institutes (e.g., National Institute of Standard and Technology (NIST), USA; PTB, National Standards Laboratory, Germany; and National Physics Laboratory (NPL), UK), is the key to allowing true intercomparability between different sensors' raw and product data sets (Fox, 2004). The chain is implemented via the use of "transfer standards" that allow traceability back to official "primary" radiometric standards using internationally agreed-upon systems of units (SI) and rigorous measurement and test protocols. Integral to the establishment of traceability is the quantification and documentation of associated uncertainties throughout the measurement chain; the fewer the number of steps in the chain, the lower the uncertainty. The advantages of maintaining traceability include a common reference base and quantitative measures of assessing the agreement of results for different sensors or measurements at different times. However, current traceability guidelines lack guidance on temporal overlap or interval length for the measurements in the unbroken chain of comparisons (Johnson, Rice, and Brown, 2004).

The successful implementation of cal/val activity needs careful planning of issues such as coordination of activities, selection and establishment of networks of sites, the development and deployment of instrumentation to support measurement campaigns, the adoption of common measurement, and data distribution/availability protocols.

8.11 Summary and Conclusion

This chapter provides a snapshot of the emerging HRS technology. Although many aspects of this promising technique are not covered herein, we hope to have provided the reader with a sense of its potential for the future, as evidenced by past accomplishments. Besides being a technology that can provide added value to the remote sensing arena, it is an expansion of the spectroscopy discipline that has been significantly developing worldwide for many years. Very soon, when sensors in air and orbit domains begin to provide SNR values that are similar to those acquired in the laboratory, all spectral techniques available today will be able to implement the HRS data and forward the applications in a generation or two. HRS technology is emerging and the general scientific community use is growing. The number of sensors is also on the rise and new companies are entering into commercial activities. The most important step in the processing of HRS data is to obtain accurate reflectance or emissivity information on every pixel in the image; at that point, a sophisticated analytical approach can be used. This means that besides the atmospheric correction method, the data has to be physically reliable and stable at the sensor level. Mixed-pixel analysis and spectral models to account for specific questions are only a few examples of what this technology can achieve. The forthcoming HRS sensors in orbit are expected to drive this technology forward by providing temporal coverage of the globe at low cost and showing decision makers that the technology can add much to other space missions. The growing sensor-development activity in the market will also permit a "sensor for all," which will also push the technology forward. As many limitations still exist, such as the TIR region not being fully covered, the information only being obtainable from a very thin layer, the time investment, high cost of data processing, and great effort needed to obtain a final product, investment in this technology is worthwhile. If the above limitations can be overcome, and other sensors' capabilities merged with it, then HRS technology can be the vehicle to real success, moving from a scientific demonstration technology to a practical commercial tool for remote sensing of the Earth.

9
LIDAR and RADAR Observations

Jacques Pelon, Gabor Vali, Gérard Ancellet, Gerhard Ehret, Pierre H. Flamant, Samuel Haimov, Gerald Heymsfield, David Leon, James B. Mead, Andrew L. Pazmany, Alain Protat, Zhien Wang, and Mengistu Wolde

9.1
Historical

Radar detection and ranging (RADAR) meteorology was born in the 1940s as a side product of the war effort to detect aircraft and other targets. Rain and clouds were RADAR disturbances originally but became subjects of scientific studies after the war. Of course, RADARs are now extensively used to provide operational weather information as well as for continued research. The technology was rapidly extended to installations of RADARs in aircraft both for severe weather avoidance and for research. Laser sources were developed in the 1960s leading to the use of LIDARs and these too were adapted to aircraft observations in the decades that followed. Because of the complementary nature of the two kinds of observations, especially for meteorology and cloud–aerosol interaction studies, making RADAR and LIDAR measurements from the same aircraft has been found to be highly valuable.

Since the early developments, technology has allowed RADAR observations at higher frequencies and the operation of reliable solid-state laser sources in LIDAR systems. Both RADAR and LIDAR systems have rapidly evolved to include wavelength and polarization diversity in addition to scanning and high data rate capabilities. New analysis methods allow more precise quantification of atmospheric parameters (boundary layer height, wind, temperature, etc.) and composition (water vapor, ozone, etc.), as well as the properties of aerosol particles, clouds, and precipitations.

9.2
Introduction

While RADARs and LIDARs share the same basic principles of measurements, because of their different wavelengths, RADARs and LIDARs are sensitive to different characteristics of the atmosphere. LIDARs are used primarily for clear air sensing of molecular and particulate species (including ice clouds). RADAR

Airborne Measurements for Environmental Research: Methods and Instruments, First Edition.
Edited by Manfred Wendisch and Jean-Louis Brenguier.

measurements are most widely used for cloud and rain sensing. However, there are overlaps in capabilities, and new applications take advantage of recently developed installations of RADAR and LIDAR on the same aircraft.

Several books exist which detail LIDAR or RADAR techniques and applications. It is not the objective of this chapter to fully describe such techniques or to delve into details about electromagnetic interaction processes. Rather, the objectives of this chapter are to (i) present, in parallel, optical and microwave measurement principles focusing on similarities and differences; (ii) highlight recent advances in observations, including synergetic ones; and (iii) to present state-of-the-art results related to the characterization of the atmospheric parameters and composition from airborne LIDAR and RADAR observations. Surface and subsurface characterization are not part of this review. This chapter starts (Section 9.3) with an overview of the fundamentals of active remote sensing and describes the main LIDAR and RADAR instrument types. Section 9.4 covers LIDAR measurements of aerosol, clouds, and gases. Sections 9.5 and 9.6 describe cloud and precipitation observations with RADAR. Section 9.7 reviews results obtained with LIDAR and RADAR combined. Additional examples of research results are given in the Supplementary Online Material in Chapter 5.

9.3 Principles of LIDAR and RADAR Remote Sensing

LIDAR and RADAR systems use active sources that emit electromagnetic waves (mostly pulse shaped, and we keep this focus throughout this chapter) to probe the atmosphere. Use is made of the information either from scattering by particles and molecules or from absorption by gases and particles in order to retrieve desired atmospheric parameters.

9.3.1 LIDAR and RADAR Equations

The flux density $F_{0,\nu}$ emitted by the pulsed sources (at optical and MW frequencies ν for the LIDAR and for the RADAR, respectively) across the surface A_{E} perpendicular to emission direction, and defined by the solid emission angle $\Omega_{\mathrm{E}} = A_{\mathrm{E}}/R^2$ linked to the emission function, can be expressed as a function of the distance R to the aircraft and the emitted power $P_{0,\nu}(t)$ as:

$$F_{0,\nu}(R,t) = \frac{P_{0,\nu}(t)}{A_{\mathrm{E}}} \cdot \exp\left[-\tau_{\mathrm{ATM}}(R,\nu)\right], \tag{9.1}$$

where R is the distance between the aircraft and the part of the atmosphere to be studied and τ is the transmission along the line of sight, taking into account the angle of observation (θ_{R} or θ_{E}. This is the same for the received power for which we will consider a slightly different solid detection angle Ω_{D} corresponding to a detection surface A_{D} (it is indeed the same if the same antennas are used both for emission and reception, see Figure 9.1). Let us consider a small element $\mathrm{d}A_{\mathrm{D}}$ in

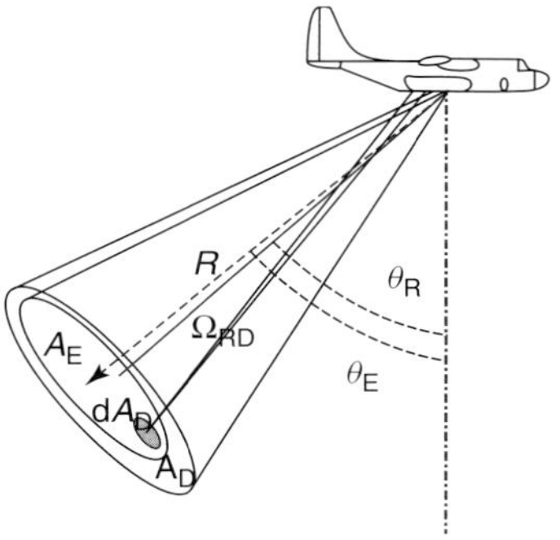

Figure 9.1 Schematic representation of airborne measurements from LIDAR or RADAR.

the surface A_D so that each element sees the surface A_R of the receiving system (antenna or telescope) with the solid angle:

$$\Omega_{RD} = \frac{A_R \cdot \mu_D}{R_D^2}, \tag{9.2}$$

with $\mu_D = \cos\theta_D$. When the scattering element dA_D is in the surface A_E (Figure 9.1), the radiance reemitted toward the emitter is equal to $B \cdot F_{0,\nu}/A_E$, where B is the atmospheric backscattering coefficient of particles and molecules (including possible molecular absorption and reemission) at the emission frequency ν'. The received power can be written as :

$$dP_R(R, \nu, \nu', t) = B(R, \nu, \nu') \cdot \frac{F_0, \nu(R, t)}{A_E} \cdot \Omega_{RD} \cdot \exp\left(-\tau_{ATM,D,\nu,\nu'}\right) dA_D \tag{9.3}$$

For LIDARs, we have to add here the solar flux scattered over the whole receiving cone (the solar emission being continuous, there is no temporal discrimination), and for both LIDAR and RADAR, the equivalent background power due to intrinsic noise in the system. Equation 9.3 needs to be integrated over the target surface to determine the full received power, using the antenna beam patterns, and accounting for the fact that the emission elementary surface may not exactly perpendicular to the receiving one. As a whole, the basic equation, considering elastic scattering only ($\nu' = \nu$), can be written as:

$$P_R(R, \nu, t) = C_\nu \cdot \frac{B(R, \nu)}{R^2} \cdot A_R \cdot P_{0,\nu}(t) \cdot \exp\left(-2\,\tau_{ATM,\nu}\right) + P_{B,\nu} \tag{9.4}$$

where C_ν is a constant linked to the system collection efficiency and $P_{B,\nu}$ is the total detected background power. The received power is further integrated over the receiver spectral bandwidth and the emission spectrum. In practice, the inverse of the above equation is used to derive the reflectance of the target from the received power.

The factors defining RADAR and LIDAR measurement capabilities and signal-to-noise ratio (SNR) optimization arise mostly from the fundamental physics of electromagnetic transmission and signal detection, as well as scattering by molecules and particles, referring in a broad sense to all kinds of gases, aerosol, and hydrometeor populations. Many other factors are related to the technology involved in constructing RADAR and LIDAR systems. It is not the purpose to detail these factors here. Good sources on general issues for LIDAR can be found in Measures (1984); Kovalev and Eichinger (2004); Weitkamp (2005). For RADAR, the reader may refer to Battan (1973); Sauvageot (1992); Rogers and Yau (1989); Doviak and Zrnic (1993), and the very detailed and relatively recent monograph of Lhermitte (2002).

9.3.2
Dependence on Atmospheric Spectral Scattering/Absorption Properties

The detected power depends, as shown above, on the scattering and the transmission of the atmosphere, through the terms B/R^2 and $T_{\mathrm{atm}} = \exp(-\tau_{\mathrm{atm}})$, respectively. This leads the two main measurement types for airborne LIDAR and RADAR systems: (i) Type I: detection of scattered energy is prevalent for the observation of cloud and aerosol particles, usually neglecting atmospheric absorption (backscatter – high spectral resolution or not – LIDAR, cloud and precipitation RADARs, Doppler LIDARs, and RADARs), (ii) Type II: analysis of the attenuation rate is applied to retrieve gas concentration using emitted wavelengths on and off the absorption lines of the gas to be measured (differential absorption LIDAR) or aerosol/cloud extinction (HSRL).

It may be noted that there is no RADAR system at present operating in the Type II mode. It should also be noted however that RADARs operating at millimeter wavelengths corrections are needed for absorption by water vapor and by oxygen.

It is evident from Eq. (9.4) and Figure 9.2 that the spectral properties of the backscattered power are controlled by the convolution of the emitted laser spectrum with the spectral distribution of the scattered flux due to Doppler broadening by the motion of the molecules and particles. The thermal motion of molecules leads to a broad Gaussian spectrum ($\Delta\nu_{\mathrm{RB}} \sim 2\,\mathrm{GHz}$ FWHM at 250 K and 532 nm) while the Brownian motion leads to a narrow spectral peak ($\sim$1–10 MHz). In a turbid atmosphere (small contribution of molecular scattering with respect to particle scattering) the scattered power spectrum is only weakly broadened. In Eq. (9.4), B represents the backscattering coefficient for LIDARs or RADAR reflectivity. It is the sum of contributions from molecules, particles or hydrometeors including precipitation, as will be discussed in next section. For particle sizes D much smaller than the wavelength $\lambda (D/\lambda \ll 1)$, the contribution to scattering is the most important as it varies as a function of λ^{-4} (or ν^4). This is known as Rayleigh scattering. It applies to scattering by molecules for LIDARs and by small cloud particles for high frequency RADARs.

It is to be noted that the propagation of an electromagnetic wave in a dense media (for which the attenuation by scattering becomes significant) also involves multiple scattering. This is the case for LIDARs in water clouds, or for RADARs in dense precipitating cloud systems. In such cases, B includes multiple order scattering. This also leads to modifications in the polarization state of the detected signal.

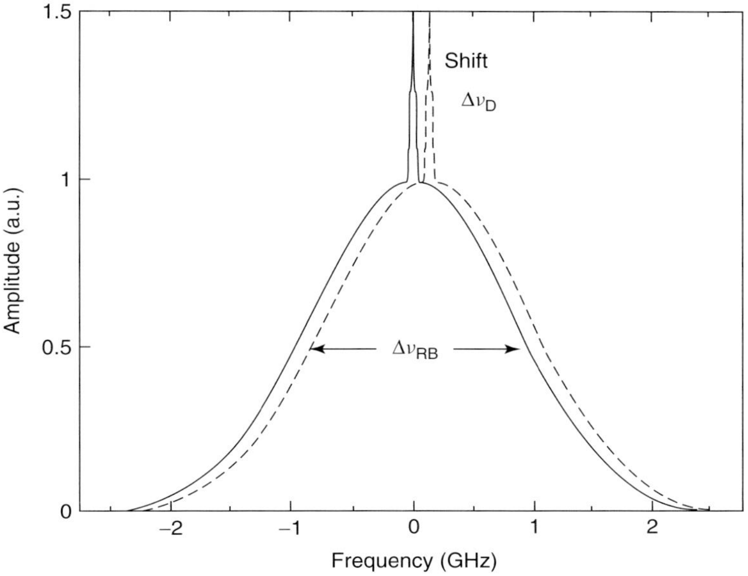

Figure 9.2 Spectral broadening of the received power P_R, arising from backscattering by molecules (broad spectrum) and by particles (narrow spectrum) and from the Doppler shift due to the line-of-sight component of the ambient wind in a weakly turbid atmosphere, for a LIDAR operating at 532 nm.

Mie theory (van de Hulst, 1981) can be applied to determine the scattering properties of spherical particles whose diameter is comparable or larger than the wavelength. The problem is more complex for ice crystals (nonspherical particles) because internal refractions and diffraction lead to significant contributions to *B* (it can be twice the value of single scattering for angular detection aperture allowing to detect the whole diffracted field). This problem is compounded by the large variety of particle shapes that are found in ice clouds (Heymsfield, 1977; Dowling and Radke, 1990). For LIDAR measurements, the wavelength is typically much smaller than the size of particles and the scattering regime can be rather well represented by geometrical optics, in a first approach. For RADAR, the measurement wavelength is usually larger than the particle size, but the size to wavelength ratio may be larger than 0.3, the upper limit for the Rayleigh scattering domain. In both these cases scattering properties are related to crystal habit.

A electromagnetic field can be represented by its Stokes quadrivector describing intensity as well as polarization. The scattering properties of scatterers (molecules, particles, droplets or ice crystals,. . .) are introduced by means of a 4×4 transformation matrix M known as the Mueller Matrix, or M–Matrix (van de Hulst, 1981; Bickel and Bailey, 1985). Obtaining such a matrix for ice particles gives access to the complete description of the scattering medium. Most computations of the

M–Matrix of ice crystals used several simplifying hypotheses, either by using basic shapes and/or assuming a random orientation of the particles.

Several methods have been used to determine scattering properties as a function of shape and size of ice particles. When the particle size is much larger than the wavelength the ray-tracing technique, based on the principles of geometric optics (Takano and Liou, 1989; Macke, 1993) can be used. This applies to most of the LIDAR measurements in ice clouds and to very high frequency RADARs. To account for smaller size parameters, other approaches are needed based on electromagnetic calculations such as Finite Domain Time Difference (Yang and Liou, 1996), or Discrete Dipole Approximation (Purcell and Pennypacker, 1973; Lemke and Quante, 1999). Other methods involving numerical solutions have also been used, e.g., the T–Matrix method (Mishchenko and Travis, 1994). This is still an important research area important for the interpretation of LIDAR and RADAR observations.

9.3.3 Basic Instrument Types and Measurement Methods

Techniques most frequently used for airborne observations in atmospheric sciences are based on the analysis of elastic scattering by particles (including hydrometeors) and molecules. With LIDARs, the scattered signal is also used as a basis for analyzing absorption by minor gaseous species such as water vapor, ozone or carbon dioxide. It is also used to analyze the Doppler frequency shift associated with the motion of particles and molecules. In most cases, quantitative analysis requires careful instrument calibration, to link directly the received power to the atmospheric parameter B in Eq. (9.4). To go further in retrieving atmospheric variables such as extinction by aerosol and cloud particles, or gas concentration, the detected signal has to be inverted using appropriate methods, as for example when scattering and extinction both contribute to the received signal; see Eq. (9.4).

9.3.3.1 Backscatter and Reflectivity

9.3.3.1.1 Broadband Spectral Analysis Operating the instruments away from any absorption line of atmospheric gases allows to treat the transmission of the atmosphere as a function of absorption and scattering by particles at first (corrections may be needed for absorption continua). Scattering is the most important contribution, but since the imaginary index of refraction for liquid water increases significantly (it is much less for ice) at high frequencies, absorption becomes important for millimeter RADARs and for IR LIDARs. Knowledge of the sizes of the scattering particles is needed to calculate scattering from molecules, aerosol particles, cloud water droplets, rain drops and ice crystals. For each of these components (identified by i in Eq. 9.5), one has to determine the scattering cross section $\sigma_{i,e}$ or efficiency factor $Q_{i,e}$ in order to determine the RADAR reflectivity Z, the LIDAR backscatter β (written as B in Eq. 9.4), or the extinction α (defined as $-1/2\mathrm{d}Ln\tau ATM/\mathrm{d}R$):

$$Z, \beta, \alpha(D, \lambda) = \int_D \sum_i \sigma_{\mathrm{i,e}} \cdot N_i(D)\mathrm{d}D$$
$$= \pi \int_D D^2 \sum_i \mathrm{N_i}(D) \cdot Q_{\mathrm{j,e}}(D, \lambda)\mathrm{d}D \tag{9.5}$$

Depending on the wavelength, e.g., when the size of particles (D) is much smaller than the wavelength (λ), the operation lies in the Rayleigh scattering domain, and the scattering efficiency varies with D^6, which means that molecules and small cloud droplets will not be detectable by RADAR. When the wavelength is much smaller than the size of particles, the approximation of geometric optics gives a constant efficiency factor $Q_{j,e}$, so that scattering and extinction coefficients are varying with D^2 and the wavelength dependence is weak. LIDAR and RADAR observations can be combined to take advantage of spectral variation of the attenuation due to scattering by particles (in the limits discussed in previous section), which can be exploited to retrieve their size. This will be discussed in Section 9.7.

9.3.3.1.2 **High Spectral Resolution Analysis** As previously emphasized, the detection of scattering by molecules introduces significant differences between LIDAR and RADAR signal formulations. This also opens specific analysis possibilities for LIDAR. As such, an important improvement in the analysis of aerosol and cloud parameters has recently been performed with the implementation of airborne high spectral resolution LIDAR (HSRL) systems in Europe at DLR (2008) and NASA in the USA (Hair *et al.* 2008). HSRL principles (Shipley *et al.* 1983) allow the spectral separation between aerosol/cloud and molecular backscattering and isolate the spectrally narrow backscattered power by particles from the broader one by molecules; see Figure 9.2, so that the term B in Eq. (9.4) is only related to one or the other species. As it still accounts for the whole atmospheric transmission, normalizing to the molecular density then allows to directly derive the total particulate attenuation, as a function of range, using the backscattered signal measured in the molecular channel. The ratio of the backscattered signals in both particulate and molecular channels reciprocally allows to directly derive the particulate backscattering coefficient after calibration. Such a determination further enables to access the ratio of the backscattering to extinction coefficients (e.g., the normalized phase function at 180°) which is of particular importance for aerosol characterization. Examples of observations from backscatter and HSR LIDARs will be presented in Section 9.4.

9.3.3.2 Doppler

A very demanding task for the recovery of Doppler velocity data from airborne LIDARs and RADARs is the removal of velocity components added by the motion of the aircraft itself. The motion of the platform introduces an additional contribution to the Doppler shift whenever the emitted beam is not exactly perpendicular to the aircraft velocity vector. Even for antennas designed to be as close to perpendicular as possible, aircraft pitch, yaw and roll lead to deviations from that ideal setting. It takes only a fraction of a degree of platform motion to make the in-beam velocity component of the forward speed roughly equal to hydrometeor and particle velocities normally found.

For all nonperpendicular beams, the forward velocity introduces a velocity component into the beams that can easily be many times the atmospheric velocity to be measured.

Removal of velocity contamination that results from platform motion requires precise measurements of the 3D velocities of the aircraft in ground coordinates at a frequency compatible with the RADAR and LIDAR sampling rates. In addition, the beam pointing direction has to be known at all times requiring knowledge of the aircraft attitude in ground coordinates. In semitransparent atmospheres, corrections can be obtained from analysis of the signal from surface return, but inertial navigation systems and GPS are the main sources of the necessary data; however, implementation of accurate corrections remains a daunting task (see Section 9.5.3).

One of the methods used to retrieve average horizontal wind speed is the use of Velocity Azimuth Display (VAD) scans.

The Doppler frequency shift along the line-of-sight (LOS) $\Delta\nu_D = -2\,V_r/\lambda$ accounts for two successive Doppler effects, forward and backward, respectively. λ is the probing wavelength, $V_r = V\cos\theta$ the radial velocity i.e., projection of the scatterers actual velocities $\vec{v} = (v_W(z) - v_a)$ on the LOS and θ the projection angle, see Figure 9.3. For full characterization of velocity vector $\vec{v}$, then 3 or 2

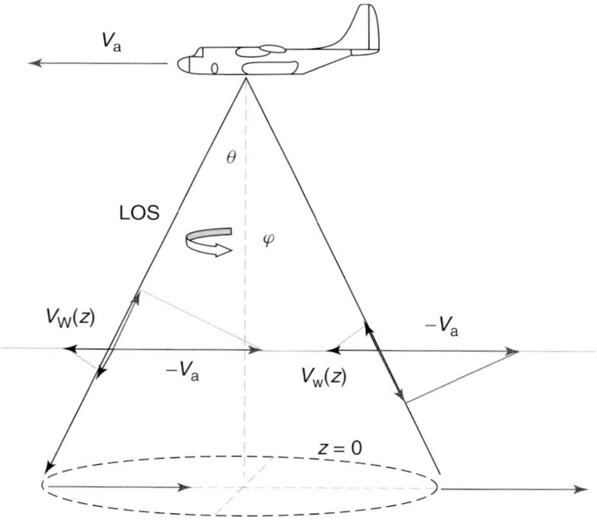

Figure 9.3 Principle of wind measurement by an airborne system using a LOS conical scan at an angle θ with respect to nadir. The aircraft velocity ($\vec{V_a}(z)$) and wind velocity ($\vec{V_W}(z)$) are in the figure plane. The resulting radial velocity in aircraft referential is ($V_r(z) = (V_W - V_a)\cos\theta\cos\phi$), where θ is fixed and the LOS rotation angle varies from $0 \leq \phi \leq 2\pi$.

independent measurements are necessary that calls for LOS scanning capability. This has been implemented for the WIND LIDAR developed in collaboration between DLR, CNES and CNRS (Werner *et al.* 2001). The fall velocity of the scatterers is assumed to be negligible in this treatment. While larger fall velocities of particles detected by RADAR impose a limitation, the technique has also been adopted for airborne RADARs by Leon and Vali (1998).

Simultaneous sampling with two antennas in the same plane, one cross-axis and one oriented at an angle in the range 20° to 45° allows dual-Doppler RADAR observations to be made. The two antennas sample the same cloud volume with a slight (range-dependent) lag in time and thus provide two components of the velocity which can then be used to solve for other desired components, assuming negligible changes in the velocity field over the lag time. The geometric overlap of the sample volumes by the two beams is never perfect, if for no other reason than for the inherent shape of the RADAR pulse volume from two different orientations. To that are added inexactitudes due to deviations of the aircraft platform from perfect alignment between the two instants of sampling. Nonetheless, these errors usually remain at tolerable levels, so that useful data can be obtained. This topic is presented in more detail in Section 9.5.3.

9.3.3.3 **Differential–Absorption**

Differential-Absorption LIDAR (DIAL) has become an indispensable tool for active remote sensing of atmospheric constituents. The DIAL technique can be used to measure the abundance of key meteorological parameters and climate related trace gases such as atmospheric water vapor water vapor (H_2O), ozone (O_3), carbon dioxide (CO_2), and methane (CH_4). In general, the DIAL technique is sensitive to the spectral feature of a gaseous constituent in the atmosphere (e.g., the absorption line), whose spectroscopic characteristics (e.g., line intensity, line position, line broadening as function of temperature and pressure) need to be precisely known, to allow optimized measurements (total absorption optical depth should be close to 1 at the measurement distance) at different latitudes and altitude (see Figure 9.4). The methodology of the DIAL technique and its related issues are described in (Schotland, 1974; Cahen, Mégie, and Flamant, 1982; Browell *et al.* 1983; Bösenberg, 1998; Bruneau *et al.* 2001b; Bruneau *et al.* 2001a; Ehret *et al.* 2008). In case the spectral feature has a narrow-band characteristics which is common to collision and Doppler broadened molecular absorption lines in the atmosphere, the frequency difference between the online and offline soundings can be made as small as ~5 GHz. Backscatter and extinction properties of the atmosphere are then approximately similar for both emitted wavelengths thus the full equation derived from Eq. (9.4) simplifies substantially and one gets the well-known DIAL equation approximation:

$$\rho_{\text{gas}}(\bar{r}) = \frac{1}{2\,(r_1 - r_2)\cdot\left[\sigma\left(\lambda_{\text{on}},\bar{r}\right) - \sigma\left(\lambda_{\text{off}},\bar{r}\right)\right]} \cdot \ln\left[\frac{P_{\text{off}}(r_2)P_{\text{on}}(r_1)}{P_{\text{on}}(r_2)P_{\text{off}}(r_1)}\right] \tag{9.6}$$

with $\bar{r} = (r_1 - r_2)/2$ is the average distance of the measured range cell of size $\Delta r = r_1 - r_2$. Unfortunately, atmospheric scattering from particles and molecules

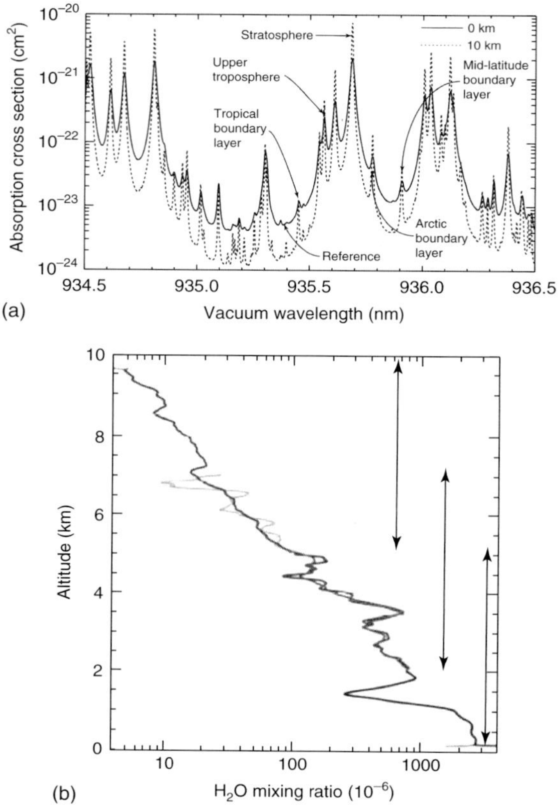

Figure 9.4 (a) Water vapor-absorption lines used for the DLR–WALES demonstrator in the NIR spectral region around 935,6 nm. Source: Taken from Wirth *et al.* (2009). (Reproduced in modified form, with permission from Springer.) (b) The profile shown has been obtained during THORPEX near 68°N, 12:50 UTC, and shows the use of the different absorption lines to cover the whole range of absorptions as identified by vertical arrows.

introduces different spectral behaviors as previously discussed. The small measurement error associated to the spectral line broadening by molecular scattering can be compensated for by applying the Rayleigh–Doppler correction algorithm which can be found in several publications; see for example Bruneau *et al.* (2001a). Thus, operating a backscatter DIAL from aircraft platforms either in nadir or zenith viewing direction yields the 2D distribution of the species concentration in vertical and horizontal direction beneath or above the aircraft flight level.

Another DIAL application using the instrument in the downward looking mode is known as Integrated-Path Differential-Absorption (IPDA) LIDAR. This type of DIAL measurement depends on the LIDAR echo from a target which can be the

Earth surface or any dense cloud top to obtain information on the path-averaged species concentration. The LIDAR signals (e.g., high SNR) resulting from hard target reflectance are substantially larger than those from the atmosphere thus enabling a high measurement sensitivity.

This technique can be used to derive integrated contents and surface fluxes of long-lived greenhouse gases such as CO_2 and CH_4 by measuring the horizontal gradient of the column-integrated mixing ratio, commonly referred to X_{gas} by applying the following equation:

$$X_{gas} = \frac{\ln\left[\frac{P_{off}(p_{surf})/E_{off}}{P_{on}(p_{surf})/E_{on}}\right]}{2\int\limits_{p_{aircraft}}^{p_{surf}} \frac{\sigma(\lambda_{on},p,T)-\sigma_{off}(\lambda_{off},p,T)}{g\cdot M_{air}}\,dp} \tag{9.7}$$

Here, the gas absorption cross section σ is expressed by the geophysical parameter p and T which denote the atmospheric pressure and temperature, respectively. p_{surf} is the surface pressure, $p_{aircraft}$ is the pressure at flight level, M_{air} is the molecular air mass, and g is the acceleration of gravity. Compared to the atmospheric backscatter DIAL, the IPDA LIDAR method is not self-calibrating. LIDAR echoes from hard target need to be calibrated by dividing the received signals by the transmitted pulse energies E_{off} and E_{on}, respectively.

9.3.4
LIDAR and RADAR Types and Configurations

This section provides mostly qualitative descriptions of the factors that characterize the varieties of airborne LIDARs and RADARs in use at this time for atmospheric measurements and the impacts of those factors on their capabilities. Desirable characteristics of airborne research active systems are, primarily high sensitivity, good spatial resolution and the least possible ambiguity in the interpretation of the received signal. In addition, there are the practical considerations of size, weight, power consumption, etc. As with almost any other instrument, compromises have to be made. The majority of these issues are manifest in the RADAR (LIDAR) equation (Section 9.3.1).

Wavelength is the most fundamental characteristics to consider, since scattering by cloud elements (or aerosol particles) is dependent on the relative magnitudes of particle size, as described in Section 9.3.1, and in a more complex fashion on particle shape (ice crystals, irregular particles) or composition (mixture). Attenuation is also wavelength dependent. Polarization may provide additional information on the shape of particles (see Sections 9.4.1.2 and 9.5.4).

Another important factor determining the performance and utility of a RADAR (or LIDAR), starting from the detection of clouds (and aerosols or gases) to the profiling of their characteristics, is the minimum detectable signal, usually expressed as the reflectivity (backscatter coefficient) that is reliably measurable at some given distance away from the RADAR (or LIDAR). This sensitivity is mainly a question of the hardware that is available, namely the power transmitted, the pulse width, antenna size, the quality of various receiver components, the transmission losses and other lesser factors. Sensitivity is also influenced by the integration time (incoherent RADAR and LIDAR systems behave the same) which is usually selectable by the user, bearing in mind that the viewed volume changes rapidly due to the speed of the aircraft.

LIDARs and RADARs frequently operate in fixed pointing directions. This is less restrictive than it sounds, because the motion of the aircraft provides a second dimension so that a fixed line of sight (LOS) actually scans a surface composed of straight lines emanating from the flight path. If the flight path is a straight line, the scanned surface is a plane, if it is a circle, the scanned surface is a cone, and so on. The simplest configuration is to have a LOS in directions perpendicular to the flight line; either up, down or sideways. This leads to simple data in Cartesian coordinates, analogous to time-height profiles from ground-based RADARs but with the important difference that the relatively slow motion of a weather system is replaced by the fast aircraft speed. The resulting data are thus quasi-instantaneous depictions of gas concentration, aerosol, clouds and precipitation.

9.3.4.1 Different Types of LIDAR Systems

LIDAR systems operate with laser sources which wavelength is chosen with respect to the measurement type aimed at. Cloud and aerosol LIDARs have no specific wavelength adjustment constraint, and more frequently solid-state laser sources using Nd–YAG as the active medium are implemented in such systems. Two to three wavelengths (355, 532 and 1064 nm, or more using non-linear frequency conversion), can be then used for aerosol particle detection, to relate the difference in spectral response to the particle microphysics (Müller *et al.* 1998; Kaufman *et al.* 2003). As previously discussed, the implementation of a high spectral resolution detection system provides independent extinction profiling capabilities. It however requires a very high separation efficiency of molecular and particle scattering signals. Airborne systems presently developed using such capabilities operate at 532 nm and are based on the use of iodine gas cell as a narrow filter to isolate the backscattered spectrum due to particles (Esselborn *et al.* 2008; Hair *et al.* 2008). Matching the emission and filter wavelengths is required. Another concept based on two-wave interferometry in the UV is in development in France (Bruneau and Pelon, 2003). All these systems are operating in direct detection (measurement of the incident power). Doppler systems may be of HSRL type or use heterodyne detection, and in this latter case, the efficiency of operation implies to operate in the IR (1 to 10 μm). For DIAL systems, the selection of the absorption bands is driving the operation wavelength. Emission in the UV (250 to 350 nm) is chosen for ozone and sulfur dioxide measurements for example, whereas for water vapor, near IR wavelengths (720 to 940 nm) are usually used. Direct detection measurements in the IR (1–10 μm) are possible but more difficult, due to more reduced SNRs.

Backscatter airborne LIDAR systems use telescope sizes which are relatively small, since the scattering efficiency is rather large (especially for clouds), but clear air reference is usually needed; see Section 9.4.6. Available laser source energy is high (although limited by eye-safety constraints). A direct signal processing is further required to retrieve usual parameters. For DIAL systems, derivative procedures are required to analyze the return signal and this leads to prefer large telescope size to increase SNR. However, in an aircraft, a limited room is available, so that telescope diameters of about 30 to 50 cm are generally used. LIDARs are mostly operated in fixed pointing directions.

Table 9.1 lists operating and planned airborne LIDARs by type and alphabetical order.

9.3.4.2 **Different Types of RADAR Systems**

Weather RADARs operate at wavelengths where absorption by oxygen and water vapor is low. These window regions, and the commonly used band designations for them, are:

$\sim$10 cm (2–4 GHz; S-band),
$\sim$5 cm (4–8 GHz; C-band),
$\sim$3 cm (8–12 GHz; X-band),
$\sim$1.5 cm (16–20 GHz; Ku-band),
$\sim$1 cm (26–40 GHz; Ka-band),
$\sim$3 mm (90–100 GHz; W-band).

Principally because the difficulties of installing RADAR antennas on research aircraft increase importantly with antenna size while to achieve a given beam width the antenna size has to increase in direct proportion to the wavelength used, the majority of airborne RADARs operate in the K or W band, with X-band used less frequently. Another influence in the choice of operating wavelength is that there has been a gradual progress over time in the development of high power transmitters operating at shorter wavelengths; for example, W-band hardware became available for civil use only in the 1980s. Airborne RADARs became important components of atmospheric research over the past quarter of a century with a great variety of systems depending on the aircraft carrying the RADAR and on the scientific goals to be achieved. Table 9.2 lists operating and planned airborne RADARs in chronological order. The version of this Table included in Table A.2 and Section A.5.1, given in the Supplementary Online Material provided on the publisher's web site, contains more complete descriptions of the systems, explains their intended uses and contains links to examples of the results obtained with them.

The use of short wavelengths mean that the simpler Rayleigh scattering calculations can be applied less frequently in the interpretation of the backscattered signal, that is, to smaller fractions of the hydrometeor populations found in clouds. This increases ambiguity in the quantitative interpretation of the RADAR signals. In addition, attenuation by water vapor becomes important for Ka and W-band wavelengths (cf. 9.5.2). The main positive impact of short wavelengths is that backscatter from hydrometeors is stronger, so there is a sensitivity gain. For this reason, the W-band RADARs are sometimes referred to as "cloud RADARs" differentiating them from RADARs operating at longer wavelengths that are better suited for observations of precipitation at long ranges.

Table 9.1 Summary of airborne LIDAR systems in operation and in planning. See text for more detail.

Unit	Aircraft	Operation Wavelength (nm)	LIDAR Type	Year
CNRS-Latmos	Falcon, ATR42	355	HSRL-depol.	In Dev.
LNG		532, 1064	Backscatter 2008	
			Scanning	
CNRS-LSCE	Falcon, ATR42	355	Backscatter -depol.	2010
DLR	Falcon, HALO	532	HSRL-depol.	2007
		1064	Backscatter-depol.	
IAO-SBRA	AN-30	532	Backscatter-depol.	2009
NASA-GSFC	ER2	532	Backscatter	2002
CPL		355, 1064	Backscatter	
NASA-GSFC	P3	540	Backscatter	2004
THOR			Multiple field-of-view	
NASA-LaRC	King Air	532	HSRL-depol.	2007
	B200	1064	Backscatter-depol.	
NCAR	C-130	532, 1064	Backscatter	1995
SABL			Scanning	
UKMO	BAE-146	355	Backscatter-depol.	2010
UWYO	King Air	532	Backscatter	2010
WCL				
CNRS-Latmos	Falcon, ATR42	720-760	DIAL H_2O	1999
LEANDRE II				
CNRS-Latmos	Falcon, ATR42	266-316	DIAL O_3	1993
ALTO				
DLR	Falcon, HALO	815-930	DIAL H_2O	2000
WALES				
NASA-LaRC	DC8, ER2	820-940	DIAL H_2O	1995
LASE				
NASA-LaRC	DC8, Electra	280-320	DIAL O_3	1983
		600, 1064	Backscatter	
DLR	Falcon, HALO	2000	Doppler	2002
DLR-Astrium	Falcon	355	Doppler	2009
A2D				
DLR-CNRS	Falcon	10600	Doppler	1999
WIND				
NASA-GSFC	DC8	355	Doppler	In Dev.
NASA-LaRC	DC8	2000	Doppler	In Dev.
DAWN				
Meteo. Serv. Canada	NRC Convair-580	532	Backscatter-depol.	2001
AERIAL		1064	Backscatter	

Table 9.2 Summary of airborne RADAR systems in operation and in planning. More complete information about the systems and about their uses is available in Appendix.

RADAR Unit	Aircraft	Band	Usable Signal Level (Best Configuration)	Year
NOAA P-3	NOAA WP-3D	C	0 dBZ at 1 km	1976
Lower Fuselage RADAR				
NOAA P3	NOAA WP-3D	X	−10 dBZ at 10 km	1976
Parabolic Antenna (Tail)				
NOAA	NOAA WP-3D	X	−10 dBZ at 10 km	1991
French Dual Flat Plate				
ELDORA	NRL P3	X	−12 dBZ at 10 km	1993
Electra Doppler RADAR				
EDOP-ER-2	NASA ER-2	X	−20 dBZ at 10 km	1993
Doppler RADAR				
WCR	U. Wyo King Air 200 T	W	−40 dBZ at 1 km	1995
Wyoming Cloud RADAR				
SPIDER	Gulfstream II	W	−30 dBZ at 5 km	1998
Super Polarimetric ...				
EC CPR	NRC Convair-580	Ka	−33 dBZ at 1 km	1999
Env. Canada Cloud				
RASTA	Falcon 20, ATR-42	W	−35 dBZ at 1 km	2000
RADAR System Airborne				
APR-2	NASA DC-8 and P-3	Ku	10 dBZ at 10 km	2001
Airborne Precip. RADAR		Ka	0 dBZ at 10 km	2001
IWRAP	NOAA WP-3D	C, Ku	0 dBZ at 1 km	2002
Imaging Wind and Rain				
CRS	NASA ER-2	W	−28 dBZ at 10 km	2002
Cloud RADAR System				
NAWX – NRC	NRC Convair-580	X	−20 dBZ at 1 km	2006
Airborne ..	.	W	−30 dBZ at 1 km	2007
HIWRAP	NASA WB-57, Global Hawk	Ku	0 dBZ at 10 km	2010
High Alt. Imaging ...		K	a−5 dBZ at 10 km	2010
G-IV	NOAA G-IV SP	X	−12 dBZ at 10 km	2010
Tail Doppler RADAR				
EXRad	NASA ER-2	X	−15 dBZ at 10 km	2010
HCR	NSF/NCAR G-V	W	−22 dBZ at 10 km	2011
HIAPER Cloud RADAR				
ACR	NASA P-3	W	n/a	In Dev.
Airborne Cloud RADAR				

The volume of cloud or precipitation sampled by the RADAR is a function of the beam width (fixed by the wavelength and the antenna size) and the pulse length (selectable within hardware limitations). Since signals can always be integrated to larger volumes and since hydrometeor populations usually vary even on scales smaller than the view volumes achievable with current RADAR units, the interpretation of the data is facilitated by having the smallest sample volume possible. However, the extent of spatial coverage (range and/or direction) possible for given hardware is inversely related to the pulse width. This leads to a tradeoff which is usually operator selectable even during flight.

For airborne applications, antenna and radome installation is the most difficult part because of structural and aerodynamic issues and because of the need to minimize wave-guide losses between the antennas and the transmitter/receiver unit. Large openings require special airframe modifications, protrusions (bubbles) beyond the skin of the aircraft are problematic, and fuselage-mounted or wing-mounted pods have serious size limitations as well as aerodynamic impacts on the aircraft. For these reasons, scanning antennas are installed on only a few aircraft. More common is the use of single or multiple fixed antennas. Movable reflector plates are added in some cases in front of the antenna thereby allowing changes in the view direction. With the use of reflector plates the cross-axis beams can be deflected, for example from sideways pointing to nadir or upwards, or can be given limited cross-path scanning capability. Cross-axis antennas can also be used for 3D sampling, as in VAD analysis (Jorgensen, Matejka, and Dugranrut, 1996; Leon and Vali, 1998) by performing prescribed flight patterns.

9.4 LIDAR Atmospheric Observations and Related Systems

9.4.1 Aerosol and Clouds

9.4.1.1 Structure

The detection of cloud and aerosol layers at high horizontal and vertical resolution is a first application of backscatter LIDARs. It allows analysis of boundary layer dynamics from airborne observations using aerosol particles as tracers (Boers, Melfi, and Palm, 1991; Flamant *et al.* 1997), and also cloud tops for entrainment in stratocumulus and derivation of microphysical parameters (Spinhirne, Hansen, and Caudill, 1982; Spinhirne, Boers, and Hart, 1989). The precise location of cloud and aerosol layers is of particular importance in radiation transfer analysis, but the retrieval of optical and microphysical parameters by LIDAR or combination of instruments (LIDAR, RADAR, radiometers, ...) is further needed.

9.4.1.2 Optical Parameters

With conventional backscatter LIDARs climatically relevant aerosol properties such as aerosol extinction can only be derived by inverting the LIDAR Eq. (9.4) under the assumption of an *a priori* known LIDAR ratio $S_p(R) = \alpha_p(R)/\beta_p(R)$, between extinction and backscatter (Klett, 1985). S_p, which is the inverse of the normalized particle phase function at 180°, is indeed a highly variable quantity which depends on particle microphysics and type (Shettle and Fenn, 1979). Uncertainties in the LIDAR ratio will consequently lead to large errors in the retrieved optical properties. An accurate determination of the extinction requires a precise assessment of the LIDAR ratio, which implies additional knowledge of aerosol/cloud particles properties or constraints on optical depth. This can be done using additional in situ and remote sensing measurements (Flamant *et al.* 1997; Pelon *et al.* 2002; Platt *et al.* 2002).

To be more efficient in terms of analysis, a backscatter LIDAR require multispectral radiometric measurements to be used at the same time. This has been shown from an analysis made during a dust outbreak event as observed during the SHADE field campaign in 2001 (Tanré *et al.* 2003). In this analysis, the dual wavelength LIDAR data obtained from the airborne backscatter LIDAR LEANDRE 1 were used as inputs to spaceborne multispectral radiometry data inversion from MODIS (Kaufman *et al.* 2003). It was shown that using the MODIS particle models and the minimization of errors between calculated and measured radiance, the retrieval of extinction profile could be achieved at the same time as the average model of particle microphysics which optimizes results. A LIDAR ratio of 63 sr was also retrieved simultaneously using a nonsphericity correction. This is however limited to a single aerosol layer.

Properties of aerosol can be better retrieved from HSRL than from a backscatter LIDAR alone, as the signal measured in the molecular channel directly gives access to the variation of the transmission term as a function of distance. The change in transmission is then due to the local extinction. An example of retrieval of such optical properties is shown in Figure 9.5 for measurements taken by DLR during a dust outbreak event over Morocco during the SAMUM campaign (Heintzenberg, 2009). The LIDAR ratio S_p which is a characteristics of the aerosol type can be directly retrieved. For desert dust values close to 55 sr^{-1} were obtained, which are very different of the ones calculated from Mie theory (close to 30 sr) as particles are not spherical. A high depolarization value (above 10 %) is an important information which helps to identify the presence of such particles, and multiple spectral depolarization combined with HSRL data inversion has proven to provide elements for aerosol classification using remote sensing only (Burton *et al.* 2012). Complementarity between HSRL and in situ measurements has been further exploited to analyze aerosol properties (Burton *et al.* 2010).

9.4.1.3 **Cloud Phase, Effective Diameter of Cloud Droplets and Ice Crystals**

Because of the density of water clouds, and the small size to wavelength ratio preventing penetration in water clouds, few observations of low clouds have been performed by LIDAR. One parameter which can be easily obtained is the cloud top altitude of stratocumulus clouds (see Section 9.4.1.1). It is however possible to retrieve the liquid water content (LWC) in the cloud, using an accurate correction for multiple scattering (Hu *et al.* 2006), or multiple field of view LIDAR (Cahalan *et al.* 2005). First characterizations of cloud top LWC were reported by Spinhirne, Boers,

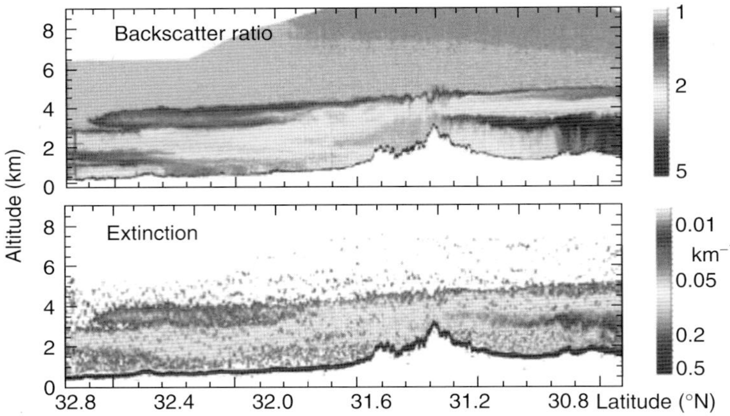

Figure 9.5 Measurement of aerosol dust scattering and extinction using DLR HSRL over Morocco during SAMUM experiment, showing a vertical extension of the dust layer up to 5 km (asl), the denser part being observed near 3 km (asl) at southern latitudes, after Esselborn *et al.* (2008).

and Hart (1989). Results obtained show that although comparable, liquid water derived from the LIDAR installed onboard the ER2 and flying at 20 km altitude was difficult to validate from in situ measurements performed from another aircraft due to non-simultaneity of measurements and cloud heterogeneity.

Recent airborne measurements revealed that supercooled water layers (which are shown to be very thin) embedded in ice clouds in frontal systems can be well evidenced by LIDAR (enhanced scattering) and RADAR (very low reflectivity) observations (Hogan *et al.* 2003b). Figure 9.6 show such an example taken during the Campaign CLARE'98. The LIDAR observations; see Figure 9.6 left, show scattering in some areas where no RADAR echoes are present. The combination of LIDAR and RADAR observations is further discussed in Section 9.7.

Despite these examples, the characterization of cloud properties by LIDAR is mostly devoted to ice clouds, which are most of the time semi-transparent to LIDAR. A lot of observations have been performed using backscatter LIDAR during previous field experiment starting in the 80's (FIRE in 1986, CEPEX in 1993, EUCREX in 1994, CRYSTAL–FACE in 2002, TC4 in 2007 ...). A large number of observations have been performed using the ER2 embarking the NASA/GSFC backscatter LIDAR system CALS (Spinhirne, Hansen, and Caudill, 1982) and then CPL (McGill *et al.* 2002). These LIDAR observations were used to infer the vertical structure of the cloud as well as the extinction profile into the cloud and the optical depth of isolated cloud layers. Clear air scattering observed above and below an isolated cloud layer can be used to determine transmission through the layer, after normalization to molecular scattering above and below it (see Section 9.4.6). Radiative properties could then be analyzed combining these observations with radiometry (Platt *et al.* 1998; Platt *et al.* 2002; Chepfer *et al.* 1999) and link optical properties to in situ observations (Heymsfield, 2007).

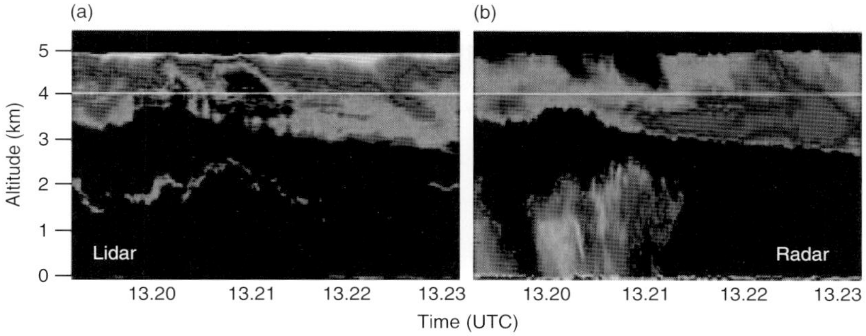

Figure 9.6 Simultaneous observations performed near Brest (France) on 10 November 2000, with combined LIDAR and RADAR ((a) backscatter LIDAR LEANDRE 1 from IPSL and (b) 94 GHz IPSL RADAR) installed in the French F27 flying at 5 km. The dashed line indicates altitude at which in situ measurements have been made with Merlin IV from Meteo–France equipped with GKSS sondes (ESA CLARE'2000 campaign). Supercooled water droplets can be identified near 13:21 UTC below the line. (*See Color Plates for color representation of the figure.*)

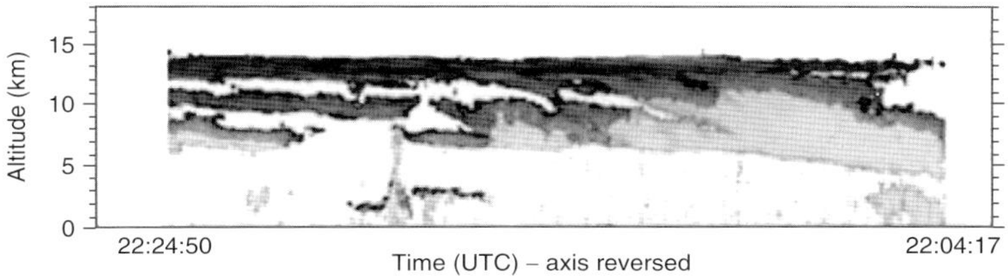

Figure 9.7 Backscatter LIDAR and cloud RADAR composite signal of a cirrus cloud observed from ER2 during TC4 (black, LIDAR only; light gray, RADAR only; mid-gray, common zone; after McGill *et al.* (2004). (Copyright 2004 American Geophysical Union (AGU). Reproduced with permission by AGU.)

In 2002, the first high altitude joint LIDAR and RADAR flights were performed with the ER2 during TC4 in ice clouds, which allowed to perform cloud classifications in terms of instrument detection sensitivity (McGill *et al.* 2004). Such an example from TC4 campaign is given in Figure 9.7. In this example, upper thin cirrus clouds are nor detected by RADAR, probably due to the small size of ice crystals at the top of te cloud (see further discussions in Section 9.7). As previously discussed, depolarization is indicative of the occurrence of irregular shape particles. As reported in Figure 9.8, it was shown from such measurements that some indication on the shape of crystals could be inferred (Noël *et al.* 2002) and that using generic types, information on their mixture could be further retrieved (Noël *et al.* 2004).

9.4.2
Winds in Cloud-Free Areas

Basically, two Doppler LIDAR techniques are used concurrently for atmospheric applications. One makes use of scattering by molecules and particles with direct

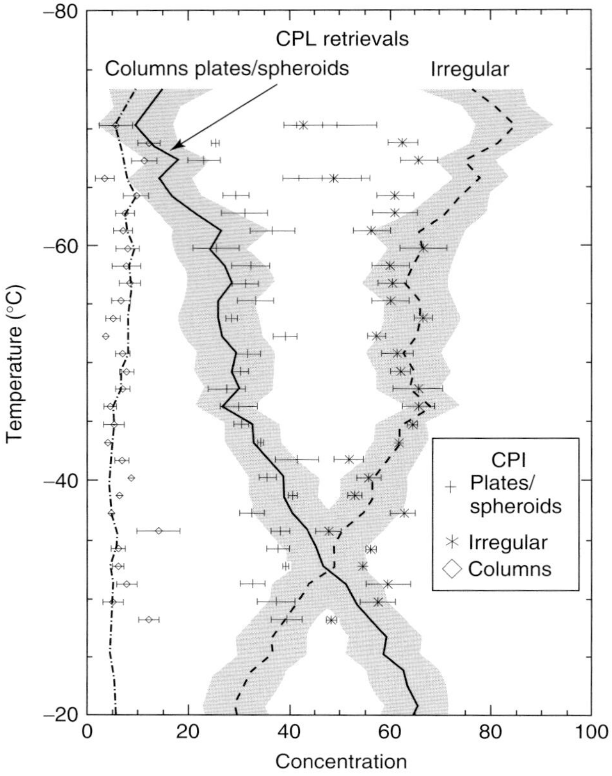

Figure 9.8 Habit percentages from CPI compared to ER2 CPL backscatter LIDAR retrievals for three cases of observations taken during CRYSTAL–Face, reported as a function of temperature, after Noel *et al.* (2004). (Copyright 2004 American Geophysical Union (AGU). Reproduced with permission by AGU.)

detection receiver, the other combines scattering by particles only with heterodyne detection receiver. Both rely on the use of a single-mode laser emission. High spectral resolution detection systems taking into account that scattering by molecules result in much broader detection bandwidth than the heterodyne systems.

9.4.2.1 Wind from Scattering by Particles

An example of LIDAR using the scattering from particles to analyze its Doppler shift due to their motion, is the WIND LIDAR. This system has been developed in collaboration between Germany and France.

Heterodyne detection LIDAR technique analyzes the ac (alternating current) signals that carry the RF (radio frequency) beat frequency after photomixing of the scattered light from atmosphere with a frequency stable CW (continuous wave) laser beam (i.e., local oscillator laser used as frequency reference). Given the sine wave form of heterodyne signals they are processed one-by-one using frequency estimators. Then the frequency estimates are accumulated for better accuracy. In heterodyne detection, efficient optical mixing requires a partial degree of coherence of the scattered light. It gives rise to speckle effects that result in more RF signal statistical fluctuations than only due to photon detection noise (i.e., shot noise). In practice SNR is limited to low values $\cong 1$ in heterodyne (or coherent) detection. Accumulation of a large number of estimates M is required for high accuracy as the improvement on $\Delta\nu_D$ goes as $M^{-0.5}$.

The best use of LIDAR instrument resources (laser transmitter mean power mostly) in heterodyne detection combines low pulse energy and high pulse repetition frequency. On the contrary incoherent light is better for direct detection LIDAR because the signal fluctuations are only due to detection noise and large SNR on single shot basis makes the best use of LIDAR instrument resources. Statistical theory enables to derive the limit of best achievable performance on Doppler frequency shift $\delta(\Delta\nu_D) \cong \Delta\nu_{sp}/\sqrt{N}$ (assuming no correlation of measurements). Basically, it is the same for direct and heterodyne detections. $\Delta\nu_{sp}$ is the scattered spectrum full width at half maximum, and N is the number of photoelectrons to be used for one single estimate. The best Doppler LIDARs reach an accuracy that is 2–3 times this statistical limit. Pulsed laser emissions enable the retrieval of V_r along the line of sight as a function of range $R = c \cdot t/2$, where t is the round trip time in the atmosphere. As for any other LIDAR instrument, the range resolution is set by the laser pulse duration and raw signal sampling capability. The combination of several (independent) LOS enables to derive 2D- or 3D-wind field.

Airborne Doppler LIDAR received a great deal of attention in the 80's for meteorological applications and as preliminary step to LIDAR deployment in space. Several groups proposed to fly airborne LIDAR combining the most advanced CO_2 laser technology (firstly CW and then pulsed laser transmitter) with heterodyne detection receiver (Woodfield and Vaughan, 1983; Bilbro *et al.* 1984b; Targ *et al.* 1991; Schwiesow and Spowart, 1996; Rothermel *et al.* 1998). However, they did not succeed to go farther beyond feasibility demonstration tests. In this respect, the airborne Doppler LIDAR "WIND" developed in Franco–German cooperation (Werner *et al.* 2001; Reitebuch *et al.* 2001; Dabas *et al.* 2003) was deployed successfully in several key atmospheric programs such as the "Mesoscale Alpine Program (MAP)" in 1999 (Reitebuch *et al.* 2003) and African Monsoon Multidisciplinary Analysis (AMMA) program in 2006 (Lebel *et al.* 2010; Bou Karam *et al.* 2008), from which results are shown in Figure 9.9. The primary science objective of WIND was a complete troposphere wind profile with $\approx$250 m vertical resolution and 10–20 km horizontal resolution at the surface. The transmitter was a pulsed CO_2–transverse excitation laser emitting 0.1 J at 10.6 μm in $\approx$2 μs, at a 10 Hz pulse repetition frequency. The receiver combines a RF excited waveguide CO_2 laser as local oscillator with a high efficiency Mercury-Cadmium-Telluride photomixer cooled at 77 K (liquid nitrogen temperature). It fits into a medium size turbo jet (DLR Falcon–20 aircraft). The germanium rotating wedge makes a revolution in 20 s to conically scans the LIDAR LOS at 30° from nadir. The slow rotation rate renders negligible the lag angle that otherwise would decrease significantly the heterodyne mixing efficiency. The heterodyne receiver measures the algebraic sum of atmospheric wind and aircraft velocities (Figure 9.2). The surface return is used to subtract the aircraft velocity from LIDAR measurements to obtain the true air speed. When the aircraft flies at $170\,m\,s^{-1}$ ($\approx 610\,km\,h^{-1}$) it moves 3.4 km during one full scan revolution (20 s). At a 10 km flying altitude the range to the ground is 11.6 km and the footprint (in diameter) varies from 5.8 km at 5 km to 11.6 km at the surface. However, despite WIND deployments in field campaign were successful during many years, the CO_2 laser technology raised issues for maintenance and operation in field campaigns (high voltage $\approx$ 30 kV, availability of spare parts). In the mean time, the solid-state 2 μm laser technology made decisive progresses (Henderson *et al.* 1993) and emerged as a reliable alternate solution for airborne heterodyne LIDAR (Targ *et al.* 1996). Along this line, the 2 μm wind LIDAR operated on board the DLR Falcon–20 has received successful applications in several meteorological field campaigns (Weissmann *et al.* 2005; Weissmann and Cardinali, 2007).

9.4.2.2 **Wind from Scattering by Molecules**

Direct detection LIDAR technique allows to analyze the scattered spectra using high spectral resolution filter, such as multiple beams interferometers i.e., Fabry–Perot (Garnier and Chanin, 1992; Korb, Gentry, and Li, 1997) or Fizeau, or two beams interferometers i.e., Mach–Zehnder (Bruneau, 2002), Michelson (Cézard *et al.* 2009). The Doppler frequency shifts and consecutive radial velocities are computed from spectrogram intensity distributions using calibrated transfer functions. As an example, for a Two channels Fabry–Perot processor it writes as $V_r = K \cdot (N_A - N_B) / (N_A + N_B) - V_{ro}$, where N_A, N_B are the signals (i.e., photo electrons) collected in channels A and B, K is an instrumental parameter and V_{ro} accounts for an offset. Signal accumulation to increase SNR for better accuracy is conducted before processing. Accumulation of frequency estimates can also be done afterward to improve accuracy.

A new airborne wind LIDAR based on such an approach is developed in the framework of the ESA's Atmospheric Dynamic Mission (ADM). The "ALADIN Airborne Demonstrator (A2D)" was successfully tested and used during airborne campaigns (Paffrath *et al.* 2009; Li *et al.* 2010). The laser transmitter is a single mode 355 nm Nd-YAG laser and the detection receiver combines a Two-channel Fabry–Perot and a Fizeau interferometers (Reitebuch *et al.* 2009). ADM–Aeolus is ESA's spaceborne LIDAR (ESA, 1999) to be launched at the end of 2013.

9.4.3 Water Vapor

In terms of specific humidity, atmospheric water vapor varies with height over four orders of magnitude. The largest values of specific humidity, of the order of $20\,g\,kg^{-1}$, occur near the Earth's surface in the tropics, while the lowest values, of the order of $0.001\,g\,kg^{-1}$ are found near and above the tropopause. A further characteristic of the tropospheric water vapor field is its high spatial and temporal variability. There are sharp vertical gradients on scales of 100 m, and horizontal

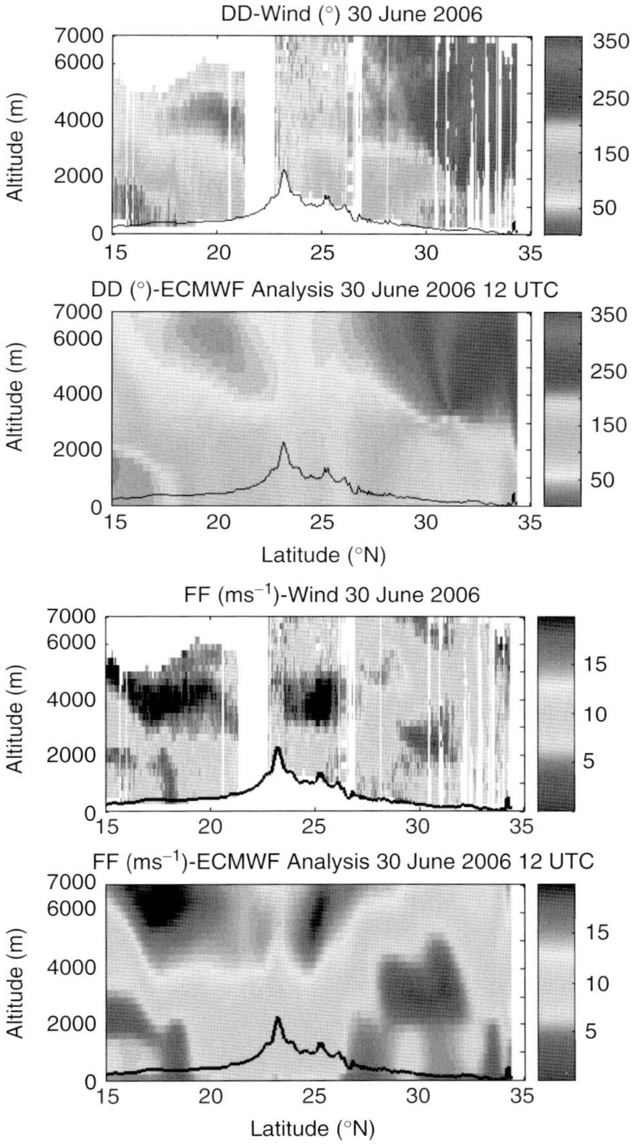

Figure 9.9 Atmospheric wind field retrieved by WIND during a flight over the Saharan desert from Djerba (Tunisia) to Niamey (Niger) on 30 June 2006. The wind field strength (FF) and direction (DD) retrieved from WIND measurements are compared to ECMWF analysis at 12.00 UTC. (Courtesy of A. Dabas.)

contrasts on scales of 10 km or less. Temporal variations range from minutes for convective processes to decades for climate change. Thus, monitoring water vapor with the required precision, resolution and temporal and spatial coverage remain important scientific and technological issues. Although Raman LIDAR has been recently tested for airborne operation (Whiteman *et al.* 2010), DIAL technique presently remains the best suited to meet these requirements.

9.4.3.1 Airborne H_2O–DIAL Instruments

A challenge of measuring atmospheric water vapor with the DIAL technique is associated to the selection of the laser transmitter to generate the online and offline wavelengths as described in Section 9.3.3. In principle the 4ν and 3ν overtone vibrational bands of H_2O in the near IR around 720 nm, 820 nm and 940 nm are suited for H_2O–DIAL application (other bands at higher wavelength can be used, but with a reduced efficiency due to the decrease of molecular scattering which provides the signal for the analysis in the free troposphere). The H_2O absorption lines found in these spectral regions are almost free from interfering absorption features arising from other trace gases and some of the lines are favorable because the corresponding absorption cross sections are only little affected by the uncertainty of the atmospheric temperature profile. Also the cross sections in the 940 nm spectral region cover more than three orders of magnitude that is a prerequisite in addressing the large dynamic range of atmospheric water vapor as can be seen in Figure 9.4. Three LIDAR groups located in the US, France, and Germany developed suitable DIAL transmitter systems in the past that can be operated on airborne platforms. The first instruments which were flown by NASA and DLR dating back to the 1980s and early 1990s. They were based on Nd:YAG-pumped dye laser technology, which is generally limited by spectral impurity caused by amplified spontaneous emission (ASE), which can significantly degrade the overall measurement accuracy. Substantial progress towards the achievement of better DIAL-transmitters was made in the 1990s where powerful solid-state lasers based on alexandrite and Ti:sapphire laser technology or nonlinear frequency converter such as optical parametric oscillator (OPO) systems became available.

One of these systems, called LEANDRE II, has been developed by the Service d'Aéronomie of CNRS in France in the mid of 1990 (Bruneau *et al.* 2001b; Bruneau *et al.* 2001a). This instrument uses a flashlamp-pumped alexandrite laser that can be operated in the 727–770 nm water vapor band for tropospheric water vapor profiling. A special feature is the double-pulse mode with a temporal separation between the online and offline pulses of 50 μs to avoid any measurement errors arising from spatial variability of atmospheric backscattering at aerosol and cloud particles. The spectral separation of both the on- and offline wavelengths is ~0.4 nm, thus the requirement for using the simple DIAL equation; see Eq. (9.6), is fulfilled by this instrument. The repetition rate of the double pulses is 10 Hz. The average power is 1 W in total and the spectral purity values of $>$ 99.99 % guarantee high measurement accuracy (Cahen and Mégie, 1981). LEANDRE II, installed on the French research aircraft, participated on the major field campaigns such as MAP in 1999, IHOP in 2002 (Weckwerth *et al.* 2003), AMMA in 2006 (Flamant *et al.* 2007), and COPS in 2007 (Richard *et al.* 2009) to measure the aerosol and water vapor fields over complex orography. The focus of these measurements was to analysis the physical processes such as convective initiation which are involved in Quantitative Precipitation Forecast (QPF). The instrument characteristics are defined for measuring the specific humidity with an accuracy of better than $0.5\,g\,kg^{-1}$ in the first 5 km at a vertical resolution of 100–300 m and a horizontal resolution of approximately 1 km. Successful comparisons were obtained between LIDAR and in situ, as well as between LIDAR observations during IHOP (Behrendt *et al.* 2007) and COPS (Bhawar *et al.* 2011) where aircraft were flown in flight formation.

The second system is called LASE (LIDAR Atmospheric Sensing Experiment) which was developed in the 1990s as a prototype for a space-based water vapor DIAL system at NASA Langley Research Center in the US. This system underwent extensive validation measurements (Moore *et al.* 1997; Browell, Ismail, and Grant, 1998). At the beginning, it has been flown almost autonomously on the high flying NASA aircraft ER2 and participated in numerous airborne campaigns since that time using the NASA DC–8 research aircraft. LASE is a Ti:Sapphire-laser-based H_2O–DIAL system operated in the 813–818 nm water vapor absorption band for tropospheric profiling. The pulse-pair separation is 300 μs and its repetition rate is 5 Hz. LASE can offer a total of 1-1.5 W laser power at a spectral purity level which is $>$ 99 %. As a special feature the instrument can be operated with two absorption wavelengths in an interleave mode in order to cope with the high dynamic range of tropospheric water vapor. LASE was found to measure specific humidity with an accuracy of better than 6 % or $0.01\,g\,kg^{-1}$ whichever is greater, across the entire troposphere.

The third system called WALES demonstrator was developed at the beginning of the new millennium (2002–2007) at the DLR Institut für Physik der Atmosphäre in Germany. The idea of this instrument was to serve as a prototype of a space-based H_2O DIAL for the WALES mission, which has been studied on Pre–Phase 0/A level at ESA in 2001–2002 (ESA, 2001). These investigations reveal that the large dynamic range of atmospheric water vapor can be addressed with an instrument looking in nadir direction using three online wavelengths with different water vapor attenuation cross-sections, and one offline wavelength as depicted in Figure 9.4. Strong lines around 935.6 nm can be used for measurements at low water vapor concentrations, e.g., in the upper troposphere and lower stratosphere, whereas weak lines are appropriate at high water vapor content and also for the entire altitude range from Arctic to tropical regions. The system concept and instrument performance of the WALES demonstrator is described in detail by (Wirth *et al.* 2009). The transmitter is based on two identical subsystems each consisting of a diode-pumped Nd:YAG laser and a nonlinear frequency conversion stage to generate pulsed radiation on the four selected wavelengths between 935 nm and 936 nm. Each transmitter module is switching between two wavelengths at a rate of 50 Hz. The time difference between the output pulses from different transmitters can be set to an arbitrary offset. A typical value is 5 ms resulting in an equidistant pulse-train of the four wavelengths at a rate of 50 Hz. The output power is 10–12 W in total which enables fast sampling or a high spatial resolution in the measured H_2O field. A spectral purity value which is $> 99.9\%$ and a frequency stability that is < 30 MHz of each wavelength guarantees a high quality data standard. The unused not converted pump radiation at 1064 nm and 532 nm is also transmitted into the atmosphere for aerosol measurements. The receiver of the WALES demonstrator uses a 48 cm Cassegrain telescope, which can be mounted either in nadir or zenith-viewing direction. The backscattering of the wavelength for the water vapor measurements are separated from backscattering at 1064 nm and 532 nm by dielectric beam splitters. To suppress the solar background radiation a 1 nm bandpass filter is used. The particle phase can be discriminated by two depolarization channels at 532 nm and 1064 nm. A special feature of the receiver subsystem is its HSRL capability through the implementation of an I2–cell to filter out the narrow-band aerosol backscattering at 532 nm. Together with the total backscattering, the spectral filtering allows to measure profiles of particle extinction and atmospheric backscatter ratios in parallel to the water vapor concentration. The latter can be used to improve the quality of the water vapor data by applying the Rayleigh–Doppler correction (determination of the effective absorption for the backscattered beam). Photomultipliers are used as detector at 532 nm, while low-noise avalanche photo diodes serve as the detection for the NIR channels. For the integration into research aircraft, like DLR's Falcon F20 or the German research aircraft HALO, a very compact setup has been realized. The system has been completed in 2007 for deployment on the Falcon 20 aircraft. Since that time, it undergoes an extensive validation program at various places of the world through participating in the field campaigns COPS_2007, SAMUM_2008, IPY-Thorpex_2008, and T–PARC_2008.

9.4.3.2 **Measurement Examples**

During long-distance ferry flights between Europe and Brazil or Australia within TROCCINOX and SCOUT-O_3, 2D distributions of tropospheric water vapor and aerosol/clouds could be measured with a precursor H_2O–DIAL installed on the Falcon aircraft (Ehret *et al.* 1999; Poberaj *et al.* 2002). Results have been obtained for mid-latitude, subtropical, and tropical regions as well as measurements over the Indian Ocean and Micronesia which are regions characterized by monsoon and intense inner tropical circulation. As an example, on the transfer between Brazil and Europe, different climatic regimes are reflected in the large-scale water vapor distribution clearly seen in the DIAL measurements (Flentje *et al.* 2007). In the tropics and subtropics part of the Hadley circulation was the dominant feature while more northerly, typical mid-latitude dynamical structures were found, resembling to those over the North Atlantic as shown in Figure 9.10. The dynamical features and the water vapor mixing ratios are quite accurately analyzed by the ECMWF at T511/L60 operational resolution. Largest deviations arise from shifts of structures/gradients subject to rapid temporal development. The averaged humidity-bias between both data fields is below 10%. While at mid-latitudes low water vapor regions are associated with filaments of enhanced particle backscatter,

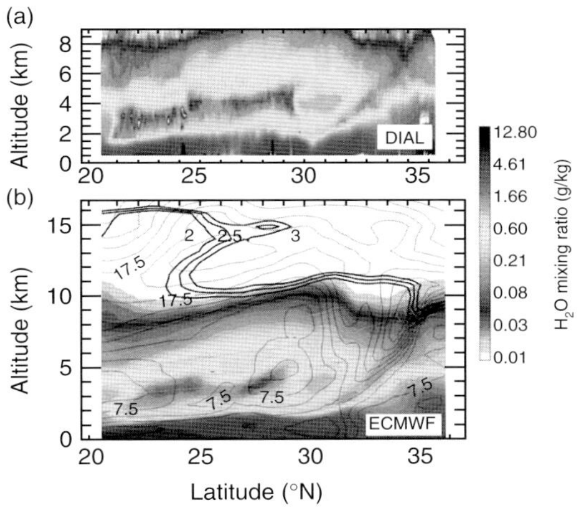

Figure 9.10 Water vapor mixing ratio *q* in g/kg along DLR DIAL flights (a) on 14 March 2004, 1630–1900 UT compared to ECMWF T511/L60 operational analysis (b). Contours are horizontal wind speed in m s^{-1} and potential vorticity at 2, 2.5, and 3 PVU. From Flentje *et al.* (2007). (*See Color Plates for color representation of the figure.*)

the H_2O signature of the tropical circulation is not correlated with the aerosol. The large H_2O gradients, small scales and the large dynamical range make the scene hardly accessible to passive sounders or sporadic soundings by radiosondes.

9.4.4
Other Gases

LIDAR profiling of atmospheric gases has been obtained by different methods. However, aircraft measurements imply high resolution (both spatial and temporal) and only the Differential Absorption method can be used to retrieve accurate gas concentration or mixing ratio from aircraft. The case of water vapor was previously described; we present here some results for other species.

9.4.4.1 Ozone

Ozone (O_3) plays a key role in the chemistry of the atmosphere both in the troposphere and stratosphere, and as a consequence of the high variability of atmospheric O_3, a relatively dense network of measurement stations in combination of ground-based, balloon- and spaceborne instruments has been set to monitor its evolution. Airborne ozone DIAL systems have the advantage of possessing the flexibility to address case studies at different locations and to provide detailed description of the spatial distribution of ozone over large areas. As many studies reveal, tropospheric ozone can be measured by the DIAL technique in the wavelength range between 270 nm and 300 nm also known as the Hartley band of ozone in the UV spectral range. However, different to the narrow-band absorption lines of the water vapor molecule in the near IR, the spectroscopy of the ozone molecule in the respective spectral range in the UV is substantial different. Ozone under atmospheric conditions exhibit a pronounced unstructured continuum absorption characteristics where the absorption cross section changes only slowly between neighboring absorption wavelengths. As a consequence, the spectral separation between the online and offline wavelengths has to be larger than 2 nm in order to achieve a sufficient optical depth in the DIAL measurement. This large wavelength separation makes it necessary to correct for the differential backscatter and extinction caused by the air molecules as depicted in Section 9.8. On the other hand, the spectral characteristics and frequency stability requirement of the DIAL transmitter is much more relaxed in an ozone DIAL experiment compared to the NIR counterpart.

After the development of first airborne systems using Nd-YAG pumped dye lasers (Browell *et al.* 1983), more compact and simpler systems were developed to avoid spectral wavelength control. For example the Service d'Aéronomie system ALTO (Airborne LIDAR for Tropospheric Ozone observation) was based on previous developments for ground-based observations (Ancellet, 1998), using a solid-state Nd:YAG laser and stimulated Raman scattering in deuterium to generate three wavelengths (266, 289, and 316 nm). Both analog and photon-counting detection methods are implemented in the detection chain to provide a measurement range up to 8 km. The system has been flown on the French Fokker 27 aircraft to perform both lower tropospheric (0.5–4 km) and upper tropospheric (4–12 km) measurements, with a 1 min temporal resolution corresponding to a 5 km spatial resolution (Ancellet and Ravetta, 2005). The vertical resolution of the ozone profile can vary from 300 to 1000 m to accommodate either a large-altitude range or optimum ozone accuracy. Comparisons with in situ ozone measurements performed by an aircraft UV photometer or ozone sondes and with ozone vertical profiles obtained by a ground-based LIDAR showed that the accuracy of the tropospheric ozone measurements is generally better than 10–15%, but that aerosol interferences necessarily need to be corrected. An example of observations performed across the mid-latitude jet (Kowol–Santen and Ancellet, 2000) showing the intrusion of stratospheric ozone rich air in the related tropopause folding as evidenced from the model analysis of potential vorticity is reported in Figure 9.11.

9.4.4.2 Carbon Dioxide

Carbon dioxide (CO_2) as the most important greenhouse gas is central in current climate change issues associated to anthropogenic activities. Then, unambiguous detection/quantification of small changes in CO_2 horizontal gradients is required at scales from tens of kilometers to continental scales. Today, optical remote sensors operated from the ground, airborne or space based platforms are seen as key players. Concurrently with passive remote sensors, active remote sensors using the DIAL technique can provide accurate CO_2 concentration measurements. The pioneering works conducted in the 80's and 90's using man made targets demonstrated the DIAL capability with convincing performance given the state of technology. Bufton *et al.* (1983), used a DIAL instrument based on two frequency-doubled pulsed TEA–CO_2 lasers emitting at 4.88 μm for atmospheric CO_2 measurements

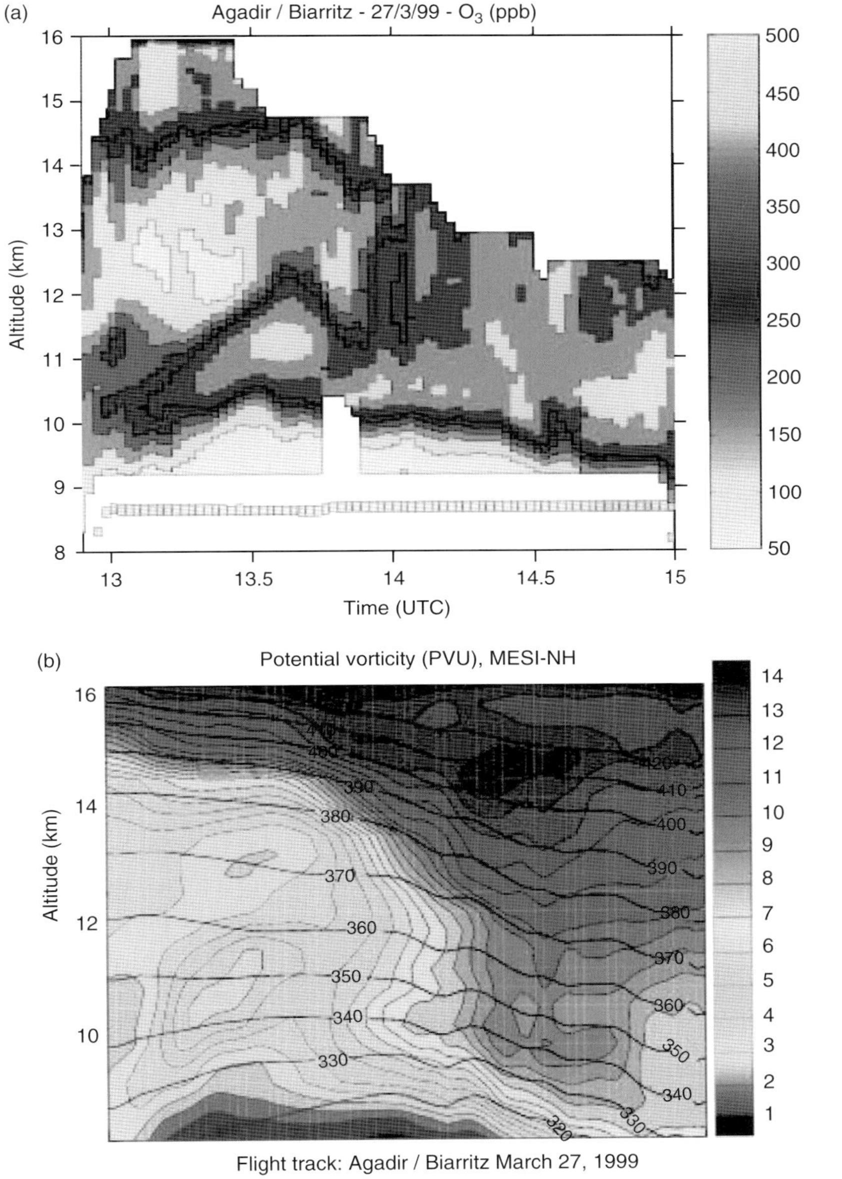

Figure 9.11 Ozone vertical cross section measured by the ALTO LIDAR (a) and comparison with a (b) high resolution PV vertical cross section from the Meso–NH model, after Kowol–Santen and Ancellet (2000). (*See Color Plates for color representation of the figure.*) (Copyright 2000 American Geophysical Union (AGU). Reproduced with permission by AGU.)

in the planetary boundary layer (PBL). Later, Sugimoto and Minato (1993), used a Nd:YAG pumped dye laser around 0.750 μm and the second Stokes radiation of a hydrogen Raman shifter to operate in the 2 μm spectral range. The wavelength of the dye laser was continuously scanned over the absorption line. The spectral bandwidth of the laser was much narrower than the bandwidth of the CO_2 absorption line at atmospheric pressure. At present, DIAL and IPDA techniques can be used considering the CO_2 absorption domain (1.6 or 2 μm), range resolved or total column content using topographic target, laser transmitter technology, direct or heterodyne detection. New airborne systems are under development in USA, France and Germany (Gibert *et al.* 2006; Ismail *et al.* 2008; Amediek *et al.* 2010).

9.4.4.3 **Methane**

First work was aiming at measuring CH_4 in the mid–IR. Ongoing actions lead to select overtone bands at 1.5–1.6 μm, to benefit from solid state laser technology. Measurement of methane is less critical than CO_2 due to relaxed requirements. A first system which takes advantage of OPO development at DLR has been developed to monitor pipeline leaks and embarked on an helicopter (Fix *et al.* 2004). An airborne demonstrator is to be built in the frame of the preparation of the new joint DLR–CNES space mission MERLIN.

9.4.5 Water Vapor Flux Measurements

Retrieval of fluxes from remote sensors onboard the aircraft may be done using the eddy-covariance technique which has been used extensively from in situ measurements (see the related chapter in this book). In general, this can only be accomplished by active remote sensing instruments such as Doppler wind measurements (either LIDAR in clear air or RADAR in clouds to derive momentum flux) or colocated Doppler and DIAL systems (for latent heat or gas fluxes). One advantage over *in situ* is that they have the ability to collect measurements simultaneously from many different altitudes. One disadvantage is the difficulty to get high SNR data and correct from aircraft motion. Operating such instruments on an airborne platform requires accurate knowledge of aircraft motion and proper alignment of the remote sensors' FOVs. It enables to measure the large-area averaged flux profile over various terrain, including water surfaces. Due to the LIDAR's cloud sensitivity, the optimal operation time is before cloud condensation occurs. In addition, information on statistical properties of turbulence and advection of gases can be derived from those measurements.

We will describe here the eddy–covariance method which has been used to derive the flux profile from airborne LIDAR measurements (Giez *et al.* 1999; Kiemle *et al.* 2007). The vertical water vapor flux F is calculated from:

$$F(L) = \sigma_{\mathrm{wq}}^2 = \overline{w' \cdot q'} \qquad (9.8)$$

where $w\prime$ and $q\prime$ are the horizontal turbulent fluctuations in the vertical wind speed w (m s^{-1}) measured by the Doppler wind LIDAR and the specific humidity q (g kg^{-1}) for each height interval measured by the H_2O–DIAL instrument. The bar represents the spatial average over the horizontal length interval L. The latent heat flux is obtained from Eq. (9.8) by multiplying with the latent heat of vaporization, which is approximately 2500 J g^{-1}. There are two main sources of random errors in the eddy flux measurement: instrumental noise and sampling error. The error contribution from instruments is given by the following relation:

$$\sigma^2_{\mathrm{F,instr}} = \frac{\Delta L}{L} \cdot \left(\sigma^2_{\mathrm{q,instr}} \cdot \sigma^2_{\mathrm{w,atmos}} + \sigma^2_{\mathrm{q,atmos}} \cdot \sigma^2_{\mathrm{w,instr}} + \sigma^2_{\mathrm{q,instr}} \cdot \sigma^2_{\mathrm{w,instr}}\right) \quad (9.9)$$

where ΔL is the horizontal spacing between two measurements for each LIDAR instrument (Giez *et al.* 1999). Typical values range between 100–150 m. The other parameters in Eq. (9.9) represent the variances of q and w which arise either from fluctuations in the atmosphere or from the measurement process itself. Since instrumental noise usually decreases with averaging at a rate of $1/\sqrt{\Delta L}, \sigma^2_{\mathrm{F,instr}}$ does not strongly depend on ΔL.

The sampling error describes the error due to the limited number of probed eddies given by the flight leg length L which is typically in the order of 50 km over inhomogeneous terrain. It can be estimated for a joint Gaussian distribution in w and q by:

$$\sigma^2_{\mathrm{F,samp}} = 2\,\frac{\lambda_{\mathrm{F}}}{L} \cdot \left(F^2 + \sigma^2_{\mathrm{w,samp}} \cdot \sigma^2_{\mathrm{q,samp}}\right) \quad (9.10)$$

with the ratio of the horizontal integral scale λ_F (a measure of the dominating eddy size in the order of the convective PBL (CBL) depth) to L determining the error (Lenschow and Stankov, 1986).

A prerequisite in measuring the latent heat flux profile using airborne LIDAR is its ability to sample the wind and water vapor fluctuations in the PBL at a frequency that is high enough to capture all flux containing eddies with sufficient accuracy and spatial resolution. From turbulence theory it ideally follows that a spatial resolution of a few tens of meters both in the vertical and horizontal domain is regarded desirable for a low sampling error. Due to the aircraft speed, this requires a sampling frequency of a few Hz or more across the PBL depth of 2 km. From an instrument point of view, the measurement of $q\prime$ in Eq. 9.6 which is in the order of 5–10% of the mean value is much more demanding than the measurement of $w\prime$. This is based on the fact that the mean vertical wind speed w being close to zero, the measurement with the Doppler wind LIDAR provides $w\prime$ in a direct manner.

Figure 9.12 displays the superposition of vertical velocity obtained with the 2 μm Doppler wind LIDAR HRDL from NOAA (Grund *et al.* 2001) and water vapor mixing ratio received with the DLR H_2O–DIAL during the IHOP2002 campaign. For these flux measurements which took place over Oklahoma/Kansas both LIDARs were deployed on the Falcon 20 aircraft from DLR, Germany for collocated soundings of the vertical wind speed and water vapor distribution with high spatial resolution (Kiemle *et al.* 2007). Figure 9.12 shows two parts of successive flight legs over the same region numbered 3 (top) and 4 (bottom). The aircraft turned back at

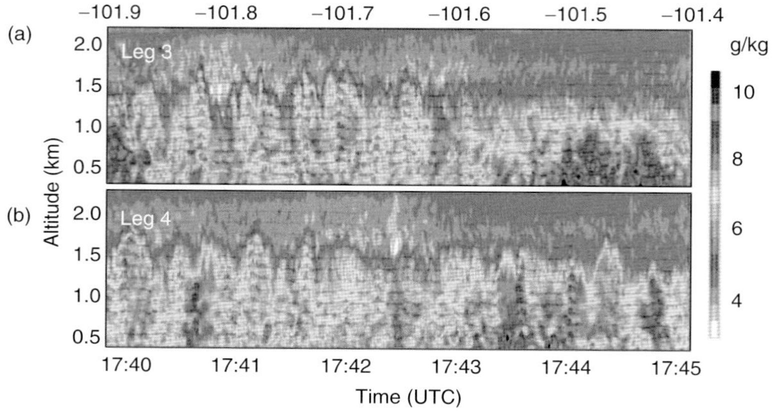

Figure 9.12 Vertical cross sections of water vapor from DLR DIAL (colors) and vertical velocity from NOAA HRDL (quasi-horizontal vertical bars) for the Falcon flight legs 3 and 4 oriented West–East at 37.4 °N over southwestern Kansas. The aircraft flew (a) leg 3 from right (East) to left (West) and turned back at 102 °W to (b) fly leg 4 on the same track. Top axis is longitude, bottom axis UTC time, 7 h ahead of LT, after Kiemle *et al.* (2007). (*See Color Plates for color representation of the figure.*) (Copyright 2007 American Meteorological Society (AMS). Reproduced with permission by AMS.)

102 °W; hence, the time interval between the two legs was 6 min at the left edge and 17 min at the right edge of Figure 9.12.

Horizontal CBL winds as measured by a Falcon dropsonde in the middle of leg 4 were 15 m s^{-1} from South-Southwest (SSW), that is, nearly perpendicular into the plane of the cross sections. This explains the absence of comparable structures in both plots.

The measurements stop at 300 m AGL because of the vertical range cell size of 150 m for both LIDARs and slight ground-level variations. The superposition of w and q gives evidence of considerable correlation between elongated humidity structures and upward motion. Strong contributions to the water vapor transport quantified by the vertical flux emanate obviously from the largest thermals that are found to extend vertically throughout most of the CBL. An inactive old thermal is visible in leg 4 at 101.42 °W, in contrast. There is also evidence of very deep entrainment, for example, in leg 4 at 101.62 °W, where dry air is transported downward.

The LIDAR–based water vapor flux profile in Figure 9.13 was derived using the method described above. The flux is zero in the free atmosphere above the tops of the strongest thermals, as expected. The flux profile shows a quasi–linear flux decrease to 200 W m^{-2} at 400 m AGL which corresponds to about one-third of the average CBL depth. This observed strong positive flux divergence is consistent with a significant upward humidity transport drying out the mid- and lower CBL by entrainment of dry air from above (see also Figure 9.12), while the entrainment zone and the part of the free atmosphere that is penetrated by the strongest thermals are humidified. The strong positive flux divergence is associated with a significant

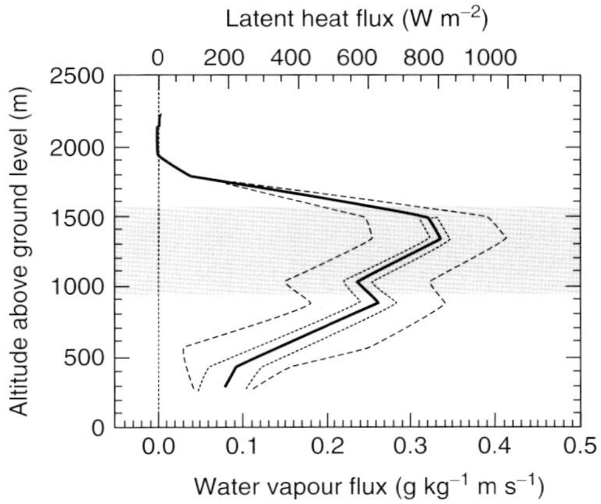

Figure 9.13 LIDAR-derived water vapor flux profile for the East-West-East oriented legs 3 and 4 (solid; around 1030 LT), with uncertainty ranges given by the thin dashed lines for the sampling error and thin dotted lines for the instrument noise taken from Kiemle *et al.* (2007). The flux peaks at the CBL top and is zero in the free atmosphere above the tops of the strongest thermals. (*See Color Plates for color representation of the figure.*) (Copyright 2007 American Meteorological Society (AMS). Reproduced with permission by AMS.)

upward humidity transport drying out the mid- and lower CBL by deep entrainment of dry air and humidifying the free atmosphere as the CBL grows vertically.This finally inhibited the growth of clouds and deep convection in this particular case.

The new airborne double-LIDAR instrumentation proves to be a valuable tool for the study of CBL processes and variability, particularly over complex terrain. Future missions will include measurements over oceans to quantify evaporation in weather-relevant regions such as the inflow domains of warm conveyor belts.

9.4.6 Calibration: Precision and Accuracy

9.4.6.1 Calibration on Molecular Scattering

LIDAR calibration is not strictly needed for all systems, as for example using DIAL or Doppler LIDARs, when focusing on gas concentrations or wind speed (see related sections). However, it is the case for backscatter LIDARs in order to quantify depolarization, and to retrieve aerosol (or cloud) optical and microphysical properties. Such a calibration has a direct impact on the accuracy of the inversion of the LIDAR signal. The propagation of the induced error is however decreasing as the distance from the reference zone and the optical depth are increasing (Klett, 1985). In all cases polarization is to be taken into account. In-lab radiometric calibration is possible, but for most backscatter LIDARs operating in the shortwave or near-IR spectral domain, calibration is usually performed so to normalize the LIDAR signal on molecular scattering. This is done using meteorological or aircraft soundings to determine temperature and pressure and thus atmospheric density vertical profiles. Referring to Eq. (9.4), this means that the received power is converted into an attenuated backscatter coefficient, equal to the product of the atmospheric backscatter and squared transmission. It can be done with a good accuracy (about 5 % in the UV to 20 % in the near IR) provided the aerosol scattering is small in the normalization altitude domain and checked to be leading to a small error at the operating wavelength (Russel, Swissler, and McCormick, 1979). In the high spectral resolution LIDAR analysis, the signals in the molecular and combined aerosol and molecular channels can be also calibrated using clear air reference (Esselborn *et al.* 2008). Russel, Swissler, and McCormick (1979) have developed a detailed analysis of error which one can refer to for further detail. This normalization method furthermore implies an accurate determination of the molecular differential scattering cross-section $d\sigma/d\Omega$, as the backscatter coefficient is equal to the product of the molecular density and $d\sigma/d\Omega$; see for example (Measures, 1984). The strength of the backscattering cross-section to consider is indeed depending on the spectral bandwidth at the reception, as it may include Raman scattering contribution or not; see (Esselborn *et al.* 2008). In the case of a larger filter bandwidth, Raman scattering is to be included and depolarization is also increased with respect to pure elastic molecular scattering (Esselborn *et al.* 2008). Dense cloud is an alternative for calibration. However, this is a much denser diffuse target, and the signal needs to account for the whole scattered photons. This means that multiple scattering is an additional factor that requires to be known. Modeling and measurements have shown that this may be possible using depolarization measurements (Hu *et al.* 2006).

9.4.6.2 Calibration Using a Hard Target

When clear air calibration is impossible, or when independent approach is preferred, a hard target scattering the whole emitted beam can be used as a known effective reflectance reference. In the case of a hard target of reflectance ρ, the effective reflectance of the target is $\rho \cdot \cos\theta/\pi$ assuming it is Lambertian. This expression has to be considered instead of the scattering coefficient B in Eq. (9.4). Painted targets have been used in the IR, allowing to meet similar accuracies (Spinhirne *et al.* 1997). Calibration of the target itself is a complex task, as it requires an accurate pointing onto the target (particularly when its size is small with respect to emission cone) for both emission and reception, and a correction of the geometrical overlap between emission beam and reception cone, as well as transmission from between the emitter and the target. Molecular calibration which does not require a complex set-up and can be performed in flight at a larger distance leading to a signal of similar amplitude which can be analyzed with the same set-up, is thus preferred when possible.

9.4.6.3 Calibration Using Sea Surface Reflectance

Another method to perform LIDAR calibration on a hard target of known reflectance at a large distance avoiding overlap correction is to use sea surface. Sea surface reflectance in the visible spectral domain was first studied using photographic analyses by Cox and Munk (1954). To provide further analysis for laser observations, airborne measurements have been performed by (Bufton *et al.* 1983), that have confirmed results previously obtained and the validity of the method for wind speed smaller than $10\,\mathrm{m\,s^{-1}}$, value at which foam begins to form and modifies reflectance mainly due to small wavelength gravity waves at the surface. Subsurface may also contribute non-negligibly to the effective surface reflectance at high incidence angle (Li *et al.* 2010). Such an approach has however only been used very rarely for airborne LIDAR calibration, due to the increased difficulty to control all experimental parameters (Bufton *et al.* 1983; Josset *et al.* 2008; Li *et al.* 2010) to perform a high absolute accuracy calibration, as compared to the molecular scattering procedure. Calibration of the depolarized signal are usually performed comparing the signals in the two polarization channels, or on molecular scattering as well, considering the receiver spectral bandwidth.

9.5 Cloud and Precipitation Observations with RADAR

9.5.1 Reflectivity from Cloud Droplets, Rain and Ice Crystals

The most frequent application of data from meteorological RADARs is to derive precipitation rate. Very extensive and varied literature exists on the Z–R (reflectivity vs rain or snowfall rate) relationships, often extended to include Z–M (reflectivity vs. mass per unit volume) relationships. Since the focus here is on airborne research applications, it is more to the point to consider the information that RADARs can provide regarding cloud composition in a more general sense.

Theoretical prediction of the reflectivity of clouds is relatively straightforward, as presented in Section 9.3, as long as the assumption of spherical drop shapes holds. Complications arise from shape distortions for drops larger than about 1 mm in diameter. For ice crystals, the combined problem arises that the calculations are complex and that actual shapes are practically impossible to predict with precision. In terms of calculations of reflectivities in cloud models, specification of the size distribution introduces considerable uncertainty for all types of hydrometeors. For all these reasons, forward modeling is possible but not precise and, in the reverse, observed reflectivities are not translatable to cloud composition with rigor. Using more RADAR parameters, in addition to reflectivity, the links between prediction and observations can be tightened considerably, as will be described in Sections 9.5.4 and 9.7.

The approach taken in this section is to present calculated reflectivities for various types of cloud composition and to thereby provide some sense of the correspondence between cloud composition and RADAR reflectivity. This can then serve as a first guide toward the interpretation of the field data.

The detection of nonprecipitating clouds is at the limit, or beyond, the sensitivities of most airborne RADARs. A cloud consisting of 10 μm diameter droplets with a liquid water content of $0.2\,\mathrm{g\,m^{-3}}$ has a reflectivity near −35 dBZ. For 20 μm droplets and the same liquid water content the reflectivity is close to −25 dBZ. These figures can be readily adjusted for other sizes and liquid water contents, linearly for LWC and with the sixth power of the sizes as in the size range of cloud droplets the Rayleigh scattering assumption holds well.

The following examples present more realistic droplet spectra, using gamma distributions. Those in Figure 9.14(a) can still be considered to be nonprecipitating, while those in Figure 9.14(b) and (c) are drizzle cases. As these spectra illustrate, as long as the mode of the spectrum is near 10 μm and even if there are drops up to 50 μm and larger, reflectivity remains considerably lower than -20 dBZ. Once the mode of the spectrum is near 20 μm or larger, reflectivity increases rapidly and for what could be considered heavy drizzle (2.6 mm h^{-1} for the broadest spectrum in Figure 9.14(c)) it may exceed +10 dBZ. These values were calculated for 94 GHz (using Mie formulas) but the results would be almost identical for lower frequencies since the contributions of droplets larger than 1 mm is negligible, so Mie resonance effects aren't significant in these cases.

For rain, the most frequent approach is to assume Marshall–Palmer size distribution and derive $Z - R$ (reflectivity vs. rainfall rate) relationships purely from calculations or from a combination of drop size and reflectivity measurements. Figure 9.15 (Figure 5.13 in Lhermitte (2002)) shows the results of such calculations for two frequencies. At 3 GHz the power–law $Z = 250R^{1.44}$ is a good description over the full range of likely values, and it is frequently used. Many similar equations are in the literature which allow for regional variations, storm characteristics and other variables. Large numbers of papers deal with the impacts of truncation of the size distribution of drops at small sizes, with the dependence of the power-law constants on the size distribution, and the impacts of deviations of drop shapes from spherical for drops larger than a few millimeters in diameter. The $Z–R$ equations (al $R(Z)$ functions) have their greatest utility for survey type of RADAR applications, and rather lesser importance for airborne RADARs. As seen in Figure 9.15, the power law is not a good fit at 35 and 94 GHz for the full range of values displayed in the graph because of the increasing influence of Mie effects for larger drops. Overestimating the concentrations of small drops by exponential distributions is also most serious at higher RADAR frequencies, as demonstrated by Kollias, Jo, and Albrecht (2005). They show that for a 1 mm h^{-1} rainfall with Marshall–Palmer size distribution drops in the 1–2 mm range contribute 66 % of the reflectivity at 94 GHz, while that proportion is only 10 % at 3 GHz, the difference being due to the larger drops having reduced scattering cross-sections at 94 GHz. The use of log–normal or gamma distributions overcomes this problem to some extent as well as providing better fits to empirical data. Reflectivity weighted fall velocity can be used to further constrain these functions, except that vertical air motions introduce large errors. Some recent papers offering specific forms of the $R(Z)$ and $M(Z)$ functions are Van Baelen, Tridon, and Pointin (2009); Tokay *et al.* (2009); Chapon *et al.* (2008); Ochou, Nzeukou, and Sauvageot (2007).

The shape distortions and oscillations of raindrops become progressively more important with increasing drop size. For example, the horizontal to vertical axis ratios of drops approach 0.9 by about 1 mm drop diameter (Beard and Chuang,

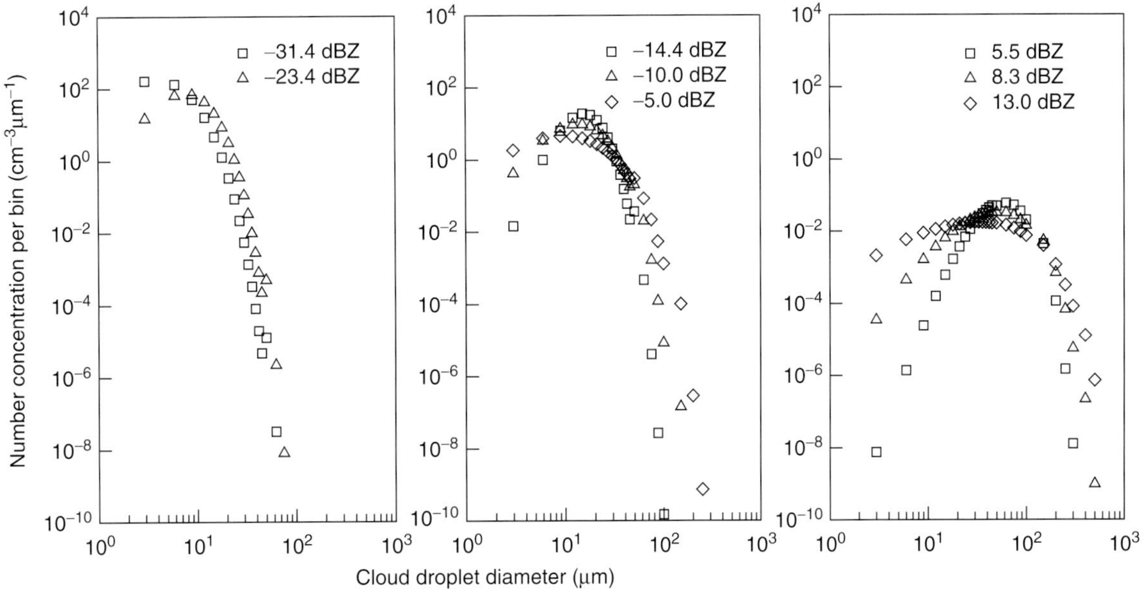

Figure 9.14 Reflectivities of cloud droplets for different size distributions defined by gamma functions. These graphs illustrate the sensitivity of reflectivity to the largest droplet sizes, or almost equivalently to the width of the size distribution. (*See Color Plates for color representation of the figure.*)

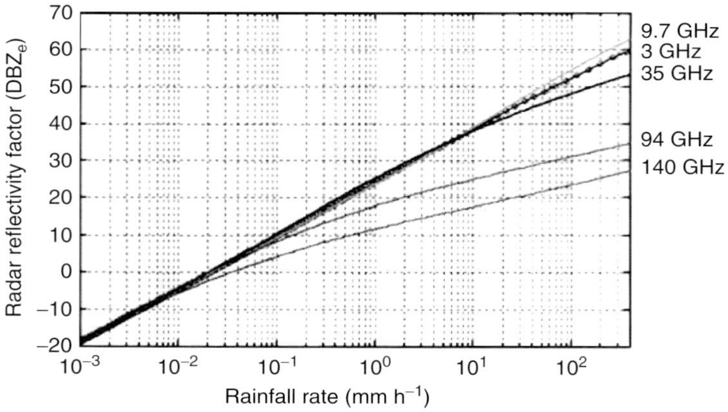

Figure 9.15 Relationship between RADAR reflectivity and rainfall rate for Marshall–Palmer size distribution. Figure 5.13 in Lhermitte (2002).

1987). This flattening leads to polarity-dependence of the backscatter cross-section. Oscillations due to turbulence or collisions diminish that dependence but become significant only for even larger drops. Both factors are exploited for improving RADAR estimation of rainfall rate via polarization and differential attenuation parameters. Essentially all work, e.g., Moisseev *et al.* (2006); Gorgucci *et al.* (2000); Gorgucci, Chandrasekar, and Baldini (2006); Thurai, Huang, and Bringi (2007), on this topic has been done for S-band and X-band RADARs, apply to rainfall rates of several tens of mm per hour (over 30 dBZ) and require integration of data over fairly large sample volumes or times. Moreover, the polarization signatures manifest themselves most strongly for horizontal beams and are lost for vertical beams. For all of these reasons, these methods have not entered so far into use for airborne RADARs.

Ice crystals constitute a large fraction of tropospheric clouds and in mixed–phase clouds grow rapidly to sizes that exceed the applicability of the Rayleigh treatment, especially for shorter wavelength RADARs. In addition, only a very small fraction of ice crystals in clouds have simple geometric shapes even when growth is just by vapor deposition. When the crystals also collect rime their shapes become quite irregular in detail, though their outlines may be close to near-spherical or conical. Because of their generally complex shapes, the backscatter cross-sections of ice particles can not be calculated from first principles except for idealized geometries and with assumed densities. Most recent solutions are based on finite-difference numerical models.

The simplest approximation to crystal shapes for which Rayleigh solutions can be obtained are oblate and prolate spheroids. Pristine ice crystals were so represented in early works dealing with RADAR signatures of ice crystals. These approximations have their place even now for the description of heavily rimed crystals (Bringi and Chandrasekar, 2001) for which the applications of more precise solutions would run into problems. When using spheroidal shapes, it is important to allow for the fact that these volumes are not fully filled with solid ice. Bohren and Battan (1980) derived the dielectric function for mixtures of ice and air and show that several methods yield nearly identical results and are corroborated by measurements. This is so, partly because the dielectric constants of air and of ice are not very different (at 3 GHz). The review by Oguchi (1983) contains further discussion of this. For water–ice mixtures such as hailstones growing at rates past the wet limit, i.e., when the heat of fusion of collected supercooled water can't all be dissipated to the ambient air and the hailstone temperature becomes 0 °C, the results are more complicated.

The application of methods based on the superposition of small dipoles arranged to fill arbitrary geometries of ice crystals became practical with advances in computer speed, though resolution of fine details is still limited by computing power. Dungey and Bohren (1993) provide of a good overview of early works and show that backscatter calculations of reasonable accuracy are possible for hexagonal shapes. In these methods, the summation of dipole resonances is clearly dependent on the polarization of the incoming radiation as well as on the relative direction of the incident beam and the axis orientation of the crystal.

Schneider and Stephens (1995) examined the extent to which the Rayleigh approximation for spheroids agrees with the discrete dipole approximation (DDA) and also the extent to which the spheroid approximation matches the DDA solution for column and plate crystals. Donovan *et al.* (2004) made a similar evaluation for four crystal shapes and for random orientations with respect to the incident beam. These analysis contain a great deal of detail and are difficult to summarize. Nonetheless, as a broad generalization, it can be said that the more elaborate methods are needed for crystal sizes above a few hundred micrometer or if specific orientations in space are to be dealt with. For example, Donovan *et al.* (2004) show that the difference between the DDA results and Mie calculations increases with size and for equivalent sphere diameters of up to 1 mm lead to 20–40 % underestimation for columns and plates, while for branched crystals there is an overestimation by about the same extent (cf. their Figure 9.9). The errors get larger for sizes above a millimeter.

Another modeling approach is the finite difference time domain (FDTD) method. Tang and Aydin (1995) applied that technique to plate, column and dendritic shapes using empirical diameter–thickness relationships for the third dimension and covering the range from 100 μm to 2 mm in diameter. Backscatter cross-sections were calculated for horizontal and vertical beams, and linear depolarization ratios were determined

Based on an extensive set of DDA calculations, Weinman and Kim (2007) show that model crystal habits made up of combinations of cylinders can be described by a single function of the phase delay parameter and thereby the scattering phase functions, extinction coefficients and asymmetry parameters obtained with greater ease for frequencies of 89, 140 and 225 GHz. Perhaps that approach will turn out to be useful at lower frequencies as well.

As much as detailed simulations of crystal shapes become more developed, there is also both a need and a reason for continued use of simpler approaches.

The need arises from expediency; the reason flows from shortcomings both of simulated shapes and of sparse observations representing the real complexities of ice particle forms and sizes. In one such work, Matrosov (2007) used the T–matrix method, which can yield solutions for spheroidal shapes similar to that of Mie (for spheres), to calculate backscatter cross-sections for snowflakes at 35 and 94 GHz. The relative ease of computation allowed comparisons to be made for different shape parameters and for the reflectivity of snowfall to be derived using empirical size distributions, densities and fall velocities. Protat *et al.* (2007) showed that the spherical approximation and Mie calculation, combined with refractive index values adjusted according to size-dependent empirical density, results in ice water contents in reasonable agreement with observations. This work, and others like Brown and Francis (1995) and Liu (2004), demonstrate that using even simplified bulk density values to account for the diversity of crystal forms leads to improvements in data evaluation.

Model calculations of RADAR backscatter for simplified ice crystal shapes have their best applicability in cirrus where vapor growth of crystals dominate, though the frequent presence of more complex shapes (bullet clusters) does inject some limits in overall usefulness. Pristine crystals are much less frequent in lower tropospheric clouds and are often mixed with irregular shapes that arise from riming and to a lesser extent from fragmentation of large crystals. This too points to the fact that, for the majority of clouds, there is little precision to be gained from assuming well-defined crystal shapes and using highly detailed calculations of scattering properties, unless detailed angle-dependence is to be studied, or polarimetric properties are to be evaluated, or, as mentioned, focus is restricted to select clouds or cloud regions of uniform crystal habits. Apart from theoretical arguments, since RADAR measurements in the atmosphere yield volume-averaged measurements, and since the most important data for practical and for modeling purposes are other bulk properties, namely ice water content (M, or IWC, in $g\,m^{-3}$) and precipitation rate (R in $mm\,h^{-1}$), much effort goes into retrieving those values from reflectivity. The desired $M-Z$ and $R-Z$ relationships can be derived from calculated backscatter for assumed size distributions and crystal habits with accompanying uncertainties as described earlier. Alternatively, coincident measurements are obtained from RADAR units on the ground or on satellites, and ice particle sampling from aircraft or falling snow at a surface sites. Since the RADAR sample volumes are 10^6-10^8 times larger than the volumes from which ice crystals are sampled the relationships derived from measurements are also subject to considerable uncertainties.

Liu and Illingworth (2000) showed that (at 94 GHz) the relationship $IWC = 0.137Z^{0.64}$ fits observations from several geographical regions within 20–30 %, while variations in ice density introduce errors up to 100 %. The latter can be reduced, they argue, by also taking into account the temperature of the observation region; see also Hogan, Mittermaier, and Illingworth (2006b); Protat *et al.* (2007). Delanoë *et al.* (2007) found that using Doppler velocity as an indicator of ice particle effective diameter, and using the measured relationship between reflectivity and Doppler velocity to yield an estimate of ice particle density, errors in retrieved IWC can be held to about 30–40 %. These papers highlight the fact that ice particle density is a key factor, density being, of course, determined by crystal shape and degree of riming. The ranges of the constants proposed for the power-law functions $Z(M)$ and $Z(R)$, or their complements $M(Z)$ and $R(Z)$, are not very large, but even small differences lead to significantly different values at the upper and lower ends of the range of values of the variables. Sato and Okamoto (2006) and Matrosov, Shupe, and Djalalova (2008) provide further examples of such analysis. Pokharel and Vali (2011) show from measurements with and imaging probe and 94 GHz RADAR co-located on the same aircraft that unknown particle density introduces up to a factor ten scatter in M(Z) and R(Z) relationships.

For ground-based RADARs with polarimetric capabilities a range of methods have been developed that improve rainfall estimates by including measures of polarization and propagation in addition to reflectivity, e.g., Giangrande and Ryzhkov (2008). These methods require averaging over periods of tens of minutes or more, and therefore are not of immediate use for airborne RADAR measurements.

Spek *et al.* (2008) developed a model to retrieve the size distributions of plate and of aggregate crystals form spectral polarimetry data collected at 45° beam angle. The method relies on shape, fall velocity, density, turbulence and other data taken from the literature, and also needs meteorological data for wind information. These generalizations of inputs, the essential role of the measured Doppler spectra, the need for averaging over tens of seconds, and the fixed beam angle exclude the use of the approach to airborne data.

9.5.2
Attenuation

Attenuation is the decrease in intensity of propagating electromagnetic waves due to absorption of energy by gases, and both absorption and scattering by liquid and solid hydrometeors. Scattering by gas molecules and by dry aerosol particles are negligible. For RADARs of cm and mm wavelengths the most significant contributions come from absorption by water vapor and oxygen and from both absorption and scattering by the same hydrometeors whose backscattered power is the RADAR signal. In general, attenuation is more pronounced the shorter the RADAR wavelength, being practically negligible at wavelengths longer than 5 cm and quite appreciable at 3 mm. Attenuation degrades measurement accuracy if not properly accounted for and raises the range-dependent minimum detectable signal level. An example of obvious attenuation and consequent distortion of the observed reflectivity field is shown in Figure 9.16.

Another illustration of attenuation is shown in Fig. 9.22 showing returned power from a fixed target over a 1 km path during a period of six months. Small drops in power due to humidity were interspersed with short intervals of large power losses due to fog events.

Since quantitative evaluation of attenuation would require knowledge of what lies along the path of the outgoing and returning RADAR beam, it is rarely possible to fully account for attenuation. There are some, but relatively few, measurements

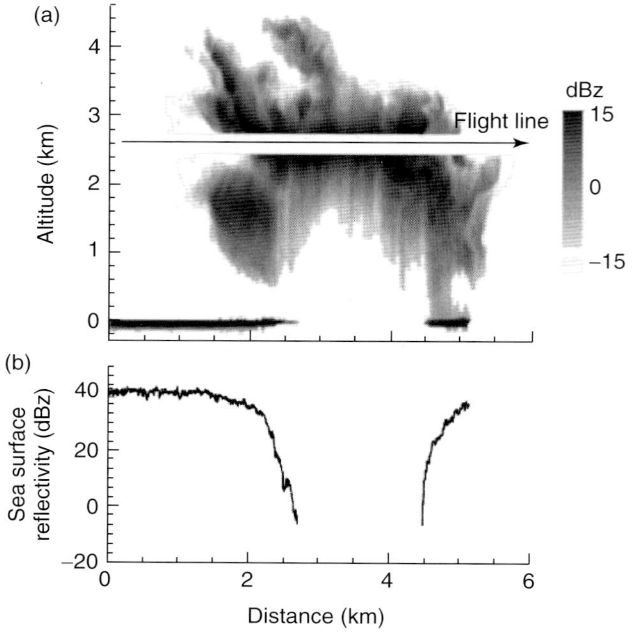

Figure 9.16 Reflectivity field (a) recorded in a precipitating marine cumulus cloud with the WCR during the RICO experiment on 17 January 2005. Strong gradients both above and below the flight line (horizontal arrow) and complete absence of data in the lower part of the cloud result from attenuation. The thick black line in the top panel at 0 km is the strong return from the ocean surface, but even this signal is lost between 2.5 and 4.5 km of flight path due to attenuation in the cloud and rain. Reflectivity values from ocean are plotted in (b). In the middle of the cloud attenuation is greater than 9 dB km^{-1}.

of attenuation rates. The following paragraphs summarize how attenuation can be estimated and what the empirical data reveal.

The rate of reduction (loss) in radiation intensity per unit distance of propagation path is proportional to the local intensity, the constant of proportionality, k being called specific attenuation (or attenuation coefficient, or extinction coefficient). Thus, the actual power received from a target at distance r is given by

$$P_r = P_{ro} \cdot \exp\left(-2\int_0^r k(r)\mathrm{d}r\right) \tag{9.11}$$

where P_{ro} is the power that would have been received from the same target in the absence of attenuation. The factor 2 accounts for the fact that both the outgoing and returned radiation passes through the intervening attenuating medium, and $k = k_g + k_h$ is the sum of the attenuation rate due to atmospheric gases (k_g) and due to hydrometeors in the air (k_h). The dimension of k is inverse distance (m^{-1}). Frequently the attenuation coefficient is given in (dB km^{-1}), with $K = 4343 \cdot k$. In the atmosphere, only oxygen and water vapor have appreciable absorption in the frequency range used for meteorological RADARs: absorption regions centered around 60 GHz, 118.8 GHz for oxygen and 22.5 GHz, 183.3 GHz for water vapor. To minimize absorption, meteorological RADARs use frequencies outside these regions. The concentration of oxygen is quite stable and attenuation is relatively minor: $< 0.01\,\mathrm{dB\,km^{-1}}$ for frequencies lower than 20 GHz and about $0.05\,\mathrm{dB\,km^{-1}}$ for 35 GHz and 94 GHz (Ulaby, Moore, and Fung, 1981).

Attenuation by water vapor is both more important and more difficult to deal with, in no small part because of the high spatial and temporal variability of this gas. The extinction coefficient can be expressed algebraically as a function of RADAR frequency, vapor density and temperature; a useful expression is given by Ulaby, Moore, and Fung (1981). The most severe degree of attenuation by water vapor is at W-band; at warm lower tropospheric conditions it can reach $1\,\mathrm{dB\,km^{-1}}$.

Attenuation by hydrometeors, and the relative contributions of absorption and scattering, are dependent on hydrometeor size, type and shape, and the RADAR frequency. The specific attenuation (in $\mathrm{dB\,km^{-1}}$) from hydrometeors in the atmosphere can be written as

$$K_{\mathrm{h}} = 4343 \int_{0}^{\infty} N(D) \cdot Q_{\mathrm{e}}(D)\mathrm{d}D \tag{9.12}$$

where $N(D)$ is the drop-size distribution and the extinction cross-section Q_{e} is the sum of the absorption Q_{a} and scattering Q_{s} cross-sections ($Q_{\mathrm{e}} = Q_{\mathrm{a}} + Q_{\mathrm{s}}$).

Because absorption and scattering by cloud droplets (e.g., smaller than $\sim 50\,\mu\mathrm{m}$) can be calculated using the Rayleigh approximation (cf. Section 9.3) it is convenient to separate the discussion of attenuation by clouds and by precipitation. Furthermore, since scattering is relatively weak by cloud droplets and since the absorption cross-section is proportional to D^3, attenuation by liquid clouds can be given as a function of liquid water content (LWC). The refractive index of water being temperature dependent, the extinction coefficient is given by the model, $K = a \cdot \theta^{b} \cdot LWC$, where $\theta = 300/T$ with T in K (Liebe, Manabe, and Hufford, 1989). Numerical values from this equation agree with values reported by Lhermitte (1990) and are shown in Figure 9.17 for various temperatures, along with a confirmation of these values by measurements in marine stratus at 95 GHz.

The data shown in Figure 9.17 were obtained in clouds with $LWC < 1\,\mathrm{g\,m^{-3}}$; for these values the predicted attenuation and the values derived from the observed rate of decrease of reflectivity with distance agree quite well. The same conclusion was drawn by Hogan, Gaussiat, and Illingworth (2005) who showed that the use of the Liebe formula led to good results in a technique that exploits the wavelength-dependence of the attenuation rate to retrieve LWC of clouds.

For attenuation by rain, the Rayleigh approximation becomes inaccurate for the RADAR frequencies used in airborne RADARs, the limit being about 0.2 mm drop diameter for W-band and 0.4 mm for Ka-band. Attenuation exceeds the Rayleigh values by up to fivefold for diameters of 0.9 mm and 2.5 mm for the two frequencies, respectively, the ratio then decreasing for even larger drops. These results can be readily applied to rain with assumed size distributions (DSD), with the resulting

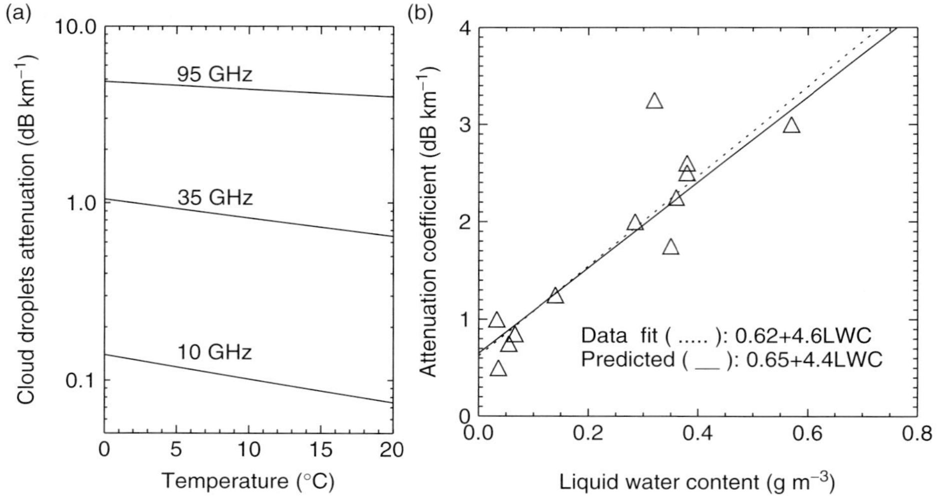

Figure 9.17 (a) One-way attenuation by cloud of LWC=1 g m^{-3} from Liebe's formula. (b) Comparison of predicted attenuation rate versus values derived from observations in marine stratus (Vali and Haimov, 2001). (Copyright 2001 IEEE. Reprinted with permission by AMS.)

accuracy depending on the degree to which the assumed size distribution holds. Kollias, Albrecht, and Marks (2003) show that calculated attenuation can differ by a factor of two or more for different assumptions of the DSD. On the positive side, measured extinction coefficients are close to the predicted values. For 10 mm h^{-1} rain the extinction rate is close to 5 dB km^{-1} at W–band, according to the paper just cited, and that value also agrees within a few dB with measurements reported in Lhermitte (1990). Attenuation varies nearly proportionally (exponent of 0.8) with rainfall rate.

Since, for hydrometeors, part of attenuation is due to scattering and even the absorption portion is dependent on many of the same parameters that determine scattering, it is reasonable to expect some relationship between the two. Hitschfeld and Bordan (1954) showed that a power–law relationship can provide useful correction for attenuation in rain. This type of formula was employed by Haddad *et al.* (1997) for rain and by Sassen and Liao (1996) for thin ice clouds. Li *et al.* (2001) showed the limitations of this approach.

Attenuation in ice clouds presents difficult theoretical problems but fortunately it is usually small in magnitude and lead to no serious practical consequences. Attenuation by falling snow is still lower than for rain but can exceed 1 dB km^{-1} at W-band and higher frequencies. T-matrix calculations for vertical incidence (no polarization dependence) of "dry" snow attenuation using an oblate spheroidal snowflake model and measured snowflake size distributions were performed by Matrosov (2007). The modeling results for a liquid-equivalent snowfall rate of less than 5 mm h^{-1} predict the attenuation coefficient to be below 0.05 dB km^{-1} and

0.5 dB km^{-1} for Ka-band and W-band, respectively. Actual W-band measurements of attenuation from falling snow were carried out by Nemarich, Wellman, and Lacombe (1988). Snow under different conditions yielded an average attenuation coefficient of 0.9 dB km^{-1} g^{-1} m^3 for W-band. Battan (1973) proposed an empirical formula for snow attenuation but this formula yield unrealistic values at the high RADAR frequencies and should not be used. Matrosov, Shupe, and Djalalova (2008) reported that wet snow produced attenuation exceeding that of rain at the same precipitation rate.

Just a brief mention will be made here of the fact that attenuation is also exploited as a useful measure of cloud and precipitation characteristics using dual-wavelength and polarization diversity setups. Application of these methods to airborne observations is hindered by the longer integration times and greater precision that is usually required.

In summary, the main facts to emphasize are that attenuation increases rapidly as the RADAR wavelength gets shorter, that attenuation depends on total LWC for clouds with small droplets, that attenuation by rain is predictable but is dependent on the raindrop size distribution, and that for ice clouds and snow estimated values for attenuation are highly tentative and lack generality; see Table A.2 (Supplementary Online Material). In all, interpretations of RADAR measurements always need to consider attenuation and very often this can only be done approximately or qualitatively. The impacts are more serious, the weaker the measured reflectivity is.

9.5.3
Doppler RADAR Measurements

The airborne Doppler RADARs listed in Table A.2 (Supplementary Online Material) range from volume-scanning systems designed to study mesoscale phenomenon, to systems intended to simulate spaceborne instruments, to systems designed to study dynamics and microphysics on a cloud scale. While the underlying physical principles remain the same regardless of application, the required accuracy and relative importance/difficulty of data analysis steps differs greatly. For example, unfolding (editing) Doppler velocities– a time consuming task for mesoscale—oriented systems such as ELDORA—is generally a minor issue for microphysics—oriented systems, which sample shorter ranges at higher elevation angles.

Regarding the removal of aircraft motion from the measured Doppler velocities (discussed in Section 9.3.3) one approach, for fixed beam systems, is to use the Doppler velocity observed for the ground or ocean surface to calculate the beam orientation with respect to the airframe. This involves performing a complete 360° turn or series of turns to generate sufficient data. Testud, Hildebrand, and Lee (1995) presented a technique using the residual Doppler velocities from the ocean surface to diagnose and correct various types of navigation errors (which increase over the course of a flight due to the Schuler oscillation of inertial navigation devices) and uncertainties in the antenna orientation. The need to apply corrections to the aircraft heading on a leg-by-leg basis (Testud, Hildebrand, and Lee, 1995; Bosart,

Lee, and Wakimoto, 2002) has been largely obviated by the adoption of GPS-based solutions, which correct the aircraft attitude as well as position and velocity.

Separating particle fallspeeds (V_p) from vertical air motion (w) remains challenging and time consuming. Some of the approaches developed to separate these components for ground-based RADAR data are difficult, though not necessarily impossible, to apply to airborne data. For example, identification of the cloud droplet population (Gossard *et al.* 1997) or the first Mie scattering minimum in the Doppler velocity spectrum to derive vertical air velocity (see Kollias *et al.* (2001); Kollias, Albrecht, and Marks (2002)) is hindered by broadening of the Doppler spectrum due to platform motion and the inherent difficulty in using long dwell times (needed for high spectral resolution) from a rapidly moving platform.

While no approach has been developed that is applicable regardless of meteorological conditions or system characteristics for separating V_p from w, a number of methods have been employed with apparent success. Since V_p ranges from $< 1\,m\,s^{-1}$ to $> 10\,m\,s^{-1}$ depending on hydrometeor phase, type, size, and density, an assessment of the predominant hydrometeor type must be made (implicitly if not explicitly). Commonly employed approaches involve determination of an empirical reflectivity–terminal velocity ($Z - V_p$) relationship for specific conditions (Marks and Houze, 1987; Black, Burpee, and Marks, 1996; Orr and Kropfli, 1999; Matrosov, Korolev, and Heymsfield, 2002; Protat, Lemaitre, and Bouniol, 2003; Heymsfield and Westbrook, 2010).

Orr and Kropfli (1999); Matrosov, Korolev, and Heymsfield (2002) and Protat, Lemaitre, and Bouniol (2003) fit Z vs. V_p for individual altitudes under the assumption that, over sufficiently large time intervals the mean of w is effectively 0 and that V_p and w are not systematically correlated. Tian *et al.* (2007) exploited differences in Doppler velocity and reflectivity at X- and W-bands resulting from different weighting of particles in the Rayleigh regime at X-band and the Mie regime at W–band to derive particle fallspeed in light stratiform rain.

Thus, far, most attempts to separate the air and particle velocity contributions for airborne RADAR data have relied on empirical reflectivity–terminal velocity ($Z - V_p$) relationships. Terminal velocities (V_p) vary widely depending on particle type, size, and density. Thus, hydrometeor type must be assessed by altitude and location so suitable $Z - V_p$ relationships for different regions can be applied (Marks and Houze, 1987; Black, Burpee, and Marks, 1996; Heymsfield and Westbrook, 2010). For RADARs sampling close to the aircraft platform, comparisons between in situ vertical wind measurements and adjacent Doppler velocities have also been used to separate fallspeeds from vertical air velocities.

For determination of horizontal winds, Mapes and Houze (1993) developed a variation of the VAD analysis method that can be applied to data from single- and dual-beam scanning airborne RADARs during 360° turns or "purls". This was subsequently extended by Protat, Lemaitre, and Scialom (1997) and Scialom, Protat, and Lemaitre (2003). The method is well-suited to retrieval of mesoscale divergence/convergence.

Airborne dual-Doppler applications exploit the motion of the aircraft to sample the same volume from two or more angles, either by using multiple antennas

(Lhermitte, 1971), (Gamache, Marks, and Roux, 1995) or by using a flight pattern designed to overlap coverage from successive legs (Jorgensen, Hildebrand, and Frush, 1983). As a result of the time lag between samples, which for multi-beam applications increases linearly with range from the aircraft (at $\sim$5 s km^{-1}) and is determined by the length of the legs and the time for the intervening turn for overlapping coverage from successive legs (a few minutes to 10 s of minutes), it is necessary to use an advection scheme to align the observations. Conceptually, an advection scheme implies a transformation from a ground-relative frame to an air- or storm-relative one. It is worth noting that the conventional definition of the drift angle infers a ground-relative frame, however the drift angle with respect to an air- or storm-relative angle is generally more relevant. This angle is typically (although not necessarily) smaller than the drift angle, as aircraft fly with a small sideslip angle. Suitable advection velocities can be determined by hand or using an optimal approach as described by Gal-Chen (1982). For a vertical-plane fixed beam system, only advection parallel to the flight track can be corrected; advection normal to the flight track results in displacements between the beams, thereby increasing uncertainties in the velocity components.

Much of the initial development of airborne multi-Doppler techniques used the single-beam, helical-scanning RADAR installed on a NOAA P3 (Jorgensen, Hildebrand, and Frush, 1983). By flying a stair step pattern, a checkerboard of squares of overlapping coverage could be obtained (Jorgensen, Hildebrand, and Frush, 1983). ELDORA, which uses a dual-beam helical-scanning antenna scanning fore and aft of the aircraft, obtains overlapping coverage during straight flight eliminating the need for complex flight patterns (Hildebrand *et al.* 1996).

Lhermitte (1971) envisioned the use of fixed, dual-beam Doppler RADAR to retrieve velocity components in a vertical plane – anticipating the configurations currently in use by systems such as the EDOP (Heymsfield *et al.* 1996), NAWX, and WCR (Leon, Vali, and Lothon, 2006; Damiani and Haimov, 2006). For fixed–beam systems, the individual beams are displaced slightly from each other and from the idealized observation plane by variations in aircraft attitude even during nominally straight and level flight. These displacements increase errors in the retrieved velocity components and motivate the use of a 3D solution in which an a priori velocity estimate is blended with the resolved velocity components so that consistent velocity components, presumably those defining the idealized observation plane, can be extracted (Leon, Vali, and Lothon, 2006; Damiani and Haimov, 2006). Since the resolved velocity components generally lie close to the idealized observation plane, this step does not tend to introduce significant errors.

Conical-scanning airborne RADARs (IWRAP and URAD; see Table A.2 (Supplementary Online Material)), which differ from helical-scanning systems (e.g., ELDORA) in that the antenna scans about an axis that is approximately perpendicular to the aircraft velocity, combine aspects of both multi-Doppler and VAD approaches. Successive scans resample the same volume from different angles (in the along-track direction) as needed for multi-Doppler analysis. However, as the angular displacement across-track increases, the difference in angle between overlapping samples decreases, effectively limiting the applicability of multi-Doppler analysis to a narrow band centered on the flight track. Since the scans occur independent of aircraft turns, this scanning-strategy is more akin to a ground-based VAD. IWRAP (Castells, Carswell, and Chang, 2001; Fernandez *et al.* 2005) the first airborne system to implement the conical-scan strategy, samples 4 elevation angles simultaneously, further increasing the similarity to ground-based extended-VAD analysis (Srivastava, Matejka, and Lorello, 1986). While it is conceptually possible to apply multi-Doppler analysis (as discussed above), this approach has yet to be demonstrated using actual data.

9.5.4
Polarization Measurements

Observations with MW RADARs can go beyond measurements of backscatter intensity by successfully exploiting the fact that scattering and to some degree absorption by cloud and precipitation particles is also sensitive to the polarization state of the incoming radiation (cf. Section 9.3). This leads to the possibility, with RADARs so equipped, to obtain additional information about the shapes of the scatterers. Transmitted RADAR power can have circular, elliptical or linear polarizations. Linear polarization is employed more frequently with pulses alternated between two orthogonal planes (designated as horizontal and vertical even if the beam direction is in some arbitrary direction). Received power is detected with those same two polarization planes. From these measurements it is possible to determine the scattering matrix and differential propagation and to derive various relatively simple quantities that describe specific properties of the particles in the sample volume: co–polar reflectivity (Z_{HH}, Z_{VV}) (Convention is for backscattered reflectivity to be designated with a first subscript indicating the received polarization and the second one indicating the transmitted polarization), cross-polar reflectivity (Z_{HV} and Z_{VH}), differential reflectivity factor (Z_{DR}), linear depolarization ratio (LDR), difference reflectivity (Z_{DP}), correlation coefficient (ρ_{HV}) and specific differential phase (K_{DP}). Detailed descriptions of data obtained from dual–polarized airborne RADARs can be found in Pazmany *et al.* (1994) and Galloway *et al.* (1997).

Of the parameters listed above, Z_{DR} and LDR are used most widely with airborne RADAR data because their interpretations is on firmer ground than the other measurements. Aydin and Walsh (1999) suggested the use of the difference reflectivity (Z_{DP}) at mm–wave to diagnose aggregates and pristine ice crystals. Galloway *et al.* (1997) reported the first airborne measurements of the correlation coefficient (ρ_{HV}) and specific differential phase (K_{DP}) at W–band from melting crystals, columns and used ρ_{HV} and LDR to infer crystal alignment due to electrification in convective clouds. Complicating factors in the interpretation of polarimetric parameters are differential attenuation and resonance effects, e.g., Evans and Vivekanandan (1990); Tang and Aydin (1995); Kropfli and Kelly (1996); Lhermitte (1990), and multiple scattering in heavy precipitation and rain (Battaglia *et al.* 2010). Here we focus on ice clouds only.

Differential reflectivity is defined as (Seliga and Bringi, 1976): $Z_{\text{DR}} = 10 \log \frac{Z_{\text{HH}}}{Z_{\text{VV}}}$ (Figure 9.18). This quantity generally depends on particle shape, fall orientation,

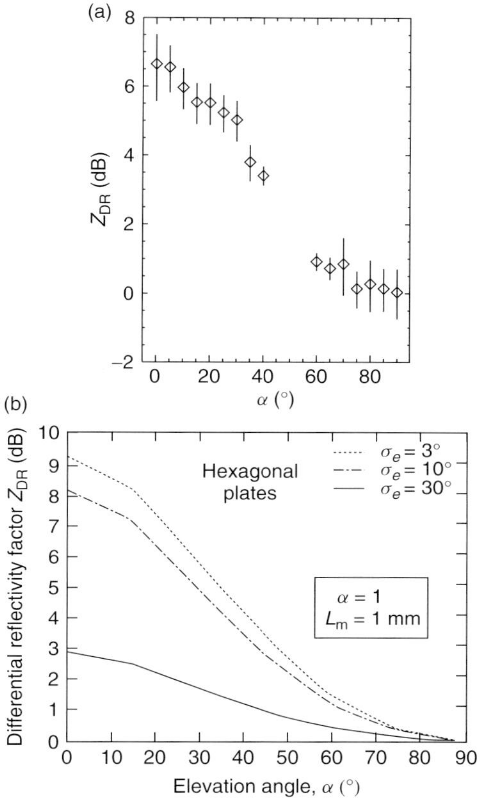

Figure 9.18 ZDR versus elevation angle (a) for hexagonal plates measured by WCR. (Source: From Wolde and Vali (2001a).) (b) Modeled for various canting angles. (Source: From Battaglia, Sturniolo, and Prodi (2001). Copyright 2001 Elsevier. Reproduced with permission.)

density (dialectic properties) and the RADAR beam orientation with respect to the hydrometeor orientation (Bader, Clough, and Cox, 1987). Both modeling and measurements show that Z_{DR} can vary from 0–9 dB for low elevation angles at mm wavelengths. Hexagonal plates and stellar type crystals have the maximum Z_{DR} values (6–9 dB) due to their high density and high aspect ratio. Similar values are also reported from ground-based RADARs (Hogan, Donovan, and Tinel, 2003a). For all ice crystals types, Z_{DR} decreases with RADAR elevation angle (α) and becomes near zero at vertical incidence. Figure 9.19 shows the results of W-band measurements obtained from cloud volumes containing hexagonal plates and stellar crystals and modeling result for the same particle types (Battaglia, Sturniolo, and Prodi, 2001).

Linear depolarization ratio (LDR) is defined as the ratio of the cross-polarized return to the co-polarized return: $LDR_{HV} = 10 \log Z_{HV}/Z_{VV}$

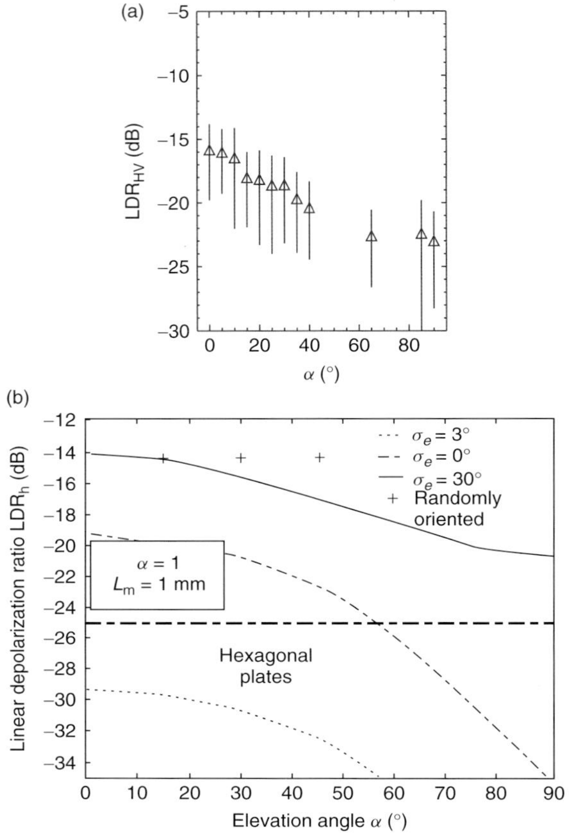

Figure 9.19 LDR versus elevation angle (a) for hexagonal plates measured by WCR. (Source: From Wolde and Vali (2001a).). (b) LDR modeled for various canting angles. (Source: From Battaglia, Sturniolo, and Prodi (2001). Copyright 2001 Elsevier. Reproduced with permission.)

or $LDR_{VH} = 10 \log Z_{VH}/Z_{HH}$. The two cross-polarized reflectivity values (Z_{HV} and Z_{VH}) can be assumed to be equal (Doviak and Zrnic, 1993; Galloway *et al.* 1997). Modeling studies, e.g., Aydin and Tang (1997); Battaglia, Sturniolo, and Prodi (2001), and observations in winter clouds Wolde and Vali (2001a) have shown that LDR can be used to discriminate between planar and columnar crystal types. The LDR decreases with beam angle for planar crystals while the opposite is the case for columnar crystals, e.g., Matrosov (1991); Aydin and Tang (1997). Figure 9.19 shows LDR observations in clouds composed of hexagonal plates and corresponding model results. Because cross-polarized signals are weak (except from melting crystals and graupel), the detectability of LDR is limited to a couple of kilometers from the aircraft even for the most sensitive airborne RADAR systems in operation today (~30 dB isolation).

Airborne RADARs can provide high spatial resolution for polarimetric measurements and the availability of near-coincident in situ measurements afford good validation of those measurements. However, polarimetric measurements put additional demands on system stability and monitoring. In addition to the polarimetric calibration issues indicated in Section 9.5.5, other factors can compromise the accuracies of the polarimetric measurements. These include ice/water deposition on the radome surface, moisture in waveguides and system instability during flights (receiver, latching circuits, transmitter, antenna beam pattern and beam pointing etc.). Cross-polarization data can suffer from leakage from the orthogonal (co-polarization) return. It is critical to include robust data quality control procedures for proper use and interpretation of airborne polarimetric measurements, including physics-based consistency checks, e.g., Ryzhkov, Giangrande, and Schurand (2005), and analysis of multiple frequency data (when more than one RADAR data available –e.g., NAWX, ER–2 RADAR).

The following discussion focuses on airborne measurements of Z, Z_{DR} and LDR in studies of cloud structure and processes. The examples cited were obtained from field projects conducted in Canada using the WCR and NAWX installed on the National Research Council of Canada Convair–580 aircraft.

Combined with in situ data, polarimetric measurements can aid the analysis of cloud organization, homogeneity and the dominant microphysical processes. This is illustrated in Figure 9.20 showing measurements in a mixed phase cloud sampled in Ontario, Canada. The in situ data reveal that clouds were of mixed phase consisting of planar crystals (hexagonal plates, stellar and dendrites) with LWC of up to $0.4\,\mathrm{g\,m^{-3}}$. The Z, Z_{DR} and in situ data suggest that wave–like activity (on the order of few hundreds of meters) resulted in systematic changes in the cloud microphysics. High Z_{DR} and weak Z were obtained in cloud volumes that consisted mainly of pristine ice crystals with no LWC. In contrast, the LWC peaks correspond to regions consisting of near zero Z_{DR} and higher Z values suggesting the presence of larger drizzle size drops. Similar observations of high Z_{DR} values in the proximity of supercooled cloud are reported by Hogan *et al.* (2003b).

Automatic classifications of RADAR polarimetric data using fuzzy-logic or similar algorithms are now routinely used in research and semi-operational cm–wavelength ground RADARs, e.g., Vivekanandan *et al.* (1999); Liu and Chandrasekar (2000). The polarimetric thresholds used for defining the membership functions for the various hydrometeor types are mostly based on computational studies. A comprehensive review of polarimetric thresholds used in image classification from cm–wavelength RADAR is given by Straka, Zrnic, and Ryzhkov (2000). Bringi and Chandrasekar (2001) also provided details on the fuzzy–logic and other techniques currently used for identifications of particle types from polarimetric measurements. As noted by Aydin and Singh (2004), airborne polarimetric RADAR measurements can provide finer classifications of ice crystals than has been reported from ground-based systems. They define polarimetric thresholds for five categories of ice crystals based on computational outputs and used fuzzy-logic based classifiers to convert polarimetric RADAR measurements into particle types with membership functions using Z, Z_{DR}, LDR and temperature.

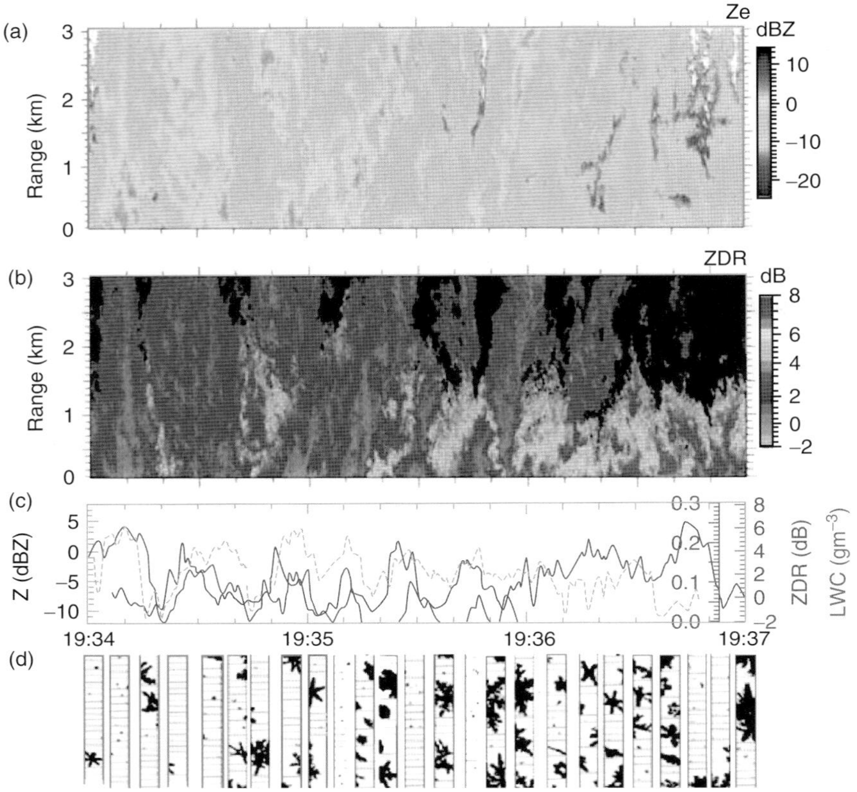

Figure 9.20 (a) and (b) Horizontal cross sections of Z_e and Z_{DR} obtained from a side-looking beam. Traces of LWC, Z_e, Temperature and Z_{DR} at 105 m from the aircraft are shown in (c). (d) Sample images from a PMS 2D–C probe. The width of each strip represents 800 μm.

Figure 9.21 shows how a fuzzy-logic based classification into ten categories (the color bar) is used to interpret flight data, and in this case be given added support by the in situ data. Such techniques can be further refined as more data become available from diverse cloud conditions, perhaps with the inclusion of more polarimetric parameters.

Another application of polarimetric data was to infer the frequency of occurrences of various particle types (pristine crystals, densely rimed crystals aggregates and graupel) in winter clouds (Wolde and Vali, 2001b).

In summary, airborne polarimetric observations can be a powerful tool in studies of cloud processes and structure. Retrieval algorithms for cloud-particle type can be best validated and improved with the near-coincident RADAR and in situ measurements from airborne systems and the findings can be transferred to ground-based and space based system.

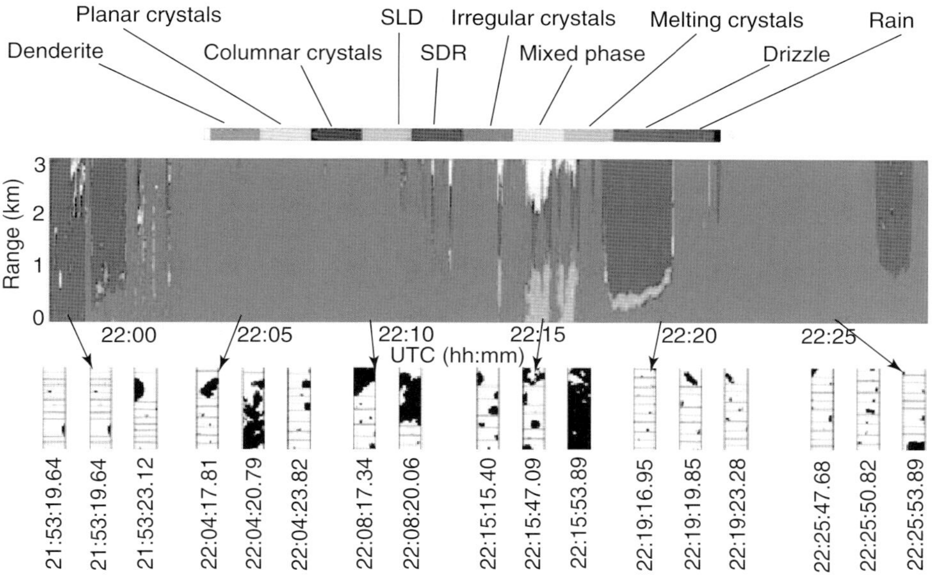

Figure 9.21 Fuzzy–logic based classification of particle from a flight in winter Ns over Ontario, Canada. Examples of particle images recorded with a PMS 2D–C probe at flight level are also shown for reference.

9.5.5
Calibration: Precision and Accuracy

There are many challenges and difficulties in calibrating RADARs and in determining the accuracies of various measured parameters. These broad challenges are compounded for aircraft installations by the very character of mobility and by the harsh conditions to which these RADARs are exposed. The following material constitutes an overview of the issues involved.

There are two basic approaches to calibration and in practice a combination of the two are needed to achieve reliable results. The two methods are the use of reference targets, and the careful measurement of relevant hardware component characteristics followed by monitoring of their performance during use. The choice of reference target (standard) is not obvious; corner reflectors and the ocean surface are two possibilities, as will be discussed later on. The bench calibration of components involves too many details to include here. A more restrictive but useful additional possibility is to explore well-known properties of hydrometeors to cross-calibrate some parameters like polarimetric channels.

In addition to calibration, which seeks to yield approximations to absolute accuracy in the numerical values of derived quantities, short-term and long-term variabilities influence the overall results. These latter factors determine what is usually called the precision of the measurements. Precision is in some ways easier to evaluate than accuracy, in some ways more difficult and less readily quantified.

In view of all this, it has to be recognized that the values quoted in Table A.2 (Supplementary Online Material) include a fair degree of subjectivity.

Modern weather RADARs measure reflectivity (Z) and Doppler velocity and new instruments are rarely developed without polarimetric capability, measuring some combination of Differential Reflectivity (Z_{DR}), Linear Depolarization Ratio (LDR), $Svv - Shh$ correlation coefficient and differential phase shift.

LDR and Z_{DR} often have to be measured to a very high degree of accuracy to detect complex particle shapes or particle asymmetry, usually requiring very large number of independent samples and a well calibrated RADAR transceiver. Gain imbalance and finite isolation between the horizontal and vertical RADAR transmitter and receiver channels, including the antenna, is the primary source of system related errors. Correcting for gain imbalance is relatively easy, because the transmitter signal source is usually common, the loss of passive components are usually stable and receiver gain imbalance variation can be minimized by temperature stabilizing the active receiver components and can be estimated from the receiver thermal noise power. Increasingly weather RADARs are being built with radiometric receiver calibration capability, equipped with hot (noise source) and warm (matched load) to monitor receiver noise figure and gain variations.

LDR measurements are often limited by insufficient sensitivity to accurately detect cross-pol signals and co-pol signal leakage due to finite polarization isolation of the antenna and the RADAR transceiver channels. The measurement of highly reflective and high LDR precipitation like hail and the melting band is relatively easy; only the gain imbalance between the co- and cross-pol transceiver channels need to be accounted for. However, the accurate measurement of low LDR, when the co-pol to cross-pol signal power ratio is greater than the antenna and RADAR polarization isolation is difficult, as it requires the characterization of the transmitter and receiver complex distortion matrix, including the antenna pattern Sarabandi, Oh, and Ulaby (1992). Moisseev *et al.* (2002) provides a good survey of how the known scattering properties of clouds and small raindrops can be used to improve polarimetric calibrations.

The measurement of mean Doppler velocity using pulse pairs is the most straightforward among all the RADAR parameters. It is a relative measurement of phase shift from pulse to pulse and so the measurement of mean Doppler velocity is unbiased with a well designed RADAR system with negligible spurious signals. However, the measured mean velocity is biased by the velocity of the largest particles and weighted by the antenna pattern, so antenna sidelobe return from ground clutter or stronger precipitation regions outside of the main antenna beam can bias the measured data. Airborne Doppler RADAR measurements are also complicated by any uncertainty in the antenna pointing angle and aircraft motion. Time synchronized aircraft Inertial Navigation System INS data can be used to correct for aircraft motion as long as the antenna pointing direction in the INS coordinate system is known. The antenna pointing direction can be calculated from ground data collected during circle flight patterns (except for zenith pointing antennas).

Many elements of calibrating the reflectivity measurements with airborne weather RADARs is similar to that of ground RADARs: (i) the transmit power is monitored using some stable power metering circuit, (ii) the antenna pattern is measured by the antenna manufacturer and the measured antenna pattern is incorporated in the data calibration algorithm, (iii) the receiver gain is measured with test equipment and any gain variation, primarily due to changes in amplifier temperature, is monitored by tracking the receiver noise power. In addition, calibration of the entire RADAR system (transmitter, antenna and receiver) is performed using external calibration targets. The most common target is a pole mounted trihedral corner reflector, which has a well characterized, bright and broad reflectivity pattern. The following discussion focuses on this method, and the use of the sea surface as a reference is discussed later.

9.5.5.1 **Calibration using Retroreflectors**

The narrow antenna beamwidth and short far-field distance of millimeter wave RADARs allow the use of a small corner reflector of just a few centimeters in size, mounted on a practical length ($\sim$6 m) non-reflective pole, placed at a clear line of sight distance of about 100 m. The best method of ensuring a minimum pole return is to tilt the pole back a few degrees (a few times the antenna beamwidth) and mount the reflector pointed downward on the pole so when the pole is tilted back, the reflector faces directly at the RADAR antenna. The pole has to be tall enough to ensure a high corner reflector signal to ground clutter signal ratio. The antenna footprint of a 0.7° antenna at a distance of 100 m is approximately 1 m, so when the beam is pointed at a 6 m pole mounted corner reflector, the ground is over five beamwidths away, resulting in better than 30 dB reflector signal to ground signal power ratio. The resulting uncertainty in the measured reflector signal power due to constructive or destructive ground signal interference is about $\pm$0.28 dB. Additional random error sources, however, like antenna pointing error, transmitter pulse shape, humidity in the air, imperfections in the corner reflector geometry make it difficult to achieve an absolute reflectivity calibration better than $\pm$3 dB with millimeter wavelength weather RADARs. The significant potential affect of atmospheric attenuation on reflector calibration is well illustrated by the data set presented in Figure 9.22. The received power from a corner reflector located 1 km from a 3 mm wavelength RADAR was sampled once every five minutes over a period of 6 months. Fog extinction (scattering and absorption) sometimes exceeded 30 dB km^{-1} and summer time humidity accounted for over 1 dB km^{-1} signal attenuation. The thin horizontal line indicates the expected corner reflector signal power level if the atmosphere was clear and dry.

In order to examine the contributions of various sources of error in the computation of reflectivity from the RADAR equation, the calculated reflectivity Z is expressed here in terms of the parameters that are subject to possible errors and by using ratios of measured values to that measured from a corner reflector:

$$Z = \frac{\Gamma \cdot \sigma_c}{V_W \cdot |K|^2} \cdot \frac{R^4}{R_c^4} \cdot \frac{P_r \cdot l_a}{P_{rc} \cdot l_{ac}} \tag{9.13}$$

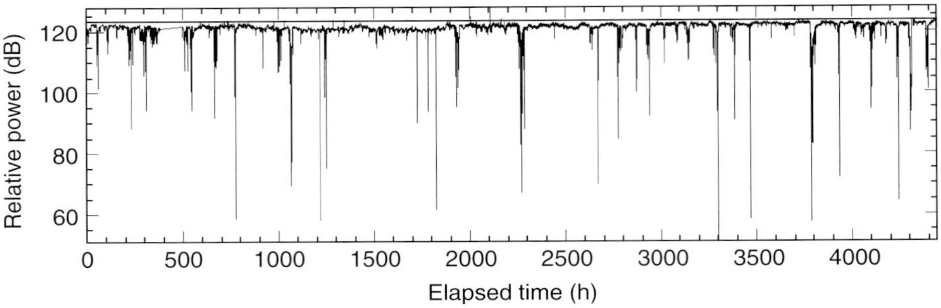

Figure 9.22 Corner reflector return signal power collected approximately every 5 minutes through a 6 month period with a ground–based 3 mm wavelength RADAR at Logan Airport. The 300 m^2 corner reflector was located 1 km from the RADAR across the Boston harbor channel. The thin horizontal line indicates the corner reflector signal power level when the atmosphere was clear and dry.

where all of the constants are lumped under the symbol Γ, V_W is the illuminated volume, σ_c is the backscatter cross-section of the corner reflector, R and R_c the range to the target and to the corner reflector, P_r and P_{rc} the received power from the target and from the corner reflector and l_a and l_{ac} the loss factors due to attenuation to the target and to the corner reflector.

Re-expressing Z in $\mathrm{d}BZ = 10 \log Z$ and differentiating with respect to the parameters yields the sensitivity of dBZ to changes in those parameters:

$$\frac{\mathrm{d}(\mathrm{d}BZ)}{\mathrm{d}(10\log\sigma_c)} = 1 \qquad \frac{\mathrm{d}(\mathrm{d}BZ)}{\mathrm{d}(10\log R)} = 4 \qquad \frac{\mathrm{d}(\mathrm{d}BZ)}{\mathrm{d}(10\log P_r)} = 1 \tag{9.14}$$

$$\frac{\mathrm{d}(\mathrm{d}BZ)}{\mathrm{d}(10\log l_a)} = 1 \qquad \frac{\mathrm{d}(\mathrm{d}BZ)}{\mathrm{d}(10\log V_W)} = -1 \qquad \frac{\mathrm{d}(\mathrm{d}BZ)}{\mathrm{d}(10\log|K|)} = -2 \tag{9.15}$$

$$\frac{\mathrm{d}(\mathrm{d}BZ)}{\mathrm{d}(10\log R_c)} = -4 \qquad \frac{\mathrm{d}(\mathrm{d}BZ)}{\mathrm{d}(10\log P_c)} = -1 \qquad \frac{\mathrm{d}(\mathrm{d}BZ)}{\mathrm{d}(10\log l_{ac})} = -1 \tag{9.16}$$

Thus, $a + x \cdot dB$ error in the corner reflector RADAR cross section will result in $a + xdB$ error in dBZ, while $a + ydB$ error in estimating the index of refraction factor, $|K|$, will map into a $-2ydB$ error in dBZ. Note that not all of these parameters vary independently. For example, the illuminated volume increases as the square of the target range and inversely with antenna gain.

The error sources in measured reflectivity are summarized below:

(i) Errors in the corner reflector RADAR cross section due to manufacturing, signal-to-clutter ratio, and switched-in receiver attenuation during calibration.
(ii) Errors in estimating the volume integral, V_W.
(iii) Errors associated with the scattering volume range and corner reflector range factors.

(iv) Uncertainty in the index of refraction of water due to temperature variation, causing errors in the assigned value of $|K|$.
(v) Errors associated with measuring the scattering volume and corner reflector power.
(vi) Propagation losses in the transceiver waveguides, between the RADAR and calibration target and between the RADAR and cloud, including loss due to water deposition on the antenna or radome.
(vii) Between calibrations, variations in receiver gain and transmit power will cause additional errors in the reported reflectivity.

The various error sources are considered separately below.

9.5.5.1.1 **Corner Reflector RADAR Cross Section** The RADAR cross-section of trihedral corner reflectors that are large compared to the RADAR wavelength is accurately modeled by the following formula (Ulaby, Moore, and Fung (1981)):

$$\sigma_c = \frac{\pi \cdot l^4}{3\,\lambda^2} \tag{9.17}$$

where l is the edge length of trihedral reflector aperture. For example, the corner reflector used for the calibration of the University Wyoming Cloud RADAR (WCR) is usually located about 100 m from the RADAR has a corner length of 3.8 cm, yielding a RADAR cross-section of 0 dB m^2.

9.5.5.1.2 **Reflector Plate Angular Errors** Manufacturing errors will result in angular errors in the alignment of the metallic plates forming the trihedral. The formula relating angular errors to errors in RADAR cross section is:

$$\frac{\sigma}{\sigma_c} = 10 \log \left(\frac{\sin q}{q} \right)^4 \text{d}B \tag{9.18}$$

with $q = 2.54\,\delta_{max} \cdot D/\lambda$ where in radians is the maximum angular deviation from 90 degrees among the plates, and $D = l/\sqrt{2}$ is the length of the inside corner of the reflector. Figure 9.23 shows the RADAR cross-section error as a function of the maximum angular error of the plates for the WACR corner reflector currently in use at SGP. This shows that the RADAR cross-section error is small for angular errors less than 0.1 degrees. Using modern manufacturing techniques, it is very likely that the RCS error due to plate misalignment is less than 0.1 dB, and therefore represents a negligible source of measurement error.

9.5.5.1.3 **Reflector Alignment Error** Trihedral corner reflectors maintain a nearly constant RADAR cross-section over a range of angles. Data presented by Robertson (1947), shows that the RADAR cross-section of an electrically large reflector varied by only 1 dB over a range of ±10 degrees. Corner reflector can easily be aligned to within ±2 degrees, so error due to reflector alignment should be negligible.

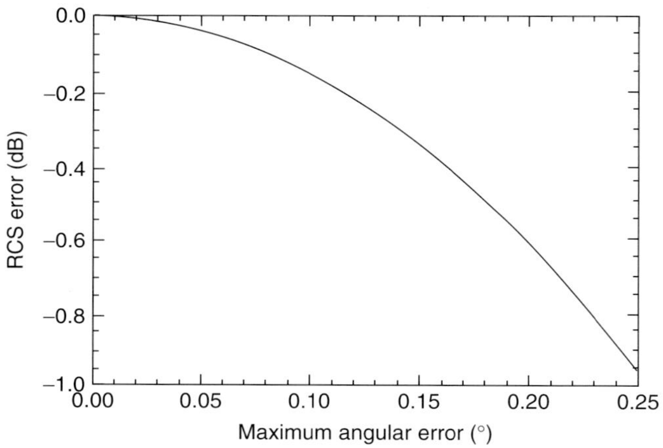

Figure 9.23 Corner reflector RADAR cross-section error as a function of maximum plate alignment error for a $D = 6.4$ inch trihedral reflector at 95.04 GHz.

9.5.5.1.4 **Reflector Signal-to-Clutter Ratio** Reflections from the mast supporting the corner reflector can bias the corner reflector RADAR cross section to a larger or smaller value, depending on the relative phase of the reflections. The maximum enhancement in RCS occurs when the reflections are in-phase and the greatest reduction in RCS when the reflections are out of phase. The maximum error, computed as a function of signal-to-clutter ratio is plotted in Figure 9.24. For a 30 dB corner reflector to ground or pole signal ratio, the maximum error is ±0.28 dB.

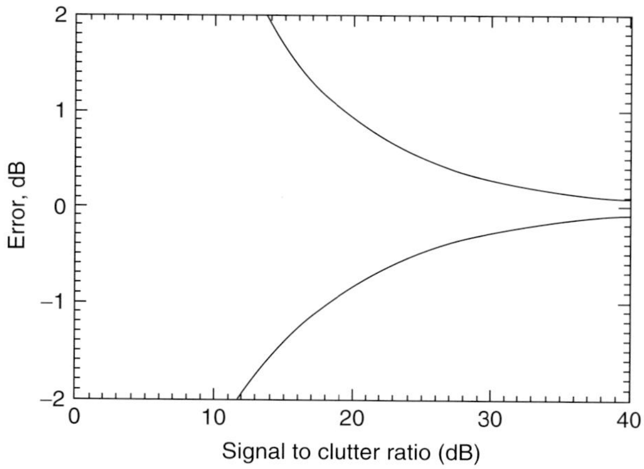

Figure 9.24 RCS error as a function of SCR. The lines show the maximum possible positive and negative error.

9.5.5.1.5 **Volume Integral Estimation** The approximation of the pulse volume assumes a Gaussian beam pattern in the cross-range direction. Gaussian beam approximations are typically very good for high gain antennas. To confirm this, the antenna pattern of a W-band RADAR system was measured by viewing the corner reflector and scanning in azimuth and elevation. These scans showed excellent agreement to a Gaussian approximation, down to the −20 dB level, as seen in Figure 9.25. The shape of the pulse at the output of the digital receiver is plotted in Figure 9.26. This shape is very close to the expected response of a square pulse passed through the digital filter which is close to a matched filter for a 300 ns pulse. The pulse volume was evaluated numerically at the corner reflector range; the analysis confirms that the pulse volume is accurately predicted by a Gaussian shape and the errors associated with this factor are negligible, even at a comparatively short range of 490 m.

9.5.5.1.6 **Temperature Dependence of the Index of Refraction of Water** The scattering cross section of small droplets is dependent on the value K, which is a function of the droplet index of refraction. The index of refraction of water is a function of frequency and temperature; for 95 GHz it varies by 0.5 dB over the temperature range 0–40 °C. Since reflectivity is dependent on $|K|^2$, uncompensated temperature variations map into a 1.0 dB reflectivity error at 95 GHz. The convention among RADAR meteorologists is to use a fixed value of $|K|$, since $|K|$ is not strongly temperature dependent at conventional weather RADAR frequencies. It may also be sufficient to assume a constant value of $|K|$ at 95 GHz, provided an assumed value is agreed upon when making comparisons between different RADAR systems. As an example, the JPL CloudSat W-band space cloud RADAR uses a value of $|K|^2 = 0.75$, or $|K| = 0.866$, which corresponds to a temperature of 8 °C.

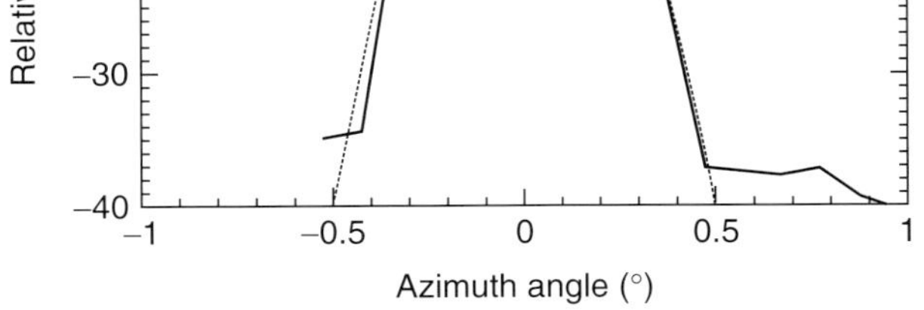

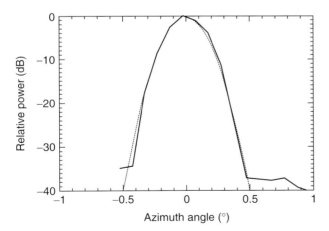

Figure 9.25 Azimuth and elevation two-way patterns of WACR SGP antenna with Gaussian fit shown as dotted line. The sidelobe in the elevation pattern at 1 ° is due to ground scattering from the main beam.

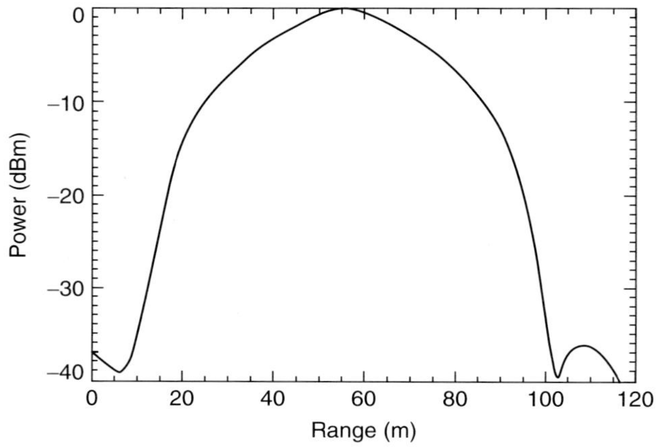

Figure 9.26 Pulse response versus range as measured at the peak of the corner reflector return.

9.5.5.1.7 **Power Measurement Errors** The following factors can potentially contribute to power measurement errors: Errors in the zeroth moment estimator that extracts signal power from the measured Doppler spectrum, errors in the transfer function used to map measured power into calibrated power.

9.5.5.2 Calibration Using Sea Surface Reflectance

As mentioned before, the ocean surface has also been widely used as a calibration target and this method has been now tested and validated for millimeter–wave frequencies, e.g., Li *et al.* (2005); Tanelli *et al.* (2008); Josset *et al.* (2008). This ocean surface calibration is being used for spaceborne RADARs such as the precipitation RADAR on board the Tropical Rainfall Measuring Mission (TRMM) (Okamoto *et al.* 2002) and the CloudSat profiling RADAR (Tanelli *et al.* 2008; Protat *et al.* 2009).

RADAR backscatter from the sea surface (generally expressed as a normalized cross section σ_0) is a function of RADAR wavelength, RADAR beam angle from the vertical, polarization, ocean surface wind speed, and surface wind direction. For incidence angles smaller than 15°, σ_0 is dominated by large-scale surface waves and at MW frequencies the quasi–specular scattering theory has been shown to work well, e.g., Li *et al.* (2005). Tanelli *et al.* (2008) recently showed that the CloudSat-derived σ_0 was in very good agreement with the Cox and Munk (1954) model modified following Li *et al.* (2005) and with the Wu (1990) model.

In using the sea surface as reference target, advantage is taken of the finding that σ_0 is least sensitive to surface wind for incidence angles near 10°, with $\sigma_0 \approx 7\,\text{dB}$, e.g., Durden *et al.* (2003). Thus, calibration involves sampling the ocean surface backscatter at that angle, though different incidence angles may also be included (by rolling the aircraft at constant altitude or during circular flight patterns at different speeds) for additional comparisons with theory. Calibration will be most accurate when: flight altitude is low and sampling is long (~1 h), water vapor concentration is low and is measured by regularly launching dropsondes (to minimize errors due to attenuation) and surface wind speed is small and virtually constant during sampling (because surface backscatter models are better for small surface wind speeds).

In summary, while it is difficult to state without qualifications, the measurement uncertainties for airborne RADARs range near 3–5 dB for reflectivity. Relative values over short data segments are more reliable than that. Doppler velocity measurements can be accurate within a few tenth of a meter per second but can also be considerably worse if turbulence causes rapid motions of the airframe and makes corrections for such motion less precise. Polarimetric measurement issues are discussed in Section 9.5.4.

9.6 Results of Airborne RADAR Observations–Some Examples

Airborne RADAR systems have played key roles in a variety of research projects and yielded a wide range of new findings. A sampling of the results in graphical form are included in Section A.5.2, given in the Supplementary Online Material provided on the publisher's web site.

Some of the features illustrated in the examples in the Supplementary Materials are listed below as a demonstration of the scope of research possible with the airborne RADARs. The specific RADARs mentioned are listed in Table 9.2; see Table A.2 and Section A.5.1, given in the Supplementary Online Material provided on the publisher's web site. Naturally, uses of each of the RADARs are not limited to the applications represented in the examples.

(i) ELDORA was used to observe a tornado, including the velocity field and the storm in which it was embedded.

(ii) WCR, with its downward pointed antenna detected the existence of a layer of snow particles lofted from the surface. The lofted crystals often merged into clouds above and changed their microphysical evolution.

(iii) EDOP/CPR observations at two different wavelengths in a stratiform cloud yielded vertical profiles of derived parameters like water content, rainfall rate, and size-distributions.

(iv) Polarization data from the NAWX RADARs reveal the variations in ice crystal types within clouds. These variations impact growth rates and indicate the degree of riming.

(v) Combined WCR RADAR and WCL LIDAR observations in wave clouds gave unprecedented depictions of the water and ice zones in those clouds and provide a basis for identifying the nucleation processes involved.

(vi) RASTA was used to provide a calibration check on CloudSat, the satelliteborne W-band RADAR.

(vii) WCR dual-beam observations were used to derive 2D Doppler velocities within maritime stratocumulus and identified cellular circulation critical to drizzle development.

9.7 Parameters Derived from Combined Use of LIDAR and RADAR

The sensitivity differences of LIDAR and RADAR to different hydrometeor size make the combined LIDAR and RADAR measurements more powerful to study cloud macro- and microphysical properties. Due to the strong cloud attenuation at LIDAR wavelength and stronger RADAR sensitivity to precipitating hydrometeor than small cloud particles, combined LIDAR and RADAR measurements are needed to provide reliable cloud boundaries (Wang and Sassen, 2001; McGill *et al.* 2004; Mace *et al.* 2009). Although ground-based LIDAR linear depolarization ratio (LDR) measurements are often enough for cloud phase determination (Sassen, 1991), combined RADAR and LIDAR measurements are often necessary for cloud phase determination for airborne and space-based measurements. But, the most important contribution of combined LIDAR and RADAR measurements is to improve the accuracy of single-instrument cloud microphysical property retrievals. In the following sub-sections, we briefly overview the general approaches to retrieve ice, water, and mixed-phase cloud microphysical properties while illustrating the benefit of combined airborne LIDAR-RADAR observations for cloud boundaries and phase determination.

9.7.1 Ice Cloud Microphysical Properties Retrieval with Airborne LIDAR and RADAR

In order to derive the microphysical and radiative properties of ice clouds and to document their vertical variability at high resolution, the RADAR-LIDAR combination and the RADAR Doppler velocity /reflectivity combination are most appropriate, e.g., Heymsfield *et al.* (2008). Broadly speaking, cloud RADARs can penetrate most ice cloud layers but will miss a portion of the thin cirrus clouds (Comstock, Ackerman, and Mace, 2002; Mace *et al.* 2006; Protat *et al.* 2006). On the other hand, LIDARs can detect these thin cirrus clouds, but the backscatter signal is often rapidly extinguished by optically thick ice clouds (optical depth $>$ 2 to 3 depending on LIDAR system and distance to clouds), e.g., Sassen and Cho (1992); Protat *et al.* (2006). There is an overlap region in which RADAR–LIDAR observations can be used to estimate some crucial ice cloud properties, from which the cloud radiative forcing can be derived. Several sophisticated retrieval methods were developed to characterize microphysical, dynamical, and radiative properties of ice clouds, e.g., Intrieri *et al.* (1993); Wang and Sassen (2002); Donovan and Lammeren (2001); Okamoto *et al.* (2003); Tinel *et al.* (2005); Delanoë and Hogan (2008); Deng, Mace, and Wang (2010). In general, existing methods using only LIDAR backscattering or extinction and RADAR reflectivity factor can be applied to airborne LIDAR and RADAR ice cloud measurements without significant modifications.

These methods all rely on the same underlying principle, but use different inversion methods (statistical relationships, iterative search, or variational framework) and make different assumptions about the ice particle habits and mass–length relationships. The common principle of all these methods is that the RADAR reflectivity and LIDAR backscatter measurements are proportional to different moments of ice crystal size distributions (6 for a Rayleigh scattering RADAR volume or about 4 for a non-Rayleigh scattering RADAR volume, and 2 for the LIDAR backscatter). These two independent measurements therefore allow for two free parameters of the ice particle size distribution to be retrieved. This framework implicitly imposes that the particle size distribution be first parameterized in those methods with a maximum of two free parameters. A lot of work has been done to derive such parameterizations for ice clouds, e.g., Austin, Heymsfield, and Stephens (2008); Delanoë *et al.* (2005); Field, Heymsfield, and Bansemer (2007); Heymsfield (2007). Finally, the ice cloud properties can then be derived by calculating other moments of retrieved particle size distribution: the ice water content (third moment), the VIS extinction (second moment), the effective diameter (ratio of the third to the second moment), the total number concentration (zeroth moment), as well as IWP and τ. For ground–based Doppler RADAR measurements, the terminal fall velocity and vertical air velocity can also be estimated (Deng and Mace, 2006; Sato *et al.* 2009), as discussed in Section 9.5.3. But large uncertainties in airborne Doppler velocity measurements limit this capability for combined airborne LIDAR–RADAR measurements.

As discussed in Protat *et al.* (2009), it is very difficult to estimate with confidence the accuracy of RADAR-LIDAR retrieval methods. There are three ways usually explored, none of them being perfect. The most popular way of evaluating cloud retrievals is to compare retrievals with airborne in situ microphysical observations by taking advantage of more direct measurements of some microphysical quantities, such as IWC, e.g., Wang and Sassen (2002); Austin, Heymsfield, and Stephens (2008); Deng, Mace, and Wang (2010). However, volume, temporal and spatial mismatch between remote sensors and in situ probes together with measurement uncertainties (e.g., shattering of ice crystals) limit the application of the approach (Wang, Heymsfield, and Li, 2005b; Heymsfield, 2007; McFarquhar *et al.* 2007). But integrated LIDAR, RADAR and in situ measurements from a single aircraft, such as the suite of instruments integrated in the Wyoming King Air, demonstrate new potentials with this approach (Wang *et al.* 2009). The second general evaluation approach is to use radiative closure by comparing calculated solar and terrestrial radiation with the retrieved ice cloud properties as inputs and measurements at the surface and/or the TOA. This approach ensures that retrieved ice cloud properties produce accurate radiative impacts; however, many other sources of errors, such as water vapor and temperature profiles, aerosol properties, and 3D cloud inhomogeneity, can cause significant discrepancy between calculations and observations. Furthermore, validation with vertically–integrated effect may mask compensating errors. The third approach uses simulated measurements based either on measured or on assumed cloud variables to evaluate algorithm performances by comparing the retrievals with inputs for the simulations (Tinel *et al.* 2005; Hogan *et al.* 2006a; Heymsfield *et al.* 2008). This approach avoids issues to collocate different measurements for evaluation; but it is hard to capture the natural variability of cloud properties in a small sample of simulated profiles.

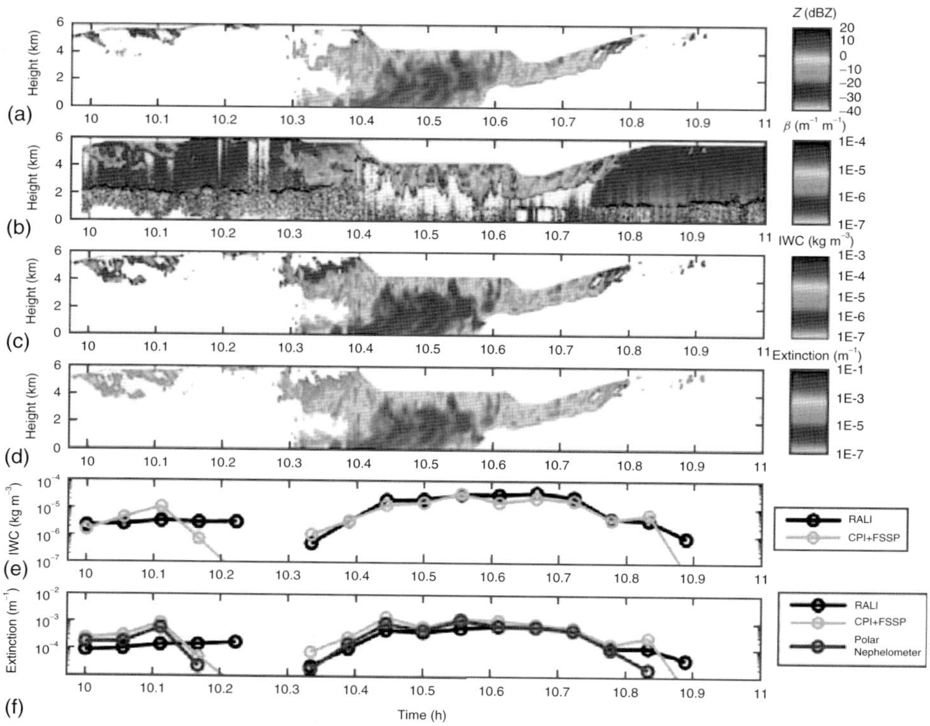

Figure 9.27 Downward looking RADAR and LIDAR observations collected in an Arctic cloud on 1 April 2008 during the POLARCAT field campaign using the RALI system on the French Falcon ATR42 aircraft. Observed RADAR reflectivity (a) and LIDAR backscatter (b) were used to retrieve the ice water content (c) and the VIS extinction (d). These parameters are compared in (e) and (f) with the values measured by *in situ* probes. (*See Color Plates for color representation of the figure.*)

It must be finally noted that the RADAR–LIDAR overlap generally corresponds to a fraction of the total ice cloud ice clouds depending on cloud type and distance from the aircraft (Illingworth *et al.* 2007). Therefore, approaches to be able to combine RADAR alone, RADAR–LIDAR, and LIDAR alone regions are needed to provide complete vertical profiles of ice cloud microphysical properties (Delanoë and Hogan, 2008), (Deng, Mace, and Wang, 2010).

The Delanoë and Hogan (2008) RADAR–LIDAR method has been adapted for airborne RADAR–LIDAR observations and an illustration of the retrieval of microphysics properties of an Arctic cloud during the POLARCAT field campaign (*http://www.polarcat.no/*) using the RALI system; see Protat *et al.* (2004), and is given in Figure 9.27. As seen in panels (e) and (f) of the figure, the retrieved and in situ measured values are in good agreement: the differences are 6 % ±12 % for ice water content and 3 %±14 % for VIS extinction. These retrieved values provide input to estimations of solar and terrestrial radiative impacts of these ice clouds.

9.7.2
Water Cloud Microphysical Properties Retrievals with Airborne Multi–Sensor Measurements

Due to the spherical shape of droplets, remotely sensing water clouds is generally simpler than ice and mixed-phase clouds (Frisch, Fairall, and Snider, 1995; Sassen *et al.* 1999). For example, in one of the earliest instances of operating a RADAR and a LIDAR on the same aircraft (Figure 9.27) the coincidence of visible cloud top (LIDAR) with RADAR echo top for a altocumulus layer was established. However, a complexity arises when combining LIDAR and RADAR measurements quantitatively for drizzling water clouds because the LIDAR signal is mainly dominated by water droplets while the RADAR signal is dominated by drizzle drops (Figure 9.28). Airborne and satellite RADAR measurements have shown the high occurrence of drizzle in stratocumulus clouds (Vali *et al.* 1995; Stevens *et al.* 2003; Leon, Wang, and Liu, 2008). Thus, combined LIDAR–RADAR measurements

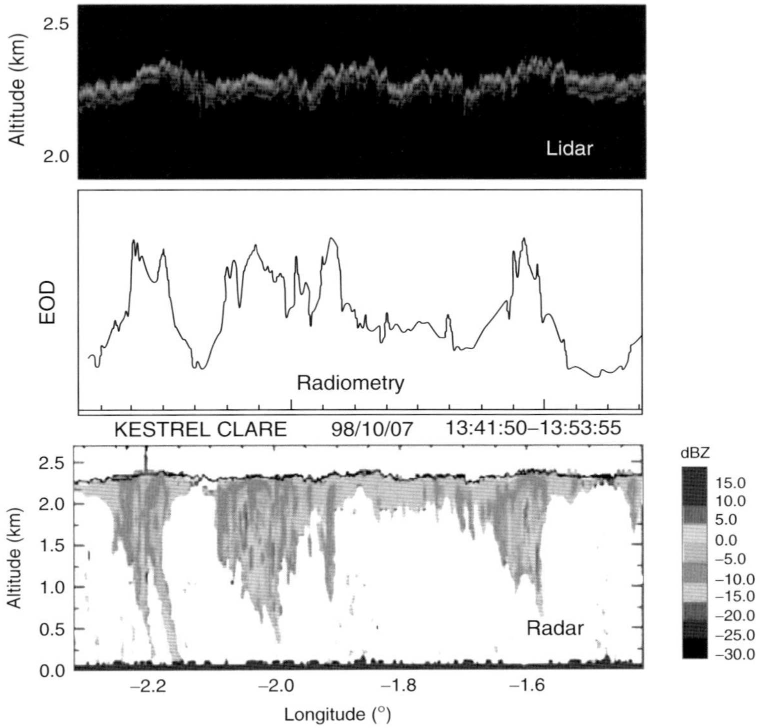

Figure 9.28 Simultaneous observations performed near Chilbolton (United Kingdom) on 7 October 1998, with combined radiometry (giving access to the Effective Optical Depth, EOD), LIDAR and RADAR (Backscatter LIDAR LEANDRE from IPSL and Wyoming Cloud RADAR WCR from University of Wyoming) installed in the French F27 flying at 3 km (ESA CLARE'1998 campaign). *(See Color Plates for color representation of the figure.)*

in marine stratocumulus don't provide enough information to constrain both cloud and drizzle properties; additional measurements (such as LWP) and/or additional assumption (such as adiabatic clouds) are needed to study drizzling water clouds.

Many observations have shown that the adiabatic cloud assumption is reasonable for most stratiform warm clouds, especially at small spatial scales and within lower part of clouds. With this assumption, LIDAR-only measurements do provide important information for water cloud number concentration and particle size, while combined LIDAR and RADAR measurements can be used to characterize drizzle properties below the cloud base (O'Connor, Hogan, and Illingworth, 2005). To fully study drizzling stratiform clouds from an aircraft, zenith pointed LIDAR and RADAR observations from below the cloud layer are needed.

Observations indicated that the size distribution of cloud droplets and drizzle drops take the lognormal form. The *kth* moment of this distribution is:

$$< D^k > = D_0^k \cdot \exp\left(\frac{k^2 \cdot \sigma_x^2}{2}\right) \tag{9.19}$$

where D_0 is the modal diameter in μm; N is the total number of particles per unit volume; and σ_x is the logarithmic width of the distribution. Thus, cloud and drizzle properties (effective diameter D_{eff} and water content q_l) and observed extinction coefficient (σ at LIDAR wavelength) and RADAR reflectivity factor (Ze) are given by:

$$D_{\mathrm{eff}} = \frac{< D^3 >}{< D^2 >} = \frac{1}{2} D_0 \cdot \exp\left(\frac{5\,\sigma_x^2}{2}\right) \tag{9.20}$$

$$q_l = \frac{\pi}{6} \rho_w N < D^3 > = \frac{\pi}{6} \cdot \rho_w \cdot N \cdot D_0^3 \exp\left(\frac{9\,\sigma_x^2}{2}\right) \tag{9.21}$$

$$\sigma = \pi\, N < D^2 > = \frac{\pi}{4} N \cdot D_0^2 \cdot \exp\left(2\,\sigma_x^2\right) \tag{9.22}$$

$$Z_e = N < D^6 > = N \cdot D_0^6 \cdot \exp\left(18\,\sigma_x^2\right) \tag{9.23}$$

where ρ_w is the density of water. By assuming σ_x as a constant, combining σ and Z_e yields N and D_0 for drizzle or non drizzling clouds. For drizzling clouds, combining σ and adiabatic q_l can determine layer mean N and D_0 (or D_{eff}) profile within the layer penetrated by the LIDAR (typically within 200 m above cloud base). For the rest of the cloud layer, cloud droplet size information can be obtained based on the following relationship of $r_{\mathrm{eff}}(z) \propto q_l(z)$ with the adiabatic assumption and cloud top determined from RADAR measurements.

Figure 9.29 shows a retrieval example of drizzling stratocumulus with this approach. WCR measurements clearly show a strong spatial variation of drizzle and most of drizzle evaporated before reach the surface. While drizzle prevents detection of cloud base from WCR measurements, WCL measurements show cloud base evidently with sharp increase in LIDAR signals (Figure 9.30). Retrieved cloud r_{eff} and N also show noticeable spatial variations. Adiabatic cloud LWPs show close

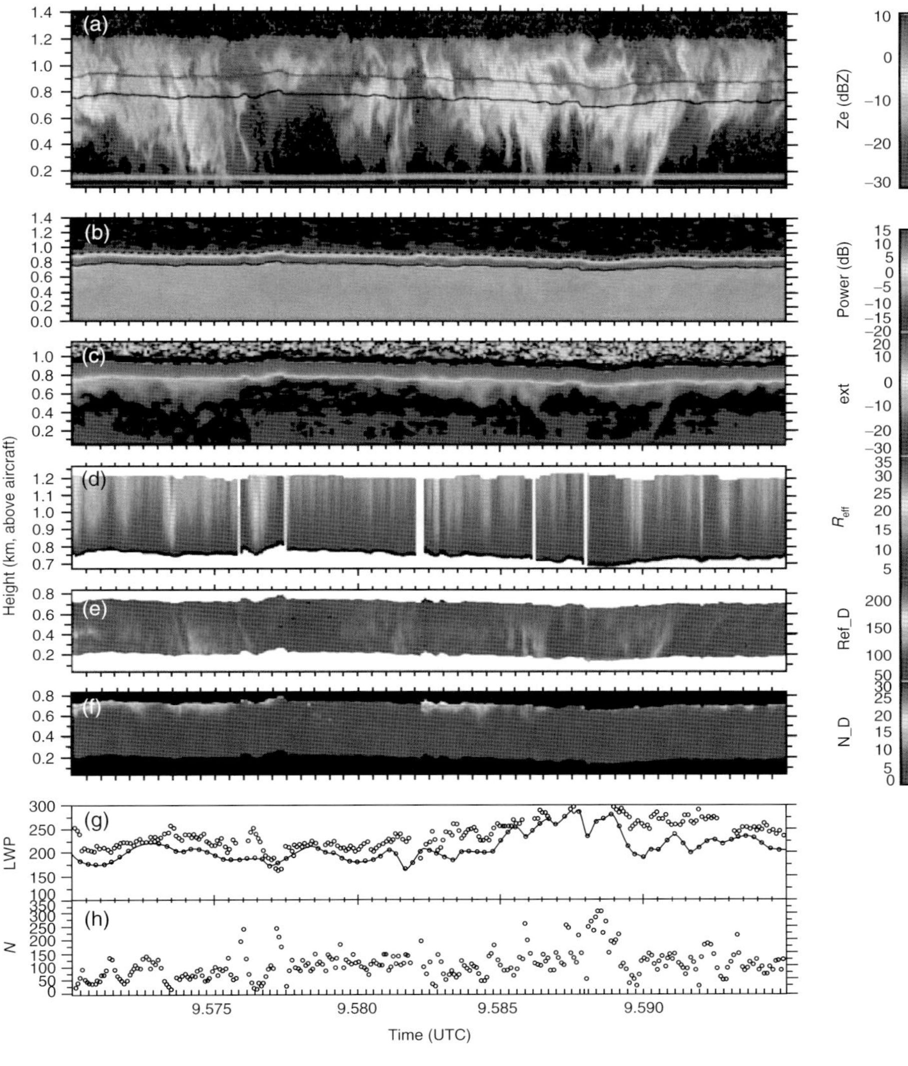

Figure 9.29 Zenith pointing WCL–WCR observations of drizzling stratocumulus clouds from NCAR/NSF C–130 on October 28, 2008 during the VOCALS field campaign: (a) WCR Z_e (b) WCL attenuated backscatter power; retrieved cloud and drizzle properties: (c) retrieved VIS extinction coefficient (d) cloud effective diameter, (e) drizzle size distribution, (f) drizzle number concentration, (g) estimated adiabatic LWP and GVR measured LWP (line with star symbols), and (e) layer mean cloud droplet number concentration. The C-130 flew at ~100 m above the surface. WCL signals provide a clear detection of cloud base (black line), but they are quickly attenuated within about 200 m. WCL cloud base and apparent top (red line) plotted in (a) shows the need of WCR measurements to provide cloud top and no information for cloud base from Z_e profiles in this case. *(See Color Plates for color representation of the figure.)*

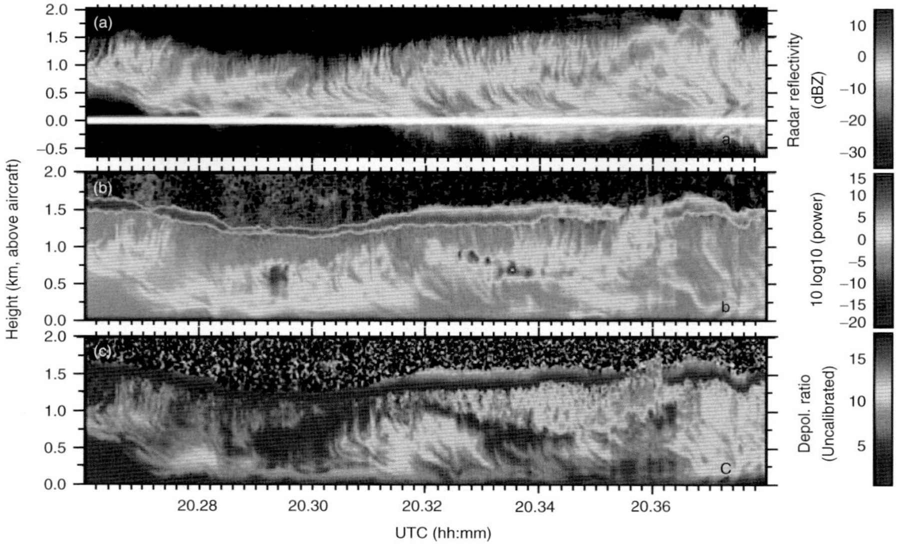

Figure 9.30 An example of WCR and WCL observations (zenith pointing) of stratiform mixed-phase clouds from the Wyoming King Air on February 20, 2008 during the Wyoming Airborne Integrated Cloud Observation (WAICO) experiment: RADAR reflectivity factor (a), LIDAR backscattering power (b), and LIDAR linear depolarization ratio (c). Water dominated mixed-phase layer is located around 1.5 km above the aircraft with cloud top temperature $\sim -25\,^{\circ}$C. Ice particles generated in the mixed-phase layer precipitate down. This is a typical example of altocumulus with ice virga observed right South of Laramie regional airport hen clouds moving from South during the WAICO. Mixed-phase stratocumulus clouds over Polar regions also have similar vertical structure. *(See Color Plates for color representation of the figure.)*

agreement with G-band MW radiometer measurements, which could be use as an important constraint to relax the adiabatic cloud assumption. Retrieved drizzle number concentration and size profiles show small drizzle evaporate quickly in short range below the base. These retrieved cloud and drizzle property together with aerosol and vertical velocity information at flight level offer a new data set to better study stratocumulus clouds.

9.7.3
Mixed-Phase Cloud Microphysical Properties Retrievals with Airborne Multi-Sensor Measurements

Due to having two-mode size distributions in mixed-phase clouds (droplets and crystals), LIDAR and RADAR measurements are necessary to characterize their macro- and microphysical properties. As the example presented in Figure 9.29, water dominated mixed-phase layer at the top (with base at $\sim$1.5 km above the aircraft) is indicated by strong LIDAR backscattering power, low LIDAR LDR and weak Z_e (Wang *et al.* 2009). Ice generation in this mixed-phase layer shows strong cell structure as indicated by the falling streaks in Z_e measurements. Due to ice supersaturation within short range below cloud base, precipitating ice crystals continue to grow. An extended precipitating ice layer is indicated with strong Ze and LIDAR LDR, and weak LIDAR backscattering power. WCL observations also show there are sub-regions (such as, ~20.29 UTC) with horizontally oriented ice crystals, which produce low LDR and strong power like water clouds, but with much weaker attenuation than water clouds.

Combined LIDAR and RADAR observations can be used to characterize these precipitating ice properties though a different ice crystal habit assumption than cirrus clouds may be necessary (Wang *et al.* 2004). For water phase microphysical properties in the mixed-phase region, a similar approach as for drizzling water cloud retrieval discussed above can be applied to stratiform mixed-phase clouds, especially with additional LWP measurements (Wang, 2007; Shupe *et al.* 2008).

9.8 Conclusion and Perspectives

Active remote sensing from aircraft with LIDARs and RADARs is a rapidly evolving set of tools primarily for observations of the spatial distributions and amounts of various atmospheric constituents. Identification of the desired components is the first task with these tools, quantitation is the next and, especially with RADARs, the interpretation of obtained signals in terms of the sizes, shapes, etc. of the scatterers is the ultimate goal. Combinations of instruments best permit the inherent ambiguities of the signals derived from one instrument to be overcome.

While the airborne instruments are just variants of devices that have been operating from the ground for many years before airborne versions could be constructed, it may be said that these airborne remote sensing tools provided resolution and acuity of observations not attainable before. The fine structures revealed by the higher resolution that results from proximity to the targets and the flexibility of selecting relevant atmospheric regions for observation are the hallmarks of these methods of study. There is also the possibility, albeit limited, to follow distinct features in time and thus overcome the snapshot limitations of the data. Having RADAR and LIDAR observations made from an airborne platform that is also making in situ measurements results in much closer coupling of these data and yield insights not otherwise obtainable.

The majority of tools and methods described in this chapter are relatively new but mature enough to be classed as parts of the standard arsenal of atmospheric research. No doubt, improvements will still be made and newer capabilities will be added, so the opportunities will continue to arise for further discoveries and for solid contributions to solving climate and weather problems.

Acknowledgments

The authors would like to thank Julien Delanoë, Christoph Kiemle, Lin Tian and Oliver Reitebuch for contributions and helpful comments.

Color Plates

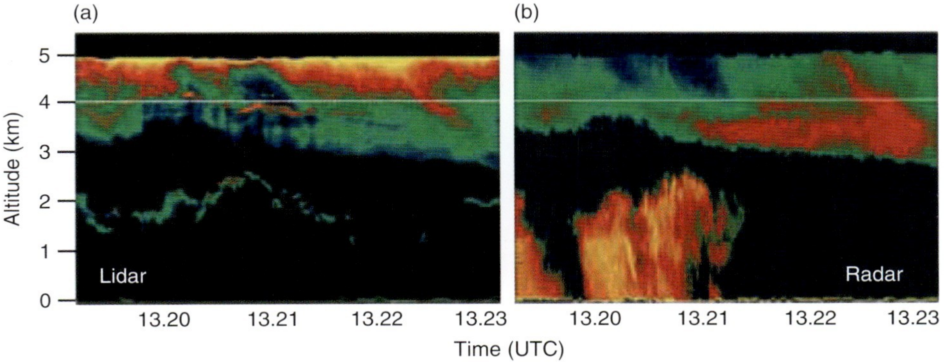

Figure 9.6 Simultaneous observations performed near Brest (France) on 10 November 2000, with combined LIDAR and RADAR ((a) backscatter LIDAR LEANDRE 1 from IPSL and (b) 94 GHz IPSL RADAR) installed in the French F27 flying at 5 km. The dashed line indicates altitude at which in situ measurements have been made with Merlin IV from Meteo-France equipped with GKSS sondes (ESA CLARE'2000 campaign). Supercooled water droplets can be identified near 13:21 UTC below the line. (Courtesy of A. Dabas.) (This figure also appears on page 475.)

Airborne Measurements for Environmental Research: Methods and Instruments, First Edition.
Edited by Manfred Wendisch and Jean-Louis Brenguier.

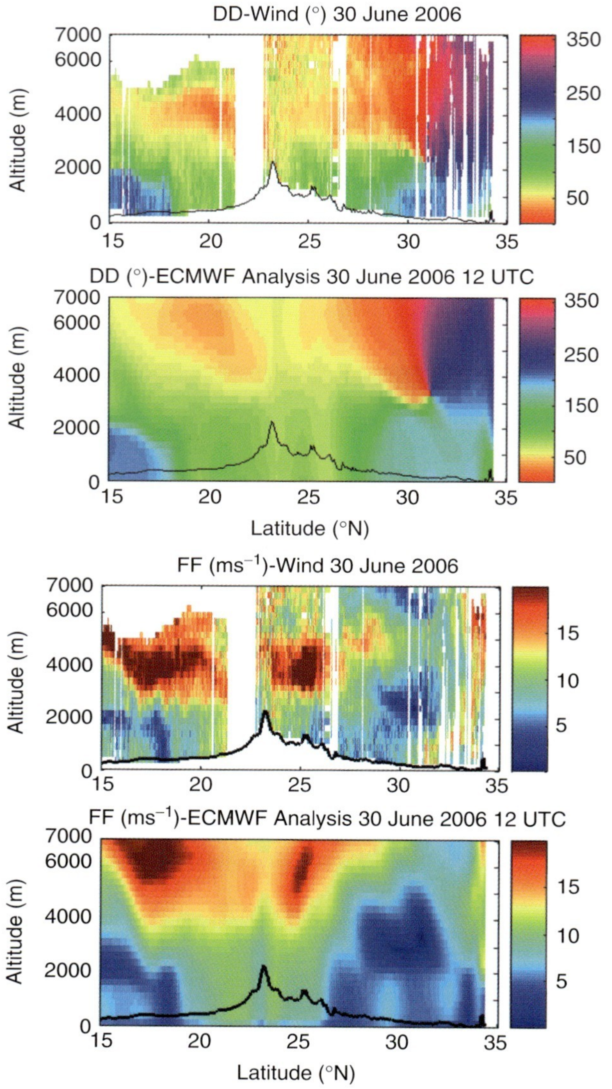

Figure 9.9 Atmospheric wind field retrieved by WIND during a flight over the Saharan desert from Djerba (Tunisia) to Niamey (Niger) on 30 June 2006. The wind field strength (FF) and direction (DD) retrieved from WIND measurements are compared to ECMWF analysis at 12.00 UTC. (Courtesy of A. Dabas.) (This figure also appears on page 479.)

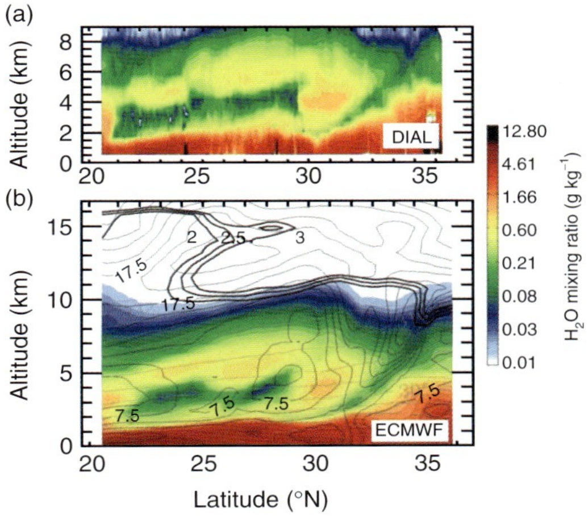

Figure 9.10 Water vapor mixing ratio q in g/kg along DIAL flights (a) on 14 March 2004, 1630–1900 UT compared to ECMWF T511/L60 operational analysis (b). Contours are horizontal wind speed in m s^{-1} and potential vorticity at 2, 2.5, and 3 PVU. From Flentje *et al.* (2007). (This figure also appears on page 483.)

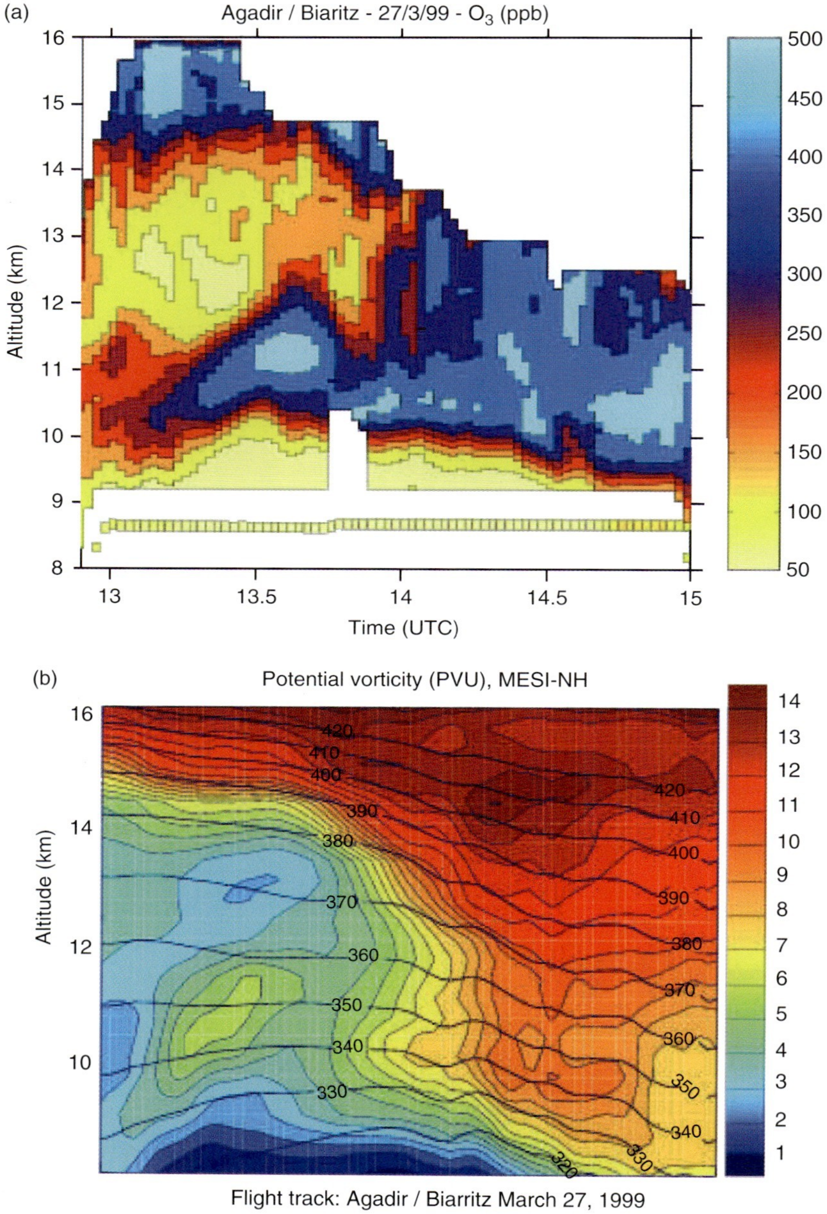

Figure 9.11 Ozone vertical cross section measured by the ALTO LIDAR (a) and comparison with a (b) high resolution PV vertical cross section from the Meso-NH model, after Kowol-Santen and Ancellet (2000). (Copyright 2000 American Geophysical Union (AGU). Reproduced with permission by AGU.) (This figure also appears on page 485.)

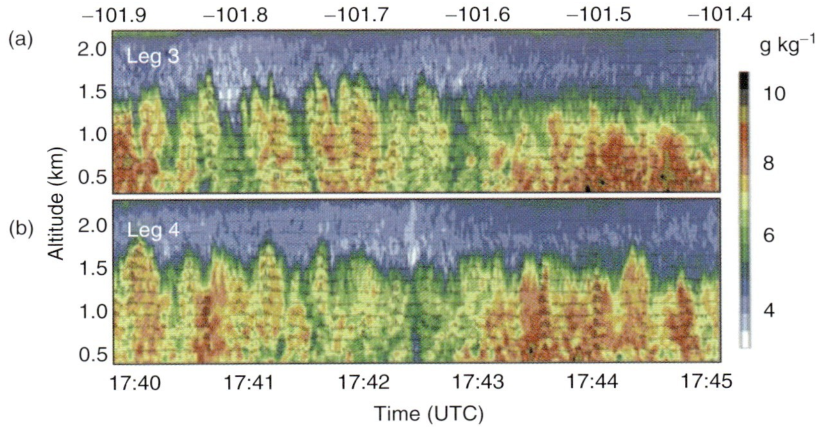

Figure 9.12 Vertical cross sections of water vapor from DLR DIAL (colors) and vertical velocity from NOAA HRDL (vertical bars) for the Falcon flight legs 3 and 4 oriented West–East at 37.4 °N over southwestern Kansas. The aircraft flew (a) leg 3 from right (East) to left (West) and turned back at 102 °W to (b) fly leg 4 on the same track. Top axis is longitude, bottom axis UTC time, 7 h ahead of LT, after Kiemle *et al.* (2007). (Copyright 2007 American Meteorological Society (AMS). Reproduced with permission by AMS.) (This figure also appears on page 488.)

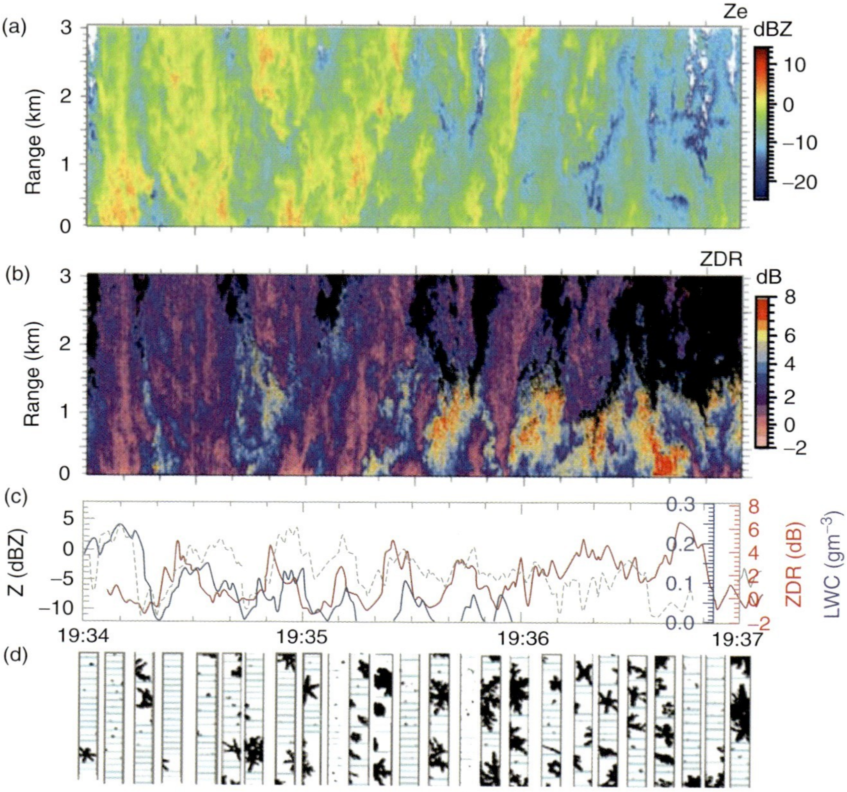

Figure 9.20 (a) and (b) Horizontal cross sections of Z_e and Z_{DR} obtained from a side-looking beam. Traces of LWC, Z_e, Temperature and Z_{DR} at 105 m from the aircraft are shown in (c). (d) Sample images from a PMS 2D-C probe. The width of each strip represents 800 μm. (This figure also appears on page 508.)

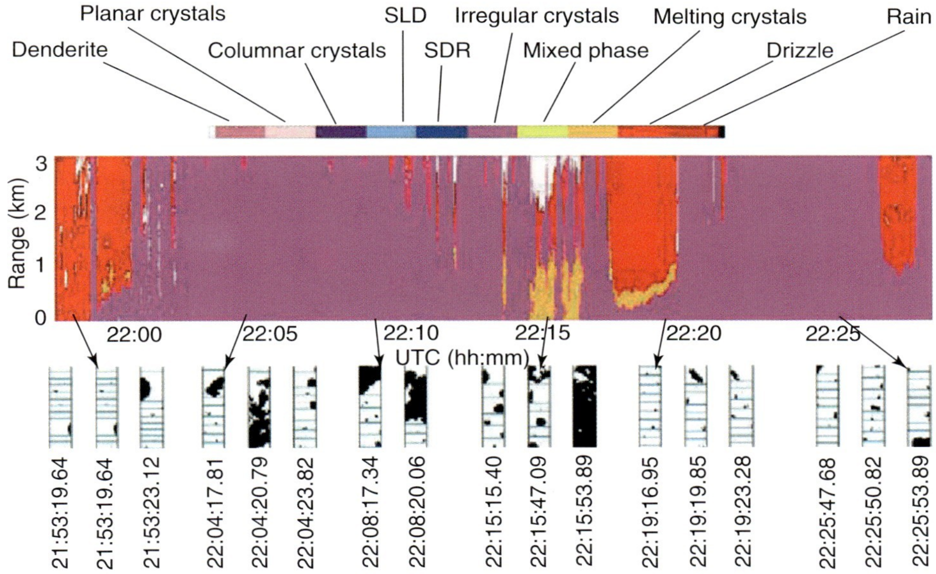

Figure 9.21 Fuzzy-logic based classification of particle from a flight in winter Ns over Ontario, Canada. Examples of particle images recorded with a PMS 2D-C probe at flight level are also shown for reference. (This figure also appears on page 509.)

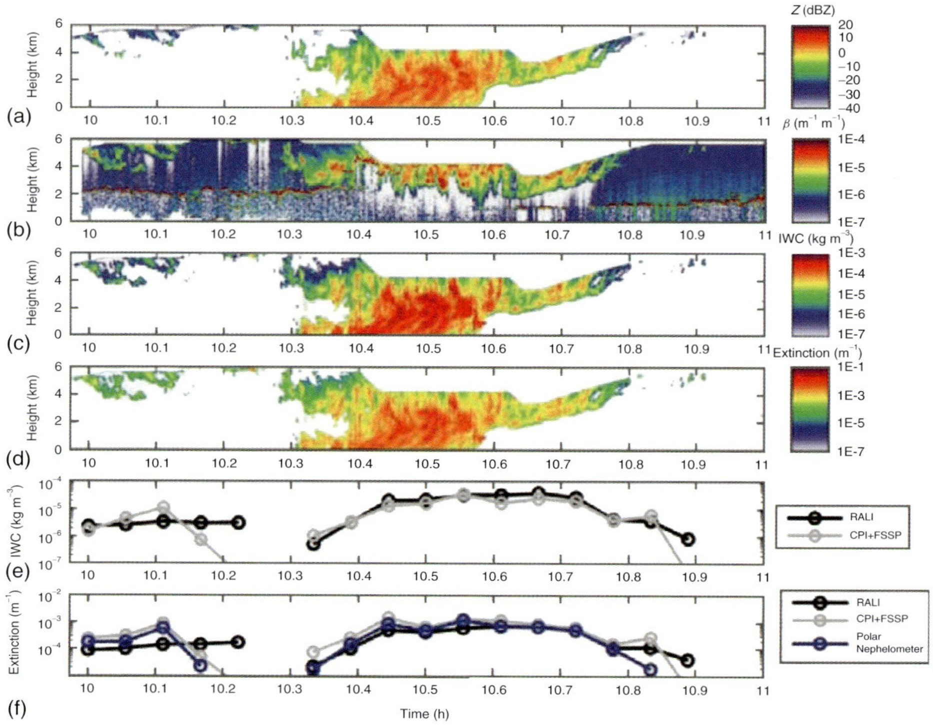

Figure 9.27 Downward looking RADAR and LIDAR observations collected in an Arctic cloud on 1 April 2008 during the POLARCAT field campaign using the RALI system on the French Falcon ATR42 aircraft. Observed RADAR reflectivity (a) and LIDAR backscatter (b) were used to retrieve the ice water content (c) and the VIS extinction (d). These parameters are compared in (e) and (f) with the values measured by *in situ* probes. (This figure also appears on page 520.)

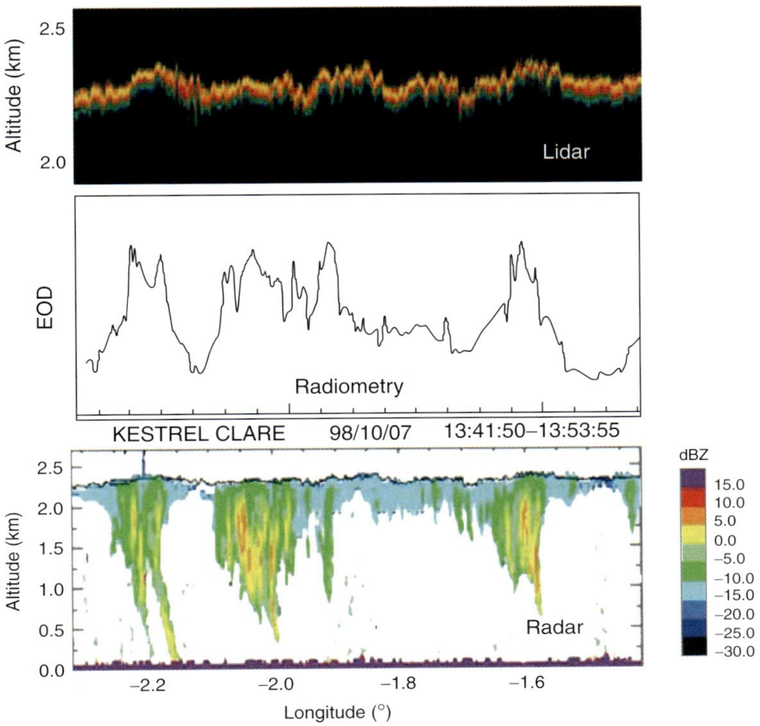

Figure 9.28 Simultaneous observations performed near Chilbolton (United Kingdom) on 7 October 1998, with combined radiometry (giving access to the Effective Optical Depth, EOD), LIDAR and RADAR (Backscatter LIDAR LEANDRE from IPSL and Wyoming Cloud RADAR WCR from University of Wyoming) installed in the French F27 flying at 3 km (ESA CLARE'1998 campaign). (This figure also appears on page 521.)

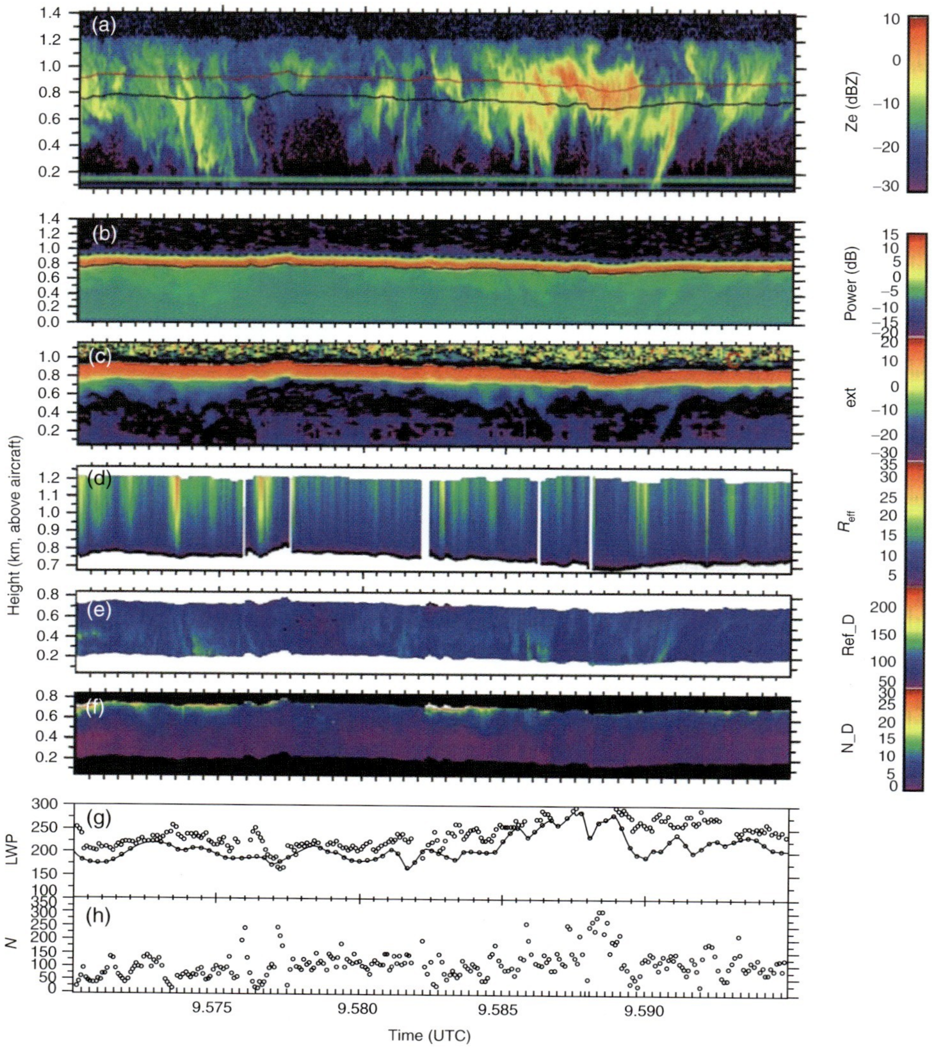

Figure 9.29 Zenith pointing WCL–WCR observations of drizzling stratocumulus clouds from NCAR/NSF C-130 on October 28, 2008 during the VOCALS field campaign: (a) WCR Z_e (b) WCL attenuated backscatter power; retrieved cloud and drizzle properties: (c) retrieved VIS extinction coefficient (d) cloud effective diameter, (e) drizzle size distribution, (f) drizzle number concentration, (g) estimated adiabatic LWP and GVR measured LWP (line with star symbols), and (e) layer mean cloud droplet number concentration. The C-130 flew at ~100 m above the surface. WCL signals provide a clear detection of cloud base (black line), but they are quickly attenuated within about 200 m. WCL cloud base and apparent top (red line) plotted in (a) shows the need of WCR measurements to provide cloud top and no information for cloud base from Z_e profiles in this case. (This figure also appears on page 523.)

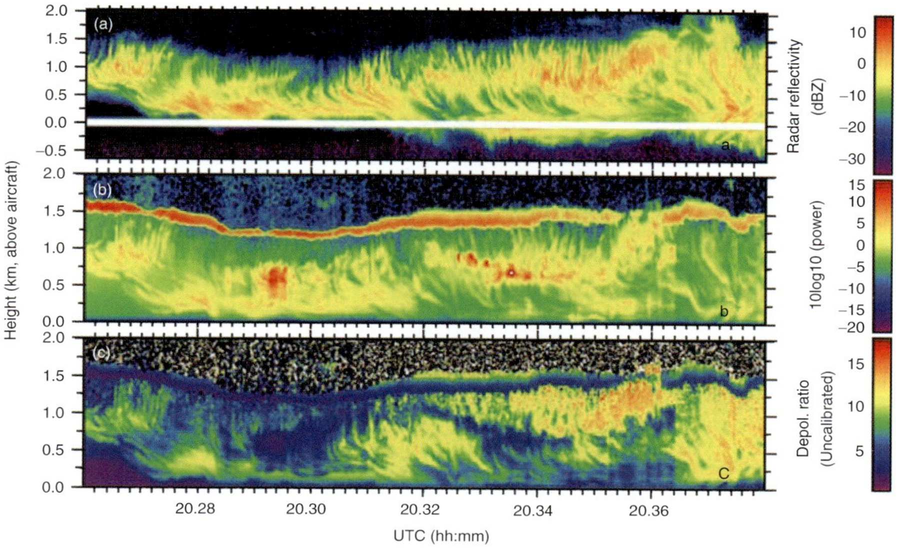

Figure 9.30 An example of WCR and WCL observations (zenith pointing) of stratiform mixed-phase clouds from the Wyoming King Air on February 20, 2008 during the Wyoming Airborne Integrated Cloud Observation (WAICO) experiment: RADAR reflectivity factor (a), LIDAR backscattering power (b), and LIDAR linear depolarization ratio (c). Water dominated mixed-phase layer is located around 1.5 km above the aircraft with cloud top temperature ∼−25 °C. Ice particles generated in the mixed-phase layer precipitate down. This is a typical example of altocumulus with ice virga observed right South of Laramie regional airport hen clouds moving from South during the WAICO. Mixed-phase stratocumulus clouds over Polar regions also have similar vertical structure. (This figure also appears on page 524.)

List of Abbreviations

1D	One-Dimensional
2D	Two-Dimensional
3D	Three-Dimensional
2D-C	2D Cloud Probe
2D-P	2D Precipitation Probe
2D-S	2D Stereo Probe
AATS	NASA Ames Airborne Tracking Sunphotometer
AC	Alternating Current
ACCENT	Atmospheric Composition Change
ACE	Aerosol Characterization Experiment
ACTOS	Airborne Cloud Turbulence Observation System
ADC	Analogue Digital Converter
AGL	Above Ground Level
AIDA	Aerosol Interaction and Dynamics in the Atmosphere
AIMMS	Aircraft-Integrated Meteorological Measurement System
AIS	Airborne Imaging Spectrometer
AISA	Airborne Imaging Spectrometer for Different Applications
AMAX-DOAS	Airborne Multiaxis Differential Optical Absorption Spectrometry
AMMA	African Monsoon Multidisciplinary Analysis
AMS	Aerosol Mass Spectrometer
AMSU	Advanced Microwave Sounding Unit
AOD	Aerosol Optical Depth
AOS	Acousto Optical Spectrometer
APD	Avalanche Photodiode
APEX	Airborne Prism Experiment
APM	Aerodynamic Particle Mass Analyzer
APS	Aerodynamic Particle Sizer
ARM	Atmospheric Radiation Measurements
ASD	Analytical Spectral Device
ASI	Airborne Spectral Imager
ASSP	Axially Scattering Spectrometer Probe
ASASP	Active Scattering Aerosol Spectrometer Probe
ASTAR	Arctic Study of Tropospheric Aerosol, Clouds, and Radiation
ATCOR	Atmospheric and Topographic Correction
ATDD	Atmospheric Turbulence and Diffusion Division
AU	Astronomical Unit

Airborne Measurements for Environmental Research: Methods and Instruments, First Edition.
Edited by Manfred Wendisch and Jean-Louis Brenguier.

AVIRIS	Airborne Visible and Infrared Imaging Spectrometer
BAT	Best Aircraft Turbulence
BBCRDS	Broadband Cavity Ring-Down Spectroscopy
BC	Black Carbon
BCP	Backscatter Cloud Probe
BHR	Bi-Hemisphere Reflection
BRDF	Bidirectional Reflectance Distribution Function
BSRN	Baseline Surface Radiation Network
CalNex	California Nexus
CARIBIC	Civil Aircraft for Regular Investigation of the Atmosphere Based on an Instrument Container
CAS	Cloud and Aerosol Spectrometer
CAS-DPOL	Cloud and Aerosol Spectrometer with Depolarization
CASI	Compact Airborne Spectrographic Imager
CBL	Convective Boundary Layer
CCD	Charge-Coupled Device
CCN	Cloud Condensation Nuclei
CCNC	Cloud Condensation Nucleus Counter
CCRF	Chemical Conversion Resonance Fluorescence
CCRS	Canadian Center for Remote Sensing
CDMA	Cylindrical Differential Mobility Analyzer
CDP	Cloud Droplet Probe
CE	Collection Efficiency
CEAS	Cavity-Enhanced Absorption Spectroscopy
CEOS	Committee of Earth Observation Satellites
CEP	Cloud Extinction Probe
CFD	Computational Fluid Dynamics
CFDC	Continuous Flow Diffusion Chamber
CFMC	Continuous Flow Mixing Chamber
CHRIS	Compact High-Resolution Imaging spectrometer
CIN	Cloud Integrating Nephelometer
CIMS	Chemical Ionization Mass Spectrometer
CIP	Cloud Imaging Probe
CIRA	Centro Italiano Ricerche Aerospaziali
CL	Chain Length
CLH	Closed Path TDL Hygrometer
CMOS	Complementary Metal Oxide Semiconductors
CNR	Council of National Research
COSSIR	Conical Scanning Millimeter-Wave Imaging Radiometer
CP	Carrier Phase
CPC	Condensation Particle Counter
CPI	Cloud Particle Imager
CPSD	Cloud Particle Spectrometer with Depolarization
CRDS	Cavity Ring-Down Spectroscopy
CRISM	Compact Reconnaissance Imaging Spectrometer
CSI	Cloud Spectrometer and Impactor
CTA	Constant Temperature Anemometer
CVI	Counterflow Virtual Impactor
CW	Continuous Wave

DAIS	Digital Airborne Imaging Spectrometer
DC	Direct Current
DC-8	Douglas DC-8 Four-Engine Jet
DDV	Dense Vegetation Approach
DEM	Digital Elevation Model
DFG	Deutsche Forschungsgemeinschaft
DGNSS	Differential GNSS
DGPS	Differential GPS
DLR	Deutsches Zentrum für Luft- und Raumfahrt
DMA	Differential Mobility Analyzer
DMPS	Differential Mobility Particle Sizer
DMS	Differential Mobility Spectrometer
DMT	Droplet Measurement Technologies
DN	Digital Number
DOE	Department of Energy
DOF	Depth of Field
DOP	Dilution of Precision
DRI	Desert Research Institute
DSD	Drop Size Distribution
DU	Dobson Units
EARSeL	European Remote Sensing Laboratories
EAS	Electrical Aerosol Spectrometer
EC	Environment Canada
eCL	effective Chain Length
ECN	Netherlands Energy Research Foundation
EEPS	Engine Exhaust Particle Sizer
EnMAP	Environmental Mapping and Analysis Program
EO	Earth Observation EO-1 Earth Orbiter 1
ESA	European Space Agency
ESG	Electrically Suspended Gyroscope
ESR	Electron Spin Resonance
EUCAARI	European Aerosol Cloud Climate and Air Quality Interactions
EUFAR	European Facility for Airborne Research
EWG	Expert Working Group
FAAM	Facility for Airborne Atmospheric Measurements
FADS	Flush Airdata Sensing
Falcon	Dassault Mystère Falcon 20 E Twin-Jet
FCDP	Fast CDP
FHP	Five-Hole Probe
FIMS	Fast Integrated Mobility Spectrometer
FINCH	Fast Ice Nucleus Chamber
FISH	Fast *In Situ* Stratospheric Hygrometer
FLAASH	Fast Line-of-sight Atmospheric Analysis of Spectral Hypercubes
FLI	Fluorescence Line Imager
FOG	Fiber Optic Gyro
FOV	Field of View
FPGA	Field Programmable Gate Array
FPA	Focal Plane Array
FSSP	Forward Scattering Spectrometer Probe

FTS	Fourier Transform Spectrometer
FWHM	Full Width at Half Maximum
GAW	Global Atmospheric Watch
GC–MS	Gas Chromatography–Mass Spectrometry
Geophysica	Myasishchev M-55 High Altitude Twin Jet
GER	Geophysical and Environmental Research Corporation
GFC	Gas Filter Correlation
GHG	Greenhouse Gases
GIFOV	Ground Instantaneous Field of View
GIS	Geographic Information System
Global Hawk UAS	Northrop Grumman RQ-4 Global Hawk Unmanned Aircraft System
GLONASS	Globalnaja Navigaziona Systema
GloPac	Global Hawk Pacific Mission
GNSS	Global Navigation Satellite Systems
GPS	Global Positioning System
GSD	Ground-Sampling Distance
Gulfstream-505	Gulfstream Aerospace Twin-Jet
HALO	High Altitude and Long Range Research Aircraft
HDRF	Hemisphere Diffuse Reflectance Function
HIAPER	High-Performance Instrumented Airborne Platform for Environmental Research
HIPPO	Pole-to-Pole Observations
HISS	Hyper Image Space Spectrometer
HOLODEC	Holographic Detector for Clouds
HPC	High Performance Computing
HPD	Hybrid Photodetector
HRS	Hyperspectral Remote Sensing
HTW	Harvard Total Water Hygrometer
HVPS	High Volume Precipitation Spectrometer
HyMAP	Hyperspectral Mapper
IAGOS	In-Service Aircraft for a Global Observing System
IC	Ion Chromatography
ICAO	International Civil Aviation Organization
ICARTT	Intercontinental Consortium for Atmospheric Research on Transport and Transformation
ICOS	Integrated Cavity Output Spectroscopy
ICPS	Isokinetic Cloud Probe System
IDI	Isokinetic Diffuser-Type Inlet
IF	Intermediate Frequency
IFOG	Interferometric Gyro
IFOV	Instantaneous Field of View
IfT	Leibniz Institute for Tropospheric Research
IKP	Isokinetic TWC Probe
ILIDS	Interferometric Laser Imaging for Droplet Sizing
ILS	Instrument Line Shape
IMU	Inertial Measurement Unit
IN	Ice Nuclei
INAA	Instrumental Neutron Activation Analysis

INDOEX	Indian Ocean Experiment
INS	Inertial Navigation System
InSb	Indium Antimonide
INSPECTRO	INfluence of Clouds on the SPectral actinic flux in the lower TROposphere
INTA	Instituto Nacional de Técnica Aeroespacial
IPCC	Intergovernmental Panel on Climate Change
IR	Infrared
IRS	Inertial Reference System
IS	Imaging Spectroscopy
ISA	International Standard Atmosphere
ISIS	International Spaceborne Imaging Spectroscopy
ISO	International Organization for Standardization
ITRES	Integral Technology for Remote Sensing
ITS	International Temperature Scale
IWC	Ice Water Content
IWV	Integrated Water Vapor
IWP	Ice Water Path
LACE	Lindenberg Aerosol Characterization Experiment
LaMP	Laboratoire de Météorologie Physique
LANDSAT	Land Satellite
LDV	Laser-Doppler Velocimetry
LED	Light Emitting Diode
LIDAR	Light Detection and Ranging
LIF	Laser-Induced Fluorescence
LIM	Leipzig Institute for Meteorology
LNA	Low Noise Amplifiers
LO	Local Oscillator
LPAS	Laser-Induced Photo-Acoustic Spectrometry
LPM	Liter per Minute
LTI	Low Turbulence Inlet
LUT	Look Up Table
LWC	Liquid Water Content
LWP	Liquid Water Path
MAAP	Multi-Angle Absorption Photometer
MARSS	Microwave Airborne Radiometer Scanning System
MAS	MODIS Airborne Simulator
MASI	Midwave Airborne Spectral Imager
MASP	Multiangle Aerosol Spectrometer
MCP	Multichannel Plate
MCS	Multichannel Spectrometer
MCT	Mercury Cadmium Telluride
MEMS	Microelectomechanical System
MIR	Mid Infrared
MIVIS	Multispectral Environment Imaging Sensor
MLS	Microwave Limb Sounder
MODTRAN	Moderate Resolution Transmission Code
MOZAIC	Measurement of Ozone and Water Vapor by Airbus In-Service Aircraft
MPI	Max Planck Institute
MSL	Mean Sea Level
MSU	Microwave Sounding Unit

MVD	Median Volume Diameter
MW	Microwave
NA	Numerical Aperture
NASA	National Aeronautics and Space Administration
NCAR	National Center for Atmospheric Research
NDI	Nested Diffuser-Type Inlet
NDIR	Nondispersive Infrared
NED	North-East Down
NEO	Norsk Elektro Optikk
NER	Noise Equivalent Radiance
NERC	Natural Environment Research Council
NIR	Near Infrared
NIST	National Institute of Standards and Technology
NMP	New Millennium Program
NOAA	National Oceanic and Atmospheric Administration
NOXAR	Swiss Nitrogen Oxides and Ozone along Air Routes Project
NPOESS	National Polar-Orbiting Operational Environmental Satellite System
NRL	Naval Research Laboratory
NRC	National Research Council
NSF	National Science Foundation
Nu	Nusselt Number
OAP	Optical Array Probe
OC	Organic Carbon
ODE	Ozone Depletion Event
OMAC	Opposed Migration Aerosol Classifier
OPC	Optical Particle Counter
PALMS	Particle Analysis by Laser Mass Spectrometer
PCA	Principle Component Analysis
PCASP	Passive Cavity Aerosol Spectrometer Probe
PC-BOSS	Particle Concentrator-Brigham Young University Organic Sampling System
PDA	Phase Doppler Analyzer
PdA	Photodiode Array
PDI	Phase Doppler Interferometer
PDPA	Phase Doppler Particle Analyzer
PeRCA	Peroxy Radical Chemical Amplification
PFA	Paraformaldehyde
PGP	Prism-Grating-Prism
PH-CPC	Pulse-Height Condensation Particle Counter
PILS	Particle-Into-Liquid Sampler
PIP	Precipitation Imaging Probe
PIXE	Particle-Induce X-Ray Emission
PM1	Particulate Matter with Particle Diameter $< 1.0\,\mu m$
PM2.5	Particulate Matter with Particle Diameter $< 2.5\,\mu m$
PMS	Particle Measuring Systems
PMT	Photomultiplier Tube
PN	Polar Nephelometer
Pr	Prandtl Number
PRISM	Processes Research for Imaging Spectrometer Mission
PROBA	Project for On-Board Autonomy

PSA	Particle Surface Area
PSAP	Particle Soot Absorption Photometer
PSD	Particle Size Distribution
PSL	Polystyrene Latex Beads
PSM	Particle Size Magnifier
PSR	Polarimetric Scanning Radiometer
PSU	Pennsylvania State University
PTFE	Polytetrafluorethylene
PToF	Particle Time-of-Flight
PTR-MS	Proton Transfer Reaction Mass Spectrometer
PVM	Particle Volume Monitor
QA	Quality Assurance
QCL	Quantum Cascade Laser
QI	Quality Indicators
QPF	Quantitative Precipitation Forecast
QUAC	Quick Atmospheric Correction
RADAR	Radar Detection and Ranging
RAOS	Reno Aerosol Optics Study
RDMA	Radial Differential Mobility Analyzer
Re	Reynolds Number
REO	Research Electro-Optics
RF	Radio Frequency
RH	Relative Humidity
RICO	Rain in Cumulus Over the Ocean
RID	Rosemount Icing Detector
RLG	Ring Laser Gyro
RONOCO	Role of Nighttime Chemistry in Controlling the Oxidizing Capacity of the Atmosphere
ROSIS	Reflective Optics System Imaging Spectrometer
RSL	Remote Sensing Laboratory
RSR	Relative Spectral Response
RT	Receiver Transmitter
SA	Selective Availability
SAGE	Stratospheric Aerosol and Gas Experiment
SAM	Spectral Angle Mapper
SAR	Synthetic Aperture Radio Detection and Ranging
SAW	Surface Acoustic Wave
SCD	Slant Column Densities
SDI	Solid Diffuser-Type Inlet
SEA	Science Engineering Associates
SEMS	Scanning Electrical Mobility Spectrometer
SFSI	Short-Wave Infrared Full Spectrum Imager
SHIVA	Stratospheric Ozone: Halogen Impacts in a Varying Atmosphere
SID	Small Ice Detector
SL	Sampling Line
SMART	Spectral Modular Airborne Radiation System
SMIRR	Shuttle Multispectral Infrared Radiometer
SMPS	Scanning Mobility Particle Sizer
SNR	Signal-to-Noise Ratio
SOLVE	SAGE III Ozone Loss and Validation Experiment

SP-2	Single Particle Soot Photometer
SPEC	Stratton Park Engineering Company
SPECIM	Spectral Imagers
SPIN	Spectrometer for Ice Nucei
SPIRIT	SPectromètre Infra Rouge In Situ
SPOT	System Probatoire d'Observation de la Terre
SSFR	Solar Spectral Flux Radiometer
SSMI/S	Special Sensor Microwave Imager/Sounder
SST	Sea Surface Temperature
SSTI	Small Satellite Technology Initiative
Stk	Stokes Number
STP	Standard Temperature and Pressure
STRAP	Stabilized Radiometer Platform
SV	Satellite Vehicle
SVM	Support Vector Machine
SWE	Snow Water Equivalent
SWIR	Short-Wave Infrared
TARFOX	Tropospheric Aerosol Radiative Forcing Observational Experiment
TAS	True Air Speed
TASI	Thermal Airborne Spectral Imager
TAU	Tel Aviv University
TDL	Tunable Diode Laser
TEC	Total Electron Content
TexAQS	Texas Air Quality Study
TLAS	Tunable Laser Absorption Spectroscopy
TIMS	Thermal Infrared Multispectral Scanner
TIR	Thermal Infrared
TM	Thematic Mapper
TOA	Top of Atmosphere
ToF	Time-of-Flight
TOPSE	Tropospheric Ozone Production about the Spring Equinox
TOR	Thermal–Optical Reflectance
TWC	Total Water Content
UAS	Unmanned Aerial Systems
UAV	Unmanned Aerial Vehicle
UHSAS	Ultrahigh Sensitivity Aerosol Spectrometer
UK	United Kingdom
UKMO	United Kingdom Meteorological Office
UK-MRF	United Kingdom Meteorological Research Flight
UNAM	Universidad Nacional Autónoma de México
US	United States
USAS	Ultrahigh Sensitivity Aerosol Spectrometer
UTC	Universal Time Coordinated
UFT	Ultra Fast Thermometer
UT/LS	Upper Troposphere/Lower Stratosphere
UV	Ultraviolet
VAD	Velocity Azimuth Display
VBA	Vibrating Beam Accelerometer
VIPS	Video Ice Particle Sampler

VIS	Visible
VITO	Flemish Institute for Technological Research
VNIR	Visible and Near Infrared
VOC	Volatile Organic Compound
VOCALS	VAMOS Ocean–Cloud–Atmosphere–Land Study
VSLS	Very Short-Lived Substances
VUV	Vacuum Ultraviolet
WAS	Whole Air Sampler
WICC	Wide Stream Impaction Cloud Water Collector
WMO	World Meteorological Organization
XRF	X-Ray Fluorescence

Constants

Greek:

$\sigma = 5.671 \times 10^{-8}\,\mathrm{W\,m^{-2}\,K^{-4}}$	Stefan–Boltzmann Constant

Latin:

$c = 2.997925 \times 10^{8}\,\mathrm{m\,s^{-1}}$		Speed of Light in a Vacuum
$c_\mathrm{p} = 1004\,\mathrm{J\,kg^{-1}\,K^{-1}}$		Specific Heat Capacities (Dry Air) at Constant Pressure
$c_\mathrm{v} = 717\,\mathrm{J\,kg^{-1}\,K^{-1}}$		Specific Heat Capacities (Dry Air) at Constant Volume
$h = 6.6262 \times 10^{-34}\,\mathrm{J\,s}$	$= 4.138 \times 10^{-15}\,\mathrm{eV\,s}$	Planck's constant
$k_\mathrm{B} = 1.3805 \times 10^{-23}\,\mathrm{J\,K^{-1}}$	$= 8.62 \times 10^{-5}\,\mathrm{eV\,K^{-1}}$	Boltzmann Constant
$k_\mathrm{W} = 2897\,\mathrm{\mu m\,K}$		Constant in Wien's Displacement Law
$\overline{R}_\mathrm{S-E} = 149.6 \times 10^{9}\,\mathrm{m}$		Average Earth–Sun distance, or 1 AU
$R_\mathrm{dry} = 287.05\,\mathrm{J\,kg^{-1}\,K^{-1}}$		Specific Gas Constant for Dry Air
$R_\mathrm{vw} = 461.7\,\mathrm{J\,kg^{-1}\,K^{-1}}$		Specific Gas Constant for Water Vapor

Airborne Measurements for Environmental Research: Methods and Instruments, First Edition.
Edited by Manfred Wendisch and Jean-Louis Brenguier.

References

Abel, S.J., Haywood, J.M., Highwood, E.J., Li, J., and Buseck, P.R. (2003) Evolution of biomass burning aerosol properties from an agricultural fire in southern Africa. *Geophys. Res. Lett.*, **30**, 1783.

Ackerman, S. and Cox, S. (1981) Aircraft observations of the shortwave fractional absorptance of non-homogeneous clouds. *J. Appl. Meteorol.*, **20**, 1510–1515.

Ackerman, T., Flynn, D., and Marchand, R. (2003) Quantifying the magnitude of anomalous solar absorption. *J. Geophys. Res.*, **108**, 674. doi: 10.1029/2002JD002.

Adams, J., Smith, M., and Johnson, P. (1986) Spectral mixture modeling: a new analysis of rock and soil types as the Viking Lander 1 site. *J. Geophys. Res.*, **91**, 8098–8112.

Adler-Golden, S.M., Matthew, M.W., Anderson, G.P., Felde, G.W., and Gardner, J.A. (2002) An algorithm for de-shadowing spectral imagery, in *Proceedings of the 11th JPL Airborne Earth Science Workshop* (ed. R. Green), 5–8 March 2002, JPL-Publication 03–04, Pasadena, U.S.A.

Adrian, R.J. (1996) Laser velocimetry, in *Fluid Mechanics Measurements* (eds E. Richard and J. Goldstein), Taylor & Francis, Washington, DC, Chapter 4, pp. 175–299.

Aiken, A.C., Decarlo, P., Kroll, J., Worsnop, D., Huffman, J., Docherty, K., Ulbrich, I., Mohr, C., Kimmel, J., Sueper, D., Sun, Y., Zhang, Q., Trimborn, A., Northway, M., Ziemann, P., Canagaratna, M., Onasch, T., Alfarra, M., Prevot, A., Dommen, J., Duplissy, J., Metzger, A., Baltensperger, U., and Jimenez, J. (2008) O/C and OM/OC ratios of primary, secondary, and ambient organic aerosol particles with high-resolution time-of-flight aerosol mass spectrometry. *Environ. Sci. Technol.*, **42**, 4478–4485.

Alakian, A., Marion, R., and Briottet, X. (2008) Retrieval of microphysical and optical properties in aerosol plumes with hyperspectral imagery: L-APOM method. *Remote Sens. Environ.*, **113**, 781–793.

Albrecht, H.-E., Borys, M., Damaschke, N., and Tropea, C. (2003) *Laser Doppler and Phase Doppler Measurement Techniques*, Springer-Verlag.

Allan, J.D., Jimenez, J., Williams, P., Alfarra, M., Bower, K., Jayne, J., Coe, H., and Worsnop, D. (2003) Quantitative sampling using an Aerodyne aerosol mass spectrometer: 1. Techniques of data interpretation and error analysis. *J. Geophys. Res.*, 108.

Allan, J.D., Deliab, A., Coea, H., Bower, K., Alfarraa, M., Jimenezc, J., Middlebrookd, A., Drewnicke, F., Onaschf, T., Canagaratnaf, M., and Worsnopf, J.J.D. (2004) A generalised method for the extraction of chemically resolved mass spectra from aerodyne aerosol mass spectrometer data. *J. Aerosol Sci.*, **35**, 909–922.

Allen, M.D. and Raabe, O.G. (1985) Slip correction measurements for aerosol particles of doublet and triangular triplet aggregates of spheres. *J. Aerosol Sci.*, **16**, 57–67.

Alonso, M. (2002) Reducing the diffusional spreading rate of a Brownian particle by

Airborne Measurements for Environmental Research: Methods and Instruments, First Edition.
Edited by Manfred Wendisch and Jean-Louis Brenguier.

an appropriate non-uniform external force field. *J. Aerosol Sci.*, **33**, 439–450.

Amediek, A., Ehret, G., Fix, A., Wirth, M., Quatrevalet, M., Büdenbender, C., and Kiemle, C. (2010) CHARM-F: airborne demonstrator for spaceborne CO_2 and CH_4 Lidars. 4th International Workshop on CO_2/CH_4 DIAL Remote Sensing, Oberpfaffenhofen, Deutschland, November 3–5, 2010.

An, W.J., Pathak, R.K., Lee, B.H., and Pandis, S.N. (2007) Aerosol volatility measurement using an improved thermodenuder: application to secondary organic aerosol. *J. Aerosol Sci.*, **38**, 305–314.

Ancellet, G. (1998) Compact Airborne Lidar for tropospheric Ozone: description and field measurements. *Appl. Opt.*, **37**, **24**, 5509–5521.

Ancellet, G. and Ravetta, F. (2005) Analysis and validation of ozone variability observed by Lidar during the ESCOMPTE-2001 campaign. *Atmos. Res.*, **74**, 435–460.

Anderson, J.G. (1975) Absolute concentration of O(p-3) in Earths stratosphere. *Geophys. Res. Lett.*, **2**, 231–234.

Anderson, J.G. (1976) Absolute concentration of $OH(x^2\pi)$ in Earths stratosphere. *Geophys. Res. Lett.*, **3**, 165–168.

Anderson, J.G., Toohey, D., and Brune, W. (1991) Free-radicals within the Antarctic vortex - the role Of CFCs in Antarctic ozone loss. *Science*, **251**, 39–46.

Anderson, T., Charlson, R., and Covert, D. (1993) Calibration of a counterflow virtual impactor at aerodynamic diameters from 1 to 15 μm. *Aerosol Sci. Technol.*, **19**, 317–329.

Anderson, T., Covert, D.S., Marshall, S.F., Laucks, M.L., Charlson, R.J., Waggoner, A.P., Ogren, J.A., Caldow, R., Holm, R.L., Quant, F.R., Sem, G.J., Wiedensohler, A., Ahlquist, N.A., and Bates, T.S. (1996) Performance characteristics of a high-sensitivity, three-wavelength, total scatter/backscatter nephelometer. *J. Atmos. Oceanic Technol.*, **13**, 967–986.

Anderson, T.L. and Ogren, J.A. (1998) Determining aerosol radiative properties using the TSI 3563 integrating nephelometer. *Aerosol Sci. Technol.*, **29**, 57–69.

Andreae, M.O., Elbert, W., Gabriel, R., Johnson, D.W., Osborne, S., and Wood, R. (2000) Soluble ion chemistry of the atmospheric aerosol and SO_2 concentrations over the eastern North Atlantic during ACE-2. *Tellus B*, **52**, 1066–1087. doi: 10.1034/j.1600-0889.2000.00 105.x.

Andrés-Hernández, M., Burkert, J., Reichert, L., Stöbener, D., Meyer-Arnek, J., Burrows, J., Dickerson, R., and Doddridge, B. (2001) Marine boundary layer peroxy radical chemistry during the aerosol particles 99 campaign: measurements and analysis. *J. Geophys. Res.*, **106**, 833–846.

Andrés-Hernández, M., Kartal, D., Reichert, L., Burrows, J., Meyer Arnek, J., Lichtenstern, M., Stock, P., and Schlager, H. (2009) Peroxy radical observations over West Africa AMMA 2006: photochemical activity in the outflow of convective systems. *Atmos. Chem. Phys.*, **9**, 3681–3695.

Andrés-Hernández, M.D., Stone, D., Brookes, D.M., Commane, R., Reeves, C.E., Huntrieser, H., Heard, D.E., Monks, P.S., Burrows, J.P., Schlager, H., Kartal, D., Evans, M.J., Floquet, C.F.A., Ingham, T., Methven, J., and Parker, A.E. (2010) Peroxy radical partitioning during the AMMA radical intercomparison exercise. *Atmos. Chem. Phys.*, **10**, 10621–10638.

Andrews, E., Sheridan, P.J., and Ogren, J.A. (2011) Seasonal differences in the vertical profiles of aerosol optical properties over rural Oklahoma. *Atmos. Chem. Phys.*, **11**, 10661–10676.

Ankilov, A., Baklanov, A., Colhoun, M., Enderle, K., Gras, J., Julanov, Y., Kaller, D., Lindner, A., Lushnikov, A., Mavliev, R., McGovern, F., O'Connor, T., Podzimek, J., Preining, O., Reischl, G., Rudolf, R., Sem, G., Szymanski, W., Vrtala, A., Wagner, P., Winklmayr, W., and Zagaynov, V. (2002a) Particle size dependent response of aerosol counters. *Atmos. Res.*, **62**, 209–237.

Ankilov, A., Baklanov, A., Colhoun, M., Enderle, K., Gras, J., Julanov, Y., Kaller, D., Lindner, A., Lushnikov, A., Mavliev, R., McGovern, F., O'Connor, T., Podzimek, J., Preining, O., Reischl, G., Rudolf, R., Sem, G., Szymanski, W.,

Vrtala, A., Wagner, P., Winklmayr, W., and Zagaynov, V. (2002b) Intercomparison of number concentration measurements by various aerosol particle counters. *Atmos. Res.*, **62**, 177–207.

Annegarn, H.J., Flanz, M., Kenntner, T., Kneen, M.A., Helas, G., and Piketh, S.J. (1996) Airborne streaker sampling for PIXE analysis. *Nucl. Instrum. Methods. Phys. Res. B*, **109**, 548–550.

ANSI (1992) *Recommended Practice for Atmospheric and Space Flight Vehicle Coordinate Systems*, American National Standards Institute. American National Standard.

Apel, E., Hills, A., Lueb, R., Zindel, S., Eisele, S., and Riemer, D. (2003) A fast-GC/MS system to measure C-2 to C-4 carbonyls and methanol aboard aircraft. *J. Geophys. Res.*, **108**, 8794. doi: 10.1029/2002JD003199.

Ardon-Dreyer, K., Levin, Z., and Lawson, R.P. (2011) Characteristics of immersion freezing nuclei at the south pole station in Antarctica. *Atmos. Chem. Phys. Discuss.*, **11**, 4015–4024.

Arijs, E., Nevejans, D., and Ingels, J. (1983) Positive ion composition measurements and acetonitrile in the upper stratosphere. *Nature*, **303**, 314–316.

Arnold, F. and Henschen, G. (1978) First mass analysis of stratospheric negative ions. *Nature*, **257**, 521–522.

Arnold, F., Krankowsky, D., and Marien, K. (1977) First mass spectrometric measurements of positive ions in the stratosphere. *Nature*, **267**, 30–32. doi: 10.1038/267030a0.

Arnold, F., Fabian, R., Henschen, G., and Joos, W. (1980) Stratospheric trace gas analysis from ions: H_2O and HNO_3. *Planet. Space Sci.*, **28**, 681–685.

Arnold, S., Methven, J., Evans, M., Chipperfield, M., Lewis, A., Hopkins, J., McQuaid, J., Watson, N., Purvis, R., Lee, J., Atlas, E., Blake, D., and Rappengluck, B. (2007) Statistical inference of OH concentrations and air mass dilution rates from successive observations of nonmethane hydrocarbons in single air masses. *J. Geophys. Res.*, **112**, D24S03. doi: 10.1029/2007JD008437.

Arnott, W.P., Dong, Y., Purcell, R., and Hallett, J. (1995) Direct airborne sampling of small ice crystals and the concentration and phase of haze particles. *Paper Presented at the 9th Symposium on Meteorological Observations and Instrumentation*, American Meteorological Society, Charlotte, NC.

Arnott, W.P., Hamasha, K., Moosmüller, H., Sheridan, P.J., and Ogren, J.A. (2005) Towards aerosol light-absorption measurements with a 7-wavelength aethalometer: evaluation with a photoacoustic instrument and 3-wavelength nephelometer. *Aerosol Sci. Technol.*, **39**, 17–29.

Arnott, W.P., Walker, J.W., Moosmüller, H., Elleman, R.A., Jonsson, H.H., Buzorius, G., Conant, W.C., Flagan, R.C., and Seinfeld, J.H. (2006) Photoacoustic insight for aerosol light absorption aloft from meteorological aircraft and comparison with particle soot absorption photometer measurements: DOE Southern Great Plains climate research facility and the coastal stratocumulus imposed pert. *J. Geophys. Res.*, **111**, D05S02. urbation experiments. doi: 10.1029/2005JD005964.

Ashbourn, S., Jenkin, M., and Clemitshaw, K. (1998) Laboratory and modelling studies of the response of a peroxy radical chemical amplifier to HO_2 and a series of organic peroxy radicals. *J. Climate*, **29**, 233.

Asner, G. and Vitousek, P. (2005) Remote analysis of biological invasion and biogeochemical change. *Proc. Natl. Acad. Sci. U.S.A.*, **102**, (12), 4383–4386.

ASTM (2005) *Standard Test Method for Calibration of a Pyranometer Using a Pyrheliometer*, ASTM, G 167-05 is the ASTM designation code for this document that contains 21 pages.

Atkinson, D.B. (2003) Solving chemical problems of environmental importance using cavity ring-down spectroscopy. *Analyst*, **128**, 117–125.

Atkinson, R. (2000) Atmospheric chemistry of VOCs and NO_x. *Atmos. Environ.*, **34**, 2063.

Atkinson, R., Winer, A., and Pitts, J.J. (1986) Estimation of night-time N_2O_5 concentrations from ambient NO_2 and NO_3 radical concentrations and the role of N_2O_5 in night-time chemistry. *Atmos. Environ.*, **20**, 331–339.

Atkinson, R., Baulch, D., Cox, R., Crowley, J., Hampson, R., Hynes, R., Jenkin, M., Rossi, M., and Troe, J. (2006) Evaluated kinetic and photochemical data for atmospheric chemistry: volume II-gas phase reactions of organic species. *Atmos. Chem. Phys.*, **6**, 3625–4055. doi: 10.5194/acp-6-3625-2006.

Auriol, F., Gayet, J.-F., Febvre, G., Jourdan, O., Labonnote, L., and Brogniez, G. (2001) In situ observations of cirrus cloud scattering phase function with 22° and 46° halos: cloud field study on 19 February 1998. *J. Atmos. Sci.*, **58**, 3376–3390.

Austin, R.J., Heymsfield, A.J., and Stephens, G.L. Retrievals of ice cloud microphysical parameters using CloudSat millimeter wave Radar and temperature. (2008) *J. Geoph. Res.* doi: 10:1029/2008JD010049.

Avissar, R., Holder, H.E., Abehserra, N., Bolch, M.A., Novick, K., Canning, P., Prince, K., Magalhaes, J., Matayoshi, N., Katul, G., Walko, R.L., and Johnson, K.M. (2009) The Duke university helicopter observation platform. *Bull. Am. Meteor. Soc.*, **90**, 939–954.

Axford, D. (1968) On the accuracy of wind measurements using an inertial platform in an aircraft, and an example of a measurement of the vertical mesostructure of the atmosphere. *J. Appl. Meteorol.*, **7**, 645–666.

Aydin, K. and Singh, J. (2004) Cloud ice crystal classification using a 95-GHz polarimetric radar. *J. Atmos. Ocean. Technol.*, **21**, 1679–1688.

Aydin, K. and Tang, C. (1997) Millimeter wave Radar scattering from model ice crystal distributions. *IEEE Trans. Geosci. Remote*, **35**, 140–146.

Aydin, K. and Walsh, T.M. (1999) Millimeter wave scattering from spatial and planar bullet rosettes. *IEEE Trans. Geosci. Remote*, **37**, 1138–1150.

Azzopardi, B. (1979) Measurement of drop sizes. *Int. J. Heat Mass Transfer*, **22**, 1245–1279.

Bachalo, W. (1980) A method for measuring the size and velocity of spheres by dual beam light scatter interferometry. *Appl. Opt.*, **19**, 363–370.

Bachalo, W. (1983) Droplet analysis techniques: their selection and applications, liquid particle size measurement techniques, in *Liquid Particle Size Measurement Techniques* (eds J.M. Tishkoff, R.D. Ingebo, and J. Kennedy), ASTM, p. 191, STP 848. doi: 10.1520/STP32613S.

Bachalo, W. (1994) Experimental methods in multiphase flows. *Int. J. Multiphase Flow*, **20**, (Suppl. 1), 261–295.

Bachalo, W. and Houser, M. (1984) Phase doppler spray analyzer for simultaneous measurements of drop size and velocity distributions. *Opt. Eng.*, 23.

Bachalo, W. and Houser, M. (1985) Spray drop size and velocity measurements using the Phase/Doppler Particle Analyzer. Proceedings of the 3rd International Conference on Liquid Atomization and Spray Systems.

Bader, M.J., Clough, S.A., and Cox, G.P. (1987) Aircraft and dual polarization Radar observations of hydrometeors in light stratiform precipitation. *Q. J. R. Meteorol. Soc.*, **113**, 491–515.

Bahreini, R., Jimenez, J.L., Wang, J., Flagan, R.C., Seinfeld, J.H., Jayne, J.T., and Worsnop, D.R. (2003) Aircraft-based aerosol size and composition measurements during ACE-Asia using an Aerodyne aerosol mass spectrometer. *J. Geophys. Res.*, 108.

Bahreini, R., Dunlea, E.J., Matthew, B.M., Simons, C., Docherty, K., de Carlo, P., Jimenez, J., Brock, C.A., and Middlebrook, A.M. (2008) Design and operation of a pressure-controlled inlet for airborne sampling with an aerodynamic aerosol lens. *Aerosol Sci. Technol.*, **42**, 465–471.

Baker, A., Rauthe-Schoch, A., Schuck, T., Brenninkmeijer, C., van Velthoven, P., Wisher, A., and Oram, D. (2011) Investigation of chlorine radical chemistry in the Eyjafjallajokull volcanic plume using observed depletions in non-methane hydrocarbons. *Geophys. Res. Lett.*, **38**. doi: 10.1029/2011GL047571.

Baker, B. and Copson, E. (1950) *The Mathematical Theory of Huygens*, Principle, Chelsea Publ. Company, New York.

Baker, B. and Lawson, R. (2006) Improvement in determination of ice water content from two-dimensional particle imagery. Part I: image-to-mass relationships. *J. Appl. Meteor. Climatol.*, **45**, 1282–1290.

Baker, B., Mo, Q., Lawson, R., O'Connor, D., and Korolev, A. (2009a) Drop size distributions and the lack of small drops in RICO rain shafts. *J. Appl. Meteor. Climatol.*, **48**, 616–623.

Baker, B., Mo, Q., Lawson, R., O'Connor, D., and Korolev, A. (2009b) The effects of precipitation on cloud droplet measurement devices. *J. Atmos. Oceanic Technol.*, **26**, 1404–1409.

Bakwin, P.S., Tans, P.P., Stephens, B.B., Wofsy, S.C., Gerbig, C., and Grainger, A. (2003) Strategies for measurement of atmospheric column means of carbon dioxide from aircraft using discrete sampling. *J. Geophys. Res.*, **108**, 4514. doi: 4510.1029/2002JD003306.

Ball, S.M. and Jones, R. (2003) Broad-band cavity ring-down spectroscopy. *Chem. Rev.*, **103**, 5239–5262.

Ball, S.M. and Jones, R. (2009) Cavity ring-down spectroscopy: techniques and applications, *Broad-band Cavity Ring-down Spectroscopy*, Blackwell Publishing Limited, pp. 57–88, Chapter 3, ISBN: 978-1-4051-7688-0.

Ball, S.M., Langridge, J.M., and Jones, R. (2004) Broadband cavity enhanced absorption using light emitting diodes. *Chem. Phys. Lett.*, **398**, 68–74.

Bange, J. (2009) *Airborne Measurement of Turbulent Energy Exchange Between the Earth Surface and the Atmosphere*, Sierke Verlag, 174 pp. -ISBN 978-3-86844-221-2.

Bange, J. and Roth, R. (1999) Helicopter-borne flux measurements in the nocturnal boundary layer over land - A case study. *Bound.-Lay. Meteorol.*, **92**, 295–325.

Bange, J., Beyrich, F., and Engelbart, D.A.M. (2002) Airborne measurements of turbulent fluxes during LITFASS-98: a case study about method and significance. *Theor. Appl. Climatol.*, **73**, 35–51.

Bange, J., Spieß, T., Herold, M., Beyrich, F., and Hennemuth, B. (2006a) Turbulent fluxes from Helipod flights above quasi-homogeneous patches within the LITFASS area. *Bound.-Lay. Meteorol.*, **121**, 127–151.

Bange, J., Zittel, P., Spieß, T., Uhlenbrock, J., and Beyrich, F. (2006b) A new method for the determination of area-averaged turbulent surface fluxes from low-level flights using inverse models. *Bound.-Lay. Meteorol.*, **119**, 527–561.

Bannehr, L. and Glover, V. (1991) Preprocessing of Airborne Pyranometer Data. NCAR *Tech. Note* NCAR/TN-364+STR, NCAR Inf. Serv., Boulder, CO, p. 35.

Bannehr, L. and Schwiesow, R. (1993) A technique to account for the misalignment of pyranometers installed on aircraft. *J. Atmos. Oceanic Technol.*, **10**, 774–777.

Barbour, N.M. (2010) Inertial navigation sensors. Lecture Series Paper 2, Low-Cost Navigation Sensors and Integration Technology, NATO Science and Technology Organisation, Rto Educational Notes. RTO-EN-SET-116(2010) AC/323(SET-116)TP/311, ISBN 978-92-837-0109-5, *ftp://ftp.rta.nato.int//PubFullText/RTO/EN/RTO-EN-SET-116(2010)///EN-SET-116(2010)-02.pdf.*

Barkey, B. and Liou, K.N. (2001) Polar nephelometer for light-scattering measurements of ice crystals. *Opt. Lett.*, **26**, 232–234.

Baron, P. and Willeke, K. (2005) *Aerosol Measurement, Principles, Techniques and Applications*, 2nd edn, John Wiley & Sons, Inc., New York, p. 1160.

Barrick, J.D.W., Ritter, J.A., Watson, C.E., Wynkoop, M.W., Quinn, J.K., and Norfolk, D.R. (1996) Calibration of NASA turbulent air motion measurement system. *Technical Paper* 3610, NASA.

Bartenbach, S., Williams, J., Plass-Dulmer, C., Berresheim, H., and Lelieveld, J. (2007) In situ measurement of reactive hydrocarbons at Hohenpeissenberg with comprehensive two-dimensional gas chromatography (GC x GC-FID): use in estimating HO and NO_3. *Atmos. Chem. Phys.*, **7**, 1–14.

Bartlett, L. and Webb, A. (2000) Changes in ultraviolet radiation in the 1990s: spectral measurements from Reading England. *J. Geophys. Res.*, **105**, 4889–4893.

Basley, B.B., Jensen, M.L., and Frehlich, R.G. (1998) The use of state-of-the-art kites for profiling the lower atmosphere. *Bound.-Lay. Meteorol.*, **87**, 1–25.

Bateson, C. and Curtiss, B. (1996) A method for manual end member selection and spectral unmixing. *Remote Sens. Environ.*, **55**, 229–243.

Battaglia, A., Sturniolo, O., and Prodi, F. (2001) Analysis of polarization Radar returns from ice clouds. *Atmos. Res.*, **59-60**, 231–250.

Battaglia, A., Tanelli, S., Kobayashi, S., Zrnic, D., Hogan, R., and Simmer, C. (2010) Multiple-scattering in Radar systems: a review. *J. Quant. Spectrosc. Radiat. Transfer*, **111**, 917–947.

Battan, L.J. (1973) *Radar Observation of the Atmosphere*, The University of Chicago Press, 325 pp.

Baumgardner, D. (1983) An analysis and comparison of five water droplet measuring instruments. *J. Appl. Meteorol.*, **22**, 891–910.

Baumgardner, D. and Korolev, A. (1997) Airspeed corrections for optical array probe sample volumes. *J. Atmos. Oceanic Technol.*, **14**, 1224–1129.

Baumgardner, D. and Rodi, A. (1989) Laboratory and wind tunnel evaluations of the Rosemount icing detector. *J. Atmos. Oceanic. Technol.*, **6**, 971–979.

Baumgardner, D. and Spowart, M. (1990) Evaluation of the forward scattering spectrometer probe. Part III: time response and laser inhomogeneity limitations. *J. Atmos. Oceanic Technol.*, **7**, 666–672.

Baumgardner, D., Strapp, W., and Dye, J.E. (1985) Evaluation of the forward scattering spectrometer probe. Part II: corrections for coincidence and dead-time losses. *J. Atmos. Oceanic Technol.*, **2**, 626–632.

Baumgardner, D., Dye, J.E., Knollenberg, R.G., and Gandrud, B. (1992) Interpretation of measurements made by the FSSP-300X during the Airborne Arctic Stratospheric Expedition. *J. Geophys. Res.*, **97**, 8035–8046.

Baumgardner, D., Jonsson, H., Dawson, W., O'Connor, D., and Newton, R. (2001) The Cloud, Aerosol and Precipitation Spectrometer (CAPS): a new instrument for cloud investigations. *Atmos. Res.*, **59-60**, 251–264.

Baumgardner, D., Kok, G., and Raga, G. (2004) Warming of the Arctic lower stratosphere by light absorbing particles. *Geophys. Res. Lett.*, **31**, L06 117. doi: 10.1029/2003GL018883.

Baumgardner, D., Chepfer, H., Raga, G., and Kok, G. (2005) The shapes of very small cirrus particles derived from in situ measurements. *Geophys. Res. Lett.*, **32**, 300. doi: 10.1029/2004GL021.

Baumgardner, D., Brenguier, J.-L., Bucholtz, A., Coe, H., DeMott, P., Garrett, T., Gayet, J.-F., Hermann, M., Heymsfield, A., Korolev, A., Krämer, M., Petzold, A., Strapp, W., Pilewskie, P., Taylor, J., Twohy, C., and Wendisch, M. (2011) Airborne instruments to measure atmospheric aerosol particles, clouds and radiation: a cook's tour of mature and emerging technology. *Atmos. Res.*, **102**, 10–29. doi: 10.1016/j.atmosres.2011.06.021.

Baumgardner, D., Avallone, L., Bansemer, A., Borrmann, S., Brown, P., Bundke, U., Chuang, P., Cziczo, D., Field, P., Gallagher, M., Gayet, J.-F., Heymsfield, A., Korolev, A., Krämer, M., McFarquhar, G., Mertes, S., Möhler, O., Lance, S., Lawson, P., Petters, M., Pratt, K., Roberts, G., Rogers, D., Stetzer, O., Stith, J., Strapp, W., Twohy, C., and Wendisch, M. (2012) In situ, airborne instrumentation: addressing and solving measurement problems in ice clouds. *Bull. Am. Meteorol. Soc.*, 29–34.

Beals, M., Fugal, J.P., Shaw, R.A., Spuler, S., Stith, J., and Campos, T. (2012) Observations of cloud droplet size distributions on microphysically relevant scales. Proceedings of the 16th International Conference on Clouds and Precip, Leipzig, Germany.

Beard, K.V. and Chuang, C. (1987) A new model for the equilibrium shape of raindrops. *J. Atmos. Sci.*, **44**, 1509.

Beaton, S. (2006) The Ophir Air Temperature Radiometer. National Center for Atmospheric Research (NCAR), Boulder, Colorado, USA. *Tech. Note*, NCAR/TN-471+EDD.

Beck, M., Hofstetter, D., Aellen, T., Faist, J., Oesterle, U., Ilegems, M., Gini, E., and Melchior, H. (2002) Continuous wave operation of a mid-infrared semiconductor laser at room temperature. *Science*, **295**, 301–305.

Behrendt, A., Wulfmeyer, V., Kiemle, C., Ehret, G., Flamant, C., Schaberl, T., Bauer, H.-S., Kooi, S., Ismail, S., Ferrare, R., and Browell, E. (2007) Intercomparison of water vapor data measured with Lidar during IHOP_2002, Part 2: airborne to airborne systems. *J. Atmos. Oceanic Technol.*, **24**, 22–39.

Belyaev, S.P. and Levin, L.M. (1974) Techniques for collection of representative aerosol samples. *J. Aerosol Sci.*, **5**, 325–338.

Ben-Dor, E. (1994) A precaution regarding cirrus cloud detection from Airborne Imaging Spectrometer data using the 1.38 μm water vapor band. *Remote Sens. Environ.*, **50**, 346–350.

Ben-Dor, E. (2001) Imaging spectroscopy, in *Imaging Spectroscopy for Urban Applications* (ed. F. van der Meer), Kluwer Academic Press, Chapter 9, pp. 243–281.

Ben-Dor, E. and Kruse, F. (1996) Detection of atmospheric gases using GER 63 channel scanner data acquired over Makhtesh Ramon Negev Israel. *Int. J. Remote Sens.*, **17**, 1215–1232.

Ben-Dor, E., Goetz, A., and Shapiro, A. (1994) Estimation of cirrus cloud and aerosol scattering in hyperspectral image data. Proceedings of the International Symposium on Spectral Sensing Research, San Diego, CA, pp. 582–593.

Ben-Dor, E., Irons, J., and Epema, A. (1999) Manual of remote sensing, in *Soil Spectroscopy*, 3rd edn (ed. A. Rencz), John Wiley & Sons, Inc., New-York, Chichester, Weinheim, Brisbane, Singapore, Toronto, pp. 111–188.

Ben-Dor, E., Patkin, K., Banin, A., and Karnieli, A. (2002) Mapping of several soil properties using DAIS-7915 hyperspectral scanner data. *Int. J. Remote Sens.*, **23**, 1043–1062.

Ben-Dor, E., Goldshalager, N., Braun, O., Kindel, B., Goetz, A.F.H., Bonfil, D., Agassi, M., Margalit, N., Binayminy, Y., and Karnieli, A. (2004) Monitoring of infiltration rate in semiarid soils using airborne hyperspectral technology. *Int. J. Remote Sens.*, **25**, 1–18.

Berden, G. and Engeln, R. (2009) *Cavity Ring-Down Spectroscopy: Techniques and Applications*, John Wiley & Sons, Inc., New York, pp. 322

Berden, G., Peeters, R., and Meijer, G. (2000) Cavity ring-down spectroscopy: experimental schemes and applications. *Int. Rev. Phys. Chem.*, **19**, 565–607.

Bergstrom, R., Pilewskie, P., Russell, P., Redemann, J., Bond, T., Quinn, P., and Sierau, B. (2007) Spectral absorption properties of atmospheric aerosol particles. *Atmos. Chem. Phys.*, **7**, 5937–5943.

Berk, A., Cooley, T.W., Anderson, G.P., Acharya, P.K., Bernstein, L.S., Muratov, L., Lee, J., Fox, M.J., Alder-Golden, S.M., Chetwynd, J.H, Hoke, M.L., Lockwood, R.B., Gardner, J.A., and Lewis, P.E. (2005) MODTRAN5: a reformulated atmospheric band model with auxiliary species and practical multiple scattering options. *Proc. SPIE*, **5655**, 88–95.

Berman, M., Kiiveri, H., Lagerstorm, R., Ernst, A., Dumme, R., and Huntington, J. (2004) ICE: a statistical approach to identifying endmembers in hyperspectral images. *IEEE Trans. Geosci. Remote*, **42**, 2085–2095.

Bernstein, L., Adler-Golden, S.M., Sundberg, R.L., and Ratkowski, A.J. (2008) In-scene-based atmospheric correction of uncalibrated VISible-SWIR (VIS-SWIR) hyper- and multi-spectral imagery. Proceedings of SPIE 7107. doi: 10.1117/12.808193.

Bertaux, J.-L. and Delan, A. (1978) Vertical distribution of H_2O in the stratosphere as determined by UV fluorescence in situ measurements. *Geophys. Res. Lett.*, **5**, 1017–1020.

Bertram, T., Perring, A., Wooldridge, P., Crounse, J., Kwan, A., Wennberg, P., Scheuer, E., Dibb, J., Avery, M., Sachse, G., Vay, S., Crawford, J., McNaughton, C., Clarke, A., Pickering, K., Fuelberg, H., Huey, G., Blake, D., Singh, H., Hall, S., Shetter, R., Fried, A., Heikes, B., and Cohen, R. (2007) Direct measurements of the convective recycling of the upper troposphere. *Science*, **315**, 816–820.

Beuttell, R.G. and Brewer, A.W. (1949) Instruments for the measurement of the visual range. *J. Sci. Instrum.*, **26**, 357–359.

Bevly, D.M. and Cobb, S. (2010) *GNSS for Vehicle Control*, Artech House, Boston, MA.

Bhawar, P.D.G., Summa, D., Flamant, C., Althausen, D., Behrendt, A., Kiemle, C., Bosser, P., Cacciani, M., Champollion, C., Iorio, T.D., Engelmann, R., Herold, C., Pal, S., Wirth, M., and Wulfmeyer, V. (2011) The water vapor intercomparison effort in the framework of the Convective and Orographically-Induced Precipitation Study: airborne-to-ground-based and airborne-to-airborne Lidar systems. *Q. J. R. Meteorol. Soc.*, **137**, 325–348, 2011.

Bianci, R., Cavalli, R., Fiumi, L., Marino, C., and Pignatti, S. (1996) CNR LARA project: evaluation of two years of airborne spectrometry. 2nd International Airborne Remote Sensing Conference and Exhibition, San Fransico, CA, pp. 534–543.

Bickel, W. and Bailey, W.M. (1985) Stokes vectors, mueller matrices, and polarized light scattering. *Am. J. Phys.*, **53**, 468–478.

Bierwirth, E., Wendisch, M., Jäkel, E., Ehrlich, A., Schmidt, K., Stark, H., Pilewskie, P., Esselborn, M., Gobbi, G., Ferrare, R., Müller, T., and Clarke, A. (2010) A new method to retrieve aerosol layer absorption coefficient from airborne flux density and actinic radiation measurements. *J. Geophys. Res.*, **115**, D14211. doi: 10.1029/2009JD013636.

Bigg, E.K. (1986) Discrepancy between observation and prediction of concentrations of cloud condensation nuclei. *Atmos. Res.*, **20**, 82–86.

Bilbro, J., Fichtl, G., Fitzjarrald, D., Krause, M., and Lee, R. (1984a) Airborne Doppler Lidar wind field measurements. *Bull. Am. Meteorol. Soc.*, **65**, 384–359.

Bilbro, J., DiMarzio, C., Fitzjarrald, D., Johnson, S., and Jones, W. (1986) Airborne Doppler Lidar measurements. *Appl. Opt.*, **25**, 3952–3960.

Bilbro, J.W., Fichtl, G., Fitzjarrald, D., and Krause, M. (1984b) Airborne Doppler Lidar wind field measurements. *Bull. Am. Meteorol. Soc.*, **65**, 348–359.

Bin Abas, M.R., Rahman, N.A., Omar, N., Maah, M.J., Samah, A.A., Orosc, D., Ottoc, A., and Simoneit, B. (2004) Organic composition of aerosol particulate matter during a haze episode in Kuala Lumpur Malaysia. *Atmos. Environ.*, **38**, 4223–4241.

Biskos, G., Reavell, K., and Collings, N. (2005) Description and theoretical analysis of a differential mobility spectrometer. *Aerosol Sci. Technol.*, **39**, 527–541.

Biter, C., Dye, J., Huffman, D., and King, W.D. (1987) The drop response of the CSIRO liquid water content. *J. Atmos. Ocean Technol.*, **4**, 359–367.

Black, M., Burpee, R., and Marks, F. (1996) Vertical motion characteristics of tropical cyclones determined with airborne Doppler radial velocities. *J. Atmos. Sci.*, **53**, 1887–1909.

Blackwell, W., Barrett, J., Chen, F., Leslie, R., Rosenkranz, P., Schwartz, M., and Staelin, D. (2001) NPOESS Aircraft Sounder Testbed-Microwave (NAST-M): instrument description and initial flight results. *IEEE Trans. Geosci. Remote*, **39**, 2444–2453.

Blake, N., Blake, D., Chen, T., Collins, J., Sachse, G., Anderson, B., and Rowland, F. (1997) Distribution and seasonality of selected hydrocarbons and halocarbons over the western Pacific basin during PEM-West A and PEM-West B. *J. Geophys. Res.*, **102**, 28315–28331.

Blanc, T.V., Plant, W.J., and Keller, W.C. (1989) The Naval research laboratory's air-sea interaction blimp experiment. *Bull. Am. Meteor. Soc.*, **70**, 354–364.

Blanchard, R. (1971) A new algorithm for computing inertial altitude and vertical velocity. *IEEE Trans. Aerosp. Elecron. Syst.*, **AES-7**, 1143–1146.

Blomquist, B., Huebert, B., Howell, S., Litchy, M., Twohy, C., Schanot, A., Baumgardner, D., Lafleur, B., Seebaugh, R., and Laucks, M. (2001) An evaluation of the community aerosol inlet for the NCAR C-130 research aircraft. *J. Atmos. Oceanic Technol.*, **18**, 1387–1397.

Blyth, A.M., Lasher-Trapp, S., Cooper, W.A., Knight, C.A., and Latham, J. (2003) The role of giant and ultragiant nuclei in the formation of early RADAR echoes in warm cumulus clouds. *J. Atmos. Sci.*, **60**, 2557–2572.

Boardman, J., Kruse, F., and Green, R. (1995) Mapping target signatures via partial unmixing of AVIRIS data. Proceedings of JPL Airborne Erath Science Workshop, pp. 23–26.

Boatman, J.F. and Auer, A.H. (1983) The role of cloud top entrainment in cumulus clouds. *J. Atmos. Sci.*, **40**, 1517–1534.

Boegel, W. and Baumann, R. (1991) Test and calibration of the DLR Falcon wind measuring system by maneuvres. *J. Atmos. Oceanic Technol.*, **8**, 5–18.

Boers, R., Melfi, S., and Palm, S.P. (1991) Cold-air outbreak during GALE: lidar observations and modeling of boundary layer dynamics. *Mon. Weather Rev.*, **119**, 1132–1147.

Boers, R., Mitchell, R., and Krummel, P. (1998) Correction of aircraft pyranometer measurements for diffuse radiance and alignment errors. *J. Geophys. Res.*, **103**, 16753–16758.

Bognar, J. and Birks, J. (1996) Miniaturized ultraviolet ozonesonde for atmospheric measurements. *Anal. Chem.*, **68**, 3059–3062.

Bohl, W. (1998) *Technische Strömungslehre*, Vogel Verlag, Würzbug, Germany, p. 137.

Bohn, B., Kraus, A., Müller, M., and Hofzumahaus, A. (2004) Measurement of atmospheric $O_3 \rightarrow O(^1D)$ photolysis frequencies using filterradiometry. *J. Geophys. Res.*, **109**, D10S90. doi: 10.1029/2003JD004319.

Bohn, B., Corlett, G.K., Gillmann, M., Sanghavi, S., Stange, G., Tensing, E., Vrekoussis, M., Bloss, W., Clapp, L., Kortner, M., Dorn, H.-P., Monks, P.S., Platt, U., Plass-Dulmer, C., Mihalopoulos, N., Heard, D.E., Clemitshaw, K.C., Meixner, F., Prevot, A.S.H., and Schmitt, R. (2008) Photolysis frequency measurement techniques: results of a comparison within the ACCENT project. *Atmos. Chem. Phys.*, **8**, 5373–5391.

Bohn, D. and Simon, H. (1975) Mehrparametrige Approximation der Eichräume und Eichflächen von Unterschall- bzw. Überschall-5-Loch-Sonden. *Arch. Tech. Mess. Ind. M.*, **3**, 31–37.

Bohren, C.F. and Battan, L.J. (1980) Radar backscattering by inhomogeneous precipitation particles. *J. Atmos. Sci.*, **37**, 1821–1827.

Bohren, C.F. and Huffman, D.R. (1983) *Absorption and Scattering of Light by Small Particles*, John Wiley & Sons, Inc., p. 544

Boiffier, J.-L. (1998) *The Dynamics of Flight - the Equations*, John Wiley & Sons, Ltd, Chichester.

Bollinger, M., Sievers, R., Fahey, D., and Fehsenfeld, F. (1983) Conversion of nitrogen dioxide, nitric acid to nitric oxide by gold-catalysed reduction with carbon monoxide. *Anal. Chem.*, **55**, 1981–1986.

Bond, T.C. and Bergstrom, R.W. (2006) Light absorption by carbonaceous particles: an investigative review. *Aerosol Sci. Technol.*, **40**, 27–67.

Bond, T.C., Anderson, T.L., and Campbell, D. (1999) Calibration and intercomparison of filter-based measurements of visible light absorption by aerosol. *Aerosol Sci. Technol.*, **30**, 582–600.

Bond, T.C., Covert, D.S., and Mueller, T. (2009) Truncation and angular-scattering corrections for absorbing aerosol in the TSI 3563 nephelometer. *Aerosol Sci. Technol.*, **43**, 866–871.

Bonsang, B., Kanakidou, M., and Lambert, G. (1990) NMHC in the marine atmosphere: preliminary results of monitoring at Amsterdam island. *J. Atmos. Chem.*, **11**, 169–178.

Born, M. and Wolf, E. (2003) *Principles of Optics*, 7th edn, *Cambridge University Press.*

Borrmann, S. and Jaenicke, R. (1993) Application of microholography for ground-based in situ measurements in stratus cloud layers: a case study. *J. Atmos. Oceanic Technol.*, **10**, 277–293.

Borrmann, S., Luot, B., and Mishchenko, M. (2000) Application of the T-matrix method to the measurement of aspherical (ellipsoidal) particles with forward scattering optical particle counters. *J. Aerosol Sci.*, **31**, 789–799.

Borstad, G.A., Edel, H.R., Gower, J.F.R., and Hollinger, A.B. (1985) *Analysis of Test and Flight Data from the Fluorescence Line Imager*, Canadian Special Publication of Fisheries and Aquatic Sciences, Ottawa

Department of Fisheries and Oceans, p. 83.

Bosart, B., Lee, W.-C., and Wakimoto, R.M. (2002) Procedures to improve the accuracy of airborne Doppler Radar data. *J. Atmos. Oceanic Technol.*, **19**, 322–339.

Bösenberg, J. (1998) Ground-based differential absorption Lidar for water-vapor and temperature profiling: methodology. *Appl. Opt.*, **37**, 3845–3860.

Bou Karam, D., Flamant, C., Knippertz, P., Reitebuch, O., Pelon, J., Chong, M., and Dabas, A. (2008) Dust emissions over Sahel associated with the West African Monsoon inter-tropical discontinuity region: a representative case study. *Q. J. R. Meteorol. Soc.*, **134**, 621–634.

Boulter, J.E., Cziczo, D., Middlebrook, A., Thomson, D., and Murphy, D. (2006) Design and performance of a pumped counterflow virtual impactor. *Aerosol Sci. Technol.*, **40**, 969–976.

Bozoki, Z., Pogány, A., and Szabó, G. (2011) Photoacoustic instruments for practical applications: present, potentials and future challenges. *Appl. Spectrosc. Rev.*, **46**, 1–37. doi: 10.1080/05704 928.2010.520 178.

Bradshaw, J.D., Rodgers, M., Sandholm, S., Kesheng, S., and Davis, D. (1985) A 2-photon laser-induced fluorescence field instrument for ground-based and airborne measurements of atmospheric NO. *J. Geophys. Res.*, **90**, 2861–2873.

Brando, V., Keen, R., Daniel, P., Baumeister, A., Nethery, M., Baumeister, H., Hawdon, A., Swan, G., Mitchell, R., Campbell, S., Schroeder, T., Park, Y.-J., Edwards, R., Steven, A., Allen, S., Clementson, L., and Dekker, A. (2010) The Lucinda Jetty Coastal Observatory's role in satellite ocean colour calibration and validation for Great Barrier Reef coastal waters. Proceedings IEEE Oceans 2010, Sydney, Australia.

Brands, M., Kamphus, M., Böttger, T., Schneider, J., Drewnick, F., Roth, A., Curtius, J., Voigt, C., Borbon, A., Beekmann, M., Bourdon, A., Perrin, T., and Borrmann, S. (2011) Characterization of a newly developed aircraft-based laser ablation aerosol mass spectrometer (ALABAMA) and first field deployment in urban pollution plumes over Paris during MEGAPOLI 2009. *Aerosol Sci. Technol.*, **45**, 46–64. doi: 10.1080/02786 826.2010.517 813.

Brandtjen, R., Klüpfel, T., Perner, D., and Knudsen, B.M. (1994) Airborne measurements during the European Arctic Stratospheric Ozone Experiment: observation of OClO. *Geophys. Res. Lett.*, **21**, 1363–1366.

Brass, M. and Röckmann, T. (2010) Continuous-flow isotope ratio mass spectrometry method for carbon and hydrogen isotope measurements on atmospheric methane. *Atmos. Meas. Technol.*, **3**, 1707–1721. doi: 10.5194/amt-3-1707-2010.

Brenguier, J. and Amodei, L. (1989a) Coincidence and dead-time corrections for particle counters. Part I: a general mathematical formalism. *J. Atmos. Ocean. Technol.*, **6**, 575–584.

Brenguier, J. and Amodei, L. (1989b) Coincidence and dead-time corrections for particle counters. Part II: high concentration measurements with an FSSP. *J. Atmos. Ocean. Technol.*, **6**, 585–598.

Brenguier, J., Rodi, A.R., Gordon, G., and Wechsler, P. (1993) Real-time detection of performance degradation of the Forward-Scattering Spectrometer Probe. *J. Atmos. Ocean. Technol.*, **10**, 27–33.

Brenguier, J., Baumgardner, D., and Baker, B. (1994) A review and discussion of processing algorithms for FSSP concentration measurements. *J. Atmos. Ocean. Technol.*, **11**, 1409–1414.

Brenguier, J., Bourrianne, T., Coelho, A., Isbert, J., Peytavi, R., Trevarin, D., and Wechsler, P. (1998) Improvements of droplet size distribution measurements with the Fast-FSSP (Forward Scattering Spectrometer Probe). *J. Atmos. Ocean. Technol.*, **15**, 1077–1090.

Brenguier, J., Chuang, P., Fouquart, Y., Johnson, D., Parol, F., Pawlowska, H., Pelon, J., Schüller, L., Schröder, F., and Snider, J. (2000a) An overview of the ACE-2 CLOUDYCOLUMN closure experiment. *Tellus B*, **52**, 815–827.

Brenguier, J.-L. (1993) Observations of cloud microstructure at the centimeter scale. *J. Appl. Meteorol.*, **32**, 783–793.

Brenguier, J.-L. and Chaumat, L. (2001b) Droplet spectra broadening in cumulus clouds. Part 1: broadening in adiabatic clouds. *J. Atmos. Sci.*, **58**, 628–641.

Brenguier, J.-L. and Chaumat, L. (2001b) Droplet spectra broadening in cumulus clouds. Part I: broadening in adiabatic cores. *J. Atmos. Sci.*, **58**, 628–641.

Brenguier, J.-L., Pawlowska, H., Schüwller.L., Preusker, R., Fischer, J., and Fouquart, Y. (2000b) Radiative properties of boundary layer clouds: droplet effective radius versus number concentration. *J. Atmos. Sci.*, **57**, 803–821.

Brenguier, J.-L., Pawlowska, H., and Schüller, L. (2003) Cloud microphysical and radiative properties for parameterization and setallite monitoring of the indirect effect of aerosol particles. *J. Geophys. Res.*, **108**, 8632. doi: 10.1029/2002JD002682.

Brenguier, J.-L., Burnet, F., and Geoffroy, O. (2011) Cloud optical thickness and liquid water path. Does the *k* coefficient vary with droplet concentration? *Atmos. Chem. Phys.*, **11**, 9771–9786. doi: 10.5194/acp-11-9771-2011.

Brenninkmeijer, C.A.M., Crutzen, P., Boumard, F., Dauer, T., Dix, B., Ebinghau, R., Filippi, D., Fischer, H., Franke, H., Friess, U., Heintzenberg, J., Helleis, F., Hermann, M., Kock, H.H., Koeppel, C., Lelieveld, J., Leuenberger, M., Martinsson, B.G., Miemczyk, S., Moret, H., Nguyen, H.N., Nyfeler, P., Oram, D., Sullivan, D.O., Penkett, S., Platt, U., Pupek, M., Ramonet, M., Randa, B., Reichelt, M., Rhee, T., Rohwer, J., Rosenfeld, K., Scharffe, D., Schlager, H., Schumann, U., Slemr, F., Sprung, D., Stock, P., Thaler, R., Valentino, F., van Velthoven, P., Waibel, A., Wandel, A., Waschitschek, K., Wiedensohler, A., Xueref-Remy, I., Zahn, A., Zech, U., and Ziereis, H. (2007) Civil Aircraft for the regular investigation of the atmosphere based on an instrumented container: the new CARIBIC system. *Atmos. Chem. Phys.*, **7**, 4953–4976.

Brewer, A.W. (1949) Evidence for a world circulation provided by the measurements of helium and water vapor distribution in the stratosphere. *Q. J. R. Meteorol. Soc.*, **75**, 351–363.

Bringi, V.N. and Chandrasekar, V. (2001) *Polarimetric Doppler Weather Radar: Principles and Applications*, Cambridge University Press, 636 pp.

Brinkmann, J., Hackenthal, E.-M., Krämer, M., Schüle, M., Schütz, L., and Sprengart-Eichel, C. (2001) Particle distribution, composition and processing during cloud, fog and rain cycles, *Dynamics and Chemistry of Hydrometeors; Final Report of the Collaborative Research Centre 233: 'Chemie und Dynamik der Hydrometeore'*, John Wiley & Sons, Ltd, ISBN 3-527-27727-7.

Brock, C.A., Schroder, F., Karcher, B., Petzold, A., Busen, R., and Fiebig, M. (2000) Ultrafine particle size distributions measured in aircraft exhaust plumes. *J. Geophys. Res.*, **105**, 26555–26567.

Brock, C.A., Cozic, J., Bahreini, R., Froyd, K.D., Middlebrook, A.M., McComiskey, A., Brioude, J., Cooper, O.R., Stohl, A., Aikin, K.C., de Gouw, J.A., Fahey, D.W., Ferrare, R.A., Gao, R.S., Gore, W., Holloway, J.S., Hübler, G., Jefferson, A., Lack, D.A., Lance, S., Moore, R.H., Murphy, D.M., Nenes, A., Novelli, P.C., Nowak, J.B., Ogren, J.A., Peischl, J., Pierce, R.B., Pilewskie, P., Quinn, P.K., Ryerson, T.B., Schmidt, K.S., Schwarz, J.P., Sodemann, H., Spackman, J.R., Stark, H., Thomson, D.S., Thornberry, T., Veres, P., Watts, L.A., Warneke, C., and Wollny, A.G. (2011) Characteristics, sources, and transport of aerosol particles measured in spring 2008 during the aerosol, radiation, and cloud processes affecting Arctic Climate (ARCPAC) Project. *Atmos. Chem. Phys.*, **11**, 2423–2453.

Brook, A. and Ben-Dor, E. (2011) Supervised vicarious calibrating of hyperspectral data. *Remote Sens. Environ.*, **115**, (6), 1543–1555.

Brookes, D. (2009) Characterisation of a PERCA instrument and measurements of peroxy radicals during the West African Monsoon 2006. PhD thesis. University of Leicester.

Brooks, B.J., McQuaid, J.B., Smith, M.H., Crosier, J., Williams, P.I., Coe, H., and Osborne, S. (2007) Intercomparison of VACC- and AMS-derived nitrate, sulphate and ammonium aerosol loadings during

ADRIEX. *Q. J. R. Meteorol. Soc.*, **133**, 77–84.

Brough, N., Reeves, C.E., Penkett, S.A., Dewey, K., Kent, J., Barjat, H., Monks, P., Ziereis, H., Stock, P., Huntrieser, H., and Schlager, H. (2003) Intercomparison of aircraft instruments on board the C-130 and Falcon 20 over southern Germany during Export 2000. *Atmos. Chem. Phys.*, **3**, 2127–2138.

Browell, E., Carter, A., Shipley, S., Allen, R., Butler, C., Mayo, M., Siviter, J. Jr., and Hall, W. (1983) NASA multipurpose airborne DIAL system and measurements of ozone and aerosol profiles. *Appl. Opt.*, **22**, 522–534.

Browell, E., Ismail, S., and Grant, W. (1998) Differential absorption Lidar (DIAL) measurements from air and space. *Appl. Phys. B*, **67**, 399–410.

Brown, E.N., Friehe, C.A., and Lenschow, D.H. (1983) The use of pressure fluctuations on the nose of aircraft for measuring air motion. *J. Clim. Appl. Meteorol.*, **22**, 171–180.

Brown, P. (1989) Use of holography for airborne cloud physics measurements. *J. Atmos. Ocean. Technol.*, **6**, 293–306.

Brown, P.R.A. and Francis, P.N. (1995) Improved measurements of the ice water content in cirrus using a total-water probe. *J. Atmos. Oceanic Technol.*, **12**, 410–414.

Brown, R.G. and Hwang, P. (1997) *Introduction to Random Signals and Applied Kalman Filtering: With MATLAB Exercises and Solutions*, John Wiley & Sons, Inc., New York.

Brown, S.S. (2003) Absorption spectroscopy in high-finesses cavities for atmospheric studies. *Chem. Rev.*, **103**, 5219–5238.

Brown, S.S., Dubé, W., Osthoff, H., Stutz, J., Ryerson, T., Wollny, A., Brock, C., Warneke, C., de Gouw, J., Atlas, E., Neuman, J., Holloway, J., Lerner, B., Williams, E., Kuster, W., Goldan, P., Angevine, W., Trainer, M., Fehsenfeld, F., and Ravishankara, A. (2007) Vertical profiles in NO_3 and N_2O_5 measured from an aircraft: results from the NOAA P-3 and surface platforms during NEAQS 2004. *J. Geophys. Res.*, **112**, doi: 10.1029/2007JD008883.

Broxmeyer, C. (1964) *Inertial Navigation Systems, McGraw-Hill Electronic Sciences Series*, McGraw-Hill, New York.

Bruegge, C., Conel, J., Green, R., Margolis, J., Holm, R., and Toon, G. (1992) Water vapor column abundance retrievals during FIFE. *J. Geophys. Res.*, **97**, 18759–18768.

Brune, W.H., Anderson, J., and Chan, K. (1989) In situ observations of BrO over Antarctica: ER-2 aircraft results from 54°S to 72°S latitude. *J. Geophys. Res.*, **94**, 16639–16647. doi: 10.1029/JD094iD14p16639.

Brune, W.H., Toohey, D.W., Anderson, J., and Chan, K. (1990) In situ observations of ClO in the Arctic Stratosphere: ER-2 aircraft results from 59°N to 80°N latitude. *Geophys. Res. Lett.*, **17**, 505–508, doi: 10.1029/GL017i004p00505.

Brune, W.H., Faloona, I., Tan, D., Weinheimer, A., Campos, T., Ridley, B., Collins, S.V.J., Sachse, G., and Jacob, L.J.D. (1998) Airborne in situ OH and HO_2 observations in the cloud-ree troposphere and lower stratosphere during SUCCESS. *Geophys. Res. Lett.*, **25**, 1701–1704.

Bruneau, D. (2002) Fringe-imaging Mach-Zehnder interferometer as a spectral analyzer for molecular Doppler wind Lidar. *Appl. Opt.*, **41**, 503–510.

Bruneau, D. and Pelon, J. (2003) Simultaneous measurements of particle backscattering and extinction coefficients and wind velocity by Lidar with a Mach-Zehnder interferometer: principle of operation and performance assessment. *Appl. Opt.*, **42**, 1101–1114.

Bruneau, D., Quaglia, P., Flamant, C., Meisonnier, M., and Pelon, J. (2001a) Airborne Lidar LEANDRE II for water-vapor profiling in the troposphere. II. First results. *Appl. Opt.*, **40**, 3462–3475.

Bruneau, D., Quaglia, P., Flamant, C., and Pelon, J. (2001b) Airborne Lidar LEANDRE II for water-vapor profiling in the troposphere. I. System description. *Appl. Opt.*, **40**, 3450–3461.

Brunelli, N.A., Flagan, R.C., and Giapis, K.P. (2009) Radial differential mobility analyzer for one nanometer particle classification. *Aerosol Sci. Technol.*, **43**, 53–59.

Bruns, M., Bühler, S., Burrows, J.P., Heue, K.-P., Platt, U., Pundt, I., Richter, A., Rozanov, A., Wagner, T., and Wang, P. (2004) Retrieval of profile information from airborne multi axis UV/visible skylight absorption measurements. *Appl. Opt.*, **43**, 4415–4426.

Bruun, H.H. (1995) *Hot-Wire Anemometry*, Oxford University Press, Oxford, 507 p.

Bruzzone, L.M.C. and Marconcini, M. (2006) Classification of remote sensing images from urban areas using fuzzy probabilistic models. *IEEE Trans. Geosci. Remote*, **3**, 40–44.

Bruzzone, L.M.C. and Marconcini, M. (2006) A novel transductive SVM for semisupervised classification of remote sensing. *IEEE Trans. Geosci. Remote*, **44**, 3363–3373.

Buchhave, P., George, W.K. Jr., and Lumley, J.L. (1979) The measurement of turbulence with the laser-Doppler anemometer. *Annu. Rev. Fluid Mech.*, **11**, 443–503.

Bucholtz, A. (1995) Rayleigh-scattering calculations for the terrestrial atmosphere. *Appl. Opt.*, **34**, 2765–2773.

Bucholtz, A., Bluth, R., Kelly, B., Taylor, S., Batson, K., Sarto, A., Tooman, T., and McCoy, R. Jr. (2008) The Stabilized Radiometer Platform (STRAP) - An actively stabilized horizontally level platform for improved aircraft irradiance measurements. *J. Atmos. Oceanic Technol.*, **25**, 2161–2175. doi: 10.1175/2008JTECHA1085.1.

Buck, A. (1985) The Lyman-alpha absorption hygrometer, in *Moisture and Humidity Symposium Washington, DC*, Instrument Society of America, Research Triangle Park, NC, pp. 411–436.

Buck, A.L. (1976) The variable-path Lyman-alpha hygrometer and its operating characteristics. *Bull. Am. Meteorol. Soc.*, **57**, 1113–1118.

Buck, A.R.C. (1991) Development of a cryogenic dew frost point hygrometer, in *Proceedings 7th Symposium Meteorological Observation Instrumentation*, American Meteorological Society, pp. 322–326.

Buckingham, R. and Staenz, K. (2008) Review of current and planned civilian space hyperpsectral sensors for EO. *Can. J. Remote Sens.*, **34**, (Suppl. 1), S187–S197.

Buehler, S., Jimenez, C., Evans, K., Eriksson, P., Rydberg, B., Heymsfield, A., Stubenrauch, C., Lohmann, U., Emde, C., John, V., Sreerekha, T., and Davis, C. (2007) A concept for a satellite mission to measure cloud ice water path, ice particle size, and cloud altitude. *Q. J. R. Meteorol. Soc.*, **133**, 109–128.

Bufton, J.L., Itabe, T., Strow, L.L., Korb, C.L., Gentry, B.M., and Weng, C.Y. (1983) Frequency-doubled CO_2 Lidar measurement and diode laser spectroscopy of atmospheric CO_2. *Appl. Opt.*, **22**, 2592–2602.

Bundke, U., Nillius, B., Jaenicke, R., Wetter, T., Klein, H., and Bingemer, H. (2008) The Fast Ice Nucleus Chamber FINCH. *Atmos. Res.*, **90**, 180–186.

Bundke, U., Reimann, B., Nillius, B., Jaenicke, R., and Bingemer, H. (2010) Development of a Bioaerosol single particle detector (BIO IN) for the Fast Ice Nucleus Chamber FINCH. *Atmos. Meas. Technol.*, **3**, 1–9.

Bunker, A.F. (1955) Turbulence and shearing stresses measured over the North Atlantic Ocean by an airplane-acceleration technique. *J. Atmos. Sci.*, **12**, 445–455.

Burnet, F. and Brenguier, J.-L. (1999) Validation of droplet spectra and water content measurements. *Phys. Chem. Earth B*, **24**, 249–254.

Burnet, F. and Brenguier, J.-L. (2002) Comparison between standard and modified forward scattering spectrometer probes during the small Cumulus microphysics studies. *J. Atmos. Oceanic Technol.*, **19**, 1516–1531.

Burnet, F. and Brenguier, J.-L. (2007) Observational study of the entrainment-mixing process in warm convective clouds. *J. Atmos. Sci.*, **64**, 1995–2011.

Burnet, F. and Brenguier, J.-L. (2010) The onset of precipitation in warm cumulus clouds: an observational case-study. *Q. J. R. Meteorol. Soc.*, **136**, 374–381.

Burton, S.P., Ferrare, R.A., Hostetler, C.A., Hair, J.W., Kittaka, C., Vaughan, M.A., Obland, M.D., Rogers, R.R., Cooks, A.L., Harper, D.B., and Remer, L.A. (2010) Using airborne high spectral resolution Lidar data to evaluate combined active plus passive retrievals of aerosol extinction profiles. *J. Geophys. Res.*, **115**, D00H15. doi: 10.1029/2009JD012130.

Burton, S.P., Ferrare, R.A., Hostetler, C.A., Hair, J.W., Rogers, R.R., Obland, M.D., Butler, C.F., Cooks, A.L., Harper, D.B., and Froyd, K.D. (2012) Aerosol classification using airborne High Spectral Resolution Lidar measurements - methodology and examples. *Atmos. Meas. Technol.*, **5**, 73–98.

Busen, R. and Buck, A. (1995) A high-performance hygrometer for air-craft use: description, installation, and flight data. *J. Atmos. Oceanic Technol.*, **12**, 73–84.

Cahalan, R., McGill, M., Kolasinski, L.J., Varnai, T., and Yetzer, K. (2005) THOR - cloud thickness from offbeam Lidar returns. *J. Atmos. Oceanic Technol.*, **22**, 605–627.

Cahen, C. and Mégie, G. (1981) A spectral limitation of the range resolved differential absorption Lidar technique. *J. Quant. Spectrosc. Radiat. Transfer*, **25**, 151–157.

Cahen, C., Mégie, G., and Flamant, P. (1982) Lidar monitoring of the water vapor cycle in the troposphere. *J. Appl. Meteorol.*, **21**, 1506–1515.

Cai, Y., Montague, D.C., Mooiweer-Bryan, W., and Deshler, T. (2008) Performance characteristics of the ultra high sensitivity aerosol spectrometer for particles between 55 and 800 nm: laboratory and field studies. *J. Aerosol Sci.*, **39**, 759–769.

Cain, S., Ram, M., and Woodward, S. (1998) Qualitative and quantitative wind tunnel measurements of the airflow through a shrouded airborne aerosol sampling probe. *J. Aerosol Sci.*, **29**, 1157–1169.

Cain, S.A. and Ram, M. (1998) Numerical simulation studies of the turbulent airflow through a shrouded airborne aerosol sampling probe and estimation of the minimum sampler transmission efficiency. *J. Aerosol Sci.*, **29**, 1145–1156.

Cairo, F., Pommereau, J.P., Law, K.S., Schlager, H., Garnier, A., Fierli, F., Ern, M., Streibel, M., Arabas, S., Borrmann, S., Berthelier, J.J., Blom, C., Christensen, T., D'Amato, F., Di Donfrancesco, G., Deshler, T., Diedhiou, A., Durry, G., Engelsen, O., Goutail, F., Harris, N.R.P., Kerstel, E.R.T., Khaykin, S., Konopka, P., Kylling, A., Larsen, N., Lebel, T., Liu, X., MacKenzie, A.R., Nielsen, J., Oulanowski, A., Parker, D.J., Pelon, J., Polcher, J., Pyle, J.A., Ravegnani, F., Riviere, E.D., Robinson, A.D., Rockmann, T., Schiller, C., Simoes, F., Stefanutti, L., Stroh, F., Some, L., Siegmund, P., Sitnikov, N., Vernier, J.P., Volk, C.M., Voigt, C., von Hobe, M., Viciani, S., and Yushkov, V. (2010) An introduction to the SCOUT-AMMA stratospheric aircraft, balloons and sondes campaign in West Africa, August 2006: rationale and roadmap. *Atmos. Chem. Phys.*, **10**, 2237–2256.

Camps-Valls, G., Gomez-Chova, L., Munoz-Mari, J., Vila-Frances, J., and Calpe-Maravilla, J. (2006) Composite kernels for hyperspectral image classification. *IEEE Trans. Geosci. Remote*, **3**, 93–97. doi: 10.1109/LGRS.2005.857031.

Canagaratna, M., Jayne, J., Jimenez, J., Allan, J., Alfarra, M., and Zhang, Q. (2007) Chemical and microphysical characterization of ambient aerosol particles with the aerodyne aerosol mass spectrometer. *Mass Spectrom. Rev.*, **26**, 185–222.

Cannon, T.W. (1970) High-speed photography of airborne atmospheric particles. *J. Appl. Meteorol.*, **9**, 104–108.

Cantrell, C., Stedman, D., and Wendel, G. (1984) Measurements of atmospheric peroxy radicals by chemical amplification. *Anal. Chem.*, **56**, 1496.

Cantrell, C., Shetter, R., Lind, J., McDaniel, A., Calvert, J., Parrish, D., Fehsenfeld, F., Buhr, M., and Trainer, M. (1993) An improved chemical amplifier technique for peroxy radical measurements. *J. Geophys. Res.*, **98**, 2897.

Cantrell, C., Edwards, G., Stephens, S., Mauldin, L., Kosciuch, E., Zondlo, M., and Eisele, F. (2003a) Peroxy radical observations using chemical ionization mass spectrometry during TOPSE. *J. Geophys. Res.*, **108**, 8371.

Cantrell, C., Edwards, G., Stephens, S., Mauldin, R.L., Zondlo, M., Kosciuch, E., Eisele, F., Shetter, R., Lefer, B., Hall, S., Flocke, F., Weinheimer, A., Fried, A., Apel, E., Kondo, Y., Blake, D., Blake, N., Simpson, I., Bandy, A., Thornton, D., Heikes, B., Singh, H., Brune, W.,

Harder, H., Martinez, M., Jacob, D., Avery, M., Barrick, J., Sachse, G., Olson, J., Crawford, J., and Clarke, A. (2003b) Peroxy radical behaviour during the Transport and Chemical Evolution over the Pacific (TRACE-P) campaign as measured aboard the NASA P-3B aircraft. *J. Geophys. Res.*, **108**, 8787. doi: 10.1029/2003JD003674.

Cantrell, C.A., Shetter, R.E., Gilpin, T.M., Calvert, J.G., Eisele, F.L., and Tanner, D.J. (1996) Peroxy radical concentrations measured and calculated from trace gas measurements in the Mauna Loa Photochemistry Experiment 2. *J. Geophys. Res.*, **101**, 14653–14664.

Cappa, C.D., Lack, D.A., Burkholder, J.B., and Ravishankara, A.R. (2008) Bias in filter-based aerosol light absorption measurements due to organic aerosol loading: evidence from laboratory measurements. *Aerosol Sci. Technol.*, **42**, 1022–1032.

Caramori, P., Schuepp, P., Desjardins, R., and MacPherson, I. (1994) Structural analysis of airborne flux estimates over a region. *J. Climate*, **7**, 627–640.

Cárdenas, L.M., Brassington, D.J., Allan, B.J., Coe, H., Alicke, B., Platt, U.K.M., Wilson, J.M.C.P., and Penkett, S.A. (2000) Intercomparison of formaldehyde measurements in clean and polluted atmospheres. *J. Atmos. Chem.*, **37**, 53–80. doi: 10.1023/A:1006383520 819.

Carrico, C.M., Kus, P., Rood, M.J., Quinn, P., and Bates, T.S. (2003) Mixtures of pollution, dust, sea salt, and volcanic aerosol during ACE-Asia: radiative properties as a function of relative humidity. *J. Geophys. Res.*, **108**, 8650. doi: 10.1029/2003JD003405.

Castells, A., Carswell, J., and Chang, P. (2001) A high-resolution imaging wind and rain airborne profiler for tropical cyclones, in *Geoscience and Remote Sensing Symposium (IGARSS)*, vol. 5, IEEE, pp. 2013–2015.

Cerni, T.A. (1983) Determination of the size and concentration of cloud drops with an FSSP. *J. Clim. Appl. Meteorol.*, **22**, 1346–1353.

Cerni, T.A. (1994) An infrared hygrometer for atmospheric research and routine monitoring. *J. Atmos. Oceanic Technol.*, **11**, 445–462.

Cess, R., Zhang, M., Minnis, P., Corsetti, L., Dutton, E., Forgan, B., Garber, D., Gates, W., Hack, J., Harrison, E., Jing, X., Kiehl, J., Long, C., Morcrette, J.-J., Potter, G., Ramanathan, V., Subasilar, B., Whitlock, C., Young, D., and Zhou, Y. (1995) Absorption of solar radiation by clouds: observations versus models. *Science*, **267**, 496–499.

Cézard, N., Dolfi-Bouteyre, A., Huignard, J.-P., and Flamant, P.H. (2009) Performance evaluation of a dual fringe-imaging Michelson interferometer for air parameter measurements with a 355 nm Rayleigh-Mie Lidar. *Appl. Opt.*, **48**, 2321–2332.

Chandra, S. and McFarland, A. (1997) Shrouded probe performance: variable flow operation and effect of free stream turbulence. *Aerosol Sci. Technol.*, **26**, 111–126.

Chang, C. (2003) *Hyperspectral Imaging: Techniques for Spectral Detection and Classification*, Kluwer Academic Plenum Publishers, New York.

Chang, C., Wu, C., Liu, W., and Ouyang, Y. (2006) A new growing method for simplex-based endmember extraction algorithms. *IEEE Trans. Geosci. Remote*, **44**, 2804–2819.

Chang, E.J. and Kailasanatha, K. (2003) Shock wave interactions with particles and liquid fuel droplets. *Shock Waves*, **12**, 333–341.

Chapon, B., Delrieu, G., Gosset, M., and Boudevillain, B. (2008) Variability of rain drop size distribution and its effect on the Z-R relationship: a case study for intense Mediterranean rainfall. *Atmos. Res.*, **87**, 52–65.

Chaumat, L. and Brenguier, J.-L. (2001a) Droplet spectra broadening in cumulus clouds. Part 2: microscale droplet concentration heterogeneities. *J. Atmos. Sci.*, **58**, 642–654.

Chaumat, L. and Brenguier, J.-L. (2001b) Droplet spectra broadening in cumulus clouds. Part II: effects of droplet concentration inhomogeneities. *J. Atmos. Sci.*, **58**, 642–654.

Chen, J., Conant, W., Rissman, T., Flagan, R., and Seinfeld, J. (2005) Effect of angle of attack on the performance of an airborne counterflow virtual impactor. *Aerosol Sci. Technol.*, **39**, 485–491.

Chepfer, H., Brogniez, G., Sauvage, L., Flamant, P.H., Trouillet, V., and Pelon, J. (1999) Remote sensing of cirrus radiative properties during EUCREX'94. Case study of 17 April 1994. Part 2: microphysical modelling. *Mon. Weather Rev.*, **127**, 486–503.

Chiu, H.-Y. and Collins, W. (1978) Spectroradiometer for airborne remote sensing. *Photogramm. Eng. Remote Sensing.*, **44**, 507–517.

Chou, C., Stetzer, O., Weingartner, E., Juranyi, Z., Kanji, A., and Lohmann, U. (2011) Ice nuclei properties within a Saharan dust event at the Jungfraujoch in the Swiss Alps. *Atmos. Chem. Phys.*, **11**, 4725–4738.

Chow, J. (1995) Measurement methods to determine compliance with ambient air-quality standards for suspended particles. *J. Air Waste Manage. Assoc.*, **45**, 320–382.

Christensen, P., Bandfield, J., Hamilton, V., Howard, D., Lane, M., Piatek, J., Ruff, S., and Stefanov, W. (2000) A thermal emission spectral library of rock-forming minerals. *J. Geophys. Res.*, **105**, 9735–9739.

Chuang, P.Y., Saw, E.W., Small, J.D., Shaw, R.A., Sipperley, C.M., Payne, G.A., and Bachalo, W.D. (2008) Airborne phase Doppler interferometry for cloud microphysical measurements. *Aerosol Sci. Technol.*, **42**, 685–703.

Chudnovsky, A., Ben-Dor, E., Kostinski, A., and Koren, I. (2009) Mineral content analysis of atmospheric dust using hyperspectral information from space. *Geophys. Res. Lett.*, **36**, 5811–5817.

Chylek, P. (1978) Extinction and liquid water content of fogs and clouds. *J. Atmos. Sci.*, **35**, 296–300.

Chýlek, P. and Wong, J.G.D. (1998) Erroneous use of the modified Köhler equation in cloud and aerosol physics applications. *J. Atmos. Sci.*, **55**, 1473–1477.

Clague, L. (1965) An improved device for obtaining cloud droplet samples. *J. Appl. Meteorol.*, **4**, 549–551.

Clark, R. (1999) Spectroscopy of rocks and minerals, and principles of spectrocopy, in *Manual of Remote Sensing, Volume 3: Remote Sensing for the Earth Sciences*, 3rd edn (ed. A.N. Rencz), John Wiley & Sons, Inc., New York, pp. 3–58.

Clark, R. and Roush, T. (1984) Reflectance spectroscopy: quantitative analysis techniques for remote sensing applications. *J. Geophys. Res.*, **89**, 6329–6340.

Clark, R., Gallagher, A., and Swayze, G. (1990) Material absorption band depth mapping of imaging spectrometer data using a complete band shape least-squares fit with library, in Proceedings of the 2nd Airborne Visible/Infrared Imaging Spectrometer (AVIRIS), Workshop, Pasadena, CA, 1990, pp. 176–186, JPL Publication 90-54.

Clarke, A.D. (1991) A thermo-optic technique for in situ analysis of size-resolved aerosol physicochemistry. *Atmos. Environ.*, **25**, 635–644.

Clarke, A.D. (1992) Atmospheric nuclei in the remote free-troposphere. *J. Climate*, **96**, 5237–5247.

Clarke, A.D. and Porter, J.N. (1991) Aerosol size distribution, composition and CO_2 backscatter at Mauna Loa observatory. *J. Geophys. Res.*, **96**, 5237–5247.

Claussen, M. (1991) Estimation of areally-averaged surface fluxes. *Bound.-Lay. Meteorol.*, **54**, 387–410.

Clemitshaw, K. (2004) A review of instrumentation and measurement techniques for ground-based and airborne field studies of gas-phase tropospheric chemistry. *Crit. Rev. Environ. Sci. Technol.*, **34**, 1–108. doi: 10.1080/10643380490265 117.

Clemitshaw, K., Carpenter, L., Penkett, S., and Jenkin, M. (1997) A calibrated peroxy radical chemical amplifier (PERCA) instrument for ground-based measurements in the troposphere. *J. Geophys. Res.*, **102**, 405.

Clevers, J.G.P.W. (1999) The use of imaging spectrometry for agricultural applications. *ISPRS J. Photogramm.*, **54**, 299–304.

Clough, P. and Thrush, B.A. (1967) Mechanism of the chemiluminescent reaction between NO and O_3. *Trans. Faraday Soc.*, **63**, 915–925.

Clyne, M., Thrush, B.A., and Wayne, R.P. (1964) Kinetics of the chemiluminescent reaction between nitric oxide and ozone. *Trans. Faraday Soc.*, **60**, 359–370.

Cober, S., Isaac, G., and Korolev, A. (2001a) Assessing the Rosemount icing detector with in situ measurements. *J. Atmos. Ocean. Technol.*, **18**, 515–528.

Cober, S., Isaac, G., Korolev, A., and Strapp, J. (2001b) Assessing cloud-phase conditions. *J. Appl. Meteorol.*, **40**, 1967–1983.

Coddington, O., Pilewskie, P., Redemann, J., Platnick, S., Russell, P., Schmidt, K., Gore, W., Livingston, J., Wind, G., and Vukicevic, T. (2010) Examining the impact of overlying aerosol particles on the retrieval of cloud optical properties from passive remote sensing. *J. Geophys. Res.* doi: 10.1029/2009JD012829.

Coddington, O.M., Pilewskie, P., and Vukicevic, T. (2012) The Shannon information content of hyperspectral shortwave cloud albedo measurements: quantification and practical applications. *J. Geophys. Res.*, **117**, D04205. doi: 10.1029/2011JD016771.

Coe, H. (2006) in *Aerosol Mass Spectrometry. Atmospheric Techniques* (eds J.D. Allan and D.E. Heard), Blackwell's Scientific Publishing.

Cohen, R. (1991) Shattering of a liquid drop due to impact. *Proc. R. Soc. London, Ser. A*, **435**, 483–503.

Collaud Coen, M., Weingartner, E., Apituley, A., Ceburnis, D., Fierz-Schmidhauser, R., Flentje, H., Henzing, J.S., Jennings, S.G., Moerman, M., Petzold, A., Schmid, O., and Baltensperger, U. (2010) Minimizing light absorption measurement artifacts of the Aethalometer: evaluation of five correction algorithms. *Atmos. Meas. Technol.*, **3**, 457–474.

Collins, D.R., Flagan, R.C., and Seinfeld, J.H. (2002) Improved inversion of scanning DMA data. *Aerosol Sci. Technol.*, **36**, 1–9.

Colomb, A., Williams, J., Crowley, J., Gros, V., Hofmann, R., Salisbury, G., Klupfel, T., Kormann, R., Stickler, A., Forster, C., and Lelieveld, J. (2006) Airborne measurements of trace organic species in the upper troposphere over Europe: the impact of deep convection. *Environ. Chem.*, **3**, 244–259.

Commane, R., Floquet, C.F.A., Ingham, T., Stone, D., Evans, M.J., and Heard, D.E. (2010) Observations of OH and HO_2 radicals over West Africa. *Atmos. Chem. Phys.*, **10**, 8783–8801.

Comstock, J.M., Ackerman, T.P., and Mace, G.G. (2002) Ground-based Lidar and Radar remote sensing of tropical cirrus clouds at Nauru Island: cloud statistics and radiative impacts. *J. Geophys. Res.*, **107**, 4714. doi: 10.1029/2002JD002203.

Comte-Bellot, G. (1976) Hot-wire anemometry. *Annu. Rev. Fluid Mech.*, **8**, 209–231.

Cooley, T., Anderson, G., Felde, G., Hoke, M., Ratkowski, A., Chetwynd, J., Gardner, J., Adler-Golden, S., Matthew, M., Berk, A., Bernstein, L., Acharya, P., Miller, D., and Lewis, P. (2002) FLAASH, a MODTRAN4-based atmospheric correction algorithm, its application and validation, Geoscience and Remote Sensing Symposium, 2002. IGARSS'202. 2002 IEEE International, pp. 1414–1418.

Cooper, W. (1978) Cloud physics investigations by the University of Wyoming in HIPLEX 1977, Final Report. University of Wyoming, vol. AS 119, Bureau of Reclamation Rep., 322 pp.

Cooper, W. (1988) Effects of coincidence on measurements with a Forward Scattering Spectrometer Probe. *J. Atmos. Oceanic Technol.*, **5**, 823–832.

Cooper, W. and Rogers, D. (1991) Effects of airflow trajectories around aircraft on measurements of scalar fluxes. *J. Atmos. Ocean. Technol.*, **8**, 66–77.

Corrigan, C.E., Roberts, G.C., Ramana, M.V., Kim, D., and Ramanathan, V. (2008) Capturing vertical profiles of aerosols and black carbon over the Indian Ocean using autonomous unmanned aerial vehicles. *Atmos. Chem. Phys.*, **8**, 737–747.

Cotton, R., Osborne, S., Ulanowski, Z., Hirst, E., Kaye, P., and Greenaway, R. (2009) The ability of the small ice detector (SID-2) to characterize cloud particle and aerosol morphologies obtained during flights of the FAAM BAe-146 research aircraft. *J. Atmos. Oceanic Technol.*, **27**, 290–303.

Coulson, K. (1975) *Solar and Terrestrial Radiation-methods and Measurements*, Academic Press, New York, San Francisco, CA, London.

Cox, C. and Munk, W. (1954) Measurements of the roughness of the sea surface from photographs of the sun's glitter. *J. Opt. Soc. Am.*, **144**, 838–850.

Cox, S. and Griffith, K. (1979) Estimates of radiative divergence during phase III of the GARP Atlantic Tropical Experiment: Part I. Methodology. *J. Atmos. Sci.*, **36**, 576–585.

Craig, R.A. (1949) Vertical eddy transfer of heat and water vapor in stable air. *J. Atmos. Sci.*, **6**, 123–133.

Craigie, J.H. (1945) Epidemiology of stem rust in western Canada. *Sci. Agric.*, **25**, 285–401.

Crawford, J., Davis, D., Chen, G., Shetter, R., Müller, M., Barrick, J., and Olson, J. (1999) An assesment of cloud effects on photolysis rate coefficients: Comparison of experimental and theoretical values. *J. Geophy. Res.*, **104**, 5725–5734.

Crawford, T., Dobosy, R., and Dumas, E. (1996) Aircraft wind measurement considering lift-induced upwash. *Bound.-Lay. Meteorol.*, **80**, 79–94.

Crawford, T.L. and Dobosy, R.J. (1992) A sensitive fast-response probe to measure turbulence and heat flux from any airplane. *Bound.-Lay. Meteorol.*, **59**, 257–278.

Crawford, T.L., McMillen, R.T., Dobosy, R.J., and MacPherson, I. (1993) Correcting airborne flux measurements for aircraft speed variation. *Bound.-Lay. Meteorol.*, **66**, 237–245.

Creasey, D., Heard, D., and Lee, J. (2000) Absorption cross-section measurements of water vapor and oxygen at 185 nm. Implications for the calibration of field instruments to measure OH, HO_2 and RO_2 radicals. *Geophys. Res. Lett.*, 27.

Cremer, M. (1999) All integrated nose boom for wind and turbulence measurement, 30th Annual Symposium, Society of Flight Test Engineers, St. Louis, MO.

Crépel, O., Gayet, J.-F., Fournol, J.-F., and Oshchepkov, S. (1997) A new airborne Polar Nephelometer for the measurements of optical and microphysical cloud properties. Part II: preliminary tests. *Ann. Geophys.*, **15**, 460–470.

Crewell, S., Fabian, R., Künzi, K., Nett, H., Wehr, T., Read, W., and Waters, J. (1994) Comparison of ClO measurements by airborne and spaceborne microwave radiometers in the Arctic winter stratosphere 1993. *Geophys. Res. Lett.*, **22**, 1489–1492.

Crosier, J., Allan, J.D., Coe, H., Bower, K.N., Formenti, P., and Williams, P.I. (2007a) Chemical composition of summertime aerosol in the Po Valley (Italy), northern Adriatic and Black Sea. *Q. J. R. Meteor. Soc.*, **133**, 61–75.

Crosier, J., Jimenez, J.L., Allan, J.D., Bower, K.N., Williams, P.I., Alfarra, M.R., Canagaratna, M.R., Jayne, J.T., Worsnop, D.R., and Coe, H. (2007b) Description and use of the new jump mass spectrum mode of operation for the aerodyne quadrupole aerosol mass spectrometers (Q-AMS). *Aerosol Sci. Technol.*, **41**, 865–872.

Crounse, J., Karena, A., McKinney, K., Kwan, A., and Wennberg, P. (2006) Measurement of gas-phase hydroperoxides by chemical ionization mass spectrometry. *Anal. Chem.*, **78**, 6726–6732.

Crowther, B. (1997) The design, construction, and calibration of a spectral diffuse/global irradiance meter. PhD thesis. University of Arizona.

Cruette, D., Marillier, A., Dufresne, J.L., Grandpeix, J.Y., Nacass, P., and Bellec, H. (2000) Fast temperature and true air speed measurements with the airborne ultrasonic anemometer-thermometer (AUSAT). *J. Atmos. Oceanic Technol.*, **17**, 1020–1039.

Crutzen, P.J., Williams, J., Pöschl, U., Hoor, P., Fischer, H., Warneke, C., Holzinger, R., Hansel, A., Lindinger, W., Scheeren, B., and Lelieveld, J. (2000) High spatial and temporal resolution measurements of primary organics and their oxidation products over the tropical forests of Surinam. *Atmos. Environ.*, **34**, 1161–1165.

Curry, M.J. and Schemenauer, R.S. (1979) The Small-particle response of an Optical Array Precipitation Probe. *J. Appl. Meteorol.*, **18**, 816–821.

Curtius, J. and Arnold, F. (2001) Measurements of aerosol sulfuric acid, 1: experimental setup, characterization, and calibration of a novel mass spectrometric system. *J. Geophys. Res.*, **106**, 31965–31974.

Curtius, J., Sierau, B., Arnold, F., de Reus, M., Strom, J., Scheeren, H., and Lelieveld, J. (2001) Measurements of aerosol sulfuric acid, 2: pronounced layering in the free troposphere during ACE-2. *J. Geophys. Res.*, **106**, 31975–31990.

Cziczo, D.J., Thomson, D.S., and Murphy, D.M. (2001) Ablation, flux, and atmospheric implications of meteors inferred from stratospheric aerosol. *Science*, **291**, 1772–1775.

Cziczo, D.J., DeMott, P.J., Brock, C., Hudson, P.K., Jesse, B., Kreidenweis, S.M., Prenni, A.J., Schreiner, J., Thomson, D.S., and Murphy, D.M. (2003) A method for single particle mass spectroscopy of ice nuclei. *Aerosol Sci. Technol.*, **37**, 460–470.

Dabas, A., Drobinski, P., Reitebuch, O., Richard, E., Delville, P., Flamant, P.H., and Werner, C. (2003) Multi-scale analysis of a straight jet streak using numerical analyses and an airborne Doppler Lidar. *Geophys. Res. Lett.*, **30**, 1049. doi: 10.1029/2001GL014040.

Dabberdt, W.R.S., Cole, H., Paukkunen, Hoerhammer, J., and Antikainen, V. (2002) Radiosondes, in *Encyclopedia of Atmospheric Sciences* (eds J. Holton, J. Pyle, and J. Curry), Academic Press, London, pp. 1469–1476.

D'Amato, F., Mazzinghi, P., and Castagnoli, F. (2002) Methane analyzer based on TDL for measurements in the lower stratosphere: design and laboratory tests. *Appl. Phys. B*, **75**, 195–202.

Damiani, R. and Haimov, S. (2006) A high-resolution dual-Doppler technique for fixed multiantenna airborne Radar. *IEEE Trans. Geosci. Remote*, **44**, 3475–3489.

Dari-Salisburgo, C., Carlo, P.D., Giammaria, F., Kajii, Y., and D'Altorio, A. (2009) Laser induced fluorescence instrument for NO_2 measurements: observations at a central Italy background site. *Atmos. Environ.*, **43**, 970–977.

Davies, J.H. and Davies, D.R. (2010) Earth's surface heat flux. *Solid Earth*, **1**, 5–24.

Davis, A., Marshak, A., Gerber, H., and Wiscombe, W. (1999) Horizontal structure of marine boundary-layer clouds from cm to km scales. *J. Geophys. Res.*, **104**, 6123–6144.

Davis, E.J. and Schweiger, G. (2002) *The Airborne Microparticle*, Springer, New York.

Davis, S., Avallone, L., Weinstock, E., Twohy, C., Smith, J., and Kok, G. (2007) Comparisons of in situ measurements of cirrus cloud ice water content. *J. Geophys. Res.*, **112**, 214. doi: 10.1029/2006JD008.

Davison, C. and MacLeod, J. (2009) Naturally aspirating isokinetic total water content probe: intake deicing and heat transfer, 1st AIAA Atmospheric and Space Environments, San Antonio, TX, June 22–25, AIAA-2009-3862, 2009.

Davison, C., MacLeod, J., and Strapp, J. (2009) Naturally aspirating isokinetic total water content probe: evaporator design and testing. 1st AIAA Atmospheric and Space Environments, San Antonio, TX, June 22–25, 2009, AIAA-2009-3861.

de Carlo, P.F., Kimmel, J.R., Trimborn, A., Northway, M.J., Jayne, J.T., Aiken, A.C., Gonin, M., Fuhrer, K., Horvath, T., Docherty, K.S., Worsnop, D.R., and Jimenez, J.L. (2006) Field-deployable high-resolution, time-of-flight aerosol mass spectrometer. *Anal. Chem.*, **78**, 8281–8289.

de Carlo, P.F., Dunlea, E.J., Kimmel, J.R., Aiken, A.C., Sueper, D., Crounse, J., Wennberg, P.O., Emmons, L., Shinozuka, Y., Clarke, A., Zhou, J., Tomlinson, J., Collins, D.R., Knapp, D., Weinheimer, A.J., Montzka, D.D., Campos, T., and Jimenez, J.L. (2008) Fast airborne aerosol size and chemistry measurements above Mexico City and Central Mexico during the MILAGRO campaign. *Atmos. Chem. Phys.*, **8**, 4027–4048.

de Gouw, J., Warneke, C., Karl, T., Eerdekens, G., van der Veen, C., and Fall, R. (2003) Sensitivity and specificity of atmospheric trace gas detection by proton-transfer-reaction mass spectrometry. *Int. J. Mass Spectrom.*, **223**, 365–382.

de Gouw, J.A., Warneke, C., Stohl, A., Wollny, A.G., Brock, C.A., Cooper, O.R., Holloway, J.S., Trainer, M., Fehsenfeld, F.C., Atlas, E.L., Donnelly, S.G., Stroud, V., and Lueb, A. (2006) Volatile organic compounds composition of merged and aged forest fire plumes from Alaska and western Canada.

J. Geophys. Res., **111**, 175. –D10303, doi: 10.1029/2005JD006.

de Leo, R. and Hagen, F. (1976) Aerodynamic Perforemance of Rosemount Model 858 AJ Air Data Sensor. *Tech. Rep.* 8767, Rosemount.

de Reus, M., Fischer, H., Arnold, F., de Gouw, J., Holzinger, R., Warneke, C., and Williams, J. (2003) On the relationship between acetone and carbon monoxide in different air masses. *Atmos. Chem. Phys.*, **3**, 1709–1723.

Dekker, A., Brando, V., Anstee, J., Pinnel, N., Kutser, T., Hoogenboom, H., Pasterkamp, R., Peters, S., Vos, R., Olbert, C., and Malthus, T. (2001) Imaging spectrometry of Water, *Imaging Spectrometry: Basic Principles and Prospective Applications: Remote Sensing and Digital Image Processing*, vol. 4, Kluwer Academic Publishers, Dordrecht, pp. 307–359.

Del Bianco, S., Carli, B., Cecchi-Pestellini, C., Dinelli, B., Gai, M., and Santurri, L. (2007) Retrieval of minor constituents in a cloudy atmosphere with remote-sensing millimetre-wave measurements. *Q. J. R. Meteorol. Soc.*, **133**, 163–170.

Delanoë, J. and Hogan, R.J. (2008) A variational scheme for retrieving ice cloud properties from combined Radar, Lidar, and infrared radiometer. *J. Geophys. Res.*, **113**, D07204. doi: 10.1029/2007JD009000.

Delanoë, J., Protat, A., Testud, J., and Bouniol, D. (2005) Statistical properties of the normalized ice particle size distribution. *J. Geophys. Res.*, **110**, D10201, doi: 10.1029/2004JD005405.

Delanoë, J., Protat, A., Bouniol, D., Heymsfield, A., Bansemer, A., and Brown, P. (2007) The characterization of ice cloud properties from Doppler Radar measurements. *J. Appl. Meteorol. Climatol.*, **46**, 1682–1698.

Delene, J.D., Deshler, T., Wechsler, P., and Vali, G. (1998) A balloon-borne cloud condensation nuclei counter. *J. Geophys. Res.*, **103**, 8927–8934.

DeMott, P.J., Petters, M.D., Prenni, A.J., Carrico, C.M., Kreidenweis, S.M., Collett, J.L. Jr., and Moosmüller, H. (2009) Ice nucleation behavior of biomass combustion particles at cirrus temperatures. *J. Geophys. Res.*, **114**, 036. doi: 10.1029/2009JD012.

DeMott, P.J., Möhler, O., Stetzer, O., Vali, G., Levin, Z., Petters, M.D., Murakami, M., Leisner, T., Bundke, U., Klein, H., Kanji, Z., Cotton, R., Jones, H., Petters, M., Prenni, A., Benz, S., Brinkmann, M., Rzesanke, D., Saathoff, H., Nicolet, M., Gallavardin, S., Saito, A., Nillius, B., Bingemer, H., Abbatt, J., Ardon, K., Ganor, E., Georgakopoulos, D.G., and Saunders, C. (2011) Resurgence in ice nucleation research. *Bull. Am. Meteorol. Soc.*, **92**, 1623–1635.

Deng, M. and Mace, G.G. (2006) Cirrus microphysical properties and air motion statistics using cloud Radar Doppler moments. Part I: algorithm description. *J. Appl. Meteorol. Climatol.*, **45**, 1690–1709.

Deng, M., Mace, G., Wang, Z. *et al.* (2010) TC4 validation for ice cloud profiling retrieval using CloudSat Radar and CALIPSO Lidar. *J. Geophys. Res.*, **115**, D00J15. doi: 10.1029/2009JD013104.

Denning, R., Guidero, S., Parks, G., and Gary, B. (1989) Instrument description of the airborne microwave temperature profiler. *J. Geophys. Res.*, **94**, 16757–16765.

Deshler, T. and Vali, G. (1992) Atmospheric concentrations of submicron contact-freezing nuclei. *J. Atmos. Sci.*, **49**, 773–784.

Deshler, T., Johnson, B.J., and Rozier, W.R. (1993) Balloonborne measurements of Pinatubo aerosol during 1991 and 1992 at 41°N-vertical profiles size distributoin and volatility. *Geophys. Res. Lett.*, **20**, 1435–1438.

Deshler, T., Hervig, M.E., Hofmann, D.J., Rosen, J.M., and Liley, J.B. (2003) Thirty years of in situ stratospheric aerosol size distribution measurements from Laramie Wyoming (41°N), using balloon-borne instruments. *J. Geophys. Res.*, **108**, 4167. doi: 10.1029/2002JD002514.

Dhaniyala, S., Flagan, R., McKinney, K., and Wennberg, P. (2003) Novel aerosol/gas inlet for aircraft-based measurements. *Aerosol Sci. Technol.*, **37**, 828–840.

Dhaniyala, S., Wennberg, P., Flagan, R., Fahey, D.W., Northway, M., Gao, R., and

Bui, T. (2004) Stratospheric aerosol sampling: effect of a blunt-body housing on inlet sampling characteristics. *Aerosol Sci. Technol.*, **38**, 1080–1090.

Di Carlo, P., Salisburgo, C.D., Aruffo, E., Biancofiore, F., Busilacchio, M., Giammaria, F., Reeves, C., Moller, S., and Lee, J. (2011) Airborne Laser Induced Fluorescence system for simultaneous measurements of nitrogen dioxide and speciated NO_y. EGU Meeting.

Dibb, J.E., Talbot, R.W., Scheuer, E.M., Seid, G., Avery, M.A., and Singh, H.B. (2003a) Aerosol chemical composition in Asian continental outflow during the TRACE-P campaign: comparison with PEM-West B. *J. Geophys. Res.*, **108**, 8815. doi: 10.1029/2002JD003111.

Dibb, J.E., Talbot, R.W., Seid, G., Jordan, C., Scheuer, E., Atlas, E., Blake, N.J., and Blake, D.R. (2003b) Airborne sampling of aerosol particles: comparison between surface sampling at Christmas Island and P-3 sampling during PEM-Tropics B. *J. Geophys. Res.*, **108**, 8230. doi: 10.1029/2001JD000408.

Dickerson, R. and Delany, A. (1988) Modification of a commercial gas filter correlation CO detector for enhanced sensitivity. *J. Atmos. Oceanic Technol.*, **5**, 424–431.

DIN (1990) Begriffe, Grössen und Formelzeichen der Flugmechanik.

Diskin, G., Podolske, J., Sachse, G., and Slate, T. (2002) Open-path airborne tunable diode laser hygrometer. *Proc. SPIE*, **4817**, 196–204.

DLR (2009) *http://www.halo.dlr.de/instrumentation/inlets.*

Dobbins, J.T. and Boone, J.M. (1998) Flat-field correction technique for digital detectors. Proceedings of SPIE, vol. 3336, Berlin.

Dobson, G.M.B.D.N.H. and Lawrence, J. (1929) Measurements of the amount of ozone in the earth's atmosphere and its relation to other geophysical conditions. *Proc. R. Soc. London, Ser. A*, **122**, 456–486.

Doebelin, E. (1990) *Measurement Systems: Application and Design*, 4th edn, McGraw-Hill, New York.

Doerffer, R., Graßl, H., Kunkel, B., and van der Piepen, H. (1989) ROSIS - an advanced imaging spectrometer for the monitoring of water color and chlorophyll fluorescence. Proceedings of the International Congress on Optical Sciences & Engineering. Conference 1129: Advanced Optical Instrumentation for Remote Sensing of the Earth's Surface from Space, Paris, April 24-28, 1989, vol. 30, pp. 117–121.

Dolson, D.A. and Klingshirn, M. (1993) Laser-initiated chemical chain-reactions-termination kinetics of the Cl_2/HBr/NO chain system. *J. Phys. Chem.*, **97**, 6645–6649.

Donovan, D.P. and van Lammeren, A. (2001) Cloud effective particle size and water content profile retrievals using combined Lidar and Radar observations-1. Theory and examples. *J. Geophys. Res.*, **106**, 27425–27448.

Donovan, D.P., Quante, M., Schlimme, I., and Macke, A. (2004) Use of equivalent spheres to model the relation between Radar reflectivity and optical extinction of ice cloud particles. *Appl. Opt.*, **43**, 4929–4940.

Dorf, M., Bosch, H., Butz, A., Camy-Peyret, C., Chipperfield, M., Engel, A., Goutail, F., Grunow, K., Hendrick, F., Hrechanyy, S., Naujokat, B., Pommereau, J., Roozendael, M.V., Sioris, C., Stroh, F., Weidner, F., and Pfeilsticker, K. (2006) Balloon-borne stratospheric BrO measurements: comparison with Envisat/SCIAMACHY BrO limb profiles. *Atmos. Chem. Phys.*, **6**, 2483–2501.

Dotan, I., Albritton, D.L., Lindinger, W., and Pahl, M. (1976) Mobilities of Co^{2+}, N_2H^+, H_3O^+, $H_3O^+.H_2O$, and $H_3O^+.(H_2O)_2$ Ions in N^{-2}. *J. Chem. Phys.*, **65**, 5028–5030.

Doviak, R.J. and Zrnic, D. (1993) *Doppler Radar and Weather Observations*, 2nd edn, Academic Press, San Diego, CA.

Dowling, D.R. and Radke, L.F. (1990) A summary of the physical properties of cirrus clouds. *J. Appl. Meteorol.*, **29**, 970–978.

Dozier, J. and Frew, J. (1981) Atmospheric corrections to satellite radiometric data over rugged terrain. *Remote Sens. Environ.*, **11**, 191–205.

Drewnick, F., Hings, S.S., DeCarlo, P.F., Jayne, J.T., Gonin, M., Fuhrer, K., Weimer, S., Jimenez, J.L., Demerjian, K.L., Borrmann, S., and Worsnop, D.R. (2005) A new time-of-flight aerosol mass spectrometer (ToF-AMS) - instrument description and first field deployment. *Aerosol Sci. Technol.*, **39**, 637–658.

Drummond, A. and MacPherson, J. (1985) Aircraft flow effects on cloud drop images and concentrations measured by the NAE Twin Otter. *J. Atmos. Oceanic Technol.*, **2**, 633–643.

Drummond, J.W., Volz, A., and Ehhalt, D.H. (1985) An optimized chemiluminescence detector for tropospheric NO measurements. *J. Atmos. Chem.*, **2**, 287–306.

Dubé, W.P., Brown, S., Osthoff, H., Nunley, M., Ciciora, S., Paris, M., McLaughlin, R., and Ravishankara, A. (2006) Aircraft instrument for simultaneous, in situ measurements of NO_3 and N_2O_5 via cavity ring-down spectroscopy. *Rev. Sci. Instrum.*, **77**. doi: 10.1063/1061.2176058.

Dubovik, O. (2004) Optimization of numerical inversion in photopolarimetric remote sensing, in *Photopolarimetry in Remote Sensing* (eds G. Videen, Y. Yatskiv, and M. Mishchenko), Kluwer Academic Publishers, Dordrecht, The Netherlands, pp. 65–106.

Dungey, C.E. and Bohren, C.F. (1993) Backscattering by nonspherical hydrometeors as calculated by the coupled-dipole method-an application in Radar meteorology. *J. Atmos. Oceanic Technol.*, **10**, 526–532.

Dupont, P. (1938) Contribution à l'étude du vol en atmosphère agitée. Rapport sur la campagne du POTEZ 540 à la Banne d'Ordanche du 19 au 30 Septembre 1936, Publications Scientifiques et Techniques du Ministère de l'Air, Gauthiers-Villard Eds.

Durant, A.J. and Shaw, R.A. (2005) Evaporation freezing by contact nucleation inside-out. *Geophys. Res. Lett.*, **32**, L20814. doi: 10.1029/2005GL024175.

Durden, S.L., Li, L., Im, E., and Yueh, S. (2003) A surface reference technique for airborne doppler Radar measurements in hurricanes. *J. Atmos. Oceanic Technol.*, **20**, 269–275.

Durry, G., Amarouche, N., Joly, L., Liu, X., Parvitte, B., and Zéninari, V. (2007) Laser diode spectroscopy of H_2O at 2.63 µm for atmospheric applications. *Appl. Phys. B*, **90**, 573. doi: 10.1007/s00340-007-2884-3.

Durst, F., Melling, A., and Whitelaw, J. (1981) *Principle and Practice of Laser-Doppler Anemometry*, Academic Press, London.

Dutton, E., DeLuisi, J., and Herbert, G. (1989) Shortwave aerosol optical depth of Arctic haze measured on board the NOAA WP-3D during AGASP-II April 1986. *J. Atmos. Chem.*, **9**, 71–79.

Dye, J.E. and Baumgardner, D. (1984) Evaluation of the forward scattering spectrometer probe: I. Electronic and optical studies. *J. Atmos. Ocean. Technol.*, **1**, 329–344.

Dyroff, C., Fuetterer, D., and Zahn, A. (2010) Compact diode-laser spectrometer ISOWAT for highly sensitive airborne measurements of water-isotope ratios. *Appl. Phys. B*, **98**, 537–548.

Eatough, D.J., Obeidi, F., Pang, Y., Ding, Y., Eatough, N.L., and Wilson, W.E. (1999) Integrated and real-time diffusion denuder sampler for PM2.5. *Atmos. Environ.*, **33**, 835–2844.

Eddy, P., Natarajan, A., and Dhaniyala, S. (2006) Subisokinetic sampling characteristics of high speed aircraft inlets: a new CFD-based correlation considering inlet geometries. *J. Aerosol Sci.*, **37**, 1853–1870.

Edwards, G.D., Cantrell, C.A., Stephens, S., Hill, B., Goyea, O., Shetter, R., Mauldin, R., Kosciuch, E., Tanner, D., and Eisele, F. (2003) Chemical ionization mass spectrometer instrument for the measurement of tropospheric HO_2 and RO_2. *Anal. Chem.*, **75**, 5217–5327.

Eerdekens, G., Ganzeveld, L., de Arellano, J.V.G., Klupfel, T., Sinha, V., Yassaa, N., Williams, J., Harder, H., Kubistin, D., Martinez, M., and Lelieveld, J. (2009) Flux estimates of isoprene, methanol and acetone from airborne PTR-MS measurements over the tropical rainforest during the GABRIEL 2005 campaig. *Atmos. Chem. Phys.*, **9**, 4207–4227.

Ehret, G., Hoinka, K.P., Stein, J., Fix, A., Kiemle, C., and Poberaj, G. (1999)

Low-stratospheric water vapor measured by an airborne DIAL. *J. Geophys. Res.*, **104**, 351–359.

Ehret, G., Kiemle, C., Wirth, M., Amediek, A., Fix, A., and Houweling, S. (2008) Space-borne remote sensing of CO_2, CH_4, and N_2O by integrated path differential absorption Lidar: a sensitivity analysis. *Appl. Phys. B*, **90**, 593–608.

Ehrlich, A., Wendisch, M., Bierwirth, E., Herber, A., and Schwarzenböck, A. (2008) Ice crystal shape effects on solar radiative properties of Arctic mixed-phase clouds - dependence on microphysical properties. *Atmos. Res.*, **88**, 266–276. doi: 10.1016/j.atmosresa.2007.11.018.

Ehrlich, A., Wendisch, M., Bierwirth, E., Gayet, J.-F., Mioche, G., Lampert, A., and Mayer, B. (2009) Evidence of ice crystals at cloud top of Arctic boundary-layer mixed-phase clouds derived from airborne remote sensing. *Atmos. Chem. Phys.*, **9**, 9401–9416. doi: 10.5194/acp-9.9401-2009.

Eidhammer, T. and Deshler, T. (2005) Technical Note: evaporation of polar stratospheric cloud particles, in situ, in a heated inlet. *Atmos. Chem. Phys.*, **5**, 97–106. SRef-ID: 1680-7324/acp/2005-5-97.

Eidhammer, T., DeMott, P.J., Rogers, D.C., Prenni, A.J., Petters, M.D., Twohy, C.H., Rogers, D.C., Stith, J., Heymsfield, A., Wang, Z., Haimov, S., French, J., Pratt, K., Prather, K., Murphy, S., Seinfeld, J., Subramanian, R., and Kreidenweis, S.M. (2010) Ice initiation by aerosol particles: measured and predicted ice nuclei concentrations versus measured ice crystal concentrations in an orographic wave cloud. *J. Atmos. Sci.*, **67**, 2417–2436. doi: http://dx.doi.org/10.1175/2010JAS3266.1.

Eisele, F., Mauldin, R., Tanner, D., Fox, J., Mouch, T., and Scully, T. (1997) An inlet/sampling duct for airborne OH and sulfuric acid measurements. *J. Geophys. Res.*, **102**, 27993–28001.

Emery, E., Miller, D., Plaskon, S., Strapp, J., and Lilie, L.E. (2004) Ice particle impact on cloud water content Instrumentation. 42nd AIAA Aerospace Sciences Meeting and Exhibit, Reno, Nevada, AIAA-2004-0731.

Engel, A., Bönisch, H., Brunner, D., Fischer, H., Franke, H., Günther, G., Gurk, C., Hegglin, M., Hoor, P., Königstedt, R., Krebsbach, M., Maser, R., Parchatka, U., Peter, T., Schell, D., Schiller, C., Schmidt, U., Spelten, N., Szabo, T., Weers, U., Wernli, H., Wetter, T., and Wirth, V. (2006) Highly resolved observations of trace gases in the lowermost stratosphere and upper troposphere from the SPURT project: an overview. *Atmos. Chem. Phys.*, **6**, 283–301. doi: 10.5194/acp-6-283-2006.

Engel, A., Möbius, T., Bönisch, H., Schmidt, U., Heinz, R., Levin, I., Atlas, E., Aoki, S., Nakazawa, T., Sugawara, S., Moore, F., Hurst, D., Elkins, J., Schauffler, S., Andrews, A., and Boering, K. (2009) Age of stratospheric air unchanged within uncertainties over the past 30 years. *Nat. Geosci.*, **2**, 28–31. doi: 10.1038/ngeo388.

Engeln, R., von Helden, G., Berden, G., and Meijer, G. (1996) Phase shift cavity ring-down absorption spectroscopy. *Chem. Phys. Lett.*, **262**, 105–109. http://dx.doi.org/10.1016/0009-2614(96)01048-2.

Engeln, R., Berden, G., Peeters, R., and Meijer, G. (1998) Cavity enhanced absorption and cavity enhanced magnetic rotation spectroscopy. *Rev. Sci. Instrum.*, **69**, 3763–3769.

English, S. (1999) Estimation of temperature and humidity profile information from microwave radiances over different surface types. *J. App. Meteorol.*, **38**, 1526–1541.

English, S. (2008) The importance of accurate skin temperature in assimilating radiances from satellite sounding instruments. *IEEE Trans. Geosci. Remote*, **46**, 403–408.

English, S., Guillou, C., Prigent, C., and Jones, D. (1994) Aircraft measurements of water vapor continuum absorption at millimetre wavelengths. *Q. J. R. Meteor. Soc.*, **120**, 603–625.

ESA (1999) The four candidate Earth Explorer core missions-Atmospheric Dynamics Mission. ESA Report for Mission Selection. *Tech. Rep.* ESA, SP-1233(4), European Space Agency, 145 pp.

ESA (2001) WALES-Water Vapor Lidar Experiment in Space. The Five Candidate Earth Explorer Core Missions, *Tech. Rep.* ESA SP-1257(2), European Space Agency, report for Assessment.

Esposito, B.M., and Marrazzo, M. (2007) *Application of PDPA System with Different Optical Configuration to the IWT Calibration*, AIAA, Reno, NV, aIAA 2007-1094, 2007.

Esselborn, M., Wirth, M., Fix, A., Tesche, M., and Ehret, G. (2008) Airborne high spectral resolution Lidar for measuring aerosol extinction and backscatter coefficients. *App. Opt.*, **47**, 346–358. http://dx.doi.org/10.1364/AO.47.000 346.

Esselborn, M., Wirth, M., Fix, A., Weinzierl, B., Rasp, K., Tesche, M., and Petzold, A. (2009) Spatial distribution and optical properties of Saharan dust observed by airborne high spectral resolution Lidar during SAMUM 2006. *Tellus B*, **61**, 131–143.

Evans, K. and Vivekanandan, J. (1990) Multiparameter Radar and microwave radiative transfer modeling of nonspherical atmospheric ice particles. *IEEE Trans. Geosci. Remote*, **28**, 423–437.

Evans, K., Wang, J., Racette, P., Heymsfield, G., and Li, L. (2005) Ice cloud retrievals and analysis with data from the compact scanning submillimeter imaging radiometer and the cloud radar system during CRYSTAL-FACE. *J. Appl. Meteorol.*, **44**, 839–859.

Facchini, M., Mircea, M., Fuzzi, S., and Charlson, R. (1999) Cloud albedo enhancement by surface-active organic solutes in growing droplets. *Nature*, **401**, 257–259.

Fahey, D., Kelly, K., Ferry, G., Poole, L., Wilson, J.C., Murphy, D., Loewenstein, M., and Chan, K. (1989) In situ measurements of total reactive nitrogen, total water, and aerosol in a polar stratospheric cloud in the Antarctic. *J. Geophys. Res.*, **94**, 299–315.

Fahey, D.W., Eubank, C.S., Hübler, G., and Fehsenfeld, F.C. (1985) Evaluation of a catalytic reduction technique for the measurement of total reactive odd-Nitrogen NO_y in the atmosphere. *J. Atmos. Chem.*, **3**, 435–468.

Faloona, I.C., Tan, D., Lesher, R., Hazen, N., Frame, C., Simpas, J., Harder, H., Martinez, M., Carlo, P.D., Ren, X., and Brune, W. (2004) A laser induced fluorescence instrument for detecting tropospheric OH and HO_2: characteristics and calibration. *J. Atmos. Chem.*, **47**, 139–167.

Febvre, G., Gayet, J.-F., Minikin, A., Schlager, H., Shcherbakov, V., Jourdan, O., Busen, R., Fiebig, M., Kärcher, B., and Schumann, U. (2009) On optical and microphysical characteristics of contrails and cirrus. *J. Geophys. Res.*, **114**, 184. doi: 10.1029/2008JD010.

Fehsenfeld, F.C., Ancellet, G., Bates, T., Goldstein, A., Hardesty, R., Honrath, R., Law, K., Lewis, A., Leaitch, R., McKeen, S., Meagher, J., Parrish, D., Pszenny, A., Russel, P., Seinfeld, H.S.J., Talbot, R., and Zbinden, R. (2006) International Consortium for Atmospheric Research on Transport and Transformation (ICARTT): North America to Europe-Overview of the 2004 summer field study. *J. Geophys. Res.*, **111**. doi: 10.1029/2006JD007829.

Feigl, C., Schlager, H., Kuhn, M., Ziereis, H., Curtius, J., Arnold, F., and Schiller, C. (1999) Observations of NO_y uptake by particles in the Arctic tropopause region at low temperatures. *Geophys. Res. Lett.*, **26**, 2215–2219.

Feingersh, T., Ben-Dor, E., and Filin, S. (2010) Correction of reflectance anisotropy: a multi-sensor approach. *Int. J. Remote Sens.*, **31**, 49–74.

Feingold, G. and Chuang, P.Y. (2002) Analysis of the influence of film-forming compounds on droplet growth: implications for cloud microphysical processes and climate. *J. Atmos. Sci.*, **59**, 2006–2018.

Feingold, G., Cotton, W., Kreidenweis, S., and Davis, J. (1999) The impact of giant cloud condensation nuclei on drizzle formation in stratocumulus: implications for cloud radiative properties. *J. Atmos. Sci.*, **56**, 4100–4117.

Feldpausch, P., Fiebig, M., Fritzsche, L., and Petzold, A. (2006) Measurement of ultrafine aerosol size distributions by a combination of diffusion screen separators and condensation particle counters. *J. Aerosol Sci.*, **37**, 577–597.

Felt, E.P. (1928) Dispersal of insects by air currents. *Bull. N.Y. State Mus.*, **274**, 59–129.

Fernandez, D., Kerr, E., Castells, A., Carswell, J., Frasier, S., Chang, P., Black, P., and Marks, F. (2005) IWRAP: the Imaging Wind and Rain Airborne Profiler for remote sensing of the ocean and the atmospheric boundary layer within tropical cyclones. *IEEE Trans. Geosci. Remote*, **43**, 1775–1787.

Fiebig, M., Stein, C., Schröder, F., Feldpausch, P., and Petzold, A. (2005) Inversion of data containing information on the aerosol particle size distribution using multiple instruments. *J. Aerosol Sci.*, **36**, 1353–1372.

Fiedler, L., Newman, S., and Bakan, S. (2005) Correction of detector nonlinearity in Fourier transform spectroscopy with a low-temperature blackbody. *App. Opt.*, **44**, 5332–5340.

Field, P. and Heymsfield, A. (2003) Aggregation and scaling of ice crystal size distributions. *J. Atmos. Sci.*, **60**, 544–560.

Field, P., Wood, R., Brown, P.R., Kaye, P., Hirst, E., Greenaway, R., and Smith, J.A. (2003) Ice particle interarrival times measured with a Fast FSSP. *J. Atmos. Oceanic Technol.*, **20**, 249–261.

Field, P.R., Heymsfield, A., and Bansemer, A. (2006) Shattering and particle interarrival times measured by Optical Array Probes in ice clouds. *J. Atmos. Oceanic Technol.*, **23**, 1357–1370.

Field, P.R., Heymsfield, A.J., and Bansemer, A. (2007) Snow size distribution parameterization for midlatitude and tropical ice clouds. *J. Atmos. Sci.*, **64**, 4346–4365.

Fierz, M., Vernooij, M.G.C., and Burtscher, H. (2007) An improved low-flow thermodenuder. *J. Aerosol Sci.*, **38**, 1163–1168.

Finlayson-Pitts and Pitts (1999) *Chemistry of the Upper and Lower Atmosphere - Theory, Experiments and Applications, International Geophysics Series*, Academic Press, ISBN: 9780122570605.

Finnegan, W.G. and Chai, S.K. (2003) A new hypothesis for the mechanism of ice nucleation on wetted AgI and Ag AgCl particulate aerosols. *J. Atmos. Sci.*, **60**, 1723–1731.

Fischer, H., Brunner, D., Harris, G.W.P., Hoor, J.L., McKenna, D.S., Rudolph, J., Scheeren, H.A., Siegmund, P., Wernli, H., Williams, J., and Wong, S. (2002) Synoptic tracer gradients in the upper troposphere over central Canada during the Stratosphere-Troposphere Experiments by Aircraft Measurements 1998 summer campaign. *J. Geophys. Res.*, **107**, 4064. doi: 10.1029/2000JD000312.

Fix, A., Ehret, G., Hoffstädt, A., Klingenberg, H.H., Lemmerz, C., Mahnke, P., Ulbricht, M., Wirth, M., Wittig, R., and Zirnig, W. (2004) CHARM: a helicopter-borne Lidar system for pipeline monitoring. 24th International Laser Radar Conference, ESA SP 561, pp. 45–48.

Flagan, R.C. (2004) Opposed Migration Aerosol Classifier (OMAC). *Aerosol Sci. Technol.*, **38**, 890–899.

Flamant, C., Pelon, J., Flamant, P.H., and Durand, P. (1997) On the determination of the entrainment zone thickness at the top of the unstable boundary layer in clear air by Lidar and in situ measurements. *Bound.-Lay. Meteorol.*, **83**, 247–284.

Flamant, C., Chaboureau, J.-P., Parker, D.J., Taylor, C.M., Cammas, J.-P., Bock, O., Timouck, F., and Pelon, J. (2007) Airborne observations of the impact of a convective system on the planetary boundary layer thermodynamics and aerosol distribution in the inter-tropical discontinuity region of the West African Monsoon. *Q. J. R. Meteorol. Soc.*, **133**, 1175–1189.

Flentje, H., Dörnbrack, A., Ehret, G., Fix, A., Kiemle, C., Poberaj, G., and Wirth, M. (2005) Water vapor heterogeneity related to tropopause folds over the North Atlantic revealed by airborne water vapor differential absorption Lidar. *J. Geophys. Res.*, **110**, D03115, 2004JD004957.

Flentje, H., Dörnbrack, A., Fix, A., Ehret, G., and Holm, E. (2007) Evaluation of ECMWF water vapor fields by airborne differential absorption Lidar measurements: A case study between Brasil and Europe. *Atmos. Chem.*

Phys., **7**, 5033–5042. www.atmos-chem-phys.net/7/5033/2007/.

Flocke, F., Atlas, E., Madronich, S., Schauffler, S.M., Aikin, K., Margitan, J.J., and Bui, T.P. (1998) Observations of methyl nitrate in the lower stratosphere during STRAT: implications for its gas phase production mechanisms. *Geophys. Res. Lett.*, **25**, 1891–1894.

Flocke, F., Herman, R.L., Salawitch, R.J., Atlas, E., Webster, C.R., Schauffler, S.M., Lueb, R.A., May, R.D., Moyer, E.J., Rosenlof, K.H., Scott, D.C., Blake, D.R., and Bui, T.P. (1999) An examination of chemistry and transport processes in the tropical lower stratosphere using observations of long-lived and short-lived compounds obtained during STRAT and POLARIS. *J. Geophys. Res.*, **104**, 26625–262664.

Fontijn, A., Sabadell, A.J., and Ronco, R.J. (1970) Homogeneous chemiluminescent measurement of nitric oxide with ozone: implications for continuous selective monitoring of gaseous air pollutants. *Anal. Chem.*, **42**, 575–579.

Foody, G. (1999) The significance of border training patterns in classification by a feed forward neural network using back propagation learning. *Int. J. Remote Sens.*, **20**, 3549–3562.

Foot, J. (1985) Improvements in pyrgeometers. UK Patent Office, No. 8515024.

Foot, J. (1986) A new prygeometer. *J. Atmos. Oceanic Technol.*, **3**, 363–370.

Forgan, B. (1994) General method for calibrating Sun photometers. *Appl. Opt.*, **33**, 4841–4850.

Formenti, P., Elbert, W., Maenhaut, W., Haywood, J., and Andreae, M.O. (2003) Chemical composition of mineral dust aerosol during the Saharan Dust Experiment (SHADE) airborne campaign in the Cape Verde region. *J. Geophys. Res.*, **108**, 8576. doi: 10.1029/2002JD002648.

Formenti, P., Schütz, L., Balkanski, Y., Desboeufs, K., Ebert, M., Kandler, K., Petzold, A., Scheuvens, D., Weinbruch, S., and Zhang, D. (2011) Recent progress in understanding physical and chemical properties of African and Asian mineral dust. *Atmos. Chem. Phys.*, **11**, 8231–8256. doi: 10.5194/acp-11-8231-2011.

Forster, P., Ramaswamy, V., Artaxo, P., Berntsen, T., Betts, R., Fahey, D., Haywood, J., Lean, J., Lowe, D., Myhre, G., Nganga, J., Prinn, R., Raga, G., Schulz, M., and Dorland, R.V. (2007) Changes in atmospheric constituents and in radiative forcing, in *Climate Change 2007: The Physical Science Basis. Contribution of Working Group I to the Fourth Assessment Report of the Intergovernmental Panel on Climate Change* (eds S. Solomon, D. Qin, M. Manning, Z. Chen, M. Marquis, K.B. Averyt, M. Tignor, and H.L. Miller), Cambridge University Press.

Fortner, E., Zhao, J., and Zhang, R. (2004) Development of ion drift-chemical ionization mass spectrometry. *Anal. Chem.*, **76**, 5436–5440.

Fortner, E.C. and Knighton, W.B. (2008) Quantitatively resolving mixtures of isobaric compounds using chemical ionization mass spectrometry by modulating the reactant ion composition. *Rapid Commun. Mass Spectrom.*, **22**, 2597–2601.

Foster, S. (2003) Accuracy requirements for airborne wind measurement in aerial application, ASAE/NAAA Technical Section. 37th Annual National Agricultural Aviation Association Convention, Reno, NV, 8 December 2003, pp. Paper Number AA03-0012.

Fox, N. (2004) Validated data and removal of bias through traceability to SI, *Post-launch Calibration of Satellite Sensors*, vol. 2, of *ISPRS Book Series*, Taylor and Francis, London, pp. 31–42.

Francis, P., Jones, A., Saunders, A., Shine, K., Slingo, A., and Zhian, S. (1994) An observational and theoretical study of the radiative properties of cirrus: some results from ICE '89. *Q. J. R. Meteor. Soc.*, **120**, 809–848.

Franz, P. and Röckmann, T. (2005) High-precision isotope measurements of $H_2{}^{16}O$, $H_2{}^{17}O$, $H_2{}^{18}O$, and the $D^{17}O$-anomaly of water vapor in the southern lowermost stratosphere. *Atmos. Chem. Phys.*, **5**, 2949–2959.

Fried, A., Lee, Y., Frost, G., Wert, B., Henry, B., Drummond, J., Hu, G., and Jobson, T. (2002) Airborne CH_2O measurements over the North Atlantic during the 1997 NARE campaign:

instrument comparisons and distributions. *J. Geophys. Res.*, **107**, 4039. doi: 10.1029/2000JD000260.

Fried, A., Crawford, J., Olson, J., Walega, J., Potter, W., Wert, B., Jordan, C., Anderson, B., Shetter, R., Lefer, B., Blake, D., Blake, N., Meinardi, S., Heikes, B., O'Sullivan, D., Snow, J., Fuelberg, H., Kiley, C.M., Sandholm, S., Tan, D., Sachse, G., Singh, H., Faloona, I., Harward, C.N., and Carmichael, G.R. (2003) Airborne tunable diode laser measurements of formaldehyde during TRACE-P: distributions and box-model comparisons. *J. Geophys. Res.*, **108**, 8798. doi: 10.1029/2003JD003451.

Friedlander, S.K. (1977) *Smoke, Dust and Haze*, John Wiley & Sons, Inc.

Friehe, C., Grossman, R., and Pann, Y. (1986) Calibration of an airborne Lyman-Alpha hygrometer and measurement of water vapor flux using a thermoelectric hygrometer. *J. Atmos. Oceanic Technol.*, **3**, 299–304.

Friehe, C.A. and Khelif, D. (1992) Fast-response aircraft temperature sensors. *J. Atmos. Oceanic Technol.*, **9**, 784–795.

Friehe, C.A., Burns, S.P., Khelif, D., and Song, X. (1996) Meteorological and flux measurements from the NOAA WP3D aircraft in TOGA COARE, *8th Conference on Air-Sea Interaction*, American Meteorological Society, pp. J42–J45.

Frisch, A.S., Fairall, C.W., and Snider, J.B. (1995) Measurement of stratus cloud and drizzle parameters in ASTEX with a K-alpha-band Doppler Radar and a microwave radiometer. *J. Atmos. Sci.*, **52**, 2788–2799.

Fritz, S. (1948) The albedo of the ground and atmosphere. *Bull. Am. Meteorol. Soc.*, **29**, 303–312.

Frost, G.J., Fried, A., Lee, Y., Wert, B., Henry, B., Drummond, J.R., Evans, M.J., Fehsenfeld, F.C., Goldan, P.D., Holloway, J.S., Hübler, G., Jakoubek, R., Jobson, B.T., Knapp, K., Kuster, W.C., Parrish, D.D., Roberts, J., Rudolph, J., Ryerson, T.B., Stohl, A., Stroud, C., Sueper, D.T., Trainer, M., and Williams, J. (2002) Comparisons of box model calculations and measurements of formaldehyde from the 1997 North Atlantic Regional Experiment. *J. Geophys. Res.*, **107**, 4060. doi: 10.1029/2001JD000896.

Froyd, K.D., Murphy, D.M., Sanford, T.J., Thomson, D.S., Wilson, J.C., Pfister, L., and Lait, L. (2009) Aerosol composition of the tropical upper troposphere. *Atmos. Chem. Phys.*, **9**, 4363–4385.

Fu, Q. (1996) An accurate parameterization of the solar radiative properties of cirrus clouds for climate models. *J. Climate*, **9**, 2058–2082.

Fuchs, H., Dubé, W., Ciciora, S., and Brown, S. (2008) Determination of inlet transmission and conversion efficiencies for in situ measurements of the nocturnal nitrogen oxides, NO_3, N_2O_5 and NO_2, via pulsed Cavity Ring-Down Spectroscopy. *Anal. Chem.*, **80**, 6010–6017. doi: 10.1021/ac8007253.

Fuchs, H., Ball, S., Bohn, B., Brauers, T., Cohen, R., Dorn, H.-P., Dubé, W., Fry, J., Häseler, R., Heitmann, U., Jones, R., Kleffmann, J., Mentel, T., Müsgen, P., Rohrer, F., Rollins, A., Ruth, A., Kiendler-Scharr, A., Schlosser, E., Shillings, A., Tillmann, R., Varma, R., Venables, D., Villena Tapia, G., Wahner, A., Wegener, R., Wooldridge, P., and Brown, S. (2010) Intercomparison of measurements of NO_2 concentrations in the atmospheric simulation chamber SAPHIR during the NO_3COMP campaign. *Atmos. Meas. Technol.*, **3**, 21–37.

Fuchs, N.A. (1963) On the stationary charge distribution of aerosol particles in bipolar ionic atmospheres. *Geofis. Pura. Appl.*, **56**, 185–193.

Fugal, J.P. and Shaw, R.A. (2009) Cloud particle size distributions measured with an airborne digital in-line holographic instrument. *Atmos. Meas. Technol.*, **2**, 259–271.

Fugal, J.P., Shaw, R.A., Saw, E.W., Schulz, T.J., and Sergeyev, A.V. (2004) Airborne digital holographic system for cloud particle measurements. *Appl. Opt.*, **43**, 5987–5995.

Fugal, J.P., Schulz, T.J., and Shaw, R.A. (2009) Practical methods for automated reconstruction and characterization of particles in digital in-line holograms. *Meas. Sci. Technol.*, **20**, 075 501. doi: 10.1088/0957-0233/20/7/075 501.

Fukuta, N. and Saxena, V. (1979) A horizontal thermal gradient cloud condensation nucleus spectrometer. *J. Appl. Meteorol.*, **18**, 1352–1362.

Gal-Chen, T. (1982) Errors in fixed and moving grame of references-applications for conventional and Doppler Radar analysis. *J. Atmos. Sci.*, **39**, 2279–2300.

Galloway, J., Pazmany, A., Mead, J., McIntosh, R., Leon, D., French, J., Kelly, R., and Vali, G. (1997) Detection of ice hydrometeror alignment using an airborne W-band polarimetric Radar. *J. Atmos. Oceanic Technol.*, **14**, 3–12.

Gamache, J., Marks, F., and Roux, F. (1995) Comparison of 3 airborne Doppler sampling techniques with airborne in situ wind observations in Hurricane Gustav (1990). *J. Atmos. Oceanic Technol.*, **12**, 171–181.

Gamero-Castano, M. and de la Mora, J.F. (2000) A condensation nucleus counter (CNC) sensitive to singly charged sub-nanometer particles. *J. Aerosol. Sci.*, **31**, 757–772.

Gao, B. and Goetz, F. (1990) Column atmospheric water vapor and vegetation liquid water retrievals from airborne imaging spectrometer data. *J. Geophys. Res.*, **95**, 3549–3564.

Gao, B. and Goetz, F. (1995) Retrieval of equivalent water thickness and information related to biochemical components of vegetation canopies from AVIRIS data CAREFUL:in word reference list the year 1994 was also given, assuming 1995 is correct. *Remote Sens. Environ.*, **52**, 155–162.

Gao, B., Goetz, A., and Wiscombe, W. (1993) Cirrus cloud detection from Airborne Imaging Spectrometer data using the 1.38 μm water vapor band. *Geophys. Res. Lett.*, **20**, 301–304.

Gao, B., Davis, C., Montes, M., and Ahmad, Z. (2000) Atmospheric correction algorithm for hyperspectral remote sensing of ocean color from space. *Appl. Opt.*, **39**, 887–896.

Gao, B., Montes, M., Davis, C., and Goetz, A. (2007) Atmospheric correction algorithms for hyperspectral remote sensing data of land and ocean. *Remote Sens. Environ.*, **113**, 17–24.

Gao, R., Keim, E., Woodbridge, E., Ciciora, S., Proffitt, M., Thompson, T., Mclaughlin, R., and Fahey, D. (1994) New photolysis system for NO_2 measurements in the lower stratosphere. *J. Geophys. Res.*, **99**, 673–681.

Gardiner, B. and Hallett, J. (1985) Degradation of in-cloud forward scattering spectrometer probe measurements in the presence of ice particles. *J. Atmos. Oceanic Technol.*, **2**, 171–180.

Garnier, A. and Chanin, M.L. (1992) Description of a Doppler Rayleigh LIDAR for measuring winds in the middle atmosphere. *Appl. Phys. B*, **55**, 35–40.

Garrett, T. (2007) Comment on "Effective radius of ice cloud particle populations derived from aircraft probes" by Heymsfield *et al. Atmos. Oceanic. Technol.*, **24**, 1495–1503. doi: 10.1175/JTECH2075.1.

Garrett, T. (2008) *Observational Quantification of the Optical Properties of Cirrus Clouds, in Light Scattering Reviews 3*, Springer, Berlin.

Garrett, T., Hobbs, P., and Gerber, H. (2001) Shortwave, single-scattering properties of Artic ice clouds. *J. Geophys. Res.*, **106**, 15155–15172.

Garrett, T., Gerber, H., Baumgardner, D., Twohy, C., and Weinstock, E. (2003) Small, highly reflective ice crystals in low-latitude cirrus. *Geophys. Res. Lett.*, **30**, 2132. doi: 10.1029/2003GL018153.

Garrett, T., Navarro, B., Twohy, C., Jensen, E., Baumgardner, D., Bui, P., Gerber, H., Herman, R., Heymsfield, A., Lawson, P., Minnis, P., Nguyien, L., Poellot, M., Pope, S., Valero, F., and Weinstock, E. (2005) Evolution of a Florida cirrus anvil. *J. Atmos. Sci.*, **62**, 2352–2372.

Garvey, D.M. and Pinnick, R.G. (1983) Response characteristics of the Particle Measuring Systems Active Scattering Aerosol Spectrometer Probe (ASASP-X). *Aerosol Sci. Technol.*, **2**, 477–488.

Gasso, S., Hegg, D.A., Covert, D.S., Collins, D., Noone, K.J., Ostrom, E., Schmid, B., Russell, P.B., Livingston, J.M., Durkee, P.A., and Jonsson, H. (2000) Influence of humidity on the aerosol scattering coefficient and its effect on the upwelling radiance during ACE-2. *Tellus B*, **52**, 546–567.

Gayet, J.-F., Brown, P., and Albers, F. (1993) A comparison of in-cloud measurements obtained with six PMS 2D-C probes. *J. Atmos. Oceanic Technol.*, **10**, 180–194.

Gayet, J.-F., Febvre, G., and Larsen, H. (1996) The reliability of the PMS FSSP in the presence of small ice crystals. *J. Atmos. Oceanic Technol.*, **13**, 1300–1310.

Gayet, J.-F., Crépel, O., Fournol, J.-F., and Oshchepkov, S. (1997) A new airborne Polar Nephelometer for the measurements of optical and microphysical cloud properties. Part I: theoretical design. *Ann. Geophys.*, **15**, 451–459.

Gayet, J.-F., Auriol, F., Oshchepkov, S., Schröder, F., Duroure, C., Febvre, G., Fournol, J.-F., Crépel, O., Personne, P., and Daugeron, D. (1998) In situ measurements of the scattering phase function of stratocumulus contrails and cirrus. *Geophys. Res. Lett.*, **25**, 971–974. doi: 10.1029/98GL00541.

Gayet, J.-F., Auriol, F., Minikin, A., Ström, J., Seifert, M., Krejci, R., Petzold, A., Febvre, G., and Schumann, U. (2002) Quantitative measurements of the microphysical and optical properties of cirrus clouds with four different in situ probes: evidence of small ice crystals. *Geophys. Res. Lett.*, **29**, 342. doi: 10.1029/2001GL014.

Gayet, J.-F., Stachlewska, I., Jourdan, O., Shcherbakov, V., Schwarzenboeck, A., and Neuber, R. (2007) Microphysical and optical properties of precipitating drizzle and ice particles obtained from alternated Lidar and in situ measurements. *Ann. Geophys.*, **25**, 1487–1497.

Gege, P., Fries, J., Haschberger, P., Schötz, P., Schwarzer, H., Strobl, P., Suhr, B., Ulbrich, G., and Vreeling, W. (2009) Calibration facility for airborne imaging spectrometers. *ISPRS J. Photogramm.*, **64**, 387–397.

Geist, J. and Blevin, W. (1973) Chopper-stabilized null radiometer based upon an electrically calibrated pyroelectic detector. *Appl. Opt.*, **12**, 2532–2535.

Geller, A.S., Rader, D., and Kempka, S. (1993) Calculation of particle concentration around aircraft-like geometries. *J. Aerosol Sci.*, **24**, 823–634.

George, I. and Abbatt, J. (2010) Chemical evolution of secondary organic aerosol from OH-initiated heterogeneous oxidation. *Atmos. Chem. Phys.*, **10**, 5551–5563.

Gerber, H. (1991a) Direct measurement of suspended particulate volume concentration and far-infrared extinction coefficient with a laser-diffraction instrument. *Appl. Opt.*, **30**, 4824–4831.

Gerber, H. (1998) Standards for measuring fog liquid water content, in *Proceedings 1st International Conference on Fog and Fog Collection* (eds R. Schemenauer and H. Bridgman), Vancouver, BC, Canada, pp. 35–39.

Gerber, H. (2007) Comment on "Effective radius of ice cloud particle populations derived from aircraft probes" by Heymsfield *et al. J. Atmos. Oceanic Technol.*, **25**, 1504–1510.

Gerber, H., Arends, B., and Ackerman, A. (1994) New microphysics sensor for aircraft use. *Atmos. Res.*, **31**, 235–252.

Gerber, H., Takano, Y., Garrett, T.J., and Hobbs, P. (2000) Nephelometer measurements of the asymmetry parameter, volume extinction coefficient and backscatter ratio in Artic clouds. *J. Atmos. Sci.*, **57**, 3021–3034.

Gerber, H., Jensen, J., Davis, A., Marshak, A., and Wiscombe, W. (2001) Spectral density of liquid water content at high frequencies. *J. Atmos. Sci.*, **58**, 497–503.

Gerber, H., Frick, G., Malinowski, S., Brenguier, J.-L., and Burnet, F. (2005) Holes and entrainment in stratocumulus. *J. Atmos. Sci.*, **62**, 443–459.

Gerber, H., Frick, G., Jensen, J., and Hudson, J. (2008) Entrainment, mixing, and microphysics in trade-wind cumulus. *J. Meteorol. Soc. Jpn.*, **86a**, 87–106.

Gerber, H.E. (1980) A saturation hygrometer for the measurement of relative humidity between 95 and 105 %. *J. Appl. Meteorol.*, **19**, 1196–1208.

Gerber, H.E. (1991b) Supersaturation and droplet spectral evolution in fog. *J. Atmos. Sci.*, **48**, 2569–2588.

Gerbig, C., Kley, D., Volz-Thomas, A., Kent, J., Dewey, K., and McKenna, D. (1996) Fast response resonance fluorescence CO measurements aboard the C-130: instrument characterization and measurements made during North

Atlantic Regional Experiment 1993. *J. Geophys. Res.*, **101**, 229–238.

Gerbig, C., Schmitgen, S., Kley, D., Volz-Thomas, A., Dewey, K., and Haaks, D. (1999) An improved fast-response vacuum-UV resonance fluorescence CO instrument. *J. Geophys. Res.*, **104**, 1699–1704.

Gersman, R., Ben-Dor, E., Beyth, M., Avigad, D., Abraha, M., and Kibreab, A. (2008) Mapping of hydrothermally altered rocks by the EO-1 Hyperion sensor, Northern Danakil Depression, Eritrea. *Int. J. Remote Sens.*, **29**, 3911–3936.

Geyer, A., Alicke, B., Mihelcic, D., Stutz, J., and Platt, U. (1999) Comparison of tropospheric NO_3 radical measurements by differential optical absorption spectroscopy and matrix isolation electron spin resonance. *J. Geophys. Res.*, **104**, 26097–26105.

Gherman, T., Venables, D., Vaughan, S., Orphal, J., and Ruth, A. (2008) Incoherent broadband cavity-enhanced absorption spectroscopy in the near-ultarviolet: application to HONO and NO_2. *Environ. Sci. Technol.*, **42**, 890–895.

Giangrande, S.E. and Ryzhkov, A.V. (2008) Estimation of rainfall based on the results of polarimetric echo classification. *J. Appl. Meteorol. Climatol.*, **47**, 2445–2462.

Gibert, F., Flamant, P.H., Bruneau, D., and Loth, C. (2006) 2 μm heterodyne differential absorption Lidar measurements of the atmospheric CO_2 mixing ratio in the boundary layer. *Appl. Opt.*, **45**, 4448–4458.

Giez, A., Ehret, G., Schwiesow, R.L., Davis, K.J., and Lenschow, D.H. (1999) Water vapor flux measurements from groundbased vertically pointed water vapor differential absorption and Doppler Lidars. *J. Atmos. Oceanic Technol.*, **16**, 237–250.

Gleason, S. and Gebre-Egziabher, D. (2009) *GNSS Applications and Methods*, Artech House, Boston, MA.

GloPac (2010) *http://www.nasa.gov/centers/dryden/research/GloPac/index.html.*

Glover, A., Skippon, S., and Boyle, R. (1995) Interferometric laser imaging for droplet sizing: a method for droplet-size measurement in sparse spray systems. *Appl. Opt.*, **34**, 8409.

Goetz, A. (1991) Imaging spectrometry for studying Earth, air, fire and water. *EARS Adv. Remote Sens.*, **1**, 3–15.

Goetz, A. (2009) Three decades of hyperspectral remote sensing of the Earth: a personal view. *Remote Sens. Environ.*, **113**, Imaging Spectroscopy Special Issue, 5–16.

Goetz, A. and Wellman, J. (1984) Airborne imaging spectrometer: a new tool for remote sensing. *IEEE Trans. Geosci. Remote*, **22**, 546–550.

Goetz, A., Preining, O., and Kallai, T. (1961) The metastability of natural and urban aerosol particles. *Pure Appl. Geophys.*, **50**, 67–80.

Goetz, A., Vane, G., Solomon, J., and Rock, B. (1985) Imaging spectrometry for Earth remote sensing. *Science*, **228**, 1147–1153.

Goetz, F. (1987) HIRS, High Resolution Imaging Spectrometer: Science Opportunities of the 1990, Earth Observation System, *Tech. Rep.* vol. 2c, Instrument Panel Report to NASA, 74 p.

Goldstein, R.J. (ed.) (1996) *Fluid Mechanics Measurements*, Taylor & Francis, Washington, DC, p. 712.

Golitzine, N. (1950) Method for Measuring the Size of Water Droplets in Clouds, Fogs and Sprays, *Tech. Rep.* Report ME-177, National Research Council, Ottawa, Canada, also published as Note 6 (1951), National Aeronautical Establishment, Ottawa, Canada.

Goodman, J. (1996) *Introduction to Fourier Optics*, 2nd edn, McGraw Hill, Boston, MA.

Gorgucci, E., Scarchilli, G., Chandrasekar, V., and Bringi, V.N. (2000) Measurement of mean raindrop shape from polarimetric Radar observations. *J. Atmos. Sci.*, **57**, 3406–3413.

Gorgucci, E., Chandrasekar, V., and Baldini, L. (2006) Rainfall estimation from X-band dual polarization Radar using reflectivity and differential reflectivity. *Atmos. Res.*, **82**, 164–172.

Gossard, E., Snider, J., Clothiaux, E.E., Martner, B., Gibson, J., Kropfli, R., and Frisch, A. (1997) The potential of 8 mm Radars for remotely sensing cloud drop size distributions. *J. Atmos. Oceanic Technol.*, **14**, 76–87.

Govind, P. (1975) Dropwindsonde instrumentation for weather reconnaissance aircraft. *J. Appl. Meteorol.*, **14**, 1512–1520.

Gower, J.F.R. and Borstad, G.A. (1989) Phytoplankton remote sensing with the FLI imaging spectrometer. *Adv. Space Res.*, **9**, 461–465.

Grace, J., Nichol, C., Disney, M., Lewis, P., Qquaife, T., and Bowyer, P. (2007) Can we measure terrestrial photosynthesis from space directly, using spectral reflectance and fluorescence? *Glob. Change Biol.*, **13**, 1484–1497.

Gracey, W. (1956) Measurement of Static Pressure On Aircraft. *Tech. Rep. NACA Report* 1364, National Advisory Committee for Aeronautics NACA.

Graham, B., Mayol-Bracero, O.L., Guyon, P., Roberts, G.C., Decesari, S., Facchini, M.C., Artaxo, P., Maenhaut, W., Köll, P., and Andreae, M.O. (2002) Water-soluble organic compounds in biomass burning aerosol particles over Amazonia. *J. Geophys. Res.*, **107**, 8047. doi: 10.1029/2001JD000336.

Green, R. (2001) Measuring the spectral expression of carbon dioxide in the solar reflected spectrum with AVIRIS. Proceedings of the 11th Annual Airborne Earth Science Workshop, Jet 442 Propulsion Laboratory, Pasadena, CA.

Green, R. and Pavri, B. (2000) AVIRIS in-flight calibration experiment, sensitivity analysis, and intraflight stability, *Proceedings of the 9th Airborne Visible/Infrared Imaging Spectrometer (AVIRIS) Workshop*, Jet Propulsion Laboratory, Pasadena, CA, JPL Publ. 00-18, pp. 207–221.

Green, R. and Shimada, M. (1997) On-orbit calibration of a multi-spectral satellite sensor using a high altitude airborne imaging spectrometer. *Adv. Space Res.*, **19**, 1387–1398.

Green, R., Estwood, M., Sarture, C., Chrien, T., Arnonsson, M., Chippendale, B., Faust, J., Pavri, B., Chovit, C., Soils, M., Olaha, M., and Williams, O. (1998) Imaging Spectroscopy and the airborne visible/infrared imaging spectrometer (AVIRIS). *Remote Sens. Environ.*, **65**, 227–248.

Green, R., Pavri, B., and Chrien, T. (2003a) On-orbit radiometric and spectral calibration characteristics of EO-1 Hyperion derived with an under flight of AVIRIS and in situ measurements at Salar de Arizaro Argentina. *IEEE Trans. Geosci. Remote*, **42**, 1194–1203.

Green, T., Brough, N., Reeves, C., Edwards, G., Monks, P., and Penkett, S. (2003b) Airborne measurements of peroxy radicals using the PERCA technique. *J. Environ. Monit.*, **5**, 75–83.

Green, T.J., Reeves, C., Fleming, Z., Brough, N., Rickard, R.A., Bandy, B., Monks, P., and Penkett, S. (2006) An improved dual channel PERCA instrument for atmospheric measurements of peroxy radicals. *J. Environ. Monit.*, **8**, 530–536. doi: 10.1039/b514630e.

Griffiths, P. and de Haseth, J. (1986) Fourier transform infrared spectrometry. *Chem. Anal.*, **83**, 9729.

Grody, N. (1991) Classification of snow cover and precipitation using the Special Sensor Microwave/Imager (SSM/I). *J. Geophys. Res.*, **96**, 7423–7435.

Grossman, R.L. (1984) Bivariate conditional sampling of moisture flux over a tropical ocean. *J. Atmos. Sci.*, **41**, 3238–3252.

Grossman, R.L. (1992) Sampling errors in the vertical fluxes of potential temperature and moisture measured by aircraft during FIFE. *J. Geophys. Res.*, **97**, 18439–18443.

Grund, C.J., Banta, M., George, J., Howell, J., Post, M., Richter, R., and Weickmann, A. (2001) High-resolution Doppler Lidar for boundary layer research. *J. Atmos. Oceanic Technol.*, **18**, 376–393.

Guan, H., Schmid, B., Bucholtz, A., and Bergstrom, R. (2010) Sensitivity of shortwave radiative flux density, forcing, and heating rate to the aerosol vertical profile. *J. Geophys. Res.*, **11**, D06209. doi:10.1029/2009JD012907.

Guanter, L., Estellés, V., and Moreno, J. (2006) Spectral calibration and atmospheric correction of ultra-fine spectral and spatial resolution remote sensing data. Application to CASI-1500 data. *Remote Sens. Environ.*, **109**, 54–65.

Guildner, L.A., Johnson, D.P., and Jones, F.E. (1976) Vapor pressure of water at its triple point. *J. Res. Natl. Bur. Stand.*, **80A**, 505–521.

Guilfoyle, K., Althouse, M.L., and Chang, C. (2001) A quantitative and comparative analysis of linear and nonlinear spectral mixture models using radial basisi function neural network. *Remote Sens. Environ.*, **39**, 2314–2318.

Guimbaud, C., Catoire, V., Gogo, S., Robert, C., Laggoun-Défarge, F., Chartier, M., Grossel, A., Albéric, P., Pomathiod, L., and Nicoullaud, B. (2011) A portable infrared laser spectrometer for flux measurements of trace gases at the geosphere-atmosphere interface. *Meas. Sci. Technol.*, **22**, D10S03.

Gunthe, S.S., King, S.M., Rose, D., Chen, Q., Roldin, P., Farmer, D.K., Jimenez, J.L., Artaxo, P., Andreae, M.O., Martin, S.T., and Pöschl, U. (2009) Cloud condensation nuclei in pristine tropical rainforest air of Amazonia: size-resolved measurements and modeling of atmospheric aerosol composition and CCN activity. *Atmos. Chem. Phys.*, **9**, 7551–7575.

Gurlit, W., Zimmermann, R., Giesemann, C., Fernholz, T., Ebert, V., Wolfrum, J., Platt, U., and Burrows, J. (2005) Lightweight diode laser spectrometer CHILD (Compact High-altitude In situ Laser Diode) for balloonborne measurements of water vapor and methane. *Appl. Opt.*, **44**, 91–102.

Haddad, Z.S., Short, D., Durden, S.L., Im, E., Hensley, S., Grable, M., and Black, R. (1997) A new parameterization of the rain drop size distribution. *IEEE Trans. Geosci. Remote*, **35**, 532–539.

Haering, E.A. (1990) Airdata Calibration of a High-Performance Aircraft for Measuring Atmospheric Wind Profiles. *Tech. Mem.* NASA-TM-101714, NASA, 24 pp.

Haering, E.A. (1995) Airdata Measurement and Calibration. *Tech. Mem.* 104316, NASA, 19 pp.

Haggerty, J. and Curry, J. (2001) Variability of sea ice emissivity estimated from airborne passive microwave measurements during FIRE SHEBA. *J. Geophys. Res.*, **106**, 15265–15277.

Hair, J.W., Hostetler, C.A., Cook, A.L., Harper, D.B., Ferrare, R.A., Mack, T.L., Welch, W., Izquierdo, L.R., and Hovis, F.E. (2008) Airborne high spectral resolution Lidar for profiling aerosol optical properties. *Appl. Opt.*, **47**, 6734–6752.

Hallett, J., Arnott, W., Purcell, R., and Schmidt, C. (1998) A technique for characterizing aerosol and cloud particles by real time processing, *Proceedings of An International Specialty Conference*, Air & Waste Management Association, Long Beach, CA, vol. 1, pp. 318–325.

Halthore, R. (1997) Sun photometric measurements of atmospheric water vapor column abundance in the 940 nm band. *J. Geophys. Res.*, **102**, 4343–4352.

Haman, K.E. and Malinowski, S.P. (1996) Temperature measurements on a centimetre scale. Preliminary results. *Atmos. Res.*, **41**, 161–175.

Haman, K.E., Makulski, A., Malinowski, S.P., and Busen, R. (1997) A new ultrafast thermometer for airborne measurements in clouds. *J. Atmos. Oceanic Technol.*, **14**, 217–227.

Haman, K.E., Malinowski, S.P., Struś, B.D., Busen, R., and Stefko, A. (2001) Two new types of ultrafast aircraft thermometer. *J. Atmos. Oceanic Technol.*, **18**, 117–134.

Haman, K.E., Malinowski, S.P., Kurowski, M.J., Gerber, H., and Brenguier, J.-L. (2007) Small-scale mixing processes at the top of a marine stratocumulus - A case-study. *Q. J. R. Meteorol. Soc.*, **133**, 213–226.

Hamburger, T., McMeeking, G., Minikin, A., Birmili, W., Dall'Osto, M., O'Dowd, C., Flentje, H., Henzing, B., Junninen, H., Kristensson, A., de Leeuw, G., Stohl, A., Burkhart, J.F., Coe, H., Krejci, R., and Petzold, A. (2003) State of the climate in 2002. *Bull. Am. Meteorol. Soc.*, **84**, S1–S68.

Hammer, P., Valero, F., and Kinne, S. (1991) The 27-28 October 1986 FIRE cirrus case study: retrieval of cloud particle sizes and optical depths from comparative analyses of aircraft and satellite-base infrared measurements. *Mon. Weather. Rev.*, **119**, 1673–1692.

Hancock, G. and Orr-Ewing, A. (2009) Applications of Cavity Ring-Down Spectroscopy in atmospheric chemistry, *Cavity Ring-Down Spectroscopy: Techniques and Applications*, Blackwell Publishing Limited, pp. 181–212, ISBN 978-1-4051-7688-0.

Hänel, G. (1976) Single-scattering albedo of atmospheric aerosol particles as a function of relative humidity. *J. Atmos. Sci.*, **33**, 1120–1124.

Hanke, M., Uecker, J., Reiner, T., and Arnold, F. (2002a) Atmospheric peroxy radicals: ROXMAS, a new mass-spectrometric methodology for speciated measurements of HO_2 and ΣRO_2 and first results. *Int. J. Mass Spectrom.*, **213**, 91–99.

Hanke, M., Uecker, J., Reiner, T., and Arnold, F. (2002b) Atmospheric peroxy radicals: ROXMAS, a new mass-spectrometric methodology for speciated measurements of HO_2 and ΣRO_2 and first results. *Int. J. Mass Spectrom.*, **213**, 91–99.

Hansel, A., Singer, W., Wisthaler, A., Schwarzmann, M., and Lindinger, W. (1997) Energy dependencies of the proton transfer reactions $H_3O^+ + CH_2O \nLeftrightarrow CH_2OH^+ + H_2O$. *Int. J. Mass Spectrom.*, **167**, 697–703.

Hansen, A.D.A., Rosen, H., and Novakov, T. (1984) The aethalometer-an instrument for the real-time measurement of optical absorption by aerosol particles. *Sci. Total Environ.*, **36**, 191–196.

Hansen, J., Sato, M., and Ruedy, R. (1997) Radiative forcing and climate response. *J. Geophys. Res.*, **102**, 6831–6864.

Hansford, G., Freshwater, R., Eden, L., Turnbull, K., Hadaway, D., Ostanin, V., and Jones, R. (2006) Lightweight dew-/frost-point hygrometer based on a surface-acoustic-wave sensor for balloon-borne atmospheric water vapor profile sounding. *Rev. Sci. Instrum.*, **77**, 014502, doi: 10.1063/1.2140275.

Hanson, D.R., Koppes, M., Stoffers, A., Harsdorf, R., and Edelen, K. (2009) Proton transfer mass spectrometry at 11 hPa with a circular glow discharge: sensitivities and applications. *Int. J. Mass Spectrom.*, **282**, 28–37.

Harlow, R. (2009) Millimeter microwave emissivities and effective temperatures of snow covered surfaces: evidence for Lambertian surface scattering. *IEEE Trans. Geosci. Remote*, **37**, 1957–1970.

Harlow, R. (2010) Can millimeter-wave data be used to improve snow water equivalent estimates? Conference Proceedings, presented at Microwave Signatures, Florence, Italy.

Harlow, R. (2011) Sea ice emissivities and effective temperatures at AMSU-B frequencies: an analysis of airborne microwave data measured during two Arctic campaigns. *IEEE Trans. Geosci. Remote*, **49**, 1223–1237.

Harrison, L., Michalsky, J., and Berndt, J. (1994) Automated multifilter rotating shadowband radiometer: an instrument for optical depth and radiation measurements. *Appl. Opt.*, **33**, 5118–5125.

Harsanyi, J. and Chang, C. (1994) Hyperspectral image classification and dimensionality reduction: an orthogonal subspace projection. *IEEE Trans. Geosci. Remote*, **32**, 779–785.

Hauf, T. (1984) Turbulenzmessungen mit dem Forschungsflugzeug Falcon. *Meteorol. Rdsch.*, **37**, 163–176.

Hayasaka, T., Kikuchi, N., and Tanaka, M. (1995) Absorption of solar radiation by stratocumulus clouds: arcraft measurements and theoretical calculations. *J. Appl. Meteorol.*, **34**, 1047–1055.

Hayward, S., Hewitt, C.N., Sartin, J.H., and Owen, S.M. (2002) Performance characteristics and applications of a proton transfer reaction-mass spectrometer for measuring volatile organic compounds in ambient air. *Environ. Sci. Technol.*, **36**, 1554–1560.

Haywood, J. and Boucher, O. (2000) Estimates of the direct and indirect radiative forcing due to tropospheric aerosol particles: A review. *Rev. Geophys.*, **38**, 513–543. doi: 10.1029/1999RG000078.

Haywood, J., Francis, P., Osborne, S., Glew, M., Loeb, N., Highwood, E., Tanré, D., Myhre, G., Formenti, P., and Hirst, E. (2003) Radiative properties and direct radiative effect of Saharan dust measured by the C-130 aircraft during SHADE: 1. Solar spectrum. *J. Geophys. Res.*, **108**, 8577. doi: 10.1029/2002JD002687.

Haywood, J., Johnson, B., Osborne, S., Baran, A., Brooks, M., Milton, S., Mulcahy, J., Walters, D., Allan, R., Woodage, M., Klaver, A., Formenti, P., Brindley, H., Christopher, S., and Gupta, P. (2011) Motivation, rationale and key results from the GERBILS Saharan dust measurement campaign. *Q. J. R. Meteorol. Soc.*, **137**, 1106–1116. doi: 10.1002/qj.797.

Haywood, J.M. and Shine, K.P. (1995) The effect of anthropogenic sulfate and soot aerosol on the clear sky planetary radiation budget. *Geophys. Res. Lett.*, **22**, 603–606.

Heard, D.E. (2006) *Analytical Techniques for Atmospheric Measurement*, Blackwell Publishing Ltd, p. 528.

Heard, D.E. and Pilling, M.J. (2003) Measurement of OH and HO_2 in the troposphere. *Chem. Rev.*, **103**, 51633–515198.

Hegg, D., Larson, T., and Yuen, P.F. (1993) A theoretical study of the effect of relative humidity on light.scattering by trospopsheric aerosol particles. *J. Geophys. Res.*, **98**, 18435–18439.

Hegg, D., Covert, D., Jonsson, H., and Covert, P. (2005) Determination of the transmission efficiency of an aircraft aerosol inlet. *Aerosol Sci. Technol.*, **39**, 966–971. doi: 10.1080/02786820500377814.

Hegg, D.A., Gao, S., Hoppel, W., Frick, G., Caffrey, P., Leaitch, W., Shantz, N., Ambrusko, J., and Albrechcinski, T. (2001) Laboratory studies of the efficiency of selected organic aerosol particles as CCN. *Atmos. Res.*, **58**, 155–166.

Heidt, L.E., Vedder, J.F., Pollock, W.H., Lueb, R.A., and Henry, B.E. (1989) Trace gases in the Antarctic atmosphere. *J. Geophys. Res.*, **94**, 11599–11611.

Heikes, B., Lee, M., Jacob, D., Talbot, R., Bradshaw, J., Singhs, H., Blake, D., Anderson, B., Fuelberg, H., and Thompson, A. (1996) Ozone, hydroperoxides, oxides of nitrogen, and hydrocarbon budgets in the marine boundary layer over the South Atlantic. *J. Geophys. Res.*, **101**, 221–234.

Heikes, B., Snow, J., Ergli, P., O'Sullivan, D., Crawford, J., Orlson, J., Chen, G., Davis, D., Blakes, N., and Blake, D. (2001) Formaldehyde over the central Pacific during PEM-Tropics B. *J. Geophys. Res.*, **106**, 717.

Heinemann, G. (2002) Aircraft-based measurements of turbulence structures in the katabatic flow over Greenland. *Bound.-Lay. Meteorol.*, **103**, 49–81.

Heintzenberg, J. (2009) The SAMUM-1 experiment over Southern Morocco: overview and introduction. *Tellus B*, **61**, 2–11. doi: 10.1111/j.1600-0889.2008.00403.x.

Heintzenberg, J., Okada, K., and Ström, J. (1996) On the composition of non-volatile material in upper tropospheric aerosols and cirrus crystals. *Atmos. Res.*, **41**, 81–88.

Heintzenberg, J., Wiedensohler, A., Tuch, T.M., Covert, D.S., Sheridan, P., Ogren, J.A., Gras, J., Nessler, R., Kleefeld, C., Kalivitis, N., Aaltonen, V., Wilhelm, R.T., and Havlicek, M. (2006) Intercomparisons and aerosol calibrations of 12 commercial integrating nephelometers of three manufacturers. *J. Atmos. Oceanic Technol.*, **23**, 902–914.

Heinz, D. and Chang, C. (2001) Fully constrained least squares linear mixture analysis for material quantification in hyperspectral imagery. *IEEE Trans. Geosci. Remote*, **39**, 529–545.

Helmig, D., Boulter, J., David, D., Cullen, J.B.N., Steffen, K., Johnson, B., and Oltmans, S. (2002) Ozone and meteorological Summit, Greenland, boundary-layer conditions at during 3–21 June 2000. *Atmos. Environ.*, **36**, 2595–2608.

Helmig, D., Tanner, D., Honrath, R., Owen, R., and Parrish, D. (2008) Nonmethane hydrocarbons at Pico Mountain, Azores: 1. Oxidation chemistry in the North Atlantic region. *J. Geophys. Res.*, **113**, D20S91.

Helten, M., Smit, H.G.J., Straeter, W., Kley, D., Nedelec, P., Zöger, M., and Busen, R. (1998) Calibration and performance of automatic compact instrumentation for the measurement of relative humidity from passenger aircraft. *J. Geophys. Res.*, **103**, 25643–25652.

Helten, M., Smit, H., Kley, D., Ovarlez, J., Schlager, H., Baumann, R., Schumann, U., Nedelec, P., and Marenco, A. (1999) In-flight intercomparison of MOZAIC and POLINAT water vapor measurements. *J. Geophys. Res.*, **104**, 26087–26096.

Henderson, S., Suni, P., Hale, C., Hannon, S., Bruns, J.M.D., and Yuen, E.

(1993) Coherent laser Radar at 2 μm using solid-state lasers. *IEEE Trans. Geosci. Remote*, **31**, 4–15.

Hendricks, J., Karcher, B., Dopelheuer, A., Feichter, J., Lohmann, U., and Baumgardner, D. (2004) Simulating the global atmospheric black carbon cycle: a revisit to the contribution of aircraft emissions. *Atmos. Chem. Phys.*, **4**, 2521–2541.

Hengstberger, F. (1989) *Absolute Radiometry: Thermal Detectors of Optical Radiation* (ed. F. Hengstberger), Academic Press, Inc., San Diego, CA.

Hering, S., Appel, B., Cheng, W., Salaymeh, F., Cadle Mulawa, S., Cahill, T., Eldred, A., Surovik, M., Fitz, D., Howes, J., Knapp, K., Stockburger, L., Turpin, B., Huntzicker, J., Zhang, X.-Q., and McMurry, P. (1990) Comparison of sampling methods for carbonaceous aerosol particles in ambient air. *Aerosol Sci. Technol.*, **12**, 200–213.

Hering, S.V. and Stolzenburg, M.R. (2005) A method for particle size amplification by water condensation in a laminar, thermally diffusive flow. *Aerosol Sci. Technol.*, **39**, 428–436.

Hermann, M. and Wiedensohler, A. (2001) Counting efficiency of condensation particle counters at low-pressures with illustrative data from the upper troposphere. *J. Aerosol Sci.*, **32**, 975–991.

Hermann, M., Stratmann, F., and Wiedensohler, A. (1999) Calibration of an aircraft-borne aerosol inlet system for fine particle measurements. *J. Aerosol Sci.*, **30**, S153–S154.

Hermann, M., Stratmann, F., Wilck, M., and Wiedensohler, A. (2001) Sampling characteristics of an aircraft-borne aerosol inlet system. *J. Atmos. Oceanic Technol.*, **18**, 7–19.

Hermann, M., Adler, S., Caldow, R., Stratmann, F., and Wiedensohler, A. (2005a) Pressure-dependent efficiency of a condensation particle counter operated with FC-43 as working fluid. *J. Aerosol Sci.*, **36**, 1322–1337.

Hermann, M., Brenninkmeijer, C., Heintzenberg, J., Martinsson, B., Nguyen, H., Reichelt, M., Slemr, F., Wiedensohler, A., and Zahn, A. (2005b) CARIBIC-Next generation: an airborne measurement platform for the global distribution of aerosol particles. Presentation at the European Aerosol Conference (EAC), Ghent, Belgium, September 28. August-2.

Herndon, S.C., Zahniser, M.S., Nelson, D.D. Jr., Shorter, J., McManus, J.B., Jimenez, R., Warneke, C., and de Gouw, J.A. (2007) Airborne measurements of HCHO and HCOOH during the New England Air Quality Study 2004 using a pulsed quantum cascade laser spectrometer. *J. Geophys. Res.*, **112**, D10S03.

Herold, M., Roberts, D., Noronha, V., and Smadi, O. (2008) *Imaging Spectrometry and Asphalt Road Surveys*, Transportation Research Part C: Emerging Technologies.

Hess, M., Koepke, P., and Schult, I. (1998) Optical properties of aerosol particles and clouds: the software package OPAC. *Bull. Am. Meteorol. Soc.*, **79**, 831–844.

Heue, K.-P., Wagner, T., Broccardo, S.P., Walter, D., Piketh, S.J., Ross, K.E., Beirle, S., and Platt, U. (2008) Direct observation of two dimensional trace gas distributions with an airborne Imaging DOAS instrument. *Atmos. Chem. Phys.*, **8**, 6707–6717.

Hewison, T. (2001) Airborne measurements of forest and agricultural land surface emissivity at millimetre wavelengths. *IEEE Trans. Geosci. Remote*, **39**, 393–400.

Hewison, T. (2006) Aircraft validation of clear air absorption models at millimeter wavelengths (89–183 GHz). *J. Geophys. Res.*, 111. doi: 10.1029/2005JD006719.

Hewison, T. and English, S. (1999) Airborne retrievals of snow and ice surface emissivity at millimetre wavelengths. *IEEE Trans. Geosci. Remote*, **37**, 1871–1879.

Hewitt, C.N., Hayward, S., and Tani, A. (2003) The application of proton transfer reaction-mass spectrometry (PTR-MS) to the monitoring and analysis of volatile organic compounds in the atmosphere. *J. Environ. Monit.*, **5**, 1–7.

Heymsfield, A. (2007) On measurements of small ice particles in clouds. *Geophys. Res. Lett.*, **34**, L23812. doi: 10.1029/2007GL030951.

Heymsfield, A. and McFarquhar, G. (1996a) High albedos of cirrus in the tropical

Pacific warm pool: microphysical interpretations from CEPEX and from Kwajalein, Marshall Islands. *J. Atmos. Sci.*, **53**, 2424–2451.

Heymsfield, A. and McFarquhar, G. (1996b) High albedos of cirrus in the tropical pacific warm pool: microphysical interpretations from CEPEX and from Kwajalein, Marshall Islands. *J. Atmos. Sci.*, **53**, 2424–2451.

Heymsfield, A. and Miloshevich, L. (1989) Evaluation of liquid water measuring instruments in cold cloud sampled during FIRE. *J. Atmos. Oceanic Technol.*, **6**, 378–388.

Heymsfield, A. and Parrish, J.L. (1978) Techniques employed in the processing of particle size spectra and state parameter data obtained with the T-28 aircraft platform. NCAT *Tech. Note* NCAR/TN-137 + 1A, National Center for Atmospheric Research, Boulder, Colorado, USA, p. 78.

Heymsfield, A., Kennedy, P., Massie, S., Schmitt, C., Wang, Z., Haimov, S., and Rangno, A. (2010) Aircraft-induced hole punch and canal clouds: inadvertent cloud seeding. *Bull. Am. Meteorol. Soc.*, 753–766. doi: 10.1175/2009BAMS2905.1.

Heymsfield, A.J. (1977) Precipitation development in stratiform ice clouds: a microphysical and dynamical study. *J. Atmos. Sci.*, **34**, 367–381.

Heymsfield, A.J. and Westbrook, C.D. (2010) Advances in the estimation of ice particle fall speeds using laboratory and field measurements. *J. Atmos. Sci.*, **67**, 2469–2482.

Heymsfield, A.J., Protat, A., Austin, R.T., Hogan, R., Delanoë, J., Okamoto, H., Sato, K., van Zadelhoff, G.-J., Donovan, D.P., and Wang, Z. (2008) Testing IWC retrieval methods using Radar and ancillary measurements with in situ data. *J. Appl. Meteorol. Climatol.*, **47**, 135–163.

Heymsfield, G., Bidwell, S., Caylor, I., Ameen, S., Nicholson, S., Boncyk, W., Miller, L., Vandemark, D., Racette, P., and Dod, L. (1996) The EDOP Radar system on the high-altitude NASA ER-2 aircraft. *J. Atmos. Oceanic Technol.*, **13**, 795–809.

Hignett, P., Taylor, J., Francis, P., and Glew, M. (1999) Comparison of observed and modeled direct aerosol forcing during TARFOX. *J. Geophys. Res.*, **104**, 2279–2287.

Hildebrand, P.H., Lee, W.-C., Walter, C.A., Frush, C., Randall, M., Loew, E., Neitzel, R., Parsons, R., Testud, J., Baudin, F., and LeCornec, A. (1996) The ELDORA-ASTRAIA airborne Doppler weather Radar: high resolution observations from TOGA-COARE. *Bull. Am. Meteorol. Soc.*, **77**, 213–232.

Hindman, E. (1987) A 'Cloud Gun' Primer. *J. Atmos. Oceanic Technol.*, **4**, 736–741.

Hinds, W.C. (1999) *Aerosol Technology, Properties, Behavior and Measurement of Airborne Particles*, 2nd edn, John Wiley & Sons, Inc., New York, p. 504.

Hirleman, E., Oechsle, V., and Chigier, N. (1984) Response characteristics of laser diffraction particle size analyzers: optical volume extent and lens effects. *Opt. Eng.*, **23**, 610–619.

Hirst, E., Kaye, P., Greenaway, R., Field, P., and Johnson, D. (2001) Discrimination of micrometre-sized ice and supercooled droplets in mixed phase clouds. *Atmos. Environ.*, **35**, 33–47.

Hitschfeld, W. and Bordan, J. (1954) Errors inherent in the Radar measurement of rainfall at attenuating wavelengths. *J. Meteorol.*, **11**, 58–67.

Hlao, N. and Wong, F. (2000) Hyperspectral Imagery Market Forcast: 2000–2005. *Tech. Rep.*, Contract No F4701-c-0009, Space and Missile System Center Air Force Materiek Comman US Air Force.

Hock, T. and Franklin, J. (1999) The NCAR GPS dropwindsonde. *Bull. Am. Meteorol. Soc.*, **80**, 407–420.

Hoffmann, T., Huang, R.-J., and Kalberer, M. (2011) Atmospheric analytical chemistry. *Anal. Chem.*, **83**, 4649–4664. doi: 10.1021/ac2010718.

Hofmann, D.J., Rosen, J.M., Pepin, T.J., and Pinnick, R.G. (1973) Particles in the polar stratospheres. *Nature*, **245**, 369–371.

Hofmann-Wellenhof, B., Lichtenegger, H., and Collins, J. (2001) *Global Positioning System: Theory and Practice*, 5th edn, Springer-Verlag, Wien.

Hofzumahaus, A. (2006) Measurement of photolysis frequencies in the atmosphere, *Analytical Techniques for Atmospheric Measurement*, Blackwell Publishing, pp. 406–500.

Hofzumahaus, A., Kraus, A., and Müller, M. (1999) Solar actinic flux spectroradiometry: a technique for measuring photolysis frequencies in the atmosphere. *Appl. Opt.*, **38**, 4443–4460.

Hofzumahaus, A., Kraus, A., Kylling, A., and Zerefos, C. (2002) Solar actinic radiation (280–420 nm) in the cloud-free troposphere between ground and 12 km altitude: measurements and model results. *J. Geophys. Res.*, 107. doi: 10.1029/2001JD900142.

Hogan, R., Donovan, D., and Tinel, C. (2003a) Independent evaluation of the ability of spaceborne Radar and Lidar to retrieve the microphysical and radiative properties of ice clouds. *J. Atmos. Oceanic Technol.*, **23**, 211–227.

Hogan, R.J., Francis, P.N., Flentje, H., Illingworth, A.J., Quante, M., and Pelon, J. (2003b) Characteristics of mixed phase clouds. I: Lidar, Radar and aircraft observations from CLARE'98. *Q. J. R. Meteorol. Soc.*, **129**, 2089–2116.

Hogan, R.J., Gaussiat, N., and Illingworth, A.J. (2005) Stratocumulus liquid water content from dual-wavelength Radar. *J. Atmos. Oceanic Technol.*, **22**, 1207–1218.

Hogan, R.J., Brooks, M., Illingworth, A., Donovan, D., Tinel, C., Bouniol, D., and Baptista, J. (2006a) Independent evaluation of the ability of spaceborne Radar and Lidar to retrieve the microphysical and radiative properties of ice clouds. *J. Atmos. Oceanic Technol.*, **23**, 211–217.

Hogan, R.J., Mittermaier, M.P., and Illingworth, A.J. (2006b) The retrieval of ice water content from Radar reflectivity factor and temperature and its use in evaluating a mesoscale model. *J. Appl. Meteorol. Climatol.*, **45**, 301–317.

Holben, B.N., Eck, T.I., Slutsker, I., Tar, D., Buis, J., Setxer, A., Vemte, E., Reagan, J., Kaufman, Y.J., Nakajima, T., Lavenu, F., Jankowiak, I., and Smirnozjt, A. (1998) AERONET-A federated instrument network and data archive for aerosol characterization. *Remote Sens. Environ.*, **66**, 1–16.

Holloway, J., Jakoubek, R., Parrish, D., Gerbig, C., Volz-Thomas, A., Schmitgen, S., Fried, A., Wert, B., Henry, B., James, R., and Drummond, J. (2000) Airborne intercomparison of vacuum ultraviolet fluorescence and tunable diode laser absorption measurements of tropospheric carbon monoxide. *J. Geophys Res.*, **105**, 251–261.

Holzapfel, E.P. (1978) Transoceanic airplane sampling for organisms and particles. *Pac. Insects*, **18**, 169–189.

Holzinger, R., Williams, J., Salisbury, G., Klupfel, T., de Reus, M., Traub, M., Crutzen, P.J., and Lelieveld, J. (2005) Oxygenated compounds in aged biomass burning plumes over the Eastern Mediterranean: evidence for strong secondary production of methanol and acetone. *Atmos. Chem. Phys.*, **5**, 39–46.

Holzwarth, S., Müller, A., Habermeyer, M., Richter, R., Hausold, A., Thiemann, S., and Strobl, P. (2003) HySens-DAIS 7915/ROSIS imaging spectrometers at DLR. Proceedings of the 3rd EARSeL Workshop on Imaging Spectroscopy, Herrsching, pp. 3–14.

Homan, C.D., Volk, C., Kuhn, A., Werner, A., Baehr, J., Viciani, S., Ulanovski, A., and Ravegnani, F. (2010) Tracer measurements in the tropical tropopause layer during the AMMA/SCOUT-O_3 aircraft campaign. *Atmos. Chem. Phys.*, **10**, 3615–3627.

Honeywell (1988) *YG1854 Inertial Reference System/Global Positioning Inertial Reference System, Installation Manual*, Publication Number 95-8537A, Rev. 1, Honeywell, Inc., Commercial Flight Systems.

Honrath, R.E., Helmig, D., Owen, R.C., Parrish, D.D., and Tanner, D.M. (2008) Nonmethane hydrocarbons at Pico Mountain, Azores: 2. Event-specific analyses of the impacts of mixing and photochemistry on hydrocarbon ratios. *J. Geophys. Res.*, **112**. doi: 10.1029/2006JD007594.

Hopkins, J., Jones, I., Lewis, A., McQuaid, J., and Seakins, P. (2002) Non-methane hydrocarbons in the Arctic boundary layer. *Atmos. Environ.*, **36**, 3217–3229.

Hornbrook, R.S., Crawford, J.H., Edwards, G.D., Goyea, O., Mauldin, R.L. III, Olson, J.S., and Cantrell, C.A. (2011) Measurements of tropospheric HO_2 and RO_2 by oxygen dilution modulation and chemical ionization mass spectrometry. *Atmos. Meas. Tech.*, **4**, 735–756. doi: 10.5194/amt-4-735-2011.

Horrak, U., Mirme, A., Salm, J., Tamm, E., and Tammet, H. (1998) Air ion measurements as a source of information about atmospheric aerosol particles. *Atmos. Res.*, **46**, 233–242.

Horvath, H. (1993) Atmospheric light absorption-A review. *Atmos. Environ.*, **27**, 293–317.

Hovenac, E. and Hirleman, E. (1991) Use of rotating pinholes and reticles for calibrations of cloud droplet instrumentation. *J. Atmos. Oceanic Technol.*, **8**, 166–171.

Howell, J.F. and Mahrt, L. (1997) Multiresolution flux decomposition. *Bound.-Lay. Meteorol.*, **83**, 117–137.

Hu, Y., Liu, Z., Winker, D., Vaughan, M., Noel, V., Bissonnette, L., Roy, G., and McGill, M. (2006) Simple relation between Lidar multiple scattering and depolarization for water clouds. *Opt. Lett.*, **31**, 1809–1811.

Hubert, K. (1932) Beobachtungen über die Verbreitung des Gelbrostes bei künstlichen Feldinfektionen. *Fortschr. Landw.*, **7**, 195–198.

Hudson, J. (1989) An instantaneous CCN spectrometer. *J. Atmos. Oceanic Technol.*, **6**, 1055–1065.

Hudson, J. and Squires, P. (1976) An improved continuous flow diffusion cloud chamber. *J. Appl. Meteorol.*, **15**, 776–782.

Hudson, J., Noble, S., and Jha, V. (2010) Stratus cloud supersaturations. *Geophys. Res. Lett.*, 37.

Hudson, J.G., Jha, V., and Noble, S. (2011) Drizzle correlations with giant nuclei. *Geophys. Res. Lett.*, **38**, L05808. doi: 10.1029/2010GL046207.

Huebert, B., Howell, S., Covert, D., Bertram, T., Clarke, A., Anderson, J., Lafleur, B., Seebaugh, W., Wilson, J.C., Gesler, D., Blomquist, B., and Fox, J. (2004) PELTI: measuring the passing efficiency of an airborne low turbulence aerosol inlet. *Aerosol Sci. Technol.*, **38**, 803–826.

Huebert, B.J., Vanbramer, S., and Tschudy, K. (1988) Liquid cloudwater collection using modified Mohnen slotted rods. *J. Atmos. Chem.*, **6**, 251–263.

Huebert, B.J., Lee, G., and Warren, W. (1990) Airborne aerosol inlet passing efficiency measurements. *J. Geophys. Res.*, **95**, 16369–16381.

Huey, L.G. (2007) Measurement of trace atmospheric species by chemical ionization mass spectrometry: speciation of reactive nitrogen and future directions. *Mass Spectrom. Rev.*, **26**, 166–184.

Huey, L.G., Hanson, D.R., and Howard, C.J. (1995) Reactions of SF^{-6} and I^- with atmospheric trace gases. *J. Phys. Chem.*, **99**, 5001–5008.

Huey, L.G., Dunlea, E., Lovejoy, E., Hanson, D., Norton, R., Fehsenfeld, F., and Howard, C. (1998) Fast time response measurements of HNO_3 in air with a chemical ionization mass spectrometer. *J. Geophys. Res.*, **103**, 3355–3360.

Huffaker, R.M. and Hardesty, R.M. (1996) Remote sensing of atmospheric wind velocities using solid-state and CO_2 coherent laser systems. *Proc. IEEE*, **84**, 181–204.

Huffman, J.A., Docherty, K.S., Aiken, A.C., Cubison, M.J., Ulbrich, I.M., de Carlo, P.F., Sueper, D., Jayne, J.T., Worsnop, D.R., Ziemann, P.J., and Jimenez, J.L. (2009) Chemically-resolved aerosol volatility measurements from two megacity field studies. *Atmos. Chem. Phys.*, **9**, 7161–7182.

Huffman, P. and Thursby, W. (1969) Light scattering by ice crystals. *J. Atmos. Sci.*, **26**, 1073–1077.

Hunt, G. and Salisbury, J. (1970) Visible and near infrared spectra of minerals and rocks-I: silicate minerals. *Mod. Geol.*, **1**, 283–300.

Hunt, G. and Salisbury, J. (1971) Visible and near infrared spectra of minerals and rocks-II: carbonates. *Mod. Geol.*, 23–30.

Hunt, G., Salisbury, J., and Lenhoff, C. (1971a) Visible and near infrared spectra of minerals and rocks-III: oxides and hydroxides. *Mod. Geol.*, **2**, 195–205.

Hunt, G., Salisbury, J., and Lenhoff, C. (1971b) Visible and near infrared of minerals and rocks: halides, phosphates, arsenates, vanates and borates. *Mod. Geol.*, **3**, 121–132.

Hunter, E.P.L. and Lias, S.G. (1998) Evaluated gas phase basicities and proton affinities of molecules: an update. *J. Phys. Chem. Ref. Data*, **27**, 413–656.

Hunton, D., Ballenthin, J.O., Borghetti, J.F., Federico, G.S., Miller, T.M., Thorn, F., Viggiano, A.A., Anderson, B.E., Cofer, W.R., McDougal, D.S., and Wey, C.C. (2000) Chemical ionization mass spectrometric measurements of SO_2 emissions from jet engines in flight and test chamber operations. *J. Geophys. Res.*, **105**, 26841–26855.

Huntrieser, H., Schlager, H., Roiger, A., Schumann, U., Höller, H., Kurz, C., Brunner, D., Schwierz, C., Richter, A., and Stohl, A. (2007) Lightning–produced NO_x over Brazil during TROCCINOX: airborne measurements in tropical and subtropical thunderstorms and the importance of mesoscale convective systems. *Atmos. Chem. Phys.*, **7**, 2987–3013.

Huntrieser, H., Schlager, H., Lichtenstern, M., Roiger, A., Stock, P., Minikin, A., Höller, H., Schmidt, K., Betz, H.-D., Allen, G., Viciani, S., Ulanovsky, A., Ravegnani, F., and Brunner, D. (2009) NO_x production by lightning in Hector: first airborne measurements during SCOUT-O_3/ACTIVE. *Atmos. Chem. Phys.*, **9**, 8377–8412.

Hyson, P. and Hicks, B. (1975) A single beam infrared hygrometer for evaporation measurement. *J. Appl. Meteorol.*, **14**, 301–307.

Iaquinta, J., Isaka, H., and Personne, P. (1995) Scattering phase function of bullet rosette ice crystals. *J. Atmos. Sci.*, **52**, 1401–1413.

Ide, R. (1999) Comparison of Liquid Water Content Measurement Techniques in an Icing Wind Tunnel. *Tech. Mem.* TM-1999-209643, NASA, aRL-TR-2134, p. 24.

Ide, R. and Oldenburg, J. (2001) Icing cloud calibration of the NASA Glenn Icing Research Tunnel. 39th Aerospace Sciences Meeting and Exhibit, aIAA-2001-0234.

Illingworth, A. and Marsh, S. (1986) Static charging of aircraft by collisions with ice crystals. *Rev. Phys. Appl.*, **21**, 803–808.

Illingworth, A.J., Hogan, R.J., O'Connor, E.J., Bouniol, D., Brooks, M.E., Delanoë, J., Donovan, D.P., Eastment, J.D., Gaussiat, N., Goddard, J.W.F., Haeffelin, M., Baltink, H.K., Krasnov, O.A., Pelon, J., Piriou, J.-M., Protat, A., Russchenberg, H.W.J., Seifert, A., Tompkins, A.M., van Zadelhoff, G.-J., Vinit, F., Willen, U., Wilson, D.R., and Wrench, C.L. (2007) Cloudnet-continuous evaluation of cloud profiles in seven operational models using ground-based observations. *Bull. Am. Meteorol. Soc.*, **88**, 883–898.

Inomata, S., Tanimoto, H., Aoki, N., Hirokawa, J., and Sadanaga, Y. (2006) A novel discharge source of hydronium ions for proton transfer reaction ionization: design, characterization, and performance. *Rapid Commun. Mass Spectrom.*, **20**, 1025–1029.

Intrieri, J.M., Stephens, G.L., Eberhard, W.L., and Uttal, T. (1993) A method for determining cirrus cloud particle sizes using Lidar and Radar backscatter technique. *J. Appl. Meteorol.*, **32**, 1074–1082.

Inverarity, G.W. (2000) Correcting airborne temperature data for lags introduced by instruments with two-time-constant responses. *J. Atmos. Oceanic Technol.*, **17**, 176–184.

IPCC (2007) *Climate Change 2007: The Scientific Basis*, Cambridge University Press, 940 pp.

Iqbal, M. (1983) *An Introduction to Solar Radiation*, Academic Press, Inc.

Iribarne, J. and Godson, W. (1981) *Atmospheric Thermodynamics*, 2nd edn, D. Reidel, Dordrecht, Holland.

Irons, J., Weismiller, R., and Petersen, G. (1989) Soil reflectance, *Theory and Application of Optical Remote Sensing, Willey Series in Remote Sensing*, John Wiley & Sons, Inc., New York, pp. 66–106.

Irshad, H., McFarland, A., Landis, M., and Stevens, R. (2004) Wind tunnel evaluation of an aircraft-borne sampling system. *Aerosol Sci. Technol.*, **38**, 311–321.

Isaac, G., Korolev, A., Strapp, J., Cober, S., Boudala, F., Marcotte, D., and Reich, V. (2006) Assessing the collection efficiency of natural cloud particles impacting the Nevzorov total water content probe. AIAA 44th Aerospace Sciences Meeting and Exhibit, Reno, Nevada, AIAA-2006-1221.

Isaac, P.R., Mcaneney, J., Leuning, R., and Hacker, J.M. (2004) Comparison of aircraft and ground-based flux measurements during OASis95. *Bound.-Lay. Meteorol.*, **110**, 39–67.

Ismail, S., Koch, G., Abedin, N., Refaat, T., Rubio, M., and Singh, U. (2008) Development of laser, detector, and receiver systems for an atmospheric CO_2 Lidar profiling system. Proceedings IEEE Aerospace Conference, pp. 1–7.

ISO (1975) *Standard Atmosphere (Identical with the ICAO and WMO Standard Atmospheres from - 2 to 32 km)*, International Organization for Standardization, Geneva, Switzerland.

ISO (1985) *Flight Dynamics, Concepts and Quantities-Part 2: Motions of Aircraft and the Atmosphere Relative to the Earth*, Second edition-1985-09-01, International Organization for Standardization, Geneva, Switzerland.

ISO (1988) *Flight Dynamics-Concepts, Quantities and Symbols-Part 1: Aircraft Motion Relative to the Air*, Ref. No. ISO 1151/1, Fourth edition-1988-04-15, International Organization for Standardization, Geneva, Switzerland.

Itten, K., Meyer, P., Staenz, K., Kellenberger, T., and Schaepman, M. (1992) Evaluation of AVIRISwiss' 91 Campaign Data. Proceedings of the 3rd Annual JPL Airborne Remote Sensing Workshop, JPL Publication 92-41, pp. 108–110.

Itten, K., Dell Endice, F., Hueni, A., Kneubühler, M., Schläpfer, D., Odermatt, D., Seidel, F., Huber, S., Schopfer, J., and Kellenberger, T. (2008) APEX-the hyperspectral ESA airborne prism experiment. *Sensors*, **8**, 6235–6259.

Itten, K.I. (2007) The emergence of imaging spectrometry in Europe Remote Sensing Laboratories. Proceedings of the 5th EARSEL SIG IS Workshop, Bruges.

Jackson, T. (1990) Droplet sizing interferometry, in *Liquid Particle Size Measurements* (eds D.E. Hirleman, W.D. Bachalo, and P.G. Felton), ASTM, Philadelphia, PA.

Jacobson, M., Hansson, H., Noone, K., and Charlson, R. (2000) Organic atmospheric aerosol particles: review and state of the science. *Rev. Geophys.*, **38**, 267–294. doi: 10.1029/1998RG000045.

Jaegle, L., Jacob, D., Brune, W., Faloona, I., Tan, D., Heikes, B., Kondo, Y., Sachse, G., Anderson, B., Gregory, G., Singh, H., Ferry, R.P.G., Blake, D., and Shetter, R. (2000) Photochemistry of HO_x in the upper troposphere at northern midlatitudes. *J. Geophys. Res.*, **105**, 3877–3892.

Jaegle, L., Jacob, D.J., Brune, W.H., and Wennberg, P.O. (2001) Chemistry of HO_x radicals in the upper troposphere. *Atmos. Environ.*, **35**, 469–489.

Jäkel, E., Wendisch, M., Kniffka, A., and Trautmann, T. (2005) Airborne system for fast measurements of upwelling and downwelling spectral actinic flux densities. *Appl. Opt.*, **44**, 434–444.

Jäkel, E., Wendisch, M., and Lefer, B. (2006) Parameterization of ozone photolysis frequency in the lower troposhere using data from photodiode array detector spectrometers. *J. Atmos. Chem.*, **54**, 67–87.

Jarecke, P. and Yokoyama, K. (2000) Radiometric calibration of the Hyperion imaging spectrometer instrument From primary standards to end-to-end calibration. Proceedings of SPIE, vol. 4135, pp. 254–263.

Jayne, J.T., Leard, D.C., Zhang, X.F., Davidovits, P., Smith, K.A., Kolb, C.E., and Worsnop, D.R. (2000) Development of an aerosol mass spectrometer for size and composition analysis of submicron particles. *Aerosol Sci. Technol.*, **33**, 49–70.

Jenkin, M. and Clemitshaw, K. (2000) Ozone and other secondary photochemical pollutants: chemical processes governing their formation in the planetary boundary layer. *Atmos. Environ.*, **34**, 2499.

Jennings, S.G. and O'Dowd, C.D. (1990) Volatility of aerosol at Mace Head, on the west coast of Ireland. *J. Geophys. Res.*, **95**, 13–937–13-948.

Jennings, S.G., O'Dowd, C.D., Cooke, W.F., Sheridan, P.J., and Cachier, H. (1994) Volatility of elemental carbon. *Geophys. Res. Lett.*, **21**, 1719–1722.

Jensen, E., Lawson, P., Baker, B., Pilson, B., Mo, Q., Heymsfield, A.J., Bansemer, A., Bui, T.P., McGill, M., Hlavka, D., Heymsfield, G., Platnick, S., Arnold, G.T., and Tanelli, S. (2009) On the importance of small ice crystals in tropical anvil cirrus. *Atmos. Chem. Phys.*, **9**, 5519–5537.

Jensen, J.B. and Granek, H. (2002) Optoelectronic simulation of the PMS 260X

Optical Array Probe and application to drizzle in a marine stratocumulus. *J. Atmos. Oceanic Technol.*, **19**, 568–585.

Jensen, J.B., Austin, P.H., Baker, M.B., and Blyth, A.M. (1985) Turbulent mixing, spectral evolution and dynamics in a warm cumulus cloud. *J. Atmos. Sci.*, **42**, 173–192.

Jeung, I.S. (1990) Response characteristics of the Knollenberg Active Scattering Aerosol Spectrometer to light-absorbing aerosol particles. *Opt. Eng.*, **29**, 247–252.

Jimenez, J.L., Jayne, J., Shi, Q., Kolb, C., Worsnop, D., Yourshaw, I., Seinfeld, J., Flagan, R., Zhang, X., Smith, K., Morris, J., and Davidovits, P. (2003) Ambient aerosol sampling using the Aerodyne aerosol mass spectrometer. *J. Geophys. Res.*, 108.

Jimenez, J.L., Canagaratna, M.R., Donahue, N.M., Prevot, A.S.H., Zhang, Q., Kroll, J.H., de Carlo, P.F., Allan, J.D., Coe, H., Ng, N.L., Aiken, A.C., Docherty, K.S., Ulbrich, I.M., Grieshop, A.P., Robinson, A.L., Duplissy, J., Smith, J.D., Wilson, K.R., Lanz, V.A., Hueglin, C., Sun, Y.L., Tian, J., Laaksonen, A., Raatikainen, T., Rautiainen, J., Vaattovaara, P., Ehn, M., Kulmala, M., Tomlinson, J.M., Collins, D.R., Cubison, M.J., Dunlea, E., Huffman, J.A., Onasch, T.B., Alfarra, M.R., Williams, P.I., Bower, K., Kondo, Y., Schneider, J., Drewnick, F., Borrmann, S., Weimer, S., Demerjian, K., Salcedo, D., Cottrell, L., Griffin, R., Takami, A., Miyoshi, T., Hatakeyama, S., Shimono, A., Sun, J.Y., Zhang, Y.M., Dzepina, K., Kimmel, J.R., Sueper, D., Jayne, J.T., Herndon, S.C., Trimborn, A.M., Williams, L.R., Wood, E.C., Middlebrook, A.M., Kolb, C.E., Baltensperger, U., and Worsnop, D.R. (2009) Evolution of organic aerosol particles in the atmosphere. *Science*, **326**, 1525–1529.

Jobson, B., Niki, H., Yokouchi, Y., Bottenheim, J., Hopper, F., and Leaitch, R. (1994) Measurements of C_2-C_6 hydrocarbons during the polar sunrise 1992 experiment: evidence for Cl atom and Br atom chemistry. *J. Geophys. Res.*, **99**, 25355–25368.

Jobson, B., Parrish, D., Goldan, P., Kuster, W., Fehsenfeld, F., Blake, D., Blake, N., and Niki, H. (1998) Spatial and temporal variability of nonmethane hydrocarbon mixing ratios and their relation to photochemical lifetime. *J. Geophys. Res.*, **103**, 13557–13567.

Jobson, B.T. and McCoskey, J.K. (2009) Sample drying to improve HCHO measurements by PTR-MS instruments: laboratory and field measurements. *Atmos. Chem. Phys. Discuss.*, **9**, 19845–19877.

Joe, P. and List, R. (1987) Testing and performance of of two-dimensional Optical Array Spectrometer with greyscale. *J. Atmos. Oceanic Technol.*, **4**, 139–150.

John, W., Hering, S., Reischl, G., Sasaki, G., and Goren, S. (1983) Characteristics of Nuclepore filters with large pore size. 2. Filtration properties. *Atmos. Environ.*, **17**, 373–382.

Johnson, C., Rice, J., and Brown, S. (2004) The establishment and verification of traceability for remote sensing radiometry, with an eye towards intercomparison of results, in *2004 CEOS/IVOS Calibration Workshop*, ESA/ESTEC, Noordwijk, The Netherlands.

Johnson, D.G., Jucks, K.W., Traub, W.A., and Chance, K.V. (2001) Isotopic composition of stratospheric water vapor: measurements and photochemistry. *Atmos. Chem. Phys.*, **106**, 12211–12217.

Johnston, P.S. and Lehmann, K. (2008) Cavity enhanced absorption spectroscopy using a broadband prism cavity and a supercontinuum source. *Opt. Express*, **16**, 15013–15023.

Jones, J. (1990) Electric charge acquired by airplanes penetrating thunderstorms. *J. Geophys. Res.*, **95**, 16589–16600.

Jonsson, H.H., Wilson, J.C., Brock, C., Knollenberg, R., Newton, R., Dye, J., Baumgardner, D., Borrmann, S., Ferry, G., Pueschel, R., Woods, D., and Pitts, M. (1995) Performance of a focused cavity aerosol spectrometer for measurements in the stratosphere of particle size in the 0.06-2.0 μm-diameter range. *J. Atmos. Oceanic Technol.*, **12**, 115–129.

Jordan, A., Haidacher, S., Hanel, G., Hartungen, E., Herbig, J., Mark, L., Schottkowsky, R., Seehauser, H., Sulzer, P., and Mark, T.D. (2009a) An online ultra-high sensitivity Proton-transfer-reaction mass-spectrometer combined with switchable reagent ion capability (PTR+SRI-MS). *Int. J. Mass Spectrom.*, **286**, 32–38.

Jordan, A., Haidacher, S., Hanel, G., Hartungen, E., Mark, L., Seehauser, H., Schottkowsky, R., Sulzer, P., and Mark, T.D. (2009b) A high resolution and high sensitivity proton-transfer-reaction time-of-flight mass spectrometer (PTR-ToF-MS). *Int. J. Mass Spectrom.*, **286**, 122–128.

Jorgensen, D., Hildebrand, P.H., and Frush, C. (1983) Feasibility test of an airborne pulse-Doppler meteorological Radar. *J. Clim. Appl. Meteorol.*, **22**, 744–757.

Jorgensen, D.P., Matejka, T., and Dugranrut, J.D. (1996) Multi-beam techniques for deriving wind fields from airborne Doppler Radars. *Meteorol. Atmos. Phys.*, **59**, 83–104.

Josset, D., Pelon, J., Protat, A., and Flamant, C. (2008) New approach to determine aerosol optical depth from combined CALIPSO and CloudSat ocean surface echoes. *Geophys. Res. Lett.*, **35**, L10805. doi: 10.1029/2008GL033442.

Jourdan, O., Oshchepkov, S., Gayet, J.-F., Shcherbakov, V., and Isaka, H. (2003) Statistical analysis of cloud light scattering and microphysical properties obtained from airborne measurements. *J. Geophys. Res.*, **108**, 4155. doi: 10.1029/2002JD002723.

Junge, C.E. (1961) Vertical profiles of condensation nuclei in the stratosphere. *J. Meteorol.*, **18**, 501–509.

Junge, C.E., Chagnon, C.W., and Manson, J.E. (1961) A world-wide stratospheric aerosol layer. *Science*, **133**, 1478–1479.

Kahle, A. and Goetz, A. (1983) Mineralogic information from a new airborne Thermal Infrared Multispectral Scanner. *Science*, **222**, 24–27.

Kaimal, J.C. and Finnigan, J.J. (1994) *Atmospheric Boundary Layer Flows-Their Structure and Measurement*, Oxford University Press, p. 289.

Kaimal, J.C., Wyngaard, J.C., and Haugen, D.A. (1968) Deriving power spectra from a three-component sonic anemometer. *J. Appl. Meteorol.*, **7**, 827–837.

Kalogiros, J. and Wang, Q. (2002) Aerodynamic effects on wind turbulence measurements with research aircraft. *J. Atmos. Oceanic Technol.*, **19**, 1567–1576.

Kampe, H. (1950) Visibility and liquid water content in clouds in the free atmosphere. *J. Meteorol.*, **7**, 54–57.

Kanaya, Y., Matsumoto, J., Kato, S., and Akimoto, H. (2001) Behaviors of OH and HO_2 radicals during the observations at a remote island of Okinawa (ORION99) field campaign. 1. Observation using a laser-induced fluorescence instrument. *J. Geophys. Res.*, **106**, 24197–24208.

Kandler, K., Benker, N., Bundke, U., Cuevas, E., Ebert, M., Knippertz, P., Rodriguez, S., Schütz, L., and Weinbruch, S. (2007) Chemical composition and complex refractive index of Saharan Mineral Dust at Izana, Tenerife (Spain) derived by electron microscopy. *Atmos. Environ.*, **108**, 8576. doi: 10.1029/2002JD002648.

Karbou, F. (2005) Two microwave land emissivity parameterizations suitable for AMSU observations. *IEEE Trans. Geosci. Remote*, **43**, 1788–1795.

Karl, T., Spirig, C., Rinne, J., Stroud, C., Prevost, P., Greenberg, J., Fall, R., and Guenther, A. (2002) Virtual disjunct dddy covariance measurements of organic compound fluxes from a subalpine forest using proton transfer reaction mass spectrometry. *Atmos. Chem. Phys*, **2**, 279–291.

Karl, T., Potosnak, M., Guenther, A., Clark, D., Walker, J., Herrick, J., and Geron, C. (2004) Exchange processes of volatile organic compounds above a tropical rain forest: implications for modeling tropospheric chemistry above dense vegetation. *J. Geophys. Res.*, **109**, D18306. doi: 10.1029/2004JD004738.

Karl, T., Guenther, A., Yokelson, R.J., Greenberg, J., Potosnak, M., Blake, D.R., and Artaxo, P. (2007) The tropical forest and fire emissions experiment: emission, chemistry, and transport of biogenic volatile organic compounds in the lower atmosphere over Amazonia. *J. Geophys. Res.*, **112**, D18302. doi: 10.1029/2007JD008539.

Karl, T., Apel, E., Hodzic, A., Riemer, D.D., Blake, D.R., and Wiedinmyer, C. (2009)

Emissions of volatile organic compounds inferred from airborne flux measurements over a megacity. *Atmos. Chem. Phys.*, **9**, 271–285.

Kartal, D., Andrés-Hernández, M., Reichert, L., Schlager, H., and Burrows, J. (2010) Technical Note: characterisation of a DUALER instrument for the airborne measurement of peroxy radicals during AMMA 2006. *Atmos. Chem. Phys.*, **10**, 3047–3062.

Kasten, F. and Young, A. (1989) Revised optical air mass tables and approximation formula. *Appl. Opt.*, **28**, 4735–4738.

Katz, J. and Mirabel, P. (1975) Calculation of supersaturation profiles in thermal diffusion cloud chambers. *J. Atmos. Sci.*, **32**, 646–652.

Kaufman, Y. and Tanré, D. (1996) Strategy for direct and indirect methods for correction the aerosol effect on remote sensing: from AVHRR to EOS-MODIS. *Remote Sens. Environ.*, **55**, 65–79.

Kaufman, Y.J., Tanré, D., Léon, J.-F., and Pelon, J. (2003) Retrievals of profiles of fine and coarse aerosol particles using Lidar and radiometric space measurements. *IEEE Trans. Geosci. Remote*, **41**, 1743–1754.

Kawamura, K., Umemoto, N., Mochida, M., Howell, T.B.S., and Huebert, B. (2003) Water-soluble dicarboxylic acids in the tropospheric aerosol particles collected over east Asia and western North Pacific by ACE-Asia C-130 aircraft. *J. Geophys. Res.*, **108**, 8639. doi: 10.1029/2002JD003256.

Kaye, P., Hirst, E., Greenaway, R.S., Ulanowski, Z., Hesse, E., DeMott, P., Saunders, C., and Connolly, P. (2008) Classifying atmospheric ice crystals by spatial light scattering. *Opt. Lett.*, **33**, 1545.

Kebabian, P., Wood, E., Herndon, S., and Freedman, A. (2008) A practical alternative to chemiluminescence-based detection of nitrogen dioxide: cavity attenuated phase shift spectroscopy. *Environ. Sci. Technol.*, **42**, 6040–6045.

Keck, L., Oeh, U., and Hoeschen, C. (2007) Corrected equation for the concentrations in the drift tube of a proton transfer reaction-mass spectrometer (PTR-MS). *Int. J. Mass Spectrom.*, **264**, 92–95.

Keck, L., Hoeschen, C., and Oeh, U. (2008) Effects of carbon dioxide in breath gas on proton transfer reaction-mass spectrometry (PTR-MS) measurements. *Int. J. Mass Spectrom.*, **270**, 156–165.

Keeler, R.J., Serafin, R.J., Schwiesow, R.L., Lenschow, D.H., Vaughan, J.M., and Woodfield, A.A. (1987) An airborne laser air motion sensing system. Part I: concept and preliminary experiment. *J. Atmos. Ocean. Technol.*, **4**, 113–127.

Kelly, K., Tuck, A., Murphy, D., Proffitt, M., Fahey, D., Jones, R., McKenna, D., Loewenstein, M., Podolske, J., Strahan, S., Ferry, G., Chan, K., Vedder, J., Gregory, G., Hypes, W., McCormick, M., Browell, E., and Heidt, L. (1989) Dehydration in the lower Antarctic stratosphere during late winter and early spring, 1987. *J. Geophys. Res.*, **94**, 11317–11357.

Kemper, T. and Sommer, S. (2003) Mapping and monitoring of residual heavy metal contamination and acidification risk after the Aznalcóllar mining accident (Andalusia, Spain) using field and airborne hyperspectral data, in *Proceedings of the 3rd EARSeL Workshop on Imaging Spectroscopy* (eds M. Habermeyer, A. Muller, and S. Holzwarth), European Association of Remote Sensing Laboratories (EARSeL) Herrsching, Germany, pp. 333–343.

Kennedy, O.J., Ouyang, B., Langridge, J.M., Daniels, M.J.S., Bauguitte, S., Freshwater, R., McLeod, M.W., Ironmonger, C., Sendall, J., Norris, O., Nightingale, R., Ball, S.M., and Jones, R.L. (2011) An aircraft based three channel broadband cavity enhanced absorption spectrometer for simultaneous measurements of NO_3, N_2O_5 and NO_2. *Atmos. Meas. Technol.*, **4**, 1759–1776.

Keramitsoglou, I., Harries, J., Colling, D., Barker, R., and Foot, J. (2002) A study of the theory and operation of a resonance fluorescence water vapor sensor for upper tropospheric humidity measurements. *Meteorol. Apps.*, **9**, 443–453.

Kerstel, E. and Gianfrani, L. (2008) Advances in laser-based isotope ratio measurements: selected applications. *Appl. Phys. B*, **92**, 439–449. doi: 10.1007/s00340-008-3128.

Kerstel, E., Iannone, R., Chenevier, M., Kassi, S., Jost, H., and Romanini, D.

(2006) A water isotope (^{2}H, ^{17}O, and ^{18}O) spectrometer based on optical feedback cavity enhanced absorption for in situ airborne applications. *Appl. Phys. B*, **85**, 397–406.

Keshava, N. and Mustard, J. (2002) Spectral unmixing. *IEEE Signal Process. Mag.*, **19**, 44–57.

Khelif, D., Burns, S.P., and Friehe, C.A. (1999) Improved wind measurements on research aircraft. *J. Atmos. Oceanic Technol.*, **16**, 860–875.

Kieffera, H., Stone, T., Barnes, R., Benderc, S., Eplee, R., Mendenha, J., and Ong, L. (2003) On-orbit radiometric calibration over time and between spacecraft using the moon. Proceedings of SPIE, vol. 4881, pp. 287–298.

Kiemle, C., Brewer, W., Ehret, G., Hardesty, R., Fix, A., Senf, C., Wirth, M., Poberaj, G., and LeMone, M. (2007) Latent heat flux profiles from collocated airborne water Vvapor and wind Lidars during IHOP_2002. *J. Atmos. Oceanic Technol.*, **24**, 627–639. doi: http://dx.doi.org/10.1175/JTECH1997.1.

Kim, Y.J. (1995) Response of the Active Scattering Aerosol Spectrometer Probe (ASASP-100X) to particles of different chemical composition. *Aerosol Sci. Technol.*, **22**, 33–42.

Kim, Y.J. and Boatman, J.F. (1990) Size calibration corrections for the Active Scattering Aerosol Spectrometer Probe (ASASP-100X). *Aerosol Sci. Technol.*, **12**, 665–672.

Kindel, B. (2010) Cloud shortwave spectral radiative properties: airborne hyperspectral measurements and modeling of irradiance. PhD dissertation. University of Colorado at Boulder, United States.

Kindel, B., Schmidt, K., Pilewskie, P., Baum, B., Yang, P., and Platnick, S. (2010) Observations and modeling of ice cloud shortwave spectral albedo during the tropical composition, cloud and climate coupling experiment. *J. Geophys. Res.*, **115**, D00J18. doi: 10.1029/2009JD013127.

Kindel, B.C., Pilewskie, P., Schmidt, K.S., Coddington, O., and King, M.D. (2011) Solar spectral absorption by marine stratus clouds: measurements and modeling. *J. Geophys. Res.*, **116**, D10203. doi: 10.1029/2010JD015071.

King, L. (1914) On the convection of heat from small cylinders in a stream of fluid: determination of the convective constants of small Platinum wires with applications to hot wire anemometry. *Proc. R. Soc. London*, **90**, 563–570.

King, M., Byrne, D., Herman, B., and Reagan, J. (1978a) Aerosol size distributions obtained by inversion of spectral optical depth measurements. *J. Atmos. Sci.*, **35**, 2153–2167.

King, W. (1984) Air flow and particle trajectories around aircraft fuselages. I: theory. *J. Atmos. Oceanic Technol.*, **1**, 5–13.

King, W. and Handsworth, R. (1979) Total droplet concentration and average droplet sizes from simultaneous liquid water content and extinction measurements. *J. Appl. Meteorol.*, **18**, 940–944.

King, W. and Turvey, D. (1986) A thermal device for aircraft measurement of the solid water content of clouds. *J. Atmos. Oceanic Technol.*, **3**, 356–362.

King, W., Parkin, D., and Handsworth, R. (1978b) A hot-wire water device having fully calculable response characteristics. *J. Appl. Meteorol.*, **17**, 1809–1813.

King, W., Turvey, D., Williams, D., and Llewellyn, D. (1984) Air flow and particle trajectories around aircraft fuselages. II: measurements. *J. Atmos. Ocean. Technol.*, **1**, 14–21.

King, W., Dye, J., Strapp, J., Baumgardner, D., and Huffman, D. (1985) Icing wind tunnel tests on the CSIRO liquid water probe. *J. Atmos. Oceanic Technol.*, **2**, 340–353.

King, M.D., Parkinson, C.L., Partington, K.C., and Williams, R.G. (Eds.) (2007) Our Changing Planet. The View from Space. Cambridge University Press, Cambridge, p. 360.

Kipp & Zonen (2010) CMP Series Pyranometer Instruction Manual, The Netherlands.

Klein, H., Haunold, W., Bundke, U., Nillius, B., Wetter, T., Schallenberg, S., and Bingemer, H. (2010a) A new method for sampling of atmospheric ice nuclei with subsequent analysis in a static diffusion chamber. *Atmos. Res.*, **96**, 218–224. doi: 10.1016/j.atmosres.2009.08.002.

Klein, H., Haunold, W., Bundke, U., Nillius, B., Wetter, T., Schallenberg, S., and Bingemer, H. (2010b) A new method for sampling of atmospheric ice nuclei with subsequent analysis in a static diffusion chamber. *Atmos. Res.*, **96**, 218–224.

Kleindienst, T.E., Hudgens, E., Smith, D., McElroy, E., and Bufalini, J. (1993) Comparison of chemiluminescence and ultraviolet ozone monitor responses in the presence of humidity and photochemical pollutants. *J. Air Waste Manage. Assoc.*, **43**, 213–222.

Kleinman, L.I., Daum, P.H., Lee, Y.-N., Senum, G.I., Springston, S.R., Wang, J., Berkowitz, C., Hubbe, J., Zaveri, R.A., Brechtel, F.J., Jayne, J., Onasch, T.B., and Worsnop, D. (2007) Aircraft observations of aerosol composition and ageing in New England and Mid-Atlantic States during the summer 2002 New England Air Quality Study field campaign. *J. Geophys. Res.*, **112**, D09310. doi: 10.1029/2006JD007786.

Klett, J.D. (1985) Lidar inversion with variable backscatter/extinction ratios. *Appl. Opt.*, **24**, 1638–1643.

Kley, D. and Stone, E. (1978) Measurement of water vapor in the stratosphere by photodissociation with Ly-α (1216 Å). *Rev. Sci. Instrum.*, **49**, 691–697.

Kley, D., Stone, E., Henderson, W., Drummond, J., Harrop, W., Schmeltekopf, A., Thompson, T., and Winkler, R. (1979) In situ measurements of the mixing ratio of water vapor in the strato-sphere. *J. Atmos. Sci.*, **36**, 2513–2524.

Kline, J., Huebert, B., Howell, S., Blomquist, B., Zhuang, J., Bertram, T., and Carrillo, J. (2004) Aerosol composition and size versus altitude measured from the C-130 during ACE-Asia. *J. Geophys. Res.*, **109**, D19S08. doi: 10.1029/2004JD004540.

Kliner, D.A.V., Daube, B.C., Burley, J.D., and Wofsy, S.C. (1997) Laboratory investigation of the catalytic reduction technique for measurement of atmospheric NO_y. *J. Geophys. Res.*, **102**, 759–776.

Klippel, T., Fischer, H., Bozem, H., Lawrence, M.G., Butler, T., Jockel, P., Tost, H., Martinez, M., Harder, H., Regelin, E., Sander, R., Schiller, C., Stickler, A., and Lelieveld, J. (2011) Distribution of hydrogen peroxide and formaldehyde over Central Europe during the HOOVER project. *Atmos. Phys. Chem.*, **11**, 4391–4410. doi: 10.5194/acp-11-4391-2011.

Klobuchar, J. (1996) Ionospheric effects on GPS, in *Global Positioning Systems: Theory and Applications* (eds B. Parkinson), American Institute of Aeronautics and Astronautics.

Knighton, W.B., Fortner, E.C., Midey, A.J., Viggiano, A.A., Herndon, S.C., Wood, E.C., and Kolb, C.E. (2009) HCN detection with a proton transfer reaction mass spectrometer. *Int. J. Mass Spectrom.*, **283**, 112–121.

Knollenberg, R. (1970) The Optical Array: an alternative to scattering or extinction for airborne particle size determination. *J. Appl. Meteorol.*, **9**, 86–103.

Knollenberg, R. (1976) Three new instruments for cloud physics measurements: the 2-D spectrometer probe, the Forward Scattering Spectrometer Probe and the Active Scattering Aerosol Spectrometer, *International Conference on Cloud Physics*, American Meteorological Society, pp. 554–561.

Knollenberg, R. (1981) Techniques for probing cloud microstructure, in Clouds, Their Formation, Optical Properties and Effects (eds P.V. Hobbs and A. Deepak), Academic Press, p. 495.

Knox, R., Green, R., Middleton, E., Turner, W., Hook, S., and Ungar, S. (2010) The Hyperspectral Infrared Imager (HyspIRI) mission: a new capability for global ecological research and applications. Proceedings of the 95th ESA Annual Meeting, PS-79-114, Pittsburg, Pennsylvania.

Kobayashi, T., Kawaguchi, T., and Maeda, M. (2000) Measurement of spray flow by an improved Interferometric Laser Imaging Droplet Sizing (ILIDS). Proceedings 10th International Symposium Applications of Laser Techniques to Fluid Mechanics, Lisbon, Portugal.

Köhler, H. (1936) The nucleus in and the growth of hygroscopic droplets. *Trans. Faraday Soc.*, **32**, 1152–1161.

Kojima, T., Buseck, P., Wilson, J.C., Reeves, J., and Mahoney, M. (2004) Aerosol particles from tropical convective systems: cloud tops and cirrus anvils. *J. Geophys. Res.*, **109**, D12201. doi: 10.1029/2003JD004504.

Kokhanovsky, A. (2007) Scattered light corrections to Sun photometry: analytical results for single and multiple scattering regimes. *J. Opt. Soc. Am. A*, **24**, 1131–1137.

Kollias, P., Albrecht, B.A., Lhermitte, R., and Savtchenko, A. (2001) Radar observations of updrafts, downdrafts, and turbulence in fair-weather cumuli. *J. Atmos. Sci.*, **58**, 1750–1766.

Kollias, P., Albrecht, B.A., and Marks, F. (2002) Why Mie? Accurate observations of vertical air velocities and raindrops using a cloud Radar. *Bull. Am. Meteorol. Soc.*, **83**, 1471–1483.

Kollias, P., Albrecht, B.A., and Marks, F.D. (2003) Cloud Radar observations of vertical drafts and microphysics in convective rain. *J. Geophys. Res.*, **108**, 4053. doi: 10.1029/2001JD002033.

Kollias, P., Jo, I., and Albrecht, B.A. (2005) High-resolution observations of mammatus in tropical anvils. *Mon. Weather Rev.*, **133**, 2105–2112.

Komabayasi, M., Gonda, T., and Isono, K. (1964) Life time of water drops before breaking and size distribution of fragment drops. *J. Meterol. Soc. Jpn.*, **42**, 330–340.

Kondo, Y., Ziereis, H., Koike, M., Kawakami, S., Gregory, G., Sachse, G., Singh, H., Davis, D., and Merrill, J. (1996) Reactive nitrogen over the Pacific Ocean during PEM-West A. *J. Geophys. Res.*, **101**, 1809–1828.

König, G., Anders, K., and Frohn, A. (1986) A new light-scattering technique to measure the diameter of periodically generated moving droplets. *J. Aerosol Sci.*, **17**, 157–167.

Korb, C.L., Gentry, B.M., and Li, S. (1997) Edge technique Doppler Lidar wind measurements with high vertical resolution. *Appl. Opt.*, **36**, 5976.

Kormann, R., Fischer, H., Gurk, C., Helleis, F., Kluepfel, T., Kowalski, K., Koenigstedt, R., Parchatka, U., and Wagner, V. (2002) Application of a multi-laser tunable diode laser absorption spectrometer for atmospheric trace gas measurements at sub-ppbv levels. *Spectrochim. Acta, Part A*, **58**, 2489–2498.

Kormann, R., Fischer, H., de Reus, M., Lawrence, M., Bruhl, C., von Kuhlmann, R., Holzinger, R., Williams, J., Lelieveld, J., Warneke, C., de Gouw, J., Heland, J., Ziereis, H., and Schlager, H. (2003) Formaldehyde over the eastern Mediterranean during MINOS: comparison of airborne in situ measurements with 3D-model results. *Atmos. Chem. Phys.*, **3**, 851–861.

Korolev, A. (2007) Reconstruction of the sizes of spherical particles from their shadow images. Part I: theoretical considerations. *J. Atmos. Oceanic Technol.*, **24**, 376–389.

Korolev, A. (2011) Probe tips for airborne instruments used to measure cloud microphysical parameters. US Patent No. 7,861,584, Issued: January 4, Owner: Her Majesty the Queen in Right of Canada, as Represented by The Minister of Environment, 2011.

Korolev, A. and Isaac, G. (2005) Shattering during sampling by OAPs and HVPS. Part 1: snow particles. *J. Atmos. Oceanic Technol.*, **22**, 528–542.

Korolev, A., Kuznetsov, S.V., Makarov, Y., and Novikov, V.S. (1991) Evaluations of measurements of particle size and sample area from optical array probes. *J. Atmos. Oceanic Technol.*, **8**, 514–522.

Korolev, A., Strapp, J., Isaac, G., and Nevzorov, A. (1998a) The Nevzorov airborne hot-wire LWC-TWC probe: principle of operation and performance characteristics. *J. Atmos. Oceanic Technol.*, **15**, 1495–1510.

Korolev, A., Isaac, G., Mazin, I., and Barker, H. (2001) Microphysical properties of continental clouds from in situ measurements. *Q. J. R. Meteorol. Soc.*, **127**, 2117–2151.

Korolev, A., Strapp, J., Isaac, G.A., and Emery, E. (2008) Improved airborne hot-wire measurements of ice water content in clouds. 15th International Conference on Clouds and Precipitation, Cancun, Mexico, CD P13.4.

Korolev, A., Emery, E., Strapp, J., Cober, S., Isaac, G., Wasey, M., and Marcotte, D.

(2011) Small ice particles in tropospheric clouds: fact or artifact? Airborne Icing Instrumentation Evaluation Experiment. *Bull. Am. Meteorol. Soc.*, **92**, 967–973.

Korolev, A.V., Strapp, J.W., and Isaac, G.A. (1998b) Evaluation of accuracy of PMS Optical Array Probes. *J. Atmos. Oceanic Technol.*, **15**, 708–720.

Korolev, A.V., Isaac, G.A., Strapp, J.W., Cober, S.G., and Barker, H.W. (1999) In situ measurements of liquid water content profiles in midlatitude stratiform clouds. *Q. J. R. Meteorol. Soc.*, **133**, 1693–1699.

Kort, E.A., Patra, P.K., Ishijima, K., Daube, B.C., Jimenez, R., Elkins, J., Hurst, D., Moore, F.L., Sweeney, C., and Wofsy, S.C. (2011) Tropospheric distribution and variability of N_2O: evidence for strong tropical emissions. *Geophys. Res. Lett.*, **38**, L15806. doi: 10.1029/2011GL047612.

Kosarev, A., Mazin, I., Nevzorov, A., and Shugaev, V. (1976) Optical density of clouds. *Trans. Cent. Aerol. Obs.*, **124**, 44–110.

Kostkowski, H. (1997) Reliable spectroradiometry, Spectroradiometry consulting, La Plata, MD, p. 609.

Kotchenruther, R.A., Hobbs, P.V., and Hegg, D.A. (1999) Humidification factors for atmospheric aerosol particles off the mid-Atlantic coast of the United States. *J. Geophys. Res.*, **104**, 2239–2251.

Kovalev, V.A. and Eichinger, W.E. (2004) *Elastic Lidar*, John Wiley & Sons, Inc., New York.

Kowol-Santen, J. and Ancellet, G. (2000) Mesoscale analysis of transport across the subtropical tropopause. *Geophys. Res. Lett.*, **27**, 3345–3348.

Kozikowska, A., Haman, K., and Supronowicz, J. (1984) Preliminary results of an investigation of the spatial distribution of fog droplets by a holographic method. *Q. J. R. Meteor. Soc.*, **110**, 65–73.

Krämer, M. and Afchine, A. (2004) Sampling characteristics of inlets operated at low U/U_0 ratios: new insights from computational fluid dynamics (CFX) modelling. *J. Aerosol Sci.*, **35**, 683–694.

Kreidenweis, S.M., Chen, Y., DeMott, P.J., and Rogers, D.C. (1998) Isolating and identifying atmospheric ice-nucleating aerosol particles: a new technique. *Atmos. Res.*, **46**, 263–278.

Kritten, L., Butz, A., Dorf, M., Deutschmann, T., Kühl, S., Prados-Roman, C., Pukite, J., Rozanov, A., Schofield, R., and Pfeilsticker, K. (2010) Balloon-borne limb measurements of the diurnal variation of UV/vis absorbing radicals-a case study on NO_2 and O_3. *Atmos. Meas. Tech. Discuss.*, **3**, 431–468. doi: 10.5194/amtd-3-431-2010.

Kropfli, R.A. and Kelly, R.D. (1996) Meteorological research applications of MM-wave Radar. *Meteorol. Atmos. Phys.*, **59**, 105–121.

Krüger, G., Erzinger, H., and Kaufmann, H. (1998) Laboratory and airborne reflectance spectrometric analyses of lignite overburden dumps. *J. Geochem. Res.*, **64**, 47–65, 4th ISEG Symposium, Vail, CO, USA.

Kruse, F., Lefkoff, A., Boardman, J., Heidebrecht, K., Sapiro, A., Barloon, P., and Goetz, A. (1993) The spectral image processing system (SIPS)-interactive visualization and analysis of imaging spectrometer data. *Remote Sens. Environ.*, **44**, 145–168.

Kruse, F.A. (1988) Use of Airborne Imaging Spectrometer data to map minerals associated with hydrothermally altered rocks in the northern Grapevine Mountains, Nevada and California. *Remote Sens. Environ.*, **24**, 31–51.

Kruse, F.A., Boardman, J., Lefkoff, A., Young, J., Kierein-Young, K., Cocks, T., Jenssen, R., and Cocks, P. (2000) HyMap: an Australian hyperspectral sensor solving global problems-results from USA HyMap data acquisitions. Proceedings of the 10th Australasian Remote Sensing and Photogrammetry Conference, Causal Productions (www.causalproductions.com), Adelaide, Australia.

Kuhn, P. and Suomi, V. (1958) Airborne observations of albedo with a beam reflector. *J. Meteorol.*, **15**, 172–174.

Kuhn, P. and Suomi, V. (2003) Passive Broadband and Spectral Radiometric Measurements Available on NSF/NCAR Research Aircraft. *NCAR RAF Technical Bulletin No. 25*, (NCAR) National Center for Atmospheric Research, Boulder, Colorado, USA.

Kulkarni, P. and Wang, J. (2006a) New fast integrated mobility spectrometer for real-time measurement of aerosol size distribution-I: concept and theory. *J. Aerosol Sci.*, **37**, 1303–1325.

Kulkarni, P. and Wang, J. (2006b) New fast integrated mobility spectrometer for real-time measurement of aerosol size distribution: II. Design, calibration, and performance characterization. *J. Aerosol Sci.*, **37**, 1326–1339.

Kumala, W., Haman, K.E., Kopec, M.K., and Malinowski, S.P. (2010) Ultrafast thermometer UFTM: high resolution temperature measurements during Physics of Stratocumulus Top (POST). 13th AMS Conference on Cloud Physics, Portland OR, 2010, http://ams.confex.com/ams/13CldPhy13AtRad/techprogram/paper_170832.htm.

Kumar, A. and Viden, I. (2007) Volatile organic compounds: sampling methods and their worldwide profile in ambient air. *Environm. Monit. Assess.*, **131**, 301–321. doi: 10.1007/s10661-006-9477-1.

Kunkel, B., Blechinger, F., Viehmann, D., van der Piepen, H., and Doerffer, R. (1991) ROSIS imaging spectrometer and its potential for ocean parameter measurements (airborne and space-borne). *Int. J. Remote Sens.*, **12**, 753–761.

Kuwata, M. and Kondo, Y. (2009) Measurements of particle masses of inorganic salt particles for calibration of cloud condensation nuclei counters. *Atmos. Chem. Phys.*, **9**, 5921–5932.

Kyle, T. (1975) The measurement of water content by an evaporator. *J. Appl. Meteorol.*, **14**, 327–332.

Kylling, A., Webb, A., Kift, R., Gobbi, G., Ammannato, L., Barnaba, F., Bais, A., Kazadzis, S., Wendisch, M., Jäkel, E., Schmidt, S., Kniffka, A., Thiel, S., Junkermann, W., Blumthaler, M., Silbernagl, R., Schallhart, B., Schmitt, R., Kjeldstad, B., Thorseth, T., Scheirer, R., and Mayer, B. (2005) Spectral actinic flux in the lower troposphere: measurement and 1-D simulations for cloudless, broken cloud and overcast situations. *Atm. Chem. Phys.*, **5**, 1975–1997.

Lack, D.A., Cappa, C.D., Lanridge, J., Richardson, M., Law, D., McLaughlin, R., and Murphy, D.M., (2012) Aircraft instrument for comprehensive characterisation of aerosol optical properties, Part 2: black and brown carbon absorption and absorption enhancement measured with photo acoustic spectroscopy. *Aerosol Sci. Technol.*, **45**, 555–568.

Lala, G.G. and Juisto, J.E. (1977) An automatic light scattering CCN counter. *J. Appl. Meteorol.*, **16**, 413–418.

Lämmerzahl, P., Röckmann, T., Brenninkmeijer, C.A.M., Krankowsky, D., and Mauersberger, K. (2002) Oxygen isotope composition of stratospheric carbon dioxide. *Atmos. Chem. Phys.*, **29**, 1582.

Lance, S., Medina, J., Smith, J.N., and Nenes, A. (2006) Mapping the operation of the DMT continuous flow CCN counter. *Aerosol Sci. Technol.*, **40**, 242–254.

Lance, S., Brock, C.A., Rogers, D., and Gordon, J.A. (2010) Water droplet calibration of the Cloud Droplet Probe (CDP) and in-flight performance in liquid, ice and mixed-phase clouds during ARCPAC. *Atmos. Meas. Tech.*, **3**, 1683–1706. doi: 10.5194/amt-3-1683-2010.

Landgrebe, D. (2003) *Signal Theory Methods in Multispectral Remote Sensing*, John Wiley and Sons, Inc., New York, 508 pp.

Langer, G. (1973) Evaluation of NCAR ice nucleus counter, Part I: basic operation. *J. Appl. Meteorol.*, **12**, 1000–1011.

Langer, G. and Rogers, J. (1975) An experimental study of the detection of ice nuclei on membrane filters and other substrata. *J. Appl. Meteorol.*, **14**, 560–570.

Langley, S. (1903) The "Solar Constant" and related problems. *Astrophys. J.*, **17**, 89–99.

Langridge, J.M., Ball, S., Shillings, A., and Jones, R. (2008a) A broadband absorption spectrometer using light emitting diodes for ultra-sensitive in situ trace gas detection. *Rev. Sci. Instrum.*, **79**, 123110.

Langridge, J.M., Laurila, T., Watt, R., Jones, R., Kaminski, C., and Hult, J. (2008b) Cavity enhanced absorption spectroscopy of multiple trace gas species using a supercontinuum radiation source. *Opt. Express*, **16**, 10178–10188.

Langridge, J.M., Richardson, M.S., Lack, D., Law, D., and Murphy, D.M. (2011) Aircraft instrument for comprehensive characterization of aerosol optical properties, Part I: wavelength-dependent optical extinction and its relative humidity dependence measured using cavity ringdown spectroscopy. *Aerosol Sci. Technol.*, **45**, 1305–1318.

Lappe, U.O. and Davidson, B. (1963) On the range of validity of Taylor's Hypothesis and the Kolmogoroff spectral law. *J. Atmos. Sci.*, **20**, 569–576.

Laube, J.C., Engel, A., Bonisch, H., Mobius, T., Sturges, W.T., Brass, M., and Röckmann, T. (2010a) Fractional release factors of long-lived halogenated organic compounds in the tropical stratosphere. *Atmos. Chem. Phys.*, **10**, 1093–1103.

Laube, J.C., Martinerie, P., Witrant, E., Blunier, T., Schwander, J., Brenninkmeijer, C.A.M., Schuck, T.J., Bolder, M., Röckmann, T., van der Veen, C., Bonisch, H., Engel, A., Mills, G.P., Newland, M.J., Oram, D.E., Reeves, C.E., and Sturges, W.T. (2010b) Accelerating growth of HFC-227ea (1,1,1,2,3,3,3-heptafluoropropane) in the atmosphere. *Atmos. Chem. Phys.*, **10**, 5903–5910.

Laucks, M.L. and Twohy, C. (1998) Size-dependent collection efficiency of an airborne counterflow virtual impactor. *Aerosol Sci. Technol.*, **28**, 40–61.

Lawson, C.L. and Hanson, R.J. (1974) *Solving Least Square Problems*, Prentice-Hall.

Lawson, R. and Cormack, R. (1995) Theoretical design and preliminary tests of two new particle spectrometers for cloud microphysics research. *Atmos. Res.*, **35**, 315–348.

Lawson, R., O'Connor, D., Zmarzly, P., Weaver, K., Baker, B., Mo, Q., and Jonsson, H. (2006) The 2D-S (Stereo) probe: design and preliminary tests of a new airborne, high speed, high-resolution particle imaging probe. *J. Atmos. Oceanic Technol.*, **23**, 1462–1477.

Lawson, R.P. (2011) Effects of ice particles shattering on the 2D-S probe. *Atmos. Meas. Tech.*, **4**, 1361–1381. doi: 10.5194/amt-4-1361-2011.

Lawson, R.P. and Cooper, W.A. (1990) Performance of some airborne thermometer in clouds. *J. Atmos. Oceanic Technol.*, **7**, 480–494.

Lawson, R.P. and Rodi, A.R. (1992) A new airborne thermometer for atmospheric and cloud physics research. Part I: design and preliminary flight tests. *J. Atmos. Oceanic Technol.*, **9**, 556–574.

Lawson, R.P., Stewart, R.E., and Angus, L.J. (1998) Observations and numerical simulations of the origin and development of very large snowflakes. *J. Atmos. Sci.*, **55**, 3209–3229. doi: http://dx.doi.org/10.1175/1520-0469(1998)055<3209:OANSOT>2.0.CO;2.

Lawson, R.P., Baker, B.A., Schmitt, C.G., and Jensen, T.L. (2001) An overview of microphysical properties of Arctic clouds observed in May and July during FIRE.ACE. *J. Geophys. Res.*, **106**, 14989–15014.

Lawson, R.P., Jensen, E., Mitchell, D.L., Baker, B., Mo, Q., and Pilson, B. (2010) Microphysical and radiative properties of tropical clouds investigated in TC4 and NAMMA. *J. Geophys. Res.*, **115**, D00J08. doi: 10.1029/2009JD013017.

Lazrus, A., Fong, K., and Lind, J. (1988) Automated fluorometric determination of formaldehyde in air. *Anal. Chem.*, **1088**, 1074–1078.

Lebel, T., Parker, D., Flamant, C., Bourles, B., Marticorena, M., Mougin, E., Peugeot, C., Diedhiou, A., Haywoodand, J., Ngaminiand, J., Polcher, J., Redelsperger, J.-L., and Thorncroft, C. (2010) The AMMA field campaigns: multiscale and multidisciplinary observations in the West African region. *Q. J. R. Meteor. Soc.*, **136**, 8–33.

Lee, M., Noone, B., O'Sullivan, D., and Heikes, B. (1995) Method for the collection and HPLC analysis of hydrogen peroxide and C1-C2 hydroperoxides in the atmosphere. *J. Atmos. Oceanic Technol.*, **12**, 1060–1070.

Lee, M., Heikes, B., Jacod, D., Sachse, G., and Anderson, B. (1997) Hydrogen peroxide, organic hydroperoxide, and formaldehyde as primary pollutants from biomass burning. *J. Geophys. Res.*, **102**, 1301–1309.

Lee, M., Heikes, B., and O'Sullivan, D. (2000) Hydrogen peroxide and organic hydroperoxide in the troposphere: a review. *Atmos. Environ.*, **34**, 3475–3494.

Lee, Y., Zhou, X., Leaitch, W., and Banic, C. (1996) An aircraft measurement technique for formaldehyde and soluble carbonyl compounds. *J. Geophys. Res.*, **101**, 75–80.

Lefer, B., Shetter, R., Hall, S., Crawford, J., and Olson, J. (2003) Impact of clouds and aerosol particles on photolysis frequencies and photochemistry during TRACE-P, Part I: analysis using radiative transfer and photochemical box models. *J. Geophys. Res.*, **108**, 8821. doi: 10.1029/2002JD003171.

Leibrock, E. and Huey, L.G. (2000) Ion chemistry for the detection of isoprene and other volatile organic compounds in ambient air. *Geophys. Res. Lett.*, **27**, 1719–1722.

Lelieveld, J., Butler, T., Crowley, J., Dillon, T., Fischer, H., Ganzeveld, L., Harder, H., Lawrence, M., Martinez, M., Taraborrelli, D., and Williams, J. (2008) Atmospheric oxidation capacity sustained by a tropical forest. *Nature*, **452**, 737–740.

Lemaitre, P., Porcheron, E., and Nuboer, A. (2007) Development of Rainbow refractometry and out-of-focus imaging to characterize heat and mass transfers in two-phase flow and aerosol collection processes by droplets. 15th International Conference on Nuclear Engineering, Nagoya.

Lemke, H. and Quante, M. (1999) Backscatter characteristics of nonspherical ice crystals: assessing the potential of polarimetric Radar measurements. *J. Geophys. Res.*, **104**, 729–751.

Lemonis, G., Schmücker, M., and Struck, H. (2002) A fast response probe system for in-flight measurements of atmospheric turbulence. *Aerospace Sci. Technol.*, **6**, 233–243.

Lenschow, D. (1972) Measurement of Air Velocity and Temperature using the NCAR Buffalo Aircraft Measurement System. *Tech. Note* NCAR/EDD-74, National Center for Atmospheric Research, Boulder, CO.

Lenschow, D. (1976) Estimating updraft velocity from an airplane response. *Mon. Weather Rev.*, **104**, 618–627.

Lenschow, D. (1986) *Probing the Atmospheric Boundary Layer*, American Meteorological Society, Boston, MA.

Lenschow, D., Savic-Jovcic, V., and Stevens, B. (2007) Divergence and vorticity from aircraft air motion measurements. *J. Atmos. Oceanic Technol.*, **24**, 2062–2072. doi: 10.1175/2007JTECHA940.1.

Lenschow, D.H. and Pennell, W.T. (1974) On the measurementof in-cloud and wet-bulb temperature from aircraft. *Mon. Weather Rev.*, **102**, 447–454.

Lenschow, D.H. and Spyers-Duran, P. (1989) *Measurement Techniques: Air Motion Sensing*, Bulletin No. 23, National Center for Atmospheric Research, *http://nldr.library.ucar.edu/repository/collections/ATD-000-000-000-062*, (accessed on 1989).

Lenschow, D.H. and Stankov, B.B. (1986) Length scales in the convective boundary layer. *J. Atmos. Sci.*, **43**, 1198–1209.

Lenschow, D.H., Friehe, C.A., and Larue, J.C. (1978) The development of an airborne hot-wire anemometer system, *Symposium on Meteorological Observations and Instrumentation, 4th Denver, CO, April 10–14, 1978, Preprints. (A79-21901 07-35)*, American Meteorological Society, Boston, MA., pp. 463–466.

Lenschow, D.H., Mann, J., and Kristensen, L. (1994) How long is long enough when measuring fluxes and other turbulence statistics. *J. Atmos. Oceanic Technol.*, **11**, 661–673.

Leon, D. and Vali, G. (1998) Retrieval of three–dimensional particle velocity from airborne Doppler Radar data. *J. Atmos. Oceanic Technol.*, **15**, 860–870.

Leon, D., Vali, G., and Lothon, M. (2006) Dual-Doppler analysis in a single plane from an airborne platform. *J. Atmos. Oceanic Technol.*, **23**, 3–22.

Leon, D.C., Wang, Z., and Liu, D. (2008) Climatology of drizzle in marine boundary layer clouds based on 1 year of data from CloudSat and Cloud-Aerosol Lidar and Infrared Pathfinder Satellite Observations (CALIPSO). *J. Geophys. Res.*, **113**. doi: 10.1029/2008JD009835.

Lewis, A., Evans, M., Methven, J., Watson, N., Lee, J., Hopkins, J., Purvis, R., Arnold, S., McQuaid, J.,

Whalley, L., Pilling, M., Heard, D., Monks, P., Parker, A., Reeves, C., Oram, D., Mills, G., Bandy, B., and Coe, H. (2007) Chemical composition observed over the mid-Atlantic and the detection of pollution signatures far from source regions. *J. Geophys. Res.*, **112**. doi: 10.1029/2006JD007584.

Lewis, J.M. (1997) The Lettau-Schwerdtfeger balloon experiment: measurement of turbulence via Austausch Theory. *Bull. Am. Meteorol. Soc.*, **78**, 2619–2635.

Lewtas, J., Pang, Y., Booth, D., Reimer, S., Eatough, D.J., and Gundel, L.A. (2001) Comparison of sampling methods for semi–volatile organic carbon associated with PM2.5. *Aerosol Sci. Technol.*, **34**, 9–22.

Lhermitte, R. (1990) Attenuation and scattering of millimeter wavelength radiation by clouds and precipitation. *J. Atmos. Oceanic Technol.*, **7**, 464–479.

Lhermitte, R. (2002) *Centimeter and Milllimeter Wavelength Radars in Meteorology*, Lhermitte Publications, Miami, FL, p. 550.

Lhermitte, R.M. (1971) Probing of atmospheric motion by airborne pulse-Doppler Radar techniques. *J. Appl. Meteorol.*, **10**, 234–246.

Li, L., Sekelsky, S.M., Reising, S.C., Swift, C.T., Durden, S.L., Sadowy, G.A., Dinardo, S., Li, F.K., Huffman, A., Stephens, G., Babb, D.M., and Rosenberger, H.W. (2001) Retrieval of atmospheric attenuation using combined ground-based and airborne 95 GHz cloud Radar measurements. *J. Atmos. Oceanic Technol.*, **18**, 1345–1353.

Li, L., Heymsfield, G.M., Tian, L., and Racette, P.E. (2005) Measurements of ocean surface backscattering using an airborne 94 GHz cloud Radar-Implication for calibration of airborne and spaceborne W-band Radars. *J. Atmos. Oceanic Technol.*, **22**, 1033–1045.

Li, Z., Ackerman, T.P., Wiscombe, W., and Stephens, G.L. (2003) Have clouds darkened. *Science*, **302**, 1150.

Li, Z., Lemmerz, C., Paffrath, U., Reitebuch, O., and Witschas, B. (2010) Airborne Doppler Lidar investigation of sea surface reflectance at a 355-nm ultraviolet wavelength. *J. Atmos. Oceanic Technol.*, **27**, 693–704.

Liebe, H.J., Manabe, T., and Hufford, G.A. (1989) Millimeter–wave attenuation and delay rates due fog/cloud conditions. *IEEE Trans. Antenn. Propag.*, **37**, 1617–1623.

Lightfoot, P., Cox, R., Crowley, J., Destriau, M., Hayman, G., Jenkin, M., Moortgat, G., and Zabel, F. (1992) Organic peroxy radicals: kinetics, spectroscopy and tropospheric chemistry. *Atmos. Environ.*, **26A**, 1805.

Lilie, L., Emery, E., Strapp, J., and Emery, J. (2004) A multiwire hot-wire device for measurement of icing severity, total water content, liquid water content, and droplet diameter. 43rd AIAA Aerospace Sciences Meeting, Reno, Nevada, AIAA-2005-0860.

Lin, H. and Heintzenberg, J. (1995) A theoretical study of the counterflow virtual impactor. *J. Aerosol Sci.*, **26**, 903–914.

Lindinger, W., Hansel, A., and Jordan, A. (1998) On-line monitoring of volatile organic compounds at pptv levels by means of proton-transfer-reaction mass spectrometry (PTR-MS)-Medical applications, food control and environmental research. *Int. J. Mass Spectrom.*, **173**, 191–241.

Lipton, A., Moncet, J.-L., and Uymin, G. (2009) Approximations of the Planck function for models and measurements into the submillimeter range. *IEEE Trans. Geosci. Remote*, **6**, 433–437.

Liu, C.-L. and Illingworth, A. (2000) Toward more accurate retrievals of ice water content from Radar measurements of clouds. *J. Appl. Meteorol.*, **39**, 1130–1146.

Liu, G.S. (2004) Approximation of single scattering properties of ice and snow particles for high microwave frequencies. *J. Atmos. Sci.*, **61**, 2441–2456.

Liu, H.W. and Chandrasekar, V. (2000) Classification of hydrometeors based on polarimetric Radar measnurements: development of fuzzy-logic and neuro-logic systems and in situ verification. *J. Atmos. Oceanic Technol.*, **7**, 140–164.

Liu, P.S.K., Leaitch, W.R., Wasey, J.W., and Wasey, M.A. (1992) The response of the particle measuring systems airborne ASAP and PCASP to NaCl and latex particles. *Aerosol Sci. Technol.*, **16**, 83–95.

Liu, Y., Morales Cueto, R., Hargrove, J., Medina, D., and Zhang, J. (2009) Measurements of peroxy radicals using chemical amplification-cavity ringdown spectroscopy. *Environ. Sci. Technol.*, **43**, 7791–7796.

Livingston, J., Schmid, B., Russell, P., Eilers, J., Kolyer, R., Redemann, J., Ramirez, S., Yee, J.-H., Swartz, W., Trepte, C., Thomason, L., Pitts, M., Avery, M., Randall, C., Lumpe, J., Bevilacqua, R., Bittner, M., Erbertseder, T., McPeters, R., Shetter, R., Browell, E., Kerr, J., and Lamb, K. (2005) Retrieval of ozone column content from airborne Sun photometer measurements during SOLVE II: comparison with coincident satellite and aircraft measurements Atmospheric Chemistry and Physics Special Issue on the SOLVE II/VINTERSOL campaign. *Atmos. Chem. Phys.*, **5**, 2035–2054.

Lobl, E., Aonashib, K., Griffith, B., Kummerow, C., Liu, G., Murakami, M., and Wilheit, T. (2007) Wakasa Bay: an AMSR precipitation validation campaign. *Bull. Am. Meteorol. Soc.*, **88**, 551–558.

Lohmann, U. and Feichter, J. (2005) Global indirect aerosol effects: a review. *Atmos. Chem. Phys.*, **5**, 715–737.

Long, C., Bucholtz, A., Jonsson, H., Schmid, B., Vogelmann, A., and Wood, J. (2010) A method of correcting for tilt from horizontal in downwelling shortwave irradiance measurements on moving platforms. *Open Atmos. Sci. J.*, **4**, 78–87.

Low, R. (1969) A generalized equation for the solution effect in droplet growth. *J. Atmos. Sci.*, **26**, 608–611.

Lu, J., Fugal, J., Nordsiek, H., Saw, E., Shaw, R., and Yang, W. (2008) Lagrangian particle tracking in three dimensions via single-camera in-line digital holography. *New J. Phys.*, **10**, 125015. doi: 10.1088/1367-2630/10/12/125013.

Luftfahrtnorm (1970) LN 9300, Blatt 1, Flugmechanik, Normenstelle Luftfahrt, Leinfelden, Germany, 42 pp.

Lumley, L. and Panofsky, H. (1964) *The Structure of Atmospheric Turbulence*, John Wiley & Sons, Inc., New York, 239 pp.

Ma, Y., Weber, R.J., Maxwell-Meier, K., Orsini, D.A., Lee, Y.N., Huebert, B.J., Howell, S.G., Bertram, T., Talbot, R.W., Dibb, J.E., and Scheuer, E. (2004) Intercomparisons of airborne measurements of aerosol ionic chemical composition during TRACE-P and ACE-Asia. *J. Geophys. Res.*, **109**, D15S06. doi: 10.1029/2003JD003673.

MacCready, P. (1964) Standardization of gustiness values from aircraft. *J. Appl. Meteorol.*, **3**, 439–449.

MacDonald, J., Ustin, S., and Schaepman, M. (2009) The contributions of Dr. Alexander F.H. Goetz to imaging spectrometry. *Remote Sens. Environ.*, **113**, S2–S4.

Mace, G.G., Deng, M., Soden, B., and Zipser, E. (2006) Association of tropical cirrus in the 10-15 km layer with deep convective sources: an observational study combining millimeter Radar data and satellite–derived trajectories. *J. Atmos. Sci.*, **63**, 480–503.

Mace, G.G., Zhang, Q., Vaughan, M., Marchand, R., Stephens, G., Trepte, C., and Winker, D. (2009) A description of hydrometeor layer occurrencestatistics derived from the first year of merged Cloudsat and CALIPSO data. *J. Geophys. Res.*, **114**, D00A26. doi: 10.1029/2007JD009755.

Macke, A. (1993) Scattering of light by polyhedral ice crystals. *Appl. Opt.*, **32**, 2780–2788.

Macke, A., Mueller, J., and Raschke, E. (1996) Single scattering properties of atmospheric ice crystals. *J. Atmos. Sci.*, **53**, 2812–2825.

MacPherson, J. and Baumgardner, D. (1988) Airflow about King Air wingtip-mounted cloud particle measurement probes. *J. Atmos. Ocean. Technol.*, **5**, 259–273.

Magnusson, L.E., Koropchak, J.A., Anisimov, M.P., Poznjakovskiy, V.M., and de la Mora, J.F. (2003) Correlations for vapor nucleating critical embryo parameters. *J. Phys. Chem. Ref. Data*, **32**, 1387–1410.

Mahrt, L. (2000) Surface heterogeneity and vertical structure of the boundary layer. *Bound.–Lay. Meteorol.*, **96**, 33–62.

Mahrt, L., Vickers, D., and Sun, J. (2001) Spatial variations of surface moisture flux from aircraft data. *Adv. Water Res.*, **24**, 1133–1141.

Malinowski, S., Haman, C., Gerber, H., and Brenguier, J.-L. (2007) Small scale mixing processes at the top of a marine stratocumulus – a case study. *Q. J. R. Meteorol. Soc.*, **133**, 213–226.

Malkus, J.S. (1954) Some results of a trade–cumulus cloud investigation. *J. Atmos. Sci.*, **11**, 220–237.

Malley, D., Martin, P., and Ben-Dor, E. (2004) Application in analysis of soils, in *Near Infrared Spectroscopy in Agriculture* (eds R. Craig, R. Windham and J. Workman), Agronomy Monograph 44, American Society of Agronomy, Crop Science Society of America, Soil Science Society of America, pp. 729–784.

Mann, J. and Lenschow, D.H. (1994) Errors in airborne flux measurements. *J. Geophys. Res.*, **99**, 14519–14526.

Mao, J., Ren, X., Brune, W.H., Olson, J.R., Crawford, J.H., Fried, A., Huey, L.G., Cohen, R.C., Heikes, B., Singh, H.B., Blake, D.R., Sachse, G.W., Diskin, G.S., Hall, S.R., and Shetter, R.E. (2009) Airborne measurement of OH reactivity during INTEX–B. *Atmos. Chem. Phys.*, **9**, 163–173.

Mapes, B. and Houze, R. Jr. (1993) Cloud clusters and superclusters over the oceanic warm pool. *Mon. Weather Rev.*, **121**, 1398–1415.

Marcy, T.P., Gao, R.S., Northway, M.J., Popp, P.J., Stark, H., and Fahey, D.W. (2005) Using chemical ionization mass spectrometry for detection of HNO_3, HOl, and $ClONO_2$ in the atmosphere. *Int. J. Mass Spectrom.*, **243**, 63–70.

Maring, H., Savoie, D., Izaguirre, M., and Custals, L. (2003) Vertical distribution of dust and sea–salt aerosol particles over Puerto Rico during PRIDE measured from a light aircraft. *J. Geophys. Res.*, **108**, 8587. doi: 10.1029/2002JD002544.

Markowski, G.R. (1987) Improving Twomey's algorithm for inversion of aerosol measurement data. *Aerosol Sci. Technol.*, **7**, 127–141.

Marks, F. and Houze, R. (1987) Inner core structure of Hurricane Alicia from airborne Doppler Radar observations. *J. Atmos. Sci.*, **44**, 1296–1317.

Marticorena, B., Chatenet, B., Traoré, J.R.S., Diallo, M.C.A., Koné, I., Maman, A., Diaye, T., and Zakou, A. (2010) Temporal variability of mineral dust concentrations over West Africa: analyses of a pluriannual monitoring from the AMMA Sahelian Dust Transect. *Atmos. Chem. Phys.*, **10**, 8899–8915.

Martinez, M., Harder, H., Kubistin, D., Rudolf, M., Bozem, H., Eerdekens, G., Fischer, H., Gurk, C., Klüpfel, T., Königstedt, R., Parchatka, U., Schiller, C., Stickler, J., Williams, J., and Lelieveld, J. (2008) Hydroxyl radicals in the tropical troposphere over the Suriname rainforest: airborne measurements. *Atmos. Chem. Phys. Discuss.*, **8**, 15491–15536.

Martinez, M., Andujar, J., and Enrique, J. (2010a) A new and inexpensive pyranometer for the visible spectral range. *Sensors*, **9**, 4615–4634. doi: 10.3390/s90604615.

Martinez, M., Harder, H., Kubistin, D., Rudolf, M., Bozem, H., Eerdekens, G., Fischer, H., Klupfel, T., Gurk, C., Konigstedt, R., Parchatka, U., Schiller, C.L., Stickler, A., Williams, J., and Lelieveld, J. (2010b) Hydroxyl radicals in the tropical troposphere over the Suriname rainforest: airborne measurements. *Atmos. Chem. Phys.*, **10**, 3759–3773.

Maser, R., Franke, H., Preiss, M., and Jaeschke, W. (1994) Methods provided and applied on a research aircraft for the study of cloud physics and chemistry. *Beitr. Phys. Atmos.*, **67**, 321–334.

Massoli, P., Murphy, D.M., Lack, D.A., Baynard, T., Brock, C.A., and Lovejoy, E.R. (2009) Uncertainty in light scattering measurements by TSI nephelometer: results from laboratory studies and implications for ambient measurements. *Aerosol Sci. Technol.*, **43**, 1064–1074.

Massoli, P., Kebabian, P.L., Onasch, T.B., Hills, F.B., and Freedman, A. (2010) Aerosol light extinction measurements by Cavity Attenuated Phase Shift (CAPS) spectroscopy: laboratory validation and field deployment of a compact aerosol particle extinction monitor. *Aerosol Sci. Technol.*, **44**, 428–435.

Mastenbrook, H. (1968) Water vapor distribution in the stratosphere and high troposphere. *J. Atmos. Sci.*, **25**, 299–311.

Matejka, T. and Lewis, S.A. (1997) Improving research aircraft navigation by incorporating INS and GPS information in a variational solution. *J. Atmos. Oceanic Technol.*, **14**, 495–511.

Matrosov, S.Y. (1991) Prospects for the measurement of ice cloud particle shape and orientation with elliptically polarized Radar signals. *Radio Sci.*, **26**, 847–856.

Matrosov, S.Y. (2007) Modeling backscatter properties of snowfall at millimeter wavelengths. *J. Atmos. Sci.*, **64**, 1727–1736.

Matrosov, S.Y., Korolev, A.V., and Heymsfield, A.J. (2002) Profiling cloud ice mass and particle characteristic size from Doppler Radar measurements. *J. Atmos. Oceanic Technol.*, **19**, 1003–1018.

Matrosov, S.Y., Shupe, M.D., and Djalalova, I.V. (2008) Snowfall retrievals using millimeter–wavelength cloud Radars. *J. Appl. Meteorol. Climatol.*, **47**, 769–777.

Matsuki, A. (2005) Heterogeneous sulfate formation on the dust surface and it dependency on the mineralogy: observational insight from the balloon-borne measurements in the surface atmosphere of Beijing, China. *Water Air Soil Pollut. Focus*, **5**, 101–132.

Matsumoto, J. and Kajii, Y. (2003) Improved analyzer for nitrogen dioxide by laser induced fluorescence technique. *Atmos. Environ.*, **37**, 4847–4851.

Matsumoto, T., Russell, P., Mina, C., van Ark, W., and Banta, V. (1987) Airborne tracking sunphotometer. *J. Atmos. Oceanic Technol.*, **4**, 336–339.

Matthew, B.M., Middlebrook, A.M., and Onasch, T.B. (2008) Collection efficiencies in an Aerodyne Aerosol Mass Spectrometer as a function of particle phase for laboratory generated aerosol particles. *Aerosol Sci. Technol.*, **42**, 884–898.

Mätzler, C. (2005) On the determination of surface emissivity from satellite observations. *IEEE Geosci. Remote Sens. Lett.*, **2**, 160–163.

Mätzler, C. (2007) Dependence of microwave brightness temperature on bistatic scattering: model functions and application to AMSU–A. *IEEE Trans. Geosci. Remote*, **45**, 2130–2138.

Mauldin, R., Tanner, D., and Eisele, F. (1998) A new chemical ionization mass spectrometric technique for the fast measurements of gas–phase nitric acid in the atmosphere. *J. Geophys. Res.*, **103**, 3361–3367.

Mauldin, R., Eisele, F., Cantrell, C., Kosciuch, E., Ridley, B., Lefer, B., Tanner, D., Nowak, J., Chen, G., Wang, L., and Davis, D. (2001) Measurements of OH aboard the NASA P–3 during PEM–Tropics B. *J. Geophys. Res.*, **106**, 32657–32666.

Mauldin, R., Kosciucha, E., Henry, B., Eiselea, F., Shetter, R., Lefer, B., Chen, G., Davis, D., Huey, G., and Tanner, D. (2004) Measurements of OH, HO_2+RO_2, H_2SO_4, and MSA at the South Pole during ISCAT 2000. *Atmos. Environ.*, **38**, 5423–5437.

May, R.D. (1998) Open–path, near–infrared tunable diode laser spectrometer for atmospheric measurements of H_2O. *J. Geophys. Res.*, **103**, 19161–19172.

Mayer, J.-C., Korbinian, H., Rummel, U., Meixner, F.X., and Foken, T. (2009) Moving measurement platforms–specific challenges and corrections. *Meteorol. Z.*, **18**, 477–488.

Mayol–Bracero, O.L., Guyon, P., Graham, B., Roberts, G., Andreae, M.O., Decesari, S., Facchini, M.C., Fuzzi, S., and Artaxo, P. (2002) Water–soluble organic compounds in biomass burning aerosol particles over Amazonia, 2, apportionment of the chemical composition and importance of the polyacidic fraction. *J. Geophys. Res.*, **107**, 8091. doi: 10.1029/2001JD000522.

Mazin, I., Korolev, A., Heymsfield, A., Isaac, G., and Cober, S. (2001) Thermodynamics of icing cylinder for measurements of liquid water content in super cooled clouds. *J. Atmos. Oceanic Technol.*, **18**, 543–558.

McBratney, A.M.B. and Rossel, R.V. (2006) Spectral soil analysis and inference system: a powerful combination for solving soil data crisis. *Geoderma*, **136**, 272–278.

McCarthy, J. (1973) A method for correcting airborne temperature data for sensor response time. *J. Appl. Meteorol.*, **12**, 211–214.

McElroy, M.B., Salawitch, R., Wofsy, S., and Logan, J. (1986) Reductions of Antarctic ozone due to synergistic interactions of chlorine and bromine. *Nature*, **32**, 759–762.

McFarquhar, G.M., Um, J., Freer, M., Baumgardner, D., Kok, G., and Mace, G. (2007) The importance of small ice crystals to cirrus properties: observations from the Tropical Warm Pool International cloud Experiment (TWP–ICE). *Geophys. Res. Lett.*, 57. doi: 10.1029/2007GL029 865.

McGill, M.J., Hlavka, D.L., Hart, W.D., Spinhirne, J.D., Scott, V.S., and Schmid, B. (2002) The Cloud Physics Lidar: instrument description and initial measurement results. *Appl. Opt.*, **41**, 3725–3734.

McGill, M.J., Li, L., Hart, W.D., Hlavka, G.H.D., Vaughan, M., and Winker, D. (2004) Combined Lidar–Radar remote sensing: initial results from CRYSTAL–FACE. *J. Geophys. Res.*, **109**, D07203. doi: 10.1029/2003JD004030.

McGrath, A. and Hewison, T. (2001) Measuring the accuracy of MARSS–an airborne microwave radiometer. *J. Atmos. Oceanic Technol.*, **18**, 2003–2012.

McKenna, D., Konopka, P., Grooss, J., Gunther, G., Muller, R., Spang, R., Offermann, D., and Orsolini, Y. (2002a) A new Chemical Lagrangian Model of the Stratosphere (CLaMS)–1. Formulation of advection and mixing. *J. Geophys. Res.*, **107**, 4309. doi: 10.1029/2000JD000114.

McKenna, D.S., Grooss, J., Gunther, G., Konopka, P., Muller, R., Carver, G., and Sasano, Y. (2002b) A new Chemical Lagrangian Model of the Stratosphere (CLaMS)–2. Formulation of chemistry scheme and initialization. *J. Geophys. Res.*, **107**, 4256. doi: 10.1029/2000JD000113.

McMeeking, G.R., Morgan, W.T., Flynn, M., Highwood, E.J., Turnbull, K., Haywood, J., and Coe, H. (2011) Black carbon aerosol mixing state, organic aerosol particles and aerosol optical properties over the United Kingdom. *Atmos. Chem. Phys.*, **11**, 9037–9052.

McMurry, P.H. (2000) The history of condensation nucleus counters. *Aerosol Sci. Technol.*, **33**, 297–322.

McNaughton, C.S., Clarke, A.D., Howell, S.G., Pinkerton, M., Anderson, B., Thornhill, L., Hudgins, C., Winstead, E., Dibb, J.E., Sceuer, E., and Maring, H. (2007) Results from the DC–8 Inlet Characterization Experiment (DICE): airborne versus surface sampling of mineral dust and sea salt aerosol particles. *Aerosol Sci. Technol.*, **41**, 136–159.

Measures, R.M. (1984) *Laser Remote Sensing*, Krieger, Malabar, FA.

Mees, J., Crewell, S., Nett, H., de Lange, G., van de Stadt, H., Kuipers, J., and Panhuyzen, R.A. (1995) ASUR–an airborne SIS–receiver for atmospheric measurements at 625 to 720 GHz. *IEEE Trans. Microw. Theory*, **43**, 2453–2458.

Merceret, F. and Schricker, T. (1975) A new hot–wire liquid water meter. *J. Appl. Meteorol.*, **14**, 319–326.

Merceret, F.J. (1976a) Measuring atmospheric turbulence with airborne hot–film anemometers. *J. Appl. Meteorol.*, **15**, 482–490.

Merceret, F.J. (1976b) Airborne hot–film measurements of the small–scale structure of atmospheric turbulence during GATE. *J. Atmos. Sci.*, 1739–1746.

Mertes, S., Schroder, F., and Wiedensohler, A. (1995) The particle–detection efficiency curve of the TSI–3010 CPC as a function of the temperature difference between saturator and condenser. *Aerosol Sci. Technol.*, **23**, 257–261.

Mertes, S., Verheggen, B., Walter, S., Connolly, P., Ebert, M., Schneider, J., Bower, K., Cozic, J., Weinbruch, S., Baltensperger, U., and Weingartner, E. (2007) Counterflow virtual impactor based collection of small ice particles in mixed–phase clouds for the physico–chemical characterization of tropospheric ice nuclei: sampler description and first case study. *Sci. Technol.*, **41**, 848–864.

Michalsky, J., Harrison, L., and Berkheiser, W. (1995) Cosine response characteristics of some radiometric and photometric sensors. *Sol. Energy*, **54**, 397–402.

Michalsky, J., Schlemmer, W., Berkheiser, E., Berndt, L., and Harrison, C. (2001) Multiyear measurements of aerosol optical depth in the Atmospheric Radiation Measurement and Quantitative Links programs. *J. Geophys. Res.*, **106**, 12099–12107.

Mie, G. (1908) Beiträge zur Optik trüber Medien, speziell kolloidaler Metallösungen. *Ann. Phys.*, **25**, 377–445.

Mielke, L.H., Erickson, D.E., McLuckey, S.A., Muller, M., Wisthaler, A.,

Hansel, A., and Shepson, P.B. (2008) Development of a proton–transfer reaction–linear ion trap mass spectrometer for quantitative determination of volatile organic compounds. *Anal. Chem.*, **80**, 8171–8177.

Mihele, C. and Hastie, D. (1998) The sensitivity of the radical amplifier to ambient water vapor. *Geophys. Res. Lett.*, **25**, 1911.

Mihele, C. and Hastie, D. (2000) Optimized operation and calibration procedures for radical amplifier–type detectors. *J. Atmos. Oceanic Technol.*, **17**, 788.

Mihele, C., Mozurkewich, M., and Hastie, D. (1999) Radical loss in a chain reaction of CO and NO in the presence of water: implications for the radical amplifier and atmospheric chemistry. *Int. J. Chem. Kinet.*, **31**, 145.

Miller, T., Ballenthin, J., Meads, R., Hunton, D., Thorn, W., Viggiano, A., Kondo, Y., Koike, M., and Zhao, Y. (2000) Chemical ionization mass spectrometric technique for the measurements of HNO_3 in air traffic corridors in the upper troposphere during the SONEX campaign. *J. Geophys. Res.*, **105**, 3701–3707.

Millet, D.B., Jacob, D.J., Custer, T.G., de Gouw, J.A., Goldstein, A.H., Karl, T., Singh, H.B., Sive, B.C., Talbot, R.W., Warneke, C., and Williams, J. (2008) New constraints on terrestrial and oceanic sources of atmospheric methanol. *Atmos. Chem. Phys.*, **8**, 6887–6905.

Mirme, S., Mirme, A., Minikin, A., Petzold, A., Horrak, U., Kerminen, V.M., and Kulmala, M. (2010) Atmospheric sub–3 nm particles at high altitudes. *Atmos. Chem. Phys.*, **10**, 437–451.

Mishchenko, M. and Travis, L. (1994) T–matrix computations of light scattering by large spheroidal particles. *Opt. Commun.*, **109**, 16–21.

Miyazaki, K., Parker, A.E., Fittschen, C., Monks, P.S., and Kajii, Y. (2010) A new technique for the selective measurement of atmospheric peroxy radical concentrations of HO_2 and RO_2 using a denuding method. *Atmos. Meas. Technol.*, **3**, 1547–1554.

Möhler, O., DeMott, P., Stetzer, O., and The ICIS-2007 Team (2008) The 4th International Ice Nucleation Workshop ICIS-2007: methods and instruments. *Proceedings of the 15th International Conference on Clouds and Precipitation*, Paper 11.1, Cancun, Mexico.

Mohnen, A. (1980) Cloud water collection from aircraft. *Atmos. Technol.*, **12**, 20–25.

Moisseev, D.N., Unal, C.M.H., Russchenberg, H.W.J., and Ligthart, L.P. (2002) Improved polarimetric calibration for atmospheric Radars. *J. Atmos. Oceanic Technol.*, **19**, 1968–1977.

Moisseev, D.N., Chandrasekar, V., Unal, C.M.H., and Russchenberg, H.W.J. (2006) Dual–polarization spectral analysis for retrieval of effective raindrop shapes. *J. Atmos. Oceanic Technol.*, **23**, 1682–1695.

Molina, L.T. and Molina, M. (1987) Production of Cl_2O_2 from the self–reaction of the ClO radical. *J. Phys. Chem.*, **91**, 433–436.

Monks, P. (2005) Gas–phase radical chemistry in the troposphere. *Chem. Soc. Rev.*, **34**, 376–395.

Montgomery, R.B. (1948) Vertical eddy flux of heat in the atmosphere. *J. Meteorol.*, **5**, 265–274.

Moore, K.G., Clarke, A.D., Kapustin, V., McNaughton, C., Anderson, B., Winstead, E., Weber, R., Ma, Y., Lee, Y., Talbot, R., Dibb, J., Anderson, T., Doherty, S., Covert, D., and Rogers, D. (2004) A comparison of similar aerosol measurements made on the NASA P3–B, DC–8, and NSF C–130 aircraft during TRACE–P and ACE–Asia. *J. Geophys. Res.*, **109**, D15S15. doi: 10.1029/2003JD003543.

Moore, R.H., Nenes, A., and Medina, J. (2010) Scanning mobility CCN analysis – a method for fast measurements of size-resolved CCN distributions and activation kinetics. *Aerosol Sci. Technol.*, **44**, 861–871. doi: 10.1080/02786826.2010.498715.

Moore, R.H., Bahreini, R., Brock, C.A., Froyd, K.D., Cozic, J., Holloway, J.S., Middlebrook, A.M., Murphy, D.M., and Nenes, A. (2011) Hygroscopicity and composition of Alaskan Arctic CCN during April 2008. *Atmos. Chem. Phys.*, **11**, 11807–11825.

Moore, A. Jr., Brown, K.E., Hall, W.M., Barnes, J.C., Edwards, W.C., Petway,

L.B., Little, A.D., Luck, W.S. Jr., Jones, I.W., Antill, C.W. Jr., Browell, E.V., and Ismail, S. (1997) Development of the Lidar Atmospheric Sensing Experiment (LASE), an advanced airborne DIAL instrument, in *Advances in Atmospheric Remote Sensing with Lidar* (eds A. Ansmann, R. Neuber, P. Rairoux, and U. Wandinger), Springer–Verlag, Berlin, pp. 281–288.

Moosmüller, H., Chakrabarty, R.K., and Arnott, W.P. (2009) Aerosol light absorption and its measurement: A review. *J. Quant. Spectrosc. Radiat. Transf.*, **110**, 844–878.

Morgan, W.T., Allan, J.D., Bower, K.N., Capes, G., Crosier, J., Williams, P.I., and Coe, H. (2009) Vertical distribution of sub–micron aerosol chemical composition from North–Western Europe and the North–East Atlantic. *Atmos. Chem. Phys.*, **9**, 5389–5401.

Moteki, N., Kondo, Y., Miyazaki, Y., Takegawa, N., Komazaki, Y., Kurata, G., Shirai, T., Blake, D.R., Miyakawa, T., and Koike, M. (2007) Evolution of mixing state of black carbon particles: aircraft measurements over the western Pacific in March 2004. *Geophys. Res. Lett.*, **34**, L11803. doi: 10.1029/2006GL028943.

Moteki, N., Kondo, Y., Takegawa, N., and Nakamura, S. (2009) Directional dependence of thermal emission from nonspherical carbon particles. *J. Aerosol Sci.*, **40**, 790–801.

Mounaïm-Rousselle, C. and Pajot, O. (1999) Droplet sizing by Mie scattering interferometry in a spark ignition engine. *Part. Part. Syst. Charact.*, **16**, 160.

Moyer, E.J., Irion, F.W., Yung, Y.L., and Gunson, M.R. (1996) ATMOS stratospheric deuterated water and implications for troposphere stratosphere transport. *Geophys. Res. Lett.*, **23**, 2385–2388.

Müller, D., Wandinger, U., Althausen, D., Mattis, I., and Ansmann, A. (1998) Retrieval of physical particle properties from Lidar observations of extinction and backscatter at multiple wavelengths. *Appl. Opt.*, **37**, 2260–2263.

Müller, S., Kämpfer, N., Feist, D., Haefele, A., Milz, M., Sitnikov, N., Schiller, C., Kiemle, C., and Urban, J. (2008) Validation of stratospheric water vapor measurements from the airborne microwave radiometer AMSOS. *Atmos. Chem. Phys.*, **8**, 3169–3183, doi:10.5194/acp-8-3169-2008.

Munson, M.S.B. and Field, F.H. (1966) Chemical ionization mass spectrometry–I. General introduction. *J. Am. Chem. Soc.*, **88**, 2621–2630.

Murphy, D. and Koop, T. (2005) Review of the vapor pressures of ice and supercooled water for atmospheric applications. *Q. J. R. Meteorol. Soc.*, **131**, 1539–1565.

Murphy, D., Cziczo, D., Hudson, P., Thomson, D., Wilson, J.C., Kojima, T., and Buseck, P. (2004) Particle generation and resuspension in aircraft inlets when flying in clouds. *Aerosol Sci. Technol.*, **38**, 400–408.

Murphy, D.M. (2007) The design of single particle mass spectrometers. *Mass Spectrom. Rev.*, **26**, 150–165.

Murphy, D.M. and Schein, M.E. (1998) Wind tunnel tests of a shrouded aircraft inlet. *Aerosol Sci. Technol.*, **28**, 33–39.

Murphy, D.M., Cziczo, D.J., Froyd, K.D., Hudson, P.K., Matthew, B.M., Middlebrook, A.M., Peltier, R.E., Sullivan, A., Thomson, D.S., and Weber, R.J. (2006) Single–particle mass spectrometry of tropospheric aerosol particles. *J. Geophys. Res.*, **111**.

Muschinski, A. and Wode, C. (1998) First in–situ evidence for co–existing sub–meter temperature and humidity sheets in the lower free troposphere. *J. Atmos. Sci.*, **55**, 2893–2906.

Muschinski, A., Frehlich, R.G., Jensen, M.L., Hugo, R., Hoff, A.M., Eaton, F., and Balsley, B.B. (2001) Fine-scale measurements of turbulence in the lower troposphere: an intercomparison between a kite– and balloon–borne and a helicopter-borne measurement system. *Bound.–Lay. Meteorol.*, **98**, 219–250.

Nagel, D., Maixner, U., Strapp, W., and Wasey, M. (2007) Advancements in techniques for calibration and characterization of in situ optical particle measuring probes, and applications to the FSSP–100 probe. *J. Atmos. Oceanic Technol.*, **24**, 745–760.

National Research Council (2007) Earth science and its applications from space:

national imperatives for the next decade and beyond. *Tech. Rep.*, NRC, The National Academies Press, 500 Fifth Street, N.W., Washington, DC 20001, 455 pp.

Nedelec, P., Cammas, J., Thouret, V., Athier, G., Cousin, J., Legrand, C., Abonnel, C., Lecoeur, F., Cayez, G., and Marizy, C. (2003) An improved infra–red carbon monoxide analyser for routine measurements aboard commercial airbus aircraft: technical validation and first scientific results of the MOZAIC III program. *Atmos. Chem. Phys.*, **3**, 1551–1564.

Neel, C.B. (1955) A Heated–Wire Liquid–Water–Content Instrument and Results of Initial Flight Tests in Icing Conditions. *Tech. Rep.*, National Advisory Committee for Aeronautics (NACA) Research Memorandum, RMA54123, 33 pp.

Neel, C. (1973) Measurement of cloud liquid–water content with a heated wire, in *Proceedings of the 19th International Aerospace Symposium* (ed. B. Washburn), Instrument Society of America, Pittsburgh, PA, p. 301.

Nemarich, J., Wellman, R.J., and Lacombe, J. (1988) Backscatter and attenuation by falling snow and rain at 96, 140, and 225 GHz. *IEEE Trans. Geosci. Remote*, **26**, 319–329.

Nenes, A., Chuang, P., Flagan, R., and Seinfeld, J. (2001a) A theoretical analysis of cloud condensation nucleus (CCN) instruments. *J. Geophys. Res.*, **106**, 3449–3474.

Nenes, A., Ghan, S., Abdul-Razzak, H., Chuang, P.Y., and Seinfeld, J.H. (2001b) Kinetic limitations on cloud droplet formation and impact on cloud albedo. *Tellus B*, **53**, 133–149.

Neuman, J., Huey, L., Ryerson, T., and Fahey, D. (1999) Study of inlet materials for sampling atmospheric nitric acid. *Environ. Sci. Technol.*, **33**, 1133–1136. doi: 10.1021/es980767f.

Neumann, A., Witzke, A., Jones, S., and Schmitt, G. (2002) Representative terrestrial solar brightness profiles. *Trans. ASME*, **124**, 198–204.

Neumann, J., Gao, R., Schein, M., Ciciora, S., Holecek, J., Thompson, T., Winkler, R., McLaughlin, R., Northway, M., Richard, E., and Fahey, D. (2000) A fast response chemical ionization mass spectrometer for in situ measurements of HNO_3 in the upper troposphere and lower stratosphere. *Rev. Sci. Instrum.*, **71**, 3886–3894.

Nevzorov, A. (1980) Aircraft cloud water content meter. Comm. a la 8eme conf int. sur la phys. des nuages, vol. II, Clermont–Ferrand, France, pp. 701–703.

Nevzorov, A. and Shugaev, V. (1972) The use of integral parameters for study of cloud microstructure (in Russian). *Trans. Cent. Aerol. Obs.*, **101**, 32–47.

Nevzorov, A. and Shugaev, V. (1974) Aircraft cloud extinction meter. *Trans. Cent. Aerol. Obs.*, **106**, 3–10.

Newman, S., Clarisse, L., Hurtmans, D., Marenco, F., Havemann, S., Johnson, B., Turnbull, K., and Haywood, J. (2012) A case study of observations of volcanic ash from the Eyjafjallajökull eruption: 2. Airborne and satellite radiative measurements. *J. Geophys. Res.*, **117**, D00U13. doi: 10.1029/2011JD016780.

Nicholls, S., Leighton, J., and Barker, R. (1990) A new fast response instrument for measuring total water content from aircraft. *J. Atmos. Oceanic Technol.*, **7**, 706–718.

Nicodemus, F., Richmond, J., Hsia, J., Ginsberg, I., and Limperis, T. (1977) Geometrical considerations and nomenclature for reflectance. *Tech. Rep.*, Institute for Basic Standards, National Bureau of Standards, Stock No. 003-003-01793.

Nieke, J., Schwarzer, H., Neumann, A., and Zimmermann, G. (1997) Imaging spaceborne and airborne sensor systems in the beginning of the next century. Presented at The European Symposium on Aerospace Remote Sensing (IEE), in Conference on Sensors, Systems and Next Generation Satellites III, vol. 3221, SPIE, London.

Nieke, J., Itten, K., Meuleman, K., Gege, P., Del'Endice, F., Hueni, A., Alberti, E., Ulbrich, G., and Meynart, R. (2008) Supporting facilities of the Airborne Imaging Spectrometer APEX, *Proceedings of IGARSS 2008*, IEEE, pp. V–502–V–505.

Nielson–Gammon, J., Powell, C., Mahoney, M., Angevine, W., Senff, C., White, A., Berkowitz, C., Doran, J., and

Knupp, K. (2008) Multi–sensor estimation of mixing heights over coastal city. *J. Appl. Meteorol. Climatol.*, **47**, 27–43.

Noël, V., Chepfer, H., Ledanois, G., Delaval, A., and Flamant, P. (2002) Classification of particle effective shape ratio in cirrus clouds based on the Lidar depolarization ratio. *Appl. Opt.*, **41**, 4245–4257.

Noël, V., Winker, D.M., McGill, M., and Lawson, P. (2004) Classification of particle shapes from Lidar depolarization ratio in convective ice clouds compared to in situ observations during CRYSTAL–FACE. *J. Geophys. Res.*, **109**, D24213. doi: 10.1029/2004JD004883.

Noomem, M., Meer, F., and Skidmore, A. (2005) Hyperspectral remote sensing for detecting the effects of three hydrocarbon gases on maize reflectance. Proceedings: 31th International Symposium of Remote Sensing of Environment, Saint Petersburg.

Noone, K.B., Noone, K.J., Heintzenberg, J., Ström, J., and Ogren, J.A. (1993) In situ observations of cirrus cloud michrophysical properties using the counterflow virtual impactor. *J. Atmos. Oceanic Technol.*, **10**, 294–303.

Noone, K.J. and Heintzenberg, J. (1991) On the determination of droplet size distributions with the counterflow virtual impactor. *Atmos. Res.*, **26**, 389–406.

Noone, K.J., Ogren, J., Heintzenberg, J., Charlson, R., and Covert, D. (1988) Design and calibration of a counterflow virtual impactor for sampling of atmospheric fog and cloud droplets. *Aerosol Sci. Technol.*, **8**, 235–244.

Norment, H. (1985) Calculation of water drop trajectories to and about arbitrary three dimensional lifting and non-lifting bodies in potential airflow. *Tech. Rep.* NTIS N87-11694/3/GAR, National Aeronautics and Space Administration.

Norment, H. (1988) Three–dimensional trajectory analysis of two drop sizing instruments: PMS OAP and PMS FSSP. *J. Atmos. Oceanic Technol.*, **5**, 743–756.

Nunnermacker, L.J., Weinstein–Lloyd, J.B., Hillery, B., Giebel, B., Kleinman, L.I., Springston, S.R., Daum, P.H., Gaffney, J., Marley, N., and Huey, G. (2008) Aircraft and ground–based measurements of hydroperoxides during the 2006 MILAGRO field campaign. *Atmos. Chem. Phys.*, **8**, 7619–7636.

Ochou, A.D., Nzeukou, A., and Sauvageot, H. (2007) Parameterization of drop size distribution with rain rate. *Atmos. Res.*, **84**, 58–66.

O'Connor, E.J., Hogan, R.J., and Illingworth, A.J. (2005) Retrieving stratocumulus drizzle parameters using Doppler Radar and Lidar. *J. Appl. Meteorol.*, **44**, 14–27.

O'Donnell, C. (1964) *Inertial Navigation: Analysis and Design*, McGraw–Hill, New York.

Oebel, A., Broch, S., Raak, D., Bohn, B., Rohrer, F., Hofzumahaus, A., Holland, F., and Wahner, A. (2010) In situ measurements of vertical profiles of chemical tracers in the PBL using the airship Zeppelin NT. ISARS Meeting.

Ogren, J., Heintzenberg, J., and Charlson, R. (1985) In situ sampling of clouds with a droplet to aerosol converter. *Geophys. Res. Lett.*, **12**, 121–124.

Ogren, J.A. (1995) A systematic approach to in situ observations of aerosol properties, in *Aerosol Forcing of Climate* (eds R.J. Charlson and J. Heintzenberg), John Wiley & Sons, pp. 215–226.

Ogren, J.A. (2010) Comment on calibration and intercomparison of filter–based measurements of visible light absorption by aerosols. *Aerosol Sci. Technol.*, **44**, 589–591.

Oguchi, T. (1983) Electromagnetic wave propagation and scattering in rain and other hydrometeors. Proceedings of the IEEE, vol. 71, pp. 1029–1078.

Ohtaki, E. and Matsui, T. (1982) Infrared device for simultaneous measurement of fluctuations of atmospheric carbon dioxide and water vapor. *Bound.–Lay. Meteorol.*, **24**, 109–119.

Okamoto, H., Iwasaki, S., Yasui, M., Horie, H., Kuroiwa, H., and Kumagai, H. (2003) An algorithm for retrieval of cloud microphysics using 95 GHz cloud Radar and Lidar. *J. Geophys. Res.*, **108**, 4226. doi: 10.1029/2001JD001225.

Okamoto, K., Kubokawa, T., Tamura, A., and Ushio, T. (2002) Long term trend ofocean surface normalized Radar cross section observed by TRMM precipitation Radar. Proceedings of the 2nd Global Precipitation Mission (GPM) International Planning Workshop, vol. 9, Shinagawa, Japan, *http://ams.confex.com/ams/pdfpapers/96443.pdf*

O'Keefe, A. and Deacon, D. (1988) Cavity ring–down optical spectrometer for absorption measurements using pulsed laser sources. *Rev. Sci. Instrum.*, **59**, 2544–2551.

Olfert, J.S. and Wang, J. (2009) Dynamic characteristics of a fast–response aerosol size spectrometer. *Aerosol Sci. Technol.*, **43**, 97–111.

Olfert, J.S., Kulkarni, P., and Wang, J. (2008) Measuring aerosol size distributions with the fast integrated mobility spectrometer. *J. Aerosol Sci.*, **39**, 940–956.

Oppelt, N. and Mauser, W. (2007) The Airborne Visible / Infrared Imaging Spectrometer AVIRIS: design, characterization and calibration. *Sensors*, **7**, 1934–1953.

Orr, B.W. and Kropfli, R.A. (1999) A method for estimating particle fall velocities from vertically pointing Doppler Radar. *J. Atmos. Oceanic Technol.*, **16**, 29–37.

Orsini, D.A., Wiedensohler, A., Stratmann, F., and Covert, D.S. (1999) A new volatility tandem differential mobility analyzer to measure the volatile sulfuric acid aerosol fraction. *J. Atmos. Oceanic Technol.*, **16**, 760–772.

Orsini, D.A., Ma, Y.L., Sullivan, A., Sierau, B., Baumann, K., and Weber, R.J. (2003) Refinements to the particle–into–liquid sampler (PILS) for ground and airborne measurements of water soluble aerosol composition. *Atmos. Environ.*, **37**, 1243–1259.

Osborne, S.R., Baran, A.J., Johnson, B.T., Haywood, J.M., Hesse, E., and Newman, S. (2001) Short–wave and long–wave radiative properties of Saharan dust aerosol. *Q. J. R. Meteor. Soc.*, **137**, 1149–1167. doi: 10.1002/qj.771.

Ovarlez, J. and van Velthoven, P. (1997) Comparison of water vapor measurements with data retrieved from ECMWF analyses during the POLINAT experiment. *J. Appl. Meteorol.*, **36**, 1329–1335.

Owens, G. (1957) Wind Tunnel Calibrations of Three Instruments Designed for Measurements of the Liquid–water Content of Clouds. Technical Note 10, U. Chicago Cloud Physics Laboratory.

Paffrath, U., Lemmerz, C., Reitebuch, O., Witschas, B., Nikolaus, I., and Freudenthaler, V. (2009) The airborne demonstrator for the direct–detection Doppler wind Lidar ALADIN on ADM–Aeolus: II. Simulations and Rayleigh radiometric performance. *J. Atmos. Oceanic Technol.*, **26**, 2516–2530.

Paldus, B.A., Harb, C., Spence, T., Wilke, B., Xie, J., and Harris, J. (1998) Cavity–locked cavity ring–down spectroscopy. *J. Appl. Phys.*, **83**, 3991–3997.

Paluch, I.R. (1979) The entrainment mechanism in Colorado cumuli. *J. Atmos. Sci.*, **36**, 2467–2478.

Pan, G. and Meng, H. (2003) Digital holography of particle fields: reconstruction by use of complex amplitude. *Appl. Opt.*, **42**, 827–833.

Parrish, D. and Fehsenfeld, F. (2000) Methods for gas–phase measurements of ozone, ozone precursors and aerosol precursors. *Atmos. Environ.*, **34**, 1921–1957.

Parrish, D.D., Allen, D.T., Bates, T.S., Estes, M., Fehsenfeld, F.C., Feingold, G., Ferrare, R., Hardesty, R.M., Meagher, J.F., Nielsen-Gammon, J.W., Pierce, R.B., Ryerson, T.B., Seinfeld, J.H., and Williams, E.J. (2009) Overview of the Second Texas Air Quality Study (TexAQS II) and the Gulf of Mexico Atmospheric Composition and Climate Study (GoMACCS). *J. Geophys. Res.*, **114**, D00F13. doi: 10.1029/2009JD011842.

Pätz, H.-W., Volz-Thomas, A., Hegglin, M.I., Brunner, D., Fischer, H., and Schmidt, U. (2006) In situ comparison of the NO_y instruments flown in MOZAIC and SPURT. *Atmos. Chem. Phys.*, **6**, 2401–2410.

Pawlowska, H. and Brenguier, J.-L. (2003) An observational study of drizzle formation in stratocumulus clouds during ACE–2 for GCM parameterizations. *J. Geophys. Res.*, **108**, 8587.

Pawlowska, H., Brenguier, J.-L., and Salut, G. (1997) Optimal non–linear estimation for cloud particle measurements. *J. Atmos. Oceanic Technol.*, **14**, 88–104.

Payne, F.R. and Lumley, J.L. (1965) One-dimensional spectra derived from an airborne hot-wire anemometer. *Q. J. R. Meteor. Soc.*, **92**, 397–401.

Pazmany, A.L., McIntosh, R.E., Kelly, R.D., and Vali, G. (1994) An airborne 95 GHz dual–polarized Radar for cloud studies. *IEEE Trans. Geosci. Remote*, **32**, 731–739.

Pearlman, J., Barry, P., Segal, C., Shepanski, J., Beiso, D., and Carman, S. (2003) Hyperion, a space–based imaging spectrometer. *IEEE Trans. Geosci. Remote*, **41**, 1160–1173.

Pelon, J., Flamant, C., Chazette, P., Léon, J.-F., Tanré, D., Sicard, M., and Sathees, S.K. (2002) Characterization of aerosol spatial distribution and optical properties over the Indian Ocean from airborne Lidar and radiometry during INDOEX'99. *J. Geophys. Res.*, **107**, 8029. doi: 10.1029/2001JD000402.

Peltier, R.E., Weber, R.J., and Sullivan, A.P. (2007) Investigating a liquid-based method for online organic carbon detection in atmospheric particles. *Aerosol Sci. Technol.*, **41**, 1117–1127.

Penkett, S., Blake, N., Lightman, P., Marsh, A., and Anwyl, P. (1993) The seasonal variation of non–methane hydrocarbons in the free troposphere over the north Atlantic ocean: possible evidence for extensive reaction of hydrocarbons with the nitrate radical. *J. Geophys. Res.*, **98**, 2865–2885.

Penner, J.E., Charlson, R.J., Schwartz, S.E., Hales, J.M., Laulainen, N.S., Travis, L., Leifer, R., Novakov, T., Ogren, J., and Radke, L.F. (1994) Quantifying and minimizing uncertainty of climate forcing by anthropogenic aerosols. *Bull. Am. Meteorol. Soc.*, **75**, 375–400.

Perner, D., Arnold, T., Crowley, J., Kluepfel, T., Martinez, M., and Seuwen, R. (1999) The measurement of active chlorine in the atmosphere by chemical amplification. *J. Atmos. Chem.*, **34**, 9–20. doi: 10.1023/A:1006208828 324.

Petropavlovskikh, I., Froidevaux, L., Shetter, R., Hall, S., Ullmann, L., Bhartia, P., Kroon, M., and Levelt, P. (2008) In–flight validation of Aura MLS ozone with CAFS partial ozone columns. *J. Geophys. Res.*, **113**, D16S41. doi: 10.1029/2007JD008690.

Petters, M.D. and Kreidenweis, S.M. (2007) A single parameter representation of hygroscopic growth and cloud condensation nucleus activity. *Atmos. Chem. Phys.*, **7**, 1961–1971.

Peturson, B. (1931) Epidemiology of cereal rusts, *Rept. Dominion botanist for the year 1930*, Div. of Botany, Canada Dept. of Agric., pp. 44–46.

Petzold, A. and Schönlinner, M. (2004) Multi-angle absorption photometry - a new method for the measurement of aerosol light absorption and atmospheric black carbon. *J. Aerosol Sci.*, **35**, 421–441.

Petzold, A., Ström, J., Ohlsson, S., and Schröder, F.P. (1998) Elemental composition and morphology of ice–crystal residual particles in cirrus clouds and contrails. *Atmos. Res.*, **49**, 21–34.

Petzold, A., Kramer, H., and Schönlinner, M. (2002) Continuous measurement of atmospheric black carbon using a multi–angle absorption photometer. *Environ. Sci. Pollut. Res.*, **4**, 78–82.

Petzold, A., Schloesser, H., Sheridan, P.J., Arnott, W.P., Ogren, J.A., and Virkkula, A. (2005a) Evaluation of multiangle absorption photometry for measuring aerosol light absorption. *Aerosol Sci. Technol.*, **39**, 40–51.

Petzold, A., Fiebig, M., Fritzsche, L., Stein, C., Schumann, U., Wilson, C.W., Hurley, C.D., Arnold, F., Katragkou, E., Baltensperger, U., Gysel, M., Nyeki, S., Hitzenberger, R., Giebl, H., Hughes, K.J., Kurtenbach, R., Wiesen, P., Madden, P., Puxbaum, H., Vrchoticky, S., and Wahl, C. (2005b) Particle emissions from aircraft engines – a survey of the European project PartEmis. *Meteorol. Z.*, **14**, 465–476.

Petzold, A., Weinzierl, B., Huntrieser, H., Stohl, A., Real, E., Cozic, J., Fiebig, M., Hendricks, J., Lauer, A., Law, K., Roiger, A., Schlager, H., and Weingartner, E. (2007) Perturbation of the European free troposphere aerosol by North American forest fire plumes during the ICARTT–ITOP Experiment in summer 2004. *Atmos. Chem. Phys.*, **7**, 105–5127.

Petzold, A., Rasp, K., Weinzierl, B., Esselborn, M., Hamburger, T.,

Dörnbrack, A., Kandler, K., Schütz, L., Knippertz, P., Fiebig, M., and Virkkula, A. (2009) Saharan dust refractive index and optical properties from aircract-based observations during SAMUM 2006. *Tellus*, **61B**, 118–130.

Petzold, A., Veira, A., Mund, S., Esselborn, M., Weinzierl, C.K., Hamburger, T., and Ehret, G. (2011) Mixing of mineral dust with urban pollution aerosol over Dakar (Senegal)–impact on dust physico–chemical and radiative properties. *Tellus B*, **63**, 619–634.

Pfau, A., Schlienger, J., Kalfas, A.I., and Abhari, R.S. (2002) Virtual Four Sensor Fast Response Aerodynamic Probe (FRAP). 16th Symposium on Measuring Techniques in Transonic and Supersonic Flow in Cascades and Turbomachines, Cambridge.

Pfeilsticker, K. and Platt, U. (1994) Airborne measurements during the Arctic stratospheric experiment: observation of O_3 and NO_2. *Geophys. Res. Lett.*, **21**, 1375.

Philipona, R., Fröhlich, C., and Betz, C. (1995) Characterisation of pyrgeometers and the accuracy of atmospheric long–wave radiation measurements. *Appl. Opt.*, **34**, 1598–1605.

Pieters, C., Boardman, J., Buratti, B., Chatterjee, A., Clark, R., Glavich, T., Green, R., Head, J., Isaacson, P., Malaret, E., McCord, T., Mustard, J., Petro, N., Runyon, C., Staid, M., Sunshine, J., Taylor, L., Varanasi, S.T.P., and White, M. (2009a) The Moon Mineralogy Mapper (M3) on Chandrayaan–1. *Curr. Sci.*, **96**, 500–505.

Pieters, C., Goswami, J., Clark, R., Annadurai, M., Boardman, J., Buratti, B., Combe, J.-P., Dyar, M., Green, R., Head, J., Hibbitts, C., Hicks, M., Isaacson, P., Klima, R., Kramer, G., Kumar, S., Livo, E., Lundeen, S., Malaret, E., McCord, T., Mustard, J., Nettles, J., Petro, N., Runyon, C., Staid, M., Sunshine, J., Taylor, L., Tompkins, S., and Varanasi, P. (2009b) Character and spatial distribution of OH/H_2O on the surface of the Moon seen by M3 ON Chandrayaan–1. *Science*, **326**, 568.

Pilch, M. and Erdman, C. (1987) Use of breakup time data and velocity history data to predict the maximum size of stable fragments for acceleration-induced breakup of a liquid drop. *Int. J. Multiphas. Flow*, **13**, 741–757.

Pilewskie, P. and Twomey, S. (1987a) Cloud phase discrimination by reflectance measurements near 1.6 and 2.2 μm. *J. Atmos. Sci.*, **44**, 3419–3420.

Pilewskie, P. and Twomey, S. (1987b) Discrimination of ice from water in clouds by optical remote sensing. *Atmos. Res.*, **21**, 113–122.

Pilewskie, P. and Valero, F. (1993) Optical depths and haze particle sizes during AGASP III. *Atmos. Environ.*, **27A**, 2895–2899.

Pilewskie, P., Pommier, J., Bergstrom, R., Gore, W., Howard, S., Rabbette, M., Schmid, B., Hobbs, P.V., and Tsay, S.C. (2003) Solar spectral radiative forcing during the Southern African Regional Science Initiative. *J. Geophys. Res.*, **108**, 8486. doi: 10.1029/2002JD002411.

Pilewskie, P., Rottman, G., and Richard, E. (2005) An overview of the disposition of solar radiation in the lower atmosphere: connections to the SORCE mission and climate change. *Sol. Phys.*, **230**, 55–69.

Pinnick, R.G. and Auvermann, H.J. (1979) Response characteristics of Knollenberg light–scattering aerosol counters. *J. Aerosol Sci.*, **10**, 55–74.

Pinnick, R.G. and Rosen, J.M. (1979) Response of Knollenberg light–scattering counters to non–spherical doublet polystyrene latex aerosol particles. *J. Aerosol Sci.*, **10**, 533–538.

Pinnick, R.G., Jennings, S.G., and Fernandez, G. (1987) Volatility of aerosol particles in the arid southwestern United States. *J. Atmos. Sci.*, **44**, 562–576.

Pinnick, R.G., Pendleton, J.D., and Videen, G. (2000) Response characteristics of the particle measuring systems active scattering aerosol spectrometer probes. *Aerosol Sci. Technol.*, **33**, 334–352.

Plant, W.J., Keller, W.C., Hesany, V., Hayes, K., Hoppel, K.W., and Blanc, T.V. (1998) Measurements of the marine boundary layer from an airship. *J. Atmos. Oceanic Technol.*, **15**, 1433–1458.

Platt, C., Young, S.A., Manson, P.J., Patterson, G.R., Marsden, S.C., Austin,

R.T., and Churnside, J.H. (1998) The optical properties of equatorial cirrus from observations in the ARM pilot radiation observation experiment. *J. Atmos. Sci.*, **55**, 1977–1996.

Platt, C., Austin, R.T., Young, S.A., and Heymsfield, A.J. (2002) Lidar observations of tropical cirrus clouds in MCTEX. Part II: optical properties and base cooling in dissipating storm anvil clouds. *J. Atmos. Sci.*, **59**, 3163–3177.

Platt, U. (1994) Differential Optical Absorption Spectroscopy (DOAS), *Air Monitoring by Spectroscopic Techniques*, John Wiley & Sons, Inc., New York, pp. 27–84.

Platt, U. and Stutz, J. (2008) *Differential Optical Absorption Spectroscopy, Principles and Applications*, Series: Physics of Earth and Space Environments, Springer–Verlag, ISBN: 3540211934.

Plaza, A. and Chang, C. (eds) (2007) *High Performance Computing in Remote Sensing*, Taylor & Francis, Chapman & Hall CRC publications, Boca Raton, FL, p. 496.

Plaza, A., Martinez, P., Perez, R., and Plaza, J. (2004) A quantitative and comparative analysis of endmember extraction algorithms from hyperspectral data. *IEEE Trans. Geosci. Remote*, **42**, 650–663.

Plaza, A., Benediktsson, A., Boardamn, J., Brazile, J., Bruzzone, L., CapsValls, G., Chanussot, J., Fauvel, M., Gampa, P., Gulatiri, J., Marconcini, M., Tilton, J., and Trianni, G. (2009) Recent advances in techniques for hyperspectral image processing. *Remote Sens. Environ.*, **113**, 110–122.

Poberaj, G., Fix, A., Assion, A., Wirth, M., Kiemle, C., and Ehret, G. (2002) Airborne all–solid–state DIAL for water vapor measurements in the tropopause region: system description and assessment of accuracy. *Appl. Phys. B*, **75**, 165–172.

Pokharel, B. and Vali, G. (2011) Evaluation of co–located measurements of Radar reflectivity and particle sizes in ice clouds. *J. Appl. Meteorol. Climatol.*, **50**, 2104–2119. doi: http://dx.doi.org/10.1175/JAMC–D–10–05 010.1.

Pollack, I.B., Lerner, B.M., and Ryerson, T.B. (2010) Evaluation of ultraviolet light–emitting diodes for detection of atmospheric NO(2) by photolysis – chemiluminescence. *J. Atmos. Chem.*, **65**, 111–125. doi: 10.1007/s10874–011–9184–3.

Pontikis, C., Rigaud, A., and Hicks, E. (1987) Entrainment and mixing as related to the microphysical properties of shallow warm cumulus clouds. *J. Atmos. Sci.*, **44**, 2150–2165.

Porcheron, E., Lemaitre, P., Nuboer, A., Rochas, V., and Wendel, J. (2007) Experimental investigation in the TOSQAN facility of heat and mass transfer in a spray for containment application. *Nucl. Eng. Des.*, **237**, 1862–1871.

Porter, J., Clarke, A., Ferry, G., and Pueschel, R. (1992) Aircraft studies of size-dependent aerosol sampling through inlets. *J. Geophys. Res.*, **97**, 3815–3824.

Porter, J., Clarke, A., Reid, J., Shaw, G., Maring, H., Reid, E., and Kress, D. (2007) Handheld Sun photometer measurements from light aircraft. *J. Atmos. Oceanic Technol.*, **24**, 1588–1597.

Pöschl, U. (2005) Atmospheric aerosols: composition, transformation, climate and health effects. *Angew. Chem. Int. Ed.*, **44**, 7520–7540.

Prados–Roman, R., Butz, A., Deutschmann, T., Dorf, M., Kritten, L., Minikin, A., Theys, N., Schlager, H., Platt, U., Roozendael, M.V., and Pfeilsticker, K. (2010) BrO profiling in the Arctic troposphere during spring 2007. *Geophys. Res. Abstr.*, **12**, EGU2010–3330–1.

Prasad, S. (2008) Overcoming the small sample size problem in hyperspectral classification and detection tasks. *IEEE Trans. Geosci. Remote*, **5**, 381–384.

Pratt, K., Mayer, A., Holecek, J.C., Moffet, R.C., Sanchez, R.O., Rebotier, T.P., Furutani, H., Gonin, M., Fuhrer, K., Su, Y.X., Guazzotti, S., and Prather, K.A. (2009a) Development and characterization of an aircraft aerosol time–of–flight mass spectrometer. *Anal. Chem.*, **81**, 1792–1800.

Pratt, K.A., DeMott, P.J., French, J.R., Wang, Z., Westphal, D.L., Heymsfield, A.J., Twohy, C.H., Prenni, A.J., and Prather, K.A. (2009b) In situ detection of biological particles in cloud ice–crystals. *Nat. Geosci.*, **2**, 397–400.

Prazeller, P., Palmer, P.T., Boscaini, E., Jobson, T., and Alexander, M. (2003)

Proton transfer reaction ion trap mass spectrometer. *Rapid Commun. Mass Spectrom.*, **17**, 1593–1599.

Prenni, A.J., Petters, M.D., Kreidenweis, S.M., Heald, C.L., Martin, S.T., Artaxo, P., Garland, R.M., Wollny, A.G., and Pöschl, U. (2009) Relative roles of biogenic emissions and Saharan dust as ice nuclei in the Amazon basin. *Nat. Geosci.*, **2**, 401–404.

Prigent, C., Chevallier, F., Karbou, F., Bauer, P., and Kelly, G. (2005) AMSU–a land surface emissivity estimation for numerical weather prediction assimilation schemes. *J. Appl. Meteorol.*, **44**, 416–426.

Pringle, K., Tost, H., Pozzer, A., Pöschl, U., and Lelieveld, J. (2010) Global distribution of the effective aerosol hygroscopicity parameter for CCN activation. *Atmos. Chem. Phys.*, **10**, 5241–5255.

Proctor, B.E. and Parker, B.W. (1938) An improved apparatus and technique for upper air investigations. *J. Bacteriol.*, **36**, 175–186.

Proffitt, M.H. and McLaughlin, R. (1983) Fast–response dual beam UV absorption ozone photometer suitable for use on stratospheric balloons. *Rev. Sci. Instrum.*, **54**, 1719–1728.

Protat, A., Lemaitre, Y., and Scialom, G. (1997) Retrieval of kinematic fields using a single–beam airborne Doppler Radar performing circular trajectories. *J. Atmos. Oceanic Technol.*, **14**, 769–791.

Protat, A., Lemaitre, Y., and Bouniol, D. (2003) Terminal fall velocity and the FASTEX cyclones. *Q. J. R. Meteor. Soc.*, **129**, 1513–1535.

Protat, A., Pelon, J., Testud, J., Grand, N., Delville, P., Laborie, P., Vinson, J.-P., Bouniol, D., Bruneau, D., Chepfer, H., Delanoë, J., Haeffelin, M., Noël, V., and Tinel, C. (2004) Etude des nuages faiblement précipitants par télédétection active. La Météorologie n°47.

Protat, A., Armstrong, A., Haeffelin, M., Morille, Y., Pelon, J., Delanoë, J., and Bouniol, D. (2006) Impact of conditional sampling and instrumental limitations on the statistics of cloud properties derived from cloud Radar and Lidar at SIRTA. *Geophys. Res. Lett.*, **33**, L11805. doi: 10.1029/2005GL025340.

Protat, A., Delanoe, J., Bouniol, D., Heymsfield, A.J., Bansemer, A., and Brown, P. (2007) Evaluation of ice water content retrievals from cloud Radar reflectivity and temperature using a large airborne in situ microphysical database. *J. Appl. Meteorol. Climatol.*, **46**, 557–572.

Protat, A., Bouniol, D., Delanoe, J., May, P.T., Plana-Fattori, A., Hasson, A., O'Connor, E., Görsdorf, U., and Heymsfield, A.J. (2009) Assessment of Cloudsat reflectivity measurements and ice cloud properties using ground–based and airborne cloud Radar observations. *J. Atmos. Oceanic Technol.*, **26**, 1717–1741.

Pruppacher, H.R. and Klett, J.D. (1997) *Microphysics of Clouds and Precipitation*, 2nd edn, Kluwer Academic Publishers, Boston, p. 954.

Pueschel, R.F., Overbeck, V.R., Snetsinger, K.G., Russell, P.B., Ferry, G.V., Wilson, J.C., Livingston, J.M., Verma, S., and Fong, W. (1990) Calibration correction of an Active Scattering Spectrometer Probe to account for refractive index of stratospheric aerosol particles: comparison of results with inertial impaction. *Aerosol Sci. Technol.*, **12**, 992–1002.

Purcell, E. and Pennypacker, C.R. (1973) Scattering and absorption of light by nonspherical dielectric grains. *Astrophys. J.*, **186**, 705–714.

Purvis, R., Lewis, A., Carney, R., McQuaid, J., Arnold, S., Methven, J., Barjat, H., Dewey, K., Kent, J., Monks, P., Carpenter, L., Brough, N., Penkett, S., and Reeves, C. (2003) Rapid uplift of nonmethane hydrocarbons in a cold front over central Europe. *J. Geophys. Res.*, **108**, 1–14. doi: 10.1029/2002JD002521.

Puxbaum, H., Caseiro, A., Sánchez–Ochoa, A., Kasper–Giebl, A., Claeys, M., Gelencsér, A., Legrand, M., Preunkert, S., and Pio, C. (2007) Levoglucosan levels at background sites in Europe for assessing the impact of biomass combustion on the European aerosol background. *J. Geophys. Res.*, **112**, D23S05. doi: 10.1029/2006JD008114.

Qu, Z., Goetz, A., and Heidbrecht, K. (2001) High accuracy atmosphere correction for hyperspectral data (HAATCH). Proceedings of the Ninth JPL Airborne Earth

Science Workshop, JPL–Pub 00-18, pp. 373–381.

Quérel, A., Lemaitre, P., Brunel, M., Porcheron, E., and Grëhan, G. (2010) Real–time global interferometric laser imaging for the droplet sizing (ILIDS) algorithm for airborne research. *Meas. Sci. Technol.*, **21**, 015 306. doi: 10.1088/0957–0233/21/1/015 306.

Quinn, P.K. and Coffman, D.J. (1998) Local closure during the First Aerosol Characterization Experiment (ACE 1): aerosol mass concentration and scattering and backscattering coefficients. *J. Geophys. Res.*, **16**, 575–596.

Racette, P., Adler, R., Gasiewski, A., Jackson, D., Wang, J., and Zacharias, D. (1996) An airborne millimeter–wave imaging radiometer for cloud, precipitation, and atmospheric water vapor studies. *J. Atmos. Oceanic Technol.*, **13**, 610–619.

Ram, M., Cain, S., and Taulbee, D. (1995) Design of a shrouded probe for airborne aerosol sampling in a high velocity airstream. *J. Aerosol Sci.*, **26**, 945–962.

Ramana, M.V., Ramanathan, V., Kim, D., Roberts, G.C., and Corrigan, C.E. (2007) Albedo, atmospheric solar absorption and heating rate measurements with stacked UAVs. *Q. J. R. Meteor. Soc.*, **133**, 1913–1931.

Ranz, W. and Wong, J. (1952) Jet impactors for determining the particle size distribution of aerosol particles. *Arch. Ind. Hyg. Occup. Med.*, **5**, 464–477.

Rapp, M., Strelnikova, I., Strelnikov, B., Hoffmann, P., Friedrich, M., Gumbel, J., Megner, L., Hoppe, U.-P., Robertson, S., Knappmiller, S., Wolff, M., and Marsh, D.R. (2010) Rocket–borne in situ measurements of meteor smoke: charging properties and implications for seasonal variation. *J. Geophys. Res.*, **115**, D00I16. doi: 10.1029/2009JD012 725.

Raupach, S. (2009) Cascaded adaptive–mask algorithm for twin–image removal and its application to digital holograms of ice crystals. *Appl. Opt.*, **42**, 287–301.

Raupach, S., Vossing, H., Curtius, J., and Borrman, S. (2006) Digital crossed–beam holography for in situ imaging of atmospheric particles. *J. Pure Appl. Opt.*, **8**, 796–806.

Rawlins, F. (1989) Aircraft measurements of the solar absorption by broken cloud fields: A case study. *Q. J. R. Meteorol. Soc.*, **115**, 365–382.

Read, K., Lewis, A., Salmon, R., Jones, A., and Bauguitte, S. (2006) OH and halogen atom influence on the variability of non–methane hydrocarbons in the Antarctic boundary layer. *Tellus*, **59**, 22–38.

Read, K., Mahajan, A., Carpenter, L., Evans, M., Faria, B., Heard, D., Hopkins, J., Lee, J., Moller, S., Lewis, A., Mendes, L., McQuaid, J., Oetjen, H., Saiz–Lopez, A., Pilling, M., and Plane, J. (2008) Extensive halogen–mediated ozone destruction over the tropical Atlantic Ocean. *Nature*, **453**, 1232–1235.

Read, W.G., Lambert, A., Bacmeister, J., Cofield, R.E., Christensen, L.E., Cuddy, D.T., Daffer, W.H., Drouin, B.J., Fetzer, E., Froidevaux, L., Fuller, R., Herman, R., Jarnot, R.F., Jiang, J.H., Jiang, Y.B., Kelly, K., Knosp, B.W., Kovalenko, L.J., Livesey, N.J., Liu, H.-C., Manney, G.L., Pickett, H.M., Pumphrey, H.C., Rosenlof, K.H., Sabounchi, X., Santee, M.L., Schwartz, M.J., Snyder, W.V., Stek, P.C., Su, H., Takacs, L.L., Thurstans, R.P., Voemel, H., Wagner, P.A., Waters, J.W., Webster, C.R., Weinstock, E.M., and Wu, D.L. (2007) Aura Microwave Limb Sounder upper tropospheric and lower stratospheric H_2O and relative humidity with respect to ice validation. *J. Geophys. Res.*, **112**, D24S35. doi: 10.1029/2007JD008 752.

Redemann, J., Schmid, B., Eilers, J., Kahn, R., Levy, R., Russell, P., Livingston, J., Hobbs, P., Smith, W. Jr.,, and Holben, B. (2005) Suborbital measurements of spectral aerosol optical depth and its variability at sub-satellite grid scales in support of CLAMS, 2001. *J. Atmos. Sci.*, **62**, 993–1007. doi: 10.1175/JAS3387.1.

Redemann, J., Pilewskie, P., Russell, P., Livingston, J., Howard, S., Schmid, B., Pommier, J., Gore, W., Eilers, J., and Wendisch, M. (2006) Airborne measurements of spectral direct aerosol radiative forcing in the Intercontinental chemical Transport Experiment/Intercontinental

Transport and Chemical Transformation of anthropogenic pollution, 2004. *J. Geophys. Res.*, **111**, D14 210. doi: 10.1029/2005JD006 812.

Reeves, C. and Penkett, S. (2003) Measurements of peroxides and what they tell us. *Chem. Rev.*, **103**, 5199–5218.

Reiche, C. and Lasher–Trapp, S. (2010) The minor importance of giant aerosol to precipitation development within small trade wind cumuli observed during RICO. *Atmos. Res.*, **95**, 386–399.

Reichert, L., Andrés Hernández, M., Stöbener, D., Burkert, J., and Burrows, J.P. (2003) Investigation of the effect of water complexes in the determination of peroxy radical ambient concentrations: implications for the atmosphere. *J. Geophys. Res.*, **9**, 4017–4032.

Reid, J.S., Jonsson, H.H., Maring, H.B., Smirnov, A., Savoie, D.L., Cliff, S.S., Reid, E.A., Livingston, J.M., Meier, M.M., Dubovik, O., and Tsay, S.C. (2003) Comparison of size and morphological measurements of coarse mode dust particles from Africa. *J. Geophys. Res.*, **108**, 8593. doi: 10.1029/2002JD002 485.

Reiner, T., Hanke, M., and Arnold, F. (1997) Atmospheric peroxy radical measurements by ion molecule reaction mass spectrometry: a novel analytical method using amplifying chemical conversion to sulfuric acid. *J. Geophys. Res.*, **102**, 1311–1326.

Reiner, T., Hanke, M., Arnold, F., Ziereis, H., and Schlager, H. (1999) Aircraft–borne measurements of peroxy radicals by chemical conversion/ion molecule reaction mass spectrometry: calibration, diagnostics and results. *J. Geophys. Res.*, **104**, 18647–18659.

Reitebuch, O., Werner, C., Leike, I., Delville, P., Flamant, P.H., Cress, A., and Engelbart, D. (2001) Experimental validation of wind profiling performed by the airborne 10 μm–heterodyne Doppler Lidar WIND. *J. Atmos. Oceanic Technol.*, **18**, 1331–1344.

Reitebuch, O., Volkert, H., Werner, C., Dabas, A., Delville, P., Drobinski, P., Flamant, P.H., and Richard, E. (2003) Determination of air flow across the alpine ridge by a combination of airborne Doppler Lidar, routine radio–sounding and numerical simulation. *Q. J. R. Meteorol. Soc.*, **128**, 715–728.

Reitebuch, O., Lemmerz, C., Nagel, E., Paffrath, U., Durand, Y., Endemann, M., Fabre, F., and Chaloupy, M. (2009) The Airborne Demonstrator for the Direct-Detection Doppler Wind Lidar ALADIN on ADM–Aeolus. Part I: instrument design and comparison to satellite instrument. *J. Atmos. Oceanic Technol.*, **26**, 2501–2515. doi: 10.1175/2009JTECHA1309.1.

Rempe, H. (1937) Untersuchungen über die Verbreitung des Blütenstaubes durch die Luftströmungen. *Planta*, **27**, 93–147.

Ren, X., Brune, W., Cantrell, C., Edwards, G., Shirley, T., Metcalf, A., and Lesher, R.L. (2005) Hydroxyl and peroxy radical chemistry in a rural area of central Pennsylvania: observations and model comparisons. *J. Atmos. Chem.*, **52**, 231–257.

Ren, X., Olson, J.R., Crawford, J.H., Brune, W.H., Mao, J., Long, R.B., Chen, Z., Chen, G., Avery, M.A., Sachse, G.W., Barrick, J.D., Diskin, G.S., Huey, L.G., Fried, A., Cohen, R.C., Heikes, B., Wennberg, P.O., Singh, H.B., Blake, D.R., and Shetter, R.E. (2008) HO_x chemistry during INTEX–A 2004: observation, model calculation, and comparison with previous studies. *J. Geophys. Res.*, **113**, D05 310. doi: 10.1029/2007JD009 166.

Reuter, A. and Bakan, S. (1998) Improvements of cloud particle sizing with a 2D–Grey probe. *J. Atmos. Oceanic Technol.*, **15**, 1196–1203.

Rice, A.L., Gotoh, A.A., Ajie, H.O., and Tyler, S.C. (2001) High–precision continuous–flow measurement of delta C–13 and delta D of atmospheric CH_4. *Anal. Chem.*, **73**, 4104–4110.

Richard, E., Flamant, C., Bouttier, F., Baelen, J.V., Champollion, C., Argence, S., Arnault, J., Barthlott, C., Behrendt, A., Bosser, P., Brousseau, P., Chaboureau, J.-P., Corsmeier, U., Cuesta, J., DiGirolamo, P., Hagen, M., Kottmeier, C., Limnaios, P., Masson, F., Pigeon, G., Pointin, Y., Seiti, Y., and Wulfmeyer, V. (2009) La campagne COPS: initiation et cycle de vie de la convection en région montagneuse. *La Météorologie*, **8**, 32–42.

Richards, J. and Jia, X. (2006) *Remote Sensing Digital Image Analysis, An Introduction*, 4th edn, Springer, Berlin, pp. 3–540–25128–6.

Richardson, M. (2009) Making real time measurements of ice nuclei concentrations at upper tropospheric temperatures: extending the capabilities of the continuous flow diffusion chamber. PhD thesis. Colorado State University, Fort Collins, 268 pp.

Richardson, M.S., DeMott, P.J., Kreidenweis, S.M., Petters, M.D., and Carrico, C.M. (2010) Observations of ice nucleation by ambient aerosol in the homogeneous freezing regime. *Geophys. Res. Lett.*, **37**, L04806.

Richter, R. and Muller, A. (2005) De–shadowing of satellite/airborne imagery. *Int. J. Remote Sens.*, **26**, 3137–3148.

Richter, R. and Schläpfer, D. (2002) Geo-atmospheric processing of airborne imaging spectrometry data. Part 2: atmospheric/topographic correction. *Int. J. Remote Sens.*, **23**, 3137–3148.

Ridley, B., Grahek, F., and Walega, J. (1992) A small, high–sensitivity, medium–response ozone detector suitable for measurements from light aircraft. *J. Atmos. Oceanic Technol.*, **9**, 142–148.

Rinne, H.J.I., Delany, A.C., Greenberg, J.-P., and Guenther, A.B. (2000) A true eddy accumulation system for trace gas fluxes using disjunct eddy sampling method. *J. Geophys. Res.*, **105**, 24791–24798.

Rinne, H.J.I., Guenther, A.B., Warneke, C., De Gouw, J.A., and Luxembourg, S.L. (2001) Disjunct dddy covariance technique for trace gas flux measurements. *Geophys. Res. Lett.*, **28**, 3139–3142.

Roach, W. (1961) Some aircraft observations of fluxes of solar radiation in the atmosphere. *Q. J. R. Meteorol. Soc.*, **87**, 346–363.

Robert, C. (2007) Simple, stable, and compact multiple-reflection optical cell for very long optical paths. *Appl. Opt.*, **46**, 5408–5418.

Roberts, D., Bradley, E., Cheung, R., Leifer, I., Dennison, P., and Margolis, J. (2010) Mapping methane emissions from a marine geological deep source using imaging spectrometry. *Remote Sens. Environ.*, **114**, 592–606.

Roberts, G., Day, D., Russell, L., Dunlea, E., Jimenez, J., Tomlinson, J., Collins, D., Shinozuka, Y., and Clarke, A. (2010) Characterization of particle cloud droplet activity and composition in the free troposphere and the boundary layer during INTEX-B. *Atmos. Chem. Phys.*, **10**, 6627–6644.

Roberts, G. and Nenes, A. (2005) A continuous–flow streamwise thermal–gradient CCN chamber for atmospheric measurements. *Aerosol Sci. Technol.*, **39**, 206–221.

Roberts, G.C., Artaxo, P., Zhou, J., Swietlicki, E., and Andreae, M.O. (2002) Sensitivity of CCN spectra on chemical and physical properties of aerosol: A case study from the Amazon Basin. *J. Geophys. Res.*, **107**, 8070. doi: 10.1029/2001JD000 583.

Roberts, J., Hutte, R., Fehsenfeld, F., Albritton, D., and Sievers, R. (1985) Measurements of anthropogenic hydrocarbon concentration ratios in the rural troposphere–discrimination between background and urban sources. *Atmos. Environ.*, **19**, 1945–1950.

Roberts, P. and Hallett, J. (1968) A laboratory study of the ice nucleating properties of some mineral particulates. *Q. J. R. Meteorol. Soc.*, **94**, 25–34.

Robertson, S.D. (1947) Targets for microwave Radar navigation. *Bell Syst. Tech. J.*, **26**, 852–869.

Robinson, J.K., Bollinger, M., and Birks, J. (1999) Luminol/H_2O_2 chemiluminescence detector for the analysis of nitric oxide in exhaled breath. *Anal. Chem.*, **71**, 5131–5136.

Rodgers, C. (2000) *Inverse Methods for Atmospheric Sounding*, World Scientific.

Rodi, A.R. and Spyers-Duran, P.A. (1972) Analysis of time response of airborne temperature sensors. *J. Appl. Meteorol.*, **11**, 554–556.

Rogers, D.C. (1994) Detecting ice nuclei with a continuous flow diffusion chamber–some exploratory tests of instrument response. *J. Atmos. Oceanic Technol.*, **11**, 1042–1047.

Rogers, D.C., DeMott, P.J., Kreidenweis, S.M., and Chen, Y. (2001)

A continuous-flow diffusion chamber for airborne measurements of ice nuclei. *J. Atmos. Oceanic Technol.*, **18**, 725–741.

Rogers, D.C. (1988) Development of a continuous flow thermal gradient diffusion chamber for ice nucleation studies. *Atmos. Res.*, **22**, 149–181.

Rogers, R. and Yau, M. (1989) *A Short Course in Cloud Physics*, 3rd edn, Pergamon Press, Oxford.

Rogge, D., Rivard, B., Zhang, J., and Feng, J. (2006) Iterative spectral unmixing for optimization per–pixel endmeber sets. *IEEE Trans. Geosci. Remote*, **44**, 3725–3736.

Roiger, A., Aufmhoff, H., Stock, P., Arnold, F., and Schlager, H. (2011a) An aircraft-borne chemical ionization–ion trap mass spectrometer (CI–ITMS) for fast PAN and PPN measurements. *Atmos. Meas. Tech.*, **4**, 173–188.

Roiger, A., Schlager, H., Schäfler, A., Huntrieser, H., Scheibe, M., Aufmhoff, H., Cooper, O.R., Sodemann, H., Stohl, A., Burkhart, J., Lazzara, M., Schiller, C., Law, K.S., and Arnold, F. (2011b) In situ observation of Asian pollution transported into the Arctic lowermost stratosphere. *Atmos. Chem. Phys.*, **11**, 10975–10994. doi: 10.5194/acp–11–10 975–2011.

Romay, F., Pui, D., Smith, T., Ngo, N., and Vincent, J. (1996) Corona discharge effects on aerosol sampling efficiency. *Atmos. Environ.*, **30**, 2607–2613.

Rosa, B., Bajer, K., Haman, K., and Szoplik, T. (2005) Theoretical and experimental characterization of the ultrafast aircraft thermometer: reduction of aerodynamic disturbances and signal processing. *J. Atmos. Oceanic Technol.*, **22**, 1727–1738.

Rose, D., Gunthe, S., Mikhailov, E., Frank, G., Dusek, U., Andreae, M., and Pöschl, U. (2008) Calibration and measurement uncertainties of a continuous–flow cloud condensation nuclei counter (DMT-CCNC): CCN activation of ammonium sulfate and sodium chloride aerosol particles in theory and experiment. *Atmos. Chem. Phys.*, 1153–1179.

Rosen, J.M. (1971) The boiling point of stratospheric aerosol particles. *J. Appl. Meteorol.*, **10**, 1044–1046.

Rosenberg, P.D., Dean, A.R., Williams, P.I., Minikin, A., Pickering, M.A., and Petzold, A. (2012) Particle sizing calibration with refractive index correction for light scattering optical particle counters and impacts upon PCASP and CDP data collected during the Fennec campaign. *Atmos. Meas. Tech. Discuss.*, **5**, 97–135. doi: 10.5194/amtd–5–97–2012.

Rothermel, J., Cutten, D.R., Hardesty, R.M., Menzies, R.T., Howell, J.N., Johnson, S.C., Tratt, D.M., Olivier, L.D., and Banta, R.M. (1998) The multi–center airborne coherent atmospheric wind sensor. *Bull. Am. Meteorol. Soc.*, **79**, 581–599.

Rotta, J.C. (1972) *Turbulente Strömungen. Eine Einführung in die Theorie und ihre Anwendung*, Teubner, Stuttgart, 267 pp.

Rudolph, J., Muller, K.P., and Koppmann, R. (1990) Sampling of organic volatiles in the atmosphere at moderate and low pollution levels. *Anal. Chim. Acta*, **236**, 197–211.

Ruehl, C.R., Chuang, P.Y., and Nenes, A. (2009) Distinct CCN activation kinetics above the marine boundary layer along the California coast. *Geophys. Res. Lett.*, **36**, L15 814. doi: 10.1029/2009GL038 839.

Ruskin, R. (1965) Measurements of water–ice budget changes at −5°C in AgI seeded tropical cumulus. *J. Appl. Meteorol.*, **6**, 72–81.

Russel, P., Swissler, T., and McCormick, M. (1979) Methodology for error analysis and simulation of Lidar aerosol measurements. *Appl. Opt.*, **18**, 3783–3797.

Russell, L.M., Flagan, R.C., and Seinfeld, J.H. (1995) Asymmetric instrument response resulting from mixing effects in accelerated DMA–CPC measurements. *Aerosol Sci. Technol.*, **23**, 491–509.

Russell, P., Livingston, J., Dutton, E., Pueschel, R., Reagan, J., DeFoor, T., Box, M., Allen, D., Pilewskie, P., Herman, B., Kinne, S., and Hofmann, D. (1993) Pinatubo and pre–Pinatubo optical–depth spectra: Mauna Loa measurements, comparisons, inferred particle size distributions, radiative effects, and relationship to Lidar data. *J. Geophys. Res.*, **98**, 22969–22985.

Russell, P., Livingston, J., Pueschel, R., Pollack, J., Brooks, S., Hamill, P., Hughes, J., Thomason, L., Stowe, L., Deshler, T., Dutton, E., and Bergstrom, R. (1996) Global to microscale evolution of the Pinatubo volcanic aerosol, derived from diverse measurements and analyses. *J. Geophys. Res.*, **101**, 18745–18763.

Russell, P.B., Bergstrom, R.W., Shinozuka, Y., Clarke, A.D., de Carlo, P.F., Jimenez, J.L., Livingston, J.M., Redemann, J., Dubovik, O., and Strawa, A. (2010) Absorption Angstrom exponent in AERONET and related data as an indicator of aerosol composition. *Atmos. Chem. Phys.*, **10**, 1155–1169. doi: 10.5194/acp–10–1155–2010.

Ryerson, T.B., Buhr, M., Frost, G., Goldan, P., Holloway, J., Hubler, G., Jobson, B., Kuster, W., McKeen, S., Parrish, D., Roberts, J., Sueper, D., Trainer, M., Williams, J., and Fehsenfeld, F. (1998) Emissions lifetimes and ozone formation in power plant plumes. *J. Geophys. Res.*, **103**, 569–584. doi: 10.1029/98JD01620.

Ryerson, T.B., Huey, L.G., Knapp, K., Neuman, J.A., Parrish, D.D., Sueper, D.T., and Fehsenfeld, F.C. (1999) Design and initial characterization of an inlet for gas–phase NO_y measurements from aircraft. *J. Geophys. Res.*, **104**, 5483–5492.

Ryerson, T.B., Williams, E.J., and Fehsenfeld, F.C. (2000) An efficient photolysis system for fast–response NO_2 measurements. *J. Geophys. Res.*, **105**, 447–461.

Ryzhkov, A.V., Giangrande, S., and Schuurand, V.M.T.J. (2005) Calibration issues of dual–polarization Radar measurements. *J. Atmos. Oceanic Technol.*, **22**, 1138–1155.

Sadanaga, Y., Yoshino, A., Watanabe, K., Yoshioka, A., Wakazono, Y., Kanaya, Y., and Kajii, Y. (2004) Development of a measurement system of peroxy radicals using a chemical amplification/laser–induced fluorescence technique. *Rev. Sci. Instrum.*, **75**, 864–872.

Sagalyn, R.C. and Faucher, G. (1955) Investigation of charged nuclei in the free atmosphere. *Geofis. Pur. Appl.*, **31**, 182.

Saito, T., Yokouchi, Y., and Kawamura, K. (2000) Distributions of C_2–C_6 hydrocarbons over the western North Pacific and eastern Indian Ocean. *Atmos. Environ.*, **34**, 4373–4381.

Salawitch, R., Canty, T., Kurosu, T., Chance, K., Liang, Q., da Silva, A., Pawson, S., Nielsen, J., Rodriguez, J., Bhartia, P., Liu, X., Huey, L., Liao, J., Stickel, R., Tanner, D., Dibb, J., Simpson, W., Donohoue, D., Weinheimer, A., Flocke, F., Knapp, D., Montzka, D., Neuman, J., Nowak, J., Ryerson, T., Oltmans, S., Blake, D., Atlas, E., Kinnison, D., Tilmes, S., Pan, L., Hendrick, F., Van Roozendael, M., Kreher, K., Johnston, P., Gao, R., Johnson, B., Bui, T., Chen, G., Pierce, R., Crawford, J., and Jacob, D. (2010) A new interpretation of total column BrO during Arctic spring. *Geophys. Res. Lett.*, **37**, L21 805. doi: 10.1029/2010GL043 798.

Salisbury, G., Monks, P.S., Bauguitte, S., Bandy, B.J., and Penkett, S.A. (2002) A seasonal comparison of the ozone photochemistry in clean and polluted air masses at Mace Head, Ireland. *J. Atmos. Chem.*, **41**, 163–187.

Sander, S., Friedl, R.R., Golden, D.M., Kurylo, M.J., Moortgat, G.K., Keller-Rudek, H., Wine, P.H., Ravishankara, A.R., Kolb, C.E., Molina, M.J., Finlayson-Pitts, B.J., Huie, R.E., and Orkin, V. (2006) Chemical kinetics and photochemical data for use in atmospheric studies, Jet Propulsion Laboratory Publ. 06–2, evaluation No. 15.

Sankar, S. and Bachalo, W. (1992) Trajectory dependent scattering in phase doppler interferometry: minimizing and eliminating sizing errors. Proceedings of the 6th Annual Symposium on Applications of Laser Techniques to Fluid Mechanics, Lisbon, Portugal.

Sankar, S., Weber, B., Kamemoto, D., and Bachalo, W. (1990) Sizing fine particles with the phase doppler interferometric technique. 2nd International Congress on Optical Particle Sizing, Arizona State University, Tempe, Arizona.

Sarabandi, K., Oh, Y., and Ulaby, F.T. (1992) Measurement and calibration of differential Muller matrix of distributed targets. *IEEE Trans. Antenn. Prop.*, **40**, 1524–1532.

Sassen, K. (1991) The polarization Lidar technique for cloud research–A review and current assessment. *Bull. Am. Meteorol. Soc.*, **72**, 1848–1866.

Sassen, K. and Cho, B.S. (1992) Subvisual thin cirrus LIDAR dataset for satellite verification and climatological research. *J. Appl. Meteorol.*, **31**, 1275–1285.

Sassen, K. and Liao, L. (1996) Estimation of cloud content by W–band Radar. *J. Appl. Meteorol.*, **35**, 932–938.

Sassen, K. and Liou, K. (1979) Scattering of polarized laser light by water droplet, mixed–phase and ice crystal clouds. Part I: angular scattering patterns. *J. Atmos. Sci.*, **5**, 838–851.

Sassen, K., Mace, G.G., Wang, Z., Poellot, M.R., Sekelsky, S.M., and McIntosh, R.E. (1999) Continental stratus clouds: A case study using coordinated remote sensing and aircraft measurements. *J. Atmos. Sci.*, **56**, 2345–2358.

Sato, K., Okamoto, H., Yamamoto, M.K., Fukao, S., Kumagai, H., Ohno, Y., Horie, H., and Abo, M. (2009) 95 GHz Doppler Radar and Lidar synergy for simultaneous ice microphysics and in–cloud vertical air motion retrieval. *J. Geophys. Res.*, 114.

Sato, K. and Okamoto, H. (2006) Characterization of Ze and LDR of nonspherical and inhomogeneous ice particles for 95 GHz cloud Radar: its implication to microphysical retrievals. *J. Geophys. Res.*, **111**, D22 213. doi: 10.1029/2005JD006 959.

Saunders, R., Brogniez, G., Buriez, J., Meerkötter, R., and Wendling, P. (1992) A comparison of measured and modeled broadband fluxes from aircraft data during the ICE '89 field experiment. *J. Atmos. Oceanic Technol.*, **9**, 391–406.

Sauvageot, H. (1992) *Radar Meteorology*, Artech House.

Savitzky, A. and Golay, M.J.E. (1964) Smoothing and differentiation of data by simplified least squares procedures. *Anal. Chem.*, **36**, 1627–1639.

Saw, E.W., Shaw, R.A., Ayyalasomayajula, S., Chuang, P.Y., and Gylfason, A. (2008) Inertial clustering of particles in high-Reynolds-number turbulence. *Phys. Rev. Lett.*, **100**, 214 501.

Saxena, V.K. and Carstens, J.C. (1971) On the operation of cylindrical thermal diffusion cloud chambers. *J. Rech. Atmospher.*, **5**, 11–23.

Sayres, D., Moyer, E.J., Hanisco, T.F., Clair, J.M.S., Keutsch, F.N., Brien, A.O., Allen, N.T., Lapson, L., Demusz, J.N., Rivero, M., Martin, T., Greenberg, M., Tuozzolo, C., Engel, G.S., Kroll, J.H., Paul, J.B., and Anderson, J.G. (2009a) A new cavity based absorption instrument for detection of water isotopologues in the upper troposphere and lower stratosphere. *Rev. Sci. Instrum.*, **80**, 044 102. doi: 10.1063/1.3117 349.

Sayres, D.S., Moyer, E.J., Hanisco, T.F., St Clair, J.M., Keutsch, F.N., O'Brien, A., Allen, N.T., Lapson, L., Demusz, J.N., Rivero, M., Martin, T., Greenberg, M., Tuozzolo, C., Engel, G.S., Kroll, J.H., Paul, J.B., and Anderson, J.G. (2009b) A new cavity based absorption instrument for detection of water isotopologues in the upper troposphere and lower stratosphere. *Rev. Sci. Instrum.*, 80.

Schaepman, M., Ustin, S., Plaza, A., Painter, T., Verrelst, J., and Liang, S. (2009) Earth system science related imaging spectroscopy–An assessment. *Remote Sens. Environ.*, **113**, 123–137.

Schauffler, S.M., Atlas, E.L., Flocke, F., Lueb, R.A., Stroud, V., and Travnicek, W. (1998) Measurements of bromine containing organic compounds at the tropical tropopause. *Geophys. Res. Lett.*, **25**, 317–320.

Scheuer, E., Talbot, R.W., Dibb, J.E., Seid, G.K., DeBell, L., and Lefer, B. (2003) Seasonal distributions of fine aerosol sulfate in the North American Arctic basin during TOPSE. *J. Geophys. Res.*, **108**, 8370. doi: 10.1029/2001JD001 364.

Schiller, C., Krämer, M., Afchine, A., Spelten, N., and Sitnikov, N. (2008) The ice water content of Arctic, midlatitude and tropical cirrus. *J. Geophys. Res.*, **113**, D24 208. doi: 10.1029/2008JD010 342.

Schkolnik, G. and Rudich, Y. (2006) Detection and quantification of levoglucosan in atmospheric aerosol particles: A review. *Anal. Bioanal. Chem.*, **385**, 26–33.

Schlager, H. and Arnold, F. (1987a) Balloon–borne composition measurements of stratospheric negative ions and inferred sulfuric acid vapor abundances during the MAP/GLOBUS campaign. *Planet. Space Sci.*, **35**, 693–701.

Schlager, H. and Arnold, F. (1987b) On stratospheric acetonitrile detection by passive chemical ionization mass spectrometry. *Planet. Space Sci.*, **35**, 715–725.

Schlager, H., Konopka, P., Schulte, P., Schumann, U., Ziereis, H., Arnold, F., Klemm, M., Hagen, D., Whitefield, P., and Ovarlez, J. (1997) In–situ observations of air traffic emission signatures in the North Atlantic flight corridor. *J. Geophys. Res.*, **102**, 10739–10750.

Schläpfer, D. and Richter, R. (2002) Geo-atmospheric processing of airborne imaging spectrometry data. Part 1: parametric ortho-rectification process. *Int. J. Remote Sens.*, **23**, 2609–2630.

Schläpfer, D., Borel, C., Keller, J., and Itten, K. (1998) Atmospheric pre–corrected differential absorption techniques to retrieve columnar water vapor. *Remote Sens. Environ.*, **65**, 353–366.

Schläpfer, D., McCubbin, I., Kindel, B., Kaiser, J., and Ben–Dor, E. (2006) Wildfire smoke analysis using the 760 nm oxygen absorption feature, in *Imaging Spectroscopy: New Quality in Environmental Studies* (eds A. Zagajewski and M. Sobczak), European Association of Remote Sensing Laboratories (EARSeL), pp. 423–435.

Schlienger, J., Pfau, A., Kalfas, A.I., and Abhari, R.S. (2002) Single pressure transducer probe for 3D flow measurements. 16th Symposium on Measuring Techniques in Transonic and Supersonic Flow in Cascades and Turbomachines, Cambridge, UK.

Schmale, J., Schneider, J., Jurkat, T., Voigt, C., Kalesse, H., Rautenhaus, M., Lichtenstern, M., Schlager, H., Ancellet, G., Arnold, F., Gerding, M., Mattis, I., Wendisch, M., and Borrmann, S. (2010) Aerosol layers from the 2008 eruptions of Mount Okmok and Mount Kasatochi: in situ upper troposphere and lower stratosphere measurements of sulfate and organics over Europe. *J. Geophys. Res.*, **115**, D00L07. doi: 10.1029/2009JD013 628.

Schmid, B. and Wehrli, C. (1995) Comparison of Sun photometer calibration by use of the Langley technique and the standard lamp. *Appl. Opt.*, **34**, 4501–4512.

Schmid, B., Thome, K., Demoulin, P., Peter, R., Matzler, C., and Sekler, J. (1996) Comparison of modeled and empirical approaches for retrieving columnar water vapor from solar transmittance measurements in the 0.94 μm region. *J. Geophys. Res.*, **101**, 9345–9358.

Schmid, B., Redemann, J., Russell, P., Hobbs, P., Hlavka, D., McGill, M., Holben, B., Welton, E., Campbell, J., Torres, O., Kahn, R., Diner, D., Helmlinger, M., Chu, D., Robles-Gonzalez, C., and de Leeuw, G. (2003) Coordinated airborne, spaceborne, and ground–based measurements of massive, thick aerosol layers during the dry season in Southern Africa. *J. Geophys. Res.*, **108**, 8496. doi: 10.1029/2002JD002 297.

Schmid, B., Ferrare, R., Flynn, C., Elleman, R., Covert, D., Strawa, A., Welton, E., Turner, D., Jonsson, H., Redemann, J., Eilers, J., Ricci, K., Hallar, A., Clayton, M., Michalsky, J., Smirnov, A., and Barnard, J. (2006) How well do state-of-the-art techniques measuring the vertical profile of tropospheric aerosol extinction compare? *J. Geophys. Res.*, **111**, 25.

Schmidt, G. (2010) INS/GPS technology trends. Paper 1. Low–Cost Navigation Sensors and Integration Technology, NATO Research and Technology Organisation, RTO–EN–SET–116 (2010) AC/323(SET–116)TP/311,ISBN 978-92-837-0109-5, *ftp://ftp.rta.nato.int//PubFullText/RTO/EN/RTO-EN-SET-116(2010)///EN-SET-116(2010)-01.pdf.*

Schmidt, G. and Phillips, R. (2010) INS/GPS Integration Architectures, and INS/GPS Integration Architecture Performance Comparisons, NATO Research and Technology Organisation, RTO–EN–SET–116, Low–Cost Navigation Sensors and Integration Technology.

Schmidt, K.S. and Pilewskie, P. (2011) Airborne measurements of spectral shortwave radiation in cloud and aerosol remote sensing and energy budget studies, in *Light Scattering Reviews*, vol. 6, (ed. A. Kokhanovsky), Springer.

Schmidt, K.S., Feingold, G., Pilewskie, P., Jiang, H., Coddington, O., and Wendisch, M. (2009) Irradiance in polluted cumulus fields: measured and modeled cloud–aerosol effects. *Geophys. Res. Lett.*, **36**, L07 804. doi: 10.1029/2008GL036 848.

Schmidt, K.S., Pilewskie, P., Bergstrom, R., Coddington, O., Redemann, J., Livingston, J.M., Russell, P., Bierwirth, E., Wendisch, M., Gore, W., Dubey, M.K., and Mazzoleni, C. (2010) A new method for deriving aerosol solar radiative forcing and its first application within MILAGRO/INTEX–B. *Atmos. Chem. Phys.*, **10**, D00J22. doi: 10.1029/2009JD013 124.

Schmidt, S., Lehmann, K., and Wendisch, M. (2004) Minimizing instrumental broadening of the drop size distribution with the M–Fast–FSSP. *J. Atmos. Ocean. Technol.*, **21**, 1855–1867.

Schmitt, C. and Heymsfield, A. (2009) The size distribution and mass–weighted terminal velocity of low–latitude tropopause cirrus crystal populations. *J. Atmos. Sci.*, **66**, 2013–2028.

Schneider, T.L. and Stephens, G.L. (1995) Theoretical aspects of modeling backscattering by Cirrus ice particles at millimeter wavelengths. *J. Atmos. Sci.*, **52**, 4367–4385.

Schnell, R.C. (1984) Arctic haze and the Arctic Gas and Aerosol Sampling Program (AGASP). *Geophys. Res. Lett.*, **11**, 361–364.

Schotland, R.M. (1974) Errors in the Lidar measurement of atmospheric gases by differential absorption. *J. Appl. Meteorol.*, **13**, 71–77.

Schott, J. (2007) *Remote Sensing: The Image Chain Approach*, 2nd edn, Oxford University Press, New York.

Schowengerdt, R. (1997) *Remote Sensing: Models and Methods for Image Processing*, 2nd edn, Academic Press, Elsevier Pub., New York, 509 pp.

Schreder, J., Grobner, J., Los, A., and Blumthaler, M. (2004) Intercomparison of monochromatic source facilities for the determination of the relative spectral response of erythemal broadband filter radiometers. *Opt. Lett.*, **13**, 1455–1457.

Schröder, F. and Ström, J. (1997) Aircraft measurements of sub micrometer aerosol particles (> 7 nm) in the midlatitude free troposphere and tropopause region. *Atmos. Res.*, **44**, 333–356.

Schröter, M., Bange, J., and Raasch, S. (2000) Simulated airborne flux measurements in a LES generated convective boundary layer. *Bound.–Lay. Meteorol.*, **95**, 437–456.

Schüller, L., Brenguier, J.-L., and Pawlowska, H. (2003) Retrieval of microphysical, geometrical, and radiative properties of marine stratocumulus from remote sensing. *J. Geophys. Res.*, 108. doi: 10.1029/2002JD002 680.

Schulte, P. and Schlager, H. (1996) In–flight measurements of cruise altitude nitric oxide emission indices of commercial jet aircraft. *Geophys. Res. Lett.*, **23**, 165–168.

Schultz, M., Heitlinger, M., Mihelcic, D., and Volz–Thomas, A. (1995) A calibration source for peroxy radicals with built–in actinometry using H_2O and O_2 photolysis at 185 nm. *J. Geophys. Res.*, **100**, 811.

Schumann, U. and Huntrieser, H. (2007) The global lightning-induced nitrogen oxides source rate. *Atmos. Chem. Phys.*, **7**, 3823–3907.

Schumann, U. (2012) Atmospheric Physics, in *Background-Methods-Tools*, Springer, Heidelberg, p. 877.

Schuster, B.G. and Knollenberg, R. (1972) Detection and sizing of small particles in an open cavity gas laser. *Appl. Opt.*, **11**, 1515–1520.

Schwab, J.J. and Anderson, J. (1982) Oscillator–strengths of Cl(I) in the vacuum ultraviolet – The 2 d–2p transitions. *J. Quant. Spectrosc. Radiat. Transfer*, **27**, 445–457.

Schwarz, J.P., Gao, R., Fahey, D., Thomson, D., Watts, L., Wilson, J.C., Reeves, J., Darbeheshti, M., Baumgardner, D., Kok, G., Chung, S., Schulz, M., Hendricks, J., Lauer, A., Kärcher, B., Slowik, J., Rosenlof, K., Thompson, T., Langford, A.,

Loewenstein, M., and Aikin, K. (2006) Single–particle measurements of midlatitude black carbon and light–scattering aerosol particles from the boundary layer to the lower stratosphere. *J. Geophys. Res.*, **111**, D16 207. doi: 10.1029/2006JD007 076.

Schwarz, J.P., Spackman, J.R., Fahey, D.W., Gao, R.S., Lohmann, U., Stier, P., Watts, L.A., Thomson, D.S., Lack, D.A., Pfister, L., Mahoney, M.J., Baumgardner, D., Wilson, J.C., and Reeves, J.M. (2008) Coatings and their enhancement of black carbon light absorption in the tropical atmosphere. *J. Geophys. Res.*, **113**, D03203.

Schwarz, J.P., Spackman, J.R., Gao, R.S., Watts, L.A., Stier, P., Schulz, M., Davis, S.M., Wofsy, S.C., and Fahey, D.W. (2010) Global–scale black carbon profiles observed in the remote atmosphere and compared to models. *Geophys. Res. Lett.*, 37. doi: 10.1029/2010GL044 372.

Schwarzenboeck, A., Mioche, G., Armetta, A., Herber, A., and Gayet, J.-F. (2009) Response of the Nevzorov hot wire probe in clouds dominated by droplet conditions in the drizzle size range. *Atmos. Meas. Tech.*, **2**, 779–788. *www.atmos–meas–tech.net/2/779/2009/.*

Schwiesow, R.L. and Spowart, M.P. (1996) The NCAR airborne infrared Lidar system: status and applications. *J. Atmos. Oceanic Technol.*, **13**, 4–15.

Scialom, G., Protat, A., and Lemaitre, Y. (2003) Retrieval of kinematic fields from dual–beam airborne Radar data gathered in circular trajectories during the FASTEX experiment. *J. Atmos. Oceanic Technol.*, **20**, 630–646.

Scott, N., Chedin, A., Armante, R., Francis, J., Stubenrauch, C., Charboureau, J.-P., Chevallier, F., Claud, C., and Cheruy, F. (1999) Characteristics of the TOVS pathfinder path–B dataset. *Bull. Am. Meteorol. Soc.*, **80**, 2679–2701.

Secker, J., Staenz, K., Gauthier, R., and Budkewitsch, P. (2001) Vicarious calibration of airborne hyperspectral sensors in operational environments. *Remote Sens. Environ.*, **76**, 81–92.

Seebaugh, W. and Wilson, J.C. (1999) High–Volume Low–Turbulence Inlet for Aerosol Sampling From Aircraft. *Tech. Rep.*, Annual Report to the National Science Foundation Division of Atmospheric Chemistry, *http://www.engr.du.edu/aerosol/ATM-9713408%20Progress%201999.pdf.*

Seifert, M., Tiede, R., Schnaiter, M., Linke, C., Möhler, O., Schurath, U., and Strom, J. (2004) Operation and performance of a differential mobility particle sizer and a TSI 3010 condensation particle counter at stratospheric temperatures and pressures. *J. Aerosol Sci.*, **35**, 981–993.

Seinfeld, J.H. and Pandis, S.N. (1998) *Atmospheric Chemistry and Physics: From Air Pollution to Climate Change*, John Wiley & Sons, Inc., 1326 pp.

Seliga, T.A. and Bringi, V.N. (1976) Potential use of Radar differential reflectivity meanurements at orthogonal polarization for measuring precipitation. *J. Appl. Meteorol.*, **39**, 1341–1372.

Sellar, R.G. and Boreman, G.D. (2005) Classification of imaging spectrometers for remote sensing applications. *Opt. Eng.*, **44**, 013602–013603.

Sellegri, K., Umann, B., Hanke, M., and Arnold, F. (2005) Deployment of a ground–based CIMS apparatus for the detection of organic gases in the boreal forest during the QUEST campaign. *Atmos. Chem. Phys.*, **5**, 357–372.

Sellers, W. (1965) *Physical Climatology*, The University of Chicago Press, Chicago, IL.

Shaw, G. (1975) The vertical distribution of tropospheric aerosol particles at Barrow, Alaska. *Tellus*, **27**, 39–49.

Shaw, G. (1983) Sun photometry. *Bull. Am. Meteorol. Soc.*, **64**, 4–10.

Sheih, C.M., Tennekes, H., and Lumley, J.L. (1971) Airborne hot–wire measurements of the small–scale structure of atmospheric turbulence. *Phys. Fluids*, **14**, 201–215.

Sheridan, P., Arnott, W., Ogren, J., Andrews, E., Atkinson, D., Covert, D., Moosmüller, H., Petzold, A., Schmid, B., Strawa, A., Varma, R., and Virkkula, A. (2005) The Reno aerosol optics study: an evaluation of aerosol absorption measurement methods. *Aerosol Sci. Technol.*, **39**, 1–16.

Shetter, R. and Müller, M. (1999) Photolysis frequency measurements using actinic flux spectroradiometry during the

PEM-Tropics mission: instrumentation description and some results. *J. Geophys. Res.*, **104**, 5647–5661.

Shetter, R., Cinquini, L., Lefer, B., and Madronich, S. (2003) Comparison of airborne measured and calculated spectral actinic flux and derived photolysis frequencies during the PEM Tropics B mission. *J. Geophys. Res.*, 108. doi: 10.1029/2001JD001 320.

Shettle, P.E. and Fenn, R.W. (1979) Models for the Aerosol Particles of the Lower Atmosphere and the Effects of Humidity Variations on Their Optical Properties. *Tech. Rep.* 79 0214, Research Papers No. 676, Air Force Cambridge Research Laboratory, Hanscom Airforce Base, MA, 100 pp.

Shifrin, K., Shifrin, Y., and Mikulinski, I. (1984) Light scattering by an ensemble of large particles of arbitrary shape. *Doklady Akad. Nauk SSSR*, **227**, 582–585.

Shimoni, M., Heremans, R., van der Meer, F., and Acheroy, M. (2007) Measuring pollutant gases using VNIR and TIR imaging. Proceedings 5th EARSeL Workshop on Imaging Spectroscopy, Bruges, Belgium, p. 12.

Shiobara, M. and Asano, S. (1994) Estimation of cirrus optical thickness from sun photometer measurements. *J. Appl. Meteorol.*, **33**, 672–681.

Shiobara, M., Spinhirne, J., Uchiyama, A., and Asano, S. (1991) Optical depth measurements of aerosol, cloud and water vapor using sun photometers during FIRE Cirrus IFO II. *J. Appl. Meteorol.*, **35**, 36–46.

Shipley, S.T., Tracy, D.H., Eloranta, E.W., Trauger, J.T., Sroga, J.T., Roesler, F.L., and Weinman, J.A. (1983) High spectral resolution Lidar to measure optical scattering properties of atmospheric aerosol particles. 1: theory and instrumentation. *Appl. Opt.*, **22**, 3716–3724.

Shiraiwa, M., Kondo, Y., Moteki, N., Takegawa, N., Sahu, L.K., Takami, A., Hatakeyama, S., Yonemura, S., and Blake, D.R. (2008) Radiative impact of mixing state of black carbon aerosol in Asian outflow. *J. Geophys. Res.*, **113**, D24 210. doi: 10.1029/2008JD010 546.

Shulman, M., Jacobson, M., Charlson, R., Synovec, R., and Young, T. (1996) Dissolution behavior and surface tension effects of organic compounds in nucleating cloud droplets. *Geophys. Res. Lett.*, **23**, 277–280.

Shupe, M.D., Daniel, J.S., de Boer, G., Eloranta, E., Kollias, P., Luke, E., Long, C., Turner, D., and Verlinde, J. (2008) A focus on mixed–phase clouds: the status of ground–based observational methods. *Bull. Am. Meteorol. Soc.*, **89**, 1549.

Siebert, H. and Muschinski, A. (2001) Relevance of a Tuning-Fork effect for temperature measurements with the Gill Solent HS ultrasonic anemometer–thermometer. *J. Atmos. Oceanic Technol.*, **18**, 1367–1376.

Siebert, H., Wendisch, M., Conrath, T., Teichmann, U., and Heintzenberg, J. (2003) A new tethered balloon–borne payload for fine-scale observations in the cloudy boundary layer. *Bound.–Lay. Meteorol.*, **106**, 461–482.

Siebert, H., Franke, H., Lehmann, K., Maser, R., Saw, E.W., Schell, D., Shaw, R.A., and Wendisch, M. (2006a) Probing finescale dynamics and microphysics of clouds with helicopter-borne measurements. *Bull. Am. Meteorol. Soc.*, **87**, 1727–1738.

Siebert, H., Lehmann, K., and Wendisch, M. (2006b) Observations of small scale turbulence and energy dissipation rates in the cloudy boundary layer. *J. Atmos. Sci.*, **63**, 1451–1466.

Siebert, H., Lehmann, K., and Shaw, R.A. (2007) On the use of a hot–wire anemometer for turbulence measurements in clouds. *J. Atmos. Oceanic Technol.*, **24**, 980–993.

Siebert, H., Shaw, R.A., and Warhaft, Z. (2010) Statistics of small–scale velocity fluctuations and internal intermittency in marine stratocumulus clouds. *J. Atmos. Sci.*, **67**, 262–273.

Simoneit, B.R.T., Kobayashi, M., Mochida, M., Kawamura, K., Lee, M., Lim, H.-J., Turpin, B.J., and Komazaki, Y. (2004) Composition and major sources of organic compounds of aerosol particulate matter sampled during the ACE-Asia campaign. *J. Geophys. Res.*, **109**, D19S10. doi: 10.1029/2004JD004 598.

Simpson, W.R., von Glasow, R., Riedel, K., Anderson, P., Ariya, P., Bottenheim, J., Burrows, J., Carpenter, L.J., Friess, U., Goodsite, M.E., Heard, D., Hutterli, M., Jacobi, H.-W., Kaleschke, L., Neff, B., Plane, J., Platt, U., Richter, A., Roscoe, H., Sander, R., Shepson, P., Sodeau, J., Steffen, A., Wagner, T., and Wolff, E. (2007) Halogens and their role in polar boundary–layer ozone depletion. *Atmos. Chem. Phys.*, **7**, 4375–4418.

Sinkevich, A.A. and Lawson, R.P. (2005) A survey of temperature measurements in convective clouds. *J. Appl. Meteorol.*, **44**, 1133–1145.

Sinnerwalla, A. and Alofs, D. (1973) A cloud nucleus counter with long available growth time. *J. Appl. Meteorol.*, **12**, 831–835.

Sipila, M., Lehtipalo, K., Kulmala, M., Petäjä, T., Junninen, H., Aalto, P.P., Manninen, H.E., Kyrö, E.-M., Asmi, E., Riipinen, I., Curtius, J., Kürten, A., Borrmann, S., and O'Dowd, C.D. (2008) Applicability of condensation particle counters to measure atmospheric clusters. *Atmos. Chem. Phys.*, **8**, 4049–4060.

Sipila, M., Lehtipalo, K., Attoui, M., Neitola, K., Petaja, T., Aalto, P.P., O'Dowd, C.D., and Kulmala, M. (2009) Laboratory verification of PH–CPC's ability to monitor atmospheric sub–3 nm clusters. *Aerosol Sci. Technol.*, **43**, 126–135.

Sipila, M., Berndt, T., Petaja, T., Brus, D., Vanhanen, J., Stratmann, F., Patokoski, J., Mauldin, R.L., Hyvarinen, A.P., Lihavainen, H., and Kulmala, M. (2010) The role of sulfuric acid in atmospheric nucleation. *Science*, **327**, 1243–1246.

Sjostedt, S., Huey, L., Tanner, D., Peischl, J., Chen, G., Dibb, J., Lefer, B., Hutterli, M., Beyersdorf, A., Blake, N., Blake, D., Sueper, D., Ryerson, T., Burkhart, J., and Stohl, A. (2007) Observations of hydroxyl and the sum of peroxy radicals at Summit, Greenland during summer 2003. *Atmos. Environ.*, **41**, 5122–5137.

Skoog, D., West, D., Holler, J., and Crouch, S. (2003) *Fundamentals of Analytical Chemistry*, vol. 8, Brooks/Cole Publishing Company.

Slusher, D., Slusher, D., Huey, L., Tanner, D., Flocke, F., and Roberts, J. (2004) A thermal dissociation–chemical ionization mass spectrometry (TD–CIMS) technique for the simultaneous measurements of peroxyacyl nitrates and dinitrogen pentoxide. *J. Geophys. Res.*, **109**, D19 315. doi: 10.1029/2004JD004 670.

Slusher, D.L., Pitteri, S.J., Haman, B.J., Tanner, D.J., and Huey, L.G. (2001) A chemical ionization technique for measurement of pernitric acid in the upper troposphere and the polar boundary layer. *Geophys. Res. Lett.*, **28**, 3875–3878.

Smit, H., Volz–Thomas, A., Helten, M., Pätz, H., and Kley, D. (2008) An in–flight calibration method for near real–time humidity measurements with the airborne MOZAIC sensor. *J. Atmos. Oceanic Technol.*, **25**, 656–666.

Smith, G.E. (2001) The invention of the CCD. *Nucl. Instrum. Methods Phys. Res.*, **471**, 1–5.

Smith, M., Roberts, D., Shipman, H., Adams, J., Willis, S., and Gillespie, A. (1987) Calibrating AIS images using the surface as a reference, in *Proceedings of the 3rd Airborne Imaging Spectrometer Workshop* (ed. R. Green), University of Washington, N88- 13fdO, pp. 63–69.

Snider, J., Petters, M., Wechsler, P., and Liu, P. (2006) Supersaturation in the Wyoming CCN instrument. *J. Atmos. Oceanic Technol.*, **23**, 1323–1339.

Snider, J.R. and Brenguier, J.-L. (2000) A comparison of cloud condensation nuclei and cloud droplet measurements obtained during ACE-2. *Tellus B*, **52**, 828–842.

Snider, J.R., Guibert, S., Brenguier, J.-L., and Putaud, J.-P. (2003) Aerosol activation in marine stratocumulus clouds: 2. Köhler and parcel theory closure studies. *J. Geophys. Res.*, **108**, 8629. doi: 10.1029/2002JD002 692.

Soderman, P., Hazen, N., and Brune, W. (1991) Aerodynamic design of gas and aerosol samplers for aircraft, *http://ntrs.nasa.gov/*, nASA Technical Memorandum 103854, Report Number: A–91110, NAS 1.15:103854.

Solberg, S., Dye, C., Schmidbauer, N., Herzog, A., and Gehrig, R. (1996a) Carbonyls and nonmethane hydrocarbons at rural European sites from the Mediterranean to the Arctic. *J. Atmos. Chem.*, **25**, 33–66.

Solberg, S., Schmidbauer, N., Semb, A., and Hov, O. (1996b) Boundary layer ozone depletion as seen in the Norwegian Arctic spring. *J. Atmos. Chem.*, **25**, 301–322.

Sonntag, D. (1994) Advancements in the field of hygrometry. *Meteorol. Z.*, **3**, 51–66.

Sorooshian, A., Brechtel, F.J., Ma, Y.L., Weber, R.J., Corless, A., Flagan, R.C., and Seinfeld, J.H. (2006) Modeling and characterization of a particle–into–liquid sampler (PILS). *Aerosol Sci. Technol.*, **40**, 396–409.

Sorooshian, A., Lu, M.L., Brechtel, F.J., Jonsson, H., Feingold, G., Flagan, R.C., and Seinfeld, J.H. (2007) On the source of organic acid aerosol layers above clouds. *Environ. Science Technol.*, **41**, 4647–4654.

Spanel, P., Dryahina, K., and Smith, D. (2007) Microwave plasma ion sources for selected ion flow tube mass spectrometry: optimizing their performance and detection limits for trace gas analysis. *Int. J. Mass Spectrom.*, **267**, 117–124.

Speidel, M., Nau, R., Arnold, F., Schlager, H., and Stohl, A. (2007) Sulfur dioxide measurenments in the lower, middle and upper troposphere: deployment of an aircraft–based chemical ionization mass spectrometer with permanent in–flight calibration. *Atmos. Environ.*, **41**, 2427–2437.

Spek, A.L.J., Unal, C.M.H., Moisseev, D.N., Russchenberg, H., Chandrasekar, V., and Dufournet, Y. (2008) A new technique to categorize and retrieve the microphysical properties of ice particles above the melting layer using Radar dual–polarization spectral analysis. *J. Atmos. Oceanic Technol.*, **25**, 482–497.

Spinhirne, J., Hansen, M., and Caudill, L. (1982) Cloud top remote sensing by airborne Lidar. *Appl. Opt.*, **21**, 1564–1571.

Spinhirne, J., Boers, R., and Hart, W.D. (1989) Cloud top liquid water from Lidar observations of marine stratocumulus. *J. Appl. Meteorol.*, **28**, 81–90.

Spinhirne, J., Hart, W., and Hlavka, D. (1996) Cirrus infrared parameters and shortwave reflectance relations from observations. *J. Atmos. Sci.*, **53**, 1438–1458.

Spinhirne, J., Chudamani, S., Cavanaugh, J., and Bufton, J. (1997) Aerosol and cloud backscatter at 1.06, 1.54, and 0.53 μm by airborne hard–target–calibrated Nd:YAG/methane Raman Lidar. *Appl. Opt.*, **36**, 3475–3490.

Sprung, D., Jost, C., Reiner, T., Hansel, A., and Wisthaler, A. (2001) Acetone and acetonitrile in the tropical Indian Ocean boundary layer and free troposphere: aircraft–based intercomparison of AP–CIMS and PTR–MS measurements. *J. Geophys. Res.*, **106**, 28511–28527.

Spuler, S., Richter, D., Spowart, M., and Rieken, K. (2011) Optical fiber–based laser remote sensor for airborne measurement of wind velocity and turbulence. *Appl. Opt.*, **50**, 842–851.

Spyers–Duran, P. (1968) Comparative measurements of cloud liquid water using heated wire and cloud replicating devices. *J. Appl. Meteorol.*, **7**, 674–678.

Spyers–Duran, P. (1976) Measuring the size, concentration, and structural properties of hydrometeors in clouds with impactor and replicating devices, *Atmospheric Technology*, No. 8, National Center for Atmospheric Research, Boulder, CO, pp. 3–9.

Spyers–Duran, P. (1991) An airborne cryogenic frost point hygrometer, *Seventh Symposium on Meteorological Observation Instrumentation*, American Meteorological Society, pp. 303–306.

Squires, P. and Twomey, S. (1966) A comparison of cloud nucleus measurements over central North America and the Caribbean Sea. *J. Atmos. Sci.*, **22**, 401–404.

Srivastava, R.C., Matejka, T.J., and Lorello, T.J. (1986) Doppler Radar study of the trailing anvil region associated with a squall line. *J. Atmos. Sci.*, **43**, 356–377.

St Clair, J.M., Hanisco, T.F., Weinstock, E., Moyer, E., Sayres, D., Keutsch, F., Kroll, J.H., Demusz, J.N., Allen, N.T., Smith, J., Spackman, J., and Anderson, J.G. (2008) A new photolysis laser–induced fluorescence instrument for the detection of H_2O and HDO in the lower stratosphere. *Rev. Sci. Instrum.*, 79, Art. No. 064 101.

Stackhouse, P. and Stephens, G. (1991) A theoretical and observational study of the radiative properties of cirrus: results from FIRE 1986. *J. Atmos. Sci.*, **48**, 2044–2059.

Staenz, K. (2009) Terrestrial imaging spectroscopy, some future perspectives, *Proceedings of the 6th EARSEL SIG IS*, Tel Aviv, p. 12.

Stallabrass, J. (1978) An Appraisal of the Single Rotating Cylinder Method of Liquid Water Content Measurement. *Tech. Rep.* LTR–LT–92, NRC Canada.

Stankov, B., Cline, D., Weber, B., Gasiewski, A., and Wick, G. (2008) High–resolution airborne polarimetric microwave imaging of snow cover during the NASA cold land processes experiment. *IEEE Trans. Geosci. Remote*, **8**, 1635–1671.

Stark, H., Lerner, B., Schmitt, R., Jakoubek, R., Williams, E., Ryerson, T., Sueper, D., Parrish, D., and Fehsenfeld, F. (2007) Atmospheric in situ measurement of nitrate radical (NO_3) and other photolysis rates using spectroradiometry and filter radiometry. *J. Geophys. Res.*, **112**, D10S04. doi: 10.1029/2006JD007 578.

Stedman, D.H., Daby, E.E., Stuhl, F., and Niki, H. (1972) Analysis of ozone and nitric oxide by a chemiluminescent method in laboratory and atmospheric studies of photochemical smog. *J. Air Pollut. Control Assoc.*, **22**, 260–263.

Steeghs, M.M.L., Sikkens, C., Crespo, E., Cristescu, S.M., and Harren, F.J.M. (2007) Development of a proton–transfer reaction ion trap mass spectrometer: online detection and analysis of volatile organic compounds. *Int. J. Mass Spectrom.*, **262**, 16–24.

Steffenson, D. and Stedman, D. (1974) Optimisation of the parameters of chemiluminescent nitric oxide detectors. *Anal. Chem.*, **46**, 1704–1709.

Steinbacher, M., Dommen, J., Ammann, C., Spirig, C., Neftel, A., and Prevot, A.S.H. (2004) Performance characteristics of a proton–transfer–reaction mass spectrometer (PTR–MS) derived from laboratory and field measurements. *Int. J. Mass Spectrom.*, **239**, 117–128.

Stephens, M., Turner, N., and Sandberg, J. (2003) Particle identification by laser–induced incandescence in a solid–state laser cavity. *Appl. Opt.*, **42**, 3726–3736.

Stetzer, O., Baschek, B., Lüönd, F., and Lohmann, U. (2008) The Zurich Ice Nucleation Chamber (ZINC) – A new instrument to investigate atmospheric ice formation. *Aerosol Sci. Technol.*, **42**, 64–74.

Stevens, A., van Wesemael, B., Bartholomeus, H., Rosillon, D., Tychon, B., and Ben-Dor, E. (2008) Laboratory, field and airborne spectroscopy for monitoring organic carbon content in agricultural soils. *Geoderma*, **144**, 395–440.

Stevens, B., Lenschow, D., Vali, G., Gerber, H., Bandy, A., Blomquist, B., Brenguier, J.-L., Bretherton, C., Burnet, F., Campos, T., Chai, S., Faloona, I., Friesen, D., Haimov, S., Laursen, K., Lilly, D., Loehrer, S., Malinowski, S., Morley, B., Petters, M., Rogers, D., Russell, L., Savic-Jovcic, V., Snider, J., Straub, D., Szumowski, M., Takagi, H., Thornton, D., Tschudi, M., Twohy, C., Wetzel, M., and van Zanten, M. (2003) Dynamics and chemistry of marine stratocumulus - DYCOMS-II. *Bull. Am. Meteorol. Soc.*, **84**, 579–593.

Stewart, D.J., Taylor, C.M., Reeves, C.E., and McQuaid, J.B. (2008) Biogenic nitrogen oxide emissions from soils: impact on NO_x and ozone over west Africa during AMMA (African Monsoon Multidisciplinary Analysis): observational study. *Atmos. Chem. Phys.*, **8**, 2285–2297.

Stickler, A., Fischer, H., Williams, J., de Reus, M., Sander, R., Lawrence, M., Crowley, J., and Lelieveld, J. (2006) Influence of summertime deep convection on formaldehyde in the middle and upper troposphere over Europe. *J. Geophys. Res.*, **111**, D14 308. doi: 10.1029/2005JD007 001.

Stickney, T., Shedlov, M., and Thompson, D. (1994) Total Temperature Sensors. *Tech. Rep.* 5755, Revision C, Rosemount Aerospace, Eagan, MN.

Stimpfle, R.M., Cohen, R., Bonne, G., Voss, P., Perkins, K., Koch, L., Anderson, J., Salawitch, R., Lloyd, S., Gao, R., Negro, L.D., Keim, E., and Bui, T. (1999) The coupling of $ClONO_2$,

ClO, and NO_2 in the lower stratosphere from in situ observations using the NASA ER–2 aircraft. *J. Geophys. Res.*, **104**, 26705–26714.

Stimpfle, R.M., Wilmouth, D., Salawitch, R., and Anderson, J. (2004) First measurements of ClOOCl in the stratosphere: the coupling of ClOOCl and ClO in the Arctic polar vortex. *J. Geophys. Res.*, **109**, D03301. doi: 10.1029/2003JD003811.

Stith, J., Fye, J., Bansemer, A., Heymsfield, A., Grainger, C., Petersen, W., and Cifelli, R. (2002) Microphysical observations of tropical clouds. *J. Appl. Meteorol.*, **41**, 97–117.

Stolzenburg, M. (1988) An ultrafine aerosol size distribution measuring system. PhD thesis. Department of Mechanical Engineering, University of Minnesota.

Stolzenburg, M.R. and McMurry, P.H. (1991) An ultrafine aerosol condensation nucleus counter. *Aerosol Sci. Technol.*, **14**, 48–65.

Stone, R., Herber, A., Vitale, V., Mazzola, M., Lupi, A., Schnell, R., Dutton, E., Liu, P., Li, S.-M., Dethloff, K., Lampert, A., Ritter, C., Stock, M., Neuber, R., and Maturilli, M. (2010) A three-dimensional characterization of Arctic aerosol particles from airborne Sun photometer observations: PAN-ARCMIP, April 2009. *J. Geophys. Res.*, D13203. doi: 10.1029/2009JD013 605.

Stoner, E. and Baumgardner, M.F. (1981) Characteristic variations in reflectance of surface soils. *Soil Sci. Soc. Am. J.*, **45**, 161–1165.

Straka, J.M., Zrnic, D.S., and Ryzhkov, A.V. (2000) Bulk hydrometeor classification and quantification using polarimetric Radar data: synthesis of relations. *J. Appl. Meteorol.*, **39**, 1341–1372.

Strapp, J., Albers, F., Reuter, A., Korolev, A., Maixner, U., Rashke, E., and Vukovic, Z. (2001) Laboratory measurements of the response of a PMS OAP–2DC. *J. Atmos. Oceanic Technol.*, **18**, 1150–1170.

Strapp, J., Oldenburg, J., Ide, R., Lilie, L., Bacic, S., Vukovic, Z., Oleskiw, M., Miller, D., Emery, E., and Leone, G. (2002) Wind tunnel measurements of the response of hot–wire liquid water content instruments to large droplets. *J. Atmos. Oceanic Technol.*, **20**, 791–806.

Strapp, J., Lilie, L., Emery, E., and Miller, D. (2005) Preliminary comparison of ice water content as measured by hot wire instruments of varying configuration. 43rd AIAA Aerospace Sciences Meeting, Reno, NV, aIAA-2005-0860.

Strapp, J., MacLeod, J., and Lilie, L. (2008) Calibration of ice water content in a wind tunnel/engine test cell facility. 15th International Conference on Cloud and Precipitation, Cancun, Mexico.

Strapp, J.W. and Schemenauer, R.S. (1982) Calibration of Johnson-Williams liquid water content meters in a high–speed icing tunnel. *J. Appl. Meteorol.*, **21**, 98–108.

Strapp, J.W., Leaitch, W.R., and Liu, P.S.K. (1992) Hydrated and dried aerosol–size–distribution measurements from the Particle Measuring Systems FSSP–300 probe and the deiced PCASP-100X probe. *J. Atmos. Oceanic Technol.*, **9**, 548–555.

Stratmann, F., Hummes, D., Kauffeldt, T., and Fissan, H. (1995) Convolution and its application to DMA transfer function measurements. *J. Aerosol Sci.*, **26**, S143–S144.

Strawa, A.W., Castaneda, R., Owano, T., Baer, D.S., and Paldus, B.A. (2003) The measurement of aerosol optical properties using continuous wave cavity ring-down techniques. *J. Atmos. Oceanic Technol.*, **20**, 454–465.

Ström, J. and Ohlsson, S. (1988) Real–time measurement of absorbing material in contrail ice using a counterflow virtual impactor. *J. Geophys. Res.*, **103**, 8737–8741.

Su, T. and Chesnavich, W.J. (1982) Parameterization of the ion–polar molecule collision rate–constant by trajectory calculations. *J. Chem. Phys.*, **76**, 5183–5185.

Sugimoto, N. and Minato, A. (1993) Long–path absorption measurement of CO_2 with a Raman–shifted tunable dye laser. *Appl. Opt.*, **32**, 6827–6833.

Sullivan, R.C., Moore, M.J.K., Petters, M.D., Kreidenweis, S.M., Roberts, G.C., and Prather, K.A. (2009) Timescale for hygroscopic conversion of calcite mineral particles through heterogeneous reaction with nitric acid. *PCCP (Phys. Chem. Chem. Phys.)*, **11**, 7826–7837.

Sumner, D. (2000) Calibration methods for a seven–hole pressure probe, in *6th Triennial International Symposium on Fluid Control, Measurement and Visualization (Flucome 2000)* (ed. A. Laneville), Sherbrooke, Canada, p. 6.

Super, A.B., Boe, B.A., Heimbach, J.A., McPartland, J.T., and Langer, G. (2010) Comparison of silver iodide outputs from two different generators and solutions measured by acoustic ice nucleus counters. *J. Weath. Modif.*, **42**, 49–60.

Taipale, R., Ruuskanen, T.M., Rinne, J., Kajos, M.K., Hakola, H., Pohja, T., and Kulmala, M. (2008) Technical Note: Quantitative long–term measurements of VOC concentrations by PTR–MS–measurement, calibration, and volume mixing ratio calculation methods. *Atmos. Chem. Phys.*, **8**, 6681–6698.

Takano, Y. and Liou, K. (1989) Solar radiative transfer in cirrus clouds. Part I: single scattering and optical properties of hexagonal ice crystals. *J. Atmos. Sci.*, **46**, 3–19.

Takano, Y. and Liou, K. (1995) Radiative transfer in cirrus clouds. *J. Atmos. Sci.*, **52**, 818–837.

Talbot, R.W., Dibbl, J.E., Leferl, B.L., Scheuer, E.M., Bradshaw, J.D., Sandholm, S.T., Smyth, S., Blake, D.R., Blake, N.J., Sachse, G.W., Collins, J.E., and Gregory, L. (1997) Large–scale distributions of tropospheric nitric, formic, and acetic acids over the western Pacific basin during wintertime. *J. Geophys. Res.*, **102**, 28303–28313.

Tammet, H., Mirme, A., and Tamm, E. (2002) Electrical aerosol spectrometer of Tartu University. *Atmos. Res.*, **62**, 315–324.

Tanaka, M., Nakajima, T., and Shiobara, M. (1986) Calibration of a sunphotometer by simultaneous measurements of direct–solar and circumsolar radiation. *Appl. Opt.*, **25**, 1170–1176.

Tanelli, S., Durden, S.L., Im, E., Pak, K., Reinke, D., Partain, P., Haynes, J., and Marchand, R. (2008) Cloudsa's Cloud Profiling Radar after 2 years in orbit: performance, external calibration, and processing. *IEEE Trans. Geosci. Remote*, **46**, 3560–3573.

Tang, C. and Aydin, K. (1995) Scattering from ice crystals at 94 and 220 GHz millimeter wave frequencies. *IEEE Trans. Geosci. Remote*, **33**, 93–99.

Tanimoto, H., Aoki, N., Inomata, S., Hirokawa, J., and Sadanaga, Y. (2007) Development of a PTR–TOFMS instrument for real–time measurements of volatile organic compounds in air. *Int. J. Mass Spectrom.*, **263**, 1–11.

Tanré, D., Herman, M., Deschamps, P., and Leffe, A. (1979) Atmospheric modeling for space measurements of ground reflectances, including bidirectional properties. *Appl. Opt.*, **18**, 3587–3594.

Tanré, D., Deschamps, P., Duhaut, P., and Herman, M. (1987) Adjacency effect produced by the atmospheric scattering in thematic mapper data. *J. Geophys. Res.*, **92**, 12000–12006.

Tanré, D., Haywood, J., Pelon, J., Léon, J., Chatenet, B., Formenti, P., Francis, P., Goloub, P., Highwood, E.J., and Myhre, G. (2003) Measurement and modeling of the Saharan dust radiative impact: overview of the Saharan Dust experiment (SHADE). *J. Geophys. Res.*, **108**, 8574. doi: 10.1029/2002JD003 273.

Targ, R., Kavaya, M.J., Huffaker, R.M., and Bowles, R.L. (1991) Coherent Lidar airborne windshear sensor: performance evaluation. *Appl. Opt.*, **30**, 2013–2026.

Targ, R., Steakley, B.C., Hawley, J.G., Ames, L.L., Fomey, P., Swanson, D., Stone, R., Otto, R.G., Zarifis, V., Brockman, P., Calloway, R.S., Klein, S.H., and Robinson, P.A. (1996) Coherent Lidar airborne wind sensor II: flight–test results at 2 and 10 μm. *Appl. Opt.*, **35**, 7117–7127.

Tarnogrodzki, A. (1993) Theoretical prediction of the critical Weber number. *Int. J. Multiphas. Flow*, **19**, 329–336.

Telford, J.W. and Warner, J. (1962) On the measurement from an aircraft of buoyancy and vertical air velocity in cloud. *J. Atmos. Sci.*, **19**, 415–423.

Testud, J., Hildebrand, P.H., and Lee, W.-C. (1995) A procedure to correct airborne Doppler Radar data for navigation errors using the echo returned from the Earth's surface. *J. Atmos. Oceanic Technol.*, **12**, 800–820.

Thiel, S., Ammannato, L., Bais, A., Bandy, B., Blumthaler, M., Bohn, B.,

Engelsen, O., Gobbi, G.P., Gröbner, J., Jäkel, E., Junkermann, W., Kazadzis, S., Kift, R., Kjeldstad, B., Kouremeti, N., Kylling, A., Mayer, B., Monks, P.S., Reeves, C.E., Schallhart, B., Scheirer, R., Schmidt, S., Schmitt, R., Schreder, J., Silbernagl, R., Topaloglou, C., Thorseth, T.M., Webb, A.R., Wendisch, M., and Werle, P. (2008) Influence of clouds on the spectral actinic flux density in the lower troposphere (INSPECTRO): overview of the field campaigns. *Atmos. Chem. Phys.*, **8**, 1789–1812. doi: 10.5194/acp–8–1789–2008.

Thomason, L., Herman, B., and Reagan, J. (1983) The effect of atmospheric attenuators with structured vertical distributions on air mass determinations and Langley plot analyses. *J. Atmos. Sci.*, **40**, 1851–1854.

Thompson, B. (1974) Holographic particle sizing techniques. *J. Phys. E*, **7**, 781–788.

Thomson, D.S., Schein, M.E., and Murphy, D.M. (2000) Particle analysis by laser mass spectrometry WB–57F instrument overview. *Aerosol Sci. Technol.*, **33**, 153–169.

Thornberry, T., Murphy, D.M., Thomson, D.S., de Gouw, J., Warneke, C., Bates, T.S., Quinn, P.K., and Coffman, D. (2009) Measurement of aerosol organic compounds using a novel collection/thermal–desorption PTR–ITMS instrument. *Aerosol Sci. Technol.*, **43**, 486–501.

Thornton, B.F., Toohey, D., Avallone, L., Harder, H., Martinez, M., Simpas, J., Brune, W., and Avery, M. (2003) In situ observations of ClO near the winter polar tropopause. *J. Geophys. Res.*, **108**, 8333.

Thornton, B.F., Toohey, D., Avallone, L., Hallar, A., Harder, H., Martinez, M., Simpas, J., Brune, W., Koike, M., Kondo, Y., Takegawa, N., Anderson, B., and Avery, M. (2005) Variability of active chlorine in the lowermost Arctic stratosphere. *J. Geophys. Res.*, **110**, D22 304.

Thornton, J., Wooldridge, P., and Cohen, R. (2000) Atmospheric NO_2: in situ laserinduced fluorescence detection at parts per trillion mixing ratios. *Anal. Chem.*, **72**, 528–539.

Thurai, M., Huang, G.J., and Bringi, V.N. (2007) Drop shapes, model comparisons, and calculations of polarimetiric Radar parameters in rain. *J. Atmos. Oceanic Technol.*, **24**, 1019–1032.

Tian, L., Heymsfield, G.M., Li, L., and Srivastava, R. (2007) Properties of light stratiform rain derived from 10 and 94 GHz airborne Doppler Radars measurements. *J. Geophys. Res.*, **112**, D11 211. doi: 10.1029/2006JD008 144.

Timko, M., Yu, Z., Kroll, J., Jayne, J., Worsnop, D., Miake–Lye, R., Onasch, T., Liscinsky, D., Kirchstetter, T., Destaillats, H., Holder, A., Smith, J., and Wilson, K. (2009) Sampling artifacts from Conductive Silicone Tubing. *Aerosol Sci. Technol.*, **43**, 855–865. doi: 10.1080/02786820902984 811.

Tinel, C., Testud, J., Pelon, J., Protat, R.H.A., Delanoë, J., and Bouniol, D. (2005) The retrieval of ice–cloud properties from cloud Radar and Lidar synergy. *J. Appl. Meteorol.*, **44**, 860–875.

Tokay, A., Hartmann, P., Battaglia, A., Gage, K.S., Clark, W.L., and Williams, C.R. (2009) A field study of reflectivity and Z–R relations using vertically pointing Radars and disdrometers. *J. Atmos. Oceanic Technol.*, **26**, 1120–1134.

Toohey, D.W., Avallone, L., Allen, N., Demusz, J., Hazen, J., Hazen, N., and Anderson, J. (1993) The performance of a new instrument for in–situ measurements of ClO in the lower stratosphere. *Geophys. Res. Lett.*, **20**, 1791–1794.

Torgeson, W.L. and Stern, S. (1966) An aircraft impactor for determining the size distributions of tropospheric aerosol. *J. Appl. Meteorol.*, **5**, 205–210.

Trenberth, K.E. and Fasullo, J.T. (2011) Tracking Earth's energy: from El Niño to global warming. *Surv. Geophys.*, 311–323. doi: 10.1007/s10 712–011–9150–2.

Trolinger, J.D. (1975) Particle field holography. *Opt. Eng.*, **14**, 383–392.

Trolinger, J.D. (1976) *Airborne Holography Technique for Particle Field Analysis*, Annals New York Academy of Sciences, pp. 448–459.

Turner, D.D., Caddedu, M.P., Löhnert, U., Crewell, S., and Vogelmann, A.M. (2009) Modifications to the water vapor continuum in the microwave suggested by ground–based 150 GHz observations. *IEEE Geosci. Remote. Sens.*, **47**, 3326–3337. doi: 10.1109/TGRS.2009.2022 262.

Turner, D.D., Knuteson, R., Revercomb, H.E., Lo, C., and Dedecker, R. (2006) Noise reduction of Atmospheric Emitted Radiance Interferometer (AERI) observations using principal component analysis. *J. Atmos. Oceanic Technol.*, **23**, 1223–1238.

Twohy, C. (1998) Model calculations and wind tunnel testing of an isokinetic shroud for high–speed sampling. *Aerosol Sci. Technol.*, **29**, 261–280.

Twohy, C. and Rogers, D. (1993) Airflow and water drop trajectories at instrument sampling points around the Beechcraft King Air and Lockheed Electra. *J. Atmos. Oceanic Technol.*, **10**, 566–578.

Twohy, C., Schanot, A., and Cooper, W. (1997) Measurement of condensed water content in liquid and ice clouds using an airborne counterflow virtual impactor. *J. Atmos. Oceanic Technol.*, **14**, 197–202.

Twohy, C., Strapp, J., and Wendisch, M. (2003) Performance of a counterflow virtual impactor in the NASA icing research tunnel. *J. Atmos. Oceanic Technol.*, **20**, 781–790.

Twohy, C.H. and Poellot, M.R. (2005) Chemical characteristics of ice residual nuclei in anvil cirrus clouds: implications for ice formation processes. *Atmos. Chem. Phys.*, 2289–2297. SRef–ID: 1680–7324/acp/2005–5–2289.

Twomey, S. (1959) The influence of cloud nucleus population on the microstructure and stability of convective clouds. *Tellus*, **11**, 408–411.

Twomey, S. (1968) On the composition of cloud nuclei in North–Eastern United States. *J. Rech. Atmos.*, **4**, 281–285.

Twomey, S. (1974) Pollution and the planetary albedo. *Atmos. Environ.*, **8**, 1251–1256.

Twomey, S. (1975) Comparison of constrained linear inversion and an iterative nonlinear algorithm applied to the indirect estimation of particle size distributions. *J. Comput. Phys.*, **18**, 188–200.

Twomey, S. and Wojciechowski, T. (1969) Observations of the geographic variations of cloud nuclei. *J. Atmos. Sci.*, **26**, 684–688.

Uchiyama, A., Yamazaki, A., Matsuse, K., and Kobayashi, E. (2007) Broadband shortwave calibration results for East Asian Regional Experiment 2005. *J. Geophys. Res.*, **112**, D22S34. doi: 10.1029/2006JD008 110.

Uhlig, E., Borrmann, S., and Jaenicke, R. (1998) Holographic in situ measurements of the spatial droplet distribution in stratiform clouds. *Tellus*, **50B**, 377–387.

Ulaby, F.T., Moore, R.K., and Fung, K. (1981) *Microwave Remote Sensing: Active and Passive*, vols 1 and 2, Addison–Wesley, Advanced Book Program, Reading, MA, p. 456.

Ungar, S., Middleton, E., Ong, L., and Campbell, L.P. (2009) EO–1 hyperion onboard performance over eight years: hyperion calibration, in *Proceedings of the 6th EARSeL SIG IS*, Tel Aviv, p. 7.

Ustin, S., Roberts, D., Gamon, J., Asner, G., and Green, R. (2008) Using imaging spectroscopy to study ecosystem processes and properties. *BioScience*, **54**, (6), 523–534.

Vaillancourt, P.A. and Yau, M.K. (2000) Review of particle-turbulence interactions and consequences for cloud physics. *Bull. Am. Meteorol. Soc.*, **81**, 285–298.

Valero, F., Gore, W., and Giver, L. (1982) Radiative flux measurements in the troposphere. *Appl. Opt.*, **21**, 831–838.

Valero, F., Ackerman, T., and Gore, W. (1989) The effects of the Arctic haze as determined from airborne radiometric measurements during AGASP II. *J. Atmos. Chem.*, **9**, 225–244.

Valero, F., Bucholtz, A., Bush, B., Pope, S., Collins, W., Flatau, P., Strawa, A., and Gore, W. (1997) Atmospheric Radiation Measurements Enhanced Shortwave Experiment (ARESE): experimental and data details. *J. Geophys. Res.*, **102**, 29929–29937.

Valero, F., Pope, S., Bush, B., Nguyen, Q., Marsden, D., Ces, R., Leitner, A., Bucholtz, A., and Udelhofen, P. (2003) Absorption of solar radiation by the clear and cloudy atmosphere during the Atmospheric Radiation Measurement Enhanced Shortwave Experiments (ARESE) I and II:

observations and models. *J. Geophys. Res.*, **108**, 4016. doi: 10.1029/2001JD001 384.

Vali, G. (1985) Nucleation terminology. *J. Aerosol Sci.*, **16**, 575–576.

Vali, G. and Haimov, S. (2001) Observed extinction by clouds at 95 GHz. *IEEE Trans. Geosci. Remote*, **39**, 190–193.

Vali, G., Kelly, R.D., Pazmany, A., and McIntosh, R.E. (1995) Airborne Radar and in situ observations of a shallow stratus with drizzle. *Atmos. Res.*, **38**, 361–380.

Van Baelen, J., Tridon, F., and Pointin, Y. (2009) Simultaneous X-band and K-band study of precipitation to derive specific $Z - R$ relationships. *Atmos. Res.*, **94**, 596–605.

van de Hulst, H. (1981) *Light Scattering by Small Particles*, Dover Publications, Mineola, NY.

van den Kroonenberg, A.C. (2009) *Airborne Measurement of Small–Scale Turbulence with special regard to the Polar Boundary Layer, No. 2009-11 in ZLR-Forschungsbericht*, Sierke Verlag, Technische Universität Braunschweig, 136 pp. ISBN 13: 978-3-86844-216-8.

van den Kroonenberg, A.C., Martin, T., Buschmann, M., Bange, J., and Vörsmann, P. (2008) Measuring the wind vector using the autonomous mini aerial vehicle M^2AV. *J. Atmos. Oceanic Technol.*, **25**, 1969–1982.

van der Meer, F. and Jong, S.D. (2001) *Imaging Spectrometry: Basic Principles and Prospective Applications*, Kluwer Academic Publishers, Dordrecht, The Netherlands, p. 451.

van Overeem, M.A. (1936) A sampling apparatus for aeroplankton, in *Proceedings*, vol. 39, No. 8, 1936. Staples a bit rusty. Folded. Book number 11949, Koninklijke Akademie van Wetenschappen te Amsterdam, pp. 1–11.

Vane, G. (ed.) (1993) *Airborne Imaging Spectrometry*, Remote Sens. Environm., Special Issue, vol. 44, 117–356.

Vane, G. and Goetz, A. (1988) Terrestrial imaging spectroscopy. *Remote Sens. Environ.*, **1**, 3–15.

Varma, R.M., Venables, D., Ruth, A., Heitmann, U., Schlosser, E., and Dixneuf, S. (2009) Long optical cavities for open–path monitoring of atmospheric trace gases and aerosol extinction. *Appl. Opt.*, **48**, B159–B171.

Vay, S.A., Anderson, B., Thornhill, H., and Hudgins, C. (2003) An assessment of aircraft-generated contamination on in situ trace gas measurements: determinations from empirical data acquired aloft. *J. Atmos. Oceanic Technol.*, **20**, 1478–1487.

Veres, P., Roberts, J.M., Warneke, C., Welsh-Bon, D., Zahniser, M., Herndon, S., Fall, R., and de Gouw, J. (2008) Development of negative–ion proton–transfer chemical–ionization mass spectrometry (NI–PT–CIMS) for the measurement of gas–phase organic acids in the atmosphere. *Int. J. Mass Spectrom.*, **274**, 48–55.

Vermote, E., Tanré, D., Deuzé, J., Herman, M., Morcette, J., and Kotchenova, S. (2006) 2nd Simulation of the Satellite Signal in the Solar Spectrum–Vector (6sv). *Tech. Rep.*, Dept. of Geography, Univ. of Maryland.

Viciani, S., D'Amato, F., Mazzinghi, P., Castagnoli, F., Toci, G., and Werle, P. (2008) A cryogenically operated laser diode spectrometer for airborne measurement of stratospheric trace gases. *Appl. Phys. B*, **12**, 581–592. doi: 10.1007/s00 340–007–2885–2.

Vidaurre, G. and Hallett, J. (2009a) Particle impact and breakup in aircraft measurements. *J. Atmos. Oceanic Technol.*, **26**, 972–983. doi: 10.1175/2008JTECHA1147.1.

Vidaurre, G. and Hallett, J. (2009b) Ice and water content of stratiform mixed phase cloud. *Q. J. R. Meteorol. Soc.*, **135**, 1292–1306.

Viggiano, A. (1993) In situ mass spectrometry and ion chemistry in the stratosphere and troposphere. *Mass Spectrom. Rev.*, **12**, 115–137.

Villani, P., Picard, D., Marchand, N., and Laj, P. (2007) Design and validation of a Volatility Tandem Differential Mobility Analyzer (VTDMA). *Aerosol Sci. Technol.*, **41**, 898–906.

Vincent, J.H. (2007) *Aerosol Sampling–Science, Standards, Instrumentation and Applications*, John Wiley & Sons, Inc., New York, p. 636.

Virkkula, A. (2010) Correction of the calibration of the 3–wavelength Particle Soot Absorption Photometer (3 PSAP). *Aerosol Sci. Technol.*, **44**, 706–712.

Virkkula, A., Ahlquist, N.C., Covert, D.S., Arnott, W.P., Sheridan, P.J., Quinn, P.K., and Coffman, D.J. (2005) Modification, calibration and a field test of an instrument for measuring light absorption by particles. *Aerosol Sci. Technol.*, **39**, 68–83.

Vivekanandan, J., Oye, R., Zrnic, D.S., Ellis, S.M., Ryzhkov, A., and Straka, J. (1999) Cloud microphysics retrival using S-band dual–polarization Radar measurements. *Bull. Am. Meteorol. Soc.*, **80**, 381–388.

Voigt, C., Schlager, H., Luo, B., Dornbrack, A., Roiger, A., Stock, P., Curtius, J., Vossing, H., Borrmann, S., Davies, S., Konopka, P., Schiller, C., Shur, G., and Peter, T. (2005) Nitric Acid Trihydrate (NAT) formation at low NAT supersaturation in Polar Stratospheric Clouds (PSCs). *Atmos. Chem. Phys.*, **5**, 1371–1380. doi: 10.5194/acp–5–1371–2005.

Volz, A. and Kley, D. (1985) A resonance fluorescence instrument for the in situ measurement of atmospheric carbon monoxide. *J. Atmos. Chem.*, **2**, 345–357.

Volz–Thomas, A., Arnold, T., Behmann, T., Borrell, P., Borrell, P., Burrows, J., Cantrell, C., Carpenter, L., Clemitshaw, K., Gilge, S., Heitlinger, M., Kluepfel, T., Kramp, F., Mihelcic, D., Muesgen, P., Paetz, H., Penkett, S., Perner, D., Schultz, M., Shetter, R., Slemr, J., and Weissenmayer, M. (1998) Peroxy radical intercomparison exercise, *A Formal Comparison of Methods for Ambient Measurements of Peroxy Radicals*, Berichte des Forschungszentrums Juelich, Institute fuer Chemie und Dynamik der Geosphaere–2, ISSN 0944–2952.

Volz–Thomas, A., Berg, M., Heil, T., Houben, N., Lerner, A., Petrick, W., Raak, D., and Pätz, H.-W. (2005) Measurements of total odd nitrogen (NO_y aboard MOZAIC in–service aircraft: instrument design, operation and performanc. *Atmos. Chem. Phys.*, **5**, 583–595.

Volz-Thomas, A., Lerner, A., Paetz, H.-W., Schultz, M., McKenna, D., Schmitt, R., Madronich, S., and Roeth, E. (1996) Airborne measurements of the photolysis of NO_2. *J. Geophys. Res.*, **101**, 18613–18627.

von der Weiden, S.-L., Drewnick, F., and Borrman, S. (2009) Particle loss calculator – a new software tool for the assessment of the performance of aerosol inlet system. *Atmos. Meas. Tech.*, **2**, 479–494.

Von Hobe, M., Grooss, J., Muller, R., Hrechanyy, S., Winkler, U., and Stroh, F. (2005) A re–evaluation of the ClO/Cl_2O_2 equilibrium constant based on stratospheric in situ observations. *Atmos. Chem. Phys.*, **5**, 693–702.

Von Hobe, M., Ulanovsky, A., Volk, C., Grooss, J., Tilmes, S., Konopka, P., Gunther, G., Werner, A., Spelten, N., Shur, G., Yushkov, V., Ravegnani, F., Schiller, C., Muller, R., and Stroh, F. (2006) Severe ozone depletion in the cold Arctic winter 2004–05. *Geophys. Res. Lett.*, **33**, L17 815. doi: 10.1029/2006GL026 945.

Voutilainen, A., Kolehmainen, V., and Kaipio, J.P. (2001) Statistical inversion of aerosol size measurement data. *Inv. Prob. Engin.*, **9**, 67–94.

Wahner, A., Callies, J., Dorn, H.-P., Platt, U., and Schiller, C. (1990) Near UV atmospheric absorption measurements of column abundances during airborne Arctic stratospheric expedition, January–February 1989: 1. Technique and NO_2 observations. *Geophys. Res. Lett.*, **17**, 497–500. doi: 10.1029/GL017i004p00 497.

Wang, J. (2009) A fast integrated mobility spectrometer with wide dynamic size range: theoretical analysis and numerical simulation. *J. Aerosol Sci.*, **40**, 890–906.

Wang, J. and Chang, C. (2006) Applications of independent component analysis in endmember extraction and aboundance quantification for hyperspetrcal imagery. *IEEE Trans. Geosci. Remote*, **44**, (9), 2601–2616.

Wang, J., Flagan, R., and Seinfeld, J. (2002a) Diffusional losses in particle sampling systems containing bends and elbows. *Aerosol Sci. Technol.*, **33**, 843–857.

Wang, J., Flagan, R.C., and Seinfeld, J.H. (2002b) Diffusional losses in particle sampling systems containing bends and elbows. *J. Aerosol Sci.*, **33**, 843–857.

Wang, J., Racette, P., Piepmeier, J., Monosmith, B., and Manning, W. (2007) Airborne CoSMIR observations between 50 and 183 GHz over snow-covered Sierra mountains. *IEEE Trans. Geosci. Remote*, 55–61. doi:10.1109/TGRS.2006.885 410.

Wang, P., Richter, A., Bruns, M., Rozanov, V.V., Burrows, J.P., Heue, K.-P., Wagner, T., Pundt, I., and Platt, U. (2005a) Measurements of tropospheric NO_2 with an air-borne multi–axis DOAS instrument. *Atmos. Chem. Phys.*, **5**, 337–343. *www.atmos–chem–phys.org/acp/5/337/.*

Wang, S.C. and Flagan, R.C. (1990) Scanning electrical mobility spectrometer. *Aerosol Sci. Technol.*, **13**, 230–240.

Wang, Z. (2007) A refined two–channel microwave radiometer liquid water path retrieval for cold regions by using multiple–sensor measurements. *IEEE Geosci. & Remote Sens. Lett.*, **4**, 591–595.

Wang, Z. and Sassen, K. (2001) Cloud type and macrophysical property retrieval using multiple remote sensors. *J. Appl. Meteorol.*, **40**, 1665–1682.

Wang, Z. and Sassen, K. (2002) Cirrus cloud microphysical property retrieval using Lidar and Radar measurements. Part I: algorithm description and comparison with in situ data. *J. Atmos. Sci.*, **41**, 218–229.

Wang, Z., Sassen, K., Whiteman, D.N., and Demoz, B. (2004) Studying altocumulus with ice virga using ground–based active and passive remote sensors. *J. Appl. Meteorol.*, **43**, 449–460.

Wang, Z., Heymsfield, G., and Li, L. (2005b) Retrieving optically thick ice cloud microphysical properties by using airborne dual-wavelength Radar measurements. *J. Geophys. Res.*, **110**, D19 201.

Wang, Z.E., Wechsler, P., Kuestner, W., French, J., Rodi, A., Glover, B., Burkhart, M., and Lukens, D. (2009) Wyoming cloud Lidar: instrument description and applications. *Opt. Express*, **17**, 13576–13587.

Warhaft, Z. (2000) Passive scalars in turbulent flows. *Annu. Rev. Fluid Mech.*, **32**, 203–240.

Warneck, P. and Williams, J. (2012) *The Atmospheric Chemist's Companion: Numerical Data for Use in the Atmospheric Sciences*, International Geophysics Sseries, Springer Books, ISBN 978-94-007-2274-3.

Warneck, P. and Zerbach, T. (1992) Synthesis of peroxyacetyl nitrate in air by acetone photlysis. *Environ. Sci. Technol.*, **26**, 74–79.

Warneke, C., van der Veen, C., Luxembourg, S., de Gouw, J.A., and Kok, A. (2001) Measurements of benzene and toluene in ambient air using proton–transfer–reaction mass spectrometry: calibration, humidity dependence, and field intercomparison. *Int. J. Mass Spectrom.*, **207**, 167–182.

Warneke, C., de Gouw, J.A., Lovejoy, E.R., Murphy, P.C., Kuster, W.C., and Fall, R. (2005a) Development of proton–transfer ion trap–mass spectrometry: on–line detection and identification of volatile organic compounds in air. *J. Am. Soc. Mass. Spectrom.*, **16**, 1316–1324.

Warneke, C., Kato, S., de Gouw, J.A., Goldan, P.D., Kuster, W.C., Shao, M., Lovejoy, E.R., Fall, R., and Fehsenfeld, F.C. (2005b) Online volatile organic compound measurements using a newly developed proton–transfer ion–trap mass spectrometry instrument during New England Air Quality Study–Intercontinental Transport and Chemical Transformation 2004: performance, intercomparison, and compound identification. *Environ. Sci. Technol.*, **39**, 5390–5397.

Warneke, C., McKeen, S.A., de Gouw, J.A., Goldan, P.D., Kuster, W.C., Holloway, J.S., Williams, E.J., Lerner, B.M., Parrish, D.D., Trainer, M., Fehsenfeld, F.C., Kato, S., Atlas, E.L., Baker, A., and Blake, D.R. (2007) Determination of urban volatile organic compound emission ratios and comparison with an emissions database. *J. Geophys. Res.*, **112**, D10S47.

Washenfelder, R.A., Langford, A., Fuchs, H., and Brown, S. (2008) Measurement of glyoxal using an incoherent broadband cavity enhanced absorption spectrometer. *Atmos. Chem. Phys.*, **8**, 7779–7793.

Wayne, R. (1999) *Chemistry of Atmospheres–Third Edition*, Oxford University Press. ISBN 0 19 850375 X.

Webb, A., Kylling, A., Wendisch, M., and Jäkel, E. (2004) Airborne measurements of ground and cloud spectral albedos under low aerosol loads. *J. Geophys. Res.*, **109**. doi: 10.1029/2004JD004 768.

Weber, R., Clarke, A., Litchy, M., Li, J., Kok, G., Schillawski, R., and McMurry, P.H. (1998) Spurious aerosol measurements when sampling from aircraft in the vicinity of clouds. *J. Geophys. Res.*, **103**, 28337–28346.

Weber, R., McMurry, P., Bates, T., Clarke, A., Covert, D., Brechtel, F., and Kok, G. (1999) Intercomparison of airborne and surface-based measurements of condensation nuclei in the remote marine troposphere during ACE 1. *J. Geophys. Res.*, **104**, 21673–21683.

Weber, R.J., Orsini, D., Daun, Y., Lee, Y.N., Klotz, P.J., and Brechtel, F. (2001) A particle–into–liquid collector for rapid measurement of aerosol bulk chemical composition. *Aerosol Sci. Technol.*, **35**, 718–727.

Webster, C. and Heymsfield, A. (2003) Water isotope ratios D/H, $^{18}O/^{16}O$, $^{17}O/^{16}O$ in and out of clouds map dehydration pathways. *Science*, **302**, 1742–1745.

Wecht, K.J., Jacob, D.J., Wofsy, S.C., Kort, E.A., Worden, J.R., Kulawik, S.S., Henze, D.K., Kopacz, M., and Payne, V.H. (2012) Validation of TES methane with HIPPO aircraft observations: implications for inverse modeling of methane sources. *Atmos. Chem. Phys.*, **12**, 1823–1832. doi: 10.5194/acp–12–1823–2012.

Weckwerth, T., Parsons, D.B., Koch, S., Moore, J.A., LeMone, M.A., Demoz, B.B., Flamant, C., Geerts, B., Wang, J., and Feltz, W.F. (2003) An overview of the International H_2O Project (IHOP_2002) and some preliminary highlights. *B. Am. Meteorol. Soc.*, **85**, 253–277. doi: 10.1175/BAMS–85–2–253.

Wehner, B., Philippin, S., and Wiedensohler, A. (2002) Design and calibration of a thermodenuder with an improved heating unit to measure the size-dependent volatile fraction of aerosol particles. *J. Aerosol Sci.*, **33**, 1087–1093.

Weickman, H. (1945) Formen und Bildung atmosphärischer Eiskristalle. *Beitr. Phys. Atmos.*, **28**, 12–52.

Weickmann, H. (1949) Die Eisphase in der Atmosphäre, Berichte des Deutschen Wetterdienstes in der US–Zone, Nr. 6.

Weickmann, H. and Kampe, H. (1953) Physical properties of cumulus clouds. *J. Meteorol.*, **10**, 204–211.

Weigel, R., Hermann, M., Curtius, J., Voigt, C., Walter, S., Bottger, T., Lepukhov, B., Belyaev, G., and Borrmann, S. (2009) Experimental characterization of the Condensation Particle counting System for high altitude aircraft–borne application. *Atmos. Meas. Tech.*, **2**, 243–258.

Weinheimer, A. and Schwiesow, R. (1992) A two–path, two wavelength ultraviolet hygrometer. *J. Atmos. Oceanic Technol.*, **9**, 407–419.

Weinman, J.A. and Kim, M.J. (2007) A simple model of the millimeter–wave scattering parameters of randomly oriented aggregates of finite cylindrical ice hydrometeors. *J. Atmos. Sci.*, **64**, 634–644.

Weinstock, E., Smith, J., Sayres, D., Spackman, J., Pittman, J., Allen, N., Demusz, J., Greenberg, M., Rivero, M., Solomon, L., and Anderson, J.G. (2006) Measurements of the total water content of cirrus clouds. Part I: instrument details and calibration. *J. Atmos. Oceanic Technol.*, **23**, 1397–1409.

Weinstock, E.M., Hintsa, E.J., Dessler, A.E., Oliver, J.F., Hazen, N.L., Demusz, J.N., Allen, N.T., Lapson, L.B., and Anderson, J.G. (1994) New fast response photofragment fluorescence hygrometer for use on the NASA ER-2 and the Perseus remotely piloted aircraft. *Rev. Sci. Instrum.*, **65**, 3544–3554.

Weinstock, E.M., Smith, J.B., Sayres, D.S., Pittman, J.V., Spackman, J.R., Hintsa, E.J., Hanisco, T.F., Moyer, E.J., Clair, J.M.S., Sargent, M.R., and Anderson, J.G. (2009) Validation of the Harvard Lyman–α in situ water vapor instrument: implications for the mechanisms that control stratospheric water vapor. *J. Geophys. Res.*, **114**, D23 301. doi: 10.1029/2009JD012 427.

Weinzierl, B., Petzold, A., Esselborn, M., Wirth, M., Rasp, K., Kandler, K., Schütz, L., Koepke, P., and Fiebig, M. (2009) Airborne measurements of dust layer properties, particle size distribution and mixing state of Saharan dust during SAMUM 2006. *Tellus*, **61B**, 96–117.

Weiß, S. (2002) Comparing three algorithms for modeling flush air data systems. AIAA Aerospace Sciences Meeting and Exhibit, 40th, AIAA–2002–535, Reno, NV.

Weiß, S. and Leißling, D. (2001) Flush Air Data System Verfahren zur Luftdatenbestimmung und Fehlererkennung. DGLR Jahrestagung, Hamburg, Germany.

Weiß, S., Thielecke, F., and Harders, H. (1999) Ein neuer Ansatz zur Modellierung von Luftdatensystemen. Deutscher Luft– und Raumfahrtkongress / DGLR Jahrestagung, Berlin, Germany, p. 7.

Weissmann, M. and Cardinali, C. (2007) Impact of airborne Doppler Lidar observations on ECMWF forecasts. *Q. J. R. Meteorol. Soc.*, **133**, 1107–116.

Weissmann, M., Busen, R., Dörnbrack, A., Rahm, S., and Reitebuch, O. (2005) Targeted observations with an airborne wind Lidar. *J. Atmos. Oceanic Technol.*, **22**, 1706–1719.

Weitkamp, C. (ed.) (2005) *Range–resolved Optical Remote Sensing of the Atmosphere*, Springer, New York.

Wendel, G., Stedman, D., Cantrell, C., and Damrauer, L. (1983) Luminol–based nitrogen dioxide detector. *Anal. Chem.*, **55**, 937.

Wendisch, M. and Keil, A. (1999) Discrepancies between measured and modeled solar and UV radiation within polluted boundary layer clouds. *J. Geophys. Res.*, **104**, 27373–27385.

Wendisch, M. and Mayer, B. (2003) Vertical distribution of spectral solar irradiance in the cloudless sky – A case study. *Geophys. Res. Lett.*, **30**, 1183–1186. doi: 10.1029/2002GL016 529.

Wendisch, M. and Yang, P. (2012) *Theory of Atmospheric Radiative Transfer – A Comprehensive Introduction*, Wiley–VCH Verlag GmbH & Co. KGaA, Weinheim, Germany.

Wendisch, M., Keil, A., and Korolev, A. (1996a) FSSP characterization with monodisperse water droplets. *J. Atmos. Oceanic Technol.*, **13**, 1152–1165.

Wendisch, M., Mertes, S., Ruggaber, A., and Nakajima, T. (1996b) Vertical profiles of aerosol and radiation and the influence of a temperature inversion: measurements and radiative transfer calculations. *J. Appl. Meteorol.*, **35**, 1703–1715.

Wendisch, M., Müller, D., Schell, D., and Heintzenberg, J. (2001) An airborne spectral albedometer with active horizontal stabilization. *J. Atmos. Oceanic Technol.*, **18**, 1856–1866.

Wendisch, M., Garrett, T., and Strapp, J. (2002) Wind tunnel tests of the airborne PVM–100A response to large droplets. *J. Atmos. Oceanic Technol.*, **19**, 1577–1584.

Wendisch, M., Coe, H., Baumgardner, D., Brenguier, J.-L., Dreiling, V., Fiebig, M., Formenti, P., Hermann, M., Krämer, M., Levin, Z., Maser, R., Mathieu, E., Nacass, P., Noone, K., Osborne, S., Schneider, J., Schütz, L., Schwarzenböck, A., Stratmann, F., and Wilson, J.C. (2004) Aircraft particle inlets: state-of-the-art and future needs. *Bull. Am. Meteorol. Soc.*, **85**, 89–91.

Wendisch, M., Pilewski, P., Pommier, J., Howard, S., Yang, P., Heymsfield, A., Schmitt, C., Baumgardner, D., and Mayer, B. (2005) Impact of cirrus crystal shape on solar spectral irradiance: a case study for subtropical cirrus. *J. Geophys. Res.*, **110**, 294. doi: 10.1029/2004JD005.

Wendisch, M., Yang, P., and Pilewskie, P. (2007) Effects of ice crystal habit on thermal infrared radiative properties and forcing of cirrus. *J. Geophys. Res.*, **112**, D03 202. doi: 10.1029/2006JD007 899.

Wennberg, P., Salawitch, R., Donaldson, D., Hanisco, T., Lanzendorf, E., Perkins, K., Lloyd, S., Vaida, V., Gao, R., Hintsa, E., Cohen, R., Swartz, W., Kusterer, T., and Anderson, D. (1999) Twilight observations suggest unknown sources of HO_x. *Geophys. Res. Lett.*, **26**, 1373–1376.

Wennberg, P.O., Cohen, R., Hazen, N., Lapson, L., Allen, N., Hanisco, T., Oliver, J., Lanham, N., Demusz, J., and Anderson, J. (1994a) Aircraft–borne, laser–induced fluorescence instrument for the in situ detection of hydroxyl and hydroperoxyl radicals. *Rev. Sci. Instrum.*, **65**, 1858–1876.

Wennberg, P.O., Cohen, R., Stimpfle, R., Koplow, J., Anderson, J., Salawitch, R., Fahey, D., Woodbridge, E., Keim, E., Gao, R., Webster, C., May, R., Toohey, D., Avallone, L., Proffitt, M., Podolske, J., Chan, K., and Wofsy, S. (1994b) Removal of stratospheric O_3 by radicals in situ measurements of OH, HO_2, NO, NO_2, CLO, and BRO. *Science*, **266**, 398–404.

Wennberg, P.O., Hanisco, T., Cohen, R., Stimpfle, R., Lapson, L., and Anderson, J. (1995) In situ measurements of OH and HO_2 in the upper troposphere and stratosphere. *J. Atmos. Sci.*, **52**, 3413–3420.

Werle, P., Slemr, F., Maurer, K., Kormann, R., Mücke, R., and Jänker, B. (2002) Near- and mid-infrared laser-optical sensors for gas analysis. *Opt. Lasers Eng.*, **37**, 101–114.

Werner, C., Flamant, P.H., Reitebuch, O., Köpp, F., Streicher, J., Rahm, S., Nagel, E., Klier, M., Hermann, H., Loth, C., Delville, P., Drobinski, P., Romand, B., Boitel, C., Oh, D., Lopez, M., Meissonnier, M., Bruneau, D., and Dabas, A.M. (2001) Wind infrared Doppler Lidar instrument. *Opt. Eng.*, **40**, 115–125.

Wert, B.P., Fried, A., Henrym, B., and Cartier, S. (2002) Evaluation of inlets used for the airborne measurement of formaldehyde. *J. Geophys. Res.*, **107**, 4163. doi: 0.1029/2001JD001 072.

Wert, B.P., Fried, A., Rauenbuehler, S., Walega, J., and Henry, B. (2003) Design and performance of a tunable diode laser absorption spectrometer for airborne formaldehyde measurements. *J. Geophys. Res.*, **108**, 4350. doi: 10.1029/2002JD002 872.

Westwater, E., Crewell, S., and Mätzler, C. (2005) Surface–based microwave and millimeter wave radiometric remote sensing of the troposphere: a tutorial. *IEEE Trans. Geosci. Remote Soc. Newslett.*, 16–33.

Whalley, L., Lewis, A., McQuaid, J., Purvis, R., Lee, J., Stemmler, K., Zellweger, C., and Ridgeon, P. (2004) Two high–speed, portable GC systems designed for the measurement of non–methane hydrocarbons and PAN: results from the Jungfraujoch High Altitude Observatory. *J. Environ. Monit.*, **6**, 234–241.

Whiteman, D.N., Rush, K., Rabenhorst, S., Welch, W., Cadirola, M., McIntire, G., Russo, F., Adam, M., Venable, D., Connell, R., Veselovskii, I., Forno, R., Mielke, B., Stein, B., Leblanc, T., McDermid, T.L., and Vömel, H. (2010) Airborne and ground–based measurements using a high–performance Raman Lidar. *J. Atmos. Oceanic Technol.*, **27**, 1781–1801. doi: 10.1175/2010JTECHA1391.1.

Whitmore, S., Lindsey, W.T., Curry, R., and Gilyard, G. (1990) Experimental Characterization of the Effects of Pneumatic Tubing on Unsteady Pressure Measurements, NASA Technical Memorandum 4171.

Wiedensohler, A. (1988) An approximation of the bipolar charge distribution for particles in the submicron size range. *J. Aerosol Sci.*, **19**, 387–389.

Wiedensohler, A., Orsini, D., Covert, D.S., Coffmann, D., Cantrell, W., Havlicek, M., Brechtel, F.J., Russell, L.M., Weber, R.J., Gras, J., Hudson, J.G., and Litchy, M. (1997) Intercomparison study of the size-dependent counting efficiency of 26 condensation particle counters. *Aerosol Sci. Technol.*, **27**, 224–242.

Wiederhold, P. (1997) *Water Vapor Measurement: Methods and Instrumentation*, Marcel Dekker Inc., New York, 357 pp.

Wielicki, B., Suttles, J., Heymsfield, A., Welch, R., Spinhirne, J., Wu, M.-L., O'Starr, D., Parker, L., and Arduini, R. (1990) The 27–28 October 1986 FIRE IFO cirrus case study: comparison of radiative transfer theory with observations by satellite and aircraft. *Mon. Weather Rev.*, **118**, 2356–2376.

Wieringa, J. (1986) Roughness–dependent geographical interpolation of surface wind speed averages. *Q. J. R. Meteorol. Soc.*, **112**, 867–889.

Williams, A. and Marcotte, D. (2000) Wind measurements on a maneuvering twin–engine turboprop aircraft accounting for flow distortion. *J. Atmos. Oceanic Technol.*, **17**, 795–810.

Williams, J. (2006) Mass spectrometric methods for atmospheric trace gases, *Analytical Technique for Atmospheric Measurements*, Blackwell Publishing Ltd., ISBN–13: 978-1-4051-2357-0.

Williams, J., Pöschl, U., Crutzen, P.J., Hansel, A., Holzinger, R., Warneke, C., Lindinger, W., and Lelieveld, J. (2001) An atmospheric chemistry interpretation of mass scans obtained from a proton transfer mass spectrometer flown over the tropical rainforest of Surinam. *J. Climate*, **38**, 133–166.

Wilson, J.C., Lafleur, B., Hilbert, H., Seebaugh, W., Fox, J., Gesler, D., Brock, C., Huebert, B., and Mullen, J. (2004) Function and performance of a low turbulence inlet for sampling supermicron particles from aircraft platforms. *Aerosol Sci. Technol.*, **38**, 790–802.

Wilson, J.C., Hyun, J.H., and Blackshear, E.D. (1983) The function and response of an improved stratospheric condensation nucleus counter. *J. Geophys. Res.*, **88**, 6781–6785. doi: 10.1029/JC088iC11p06781.

Wilson, J.C., Stolzenburg, M., Loewenstein, W.C.M., Ferry, G., Chan, K.R., and Kelly, K.K. (1992) Stratospheric sulfate aerosol in and near the northern—hemisphere polar vortex—the Morphology of the sulfate layer, multimodal size distributions, and the effect of denitrification. *J. Geophys. Res.*, **97**, 7997–8013.

Wilson, K. and Birks, J. (2006) Mechanism and elimination of a water vapor interference in the measurement of ozone by UV absorbance. *Environ. Sci. Technol.*, **40**, 6361–6367.

Wilson, S., Atkinson, N., and Smith, J. (1999) The development of an airborne infrared interferometer for meteorological sounding studies. *J. Atmos. Oceanic Technol.*, **16**, 1912–1927.

Wirth, M., Fix, A., Mahnke, P., Schwarzer, H., Schrandt, F., and Ehret, G. (2009) The airborne multi-wavelength water vapor differential absorption Lidar WALES: system design and performance. *Appl. Phys. B.*, **96**, 201–213.

Wisthaler, A., Hansel, A., Dickerson, R.R., and Crutzen, P.J. (2002) Organic trace gas measurements by PTR–MS during INDOEX 1999. *J. Geophys. Res.*, **107**, 8024. doi: 10.1029/2001JD000 576.

Wofsy, S.C. The HIPPO Science Team and Cooperating Modellers and Satellite Teams (2011) HIAPER Pole-to-Pole Observations (HIPPO): fine-grained, global-scale measurements of climatically important atmospheric gases and aerosol particles. *Philos. Trans. R. Soc. London, Ser. A*, **369**, 2073–2086. doi: 10.1098/rsta.2010.0313.

Wolde, M. and Vali, G. (2001a) Polarimetric signatures from ice crystals observed at 95 GHz in winter clouds: Part I. Dependence on crystal type. *J. Atmos. Sci.*, **58**, 828–841.

Wolde, M. and Vali, G. (2001b) Polarimetric signatures from ice crystals observed at 95 GHz in winter clouds: Part II. Frequencies of occurrence. *J. Atmos. Sci.*, **58**, 842–849.

Woodcock, A. (1952) Atmospheric salt particles and raindrops. *J. Meteorol.*, **9**, 200–212.

Woodfield, A.A. and Vaughan, J.M. (1983) Airspeed and wind shear measurements with an airborne CO_2 CW laser. *Int. J. Aviat. Saf.*, **1**, 207–224.

SPARC (2000) World Meteorological Organization: SPARC Assessment of Upper Tropospheric and Stratospheric Water Vapor. *Tech. Rep.* WMO/TD–No. 1043, World Climate Research Programme, Geneva.

World Meteorological Organization (2008) Measurements of humidity, *Guide to Meteorological Instruments and Methods of Observation*, 7th edn, World Meteorological Organization, Geneva, Chapter 12. WMO No.8, *http://www.wmo.int/pages/prog/www/IMOP/publications/CIMO-Guide/CIMO_Guide-7th_Edition-2008.html* (accessed on 2008).

Wörrlein, K. (1990) Kalibrierung von Fünflochsonden. GFA *Tech. Rep.*, Technische Universität Darmstadt.

Wu, J. (1990) Mean square slopes of the wind-disturbed water surface, their magnitude, directionality, and composition. *Radio Sci.*, **25**, 37–48.

Wunch, D., Toon, C., Wennberg, P.O., and Wofsy, S.C. (2010) Calibration of the total carbon column observing network using aircraft profile data. *Atmos. Meas. Tech.*, **3**, 1351–1362.

Wyche, K.P., Blake, R.S., Willis, K.A., Monks, P.S., and Ellis, A.M. (2005) Differentiation of isobaric compounds using chemical ionization reaction mass spectrometry. *Rapid Commun. Mass Spectrom.*, **19**, 3356–3362.

Wyche, K.P., Blake, R.S., Ellis, A.M., Monks, P.S., Brauers, T., Koppmann, R., and Apel, E.C. (2007) Technical note: performance of Chemical Ionization Reaction Time–of–Flight Mass Spectrometry (CIR–ToF–MS) for the measurement of atmospherically significant oxygenated volatile organic compounds. *Atmos. Chem. Phys.*, **7**, 609–620.

Wylie, R., Davies, D., and Caw, W. (1965) *The Basic Process of the Dew–Point Hygrometer, Humidity and Moisture*, vol. 1, (ed. R.E. Ruskin), Reinhold, pp. 123–134.

Wyngaard, J.C. (2010) *Turbulence in the Atmosphere*, Cambridge University Press, 408 p.

Wyngaard, J.T. and Zhang, S.-F. (1985) Transducer–shadow effects on turbulence spectra measured by sonic anemometers. *J. Atmos. Oceanic Technol.*, **2**, 548–558.

Yang, J. and Jaenicke, R. (1999) The condensational growth of aerosol particle and its effect in aerosol measurements. *J. Aerosol Sci.*, **30**, S69–S70.

Yang, P. and Liou, K. (1996) Geometric–optics–integral–equation method for light scattering by nonspherical ice crystals. *Appl. Opt.*, **35**, 6568–6584.

Yang, P. and Liou, K.-N. (1998) Single scattering properties of complex ice crystals in terrestrial atmosphere. *Contrib. Atmos. Phys.*, **71**, 223–248.

Yarin, A. (2006) Drop impact dynamics: splashing, spreading, receding, bouncing .... *Annu. Rev. Fluid Mech.*, **38**, 159–192. doi: 10.1146/annurev.fluid.38.050 304.092 144.

Ye, J. and Hall, J. (1998) Ultrasensitive detection in atomic and molecular physics: demonstration in molecular overtone spectroscopy. *J. Opt. Soc. Am. B: Opt. Phys.*, **15**, 6–15.

Yu, Y., Alexander, M., Perraud, V., Bruns, E., Johnson, S., Ezell, M., and Finlayson-Pitts, B. (2009) Contamination from electrically conductive silicone tubing during aerosol chemical analysis. *Atmos. Environ.*, **43**, 2836–2839. doi: 10.1016/j.atmosenv. 2009.02.014.

Yushkov, V., Oulanovsky, A., Lechenuk, N., Roudakov, I., Arshinov, K., Tikhonov, F., Stefanutti, L., Ravegnani, F., Bonafe, U., and Georgiadis, T. (1999) A chemiluminescent analyzer for stratospheric measurements of the ozone concentration (FOZAN). *J. Atmos. Oceanic Technol.*, **16**, 1345–1350.

Zabrodsky, G. (1957) Measurements and some results of study of visibility in clouds, *Study of Clouds, Precipitation and Cloud Electrification*, Gidrometeizdat.

Zaer, A. and Gader, P. (2008) Hyperspectral band selection and endmember detection using sparsity promoting priors. *IEEE Trans. Geosci. Remote Sens. Lett.*, **5**, 256–260.

Zahn, A. (2001) Constraints on 2–way transport across the Arctic tropopause based on O_3, stratospheric tracer (SF_6) ages, and water vapor isotope (D, T) tracers. *J. Atmos. Chem.*, **39**, 303–325.

Zahn, A., Barth, V., Pfeilsticker, K., and Platt, U. (1998) Deuterium, oxygen–18, and tritium as tracers for water vapor transport in the lower stratosphere and tropopause region. *J. Atmos. Chem.*, **30**, 25–47.

Zahn, A., Brenninkmeijer, C., Crutzen, P., Parrish, D., Sueper, D., Heinrich, G., Güsten, H., Fischer, H., Hermann, M., and Heintzenberg, J. (2002) Electrical discharge source for tropospheric 'ozone–rich transients'. *J. Geophys. Res.*, **107**, 4638. doi: 10.1029/2002JD002 345.

Zelenyuk, A. and Imre, D.G. (2009) Beyond single particle mass spectrometry: multidimensional characterisation of individual aerosol particles. *Int. Rev. Phys. Chem.*, **28**, 309–358.

Zelenyuk, A., Yang, J., Choi, E.Y., and Imre, D.G. (2009) SPLAT II: an aircraft compatible, ultra–sensitive, high precision instrument for in–situ characterization of the size and composition of fine and ultrafine particles. *Aerosol Sci. Technol.*, **43**, 411–424.

Zhang, N., Zhou, X., Shepson, P., Gao, H., Alaghmand, M., and Stirm, B. (2009) Aircraft measurement of HONO vertical profiles over a forested region. *Geophys. Res. Lett.*, **36**, L15 820. doi: 10.1029/2009GL038 999.

Zhang, S.H., Akutsu, Y., Russell, L.M., Flagan, R.C., and Seinfeld, J.H. (1995) Radial differential mobility analyzer. *Aerosol Sci. Technol.*, **23**, 357–372.

Zhang, X.F., Smith, K.A., Worsnop, D.R., Jimenez, J., Jayne, J.T., and Kolb, C.E. (2002) A numerical characterization of particle beam collimation by an aerodynamic lens–nozzle system: Part I. An individual lens or nozzle. *Aerosol Sci. Technol.*, **36**, 617–631.

Zhang, Z.Q. and Liu, B.Y.H. (1991) Performance of TSI 3760 condensation nuclei counter at reduced pressures and flow–rates. *Aerosol Sci. Technol.*, **15**, 228–238.

Zhao, J. and Zhang, R.Y. (2004) Proton transfer reaction rate constants between hydronium ion (H_3O_+) and volatile organic compounds. *Atmos. Environ.*, **38**, 2177–2185.

Ziereis, H., Schlager, H., Schulte, P., Köhler, I., Marquardt, R., and Feigl, C. (1999) In situ observations of the NO_x distribution and variability over the eastern North Atlantic. *J. Geophys. Res.*, **104**, 16021–16032.

Ziereis, H., Schlager, H., Schulte, P., van Velthoven, P.F.J., and Slemr, F. (2000) Distributions of NO, NO_x, and NO_y in the upper troposphere and lower stratospere between 28° and 61°N during POLINAT 2. *J. Geophys. Res.*, **105**, 3653–3664.

Ziereis, H., Minikin, A., Schlager, H., Gayet, J.-F., Auriol, F., Stock, P., Baehr, J., Petzold, A., Schumann, U., Weinheimer, A., Ridley, B., and Ström, J. (2004) Uptake of reactive nitrogen on cirrus cloud particles during INCA. *Geophys. Res. Lett.*, **31**, L05 115. doi: 10.1029/2003GL018 794.

Zimmermann, F., Ebert, M., Worringen, A., Schütz, L., and Weinbruch, S. (2007) Environmental scanning electron microscopy (ESEM) as a new technique to determine the ice nucleation capability of individual atmospheric aerosol particles. *Atmos. Environ.*, **41**, 8219–8227.

Ziskind, G. (2006) Particle resuspension from surfaces: revisited and re–evaluated. *Rev. Chem. Eng.*, **22**, 1–123.

Zmarzly, P. and Lawson, R. (2000) An Optical Extinctiometer for Cloud Radiation Measurements and Planetary Exploration. Final Report in fulfillment of Contract NAS5–98032, *Tech. Rep.*, NASA Goddard Space Flight Center, 131 pp. *http://www.specinc.com/publications/Extinctiometer_Report.pdf* (accessed on 2000).

Zöger, M., Afchine, A., Eicke, N., Gerhards, M.-T., Klein, E., McKenna, D., Mörschel, U., Schmidt, U., Tan, V., Tuitjer, F., Woyke, T., and Schiller, C. (1999) Fast in situ stratospheric hygrometers: a new family of balloonborne and airborne Lyman–α photofragment fluorescence hygrometers. *J. Geophys. Res.*, **104**, 1807–1816. doi: 10.1029/1998JD100 025.

Zondlo, M., Mauldin, R., Kosciuch, E., Cantrell, C., and Eisele, F. (2003) Development and characterization of an airborne–based instrument to measure nitric acid during the NASA Transport and Chemical Evolution over the Pacific field experiment. *J. Geophys. Res.*, **108**, 8793. doi: 10.1029/2002JD003 234.

Zondlo, M., Paige, M., Massick, S., and Silver, J. (2010) Vertical cavity laser hygrometer for the National Science Foundation Gulfstream-V aircraft. *J. Geophys. Res.*, **115**, D20 309. doi: 10.1029/2010JD014 445.

Zuidema, P., Leon, D., Pazmany, A., and Cadeddu, M. (2012) Aircraft millimeter-wave passive sensing of cloud liquid water and water vapor during VOCALS-REx. *Atmos. Chem. Phys.*, **12**, 355–369.

Zukauskas, A. and Ziugzda, J. (1985) *Heat Transfer of a Cylinder in a Crossflow* (ed. G.F. Hewitt), Hemisphere Publishing Corp., Washington, DC.

Index

Airborne Measurements for Environmental Research: Methods and Instruments, First Edition.
Edited by Manfred Wendisch and Jean-Louis Brenguier.

b

d

e

f

g

h

t

u

v

w